ELECTROPHYSIOLOGICAL
METHODS
IN BIOLOGICAL
RESEARCH

Scientific Editors
Dr. JOSEF HOLUBÁŘ and Dr. JOSEF IPSER

Translated by
Dr. PETR HAHN

Dr. Jan Bureš, Dr. Mojmír Petráň,
Dr. Jozef Zachar

ELECTROPHYSIOLOGICAL METHODS IN BIOLOGICAL RESEARCH

Third revised edition

Academia
Publishing House of the
Czechoslovak Academy of Sciences
Prague

Academic Press
New York
and London

1967

NMU LIBRARY

Academic Press Inc., 111 Fifth Avenue, New York, New York 10003
Library of Congress Catalog Card Number 66-16447

CONTENTS

6

CHAPTER III

General electrophysiology of cells and tissues (by J. Bureš) 271

CHAPTER IV

Electrophysiology of isolated excitable structures in vitro (by J. Zachar). 304

Five years that have elapsed since this book was first published have shown up its good and bad points. On the whole it appears that the lay-out was rightly chosen. The book has been translated into Russian (1962) and Chinese (1963) and has become a handbook for young adepts of electrophysiology at many universities. On the other hand, the rapid development of electrophysiological techniques has led to the rapid aging of some technical informations and has considerably favoured the use of some methods that not so very long ago were considered as too special. This third edition attempts to fill in those gaps. In the technical part much attention has been paid to transistorisation of electrophysiological apparatus and to the mathematical analysis of electrophysological data. Appendix II may be used in order to refresh the mathematical knowledge necessary for reading especially the additions to chapter II. In the experimental part the number of experiments using intra- and extracellular microelectrodes has been several times increased but the experimental objects have been chosen in such a way that even beginners can successfully solve the problems with simple means. New references added to all chapters have been mainly used for the interpretation of the experimental results. Long experience with the stereotaxic atlases in the Appendix I resulted in some corrections and extensions. Finally new experiments have been included, illustrating some recent electrophysiological findings and methods (electric synapses, presynaptic inhibition, microelectrophoresis etc.).

The aim of the book remains to give a more or less complete review of the principles and applications of the most important electrophysiological methods. More detailed information may be found in special technical books and particularly in the original articles, which are amply quoted throughout.

In conclusion we should like to extend our thanks to all readers of our book, who in reviews, letters and discussion have pointed out errors and drawbacks and have given us an impetus for further work. Our thanks are also due to our colleagues at the Institute of Physiology of the Czechoslovak Academy of Sciences, Prague, at the Institute of Biophysics, Medical Faculty, Plzeň and at the Institute of Normal and Pathological Physiology, Slovak Academy of Sciences, Bratislava for all their help in preparing this 3rd edition. We are particularly grateful for valuable advice to the scientific editors Dr. J. Holubář and Dr. J. Ipser, to Dr. P. Hahn for the English translation and proof reading, to Dr. R. G. W. Prescott for editing the translated manuscript and to Mrs. M. Betková for preparing the subject and author indexes.

Prague, January, 1967.   *J. Bureš, M. Petráň, J. Zachar*

Preface

During the last twenty years electrophysiological methods have been used to an ever-increasing extent in biological and medical research. They represent today the most perfect analytical research tool giving a dynamic picture of processes that occur in living matter from individual cells to mammalian brain. Nevertheless, electrophysiology is still used insufficiently in many disciplines. Its broader application is hindered primarily by lack of qualified specialists. The need for electrophysiologists is large and their training difficult.

In the present book an attempt has been made to combine theoretical and technical information concerning electrophysiology with practical instructions for performing fundamental electrophysiological experiments. The authors have worked in different fields of electrophysiology since 1950. They had not been trained in any renowned laboratory and had therefore to solve various methodical problems, making many mistakes and overcoming many obstacles. It is hoped that their experience will help scientists who themselves intend to commence electrophysiological research and for whom the following chapters are mainly intended.

The authors take this opportunity of thanking all their colleagues from the Institute of Physiology of the Czechoslovak Academy of Sciences in Prague and from the Institute of Experimental Medicine of the Slovak Academy of Sciences in Bratislava, who helped them so much in preparing this book. The authors are much indebted for valuable advice and stimulating criticism of various parts of the manuscript to Dr M. Brazier, Dr H. Grundfest, Dr P. G. Kostyuk, Dr H. W. Magoun, Dr G. Moruzzi and Dr G. D. Smirnov. They are especially grateful to Dr J. Holubář and Dr J. Ipser, the scientific editors who carefully read the whole manuscript.

J. Bureš, M. Petráň, J. Zachar

Prague, April 6, 1960.

Introduction

Electrophysiology is concerned with electrical phenomena occurring in living matter and with the effect of electrical currents on living matter.

That part of electrophysiology which is concerned with the basis of bioelectrical phenomena might be termed general electrophysiology. The subject matter can be divided as follows:

1) A study of electrical states, e. g. polarity of tissues, cells, organisms etc.

2) A study of changes in voltage or current in these structures.

3) A study of the effect of electric current on living matter.

4) A study of the electrical characteristics of cells and tissues in general, e. g. resistance and capacitance (Walter 1956).

Applied electrophysiology, on the other hand, makes use of certain electrical manifestations during activity of some system for analysing this system itself. The use of electrophysiological technique for solving specific problems concerning the function of excitable structures may serve as an example.

The object of such research, however, is not the problem of electrogenesis or the functional significance of electrical potentials, but that of analysing relationships between certain electrical phenomena and functions, between functional elements, between the stimulus and the response etc.

In such a case the action potential is only an expression of the nerve impulse, a fundamental unit of information by which various excitatory structures are mutually connected. Studying the movements of such signals along parts of the nervous system makes it possible to study its organisation, to clarify the interaction of individual elements, the formation and transmission of more complex information. The same is valid not only for action potentials, but also for more complex electrical phenomena (e. g. spontaneous EEG) which represent the sum of activity from extensive synaptic areas.

Used in such a way electrophysiology is nothing but a research tool (often the only one available), sometimes merely completing information obtained otherwise.

According to the system studied, or to other criteria, electrophysiology may be subdivided, e. g. into electrophysiology of plants and animals, skin, muscle, nervous system etc.

Historically the roots of electrophysiology are found in the controversy between Galvani and Volta concerning the interpretation of the experiment of Galvani (1791). Galvani explained his findings by the concept of animal electricity and Volta by postulating currents due to different metals connected by salt solutions.

This question could only be settled definitely when adequate measuring instruments were available. In 1819 Oersted found that a magnetic needle is deflected under the influence of a galvanic current, and this led to the construction of galvanometers. In 1825 Nobili demonstrated the existence of muscle currents by means of a galvanometer of his own design. In 1845 Carlo Matteucci used a nerve-muscle preparation as a sensitive biological indicator of the presence of currents in another contracting muscle.

It is due to the book: "Untersuchungen über thierische Elektricität" by du Bois-Reymond (1841, 1849) that electrophysiology became an independent discipline. This author developed the technique of stimulation (inductorium) and recording using a galvanometer. He demonstrated two fundamental types of bioelectrical potentials — the resting and action potentials. Thus the 60-year-old controversy between Galvani and Volta was finally resolved.

Further knowledge concerning living matter, and especially nerves and muscles, increased with the development of electrophysiological techniques — particularly of recording apparatus. Sensitive mirror galvanometers and the telephone came into use in the 19th century (Wedensky 1883). In 1873 Lippman described the capillary electrometer, which permitted a more detailed study of electrical phenomena in living tissues.

At the beginning of the 20th century electrophysiology began to develop rapidly. Einthoven invented the string galvanometer and used it to record the ECG (1899). Electrophysiology thus moved from the laboratory into the wards and interest in this subject increased.

Further development of electrophysiology was due to the introduction of radiotechniques. The thermionic vacuum valve was invented, making it possible to amplify very small electrical voltages recorded from different tissues.

In 1922 the cathode-ray oscilloscope was introduced by Erlanger and Gasser. This marked the beginning of modern electrophysiology. Between the twenties and forties electrophysiology developed rapidly. Although the oscilloscope was already in use (it reached Europe in 1929 — Rijlant), the string galvanometer made possible the discovery of the electroencephalogram by Pravdicz-Neminski (1925) and Berger (1929, 1930). After Gibbs, Davis and Lennox (1935) discovered epileptic rhythms in the EEG and Walter (1936) applied this technique to localising tumours in the brain, electroencephalography became an independent applied discipline.

During that period electrophysiology was mainly applied to a study of the nervous system and soon became the dominant technique in this field. Muscular contractions as indicators of nervous activity were used less and less. Many physiologists were attracted by the possibility of studying an impulse wave from its origin in receptors to the central nervous system. They improved the methods used and made important discoveries.

Just before World War II, papers appeared indicating the beginning of a new phase in electrophysiology. Hodgkin and Huxley (1939), Curtis and and Cole (1940, 1942) measured membrane potentials with longitudinal micro-electrodes introduced into giant fibres. They found that the action potential measured with the microelectrode was larger than the membrane potential. At the same time the discovery of depolarisation at the myoneural junction (Schaefer and Haas 1939, Eccles and O'Connor 1939) opened up a new area of electrophysiology — synaptic potentials as a specific electrical manifestation of activity at the synapses. In 1940—1942 Schaefer systematically described electrophysiology.

After 1945 the latest phase of electrophysiology appeared. Microelectrode technique was enriched by transmembrane intracellular recording (Graham and Gerard 1946, Ling and Gerard 1949). The discovery of overshoot of the action potential over the membrane potential and thus the fall of Bernstein's hypothesis (1902, 1912; Lillie 1923), resulted in the formulation of a new ionic hypothesis (Hodgkin and Katz 1949; Hodgkin, Huxley and Katz 1949, 1952) and this in turn stimulated further development of intracellular recording techniques and other methods.

The last decade was an era of microelectrodes. Application of intra- and extracellular microelectrodes considerably increased our knowledge of the post-synaptic part of the neuro-neuronal and neuromuscular synaptic contacts. Although the presynaptic part of a synapse can be examined by intracellular microelectrodes only in a few special structures, analysis of the postsynaptic events revealed many significant features of the presynaptic processes The fundamental controversy between the chemical and electrical hypotheses of synaptic transmission, recalling in so many respects the ancient Galvani-Volta discourse, was finally resolved. It also ended in a compromise, after the existence of both chemical and electrical synapses had been convincingly de-monstrated. The discovery of quantal release of mediators underlying the miniature end-plate potentials (Fatt and Katz 1952b) and the detailed de-scription of the ionic mechanisms of the excitatory and inhibitory postsynaptic potentials (Eccles 1953, 1957) moved the investigation of synapses to a molecu-lar level. The perhaps most striking success was attained in resolving one of the great neurophysiological mysteries, the mechanism of inhibition. Shortly after the hyperpolarizing postsynaptic potential was recognized as a sign of postsynaptic inhibition in spinal motoneurons (Brock et al. 1952), another form of inhibition, so called presynaptic inhibition was found to be prevalent at this level (Frank and Fuortes 1957, Eccles, Eccles and Magni 1961, Eccles, Magni and Willis 1962).

The possibility of a direct examination of synapses influenced also the subject matter of general neurophysiology which was used to explain the functions of the inaccessible central synapses by analogies. The models of synapses in peripheral nerve fibres lost their significance and direct investigation of the once poorly defined "central excitatory states" put an end to the hegemony of spike in interpretation of the integrative activity of the nervous system.

The microelectrode technique also considerably enlarged the list of structures available to electrophysiological experimentation. Soon after impalement of the spinal motoneuron excitatory and inhibitory postsynaptic potentials were found in Betz cells of the cerebral cortex (Phillips 1956a,b, 1964). Comparative physiology continuously offers to the electrophysiologists advantageous types of nerve cells and synapses designed by evolution. Such objects helped to explain the functions of those parts of neuron or synapse which are still beyond the possibilities of the present electrophysiological technique in more common objects.

Fundamental progress was also reached in the electrophysiology of the muscle membrane, both resting and active. This knowledge contributed to a better understanding of the impulse transmission from the surface membrane to the intracellular structures (Huxley 1957a) and started thus the research into the mechanisms of intracellular coordination.

The theoretical basis for all above research was the analysis of the resting and action potentials in the giant axons of *Loligo*, which was completed by mathematical formulation of the underlying mechanisms. The mathematical model by Hodgkin and Huxley (1952) was in good very agreement with the experimental findings not only for the giant axons but also for the differently shaped action potentials of other excitable structures. The use of digital computers (Cole et al. 1955, Huxley 1959) was a prerequisite for rapid progress in this field. A great achievement of the ionic theory of nerve impulse was recording of action potentials in axons, the axoplasma of which had been replaced by arteficial washing fluid (Baker, Hodgkin and Shaw 1961).

The recent physiology of the central nervous system can be traced back to two important discoveries, made soon after the EEG technique became available. Activation of EEG during arousal was described by Rheinberger and Jasper in 1937 and at the same time the classical preparations of waking brain — "encéphale isolé" and of sleeping brain — "cerveau isolé" were introduced by Bremer (1935, 1936a,b, 1937). The epochmaking significance of microelectrodes for general neurophysiology was paralleled in brain physiology by the application of the stereotaxic technique for inserting electrodes into the deep brain structures. The extremely fruitful research on the functional organization of brain (Magoun 1958) was opened by the discovery of the desynchronizing effects of subcortical stimulation (Moruzzi and Magoun 1949). Electrophysiology became an important tool in studying even such complex phenomena as sleep, attention, motivation, conditioning etc. Chronically implanted electrodes made it possible to record EEG in freely moving animals in order to study the electrical correlates of behavioral phenomena. Macroelectrodes were soon joined by microelectrodes in brain research. The volume of nerve tissue covered by a single macroelectrode was gradually decreased until activity of one neuron could be reliably recorded by an extracellular microelectrode.

Attempts to process the vast amount of data obtained by simultaneous recording from many electrodes and the problems of detecting low signals from the noisy background caused the recent invasion of computers into the electrophysiological laboratories. This last development was closely connected with the growth of a new science, cybernetics, the birth of which had again been considerably influenced by neurophysiology (Wiener 1948). At present we can only guess what this science about control and communication in animals and machines will return to neurophysiology besides what it has already contributed to electrophysiological techniques (computers) and to neurophysiological theory (application of the information theory to nerve function-Ashby 1956, 1960).

Two characteristic trends can clearly be seen in contemporary electrophysiology: a) a more profound and exact analysis of individual data and b) an increase in the number of simultaneously obtained data. For practical purposes the two trends oppose each other to a certain extent and hence one or the other predominates in any given experimental work. In the future, their further development will require the construction of special apparatus permitting quantitative treatment of the data obtained with the help of statistical theory.

It can be seen that electrophysiology has been primarily employed in the study of the nervous system, and this is true even today. It is most frequently used in research into the following problems:

1) The relationship between stimulus and response in sense organs.

2) The afferent course of peripheral signals, their connections in the brain and reflex efflux.

3) Definition of functional relationships between different brain structures by stimulation and extirpation with simultaneous recording.

4) Correlation of spontaneous and evoked electrical activity in different nerve structures with metabolism.

5) Correlation of spontaneous and evoked electrical activity in nerve structures with the animal's behaviour.

6) The study of the space-time dynamics of physiological and pathological nerve processes with clear cut electrophysiological manifestations (e. g. an epileptic seizure).

Concrete examples for the majority of these applications are given in chapters III—X. The technique as well as the character of the data obtained are described.

The above, of course, does not exhaust by far the fields in which electrophysiological methods may be used. Its more extensive use today is not hindered so much by an isufficiency of suitable apparatus as by a lack of qualified investigators.

Great demands are made on a modern electrophysiologist. In addition to being acquainted with the physiology of the system studied, he must know the principles on which his instruments and apparatus work and the latter's capabilities and limitations. In the past electrophysiologists themselves constructed and designed their own apparatus. Today this is no longer the case. Good electrophysiological laboratories have engineers on their staff or are supplied with suitable apparatus from factories. Yet the investigator must still know his equipment. This is all the more important if fundamental problems are being studied. In such cases the scientist must seek new approaches and must thus be well acquainted with electronics. The words of I. P. Pavlov are still not out of date in this connection: "Science progresses in steps depending upon the success of techniques. Each improvement in technique raises us a stage upward, from which a new horizon is uncovered containing phenomena not known before".

I

Theoretical basis of electrophysiological phenomena

A. Some fundamentals of electrochemistry

In order to understand the theories concerning the nature of bioelectrical potentials it is essential to have at least an elementary knowledge of electrochemistry. Only a short survey of the subject is given here, for detailed information the reader is asked to consult any textbook on physical chemistry (Mac Innes 1939, Höber 1945, Glasstone 1949, Kireyev 1951, Brdička 1952, Monk 1961, Dvořák et al. 1966).

Electrode potential

A potential difference arises between two different metal conductors immersed in a solution. The value of the potential depends upon the nature of the metals and the composition of the solution. Nernst (1908) attributed the existence of this potential to a tendency of the metals to discharge cations into the solution. This "dissolving pressure" is the greater, the less firmly electrons are bound to metal atoms, i. e. the less "noble" the metal. Metal cations of the solution, on the other hand, show a tendency to transmit their positive charge to the electrode. This is the stronger, the higher their concentration and thus their osmotic pressure. Both these processes result in the formation of an electric double layer at the phase boundary between the metal and the solution. The orientation of this double layer depends upon the mutual relations of "dissolving" and osmotic pressures.

If the former preponderates, the electrode is negative to the solution (the electrode loses positive charges to the solution); if the latter predominates, the electrode is positive (the electrode receives positive charges from the solution).

The theoretical value of the potential difference produced under such conditions at a single phase boundary is defined by the relation

$$V = V_0 + \frac{RT}{nF} \ln a \tag{1}$$

where V_0 is a constant for the given electrode, temperature and unit activity of cations.

R is the gas constant (8·314 Voltcoulomb. mol^{-1} degree Kelvin^{-1})

T the absolute temperature ($t + 273°C$)

n the number of elementary charges on one cation

F Faraday constant (96 500 coulombs)

a the activity of cations in the solution (product of molar concentration c and activity coefficient f).

Since in dilute solutions the activity coefficient is nearly one, concentration (c) may replace activity (a) in equation (1).

After changing natural logarithms to Briggsian ones we obtain for V:

$$V = V_0 + 2·303 \frac{RT}{nF} \log a \tag{2}$$

and for $n = 1$, $t = 20°C$

$$V = V_0 + 0·0582 \log a$$

or for $n = 1$, $t = 25°C$

$$V = V_0 + 0·0591 \log a$$

This potential difference cannot be measured directly, since at least two such phase boundaries must be present in a galvanic cell. The hydrogen electrode is used as standard reference in most cases (c. f. page 26).

If the electrode is such that quantitative, but not qualitative changes occur when a current is passing and if those changes are completely reversible on reversing the current, then it belongs to the type of reversible (non- polarisable) electrodes. These are:

1) Metal electrodes immersed in a solution containing cations of the same metal (e. g. Zn in $ZnSO_4$, Cu in $CuSO_4$ etc.).

2) Metal electrodes covered with a layer of a poorly soluble salt of this metal, in a solution containing the anion of that salt (e. g. Ag covered with a layer of AgCl or Hg covered with Hg_2Cl_2 in contact with a solution of some chloride). This latter kind of electrodes is often used in elctrophysiology. A current passing in one direction carries ions from the electrode into the salt layer. If the current is reversed ions accumulate at the electrode. The system thus remains qualitatively unchanged and only small changes in ion concentration occur in the vicinity of the electrode. These do not have a significant effect on the e. m. f. of the cell.

Two electrodes immersed in a solution form a galvanic cell, characterised by potential differences produced between its individual parts.

The concentration cell

If two identical metal electrodes are immersed in a solution of the salt of the same metal, the potential arising at both phase boundaries will be the same so that no potential difference can be registered. If, however, the concentration of the solution is higher at one electrode than at the other, a potential difference E_c arises between them, according to equation (1)

$$E_c = V_1 - V_2 = \left[V_0 + \frac{RT}{nF} \ln c_1 \right] - \left[V_0 + \frac{RT}{nF} \ln c_2 \right] \qquad (3)$$

$$E_c = \frac{RT}{nF} \ln \frac{c_1}{c_2}$$

depending mainly on the relative concentrations of the corresponding cations or, more exactly, on their relative activities

$$E_c = \frac{RT}{nF} \ln \frac{c_1 f_1}{c_2 f_2} \qquad (4)$$

The electrode immersed in the more dilute solution is negative.

Liquid junction potential

Different concentrations at the electrodes may be obtained by separating the two solutions by a porous membrane, an agar bridge or by other means, slowing down equilibration of the concentration gradients but permitting free ion exchange.

If the mobilities of cations and anions in the solution differ, the more mobile ions will diffuse more rapidly into the dilute solution than the less mobile ones. In this way a certain separation of more and less mobile ions occurs at the junction with a corresponding potential difference, the so called diffusion potential or liquid junction potential. Electrostatic forces, of course, prevent any analytically detectable separation of ions with opposite charges.

The size of the diffusion potential E_d at the junction between two concentrations of the same electrolyte is given by the equation

$$E_d = \frac{RT}{nF} \frac{u - v}{u + v} \ln \frac{c_1}{c_2} \qquad (5)$$

where u is the mobility of the cation and v the mobility of the anion. The mobilities of the biologically most important anions and cations are given in Tab. 1. If u is larger than v, the dilute solution (c_2) is positive, since the con-

centrated solution (c_1) passes the positive charges to the dilute solution more rapidly than the negative ones. The electromotive force of a concentration cell is either increased or decreased by the liquid junction potential. Thus, for

TABLE 1

Mobilities of some physiologically important ions at 18°C
(10^{-4} cm². Volt^{-1}. sec^{-1})

Cations	H+ 32·7	K+ 6·7	Na+ 4·5	NH$_4$+ 6·7	½Mg++ 4·7	½Ca++ 5·3
Anions	OH− 18·0	Cl− 6·8	HCO$_3$− 4·6	½SO$_4$−− 7·1		

instance, a potential difference is produced between two hydrogen electrodes (platinum covered with platinum black and partly immersed in a solution containing H ions, partly in an hydrogen atmosphere) immersed in 0·1N and 0·01N-HCl, connected by a liquid junction. This potential consists of:

a) a concentration potential

$$E_c = \frac{RT}{nF} \ln \frac{c_1}{c_2}$$

which at 20°C and the given concentration of both solutions (10 : 1) is equal to $E_c = 0·058$ V (the more dilute solution negative)

b) a diffusion potential

$$E_d = \frac{RT}{nF} \frac{u - v}{u + v} \ln \frac{c_1}{c_2}$$

which after substitution of the corresponding mobilities ($u = 32·7 . 10^{-4}$, $v = 6·75 . 10^{-4}$ cm² . sec^{-1} V^{-1}) at 20°C is equal to

$$E_d = 0·058 \frac{25·9}{39·5} = 0·038 \text{ V}$$

(the more dilute solution positive).

Since in the given case the polarity of the diffusion potential is the opposite to that of the concentration potential, the resulting potential will be determined by the difference between the two,

$$E = E_c - E_d = 0·020 \text{ V}$$

(the more dilute solution is negative).

If, however, calomel electrodes are immersed in these solutions, a different result is obtained. The diffusion potential does not change and (assuming equal activity of H^+ and Cl^- ions) the size of the concentration potential also remains constant. Its polarity, however, is reversed: with calomel electrodes which are in equilibrium with the anions of the system the more dilute solution is positive. As a result of this the polarity of the concentration and diffusion potentials is the same and the resultant potential difference is

$$E = E_c + E_d = 0.096 \text{ V}.$$

Practically, of course, diffusion potentials rather than concentration potentials are measured in biological experiments. That is so because we prefer to use nonpolarisable Ag-AgCl or calomel electrodes for the determination of potential differences. Their electrode vessels are filled with a solution of exactly defined concentration and composition. No potential difference is produced between such two equal electrodes, but diffusion potentials can arise at the junction between the electrode solution and tissue fluid or different solutions applied to the tissue.

Diffusion potentials also occur between two solutions of equal concentration but different composition. If, in the simpler case, we have two equally concentrated solutions with the same anion but different cations (e. g. 0·1 M-NaCl in the electrode vessel and 0·1M-KCl at the treated part of the nerve) then the size of the diffusion potential is given by the equation

$$E_d = \frac{RT}{F} \ln \frac{l_{Na} + l_{Cl}}{l_K + l_{Cl}} \tag{6}$$

or in general

$$E_d = \frac{RT}{F} \ln \frac{l_{k_1} + l_a}{l_{k_2} + l_a} \tag{7}$$

where l_{Na}, l_K, l_{Cl} are the limiting equivalent conductances (cm^2 . Ω^{-1}) of the corresponding ions (the mobilities multiplied by Faraday constant) at 18°C. After substituting we obtain:

$$E_d = 0.058 \log \frac{43.3 + 65.3}{64.5 + 65.3} = 0.058 \log 0.836$$

$$E_d = 0.058 . (0.922 - 1) = - 0.005 \text{ V}$$

(NaCl solution positive).

A more complex relationship occurs with two solutions of different concentrations with a common anion and different cations. The following applies for the diffusion potential

$$E_d = \frac{(l_{k_1}c_1 - l_a c_1) - (l_{k_2}c_2 - l_a c_2)}{(l_{k_1}c_1 + l_a c_1) - (l_{k_2}c_2 + l_a c_2)} \; 0\cdot058 \log \frac{l_{k_1}c_1 + l_a c_1}{l_{k_2}c_2 + l_a c_2} \tag{8}$$

where l_{k_1}, l_{k_2} and l_a are the limiting equivalent conductances and c_1 and c_2 the concentrations of ions on either side of the junction. Thus, for instance, a capillary electrode filled with 3M-KCl gives the following diffusion potential against 0·1M-NaCl:

$$E_d = \frac{43\cdot3 \cdot 0\cdot1 - 65\cdot3 \cdot 0\cdot1 - 64\cdot5 \cdot 3\cdot0 + 65\cdot3 \cdot 3\cdot0}{43\cdot3 \cdot 0\cdot1 + 65\cdot3 \cdot 0\cdot1 - 64\cdot5 \cdot 3\cdot0 - 65\cdot3 \cdot 3\cdot0} \; 0\cdot058$$

$$. \log \frac{43\cdot3 \cdot 0\cdot1 + 65\cdot3 \cdot 0\cdot1}{64\cdot5 \cdot 3\cdot0 + 65\cdot3 \cdot 3\cdot0}$$

$$E_d = \frac{0\cdot2}{-400\cdot2} \; 0\cdot058 \log \frac{10\cdot8}{384\cdot4}$$

$$E_d = (-0\cdot0005) \cdot 0\cdot058 (-1\cdot57)) = 0\cdot00004 \text{ V}$$

(NaCl solution negative).

The low value of the diffusion potential in this case is due to the high KCl concentration, the anions and cations of which have nearly the same mobility. This considerably decreases the numerator and increases the denominator of the fraction in the equation for E_d.

If, instead of KCl, 3M-NaCl is used in the same electrode vessel, the diffusion potential against 0·1M-NaCl is given by equation (5)

$$E_d = \frac{43\cdot3 - 65\cdot3}{43\cdot3 + 65\cdot3} \; 0\cdot058 \log \frac{3\cdot0}{0\cdot1}$$

$$E_d = \frac{-22\cdot0}{108\cdot6} \; 0\cdot058 \cdot 1\cdot477 \doteq 0\cdot0174 \text{ V}$$

(the more dilute solution negative).

When the microelectrode filled with 3M-KCl is moved from 0·1M-NaCl into 0·1M-KCl (this approximately corresponds to passing from extracellular into intracellular space), the diffusion potential changes only slightly from 0·04 mV to 0·53 mV (microelectrode positive).

The diffusion potential of the microelectrode filled with 0·1M-NaCl would under the same circumstances change by 4·4 mV (the microelectrode positive) and that of a microelectrode filled with 3M-NaCl by 0·35 mV (microelectrode positive):

$$E_d = \frac{64\cdot5 \cdot 0\cdot1 - 65\cdot3 \cdot 0\cdot1 - 43\cdot3 \cdot 3\cdot0 + 65\cdot3 \cdot 3\cdot0}{64\cdot5 \cdot 0\cdot1 + 65\cdot3 \; 0\cdot1 - 43\cdot3 \; 3\cdot0 - 54\cdot3 \cdot 3\cdot0}$$

$$0.058 \log \frac{64.5 \cdot 0.1 + 65.3 \cdot 0.1}{43.3 \cdot 3.0 + 65.3 \cdot 3.0}$$

$$E_d = \frac{-0.1 + 66.0}{13.0 - 325.8} \, 0.058 \log \frac{13}{325.8} = (-0.21) \cdot 0.058 \, (-\log 25.1)$$

$$E_d = (-0.21) \quad 0.058 \, (-1.40) = 0.0170 \text{ V}$$

(the more dilute solution negative).

As can be seen, the diffusion potential is directly proportional to the difference in mobilities of the cations and anions of the solution used for filling the electrodes. If different ions are present at the junction, the diffusion potential is determined chiefly by the more concentrated ones. In order to reduce the diffusion potentials occurring between the electrode and the test solution or tissue, 3M-KCl is used for filling the microelectrodes or saturated KCl for filling very stable reference electrodes (e. g. for measuring pH). Agar bridges with 3M-KCl are used for connecting solutions of different concentrations and for reducing the liquid junction potential. When working with nonpolarisable macro-electrodes (with Ag-AgCl or calomel electrodes), diffusion of KCl from the electrode vessel might act on tissue which reacts very sensitively to even small increases in potassium ion concentration in the extracellular space. Such electrodes are preferably filled with solutions similar in composition to extracellular fluid (physiological saline, Ringer solution). Even in such cases, however, large diffusion potentials are usually not produced as long as the composition of the extracellular fluid remains within physiological limits.

The theoretically predicted diffusion potential often considerably differs from the real values found with KCl filled microelectrodes. This is due to a selective reduction of the mobility of some ions (particularly anions) in the microelectrode tip caused by the negative charge of the glass walls and of the protein material which sometimes clogs the microcapillary orifice. Due to a higher mobility of cations the concentrated solution in the microelectrode vessel tends to lose positive charge and becomes, therefore, up to 70 mV negative. Potentials generated by this mechanism are called "tip potentials". Their magnitude can be estimated by comparing the potential of the same electrochemical cells when the two solutions are connected with a semi-microelectrode (tip 20-40μ) and with a microelectrode. Only microelectrodes with a tip potential below 5 mV are suitable for microelectrode work. The tip potential may also be influenced by the composition of the external medium. It is therefore important to check the voltage difference obtained when the microelectrode is immersed in 0.1M-KCl and 0.1M-NaCl (Adrian 1956). For further details see p. 217.

Membrane potential

A diffusion potential occurs at any junction of two solutions of different composition or concentration. Somewhat different conditions prevail if such solutions are separated by a collodion or protein membrane. If the latter's pores are large, only the corresponding diffusion potential is produced. In membranes with smaller pores, the difference in permeability to cations and anions gradually increases. This is due to the fact that protein molecules forming the membrane behave as acids at a normal pH and are therefore charged negatively. For that reason they repel negatively charged anions, which are unable to pass through the pores of the membrane. It is sufficient, however, to decrease the pH of the solution and the amphoteric proteins change their negative charge for a positive one, and the membrane ceases to be permeable for cations and becomes permeable for anions. Membranes selectively permeable to anions even at normal pH can be produced by impregnating collodion membranes with basic dyes or alkaloids.

The potential arising between two concentrations of the same solution separated by a collodion membrane which is nearly impermeable to anions can be calculated from the equation for the diffusion potential

$$E_d = \frac{u - v}{u + v} \, 0{\cdot}058 \log \frac{c_1}{c_2}$$

in which $v = 0$ so that

$$E_m = \frac{u}{u} \, 0{\cdot}058 \log \frac{c_1}{c_2} = + \, 0{\cdot}058 \log \frac{c_1}{c_2}$$

For a membrane selectively permeable to anions in a similar way $u = 0$ and

$$E_m = \frac{-v}{v} \, 0{\cdot}058 \log \frac{c_1}{c_2} = - \, 0{\cdot}058 \log \frac{c_1}{c_2}$$

In both cases the sign of the potential will be that of the more dilute solution. Thus, for instance, in a cell with a negative membrane

$$\text{Hg.Hg}_2\text{Cl}_2.\text{KCl (sat).0}{\cdot}\text{1M-KCl.membrane.0}{\cdot}\text{01M-KCl.KCl (sat).Hg}_2\text{Cl}_2.\text{Hg,}$$

in which we can neglect the diffusion potentials produced between the saturated KCl in the calomel electrodes and the KCl solution on both sides of the membrane, the potential will be near the value 58 mV predicted by equation (9).

If Ag-AgCl electrodes were directly immersed in the solution on both sides of the membrane, we should measure in the cell

$$\text{Ag.AgCl(solid).0}{\cdot}\text{1M-KCl.membrane.0}{\cdot}\text{01M-KCl.AgCl(solid).Ag}$$

not only the membrane potential, but also the concentration potential corresponding to different concentrations of Cl ions at each electrode. This potential, the size of which is determined by equation (3) has the same polarity as the membrane potential (the more dilute solution is positive), so that the resultant potential will approximate 116 mV.

If the collodion membrane separates the same concentrations of two salts with different cations and a common anion, then the potential across the membrane will mainly depend upon relative mobilities of the cations in the membrane. If they remain the same as in aqueous solutions, the membrane potential E_m would be determined by equation (7), in which $l_a = 0$.

$$E_m = 0{\cdot}058 \log \frac{l_{k_1}}{l_{k_2}}$$

For a $0{\cdot}1$M-NaCl and $0{\cdot}1$M-KCl solutions thus would result in an increase in the diffusion potential from 5 mV to

$$E_m = 0{\cdot}058 \log \frac{64{\cdot}5}{43{\cdot}3} = 0{\cdot}058 \log 1{\cdot}49 = 0{\cdot}010 \text{ V}$$

The potentials actually recorded, however, are much greater. This indicates that the mobility of one kind of cations is decreased more than that of another kind. Thus, while in water the proportion of mobilities of K^+ and Na^+ is $1{\cdot}49$, this same proportion in a collodion membrane is $7{\cdot}15$. Conditions are even more complex on membranes surrounded by solutions of different ionic composition and having a different and variable permeability for anions and cations of the system. Such conditions prevail on biological membranes of cells and tissues. Since biological membranes are part of the living system of the cell their characteristics are determined not only by passive physico-chemical processes but also by metabolic activity. So-called active transport of ions (the ability of living tissue to transfer a certain kind of ions against electrochemical gradients) is particularly important for electrogenesis. These problems will be discussed in more detail later.

B. The membrane theory of bioelectric phenomena

At the beginning of this century the first attempts were made to give a quantitative explanation of demarcation potential — potential difference between the intact surface of a muscle or nerve fibre and a cut surface — using physicochemical laws and equations derived by Nernst. Bernstein (1902) following the ideas of du Bois-Reymond (1849), suggested that the source of this potential is not the injured site but the undisturbed surface of the tissue. The potential difference is produced on a hypothetical membrane between the external medium and the inside of the fibre. The electrode placed onto the injured site is actually in contact with the inside of the fibre and thus with the inner side of the membrane. The electrical potential is due to a tendency of potassium ions, accumulated within the fibre, to leave the latter through the membrane, which is selectively permeable to K^+. Quantitatively the relation between the potential and the potassium concentration in the external medium (K_0) and the internal medium (K_i) is given by Nernst's relation.

Overton (1902) made another important discovery nearly at the same time. He found that the muscle loses its excitability in solutions devoid of Na ions, and put forward the hypothesis according to which muscular activity is accompanied by an exchange of intracellular K^+ for extracellular Na^+. Despite the fact that much study has been devoted to these problems, only technical progress of the last 25 years (microelectrode and micro-injection techniques, radiactive tracers, chemical micro-analysis etc.) and the choice of suitable experimental objects, especially the giant fibres of the squid and *Loligo* have made further advance possible.

The first experiments with intracellular electrodes yielded unexpected results. According to Bernstein's membrane theory the action potential was considered to be an expression of membrane depolarisation. It was shown experimentally, however, that more than depolarisation occurs during the action potential, since the polarity of the membrane is not destroyed but reversed. It was therefore necessary to supplement and alter the original hypotheses. It is mainly to Hodgkin, Huxley and Katz that we owe the elaboration of the modern membrane theory (Hodgkin 1951, 1958).

The axoplasma of nerve or muscle fibres contains K^+ as the main cation. Organic acids, chiefly amino acids, are the main anions. The surface of the fibres is formed by a membrane $50-100$ Å thick (according to data obtained by electron microscopy) with a specific transverse resistance $1000 \ \Omega.cm^2$ in the squid (Cole and Hodgkin 1939) and a capacitance of about $1 \ \mu F.cm^{-2}$ (Curtis and Cole 1938). This membrane forms a diffusion barrier between the interior and exterior. As it is practically impermeable to proteins, amino acids and Na-ions, a Donnan equilibrium is formed across it. Less Cl^- will be present on the side of the protein anions. For the majority of excitable tissues the following ratio was found:

$$\frac{[K_i]}{[K_o]} \doteq \frac{[Cl_o]}{[Cl_i]} = 20 \text{ to } 50$$

On the basis of this distribution, the equilibrium potential produced on a membrane selectively permeable to K^+ can be calculated from the Nernst relation

$$E_{\mathrm{K}} = 0.058 \log \frac{[\mathrm{K_i}]}{[\mathrm{K_o}]} = 0.070 \div 0.099 \text{ V} \tag{1}$$

the inside of the cell being negative.

This potential was actually measured with a microelectrode introduced into the inside of a nerve or muscle fibre. Its value changes according to equation (1) with changing external potassium concentrations. When decreasing $\mathrm{K_o}$ below the physiological range, however, the membrane potential does not increase beyond a certain level.

Sodium has a completely different distribution. It is the main extracellular cation and occurs in only small amounts intracellularly (usually $\mathrm{Na_i} : \mathrm{Na_o} = = 1 : 10$). Such a distribution can be explained by an active metabolic process removing $\mathrm{Na^+}$ from the cell against a concentration gradient. The distribution of $\mathrm{Na^+}$ corresponds to an equilibrium potential

$$E_{\mathrm{Na}} = 0.058 \log \frac{[\mathrm{Na_i}]}{[\mathrm{Na_o}]} = -0.058 \text{ V} \tag{2}$$

The negative sign shows that the inside of the cell in this case is positive. This potential cannot be normally observed because of the low permeability of the resting membrane for sodium. Different conditions, however, occur when a nerve impulse passes. The permeability of the active membrane for $\mathrm{Na^+}$ increases about 500 times so that it is now several times larger than that for $\mathrm{K^+}$. This results in the production of a new equilibrium corresponding to E_{Na} with a sign opposite to that of the resting membrane potential. For that reason the action potential is higher than the membrane potential. It is actually the sum of E_{K} and E_{Na}. This assumption was confirmed by experiments showing that the amplitude of the action potential depends on the extracellular concentration of $\mathrm{Na^+}$ roughly according to equation (2).

The voltage clamp method permitted a more detailed analysis of ionic currents in the membrane (Hodgkin, Huxley and Katz 1952). This method utilizes sudden displacement of the membrane potential by an imposed potential difference put between a microelectrode introduced longitudinally into the inside of the axon and the external solution. If the displaced membrane potential is held at a fixed value by a feed-back amplifier, changes in membrane conductivity can be studied by recording the current flowing through the membrane under these conditions. It was shown that depolarization of the membrane by 15 to 100 mV at first evokes an inward current, flowing into the axon, although the electric pulse resulted in increased interior positivity and should therefore initiate an outwardly directed current. Within 0.5 to 1 msec. this inward current is changed to an outward current, which attains a constant value, pursuing a sigmoid curve. This second component corresponds to a potassium current carried by K^+ efflux from the depolarised fibre while the first component might be explained only by the assumption that the depolarised membrane changes its permeability to sodium. This explanation is supported by the fact that equilibration of the electrical (depolarisation by $E_K + E_{Na}$) or chemical ($Na_i = Na_o$) sodium gradient between the interior and exterior of the axon causes the initial current component to disappear.

The use of the same depolarisation in a normal Ringer solution and immediately afterwards in Ringer solution in which 90% of sodium has been replaced by choline makes it possible to obtain a summation curve $I_{Na} + I_K$ in the first case, a curve for I_K alone in the second case and by subtraction the isolated I_{Na} curve. The sodium current rapidly reaches a maximum but then immediately begins to drop towards zero despite the continued depolarisation.

On the basis of these data it is now possible to describe the events occurring in a nerve fibre during a nerve impulse. Currents flowing through the interior of the axon in one direction and through the external fluid in the reverse direction form local circuits between the resting and active regions of the nerve fibre. These currents displace the resting potential of the membrane just ahead of the impulse. When a certain degree of depolarisation is reached (20 mV), the permeability of the membrane for Na^+ increases. Due to the entry of Na^+, depolarisation is enhanced until, at the peak of the action potential, an equilibrium potential for sodium is attained. This prevents further increase of Na^+ influx and the sodium permeability of the membrane begins to decline. At the same time permeability for K^+ rises. Increased K^+ efflux returns the membrane potential to the original value and accelerates the decrease in sodium permeability. The raised potassium permeability persists for a further few msec. For the period corresponding to the refractory period, the mechanism increasing sodium permeability is blocked. Only when the membrane is again capable of increasing its permeability for Na^+ can a new action potential be conducted. During impulse conduction the axoplasm has gained an insignificant amount of Na^+ and has lost about the same amount of K^+.

The movements of potassium and sodium during impulse activity can be expressed quantitatively on the basis of theoretical considerations and direct measurements. If we assume that sodium enters the inside of the axon only up to the time when depolarisation is completed, then according to Keynes and Lewis (1951), the minimum necessary amount of Na^+ can be calculated from the capacity of the membrane ($C = 1.5 \, \mu F . cm^{-2}$ for *Loligo*) and the amplitude of the action potential ($V = 100 \, mV$). The necessary quantitity of electricity is thus

$$C . V = 1.5 \, \mu F . cm^{-2} . 100 \, mV = 150 \, m\mu coulomb . cm^{-2}$$

which may be transferred to the interior of the axon by $(C . V)/F$ gramequivalents of Na$^+$. This is $(150 . 10^{-9})/96500 = 1.5 . 10^{-12}$ grameq.cm^{-2}.imp^{-1}

Experimentally (using accumulation of Na^{24} in the axon, K^{42} efflux from a previously saturated axon and activation analysis) a sodium influx of $3.8 . 10^{-12}$ was found, i. e. about $2-3$ times more than the theoretical assumption. This can be explained if increased Na$^+$ influx outlasts the crest of the action potential. The same quantity of electricity is necessary for repolarisation of the membrane. This is now supplied by a K$^+$ current. Experimentally values of $3.6 . 10^{-12}$ grameq.cm^{-2}.imp^{-1} were found.

Despite general agreement between theoretically assumed and experimentally determined values of membrane and action potentials, there are some cases in which the calculated and determined values differ considerably. Thus, for instance, the resting membrane potential does not follow equation (1) if K$_0$ falls below 3.5 meq./l. This is due to the fact that the selective permeability of the membrane for K and Na is only relative. Goldman's equation takes relative permeabilities into account

$$E_m = 58 \log \frac{P_K [K_i] + P_{Na} [Na_i] + P_{Cl} [Cl_o]}{P_K [K_o] + P_{Na} [Na_o] + P_{Cl} [Cl_o]} \tag{3}$$

For the resting membrane $P_K : P_{Na} : P_{Cl} = 1 : 0.04 : 0.45$.
For the active membrane $P_K : P_{Na} : P_{Cl} = 1 : 20 : 0.45$.

It follows from equation (3) that after removal of potassium from the external medium the value E_m is determined by further members of the denominator, especially by the value $P_{Na} [Na_o]$. It appears that an estimation of the relative permeabilities $P_K : P_{Na}$ ought rather to be $1 : 0.01$ for the resting and $1 : 30$ for the active membrane.

Another cause of the discrepancy between theoretical and measured values is the unstable state of isolated excitable structures, which lose K$^+$ and gain Na$^+$ apparently due to injury during dissection. Under such conditions, resting membrane potentials of only about 50 or 60 mV were recorded instead of 90 mV. Equations (1), (2) and (3) characterise steady state systems in which the efflux (m_o) of a certain ion is compensated by an equally strong influx (m_i). They can therefore only be applied if $m_i/m_o = 1$.

The relation between influx and outflux, internal and external concentrations and the steady state potential for a certain ion is given by the equation derived by Ussing (1949):

$$\frac{m_i}{m_o} = \frac{c_o}{c_i} e^{EF/RT} \tag{4}$$

35

or after converting to logarithms

$$\ln \frac{m_i}{m_o} = \ln \frac{c_o}{c_i} + EF/RT \tag{5}$$

If we substitute for E the membrane potential calculated according to equation (1)

$$E = \frac{RT}{F} \ln \frac{c_i}{c_o}$$

we obtain

$$\ln \frac{m_i}{m_o} = 0$$

If, however, the value of E in equation (5) differs from the theoretical value, $\frac{m_i}{m_o}$ is not equal to 1. Thus in the nerve fibre of *Loligo*, $\frac{m_i}{m_o} = 0.35$ was obtained for $E = 62$ mV, $K_i = 300$ mM. Experimentally, using labelled potassium, it was found in the same fibre

$m_i = 10.6 \cdot 10^{-12}$ mol cm^{-2} sec^{-1}

$m_o = 27.5 \cdot 10^{-12}$ mol cm^{-2} sec^{-1} and thus

$\frac{m_i}{m_o} = 0.39$, which agrees well with the calculated ratio.

The recovery mechanism maintaining a stable internal medium in nerve and muscle fibres by removing entering Na$^+$ and reabsorbing outflowing K$^+$, particularly during and after intensive impulse activity, is a necessary supplement to passive ionic movements. During the action potential ions move along existing concentration gradients, which provide the immediate source of energy for conduction. The recovery process, on the other hand, requires movements of ions against electrochemical gradients, making participation of metabolic energy necessary. A number of poisons affecting cellular metabolism (2,4-dinitrophenol, NaCN, Na$_3$N) retard the excretion of Na^{24} from stimulated nerve previously loaded with Na^{24}. The membrane and action potentials do not change. Chemical analysis has shown a close relationship between the mechanism excreting sodium, the so called sodium pump, and high energy phosphate compounds (ATP, arginine phosphate, creatine phosphate etc.). The movement of K$^+$ across the membrane probably depends on the action of the sodium pump not only passively (removal of one cation from inside the cell must be followed by entry of another in order to fulfil the conditions of electroneutrality in the internal fluid and of osmotic balance between internal and external fluid) but also actively. This is demonstrated by the fact (Hodgkin and Keynes 1955) that removal of K^+ from the external medium decreases the Na-outflux. According to some theories the membrane contains a carrier that combines with Na^+ on the inner side of the membrane to form a neutral molecule X. This passes through the membrane and at its outer surface Na^+ is released and K^+ accepted, thus changing the carrier into a neutral molecule Y, which diffuses back through the membrane to release K^+ inside the cell.

A new important advance in this field was the perfusion of the isolated giant axon with isotonic solutions (Baker et al. 1961, 1962 a,b, Oikawa et al. 1961, Tasaki et al. 1962). This makes it possible to achieve within a few seconds any desirable changes in the ionic composition of the medium on both sides of the excitable membrane and to study resultant changes in resting and action potentials. Results confirmed the assumption that the axoplasm has no decisive significance for the conduction of impulses since normal action potentials can be recorded even after substituting isotonic K_2SO_4 for 95% of the axoplasmatic volume. Equation (3) is valid also for changes in K_i and Na_i. The resting potential approaches zero at

$K_i = K_o$ and is inversed (inside of axon is positive) when
$K_i < K_o$.

With increasing Na_i the action potential decreases and eventually disappears when the membrane is depolarized to below -30 mV, probably because of inactivation of the "Na carrier" system. Principally new results are obtained with perfusion of the axon with solutions having a low electrolyte content. Thus if sea water without potassium is in the external environment of the axon and isotonic saccharose with 6 mM-NaCl inside, the resting potential is zero. Stimulation of the axon, however, evokes an action potential of normal amplitude and very long duration (plateau about 1 sec), evidently because the inactivation of the Na carrier is slowed down in the absence of potassium ions.

It remains, of course, an open question how far results obtained with giant fibres of the squid may be applied to other excitable structures, especially nerve cells of the mammalian brain. It has been possible in recent years to demonstrate convincingly that the fundamental characteristics of the membrane of the neurone soma and that of the giant fibre are similar. This was made possible by the use of intracellular recording of potentials in motoneurones of spinal cord (Brock et al., 1952; see also Eccles 1957) and other nerve cells (pyramidal cells of the cerebral cortex — Phillips 1956a, b, Renshaw cells — Frank and Fuortes 1956a, sympathetic ganglion cells — Eccles R. M. 1955, hippocampal pyramids — Kandel et al. 1961; for review of literature see Eccles 1964a).

According to Eccles (1957), the average capacity of the motoneurone membrane is $3 \cdot 10^{-9}$ F, its potential -70 mV and its resistance 800 KΩ. The membrane potential corresponds to the equilibrium potential of chloride ions, but is somewhat lower than would correspond to the electrochemical equilibrium for K^+ (-90 mV). Depolarisation of the membrane by 10 to 30 mV evokes a selfregenerative process of increased sodium permeability. Thus the equilibrium potential for Na^+ is approached and a spike attaining 80 to 110 mV produced. Membrane potential changes are initiated at the synapses — points of contact between the axon of one neurone and the cell body or dendrites of the other. The synaptic change in membrane potential is usually not due to direct electrical action of presynaptic activity. This follows chiefly from the fact that the postsynaptic potential starts at a time when the pre-synaptic spike has nearly dissappeared. It has been demonstrated that a number of synapses cannot be stimulated electrically (Grundfest 1957b). It seems that as a result of the arrival of an impulse a transmitter substance is released which diffuses through the synaptic space and characteristically affects the pro-perties of the subsynaptic membrane (Eccles 1957, 1964a).

Synaptic changes of the membrane potential are of two kinds: excitatory (EPSP-excitatory postsynaptic potential) and inhibitory (IPSP-inhibitory postsynaptic potential). While the EPSP is similar, to a certain extent, to local changes in a nerve fibre and is characterised by depolarisation of the membrane, the IPSP manifests itself by hyperpolarisation of the membrane or by an increase in the latter's resistance to depolarising effects. The EPSP is due to increased membrane permeability to all ions, the IPSP, on the other hand, to increased membrane permeability to K^+ and Cl^- with the low Na^+ permeability preserved.

Since a number of synapses may be active on one neurone, the resultant reaction is determined by complex mutual interaction. The contrary changes in membrane potential produced by IPSP and EPSP summate algebraically and thus may cancel each other under certain conditions. Excitatory synaptic effects sufficient to evoke a spike become ineffective if they coincide in time with inhibitory effects — inhibition occurs.

Even though the above picture of chemical synaptic mechanisms is in agreement with numerous results of experiments performed in many labora-tories it is not quite universally valid. An alternative hypothesis of electric interaction between pre- and postsynaptic elements has been demonstrated experimentally for some special synapses of invertebrates (Furshpan and Potter 1959a, b, Watanabe and Grundfest 1961 — see also p. 436) and also verte-brates (Mauthner's cells in fish — Furukawa and Furshpan 1963). Electric synaptic transmission is realized particularly in those cases in which most of the current generated by the presynaptic element passes through the post-synaptic membrane (i. e. the synaptic cleft is absent and the resistance of the pre- and subsynaptic membrane is low). Electric synapses are evidently developmentally the older type of mutual connection of nerve cells. Their possible significance in the nervous system of higher organisms is being studied.

Another important contribution to the physiology of synapses was the discovery of axo-axonal contacts (Gray 1962) mediating particularly so called presynaptic inhibition (Frank and Fuortes 1957, Eccles 1964b). This type of inhibition manifests itself by a decrease in EPSP which occurs without any change in the characteristics of the post-synaptic membrane and is thus evidently due to a decrease in the excitatory effect of the presynaptic signal. A detailed analysis of this phenomenon by Eccles et al. showed that in spinal motoneurons the decrease in EPSP is related to depolarization of presynaptic nerve fibres (primary afferent depolarization — PAD). Depolarization is the result of activation of interneurons, the axons of which form synaptic contacts with the terminal branches of afferent fibres. Here the mediator is released which increases the permeability of the membrane of the afferent terminals and thus induces a fall in their resting potential by about 30 mV. The relatively small decrease in the amplitude of the afferent action potentials in the depolarized region leads to a considerable fall in the amount of mediator released at the corresponding axodendritic or axosomatic synapses and thus also to a decrease in EPSP. Presynaptic inhibition and probably also presynaptic facilitation must thus be considered as a further important mutual connection of nervous elements in all parts of the brain.

The membrane theory thus gives a satisfactory explanation of processes occurring in excitable tissues and makes even possible their quantitative description. Other hypotheses have been proposed, however, to explain the nerve impulse by changes occurring in the whole axoplasma, especially in its protein structures. The founder of this theory (so called alteration theory) was L. Hermann (1879), its modern followers were Nasonov and Aleksandrov (1940), Segal (1956), Heilbrunn (1956), Ungar (1957) and Troshin (1956).

Although the alteration theory is of merely historical significance now, problems of the protein metabolism of nerve cells continue to attract attention of scientists in connection with the synthesis of mediators (Nachmansohn 1961), the degeneration and regeneration of nerve fibers (Gutmann 1962) and plastic changes of the nervous networks in learning and memory (Hydén 1959).

II

Electrophysiological apparatus and technique

In contrast to many other biological disciplines, electrophysiology requires a fairly good knowledge of several branches of physics, particularly electricity (electrotechniques and electronics), both theoretical and practical.

This second chapter is in no way an exhaustive account, and is only intended to inform the reader about those parts of physics and electronics that are important for electrophysiology.

Instruction for the design and construction of apparatus, although important didactically, are not given here.

The following literature may be recommended to electrophysiologists: Any good textbook of physics for university students, books giving instructions for construction and calculations of the most important circuits and their components, such as the books by Terman (1943) translated and extended by Smirenin (1950) or by Rindt and Kretzer (1954—57). The following are important references for constructing amplifiers: Bayda and Semenkovich (1953), Valley and Wallman (1948), Haapanen (1952), Haapanen et al. (1952), Kozhevnikov (1956), Kaminir (1956), Nüsslein (1952), Korn and Korn (1955), Johnston (1947), Grundfest (1950). The theory and practice of electrophysiological techniques is described in detail in the work of Whitfield (1959), Donaldson (1958), Nastuk (1963, 1964), Dickinson (1950), Gulyajev and Zhukov (1948), Mikhaylov (1958), Vodolazsky (1952).

As already stated, the fundamental methods of electrophysiological research are the following: stimulation of tissues and registration of potential differences between different points in tissue, cells or fibres. Time changes of those potential differences, occurring usually in response to electric stimulation, but also, as the result of natural stimuli, or metabolic processes or even without any evident stimulus, are recorded. In general, the experimental system — electrical stimulation — tissue — reccording of potential changes — is the most common to day, but by no means the only one possible. Technically simpler and older methods of stimulation (electrical or other) and observation of mechanical (or other non-electrical) responses of tissues or animals are still important. Beginners are advised to follow the well tested path from the more simple to the more complex, especially if their practical experience with electro-technical apparatus is limited.

A. Stimulation technique

The purpose of stimulation is generally to produce a change in vital tissue processes, e. g. local or spreading excitation (an impulse) in excitable tissue. Usually one of the important requirements is the possibility of using a quantitatively variable stimulus. The size of the stimulus is defined either absolutely, so that results obtained from completely different experiments may be compared, or relatively, so that only results obtained under the same conditions with the same apparatus and in the same tissue can be compared. If only electrical stimulation is taken into account, the following points must be considered in order to evaluate the possibility of quantitative determination of stimulus parameters: 1) which physical parameter (whether voltage, current etc.) determines the extent of the effect of the stimulus on the tissue, 2) how does this value, which is usually only measured outside the tissue, change within the tissue using various kinds of stimulation.

a) The physical effect of the electrical stimulus on the tissue

Ions and molecules with a large dipole moment are the most important components of living matter subject to the action of electricity. The change caused by an electrical stimulus of medium duration would then be a change in ion concentration at a certain point (as the result of different ion mobilities) and/or a change in structure due to changed orientation of linked dipoles.

In both cases the action of an electrical stimulus is determined not only by duration, but also by the intensity of the field produced by the stimulus in the tissue.

In this section only the action of the electrical stimulus on ionic movements will be considered. This evidently is more important and simpler. It must be kept in mind, however, that difficulties encountered when defining the stimulus effect on ionic movements in tissues are even greater for dipole movements.

The production of an impulse in an isolated nerve fibre lying on metal electrodes in a non-conductive medium (paraffin oil) will be considered first (Fig. 1).

41

From a purely physical aspect the field intensity may be defined (and this is better for clarity's sake) as the potential difference per unit length $E = V/s$. This means that the intensity of the field will be the greater the larger the potential difference between the electrodes and the smaller the distance between them. This is true only if the medium is homogeneous and the electrodes are large. In non-homogeneous, geometrically complex media, with electrodes

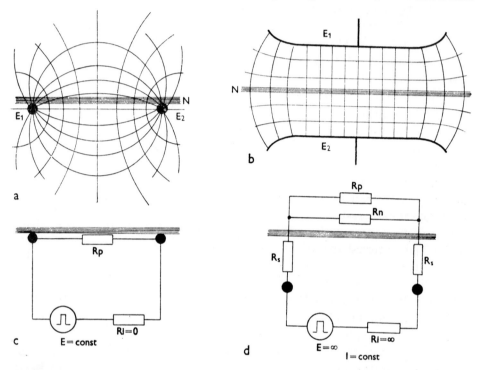

Fig. 1. Electrical stimulation of a nerve. 1a, b: electric field between stimulating electrodes of small dimensions (a) and between large flat electrodes (b). *N*-nerve. E_1, E_2—electrodes. c: stimulation with constant voltage. d: stimulation with constant current.

of small dimensions, the intensity of the field must be defined according to the equation $E = dv/ds$, which states that the field intensity is given by the space increase in potential between two points unlimited close to each other.

Our objective is 1) to find such an arrangement of electrodes and fibre that the field intensity at the stimulated point is least dependent on small changes in the geometric arrangement (especially on changing the position of the nerve fibre with respect to the electrodes), 2) to have an intensity of the field at the stimulated point that is least dependent on tissue and medium resistance and 3) to define exactly the point at which the impulse originates and to keep this point constant.

The first condition is fulfilled most satisfactorily with large plane electrodes situated as far as possible from one another and parallel to the fibre, which is situated half way between the electrodes (Fig. 1b). This arrangement, however, complies poorly with the third condition, since the field intensity will be nearly constant over a relatively large area of the fibre and thus the point of origin of the impulse will be more or less left to chance. The third condition is fulfilled the more exactly the smaller the electrodes (or the smaller the different electrode), the closer the electrodes are to each other and, most important, the closer they are to the nerve fibre (or the closer the different electrode is to the fibre) (Fig. 1a). The second condition can be complied with only in two special cases:

a) The variable leakage resistance (R_p) in parallel with the stimulated element (which need not have a constant resistance) is connected directly to the electrodes. The independent field intensity can be fulfilled by using a supply with a zero internal resistance (R_i), or, as it is usually called, a constant voltage output (Fig. 1c).

b) The variable leakage resistances (R_s) are connected in series with the stimulated element. The resistance of the latter (R_n) or that including a further parasitic resistance (R_p) connected in parallel, is constant (Fig. 1d). The above condition can then be fulfilled using a high electromotive force with a high internal resistance, so that the current does not depend upon the external resistance. This is usually called constant current output.

Even this very rough analysis shows that the three above conditions cannot be fulfilled at the same time. Under certain circumstances it is possible to ensure approximately that the actual effective stimulus is proportional to the stimulus applied at the electrodes. It is, however quite impossible to ensure identity of the effective stimulus for different tissues, not even with the same apparatus and equipment, and it cannot be excluded, even with the same kind of preparation, that in different experiments the actually effective stimuli are very different from the applied stimuli of the same characteristics. These difficulties may be overcome to a great extent using intracellular stimulation.

b) The physical characteristics of the stimulus

As explained above, the effectiveness of a stimulus is given by transfer of ions and changes in their concentration and hence it is determined not only by the intensity of the field but also by its time course, which may be, however, infinitely variable. If only periodic stimuli are considered (single stimuli being a special case) the harmonic stimulus is determined by three physical parameters: the amplitude (of voltage or current), the frequency and the duration (or number of periods). The so called rectangular pulse is determined by four parameters: the amplitude, the width of the pulse, the repetition frequency and the duration of the pulse train (or the number of pulses). The amplitude can, of course, be either positive or negative. If these stimuli have a D. C. component, a further parameter is added. More complex stimuli are determined by a correspondingly larger number of parameters so that it is often impossible to characterise them exactly.

In electrophysiological experiments it is often important to isolate one variable factor while maintaining the other factors constant, so that complex relationships between stimulus and response may be understood. If different stimuli are considered from this aspect, the most essential question is how far it is possible to change only one parameter, i. e. how far a change in one parameter results in an alteration of another parameter or of other functions determining the effectiveness of the stimulus.

With a harmonic stimulus, a change in amplitude also changes the slope of the stimulus, i. e. the rate at which the voltage increases and decreases at the electrodes. A change in frequency also means a change in the slope of the stimulus. In addition, the effect of the stimulus may depend to a large extent on the phase of the wave at which the stimulus is switched on or off.

Rectangular pulse (ideal). The slope of front and trailing edges is infinite and thus not dependent upon the amplitude and frequency. Even though the slope cannot be infinite in reality, it is very large, nevertheless, so that its change with the amplitude is of little significance and with changing frequency the slope usually changes hardly. Consequently rectangular stimuli are now preferred to harmonic or more complex or undefinable stimuli.

It must further be pointed out that, under certain conditions, unpredicted effects may occur when a stimulus is applied, which are not considered when interpreting the results, and may lead to incorrect conclusions. These may be heat effects, electrolysis of the medium, entry of the electrode metals into the tissue (particularly copper, but other metals as well), and polarisation of the electrodes (Rohlíček 1964).

It must further be borne in mind that the stimulus may also act at a point other than that assumed, e. g. at the earthing plate, at different clamps etc. (see p. 196). In general the danger is the greater the greater the slope of the front and trailing edges of the pulse and the greater the amplitude of the stimulus.

Finally it must be remembered that the shape of the stimulus at the excitable elements themselves may considerably differ from the shape which can be registered at the stimulating electrodes. Distortion may be smaller when using a low resistance output. In general it is the larger the closer the shape of the stimulus approaches the ideal rectangular pulse and the shorter its duration.

c) The main requirements for a stimulator

Practical requirements for a good stimulator are thus as follows: a well defined shape of the stimulus, independent of other stimulus parameters (amplitude, duration etc.); a low internal impedance; easy, rapid and sufficiently fine control of individual stimulus parameters over a wide range. In addition to the above fundamental requirements there are others, that are of prime importance when recording responses: insulation (resistive and capacitive) of the stimulator output from earth and also from other parts of the apparatus, the possibility of synchronising the stimulus with the time base of the oscilloscope, with the time marker, with another stimulator etc., as will be discussed later on (see p. 174 and 184).

Different kinds of stimulators possess these features to different extents, and in choosing a stimulator for a definite purpose, the important stimulus parameters must be taken into consideration. Different kinds of stimuli will be described below, together with their electrophysiologically important characteristics.

The stimulator thus produces at the electrodes a potential difference with an exactly defined shape. The potential may either be determined in absolute units or may be automatically adapted to tissue resistance between the electrodes in such a way that the current passing between them is exactly defined.

d) Kinds of stimuli

Some physical aspects of the stimulus parameters will be considered now, together with their relation to the electric circuits of the stimulator and the tissue.

1. Periodic voltage — Harmonic voltage

The most simple type is the harmonic or sine-wave stimulus. Here the potential e on the electrodes is defined by equation

$$e = E_0 \sin \omega t \quad \text{where} \quad \omega = \frac{2\pi}{T} = 2\pi f$$

E_0 is the maximum potential value, ω is the angular frequency, T the period (repetition interval), f the frequency (number of periods per sec).

The current passing between the electrodes is defined by equation

$$i = I_0 \sin (\omega t + \varphi) = I_0 \sin \omega (t + t_0)$$

where I_0 is the maximal current, φ is the phase shift between potential and current, t_0 is the time shift between potential and current. The following relation is valid between constants φ and t_0

$$\varphi = \omega t_0 = 2\pi f t_0 = 2\pi \frac{t_0}{T}$$

It is evident, especially from the last equation, that φ defines the relative and t_0 the absolute shift between potential and current.

The potential and current are generally not in phase (i. e. they are shifted) because during the passage of current through some conductors (coils, condensers, electrolytes) new potentials arise which are added to the original

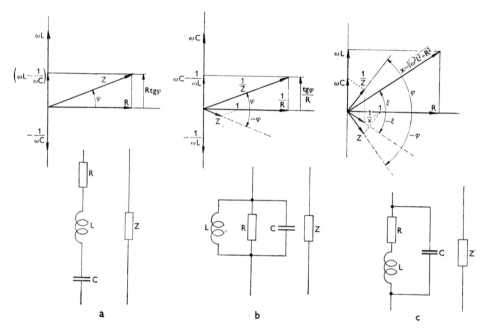

Fig. 2. Graphical determination of impedance in complicated circuits. Bottom: diagram of the circuit. Top: the corresponding vector diagram. The inverse values of some vectors necessary for further calculation are constructed in such a way that a parallel line is drawn to the given vector passing through point $(+1;0)$. The point of intersection of this line with the side of the negative angle of the original vector is the end point of the required inverse value.

potential e. During passage through a condenser the current leads the voltage by 90^0, or voltage lags behind the current by 90^0. When passing through an ideal coil, on the other hand, the current lags behind the voltage by 90^0. In a resistance the current and potential remain in phase. In more complex circuits containing resistances, capacitances and inductances the current may be shifted from $+ 90^0$ to $- 90^0$ in respect to the voltage according to which component is more important. Their relative significance is best analysed using a vector diagram as shown in Fig. 2. The ohmic resistance does not depend on f (at least for low frequencies); thec oil and condenser form "resistances" dependent on the frequency, so called reactances; inductive reactance $X_L = \omega L$, capacitive reactance $X_C = 1/\omega C$; L is the inductance of the coil, in henrys (H), $C =$ capacity of the condenser in farads (F). In networks containing the components R, L and C in series, the total impedance of the circuit and the phase shift are

$$Z = \left[R^2 + \left(\omega L - \frac{1}{\omega C} \right)^2 \right]^{\frac{1}{2}} \quad \text{and} \quad \tan \varphi = \frac{\omega L - (1/\omega C)}{R}$$

For R, L and C in parallel the admittance is

$$\frac{1}{Z} = \left[\frac{1}{R^2} + \left(\frac{1}{\omega L} - \omega C \right)^2 \right]^{\frac{1}{2}} \quad \text{and} \quad \tan \varphi = \frac{- (1/\omega L) + \omega C}{1/R} =$$

$$= R \left(\omega C - \frac{1}{\omega L} \right)$$

One of the more complex cases (resistance and inductance in series, capacitance in parallel) is shown and analysed in Fig. 2c. Other complex examples may be solved in an analogous way. Further details must be sought in electrotechnical textbooks.

The power of alternating current

The instantaneous power of a sinusoid current is

$$n = e.\, i. = E_0 I_0 \sin \omega t \sin (\omega t + \varphi)$$

It is evident that during one period T, n is zero four times (twice when $t = 0$, twice when $\omega t + \varphi = 0$) and attains extreme values also four times. The average power of the sinusoid current is then:

$$P = \frac{1}{T} \int_0^T ei \, dt = \frac{1}{T} \int_0^T E_0 I_0 \sin \omega t \sin (\omega t + \varphi) \, dt$$

$$P = \frac{1}{2} E_0 I_0 \cos \varphi$$

47

In order to obtain formal agreement of this equation with the analogous equation for the power of D. C. current (which is $P = EI$ or $P = E^2/R$ or $P = I^2R$) we introduce

$$E_{eff} = \frac{E_0}{\sqrt{2}} \quad \text{and} \quad I_{eff} = \frac{I_0}{\sqrt{2}} \text{ and obtain}$$

$$P = E_{eff}I_{eff} \cos \varphi = \frac{E_{eff}^2}{Z} \cos \varphi = I_{eff}^2 Z \cos \varphi$$

E_{eff} and I_{eff} are the r. m. s. (root mean square) values of current and voltage and after substituting for $\sqrt{2}$

$$E_{eff} = 0.707E_0 \text{ and } E_0 = 1.414E_{eff} \text{ and similarly for } I_{eff} \text{ and } I_0.$$

Mean current. As shown, the r. m. s. values are defined on the basis of power. Values of alternating current (or voltage) may, however be defined also in another way as integrated current (or voltage) corresponding to the area limited by the time axis and the curve of the current (or voltage).

$$I_m = 2\frac{I_0}{T} \int_0^{T/2} \sin \omega t \, dt = \frac{2}{\pi} I_0$$

$$E_m = 2\frac{E_0}{T} \int_0^{T/2} \sin \omega t \, dt = \frac{2}{\pi} E_0$$

These values are termed mean values and thus

$$E_m = 0.636E_0 \quad \text{and}$$

$$E_0 = 1.572E_m.$$

In electronics and physics E_0 and I_0 are usually used. In engineering and usual calculations r. m. s. values are used. Mean values are applied only rarely, especially when determining the quantity of electricity (e. g. in integrators). In this case $Q = I_m t$ (if $t = nT$).

The sensitivity of different measuring apparatus must also be considered.

Measuring A. C. voltage and current

a) Apparatus with a rotating coil measure the D. C. component only.

b) If a rectifier is added, the apparatus measures mean current values, but as an A. C. apparatus it is usually calibrated in r. m. s. values. Such apparatuses show r. m. s. values although their deviation is due to mean current (they automatically recalculate mean current to r. m. s. current). If currents and potentials of non-sinusoidal shape are measured, such instruments do not show correct values. They can then be only used for measuring mean currents, which must, however, be calculated from the figures on the scale according to

$$I_m = 0.900 I_{eff}$$

where I_m is the actual mean current and I_{eff} the value shown by the apparatus.

c) Dynamic apparatus always measure exact r. m. s. values (and if connected as wattmeters, the power).

d) Electrostatic apparatus show the r. m. s. voltage (only in some special connections do they show other, e. g. maximal, values).

f) Electronic voltmeters show either mean or maximal values according to how they are connected.

g) Oscilloscopes show the whole time course of the current or voltage.

h) Coulombmeters and ampere-hour meters show the amount of electricity that has passed (the charge)

$$Q = \int_0^t i \, dt$$

i) Watt-hour meters show the amount of energy consumed by the circuit

$$W = \int_0^t e \, i \, dt$$

2. Nonharmonic voltage

Periodic nonharmonic voltage.

This can always be transformed to a sum of harmonic potentials, the frequencies of which are a multiple of the basic frequency (so called second, third etc. harmonics), the phases of which are generally shifted. This analysis into a sum of higher harmonic components is termed harmonic (also Fourier) analysis. This may be performed mathematically, graphically, mechanically or electrically (using so called harmonic analysers) (Smirenin 1950).

While the harmonic potential remains harmonic in all circuits not having any nonlinear elements*) and only its phase changes, the shape of the non-harmonic potential is preserved only in circuits containing no other elements but ohmic resistances.

Inductance and capacitance produce changes in the shape of the stimulus by changing the amplitude of its components (especially of higher harmonics) and by affecting their phase relations. These changes may be used to adjust the shape of the pulse for which purpose nonlinear elements are often used. They also cause serious dificulties, however, because the shape, particularly of sharp (rectangular or triangular) pulses, is distorted. In fundamentally the same way these components act on individual stimuli and hence their influence will be analysed systematically.

Aperiodic voltage; Voltage step

A D.C. voltage step (positive or negative). Fig. 3.

The voltage across the electrodes rises suddenly to a constant value E (Fig. 3a). Obviously this ideal time course can only be approximated. Depending upon how close an approximation is required, the appropriate apparatus is chosen. Let us consider an ideal voltage generator (Fig. 3b) composed of a galvanic cell E, a switch S and a resistance in series R_i (this is really the internal resistance of the source). To the generator a capacitance C_p (the capacitance of the output and the electrodes) and a resistance R_e is connected in parallel. If S is switched on, the current begins to flow with an initial value of $i_0 = E/R_i$ which, on charging C_p, decreases asymptotically to a value

$$i_\infty = \frac{E}{R_i + R_e}$$

The charge of C_p increases with time until its voltage attains a value

$$e_\infty = E \, \frac{R_e}{R_i + R_e}$$

---

*) Nonlinear elements change their characteristics with the size of the voltage or current. For instance, ohmic resistance independent of the temperature, the capacitance of a condenser and inductance of an air cored coil are all linear elements since their values are constant and independent of the voltage and current. The resistance of a bulb increases on heating (and thus with the current); inductance of an iron cored coil first increases with the current and then (after saturation) decreases; the resistance between electrodes of a valve depends on the voltage across these and other electrodes. The resistance of a rectifier (diode, germanium diode, selenium rectifier) depends on the voltage and current. Thus valves, rectifiers, bulbs, iron cored coils and transformers are examples of nonlinear elements.

The instantaneous voltage on it is

$$e = e_\infty \left(1 - e^{-t/\tau}\right) \quad \text{(Fig. 3c)}$$

where e is the base of natural logarithms (e $\doteq$ 2·7); $\tau = R_i C_p$ is the so called time constant. This is the time during which C_p is charged to a voltage of $0·63e_\infty$,

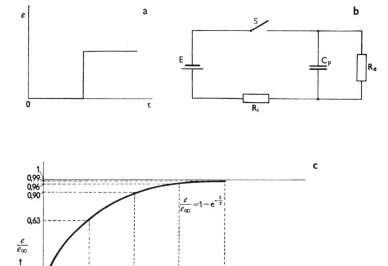

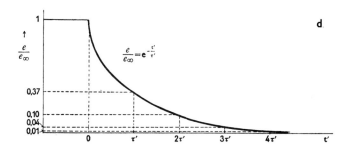

Fig. 3. Distortion of the voltage step by parasitic capacitance C_p. a) ideal step, b) equivalent circuit diagram. Distortion when switching the switch S on (c) and off (d).

i. e. approximately $\frac{2}{3}$ of the maximum voltage. The time constant is an important characteristic of the circuit. During 2τ, C_p is charged to $0·9\ e_\infty$, during 3τ to $0·96e_\infty$ etc. It may, therefore, be generally assumed that the maximum value is attained after 2 to 4τ (with an error of 10 to 1%.)

It can be seen that the "parasite capacitance" C_p prolongs the rising time of the front of the D. C. pulse and this prolongation depends not only on C_p but also on the resistance R_i.

Rectangular pulses

All that was said in the preceding paragraph is also valid for the front edge of a rectangular pulse. Its trailing edge will also be distorted by the parasitic capacitance C_p. If the key S is opened in time ∞ (i. e. practically in time $t > 2\tau$) the voltage at R_e does not drop to zero immediately but falls as C_p is discharged across R_e (Fig. 3d). The voltage across R_e is:

$$e = e_\infty \, e^{-t'/\tau'}$$

where t' is the time from the moment when S is opened and $\tau' = R_e C_p$ is the time constant of the trailing edge of the pulse. The shape of a rectangular pulse with the width t_p is shown in Fig. 4 for a) $t_p \gg \tau$, b) $t_p \approx \tau'$ and c) $\tau > t_p$. (In all cases $\tau' \doteq 3\tau$.) It is evident that in the first case the parasite capacitance only distorts the shape of the pulse but its amplitude is nearly independent of small changes in t_p. Such a shape as shown in Fig. 4a, may still be used in stimulators for the shortest pulses available. The conventional condition is approximately $t_p > 4\tau$ ($\tau > \tau'$), or $t_p > 4\tau'$ (if $\tau < \tau'$).

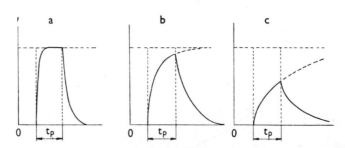

Fig. 4. Distortion of the pulse by $RC = \tau$, $R'C' = \tau'$

a) $t_p \gg \tau$; $\tau' = 3\tau$,   b) $t_p = 2\tau$; $\tau' = 3\tau$,   c) $t_p < \tau$; $\tau' = 3\tau$
For details see text.

The impulse shown in Fig. 4b, on the other hand, will be undesirable in a good stimulator, since for instance, a slight change in the length of the pulse (or the time constant RC) also results in a considerable change in amplitude. The shape of the pulse in Fig. 4c scarcely resembles the required rectangular shape.

Damped oscillations

So far only a circuit containing resistance and capacitance has been considered. The course of the voltage as was found in this case is called aperiodic. A similar course would also be seen with resistance and inductance, but this case hardly ever occurs in electrophysiology. If however, inductances are connected in the circuit in addition to resistance and capacitance, much more

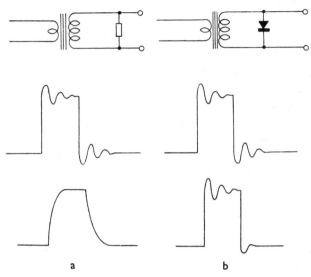

a   b

Fig. 5. Damping of undesirable oscillations in the transformer by resistance (a) or a germanium diode (b). Top: circuit diagram. Middle: pulse shape before damping. Bottom: after connecting a resistance or diode.

complex relations are generally obtained. If the resistance (or conductance) predominates, the shape of the distorted pulse is again aperiodic. If the effect of reactances is higher, damped oscillations appear with the voltage step. If the resistance is neglected at a first approximation, the oscillation frequency is

$$f = \frac{1}{2\pi\sqrt{LC}} \quad \text{or} \quad \omega = \frac{1}{\sqrt{LC}} \quad \text{i. e.} \quad T = 2\pi\sqrt{LC}$$

The effect of resistance is to decrease the oscillation frequency somewhat and chiefly to damp the oscillations*). Since parameters may differ when switching the current on and off, the oscillations at the front and trailing edge of the pulse may differ considerably.

---

*) In a parallel circuit (Fig. 2b) critical damping is attained if $R = \frac{1}{2}\sqrt{(L/C)}$; if the by-pass resistance R is greater, the circuit oscillates, if it is smaller, the distortion of the edges is too large. Since we usually do not know the inductance L and especially the parasitic capacitance C of the transformer with sufficient accuracy, it is best to find the damping resistance by trial and error.

If a transformer is used in the stimulator output it is usually necessary to eliminate oscillation phenomena. Oscillations can be reduced (Fig. 5):

1) By damping resistance: An appropriate resistance is connected in parallel to the coil so that the voltage course becomes aperiodic. This, however, prolongs the front and trailing edge of the wave so that only specially designed transformers ought to be used having small capacitance between the turns, layers of turns and coils on a common core. If the frequency of the oscillations before connecting the aperiodic resistance is sufficiently high, the prolongation of the front and trailing edge caused by damping is not significant. The amplitude is, however, decreased considerably.

2) By damping with a valve, diode or germanium rectifier connected as the damping resistor in case 1 (Fig. 5b). The use of a germanium diode is particularly simple, cheap and effective. With it the second half of the wave is strongly damped while the first half remains practically unchanged. Thus most of the oscillation energy is transformed to heat and, the third half-wave is less than $\frac{1}{10}$ of the first. It is important that the slope of the front and trailing edge is much steeper than when damping with resistances.

"Sawtooth" oscillations

Sawtooth waves are of two types: with a steep front or a steep trailing edge. What has been said above also holds true for the steep side of this wave form. It is usually not difficult to lead the gradual part through the circuit with considerably less distortion than the steep side.

Triangular pulses are distorted mainly at their peaks where aperiodic flattening or oscillations due to the same causes as in rectangular impulses occur.

Repetitive stimuli

Multiple stimuli, especially paired rectangular pulses. These are used more and more frequently. In addition to what has already been stated concerning single rectangular impulses, the intervals between them must also be considered. A low time constant must be maintained, particularly if the first impulse is much higher than the second.

e) Transformers

Transformers are often used in electrophysiological instrumentation, especially in connection with stimulators, both electromechanical and electronic, e. g. as mixing, derivative and isolation transformers. They are also used for isolation of stimuli from earth, for separating off the A. C. component or for simultaneous stimulation at several pairs of electrodes which must not be connected directly etc. In all these cases the lowest and highest frequencies transferred by the transformer must be lower and higher than the frequencies*) used for stimulation. Lower frequencies are distorted by insufficient inductance (the transformer acts as a differentiating coil), higher frequencies by stray capacitance (capacitances between turns and coils). In order to emphasize a certain frequency a resonant circuit**) or a tuned transformer is used.

---

*) This applies not only to the basic frequency, but also to all higher harmonics necessary for maintaining the given shape of the stimulus.

**) The networks shown in Fig. 2 are essential for explaining resonant circuits. We know (Fig. 2a) that

$$Z = \left[R^2 + \left(\omega L - \frac{1}{\omega C} \right)^2 \right]^{\frac{1}{2}}$$

It is evident that there is a certain frequency ω_0 for which $\omega_0 L - 1/\omega_0 C = 0$ and thus $Z_0 = R$. This value Z_0 is evidently the least of all Z in a given circuit. Thus for $Z_0 = Z_{minim}$ the resonance frequency

$$\omega_0 L = \frac{1}{\omega_0 C}$$

or

$$\omega_0 = \frac{1}{\sqrt{(LC)}} ; \quad f_0 = \frac{1}{2\pi\sqrt{(LC)}} ; \quad T = 2\pi\sqrt{(LC)}$$

i. e. Thomson's equation.

This connection (R, C, L in series) is termed a *series resonant circuit*.

It follows from: $R = 0$ that
          $Z = 0$

the ideal series resonant circuit (with an ideal coil with zero resistance) has a *zero impedance* for the resonance frequency.

In an analogous way for a network shown in Fig. 2c we have

$$\frac{1}{Z} = \left[\left(\frac{R}{R^2 + \omega^2 L^2} \right)^2 + \left(\frac{\omega L}{R^2 + \omega^2 L^2} - \omega C \right)^2 \right]^{\frac{1}{2}}$$

Here too there is evidently a resonance frequency for which $1/Z$ is minimal. This occurs if $\omega_0 L/(R^2 + \omega_0^2 L^2) = \omega_0 C$
hence

$$\omega_0 = \frac{1}{\sqrt{(LC)}} \left(1 - \frac{R^2 C}{L} \right)^{\frac{1}{2}}$$

Usually, however $R \ll \omega.L$ and thus $\omega_0 \approx 1/\sqrt{(LC)}$. For $R = 0$ in the preceding equations $1/Z = 0$ and therefore $Z = \infty$. For the resonance frequency (ω_0) an ideal *parallel resonant circuit* has an *infinite impedance*. It is important to bear in mind that for a resonance frequency the parallel as well as the series resonant circuits represent pure resistances since $\tan \varphi = 0$.

For close coupling between the coils (the mutual inductance M is only slightly smaller than the geometric mean of inductances of the primary coil L_1 and the secondary coil L_2) only one coil is tuned. With loose coupling ($M \ll \ll \sqrt{L_1 L_2}$) often both coils are tuned and thus not only the transferred frequency can be controlled but also the transferred band width. For calculations and further details see Terman (1943).

The inductance of coils is increased by the use of laminated iron cores. As losses due to eddy currents rise with the frequency the lamination is the finer the higher the transferred frequency. For the same reason powdered-iron cores or special alloys (e. g. permalloy) or ferrites are used.

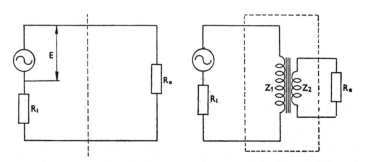

Fig. 6. Matching of the source resistance (Re) to the load resistance (Re) using a transformer.

Transformers are also often used for matching the load impedance to the source impedance. It can easily be demonstrated that a given power supply with impedance R_i will have its maximum possible output when the impedance of the load R_e is equal to R_i ($R_e = R_i$). This will be explained in an example (Fig. 6a)*). The resistance of the source $R_i = 100\Omega$ and its e. m. f. $E = 100$ V.

If a load having an impedance $R_e = 1\ \Omega$ is connected to this source the power will be

$$P = 1\ \text{W} \left(I = \frac{100\ \text{V}}{101\ \Omega} \doteq 1\ \text{A}\ ;\quad P = I^2 R \right)$$

for $R_e = $	10 Ω	$P = \ 8$ W	for $R_e = $ 1000 Ω	$P = \ 8$ W
for $R_e = $	100 Ω	$P = 25$ W	for $R_e = $ 10000 Ω	$P = \ 1$ W

---

*) The abridged proof is as follows

$$P = RI^2 \qquad I = \frac{E}{R_i + R_e} \qquad P = \frac{R_e}{(R_i + R_e)^2} \cdot E^2$$

for $P = $ max the differential quotient dP/dR_e must be zero.

$$\frac{dP}{dR_e} = E^2 \frac{(R_i + R_e)^2 - 2R_e(R_i + R_e)}{(R_i + R_e)^4} = 0$$

and thus $R_e = R_i$.

If $R_e \neq R_i$ and if A. C. current is used, the load impedance can be adjusted to the source impedance using a matching transformer (Fig. 6b). If z_1 is the number of turns on the primary coil, z_2 the number of turns on the secondary coil, V_1, V_2 the primary and secondary voltages, I_1 and I_2 the primary and secondary currents and $p = z_1/z_2$ the turns ratio, then the following relation is approximately valid:

$$\frac{z_1}{z_2} = p = \frac{V_1}{V_2} = \frac{I_2}{I_1}$$

From the above we obtain:

$$p^2 = \frac{V_1}{V_2} \cdot \frac{I_2}{I_1} = \frac{V_1}{I_1} : \frac{V_2}{I_2}$$

As $V_1/I_1 = Z_1$ is the impedance of the primary coil and $V_2/I_2 = Z_2$ the impedance of the secondary coil, $Z_1/Z_2 = p^2 = z_1^2/z_2^2$. The values in the figure must be chosen in such a way that $Z_1 = R_i$; $Z_2 = R_e$ and the primary to secondary turn ratio

$$p = \frac{z_1}{z_2} = \sqrt{\frac{R_i}{R_e}}$$

With complex circuits containing resistances, capacitances and inductances even more perfect adjustment may be obtained by matching not only the source and load impedance but also the phase angles of the primary and the secondary current so that they are of equal value, but of opposite sign.

The elements described above were used in different mutual connections and particularly in connection with various types of mechanical switches (of which only a few could fulfill the great demands placed on them by electrophysiologists) to construct different types of electro-mechanical stimulators.

f) Electromechanical stimulators

Mechanical stimulators are rarely used at present although the best of them may be more satisfactory than the average electronic apparatus. In addition to simplicity (see e. g. Peters 1859), their main advantage lies in the fact that their parameters do not change in time, are not influenced by the instability, interference and hum of the mains supply and, last but not least, that perfect resistive and capacitative insulation of their output voltage from earth and other parts of the apparatus particularly from the amplifier, can be attained. Their main disadvantage is that they cannot be used universally and that sometimes it is impossible or difficult to change important parameters of the stimuli.

Only one device of this group can preserve its position also for the future: the galvanic forceps (Fig. 7). Two arms, one of zinc, the other of copper, are soldered together with tin solder. Actually these are two electrodes of the original Volta cell which are short circuited. The stimulated tissue is the electrolyte. During dissection, the galvanic forceps is applied on nerves in order to determine which branch leads to which muscle.

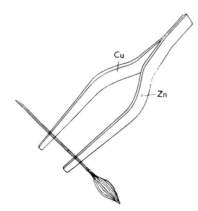

Fig. 7. Stimulation of a nerve-muscle preparation with a galvanic forceps.

Practically all stimulators used today have electronic elements as switches. Hence it is necessary to insert here a more detailed description of electronics which will serve not only as an introduction to the section on stimulators but also on amplifiers, logical circuits and computers.

B. Basic principles of valve and semiconductor function

Electronics, approximately speaking, deals with the laws of current conduction in a range where Ohm's law does not apply, i. e. particularly in a vacuum, in gases (partly also in electrolytes) and in semiconductors, in other words in media and devices the voltamper characteristics of which are not linear. Hence such elements are often termed *nonlinear elements*. These elements can be connected into electric circuits as two poles (e. g. rectifier) or four poles (e. g. triode with an amplifier function). Examples of voltamper characteristics are shown in Fig. 9. For modern trends see also Nekrasov (1965).

Electronic elements are most frequently used as four poles, connected simultaneously into two circuits: on one side is the input circuit, which via the nonlinear four pole affects in a defined way the output circuit connected from the other side. (Often one input pole is also an output pole.) The purpose of such a four pole is then to control the output circuit by the input one. With a certain, usually very small and often negligible energy taken from the input circuit the much greater power of the output is regulated. This, however, is an ideal example which can only be approximated, since in reality the output circuit in turn acts on the input circuit and changes conditions prevalent in the latter. This interaction arises in the electronic four-pole itself and also in subsidiary elements and is termed feedback. Its significance will be discussed below. The part of the feedback that is inherent to the electronic element itself is very important not only for the construction of circuits but also for understanding the basic functions of individual electronic elements. Since this feedback is smallest in valves these will be discussed first.

a) Valves

The diode

The perfectly exhausted glass vessel contains a heated cathode and a cold anode. Electrons leaving the cathode either remain concentrated around the cathode in the empty valve space (space charge) or reach the anode and can be led off from it. The number of electrons reaching the anode during unit time (anode current) depends on the composition and temperature of the

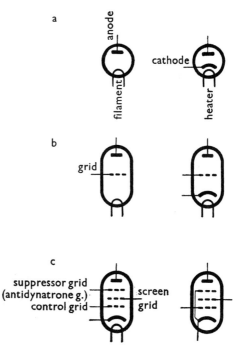

Fig. 8. Symbols for a) diode, b) triode (left — directly, right — indirectly heated), c) pentode: Left — all grid terminals independent. Right — third grid connected to cathode already inside the envelope and the filament of the tube is left out in the symbol, as is usual for simplification of diagrams.

cathode and on the potential difference between the cathode and the anode. If the anode is negative to the cathode, no anode current flows. If the anode voltage is raised, a very small current begins to flow already with a small negative voltage and becomes definite with zero voltage. As the anode voltage continues to rise the anode current rapidly increases until all electrons released from the heated cathode reach the anode — the region of saturated current is attained. The current cannot rise any further (there being no further electrons available). The graphic representation of this relationship is termed the anode characteristic. The saturated current can only be increased by increasing the temperature of the cathode. The cathode is heated electrically. The cathode is either formed by the heated filament itself (directly heated cathode) or the cathode, in the form of a cylinder, is heated by an insulated spiral inside it (indirectly heated cathode). In electrophysiological instruments indirectly heated valves are used nearly exclusively. The saturated current depends: 1) On the material of the cathode surface. Consequently the majority of indirectly heated cathodes are covered by a layer of the oxides of Ca, Sr or Ba. These emit electrons at relatively low temperatures, several hundreds of degrees Centigrade. They cannot, however, be loaded with a saturated current, which destroys them within a short time. 2) On the temperature. (Thermionic emission rises rapidly with increasing cathode temperature).

Diodes are used to rectify A. C. currents. Today selenium, germanium or silicon rectifiers often replace them (unless a practically infinity resistance for the inverse voltage or a very high inverse voltage value is required).

Triode

This has a third electrode — the control grid — between the cathode and anode. The anode current does not only depend on the temperature of the cathode and the anode voltage, but also on the grid voltage, i. e. on the voltage difference between the grid and cathode. This is shown in Fig. 9b, d. The most important relationship is that between the anode current and the grid voltage, the so-called grid-anode characteristic (Fig. 9b). It can be seen that a change of the grid voltage has a much greater effect upon the anode current than the same change of the anode voltage. This is the key to understanding the function of a triode as an amplifier: a small change in grid voltage causes a large change in anode current. In series with the anode is an anode load across which the anode current evokes a voltage drop having a D. C. and an A. C. component. The A. C. component corresponds to the A. C. component of the grid voltage, but is larger and of the opposite direction (the phase of the anode voltage is shifted 180 degrees from that of the grid voltage). The suitable size of the anode load must be chosen according to the valve properties. For that purpose some of the triode parameters must be defined: the penetration coefficient

$$D = \frac{\Delta V_g}{\Delta V_a} \quad (I_a = \mathrm{const})$$

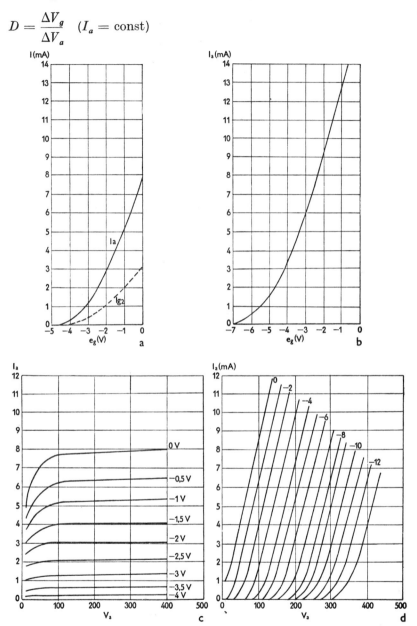

Fig. 9. Characteristics of pentode EF6. a, c — connected as pentode, b, d — connected as triode (screen and suppressor grids connected to anode). a, b — grid characteristics. Anode voltage 150 V, screen grid voltage 100 V. c, d — anode characteristics (control grid voltage is a parameter and is given for each curve). The difference between a triode and pentode is best seen when comparing c and d.

states by how large a factor the anode current is affected more by the anode voltage than by the grid voltage. $1/D = \mu$ is called the amplification factor and gives the theoretically maximum possible amplification.

The mutual conductance or transconductance

$$g_m = \frac{\Delta I_a}{\Delta V_g} \quad (V_a = \text{const})$$

indicates the extent to which changes in grid voltage evoke changes in anode current. The anode resistance

$$R_i = \frac{\Delta V_a}{\Delta I_a} \quad (V_g = \text{const})$$

shows how a change in anode voltage influences the anode current.

It follows from these definitions that

$$D \cdot g_m \cdot R_i = \frac{\Delta V_g}{\Delta V_a} \cdot \frac{\Delta I_a}{\Delta V_g} \cdot \frac{\Delta V_a}{\Delta I_a} = 1 \quad \text{(Barkhausen's equation)}$$

It must be underlined that, except for the penetration coefficient (D), these parameters are not constant, but depend on the conditions under which the valve is working (Fig. 10).

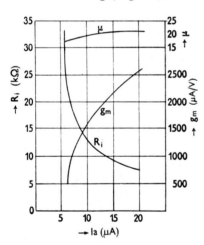

Fig. 10. Dependence of triode coefficients on anode current. Abscissa: anode current. Ordinate: scale for anode resistance in kΩ (left), transconductance in µA/V or micromho (right), amplification factor (top right). Curve R_i — anode resistance; g_m — transconductance; μ — amplication factor.

If a small anode load R_a is chosen, even a large change in anode current causes only a small change in voltage drop across it

$$\Delta V_a \approx R_a \Delta I_a, \quad \text{when} \quad \Delta I_a = g_m \Delta V_g \text{ and thus } \Delta V_a \approx R_a g_m \Delta V_g$$

In this case the amplification will be small. If, on the other hand, a very large R_a is chosen, the anode voltage will approach zero and the amplification will again be small, since for small anode voltage the anode current and the trans-conductance, which is its function, fall considerably. It can be demonstrated that maximum voltage amplification is obtained for $R_a = R_i$, which for the majority of triodes is equal to several tens of kΩ.

Pentode

It is evident from the above that the amplification could be increased considerably by increasing R_i (the anode resistance of the valve), i.e. by decreasing the effect of changes in anode voltage on changes in anode current. This is achieved by introducing a further grid, a so-called screen grid, between the control grid and the anode. The screen grid has a voltage near to that of the anode and acts, therefore, as a kind of first anode. Since it has a constant potential, the electron current between the control and screen grid

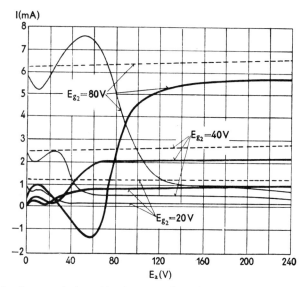

Fig. 11. Tetrode characteristics. Abscissa: anode voltage. Ordinate: current. Full heavy line — anode current curve (I_a), full thin line — screen grid current curve (I_{g_2}), interrupted line — cathode current curve (I_k). All curves for 3 different screen grid voltage values (20, 40, 80 V). The initial waves (left) in curves I_a and I_{g_2} are due to secondary emission. A negative anode current with a minimum for E_a 60 V is seen on the curve I_a for E_{g_2} 80 V.

is not affected by changes in the anode voltage (Fig. 9c). (This is the case for triodes, where the drop across the anode load R_a causes variations of anode voltage and decreases the effect of the control grid on the anode current.) The screen grid accelerates electrons, but intercepts only about 10% to 20% of them. The remainder pass through its openings and thus reach the anode nearly independently of the anode voltage, i. e. also independently of the voltage drop across R_a. The latter can thus be chosen to have very high values, up to several MΩ. In other words, the anode resistance is high. Such a valve is termed *tetrode* or screen-grid valve. In addition, however, to the above advantages, tetrodes have a considerable drawback due to secondary emission of electrons from the anode (Fig.11). The electrons falling onto the anode release so-called secondary electrons because of their high kinetic energy. There may even be several secondary electrons to a single primary one. These secondary electrons enter the space between the screen grid and the anode and if the former has a larger voltage than the latter, they may return to the screen grid. Thus a negative anode current is produced: more (secondary) electrons leave the anode than are trapped to it. This is called the dynatron effect and is usually very unwelcome. It can be eliminated either by a special arrangement and shape of the electrodes in so-called beam tetrodes, or by introducing a further, so-called suppressor or antidynatron grid. This has a negative voltage with respect to the anode (it is usually connected with the cathode). Rapid primary electrons pass through it without difficulty while slow secondary ones are returned to the anode by negative voltage of the suppressor grid. They do not release further "tertiary" electrons at the anode on their return, since their kinetic energy is small. Such valves are termed *pentodes*.

Some important interference phenomena in valves

When designing amplifiers the following phenomena have to be considered:

Noise

A voltage may be found at the output of every valve which does not correspond to any voltage applied to the input. This is called noise voltage. It has several causes and differs according to them.

Thermal noise

Every conductor contains free electrons which have thermal movement. At a certain moment more electrons may move in one than in another direction. Thus a voltage due to an instantaneous excess of electrons arises at one end of the conductor as compared to the other end. This noise voltage continually varies and changes its polarity. Theoretically it contains voltages of all frequencies from 0 to ∞. Hence the noise voltage is usually determined only for a certain frequency bandwidth. If the resistance of the conductor does not depend on the frequency, then the r. m. s. thermal noise voltage (in volts) is

$$V = 2\sqrt{(kTR\Delta f)}$$

where k is Boltzmann's constant (k $= 1\cdot38 \cdot 10^{-23}$ Joule . degree^{-1}), T is the absolute temperature of the conductor (in degrees Kelvin), R is the resistance of the conductor in ohms, Δf the bandwidth of the measured frequency in c/sec.*) Thermal noise is found not only in all valves, but also in all circuits. Noise can never be reduced below values calculated from the above equation. This equation mainly serves to give a rough indication of demands on the recording apparatus and to evaluate the final product.

The equation shows that the only practical possibility of reducing thermal noise is to decrease load resistance and to narrow the passband of the amplifier. In addition to thermal noise, there are other noise forms in resistances, caused by changes in the quality of the resistor. This resistor component is the largest in graphite and carbon resistors, the smallest in wire-wound ones.

Valve noise

The current in a valve is produced by electron emission from the cathode. The number of emitted electrons is determined statistically. It varies and thus causes fluctuation in mean anode current, which produces so-called shot noise voltage across the anode load. The spectral density of the shot noise is similar to that of thermal noise. The noise current is (in amperes)

$$i = 5\cdot64 \cdot 10^{-9}\sqrt{(I_{ao}\Delta f)} \quad (A; A; c/sec)$$

(where I_{ao} is the mean anode current). This holds only if all electrons are immediately taken over by further (positive) electrodes. Usually, i. e. with a negative grid voltage, electrons accumulate at the cathode and thus form a so-called space charge. This evidently considerably decreases the shot noise. The equivalent noise resistance of the valve equals an equivalent grid leak, which would produce at the output of an ideal valve a noise equal to the noise of the valve at room temperature (300 degrees Kelvin) and for the same bandwidth Δf. For a diode with a saturated current

$$R_{eq.} = \frac{1}{20\,I_{ao}} \; ; \text{for a diode with a space charge}$$

$$R_{eq.} = \frac{1}{30\,I_{ao}} \; ; \text{for a triode with a space charge}$$

---

*) After substitution (for 20°C)

$V = 0\cdot13\sqrt{(R\,\Delta f)} \quad (\mu V_{eff}; k\Omega; kc)$

$$R_{eq.} = \frac{2 \cdot 5}{g_m}, \text{ where } g_m \text{ is mutual conductance in ampers per volt.}$$

The noise of pentodes is considerably higher than that of triodes because of the variation in the distribution of the cathode current between the anode and the screen grid. The equivalent noise resistance is equal to:

$$R_{eq.} = \frac{I_{ao}}{I_{ao} + I_{go}} \left(\frac{2 \cdot 5}{g_m} + \frac{20 \, I_{go}}{g_m^2} \right)$$

where I_{go} refers to the mean screen grid current.

Other less important sources of noise are irregularities in the oxide layers of the cathode and their changes (flicker noise), ionic cathode and grid current (for imperfect vacua), thermal and secondary electron emission of "cold" electrodes, etc.

Valve hum

Valve hum is due to imperfect filtration of anode voltage on the one hand and the magnetic field of the heating filament on the other. In order to suppress it completely in the first stages of the most sensitive amplifiers, valves must be heated by D. C. current from a battery or a regulated rectifier.

Microphony

Microphony is due to mechanical vibrations of the valve. The vibration of the electrodes (especially of the grid) considerably changes the characteristics which depend upon the electrode distances. In some of the newest valves microphony has been nearly completely eliminated. Others must be mounted resiliently.

Other interferences

In addition to the above mentioned, most obvious disturbing phenomena, attention must also be paid to other frequent sources of interference, which are most undesirable in electrometric amplifiers. In most cases they are due to the following: Secondary electron emission from the anode and the screen grid (see above). Thermal emission from the control grid especially in valves having a high transconductance, where the grid is very close to the heated cathode.

Leaks between the electrodes inside the glass envelope (these are often due to getter — a layer of reactive metals of groups I and II which are introduced into the envelope to remove the last traces of gases) and also outside (dirt, moisture) or even in the glass (the leak is the larger, for the same glass properties, the larger the metal-to-glass seals).

An imperfect vacuum results in frequent collisions of the electrons coming from the cathode and going to the positive electrodes with gas molecules. In this way other electrons and especially positive ions flowing to the negative electrodes (the cathode and the grid) are formed. The ion current flows in a direction opposite to that of the electrons. Ions cause most disturbances at the grid. They produce a negative grid current. With a high grid leak resistance, the grid is charged positively. This results in an increase in anode current and may also cause an electron (positive) grid current and secondary emission. Consequently the grid resistance for the usual circuits and valves cannot be increased above 2 MΩ (0·5 MΩ for the power valves). The grid currents (positive and negative, i. e. electron and ionic) may considerably affect the tissue between the electrodes and hence must be reduced to a minimum, particularly when working with single fibres or single cells. This will be discussed later. A strong ionic current also rapidly destroys oxide cathodes.

In summary we may say that in electronic valves electrons are released from the cathode into the vaccum (most usually by heat, more rarely, e. g. in vacuum photocells, by light) where their path is determined by the shape of the electric or also magnetic field. Amplification is usually achieved by altering the number of electrons falling onto one electrode with the aid of changes in the field, usually by changing the voltage of other electrodes, most frequently that of the control grid. In some special cases amplification is attained by increasing the number of electrons with the aid of secondary emission — see p. 64. Such valves are termed electron multipliers and the photomultiplier is the best known of these. In it a very weak flow of electrons arises after illumination of the photocathode by a weak light source and is then amplified up to 10^7 times.

b) Semiconductors

Free electrons can exist not only in a vacuum but also in substances called semiconductors (and also in metals, but this will not be discussed for the present). The most important semiconductors are pure elements of group IV of Mendeleyev's periodic table, particularly silicon and germanium.

Free electrons in them are generated e. g. by thermal oscillations of atoms when some valency electrons gain so much energy that they are no longer under the influence of neighbouring atoms about which they had been orbiting. An electron thus released no longer belongs to several atoms but to the whole crystal. It can fairly easily change its location in the crystal under the influence of an electric field and thus can carry a current. The number of free electrons (and thus the conductivity of the crystal) rapidly increases with increasing temperature. Hence such a semiconductor can be used to measure changes in temperature. It is termed thermistor and is being used for measurements of body temperature, expired air temperature, tissue and blood temperature etc. Germanium is more suitable for lower temperature, silicon for higher temperature.

The number of free electrons depends also to a considerable extent on the purity of the substance. Ideally pure semiconductor elements are termed pure semiconductors. They are characterized by having the same number of free electrons and holes. A hole is the space left after an excited electron has left its position in the crystal lattice. While the energy required to release an electron is large, about 1 eV*), the energy necessary for a bound electron to jump from one place in the crystal lattice to a free site (hole) is very low, about 10^{-20}eV. This extremely low energy is also required to induce the migration of free electrons under the influence of thermal agitation. The free electron moves randomly from one site to another and in the same random way bound electrons move into the free holes, leaving behind them new free holes. Thus this movement appears as a movement of holes. Hence we speak of empty holes and of their movement as a movement of positive charges (about as mobile as negative free electrons), although in reality no real mobile positive particles exist in the crystal. This random movement which is subject to the same laws as diffusion of gases is termed diffusion of free electrons and holes.

If an electric field is applied on the semiconducting crystal then the movement of holes in the direction of the field and of free electrons in the opposite direction is superimposed upon diffuse thermal movement of the above. Holes and electrons thus carry a charge under the influence of the potential gradient in the same direction (i. e. in the conventional sense in the direction of the field). The current in the semiconductor is proportional to the number of free carriers present and is also dependent on the intensity of the field. The number of free carriers, of course, is not stable but represents an equilibrum. Free electrons constantly arise, leaving free holes behind them, but also constantly disappear combining with (or, if you like, falling into) the holes. This is termed recombination of electrons and holes.

If a slight amount of an element of group V e. g. arsenic (As) or antimony (Sb) (having 5 valency electrons instead of 4 as have Ge and Si) is added to a pure semiconductor of group IV, then a certain number of electrons is bound to the crystal lattice much less firmly than the others, i. e. only by electrostatic forces of the nuclei of group V. Considerably less energy is required for the release of such electrons and no holes remain in the lattice after their displacement. Such a semiconductor has more free electrons than holes and consequently is termed a type N semiconductor (negative). If, on the other

---

*) Substances in which releasing an electron needs much more energy than this are termed insulators.

hand, an element of group III (e. g. indium) is added to germanium or silicon a deficit of bound electrons will arise. As there will be more holes than free electrons, such a semiconductor is termed type P (positive). Both P and N semiconductors are named impurity, or extrinsic semiconductors.

If two types of impurity semiconductors are connected, e. g. type P on the left and type N on the right (Fig. 12) excess holes diffuse from the left half

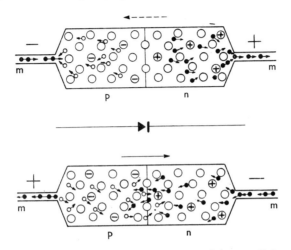

Fig. 12. The *p-n* junction in semiconductor and its rectifying ability.

a) Free electrons (black discs) are sucked in the direction of the field from the junction region, holes (small circles) in the opposite direction and in this way the junction is made free from charge carriers, i. e. nonconducting.

b) Electrons and holes freely diffuse across the junction. The electric field leads free electrons from the metal lead (*m*) to the left and electrons from the *p*-type semiconductor take the place of the electrons carried away. Thus new holes are formed diffusing to the right. Electrons entering the semiconductor from the right conductor because of the effect of the field, substitute for the electrons diffusing across the junction to the left. The holes that have moved to the right gradually recombine with the free electrons present there and electrons that have moved to the left recombine with holes there. Thus free carriers that disappear (electrons and holes) are constantly renewed: on the left new holes are formed and on the right new free electrons.

into the right one and excess electrons migrate in the opposite direction, until the left part is charged negatively (it has lost positive holes and gained negative electrons) and the right one positively (the number of holes has increased and that of electrons decreased). In other words: By diffusion the holes and electrons are more or less equally distributed but the charges of the nuclei are bound in the crystal lattice and hence on the right (where there are nuclei with a higher positive charge) the positive charge predominates, while on the left the negative charge is prevalent (electrons). The electric field thus formed induces an equilibrium: the negative charge on the left prevents further influx of electrons and further outflux of holes, while the opposite applies to the right part. This phenomenon can be utilized for constructing a power thermoelectric couple: The junction is heated and the two ends are connected to a load. In a semiconductor photoelement carriers are released at the junction by light and the charge is carried through it only in one direction as described above.

Semiconductor diode

If a voltage source is connected to the semiconductors in such a way that the carriers diffusing across the junction do not accumulate beyond it, but are continuously removed, i. e. if the left half is connected to a positive pole the right one to a negative pole, then the carriers migrate continuously across the junction. If, on the other hand, the left half is connected to a negative pole and the right one to a positive one, holes will move to the left and electrons to the right and, therefore no current will flow across the junction. This is the principle of germanium and silicon rectifiers which in recent years have been used to an ever increasing extent in science and industry.

If an A. C. voltage is applied to the rectifier, current is allowed to pass only in one direction, similarly as in a diode. In contrast to the vacuum diode, however, resistance in the conducting direction (forward resistance) is much smaller and no heating is required. Such a rectifier is also cheaper, lasts longer and is more robust. On the other hand, its disadvantage rests in the fact that a measurable current flows when a reverse voltage is applied. This current is due to "minority carriers", i. e. to holes in N-type semiconductor and electrons in P-type semiconductor belonging to the atoms of germanium or silicon, not to the admixtures. Their number rapidly increases with increasing temperature. Also the reverse voltage which a semiconductor rectifier can withstand is smaller. The greater the voltage applied in the inverse direction, the more the semiconductor is heated due to the inverse current. Thus the temperature is raised and so is the current and this may cause a further rise in both until the semiconductor is destroyed by heat.

If such a semiconductor rectifier is connected to a D. C. voltage in the inverse direction, only a very weak current will flow. This current, however, can be considerably increased by (a) temperature, (b) light, (c) ionizing radiation, (d) too high a voltage, (e) "injection" of electrons or holes into one semiconductor region. All these mechanisms can be utilized practically for (a) temperature determination, (b) light measurement (so called photodiodes), (c) measurement and indication of radiation, (d) regulation and control of voltage by so called Zenner diodes and for (e) amplification of current, voltage and power using "crystal triodes" or transistors (Fig. 13.).

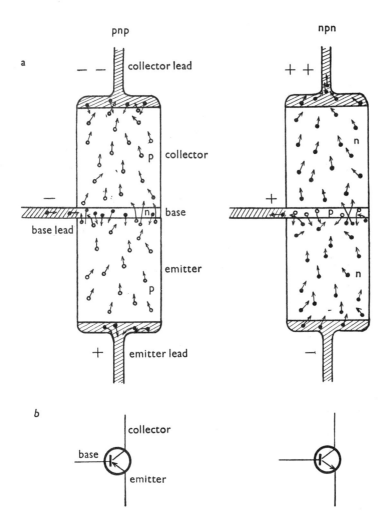

Fig. 13. *pnp* tansistor (left) a) principle, b) scheme. The junction between the base (middle part) and collector (upper part) is connected in non conducting direction and the junction between the base and emitter is connected in the conducting direction, free electrons (dots) pass from the base into the emitter and holes (circles) the other way round. Since the upper part of the semiconductor (type *p*), however, has no holes — the electric field between the base and the collector continuously removes them and cleans the upper junction — only a smaller part of the holes passing from the emitter through the lower junction ends in the outer lead to the base but most holes diffuse also through the upper junction and are sucked off by the collector. Their number is so proportional to the current through the lower junction. Onto the right: *npn* type transistor and its scheme.

Semiconductor triodes

A simplified explanation of a transistor is the following: 3 different semiconductors are connected in series, e. g. P-N-P. These three parts are called emitter, base and collector. A fairly large voltage (1-20 V) is applied between the base and the collector in such a way that no current flows through the base-collector junction (i. e. in the inverse direction). The source of the measured voltage is connected between emitter and the base with such a polarity (it may also be biassed) that current flows. The emitter-base junction is thus connected in the conducting direction. Thus free carriers flow across the emitter-base junction their number being determined by the emitter voltage. Since the semiconductor forming the base is very thin (sometimes less than 10^{-3} mm), many carriers (mostly much more than 90%) diffuse to the base-collector junction which is connected in the nonconducting direction. The free carriers "emitted" to this junction by the emitter (hence its name) are collected by the potential difference between the base and the collector, i. e. the collector collects them (hence its name) and thus an electric current appears in the collector-base circuit which, of course, is somewhat smaller than the current flowing across the emitter-base junction. Since the voltage used in the collector-base circuit is rather high, the load resistance in this circuit may be considerably greater than the internal resistance of the source supplying current to the emitter-base circuit. Since the power $P = I^2 R$ and the currents are practically the same in both circuits, the power is amplified in proportion to the ratio of the internal resistance of the collector to the internal resistance

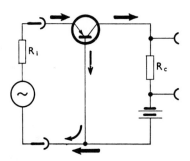

Fig. 14. Common base transistor amplifier. The emitter current is about the same as the collector current and the current amplification is about minus one. The input resistance is very low — tenths of ohms, the output resistance very high (about 1 megaohm). Amplification of both voltage and power is the same, several hundred times. Strong arrows: strong current, fine arrows: weak current.

of the emitter. Hence this is a typical impedance transformer which represents a 'mirror' function of the cathode follower. In the latter the cathode voltage is nearly the same as the grid voltage, but the cathode resistance is smaller. In the above transistor amplifier the current is the same in the input and output circuits but the output resistance is greater. Such a circuit is termed a common base amplifier since the base is common for both input and output (Fig. 14).

The amplifier described above differs essentially from the coresponding valve circuit because a considerable current is drawn from the measured source. Hence a different connection is more frequently used which draws a much smaller input current (though more than the grid circuit of a triode). This connection has a common emitter: The input circuit is connected between the base and the emitter and the polarity of the applied voltage is such that

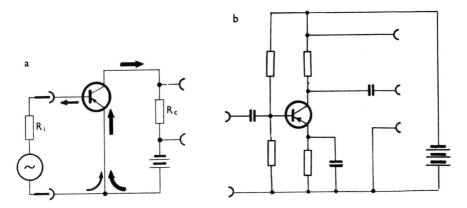

Fig. 15. Transistor amplifier with common emitter. The current of the base is relatively low — it is the difference between the emitter and collector current. Hence a higher input resistance (in the range of kiloohms). High amplification of current ($100\times$) and voltage ($100\times$, with inverted phase) and hence the greatest power amplification, about 10 000.
a) Scheme. b) One of used circuits.

the emitter-base junction is again conducting (Fig. 15.) The output circuit between the emitter and the collector is connected to the supply voltage in such a way that the second junction (between the base and the collector) is nonconducting. Current flowing between the base and the emitter is the same,

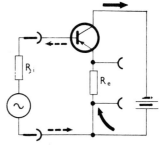

Fig. 16. Transistor amplifier with common collector (emitter follower).The load resistance is common for the input and output circuits and a strong negative feedback occurs at this resistance so that the voltage amplification is nearly exactly plus one. Current amplification is considerable (tens) since again only the difference in currents passes through the outer base connection. The voltage negative feedback, however, increases the input resistance to hundreds of kiloohms and decreases the output resistance to hundreds of ohms. Amplification of power about $10\times$. It is suitable only as impedance transformer when measuring sources with high internal resistance, in the same way as a cathode follower.

as in the previous case, and carriers, say 95% of them, diffuse to the second junction. This, however, means that only about 5% of the current flows from the base via the external lead, while 95% of the current passing through the emitter (which is the common electrode for both circuits) is supplied by the output circuit (i. e. from the supply battery). The base only removes those carriers that did not diffuse. This amplifier is not only an amplifier of power but also of voltage and amplifies the latter while drawing much less current from the source. The third possible connection with a common collector is only rarely used (Fig. 16.).

The main advantages of a transistor over a valve are small dimensions, low current consumption and low supply voltage; no special heating circuit is needed. This also makes stabilization of the supply voltage much easier — mostly a battery with a large capacity is sufficient.

The disadvantages compared to valve are still large, however: An important difference which is particularly disadvantageous in the input stage of amplifiers is the non-negligible control power (which is of fundamental importance in electrophysiology — see paragraph on grid current in cathode followers, p. 153) which has only recently been reduced to a sometimes bearable degree. This is related to the presence of feedbacks: in the valve the feedback effect of the output on the input can be neglected and if a feedback is required (positive or negative) special measures have to be taken. In transistors not only does the input circuit act on the output, but also vice versa the output circuit has a considerable effect on the input. In other words there is always a feedback in the transistor itself and to decrease it to within reasonable limits is a considerable problem.

Prevention of the temperature effect is another problem which cannot be solved by only stabilizing the temperature (e. g. by placing the transistor in a thermostat) since it is heated during passage of the current. The difficulties increase with increasing power demands and increasing ambient temperature. In this respect silicon transistors are much better than the germanium ones.

The last main problem is noise. Even though this has been decreased by several orders of magnitude with the development of transistor technology, it is still larger than in good valves. This prevents the use of transistors in the input stages of amplifiers so that transistorized EEG, ECG and similar apparatus mostly still have valves in their inputs. A further disadvantage is related to the frequency which the transistor can amplify. This is usually lower than for valves even though transistors for hundreds of megacycles have already been constructed. These, however, are expensive and rare. Fortunately this disadvantage is usually of no significance for electrophysiologists.

It is very easy to use transistors as switches: if a current flows through the emitter-base circuit, carriers diffuse into the collector circuit (as described above) and this appears as a considerable decrease in its resistance. Let us

74

call this resistance $R_{kz} = U_{kz}/i_{kz}$. As long as no current flows through the emitter, the collector-base junction is closed and practically no current flows, the resistance of the collector being very high. We may call it $R_{ko} = U_{ko}/i_{ko}$. A large load resistor R_a is placed in the collector circuit so that $R_{kz} < R_a < < R_{ko}$. If the transistor is made conducting (by the emitter-base current), the voltage drop on resistor R_a is nearly equal to the supply voltage while it is

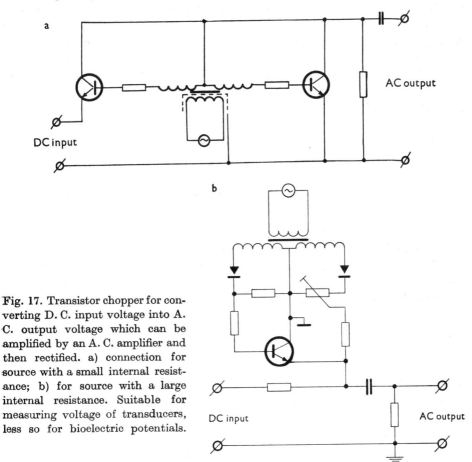

Fig. 17. Transistor chopper for converting D. C. input voltage into A. C. output voltage which can be amplified by an A. C. amplifier and then rectified. a) connection for source with a small internal resistance; b) for source with a large internal resistance. Suitable for measuring voltage of transducers, less so for bioelectric potentials.

almost zero when the transistor is closed. What is most important, the voltage at R_a is nearly independent of the emitter-base current as long as the latter does not exceed certain limits.

This principle can be used for two purposes:

1) For building amplifiers that change a D. C. voltage to an A. C. voltage (see also pp. 163 and 636) of a certain constant frequency which is then further amplified by an A. C. amplifier, usually tuned to a narrow frequency band in order to limit the noise. Thus it is possible to use, instead of a mechanical

chopper, a common base transistor circuit, the emitter of which is supplied from an A. C. source while the measured voltage is connected via resistor R_a to the collector. (Fig. 14). The measured voltage will appear and disappear on the collector with the frequency of the current applied to the emitter. The A. C. voltage thus formed is further amplified by a tuned transistor amplifier, synchronous detector etc. In a similar way more complex choppers can be constructed (Fig. 17a, b). The current which switches over the transistor can be produced, e. g. in the system described below (2), in a tone generator, by transformed voltage of the mains etc. The advantage of a transistor chopper over a mechanical one is in its greater robustness. There are no moving contacts which could be oxidized and the switching frequency can be relatively high — hundreds or even thousands of cycles — so that the upper limiting frequency of such a "D. C." amplifier (which is usually lower by one order of magnitude than the switching frequency) can be much greater than in a mechanical chopper.

2) If a feedback is introduced into a transistor switch a system similar to a multivibrator is formed (astable, monostable or bistable). Thus circuits important for stimulators can very easily be constructed of transistors instead of valves and the apparatus is then much cheaper. Transistor stimulators are frequently described in the literature and undoubtedly will be further developed. These circuits are the more advantageous than valve circuits, the more complex is the system and the more frequently its basic elements are repeated, as is the case particularly in computers, logical circuits of cybernetic machines, medical diagnostic apparatus etc. (Shats 1963, Tishenko et al. 1964, Yakovchuk et al. 1961).

Unipolar transistors (FET and MOS).

The movement of free carriers in semiconductors can be modified by the spatial configuration of an imposed electrostatic field in a similar way as the movement of electrons in a vacuum, i. e. without passing current through the field control electrodes. Already the pioneers of semiconductor electronics developed components that resembled valves more than the commonly used planar transistors (Shockley 1952), but because of technological difficulties transistors of this type became commercially available only recently.

In unipolar transistors the main current (analogous to the cathode current in valves) passes through a single channel of a P or N-type semiconductor. The current, carried by majority carriers, enters this channel through a source electrode S (analogous to the cathode) and leaves through the drain electrode D (analogous to the anode). The current increases with the source — drain voltage until a certain "saturation" level is attained (compared with the anode characteristics of a triode and pentode). The third electrode G, termed gate and analogous to the control grid of a valve, is connected to a thin section of the conduction channel in such a way that the applied voltage controls the width of the underlying section of the channel and thus the passing current.

Two main types can be distinguished according to the gate construction.

a) The field effect transistor (FET) consists of a channel of N-type semiconductor with metal S (−) and D (+) terminals. The channel is formed on a larger piece of P-type material. The gate represents a small zone of P-type semiconductor forming a P-N junction with the channel. The negative bias of the gate (analogous to the negative grid bias in valves) makes this junction non-conducting: While the free carriers (holes) are removed from the junction to G, the negative charge of the junction increases and restricts the movement of the majority carriers (electrons) to a narrow zone of the channel. A small A. C. voltage superimposed on the gate bias elicits changes of the S-D current closely resembling modifications of valve current by the control grid voltage. The S-D current can also be modified by the potential of the P-type basis (equivalent to the second grid). The input resistance of a FET can be higher than that of the best valves. Its input capacitance does not exceed several $\mu\mu$F, but its output capacitance is higher by several orders of magnitude than that of a pentode. With a mutual transconductance of approximately 1 milimho the noise is lower than in common transistors (a few dB in the $0-50$ kc range).

b) Metal oxide semiconductor (MOS). In the FET described above, the gate P-N junction represents not only a capacitance charged or discharged through the input circuit, but also a small leak caused by the minority carriers (holes) in the channel. This leakage current can be reduced by several orders of magnitude, when the channel current is controlled capacitatively: the narrow portion of the channel is influenced by the metal gate electrode through a dielectric layer (usually SiO_2). The advantages are high impedance and low capacitance of the input. No gate bias is necessary, and even positive gate voltage can be used. The noise is somewhat higher than in the FET, however.

Unipolar transistor circuits are essentially similar to those using valves. More attention must be paid to the negative feedback used for compensating the high output capacitance. Basic circuits were described by Butler (1965), Gosling (1964), Büttner (1966), Grisswold (1964) and Webb (1965).

Semiconductors and semiconductor devices are important for electrophysiology also for another reason. Indirect evidence indicates increasingly that the protoplasm has some semiconductor characteristics (due to conduction of ions or even of electrons). The elements of a semiconductor theory of the function of excitable structures are being laid. The recent physical and technical literature contains reports on liquid semiconductor elements — diodes and transistors — and these must be confronted with the unidirectional transfer (or rectification) at the synapses and with the amplification of nerve impulses into muscle impulses which occurs at the muscle endplate (Digby 1965).

The literature on semiconductors is rapidly increasing in size, hence only the basic monographs by Yoffe (1957) and Smith (1959) are mentioned here containing a large number of references. The newer literature can be found in encyclopedic dictionaries of physics (e. g. Thewlins 1960) under the corresponding headings (semiconductors, transistors, rectifiers, photoelements, thermoelements etc.).

c) Gas filled tubes

Electronic components using conduction of electricity in gases are being used less and less in electrophysiology. Hence they will be described only briefly.

The basis of all gas elements is generation of an electric discharge or a change of its character. Gas becomes conducting only if acted on by ionizing factors. These are either photons (infrared, visible, ultraviolet, X and gamma radiation) or charged particles entering the gas. Ions or electrons thus formed move in the electric field between the electrodes and produce a current that for a small field intensity is proportional to the number of newly formed charged particles. With increasing field intensity, i. e. with increasing potential difference between the electrodes, the current is raised only slightly since the probability of recombinations decreases. For large field intensities particularly the free electrons are accelerated to such an extent that their energy suffices to induce further ionizations, i. e. the number of ions increases explosively and within 10^{-6} or even 10^{-9} sec all the gas in the tube is ionized. The "resistance" between the electrodes falls to a value approaching that of metal. Thus gas-filled tubes usually serve as switches because their resistance can be lowered within fractions of a second from values close to infinity to values close to zero. Their use in stimulators will be discussed below.

d) The basic electronic circuits

The circuits in which the above mentioned electronic components are used can be discussed from two different aspects: Either from the aspect of the classical theory of electrical circuits, i. e. as it is necessary for designing and building such circuits and also for those who want to understand them perfectly; or from the point of view of the theory of information, or cybernetics, i. e. from the aspect of the consumer who wants to use these circuits for the transmission, transformation and processing of information. The more complex this information processing the more important is this new theoretical approach which has already created its own symbolic mathematical apparatus, a kind of "algebraic logic" or "logical algebra" which helps to find the optimum circuitry for information processing irrespective of the realization of the individual elementary circuits (i. e. whether valves, semiconductor, ionic or even electromechanical relays are used) and without taking into account the problems of energy.

The character of this book leads us to deal with simple electronic circuits with an eye on both aspects; of course not always to the same extent.

The majority of circuits used in electronics is derived from two main types: the *rectifier* and the *amplifier* which have been dealt with when describing diodes (vacuum and semiconductor), triodes, pentodes and transistors. In order to understand the different apparatus sufficiently, we must first of all grasp the difference between so called differential characteristics (resistance, conductance, mutual conductance, penetration coefficient, etc.) and the so called D. C. equivalent resistance or conductance. If, for example, we connect the rectifier to some D. C. or A. C. voltage, then the ratio $v/i = R$ (where v is the instantaneous value of voltage and i that of the current) is called the equivalent resistance. This is significant particularly if the component (i. e. the rectifier) is used as a switch. It must be realized, of course, that this is not a constant characteristic of the component but that it depends on a number of variables (e. g. voltage, current, temperature, illumination, magnetic field etc.) If now $v_1/i_1 = R_1$ and $v_2/i_2 = R_2$, then

$$\lim_{(i_2 - i_1) \to 0} \frac{v_2 - v_1}{i_2 - i_1} = \lim_{\Delta i \to 0} \frac{\Delta v}{\Delta i} = \frac{dv}{di} = r$$

The quantity r is termed the internal or differential or (for valves) the anode resistance. It is of particular significance in constructing amplifiers and oscillators. The difference between R and r is best understood if we consider that R is always positive while r may be either positive or negative. For this it is sufficient for the component to have a decreasing characteristic, i. e. the voltage across it declines with rising current. Such components with negative (differential) resistance (within a certain range of voltages and currents) play an important role in electronic circuits, particularly in all kinds of oscillators.

If the element is used as a dipole, it is more or less fully characterized at a certain working point by its voltage, current and differential resistance and sometimes also by its dependence on the rate of change of voltage or current, i. e. by its frequency dependence. In a fourpole the situation is more complex since here we have two equivalent and two differential resistances in the input and output and also feed back "resistances" between the input and output. None of these resistances need be pure resistance but usually also has a capacitative and rarely also an inductive component, the significance of which is the greater the higher the transmitted frequency.

We shall now discuss the most fundamental circuits used in electrophysiology. Some special problems of amplifiers will be mentioned later (p. 132).

1. Linear amplifiers

Triodes, pentodes and transistors are mostly used as amplifying elements. The basic characteristic of amplifiers is to increase power. This is attained either by amplifying the voltage (such an amplifier is termed a voltage amplifier), or by decreasing the differential resistance (the amplifier is then termed an impedance transformer) or by both procedures (most of the so called power amplifiers).

Triode voltage amplifier (Fig. 18)

This has two circuits: (1) the input or grid circuit connected between the cathode and the grid, to which the input voltage (e_i) is applied and (2) the output or anode circuit from where the output voltage (e_o) is taken and where the power supply is connected (for details see p. 130).

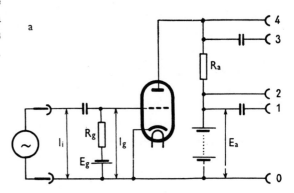

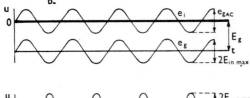

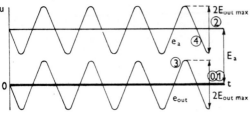

Fig. 18. Basic connection of triode amplifier with common cathode. a) scheme, b) voltage shapes. e_i — input voltage, e_g — total grid voltage, E_g — grid bias, E_a — anode supply voltage. The numerals show voltages measured at the corresponding points. All voltages measured relative to the cathode.

Quite generally we may write $e_o = \varphi(e_i)$. According to the "shape" of this function amplifiers are divided into several categories. If within a certain range of values of e_i the equation $e_o = k \cdot e_i +$ const. is valid, the amplifier is termed a linear voltage amplifier. For recording bioelectric potentials we try to approach such a linear amplifier as far as possible. Its function has already been described (see "triode" and "pentode") and will be discussed in more detail later (p. 132). The linear current amplifier is described by an analogous equation $i_o = kt_i +$ const. It is easiest to approximate to it with a transistor.

2. Wide band amplifiers

They have $e_o = \varphi(e_i)$, i. e. the output voltage e_o is only a function of the input voltage e_i and does not depend (at least within certain limits) on anything else, particularly not on the shape of the input voltage.

3. Tuned amplifiers

They have $e_o = \varphi(e_i, f)$, where f denotes frequency. For $e_i =$ const, e_o attains a maximum or minimum for one or several frequencies called the resonance frequency. They can be linear or nonlinear. The frequency dependence is attained either by resonant circuits connected to the input or output circuit of the amplifier or only by the choice of RC or L/R time constants. As a whole, amplifiers with resonant circuits function mostly as linear amplifiers.

Adding a positive feed back to tuned amplifiers or adding a frequency dependent and time delayed feed back to wide band amplifiers leads to oscillations which are a frequent disturbance in common amplifiers, often very difficult to remove. Oscillations are sometimes also achieved on purpose (e. g. in oscillators).

Amplifiers in the narrow sense of the word will be dealt with in the part on recording. Nonlinear amplifiers are used in circuits shaping the stimulation voltage and also in the so called logical circuits and in circuits used in computers for transforming information contained in the input voltage. Hence we shall first discuss the production of voltage of the required properties, particularly of the required frequency, and then the means of its transformation (shaping).

C. Electronic stimulators

These, according to the shape of the stimulus, are divided into: 1) generators of harmonic oscillations, 2) generators of rectangular pulses, 3) generators of special wave forms. It must be pointed out right at the beginning that any problem may usually be solved by different types of apparatus and that the problem must be thoroughly analysed at first in order that the most suitable approach for the required purpose be chosen. Criteria important for this choice will be given for all examples mentioned.

In general the same purpose may be achieved in two fundamentally different ways: by using a simple single purpose apparatus which produces impulses of a given frequency, duration and amplitude, or by using a universal apparatus consisting of several parts, the first of which may initiate oscillations of a given frequency, the second may change them to rectangular pulses, the following adjust their duration and the last one gives them the required size. Although this second way may seem needlessly complex, it is much more suitable in practice since 1) it permits the variation of any parameter independently of all others, 2) it permits the use of the apparatus for various purposes according to immediate requirements, 3) developing and producing individual standardized parts to be combined as necessary is more economical, 4) failures and disturbances are more easily found and repaired in such apparatus.

In electronics, one of the most important criteria for impulse generators is the pulse width and the slope of its leading edge. This is not very important for our purposes, since it is relatively easy to attain values that are quite suitable for electrophysiology. On the contrary, however, requirements concerning the flatness of the top of pulses or the slope of the trailing edge are sometimes stricter than in electronics. The individual functional elements and the single purpose apparatus will be dealt with systematically below. In the next part the connection of these elements into complete apparatus will be described (see p. 123). It is assumed that the reader is acquainted with the fundamental electrical laws and the principles of valves and semiconductors (see also Nelepets and Nelepets 1960 and Neeteson 1955).

a) Generators of harmonic oscillations

1. High frequency oscillators

High frequency oscillators are amplifiers whose grid or anode impedance or both are frequency dependent (they have parallel resonant circuits, see p. 55) so that amplification depends on frequency and is maximal for the resonance frequency. In addition there is a feedback circuit between the anode and the grid by which a fraction of the output is returned to the input. The energy fed back should be only slightly larger than losses in the circuit. The phase or polarity of the feedback is also important. For a positive (regenerative) feedback (which increases the amplification), the voltage transferred from the anode to the grid must be reversed in phase, i. e. if the anode voltage decreases, the grid voltage must be increased by the feedback and vice versa.*)

The oscillator works as follows: a small change in voltage is amplified, and returned to the grid by the positive feedback. This continues until the

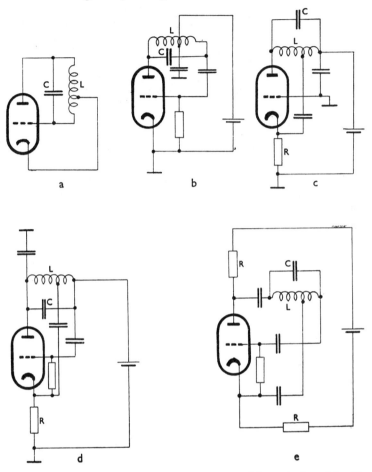

Fig. 19. Hartley's oscillator. a) basic diagram; b—e possible variations; b) with an earthed cathode; c) with an earthed grid; d) with an earthed anode; e) with an oscillation circuit not connected galvanically with any electrode. In place of *R* a choke is more suitable.

*) In other words: if a positive or negative voltage change appears accidentally or is induced across the input, the absolute value of this change is increased by the positive feedback. A negative (degenerative) feedback, on the contrary, results in decreased changes in input voltage because the output is fed back in such a phase as to oppose the input (i. e. plate voltage is returned to the grid of the same valve unchanged in phase). It may of course happen that the feedback is negative for a certain frequency and positive for another. Negative feedback is used for improving the frequency-response characteristics of low frequency amplifiers. It may then happen that such an amplifier becomes an oscillator at a high frequency because the negative feedback has changed to a positive one for this high frequency.

feedback is no longer capable of compensating for the decrease in amplification due to the shift of the grid voltage to the curved part of the tube characteristic. The plate and the grid voltages (their absolute value) now begin to
fall. The rate of decrease is regulated by the resonant circuit. The grid and the
plate voltages are again increased by the positive feedback etc. It is evident
that the frequency of the oscillator depends on the resonance frequency ω_0
of the frequency control unit and on the amount of the feed back. The smaller
the feedback, the more exactly the frequency of the oscillator approaches the
resonance frequency ω_0 and the sinusoidal wave form.

When constructing an apparatus a certain circuit must be chosen (for
details see Smirenin 1950, Terman 1943). Only one of the most usual and most
reliable circuits, the Hartley oscillator, will be described here (Fig. 19). The
resonant circuit is connected between the plate and the grid (Fig. 19a). The
feedback voltage is obtained by connecting the tap of the coil to the cathode
(the coil thus acts as an autotransformer). This connection is also termed
a three-point circuit since the coil is connected to the valve at three points.
One of these points must be earthed, either directly or across a condenser.
Several circuit variants may be used, e. g. the coil may have the D. C. voltage
of the plate, cathode or grid. This is best seen in Figures 19b-c. Similar
combinations are possible for all other types of oscillators. A Hartley oscillator

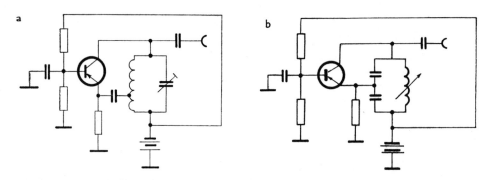

Fig. 20. Transistor three-point oscillators. a) with a tap on coil (Hartley), b) with a "tap"
on condenser (Colpits).

may be constructed with conventional valves and air- or iron-cored coils for
frequencies of several kc/sec. to several tens or hundreds of megacycles.
The same circuit can be constructed also in a transistorized version (e. g.
as a common emitter circuit, Fig. 20). The resonant circuit is usually connected
between the base and the collector, the tap to the emitter. Its position,
just as that of the cathode tap in the valve version is determined by the
amplification of the transistor, by the voltage and power consumption, by
the permissible distortion etc. In general it will be close to the base lead.

2. Low frequency (audio frequency) generators

α) *LC* oscillators are analogous to the high frequency generator described above. Hartley's circuit is usually not employed. Instead a circuit with separate oscillation and feedback coils on the same laminated iron core is used. The oscillators are very sensitive to the amount of the feedback. It is difficult to construct them in such a way that their frequency can be regulated

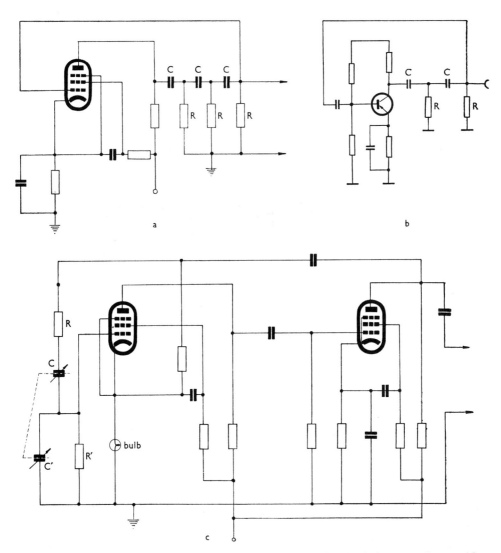

Fig. 21. *RC* oscillators a) with *RC* phase inverter in the feedback from anode to grid; b) transistor version of the same; c) tuned by either a change in *C* or *R* forming Wien's bridge. The bulb serves to stabilise the amplitude.

continuously and they are therefore used only as a source of fixed frequencies. The frequency selector simultaneously switches on the oscillation condensers (fixed) and corresponding resistors controlling the amount of the feedback. They are little used, since large coils are expensive.

β) Two-point oscillators. For detail see Terman (1943), Smirenin (1950). Their description is beyond the scope of this book.

γ) Beat-frequency oscillators were the most exact laboratory oscillators up to recent times. Low frequency is produced as the frequency difference between two high frequencies (about 100 kc/sec) one of which changes over a small range. Thus the frequency difference is changed e. g. from 0 to 16 000 c/sec (Philips GM 2307 oscillator).

δ) They have been replaced by much simpler and more reliable *RC* generators. Fundamentally these are amplifiers with a feedback having variable phase shift. There is a certain frequency for which the phase shift is just 180 degrees. Because of this frequency dependent feedback the amplifier works as an oscillator with a frequency for which the amplification is maximal, i. e. a frequency for which the phase shift is 180 degrees. The circuit shown in Fig. 21a,b is suitable as a fixed frequency source, that shown in Fig. 21c as a variable frequency oscillator controlled by variable resistors and capacitors.

Such an oscillator may also be used as a filter, if the feedback is adjusted so that the oscillator just ceases oscillating. If now any low and distorted voltage of a frequency to which the oscillator is tuned is inserted into the grid circuit, a sinusoidal voltage of the same frequency is obtained at the output.

Using a somewhat higher exciting voltage, this "filter" may oscillate with its fundamental frequency even if a higher harmonic or a subharmonic frequency is used at the input.*) It can thus also be used as a frequency multiplier or divider with the advantage that it is often possible to obtain a nearly sinusoidal voltage by using a single stage apparatus.

b) Generators of nonharmonic oscillations

These produce a voltage having a shape that differs considerably from the sine wave. In other words, their oscillations have many higher harmonic components.

For physiology, a rectangular pulse is most important. Its main parameters are the amplitude E, the repetition interval T or the repetition frequency $f = 1/T$, the pulse width τ and the quiescent interval $T-\tau$, the ratio of the pulse width to the repetition interval, so called duty cycle $\tau/T = \tau f$.

---

*) If the frequency of the filter is F, the n-th harmonic frequency is $f_n = nF$ and the n-th subharmonic (submultiple) one $f_s = F/n$.

Another important waveform are the sawtooth waves (ascending and descending) where in addition to E, T and f, the duration (absolute and even more relative) of the steep edge of the wave is essential (usually the shortest possible is required). It must be borne in mind that the rising part of the sawtooth wave is never linear with time but always curved. The voltage-rise curve is mostly exponential but it may also be the middle, almost straight, portion of a sine wave (passing through zero) etc.

The majority of oscillators that are described here function according to one of the following principles: a) they have a very large positive feedback which results in an unusually rapid rise and fall of voltage. Flattening of the pulse top then occurs in the curved part of the valve characteristic. These include multivibrators, blocking oscillators and transitrons. b) An important characteristic of a discharge in gases is made use of, i. e. that the ignition voltage is usually higher than the extinction voltage.

The majority of oscillators can be driven with an external A. C. voltage close to the fundamental frequency of the oscillator or to harmonic or sub-harmonic frequencies. The majority of oscillators can be changed to shaping circuits, i. e. they can be blocked so that they do not oscillate unless an external signal is applied . They then produce only a single wave.

1. Multivibrators

A multivibrator is a two-stage resistance-coupled amplifier with a tight feedback. The output of the second tube is returned to the grid of the first, and the plate of the first valve is connected to the grid of the second. Both couplings may be capacitative (astable multivibrator), resistive (bistable multivibrator) or mixed.

Astable (free-running) multivibrator

The circuit is usually more or less symmetrical, the valves and the corresponding capacitors and resistors being equal. The apparatus works as follows (Fig. 22): The positive voltage appearing at the grid of the first valve is amplified in this valve and reversed in phase 180 degrees. In the second valve it is further amplified and again reversed in phase. Thus the plate voltage from valve E_2, which is fed back across C_{g_2} to grid E_1, is added to the original signal voltage, returning along the same path through the feedback. As a result, the plate voltage in E_1 rapidly drops and that in E_2 rapidly increases, so that the first valve is rapidly "opened" to the highest possible plate current (limited by the load resistance) and the second is closed so that practically no current flows through it. The current in E_1 can rise no longer nor can it continue to fall in E_2 and despite the feedback, the valves can no longer amplify. Under these conditions the grid E_1 has a maximum positive voltage (i.e. approximately zero), E_2 a maximum negative voltage (about -10 to -50 V). The condenser

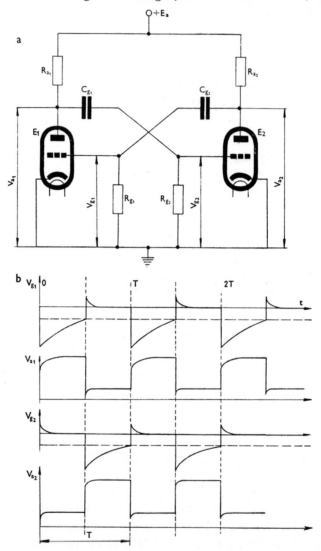

Fig. 22. Multivibrator. a) circuit diagram; b) shape of voltages at electrodes.

C_{g_1} is discharged across R_{g_2} with a time constant $R_{g_2}C_{g_1}$. When the voltage across C_{g_1} has fallen (the voltage of E_2 has risen) sufficiently for a plate current to flow through valve E_2, the plate voltage at anode E_2 decreases somewhat (by the voltage drop caused by the plate current flowing through R_{a_2}). This decrease is carried by the feedback condenser C_{g_2} to the grid of E_1 through which the maximal possible current has been flowing. This current is somewhat decreased and thus the plate voltage of E_1 rises to a certain extent. This increase is transferred across C_{g_1} to grid E_2 and, consequently, the rate of current increase in E_2 is accelerated. Thus a decrease in plate voltage of E_2, and a corresponding plate current decrease in E_1 is rapidly built up. In this

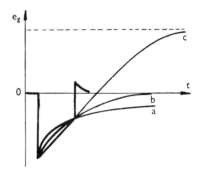

Fig. 23. Different pulse [shapes at multivibrator grid. a) with negative grid bias; b) with zero bias; c) with positive bias.

way the current is interrupted in E_1 and attains maximal values in E_2 during a very short time. This state lasts until the large voltage across C_{g_2} is discharged through R_{g_1} to such an extent that a plate current begins to flow in E_1. This results again in a rapid "reverting" of the multivibrator and conditions as described at the beginning are again restored. From this description the time curves of voltage at the individual electrodes may be derived (Fig. 22c): The time of reverting is exceptionally short in comparison with repetition interval T, so that for the large majority of the operation cycle either a maximal or zero current flows through one or the other valve. Thus the voltage at the anodes is constant — either minimum (about 10 V) or maximum (the full voltage of the anode supply). The plate voltage changes in steps. The plate voltage of one valve is a mirror image of that of the other. The grid voltage, however, changes according to another time curve. During change-over, the grid of the valve which has become non-conductive (and has a maximal plate voltage) is charged to a minimum voltage (i. e. highly negative). Thus the grid and plate voltages are 180 degrees out of phase, as in any amplifier. In distinction to the plate voltage, this grid voltage does not remain constant but gradually increases (becomes less negative) as the C_{g_2} is being charged (or "discharged", if the absolute voltage value across C_{g_2} is considered). It is evident that the voltage across the grid has an exponential shape and that again both the grid voltages are approximately mirror images of each other.

The width of the positive pulse at plate E_1 thus depends upon the time constant, given by the product R_{g_1}, C_{g_2} and on the "cut-off" voltage of valve E_1, i. e. on the position of the cut-off point on the valve characteristic curve. Thus the kind of the valve, the plate voltage and also the voltage at the lower end of R_{g_1} are important (Fig. 23b). If R_{g_1} is connected directly to the cathode, the plate current will begin to flow when C_{g_1} is already considerably discharged and when consequently the discharge rate is already slow. If R_{g_1} is connected to a slightly negative bias voltage, the voltage across E_{g_2} attains the "cut-off" grid voltage even later and more slowly. This results in a decreased frequency, an increased pulse width and also in a deterioration of the stability of the oscillator. The curve of the voltage drop across C_{g_2} intersects the cut-off bias level at a very sharp angle so that the point of transection is not determined very accurately (Fig. 23a). Such a connection is of advantage only rarely, e. g. if the multivibrator is to be synchronised within a relatively large range of frequencies. This is really a transition to a monostable multivibrator. If, on the contrary, R_{g_1} has a positive bias (Fig. 24a), the frequency increases several times, the pulses are shorter and the moment when reversal occurs is determined much more accurately. Therefore the frequency and pulse width are much more stable and variations drop to only 1%. In addition, the grid voltage curve is formed by shorter exponential sections than in the above examples (Fig. 23c). Such a multivibrator may thus be used also as a source of voltage having a sawtooth wave form, but the output voltage must then be taken from the grid and not from the anode.

Frequency and pulse width controls: If $C_{g_1} = C_{g_2}$, $R_{a_1} = R_{a_2}$ and both valves have identical properties, as far as possible, then the pulse width and the quiescent interval are equal. In that case the frequency can be changed while maintaining a stable ratio $1 : 2$ of pulse width to repetition interval (so called duty cycle) by changing simultaneously and to the same extent R_{g_1} and R_{g_2} or C_{g_1} and C_{g_2}. In a non-symmetrical circuit ($C_{g_1} \neq C_{g_2}$, $R_{g_1} \neq R_{g_2}$) pulses longer or shorter than the quiescent interval may be obtained, depending on the valve from which the output voltage is taken. It is, however, impossible to obtain a pulse width independent of repetition frequency and vice versa. (Thus the multivibrator is not very generally useful as a stimulator and can be used rather as a fundamental oscillator, the pulses of which are given the required width and amplitude later on). Neither frequency nor pulse width change linearly with a change in the control elements. The multivibrator can be used for frequencies from about $0 \cdot 001$ c/sec to 100 kc/sec and more, and for duty cycles ranging from $1 : 100$ to $99 : 100$. If we want to attain very steep front and trailing edges, valves with a large transconductance must be used, having a high plate and screen dissipation, small internal capacitance and small anode load. Thus pulse widths of several tenths of a μsec. may be obtained.

Special types of multivibrators

The following special types of multivibrators may be mentioned:

Electronically coupled multivibrator: The multivibrator is connected to triode parts of two pentodes (the screen grids are used as anodes). The anode circuits contain loads from which the output is taken. In such a circuit the frequency of the multivibrator depends only very little on the load (Fig. 24b).

In a multivibrator giving an improved shape of oscillations the coupling condensers C_{g_1} and C_{g_2} are connected to the grids not directly but across

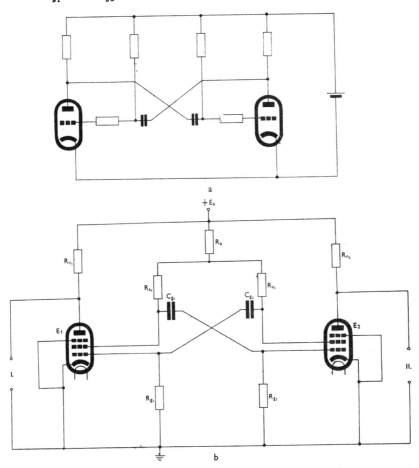

Fig. 24. a) Multivibrator with improved pulse shape and better frequency stability. b) Electronically coupled multivibrator.

a series resistor of 100kΩ or more, which may be blocked by a condenser with several tens pF. In such a way the grid circuit also acts as a grid limiter. The pulses are more rectangular, but the edges are not so steep as in the usual circuit (Fig. 24a).

A multivibrator with a positive bias: The common joint of the grid resistors is blocked with a large condenser (0·1 μF) to the frame and connected to the tap of a potentiometer which is situated between the poles of the D. C. source. By changing the bias voltage the frequency is regulated within a very wide range (up to 1 : 40).

There are many more modifications of multivibrators. The multivibrator with a cathode coupling, the multivibrator producing trains of pulses and Antipov's series multivibrator are worth mentioning. They are described in the literature (Petrovich and Kozyrev 1954, Antipov 1951, Meyerovich and Zelichenko 1953).

The multivibrator can easily be converted into a shaping circuit which either prolongs a short pulse to a constant width and definite shape (rectangular) or changes it into a positive or a negative voltage step (either positive or negative depending on the preceding state).

Monostable (one-shot, single-kick) multivibrator

The monostable multivibrator is an asymmetrical multivibrator in which one valve is constantly non-conducting because of a large negative grid bias while the other one is constantly conducting (Fig. 25). The arrival of the trigger pulse (positive at grid E_1 or anode E_2, negative at grid E_2 or anode E_1) results

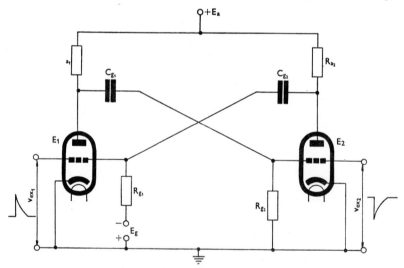

Fig. 25. Monostable multivibrator.

in change-over of the multivibrator which remains in this reversed state (or produces a pulse) so long as the grid condenser is not discharged through the grid resistance. It then reverts to the original state, which lasts until the next triggering pulse arrives. Thus the monostable multivibrator has one stable and one unstable equilibrium position.

Bistable multivibrator

The bistable multivibrator (Eccles-Jordan circuit, resistance coupled trigger circuit, binary reductor, flip-flop) is a multivibrator with two stable positions. The circuit (Fig. 26a) is completely symmetrical. Both grids receive D. C. voltage from a voltage divider connected between the large negative

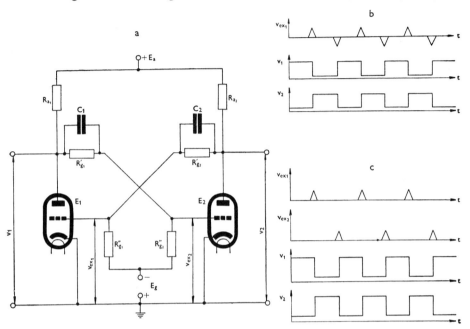

Fig. 26. Bistable multivibrator. a) circuit diagram; b) supplied with pulses of alternating polarity on one gid; c) supplied with two trains of positive pulses.

bias and the anode of the other valve. The anode section of the divider is by-passed by a small condenser (10 — 100 pF). Let it be assumed that the valve E_1 carries anode current. Then its anode voltage is very low and the potential on the grid E_2 is below the cut-off value, so that E_2 is nonconducting. The voltage at its anode is very high (since no anode current is flowing). This high voltage is communicated to the grid E_1 and a maximal anode current is therefore flowing through E_1. The grid voltage is near zero, however, because of the voltage drop caused by the grid current on the series grid resistance. This state is evidently stable. If a negative pulse is applied to the grid E_1 (or to anode E_2 from where it is transmitted to grid E_1 through the condenser), then E_1 is blocked, the anode current ceases to flow and the anode voltage rises up to the values of the anode supply.

This positive pulse is transferred to grid E_2 through the condenser C_1 and resistor R'_{g_1}, resulting in the appearance of an anode current at E_2. The voltage at the anode of this valve decreases and this decrease is transferred through the condenser C_2 and the resistor R'_{g_2} to grid E_1. Thus this new state is fixed even when the triggering pulse has passed. This second equilibrium position is also stable. Change-over of the circuit to the original state occurs only when a further triggering pulse of suitable polarity reaches an appropriate point of the circuit. In order to produce one rectangular pulse two triggering pulses are necessary. Thus a bistable multivibrator can be used for frequency division (to half the frequency of the signal).

Synchronization of multivibrators

The synchronisation of a multivibrator, triggering of a monostable multivibrator and change-over of a bistable multivibrator must be considered.

As has already been mentioned, a multivibrator can be made to lock, in with a frequency close to its own frequency (somewhat higher so that the change-over due to the injected voltage occurs sooner than automatic change-over would occur) or close to its higher harmonic or even submultiple (subharmonic) frequency. In the latter case the pulses of the multivibrator are of unequal width, since those triggered by the submultiple frequency are somewhat shorter than the others. A synchronised multivibrator is most frequently used for frequency division. i. e. it is synchronised by a frequency several times higher than its own. The higher the synchronising voltage the greater the acceleration of the multivibrator's oscillations, i. e. the further it deviates from its own frequency, until for a sufficiently large synchronising voltage, it is driven so hard.that it is locked in with the injected signal. Thus it is possible for instance, to synchronise a multivibrator using a voltage of about 1 V to a frequency 12 times lower than the synchronising frequency. For a higher synchronising voltage, the frequency of the multivibrator increases so that it is 11, then 10, 9, 8 etc. times lower than the synchronising frequency until for a synchronising voltage of about 40 V the frequency of the multivibrator reaches the synchronising frequency.

Since it is possible to inject the synchronising voltage into both valves of the multivibrator in the same or opposite phase, the multivibrator can be synchronised either with even or odd submultiples of the synchronising frequency. If the signal voltage is directed to both grids or both anodes in the same phase, synchronisation for even submultiples is accentuated. If the potentials applied to both grids or anodes are reversed in phase, synchronisation for odd submultiples is easier. If neither odd nor even submultiples are preferred, the synchronising voltage is applied only to one grid or anode. The above applies to the symmetrical multivibrator. In an asymmetrical multivibrator, it is even possible to make it several times more sensitive for a certain submultiple than for all others.

Trigger wave form

Fundamentally it is possible to synchronise multivibrators using any periodic voltage, i. e. also harmonic (sine wave) voltage. Synchronising voltage peaks with a steep wave front and an exponential trailing edge, such as are obtained e. g. by differentiation of a rectangular voltage, are most effective and also determine the moment of reversal most accurately. Pulses of the same shape (derivative spikes) are also most suitable for triggering monostable and especially for reversing bistable multivibrators. For this purpose, however, they must have a sufficiently large amplitude, exceeding the negative grid voltage (usually $10-100$ V), and a sufficiently long duration (usually $0 \cdot 1 - 10$ μsec), so that after their passage the decrease of bias voltage will not again cut off the valve which has just started conducting. An exponential trailing edge is desirable, since the exciting voltage gradually falls while the internal feed-back reversing voltage gradually rises. After the first voltage step produced by the triggering pulse front, no further abrupt potential changes occur which might either interrupt the process of change-over or might produce, *sit venia verbo*, an extrasystolic change-over after the first one has terminated.

Triggering of monostable and bistable multivibrators

Triggering of a monostable multivibrator is simple, since the triggering pulse is applied to a single point of the circuit. It is more complex in a bistable multivibrator and consequently will be discussed in more detail.

a) Bistable multivibrator may be reversed in a similar way to monostable multivibrators: Pulses of both polarities are directed to one point (the grid or the anode — Fig. 26b). Then a negative pulse applied to the grid of the first valve, which is conducting, blocks it, while a positive pulse arriving under the same conditions has no effect. If, on the other hand, the valve is not conducting, only a positive pulse can swing its grid above the cut-off point while a negative one has no effect.

b) Two trains of impulses of the same polarity are applied, one to the grid of the first valve and the second to the grid of the second valve (Fig. 26c). After the bistable multivibrator has been reversed by a certain impulse of one train, further impulses of the same train applied to the same point remain without effect so long as the next impulse of the second train does not revert the circuit back again. The same would hold good for impulses directed to the anode.

c) The third method of reversal is fundamentally different. It is characterised by complete symmetry and formal uncertainty as to which valve will be reversed by any impulse. This means that every impulse reverses the circuit (in other words all impulses are effective) and that the direction of change-over is independent of the circuit and only depends upon the actual state of each valve (if it does or does not conduct). The triggering impulses are applied

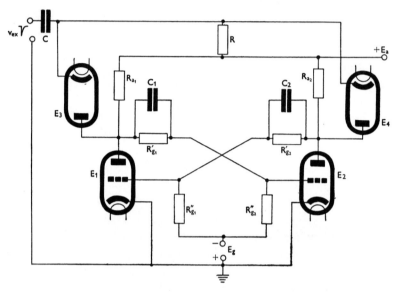

Fig. 27. Bistable multivibrator with electronic switching of pulses. (Instead of vacuum diodes semiconductor rectifiers may be used.)

simultaneously in both valves (either to both grids or to both anodes, or — in a pentode circuit — to both suppressor grids). It is evident that the impulse at one valve tends to reverse the circuit, while the same impulse applied simultaneously to the second valve counteracts the change-over and makes it more difficult. Yet it is possible to find a certain size and width of the pulse which will regularly produce the change-over, since the conducting and nonconducting valves differ in their sensitivity to the pulse. Driving would be much more reliable, however, and the size and duration of the triggering pulses much less critical, if they could be directed only into that valve in which they can produce a change-over and if they could be prevented from reaching the valve in which they will oppose the turning over. This is best attained by a circuit shown in Fig. 27. The negative triggering pulse is directed to the anodes of both valves through germanium rectifiers (or vacuum diodes). Impulses are applied simultaneously to the cathodes of both rectifiers having the D. C. voltage of the anode supply (through R), i.e. the plate voltage of a nonconducting valve. The anodes of both rectifiers, however, have different voltages, since they are connected to the anodes of the valves, one of which is conducting, and thus its anode voltage is very low (10—50 V), the other is cut off and has the full plate voltage equal to anode supply. In other words: the rectifier connected to the blocked valve has a relative anode voltage (i. e. a potential difference between its cathode and anode) of about 0 V, while in the rectifier connected to the conducting valve the anode is about 100 to 300 V more negative than the cathode (according to the circuit and the anode supply voltage). Negative pulses having an amplitude of 50—100 V arriving at the cathode of both rectifiers can then only pass through the rectifier having zero relative anode voltage and cannot overcome the negative anode-to-cathode voltage of the other rectifier. Thus negative trigger impulses only reach the anode of the nonconducting valve, are transmitted through the coupling unit to the grid of the conducting valve which is thus cut off. This electronic switching of trigger impulses makes change-over considerably more accurate and less sensitive to correct adjusting of the width, shape and amplitude of the triggering impulse. If the circuit is to be reversed by positive impulses, it is usually sufficient to change the polarity of both properly biassed rectifiers.

In addition to these completely symmetrical bistable multivibrators there are also slightly asymmetrical circuits, e. g. with a cathode coupling. They do not require a source of high negative grid-bias, but calculations and adjustment are more difficult.

Since all calculations are only approximate, it is best to test them by building a "bread-board" model in which special care is taken to achieve symmetry. We do not check the exact absolute values of every component, but the symmetry of both R_a, R_g etc., and particularly the symmetry of valves. We choose valves that have the same characteristics — chiefly the cut-off grid voltage and the same D. C. resistance for zero grid bias (i. e. for the grid connected to the cathode).

2. Transistor multivibrators and triggers

The transistor multivibrator basically works on the same principles as the valve multivibrator. It consists of two cross connected (i. e. quite symmetrically connected) common emitter amplifiers (Fig. 28). Let us assume that transistor T_1 has just started to conduct, i. e. the voltage at the collector has risen practically to zero. This voltage step is transferred to base T_2 by condenser C_1 and causes the decrease or disappearance of the current between the emitter and base T_2. Before the jump the base current was limited by resistance R_4, the voltage drop on which ensured that the voltage of the base T_2 was practically the same as that of the emitter. The transferred voltage step leads to considerable rise of the base voltage so that the emitter-base junction becomes nonconducting. Transistor T_2 does not conduct and this lasts as long as the charge induced by the voltage step through C_1 to base T_2 is not discharged through the resistance R_4. Then the voltage at the base again falls to that of the emitter and at that moment a weak current again starts to flow through the emitter and thus also through the collector. The collector voltage rises and this rise is transmitted by the condenser C_2 to base T_1 decreasing the current in the latter

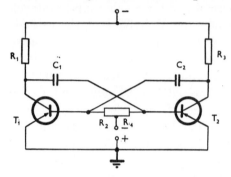

Fig. 28. Astable transistor multivibrator.

transistor. The decrease of its collector voltage accelerates the rise in T_2 current. Thus the voltage rise at collector T_2 and the voltage fall at collector T_1 occur very rapidly and only cease when T_1 is fully blocked and T_2 fully open. At that moment, however, the voltage of base T_1 is very high and thus this state is maintained until the charge of condenser C_2 flows away through R_2. Then there is rapid change-over to the initial state and the whole process is repeated. Hence the duration of each "half" of the cycle is determined by the time con-

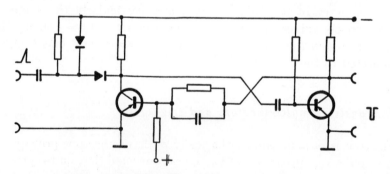

Fig. 29. Monostable transistor multivibrator.

stant of the coupling condenser and of the base resistor, i. e. C_2R_2 and C_1R_4 and $f \approx (C_2R_2 + C_1R_4)^{-1}$.

A monostable multivibrator in transistor version may be obtained from an astable one by introducing a blocking bias (Fig. 29).

It is also possible to derive a monostable multivibrator from a bistable one (see below) by introducing a purely capacitative coupling instead of one of the resistive (or resistive-capacitive) couplings. Then the transistor to the base of which this coupling is applied cannot remain indefinitely in the blocked state but returns to the conducting state after the condenser is discharged.

The bistable transistor multivibrator is also an analogy of the valve one (Fig. 30) and is used for the same purposes. This circuit is very common, particularly in counters and computers and hence there are many types developed

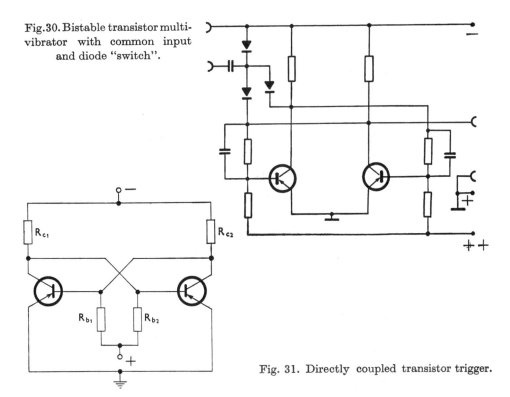

Fig.30. Bistable transistor multivibrator with common input and diode "switch".

Fig. 31. Directly coupled transistor trigger.

for different applications. The simplest are circuits working with the saturated current of an open transistor (Fig. 30). In this case the open transistor can be considered as a point in which the emitter, base and collector are galvanically joined. The main disadvantage of this circuit is the necessity of individual calculations or adaptations for each individual transistor used. For serial production nonsaturated circuits are therefore most frequently used (Kononov 1960).

Using transistors, flip-flops may also be constructed which have no valve analogies, e. g. direct coupled triggers (Fig. 31) which are particularly simple.

Schmitt's circuits (monostable or bistable) are fairly frequently used instead of multivibrators. They are particularly useful as limiters (p. 118) since their turn-over occurs suddenly as soon as the input voltage attains a certain value and the output voltage has two fixed values (two states). (Fig. 32.) (An analogous circuit can, of course, also be constructed with valves.)

The triggering of transistor flip-flops is the same as described for valve circuits. It may be achieved by applying impulses to the bases (one or both, each separately, or both simultaneously), to the emitters or to the collectors. Very often diodes (or transistors) are used analogously as described above in

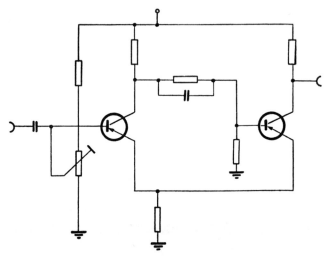

Fig. 32. Transistorized Schmitt trigger.

Fig. 27. The point of connection of the triggering impulses, their polarity and thus also the diode orientation must be selected in such a way that the "triggering" impulse closes the open transistor, not the other way round, since the closing is more rapid and requires less energy.

In general it must be mentioned that for transistor multivibrators of all kinds the possibilities of varying the circuits are much greater than in the valve versions since e. g. polarity can be chosen in such a way that we take either PNP or NPN transistors. It is also possible to construct even simpler circuits using special semiconductor parts such as tunnel diodes, four layer diodes, controlled rectifiers (semiconductor analogues of thyratrons), unijunction transistors etc.

For the sake of interest only, the relaxation generator with a unijunction transistor (Fig. 33) is described here. The unijunction transistor has one diode PN junction between the P-type emitter and the semiconductor of the N-type. On the other side of the N-semiconductor disc there are two leads with an ohmic contact, called base-one (usually earthed) and base-two (lying closer to the emitter and connected to the positive pole of the supply). The internal resistance between the bases is of the order of 10 kΩ, its size remaining more or less constant as long as no current flows through the emitter.

If a positive voltage V_E connected between the base-one and the emitter is increased, only a very small leakage current flows as long as the voltage on the emittter does not exceed the voltage at the internal resistance between the bases at the site where the rectifying emitter PN junction is situated (this is connected in the nonconducting direction). On further increasing V_E to V_p the junction is connected in the conducting direction and current I_E begins to flow

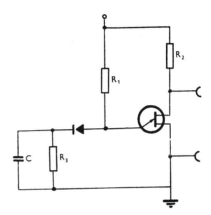

Fig. 33. Relaxation oscillator using unijunction transistor.

carried by the holes from the emitter into base one. This increases the carrier concentration in a large part of the semiconductor volume and thus decreases its resistance, particularly between the emitter and base-one. A fall in resistance means a fall in voltage both between bases (V_{BB}) and between the emitter and base-one (V_E). In other words: although the voltage on the emitter V_E falls, its junction remains connected in the conducting direction since the voltage on the semiconductor portion between the emitter and base-one falls just as rapidly. Hence the section emitter — base-one now has a negative differential resistance $r_E = dV_E/d\,I_E$ and can be used for generating undamped oscillations, e. g. of the relaxation type (Fig. 33). Let us assume that condenser C is discharged. When the circuit is connected to the voltage supply, C begins to be charged with a time constant $\tau_n = R_1 C$; as soon as the voltage on the emitter V_E rises to the "ignition" voltage V_p, emitter current begins to flow, V_E falls below the condenser voltage, the diode becomes nonconducting and the condenser C begins to discharge with a time constant $\tau_d = R_3 C$. As soon as the condenser voltage falls below the emitter voltage, the diode is again opened, the emitter current falls and the emitter resistance rises progressively. Thus the emitter-base-one junction is closed and the charging of C through R_1 starts again.

Unijunction transistors are a great simplification of many circuits used in electrophysiology and function very reliably.

3. Blocking oscillator (blocking generator, generator of impulses with transformer coupling)

Fig. 34 shows a blocking oscillator consisting of a single triode and a transformer coupled feedback circuit. The blocking oscillator somewhat resembles a harmonic oscillator with an inductive feedback. It differs from the

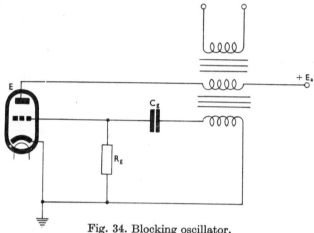

Fig. 34. Blocking oscillator.

latter in the fact that neither the anode nor the grid windings form a part of the resonant circuit and the transformer has a minimum stray magnetic field. Let it be assumed that condenser C_g is charged to a high negative voltage. The valve is thus cut off and the condenser is slowly discharged through R_g. As soon as discharge has proceeded sufficiently so that the anode current starts in the valve, this small initial current causes a further increase of the grid voltage communicated to the grid by the transformer feedback. Since this feedback is very tight, the grid voltage, and thus the anode current, increases rapidly up to limiting values. This state lasts for a short time. Since the grid voltage has attained positive values a grid current flows and the cathode current is thus divided into two parts — the anode current flowing through the transformer and producing the feedback — and the grid current charging grid condenser C_g. As voltage across C_g increases, the grid current decreases, but the anode current continues to flow, since the grid voltage has not yet dropped below zero. As soon as the grid condenser C_g is charged to such an extent that the anode current slightly decreases, this decrease in the primary (anode) coil of the transformer produces a voltage drop in the secondary (grid) coil and this is transferred to the grid through the grid condenser C_g (which a moment sooner had a high positive voltage at the condenser plate nearer to the transformer and a small positive voltage decreased by the grid current at the plate nearer to the grid). In this way the grid voltage is further reduced.

Consequently the grid and anode currents continue to fall rapidly; this drop is again fed back through the transformer to the grid and so on. The anode current continues to fall very rapidly. After this current and the voltage across the secondary winding of the transformer have disappeared, a large negative charge remains on the grid condenser. This maintains the whole circuit at rest (in the blocked state) so long as the grid condenser is not discharged through the grid leak. Thus C_g discharges through R_g with a relatively long time constant $C_g R_g$ and it is charged through the grid resistance of the valve, which is several orders of magnitude smaller. Charging corresponds to the pulse width, discharging to the quiescent interval. The transformer increases the steepness of the rise and fall of the anode current, i. e. the slope of the leading and trailing edges of the pulse. Exact calculations of the blocking oscillator are very difficult and time consuming. Meyerovich and Zelichenko were the first to describe this accurately (1953).

Changing of an individual element usually affects all the characteristics of the oscillator simultaneously (as in the multivibrator), but some more than others. Thus the frequency is mainly influenced by R_g, but also by C_g. C_g acts mainly on the pulse width, but also on the repetition frequency. With increasing R_g (and also C_g) the frequency falls. The pulse width increases with increasing C_g. The pulse width of a conventional blocking generator may be decreased down to $0 \cdot 1\ \mu$sec, but its upper limit is also relatively low. If the pulse width is prolonged above several tens of μsec, the shape of the pulse is maintained with great difficulty. The frequency of a blocking oscillator, on the other hand, can be controlled within a wide range from several c/sec to tens and sometimes also hundreds of kc/sec. In every case, however, the quiescent interval is at least several tens (up to several ten thousands) times longer than the pulse width. This permits the accumulation of energy (which will be used up in the production of the pulse) in the condenser C_e (Fig. 35) separated from the anode supply and thus from the other parts of the circuit by the charging resistance R_e. Thus even for a large pulse energy a relatively small anode supply can be used (it is calculated not for maximum but for average power which is T/τ times smaller). The effect of the blocking oscillator on the other parts of the apparatus is also reduced. C_e is calculated so that the voltage across it does not fall during the pulse passage by more than $10-30\%$. R_e is chosen in such a way that C_e can be charged to nearly the full anode supply voltage during the quiescent interval. At the same time R_e must be large enough to avoid more than a minimal drop in supply voltage during the pulse passage. Let us now consider the most important part of the blocking oscillator, the pulse transformer. Very exacting demands are made on it. Its own resonant frequency must be as high as possible, i. e. the winding capacitance, leakage inductance and core losses must be as small as possible. Hence it is usualy constructed as follows:

a) The core material has the highest possible permeability and the least possible losses. Usually a very thin band of permalloy or other alloy is rolled into a small torroidal core. For instance, a band $0 \cdot 1 \times 5$ mm, internal core diameter 15 mm, a total of 30 turns of the band (insulated with varnish).

b) Primary, secondary (and if necessary also tertiary) coils of approximately the same size are wound with a thin and a thickly insulated wire (for decreasing the capacitances). For a core described above, each coil has 85 turns of copper wire, $\varnothing$ $0 \cdot 15$ mm, EDS (enamel, double silk).

c) The whole transformer is covered with a good insulator having small dielectric losses and a small dielectric constant (e. g. ozokerite) or with bee's wax or paraffin.

Even the best transformer produces a small backswing at the end of the pulse. A bad transformer may give very large and long lasting damped oscillations. These can be considerably reduced in a good transformer by shunting one winding of the transformer with a resistance of about 5 kΩ—10 kΩ. This, however, is of no avail in a bad transformer. A still smaller resistance would

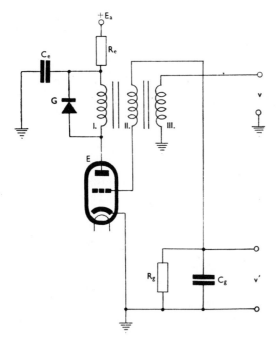

Fig. 35. Blocking oscillator with separation filter $(C_e R_e)$, damping diode (G) and output coil in transformer (III).

help, but this would considerably distort the shape of the pulse and decrease its amplitude. Even in such a bad instance shunting of the anode coil with a germanium rectifier which has its anode connected to the anode of the triode and its cathode to the opposite end of the primary coil, usually improves things (Fig. 35). Its function is the following: it has no effect during the main pulse itself since no current can flow through it. As soon as the main pulse has passed, however, and the backswing of opposite polarity is due to appear, i. e. when the triode plate becomes more positive (at the expense of energy accumulated in the transformer) than the anode supply voltage, the stored energy is destroyed through the rectifier. Thus no further oscillations occur. This arrangement sometimes makes it possible to use a conventional low frequency transformer if demands on the steepness of the front and trailing edges of the pulse are small.

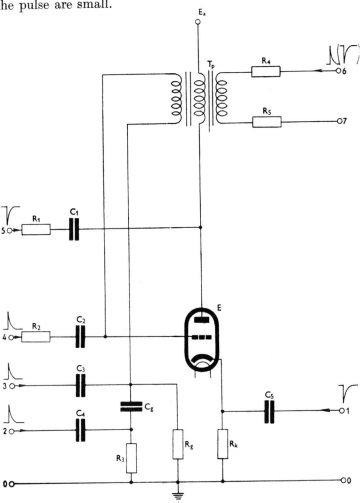

Fig. 36. Different methods of synchronising a blocking oscillator.

It must be stressed that a blocking oscillator can oscillate only if the transformer coils have the correct polarity. If it does not oscillate, the error is usually here and it is sufficient to exchange both ends of one coil (either the anode or the grid coil).

The stability of the pulse width and repetition frequency is rather poor (5—10%) for the conventional blocking oscillator circuit. It is most easily increased, as in the multivibrator, by connecting the grid lead to a positive potential. For this purpose the potential of the anode supply is best. The repetition frequency, of course, increases approximately 10 times. The frequency can be stabilised further by synchronising it with a frequency about 15% higher. The synchronising sine wave or peak voltage can be directed to any more or less arbitrary point (Fig. 36).

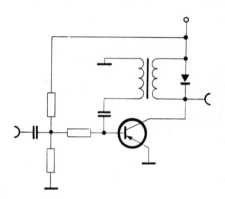

Fig. 37. Transistor blocking oscillator.

The basic circuit of a blocking oscillator can be changed considerably. For instance, C_g and R_g can either be connected directly to the grid (Fig. 34) or to the other end of the secondary coil (Fig. 35). The anode can be connected either in series with the primary coil or in parallel.

The output voltage may be taken from the blocking oscillator either from the anode, from the tertiary coil of the transformer (v in Fig. 35) (more or less rectangular pulses), from the grid or from the other end of the grid coil (if it is not earthed) (v' in Fig. 35). The shape of this voltage approximates a sawtooth wave.

As in the monostable multivibrator, blocking oscillators may be blocked by a negative grid bias and triggered off by external triggering pulses. Since this is used only rarely in electrophysiology, it will not be discussed in detail (cf. Petrovich and Kozyrev 1954).

A blocking oscillator can also be easily constructed using a simple transistor circuit with a common emitter (Fig. 37). If the base is connected via a resistance to the emitter terminal of the supply, the oscillator is blocked. It is triggered by an impulse led via a condenser to the base (negative for PNP and positive for NPN transistors). If the base is connected via the resistance to the collector terminal of the supply, the oscillator oscillates freely.

Glow-discharge and thyratron oscillators

The basic component is a two to four electrode valve filled with gas at low pressure (H_2, He, Ar, Ne, Hg, Hg + Ar) with a cold or hot cathode, a so-called gas tube. If a voltage is placed across its main electrodes (cathode and anode), a current will pass through it so that at first the current is small and rises linearly with the voltage. At a certain voltage, termed breakdown voltage, the current rises suddenly to such a high value that, if the valve were not protected by a series resistance, it would be destroyed by heat dissipation. With the series resistance the anode-to-cathode voltage falls during conduction, independently of the supply voltage, to a certain constant value. If now the source voltage is increased, the voltage across the valve remains constant and only the current increases (and also of course, the voltage drop across the series resistance). The circuit behaves as if the valve had a zero resistance. If, on the contrary, the anode supply voltage is decreased, the current also decreases, but the voltage across the valve again remains constant. Only after the supply voltage falls to that of the valve is the current flow interrupted and the valve switched off. It is evident that this extinction voltage is only slightly lower than the anode-to-cathode voltage during conduction and is considerably lower than the ignition voltage for which this large current begins to flow. This part of the ionic discharge in a gas is termed *glow discharge*. Cold cathode valves in which it is utilised for electronic or lighting puposes are called glow discharge tubes. The extinction voltage is usually between 70 and 200 V, according to the type, and the ignition voltage is by several tens of V higher. This difference between the extinction and ignition voltages can be reduced to several volts or less (e. g. in voltage regulators). Accurate measurement of the valve resistance during the glow discharge shows that it is not zero but $100 - 10\ 000\Omega$.

The ignition voltage is not stable but depends considerably upon the number of free electrons in the gas. This means that it will decrease with external temperature, with ionising radiation etc.

The glow discharge is only maintained if the current is not too large. If, by increasing the supply voltage the current is raised considerably, the voltage across the valve begins to rise again until it reaches the second critical value (analogous to the ignition voltage attained earlier). This occurs when the temperature of the cathode attains such a value that considerable thermionic emission sets in, usually with a current of several amperes. Then the character of the discharge changes, its resistance falls to $1-100\ \Omega$ and the voltage across the electrodes (if there is a limiting series resistance) drops to about 10 V. Evidently analogous phenomena as in the glow discharge occur here. This part of the discharge is termed the *arc*. (Another kind of discharge — the spark is of similar character.) Thus a hot cathode (or at least a hot point on the cathode) is an important condition for the arc to occur. This heating is attained either by Joule heat at the electrode contact (the carbon lighting arc), by bombarding a metallic surface (usually mercury) with positive ions (ignitrons etc.) or finally by heating the cathode with an auxiliary current (thyratrons). Only thyratrons will be considered here. Further details concerning discharges in gases can be found in Kaptsov (1950, 1956).

In the oscillators considered in this paragraph the fact is utilised that the ignition voltages of the glow and arc discharge are higher than the respective extinction voltages. In glow discharge tubes this difference is about 20% of the ignition voltage, in thyratrons up to about 99%. Thyratrons, in addition to the anode and directly or indirectly heated cathode, contain a third electrode, the grid, which does not, however, control the anode current, but only the ignition voltage. The grid voltage is usually negative with respect to the cathode by tens to hundreds of volts and since the grid completely surrounds the cathode, it prevents the emitted electrons from leaving. The arc is started as follows: a) either the anode voltage rises sufficiently so that enough electrons leave the cathode through the opening in the grid. If they are accelerated enough as to evoke mass ionisation in the space between the grid and the anode, the discharge is initiated. The grid loses all influence on the discharge since it is surrounded by positive ions and thus its field is screened. The discharge is interrupted only when the anode-to-cathode voltage falls to the value of the extinction voltage and is mantained at this low value for some time (10—200 μsec.) necessary for clearing the anode-cathode space of the positive ions, which are neutralised by free electrons (deionisation time). b) The grid voltage rises (i. e. the grid bias decreases) to such an extent that with the same anode voltage, electrons begin to force their way to the anode. The further course is the same as in (a). It can be seen that for a certain anode voltage there is a definite maximum grid voltage (i. e. a minimum bias) for which the arc discharge just commences. The graphic representation of this relationship is termed the grid characteristic of the thyratron (this evidently is in no way similar to the grid characteristic of a vacuum triode, see page 61). The following characteristics must be maintained if the thyratron is to be operated properly: a) The heater voltage or current. Over- or underheating considerably decreases the valve life. On underheating, the discharge current does not flow over the whole of the cathode's surface but only through one point which is then overheated and may burn, or in indirectly heated tubes it may lose emission because the superficial oxide layers have been destroyed. b) The grid characteristic. c) The grid leak. Too large a resistance decreases the grid control, a small resistance may damage the grid. d) Maximum anode voltage. If it is exceeded, the grid loses its ability to maintain the thyratron in the blocked state. e) Maximum inverse anode voltage. If it is exceeded, the glow and then also the arc discharge is started between a positive heated cathode and a negative anode. f) Maximum average anode current. This determines the maximum permissible heating of various parts of the valve; if cooling is insufficient, the anode current must be decreased below catalogue values. g) Maximum peak current. If it is exceeded, the cathode is destroyed in the same way as when underheated. In its stead the minimum anode resistance is sometimes given or the maximum capacitance that can be discharged directly without a pro-

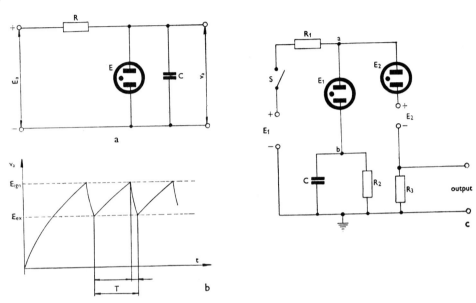

Fig. 38. a) glow tube oscillator; b) the shape of its output voltage; c) glow tube oscillator giving rectangular pulses.

tective resistance. h) Maximum repetition frequency or minimum repetition interval depends on the deionisation time; if it is exceeded, the thyratron may conduct even though the grid is at negative potential and does not regain its proper function so long as the anode circuit is not interrupted or the anode voltage supply cut off. This failure may occasionally occur even if the thyratron is operated near the maximum frequency limit.

Gas tube oscillators are synchronised in a similar way to the types mentioned earlier. The glow discharge tube is difficult to synchronise. This is easier for the thyratron. By changing the grid bias, any thyratron oscillator may be changed to an autooscillator or to a triggered pulse generator.

The oscillations are either sawtooth (exponential) or rectangular waves depending on the circuit. The steepness of the front (up to 10^{-7} sec) with a large pulse amplitude (up to several hundreds or thousands volt) and a small internal resistance may be of advantage. The difficulty, however, of obtaining a steep trailing edge (i. e. rectangular pulses) and controlling the pulse width, low maximum frequency (several 100 to 1000 c/sec, except for hydrogen thyratrons), poor frequency stability and limited life span are other drawbacks. Thyratrons are delicate instruments and sometimes difficult to set for an amateur. They were used often in physiology in earlier days.

The fundamental circuit of a glow discharge tube oscillator (Fig. 38a, b) needs no further comment.

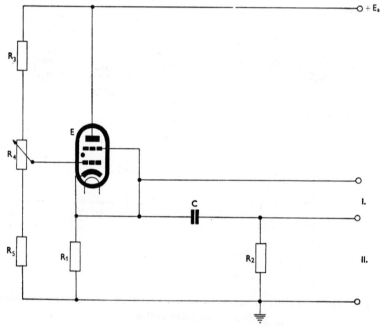

Fig. 39. Thyratron oscillator with high frequency stability. Sawtooth pulses at output I, rectangular pulse at output II. $R_1 \gg R_2$.

The circuit of a thyratron oscillator is given in Fig. 39.

5. Transitron generators and some related oscillators

These are valve relaxation oscillators with a negative resistance.*) The transitron is the most important and works as follows (Fig. 40a): Let us start from the moment the valve is blocked. The anode voltage is the same as the supply voltage and no anode current flows because of the negative voltage of g_3. All the cathode current thus flows through the screen grid g_2, the voltage of which is very low. The negative charge from C_2 (and thus g_3) leaks through

---

*) In a negative resistance, an increase in current results in a decrease in the voltage drop across it, while in a positive resistance, the voltage rises with the current.

R_4. As soon as the voltage at g_3 begins to come close to zero, an anode current begins to flow, i. e. part of the cathode current ceases to flow to g_2 and begins to flow to the anode. Consequently the g_2 voltage rises somewhat, this rise is transferred to g_3 by condenser C_2. This results in a further increase in the potential of g_3, and the anode current is further increased until a strong anode current flows with a positive voltage at g_2 and g_3. Since the anode current

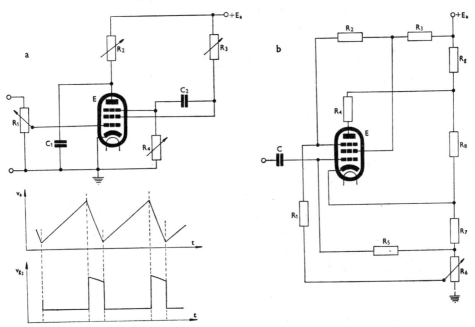

Fig. 40. Transitron. a) astable transitron and the voltages at its electrodes. b) bistable transitron.

discharge C_1, the anode voltage and then also the anode current decrease. In addition the positive charge from g_3 disappears both through the grid leak R_4 and, chiefly, at the expense of the grid current to g_3. As the anode current decreases, the grid current g_2 begins to flow. Its voltage falls and this drop is transferred through C_2 to g_3. This accelerates the rapid fall of anode current. The grid current g_2 is further increased and the grid voltage g_2 decreased. The valve is blocked, the anode current falls to zero, the anode voltage begins to rise as C_1 is being charged, the g_2 voltage falls considerably and the g_3 voltage reaches even larger negative values. The negative charge now leaves C_2 only through R_4 i.e. at a much slower rate than the earlier positive charge. As soon as the condenser C_2 is discharged (and C_1 charged), the g_3 potential rises to zero, an anode current appears, the screen grid current falls etc.

It is evident that during the longer part of the repetition interval, C_1 is charged through R_2 and C_2 discharged through R_4. These elements thus mainly affect the frequency. The shorter part of the repetition interval is determined by discharging of C_1 and by the conductance of the valve, i. e. its characteristic, the voltages at all three grids and the grid leaks, especially R_3, and also by the cathode resistance, if it is used in the circuit. Transitrons are often used to obtain sawtooth voltages from the anode for the time base of the cathode-ray oscilloscope, and then we try to obtain very rapid discharge. They may also be used for producing rectangular voltage (from g_2), and in this case a duty cycle of $\frac{1}{2}$ or less may be obtained by a suitable choice of resistances. The synchronising voltage is usually applied at g_1.

The transitron may be blocked and changed to a triggered transitron with one or two stable positions by a suitable choice of g_1 and g_3 bias (Fig. 40b). It then fully corresponds functionally to a mono- or bistable multivibrator. In comparison to multivibrators it is simpler and can be triggered off by both positive and negative pulses. Its disadvantages are asymmetry, less sensitivity (larger and longer triggering pulses are necessary), a smaller amplitude of the output voltage (50 V or less), more difficult calculation and setting. Particularly for pulse width near the quiescent interval a large dependence of the actual pulse width not only on the preceding interval but also on the width of the preceding pulse may be observed. This may even result in even pulses being about twice as long as odd pulses, or vice versa. Triggering pulses are usually injected into the control grid circuit, which has the highest sensitivity and is not used in the feedback circuit itself. It is, however, possible to direct them to any other electrode.

In monostable transitrons the coupling between g_2 and g_3 is achieved with a condenser in the same way as in transitron auto-oscillators. In bistable transitrons a resistance coupling must be used (as in bistable multivibrators — Fig. 40b).

Dynatrons are somewhat similar to transitrons. A negative resistance is formed in them as a result of secondary electron emission from the anode (see page 64). Since the dynatron effect depends to a large extent on the voltage, it is unstable and, in modern valves, is purposely avoided. Dynatrons are rarely used today. A negative resistance and thus generation of rectangular or sine wave oscillations may also be obtained with a crystal of pyrite or galenite, with a germanium diode, with a transistor, a tunnel diode, etc.

c) Pulse shaping circuits

The pulses described above can rarely be used directly. Usually it is necessary to shape them, i. e. to adjust their shape in such a way that it approximates the ideal course with the accuracy chosen. Shaping involves three problems: a) determination of the shape of the curve (harmonic, damped oscillations, rectangular, sawtooth, exponential etc.), b) adjustment of the time scale (width of the rectangular or sawtooth pulses, time constant of the exponential voltage drop, frequency or duration of the harmonic and damped oscillations), c) determination of the amplitude.

1. Determination of the shape

1) Harmonic oscillations may be produced: α) Directly in harmonic oscillators (see p. 82). They are then further amplified only by linear amplifiers or tuned amplifiers which do not change their shape. β) Nonharmonic oscillations are changed to harmonic ones by passing them through tuned filters (either through resonant circuits, especially for higher frequencies, or through more complex RC filters containing often also amplifying valves, see e. g. p. 86). γ) Damped oscillations are damped so little (or damping is artificially reduced) that for certain purposes they are equivalent to undamped ones (cf. the next paragraph and also p. 184).

2) Damped oscillations are nearly exclusively produced by injecting either a potential step or a short pulse into the parallel resonant LRC circuit. Damping is controlled by a resistance connected either in parallel to the whole LC circuit (e. g. a valve, germanium rectifier) or in series with L (resistance of the coil windings and losses in the iron core). The first stage of the apparatus in Fig. 67b shows the circuit most frequently used. The resonant circuit in the cathode of a triode is highly damped by the grid current, which is always flowing for a positive grid bias. The negative voltage step applied to the grid through C_1 blocks the cathode current. Disappearance of the cathode current (and thus of magnetic induction) in coil L results in oscillations in the circuit LC, which is only slightly damped since it is isolated by the blocked valve. When the negativity of the grid ends, the positive grid bias produces a grid and anode current which very rapidly damps the circuit. The damping of the circuit at a time when it should operate can further be decreased using a positive feedback (see p. 185). Sometimes, on the other hand, damping ought to be increased.

2a) By using a parallel resistance or, still better, a parallel diode (see p. 53) an aperiodic course is achieved, which may also be obtained by critical damping for

$$R = \frac{1}{2} \sqrt{\frac{L}{C}}$$

3) For producing square waves many different circuits may be used. These were already described or will be discussed later (p. 119).

4) The same applies to sawtooth waveforms.

5) An exponential curve is most frequently obtained by charging or discharging a condenser across a resistor. Such a network is called an *RC* differentiating circuit. Quantitatively it is characterised by its time constant

$$\tau = RC$$

which determines the shape of the voltage rise or decay at the condenser

$$v_1 = E_0(1 - e^{-t/\tau}) \quad \text{(on charging)}$$

$$v_2 = E_0 e^{-t/\tau} \quad \text{(on discharging)}$$

Evidently for $t = \tau$ the voltage drops to E/e, i. e. to about $0 \cdot 37 E_0$ (when discharging) or rises to $E_0(1 - 1/e)$, i. e. to $0 \cdot 63 E_0$ (when charging). See also p. 50.

Such derivative peaks are very often used for triggering monostable and bistable multivibrators or as standard pulses, as in counters, for synchronisation, or also directly as stimuli (in the older chronaximeters).

2. Time relations

The second parameter of shaping—time—is either included implicitly in the first, and controlled together with it (the frequency of harmonic and damped oscillations, time constant of exponential pulses) or controlled separately, as in the case of rectangular waveforms if the frequency, pulse width and amplitude are controlled independently. The width of a rectangular pulse may be controlled by the following means:

1) Using an astable multivibrator. Here the pulse width is not fully independent of the frequency, however. See p. 87.

2) Using a monostable multivibrator which is driven by short pulses of the chosen frequency. The leading and trailing edges are steep, but the pulse width is slightly frequency dependent and cannot be controlled accurately (variations of several %). It is difficult to control the pulse width within a wider range.

3) Using bistable multivibrators, as in Fig. 26. A sinusoidal voltage is shifted in phase by a *RC* circuit and both the original and the phase shifted voltage are given a symmetrically rectangular shape by limiters (see p. 118). This voltage is differentiated and the bistable multivibrator is reversed by the derivative peaks. Since the phase can be controlled very exactly (to a fraction of $1^0/_{00}$), this method of controlling the pulse width is very exact, but rather complicated. This, however, is the only method by which the pulse width can be controlled relatively, i. e. the frequency can be changed as desired without changing the duty cycle, i. e., the constant ratio of pulse width to the

repetition interval is preserved. The method may also be modified in such a way that both rectangular voltages, shifted in time, are directed into a suitable selector circuit in which they are summed.

An analogous transitron connection may be used instead of the bistable, monostable or astable multivibrator. Its advantages and disadvantages are discussed on p. 110).

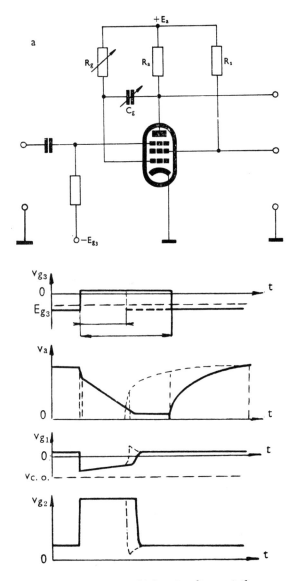

Fig. 41. Phantastron: a) circuit diagram; b) input voltage at the suppressor grid, voltages at the anode, at the control and screen grids; $V_{c.o.}$ — cut-off voltage.

4) For our purposes it is usually advantageous to use more complicated integrating circuits known as phantastrons. The basic circuit (always termed Miller's transitron) is shown in Fig. 41.

In the blocked state the cathode current divides into that of the control grid (which is connected to the positive supply pole through the leak R_g) and that of the screen grid, which thus has a very low voltage. The anode current does not flow, since the suppresor grid has a large negative bias. Thus the anode has the full supply voltage. If a positive voltage (a long pulse) is applied to the suppressor grid, the anode current suddenly begins to flow. This results in a fall in anode voltage. This fall is transferred to the control grid through a negative feedback condenser C_g. In this way the rate of the cathode current increase is slowed. (Since the anode current begins to flow, the screen grid current decreases and its voltage rises.) The negative charge transmitted to the control grid from the anode leaks through resistance R_g, however, so that the cathode and anode current continue to rise. This rise is, of course, very slow, since the fall in anode voltage due to anode current increase is transferred through C_g to the control grid and suppresses the anode current, which is affected more by the grid voltage than by the anode voltage (this ratio is determined by the amplification factor). This gradual and linear rise in anode current or fall in anode voltage may terminate in two ways: a) the voltage step at the suppressor grid is terminated — the valve is again blocked; b) the anode current cannot rise and the voltage cannot continue to fall, since the curved part of the characteristic is reached. Here the condenser C_g ceases to have an effect, the grid voltage again rises to positive values and thus the cathode current continues to rise rapidly. This, of course, cannot further increase the anode current, as this is saturated, but causes a rise in the control grid current and thus a rapid fall in its voltage. The anode voltage remains at the low value until the suppressor grid is again blocked.

In practice either the linearly falling part of the anode voltage (a very good time base for the oscilloscope) or the rectangular pulse at the screen grid is used. Its width depends upon the electrode voltages and on the values of R_a, R_g and C_g. Calculations are complicated and unnecessary since pulse width depends only on R_g and C_g very closely according to the equation:

$$\tau = kR_gC_g$$

where the constant k includes the effect of different parameters (the other resistances, the valve, the electrode voltages). This means, for practical purposes, that it is sufficient to determine the pulse width τ_0 in one special case and then calculate either C_g or R_g necessary for a certain different pulse width τ or determine the pulse width for any R_g and C_g. Then

$$\tau = \tau_0 \frac{C_gR_g}{C_{go}R_{go}} ; \quad C_g = \frac{\tau}{\tau_0} \cdot \frac{C_{go}R_{go}}{R_g} ; \quad R_g = \frac{\tau}{\tau_0} \cdot \frac{C_{go}R_{go}}{C_g} .$$

The pulse width can, of course, also be controlled otherwise, by a voltage applied to the anode through a so-called fixing diode. When the anode current rises the anode voltage cannot fall below the potential of the cathode of the fixing diode. The end of the pulse is thus formed at this "fixed" voltage.

A further advantage of the phantastron circuit is the possibility of obtaining any desired scale (as follows from the above equations) using variable resistors or condensers, and numerically the scales for multiple ranges agree and a unipolar switch is sufficient for the range selector. What is most important, the circuit is very accurate. Variations in the anode voltage and the heating current supplies by $\pm 10\%$ result in variations of pulse width by about only $\pm 1°/_{00}$. The pulse width depends on the amplification factor of the valve and this is probably the only drawback. This must, therefore, be checked from time to time, and calibration must be performed only after longer use (50 to 100 hrs.). If an accurate and equal scale is to be preserved even as the valve ages or after exchanging it, a variable resistor or condenser (trimmer) must be used for adjusting the scale to the dial. If a variable resistor serves as the main tuning element, a trimmer is used for scale adjusting while a variable resistor is used if the tuning element is a condenser.

In addition to this simplest, most reliable and accurate circuit of the phantastron, more complex circuits are used frequently. These are especially suitable if the phantastron is used for shifting short pulses in time, i. e. for producing delayed pulses. This might be obtained as follows: The input pulses trigger a monostable multivibrator or transitron and this starts the phantastron. This may be achieved in a more simple way with a single valve phantastron circuit using a pentode or pentagrid. This operates simultaneously as a triggered mono- or bistable transitron and phantastron. Such connections are simpler as far as the number of valves and components are concerned, but need more careful adjustment and are usually less accurate. It is better not to use them even though more valves are then required. Reliable circuits may be found in Petrovich and Kozyrev (1954), Meyerovich and Zelichenko (1953), Markus and Zeluff (1948).

So-called transmission lines are used especially for shorter pulses of about 1 μsec both for pulse shaping and delaying. They have the common property that pulses pass through them or are reflected by them, the pulse propagation rate in them being relatively low (i. e. much less than the speed of light in a vacuum). Their main advantage lies in their accuracy and complete independence of the input voltage etc. For our purposes they are unsuitable, since because of long pulses and delays, their dimensions are tremendous, especially if steep front and trailing edges are to be preserved. (For details see Meyerovich and Zelichenko, 1954). Here it need only be said that usually they consist of a number of coils and condensers (T or π networks) with more complicated coupling between the individual elements.

Using transistor circuits, it is unfortunately not possible to control time relationships as exactly and simply as with the phantastron. (Transistorized phantastrons are not very exact — see e. g. Shterk 1964, Parris and Staar 1960). The simplicity, small dimensions and low consumption of transistor circuits however, allow us to use for this purpose more complex circuits, most frequently monostable multivibrators. The best means is to trigger with a negative

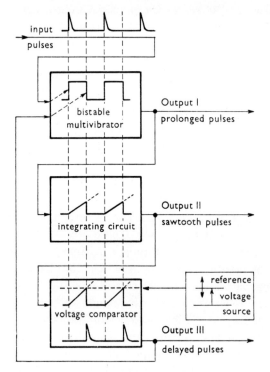

Fig. 42. Block diagram of more complex transistor circuits for prolonging or delaying impulses used instead of phantasthron.

impulse a bistable multivibrator which feeds an integrating generator of saw-tooth waves. A comparator is used to establish the equality of the saw-tooth voltage and of the preset fixed voltage (Fig. 42). At this moment the comparator forms a pulse which is shifted in time against the triggering pulse. This pulse is also led into the second half of the bistable multivibrator and resets it. By further shaping the output impulses it is possible to obtain rectangular pulses of exactly controlled duration, delayed pulses etc. (Shterk 1964, Merril and Slater 1956).

3. Limiters

Pulse amplitudes are usually determined in voltage limiters. These are nonlinear elements distorting sinusoidal or other input voltage in such a way that the output voltage has an approximately rectangular course. In other words, a limiter is an amplifier, the amplification of which changes in relation to the input voltage in such a way that the output voltage attains only two values, one of which is usually zero. The nonlinear element of the limiter is usually a vacuum valve (diode, triode, pentode), a germanium or selenium rectifier, more rarely a diode with a pure tungsten filament working in the region of saturated emission current, an iron cored coil, a thyratron, a gas discharge tube etc.

The series diode limiter

The diode permits current from the source to pass in one direction only, so that a pulsating voltage is obtained across resistance R as shown in Fig. 43a. If a positive or negative bias is given to the diode, the zero line is shifted above or below the symmetry axis of the input sine wave voltage (Fig. 43b). If the diode is reversed, a pulsating voltage of opposite polarity is obtained (Fig. 43b). A germanium or selenium rectifier may be used instead of the diode.

By using two biased limiters the sine wave voltage may be limited from both sides, so that a voltage having a trapezoidal shape is obtained. The voltage approaches a rectangular shape the more closely, the larger the limiting, and thus the smaller the output voltage. This is a disadvantage of all limiters working without amplification.

The parallel diode limiter (Fig. 43c)

The diode connected in parallel to the load R is a resistance which attains very large (infinite) values for voltage of one polarity, and very small ones (R_s — the so-called D. C. resistance of the diode, smallest in germanium rectifiers) for voltage of the opposite polarity. Thus during one half cycle the input is shunted. Good suppression is obtained for $R_s \ll R$ and $R_{in} \gg R$ (R_{in} is the internal resistance of the input signal source). The polarity in this limiter may be changed by reversing the diode and the level of suppression may be regulated by the D. C. bias (Fig. 44a). With a parallel diode limiter symmetrical limiting may be also obtained (Fig. 44b).

The triode limiter using grid current (Fig. 45a)

The circuit cathode — grid is really a parallel diode limiter. Due to the grid current and the drop across resistance R_g, the grid voltage and thus the anode

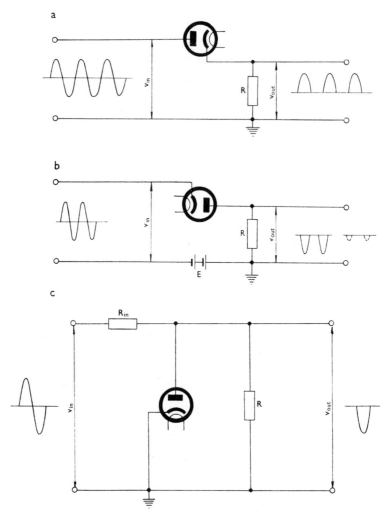

Fig. 43. Diode limiter. a) series negative limiter; b) series positive limiter with bias; c) parallel positive limiter.

current are constant during the positive halfwave. The anode voltage fall
to minimum values. The negative half wave is amplified. R_g is 10^4 to 10^6 Ω.

The triode limiter using the anode current cut-off

This limits the input voltage from the opposite side as the limiter described above. The negative grid bias is chosen in such a way that the anode
current disappears during the negative half waves of the input voltage and the
positive half waves are amplified (Fig. 45b).

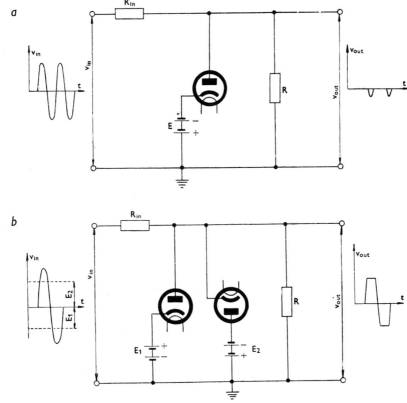

Fig. 44. Biased diode limiters. a) parallel positive limiter with bias; b) parallel symmetrical limiter with bias.

A limiter using shift of cathode current

In the region of zero and positive grid voltages, the valve (triode and pentode) having a large anode resistance ceases to amplify even if the source of input voltage has an internal resistance sufficiently small to compensate the grid current leakage.

The symmetrical triode limiter using grid current and the cut-off of the anode current (Fig. 46)

This combination of two types described above is used more often. In a similar way any two opposite types of limiters may be combined into a symmetrical limiter. By controlling the bias, the proportion of pulse width to the quiescent interval is also regulated. All types of limiters may also be designed with a pentode. The pentode limiters are mostly more efficient. Two stages

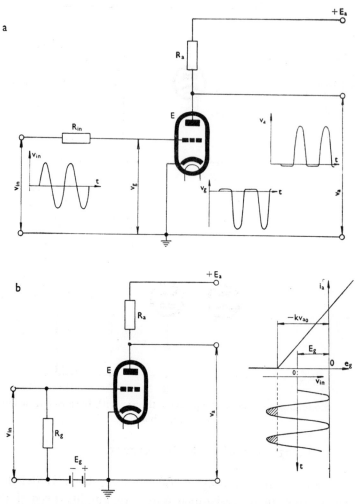

Fig. 45. Triode limiter. a) grid current negative limiter; b) anode current cut-off limiter with A. C. input and output.

of such a symmetrical pentode limiter normally are sufficient for producing a very rectangular voltage from a sine wave voltage. For limiting a voltage with a steeper course, e. g. for topping a multivibrator (or a transitron) pulse one stage is sufficient.

Limiters, especially non-symmetrical limiters, are often used, after differentiating the rectangular pulses, for removing peaks of one or the other polarity. Often limiters are also used where they are not absolutely necessary, i. e. as isolating amplifiers, phase inverters etc. This lessens the danger of interference, accidental feedbacks etc.

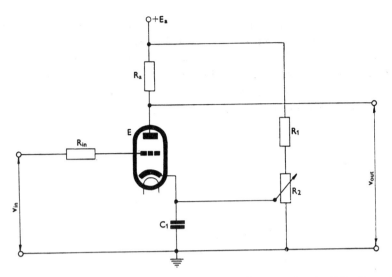

Fig. 46. Triode grid current negative limiter and anode current cut-off limiter with controlled level of limiting.

Transistorized limiters

Transistorized limiters can easily be constructed according to the principles described for triode circuits, but such are used very rarely since in transistor circuits it is not necessary to limit the number of components so carefully as in valves. As long as our demands on the slope of the leading and trailing edges of the pulse and on its amplitude are not too great or if an input pulse of sufficient amplitude or with sufficiently sharp edges is available, diode limiters with semiconductor diodes are most widely used. If the slope of the leading and trailing edges have to be considerably increased a modification of the bistable multivibrator or (more often) a Schmitt trigger is used (see p. 99).

d) Assembling a stimulator

We must now consider the main principles for synthesis of such elementary circuits into an electronic stimulator. First, these circuits are connected in series according to our general requirements, i. e. a so-called block diagram is made (Fig. 47). According to the stimulus parameters required and to the accuracy with which they are to be regulated, the individual circuits are now designed in more detail. This is best illustrated by an example.

A stimulator giving paired rectangular pulses is required. The absolute width of each pulse is to be regulated from 10^{-5} to 10^{-2} sec, their delay from

10^{-4} to 10^{-1} sec , their amplitude from 10^{-2} to 10^2 V. Both pulses must be preceded, with a controllable delay of 10^{-4} to 10^{-1} sec , by the impulse for synchronising the time base of the cathode-ray oscilloscope, which has a duration of about 10^{-3} to 10^{-4} sec , and an amplitude of about 20 V or more. The accuracy of all time controls is about $1^o/_{oo}$, that of the amplitude control 2%. The time constants of the front and trailing edges of both stimulating

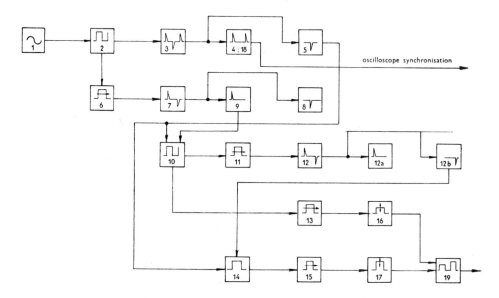

Fig. 47. Block diagram of complex stimulator.

pulses are 2μsec for the shortest pulses and somewhat longer for the other pulses. The output pulses may be applied to the same or to different electrode pairs. This system of pulses may be repeated with a frequency that can be controlled from 10^{-1} to 10^3 c/sec (accuracy $1^o/_{oo}$) or is produced only once (by pressing a key). To satisfy these requirements, the following principles are to be observed.

A) In view of the required accuracy, the generator of the fundamental frequency must be harmonic. In view of the large frequency range and the very low lower limit, a symmetrical *RC* oscillator will be best (see p. 86).

B) In view of the requirement of triggering the stimulus with a key, it is necessary that the whole cycle occur, after triggering with one potential step. The return to the original state can be attained either by a step of opposite polarity or automatically.

C) In view of the required accuracy of time relations, all these relations must be controlled by phantastrons.

D) In order to attain simple graduation of dials (readings without calculations, graphs etc.), the corresponding times must be directly determined by phantastrons. Now the diagrams of the time relations for a frequent stimulus are drawn (Fig. 48).

The sine wave voltage (1) of the oscillator is changed to a rectangular one (2). Using differentiating circuit (3) and limiters, peaks for synchronisation

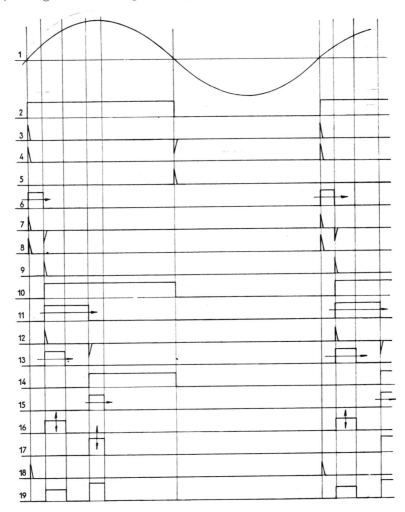

Fig. 48. Time diagram of stimulator shown in Fig. 47. The row numbers indicate the block diagram stages (for details see text).

of the cathode-ray oscilloscope (4) and bistable multivibrators (5) are obtained. A variable delay of the first pulse, i. e. the time by which the synchronising pulse precedes the first stimulating pulse, is produced by the phantastron

(6). The width of the first stimulating pulse must be controlled independently of this delay. Consequently we use derivative peaks of the trailing edge of the phantastron pulse (9) and the derivative peaks (5) of the basic rectangular waves (2) for the formation (in the bistable multivibrator) of the first main pulse (10). This is shortened by the phantastron to the required width of the first stimulating pulse (13). The same main pulse (10) is used to drive a further phantastron, the pulse of which (11) determines the delay of the front edge of the second main pulse (14) from that of the first one (10). By differentiating the phantastron pulses, peaks (12) are produced, the second of which is used for triggering a further bistable multivibrator (14). The end of the pulse of this is determined either by the peak terminating the first half cycle of the basic frequency (5) (if the whole second pulse lies in the first half cycle), or by the peak terminating the second half cycle (4) (for long or very much delayed second pulses). Thus the second main pulse (14) is obtained, from which the second stimulating pulse width (15) is formed by a phantastron.

This ends the first part of the rough design, the part determining the time relations. Now both pulses are formed and their amplitude, polarity and mutual relationship determined. Each pulse is directed separately into a separating limiter and then into an amplifier, where their size and polarity (16, 17) are determined. If both pulses are to be supplied to the same electrodes, they must be lead into a linear mixer (19). Now the number of the oscillator ranges is determined. Because of the high accuracy required, an oscillator with several, not to wide ranges must be used, e. g. 0·1 c/sec to to 0·3 c/sec , 0·3 c/sec to 1 c/sec , 1 c/sec to 3 c/sec , 3 c/sec to 10 c/sec , 10 to 30 c/sec , 30 to 100 c/sec , 100 to 300 c/sec , 300 to 1000 c/sec , i. e. altogether 8 ranges. Phantastrons must be made in a similar way. They must have about 6 ranges (for an accuracy of 1% 2—3 ranges would suffice). Time constants of the circuits are then chosen according to the required slope of the front and trailing edges of the pulses, etc.

The requirements of this example can, of course, also be met in many different ways. For less strict demands another example is given in Fig. 49.

Several simple commercial stimulators are constructed in such a way that they can be connected in series so that one stimulator triggers the next and thus the stimulating apparatus may be adapted to even the most complicated requirements, e. g. trains of paired pulses etc. If requirements are not so strict it is, of course, possible to use each of these units as a separate apparatus.

A transistor stimulator can be constructed in a similar way. The main difference lies in the use of other circuits than phantastrons for controlling the width and delay of the impulse (p. 118). It folows that the stimulator will be more complex and will contain more elements.

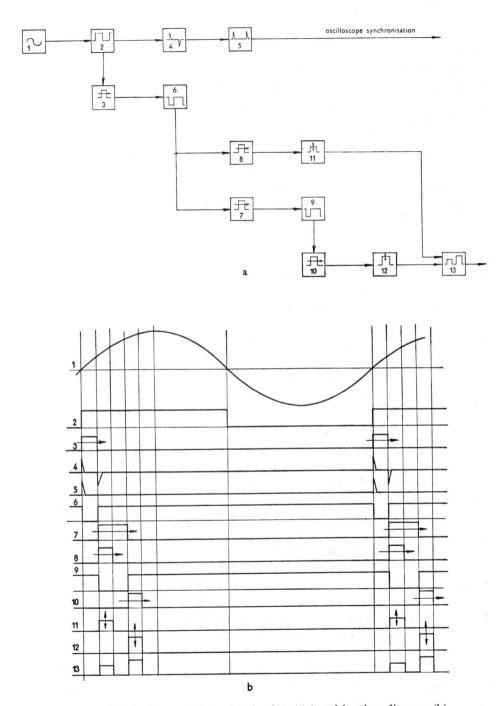

Fig. 49. Block diagram of simple stimulator (a) and its time diagram (b).

e) *Stimulus isolation*

Another very important auxiliary arrangement must be described when dealing with stimulators: radiofrequency stimulator, radiofrequency transformer or radiofrequency output unit (Schmitt and Dubert 1949) (Fig. 50). This is an arrangement permitting complete resistive and capacitative isolation of the stimulating electrodes from earth (cf. p. 196). It is based on the fact that the stimulus is transmitted to the electrodes without wires (as in broadcasting). The transmitter is an anode modulated Hartley oscillator (cf. p. 83). Its anode receives positive impulses from any stimulator. If such an impulse reaches it, the oscillator begins to oscillate for a time corresponding to the pulse duration with a frequency of about 10 Mc/sec. The receiver consists of a coil (at a distance of 1—2 cm from the oscillator coil) and a condenser. It is tuned by moving the powdered iron core or with a trimmer. A germanium diode forms the detector. After filtration to remove the remaining high frequency voltage, an impulse of the same duration as that modulating the oscillator is applied at the stimulating electrodes. Its amplitude, however, is only a fraction of that of the modulating pulse, but is proportional to the latter (from a certain minimum value). The minimum amplitude of the modulating pulse necessary for oscillating the oscillator is 1—10 V, depending upon the type of valve and the design of the apparatus. It may happen if modulating voltage is reduced by radiofrequency transmission to one third, that the amplitude of the output pulse is still too high. It is then necessary to connect a voltage divider to the output. The heating voltage for the valve is either taken from the modulating stimulator or from a battery.

Even though the isolation of the stimulating impulse from earth is still exceptionally important, yet the above circuit is out of date to some extent. Transistor stimulators can work without an isolation transformer, particularly if the apparatus is physically small and has a built in energy supply (1). If more than one pulse and several pairs of electrodes are used the radiofrequency transformers often affect each other even though they are tuned to different frequencies (2). The output voltage of the radiofrequency transformer is fairly soft (3). The slope of the leading edge is limited by the time constant of the filter, which must be several times the inverse value of the carrier frequency (4). For all those reasons isolation units have been developed working on different principles. Dawson et al. (1960), for instance, used photoelectric transmission: The output of the conventional stimulator is connected to a glow discharge tube, to an optic voltage indicator or to a similar light source that can be modulated by the pulses. The modulated light is optically focussed and received by a photodiode, the output voltage of which controls a Schmitt trigger. Its pulses are amplified and after amplitude adjustment led to the stimulating electrodes. The main advantage is the possibility of using an unlimited number of channels without their mutual interference, transmission of pulses of any desidered slope for any distance (e. g. across the screening of a Faraday cage) and a low output resistance. The main disadvantage is the necessity to control the amplitude of the stimulating pulse at the electrodes (not in the stimulator). On the same principles systems for distant stimulation of freely moving animals can also be developed. See also Cerf and Libert (1955).

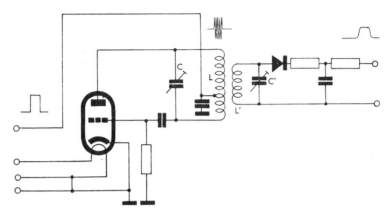

Fig. 50. A radiofrequency-coupled output stage of a stimulator.

D. Recording technique

The second cardinal task of electrophysiology is to study electrical signs of vital processes either in cells or their parts or in tissues, organs or whole organisms. This is done by using recording apparatus. Before describing the individual typical components of such apparatus, demands made upon them in general, and how far these can be realised, will be discussed.

a) The fundamental characteristics of signals and their sources

Usually a change in the potential difference between two parts of a living object is considered as an electrical expression of vital functions. It must, of course, be remembered that a change in potential difference is always accompanied by a number of changes in other electrical properties (resistance, capacitance etc.) and often also of non-electrical characteristics (optical, mechanical etc.) that have not yet been studied sufficiently. Such a change is called an electrical impulse in the theory of communication (which forms the basis of applied electrophysiology). A single electrical impulse is characterised by its shape, amplitude and, further, by characteristics of its source, particularly its internal resistance. The amplitude of impulses recorded in electrophysiology lies between 0 (practically several μV) and 0·1 V for nerve and sensory cells, nerve and muscle fibres (excepting the electric organs of some fish).

The time course of the amplitude changes can be developed into the so-called frequency spectrum according to the principles of harmonic analysis, which are described in most radiotechnical and physical textbooks and the practical procedure is given in textbooks of higher mathematics. Here it is sufficient to state the qualitative rule that each electric impulse is formed also by oscillations of a much higher frequency than is the repetition frequency and the inverse value of the pulse width. The participation of these "higher harmonics" is the larger the more square or pointed the impulse shape. How far the whole of this "frequency spectrum" is to be respected is often an important question in electrophysiology. A further important characteristic of impulse is the internal resistance of their source, i. e. the internal resistance of the tissue. Its significance can be illustrated by an example: If the size of an impulse recorded from the whole nerve is measured with an apparatus (e. g. cathode-ray oscilloscope) having an internal resistance of 100 000 Ω the result will be roughly the same as with an apparatus having a resistance of 1 000 000 Ω. It follows that the internal resistance of the impulse source (of the nerve) is considerably smaller than 100 000 Ω. If, however, microelectrodes are used to record impulses from a single fibre, the first apparatus will show nothing and the second only very low values. An apparatus having an internal resistance of 10 MΩ or more will show considerably larger amplitude. Thus the internal resistance of a single fibre is about 10 MΩ or more. It is evident that the internal resistance of the impulse source does not only depend upon its nature, but also considerably upon the external medium in which the source is situated (conductive, nonconductive), on the size and shape of the electrodes, etc. It even seems that the tissues may have a different internal resistance for different frequency components of the impulse. It follows that it is sometimes necessary to increase the internal resistance of the input circuit of the recording apparatus to extremely high values.

b) Demands on the amplifier stages

The recording apparatus consists of several components with different functions. Impulses from the tissue are lead into the input circuit, which must not distort them. This is followed by one or more (or no) middle stages which change the impulse (usually only its amplitude) in such a way that it can be transferred to the output stage connected with the actual recording or measuring instrument. Demands on the individual components differ according to these different functions.

Input circuits

The following are the most important requirements:

1) The input circuit must not affect the studied substrate. It is especially important that it must not take up too much energy nor supply energy to the substrate.

2) The input circuit must pass on impulses without distortion, i. e. the output amplitude must be exactly proportional to the input amplitude. This means that no frequency component must be suppressed or accentuated and no voltage not corresponding to the input voltage must be transmitted. For details see below.

3) The input circuit must amplify the energy of the input signal. In some cases this is achieved by the output impulse having a considerably larger amplitude than the input impulse. At other times the amplitude remains practically unchanged or even decreases, but the output resistance is much smaller than the input resistance. Since power is e^2/R, power is also amplified in this case.

The first two requirements can be fulfilled only in part and the corresponding compromise must be considered for each case separately.

Middle stages

Requirements are somewhat less strict:

1) undistorted transmission (see above),
2) the greatest possible amplification (usually voltage amplification).

The output (power) stage

1) must have the necessary amplification (of voltage or power),
2) must compensate for the distortion occurring in the recording system.

The recording or measuring system

1) must permit registration or measurement within the whole required frequency and amplitude range,

2) must either work without distortion, or its distortion must be compensated for by the preceding amplifier.

The following are the most commonly used instruments: The rotating coil instrument (Deprez-d'Arsonval), mechanical recording system with a moving coil, magnet or iron, a cathode-ray tube. For details see p. 169.

c) Biological amplifiers

All stages except the last one are usually constructed as valve amplifiers. They will now be discussed and the types which can be used for the different stages will be described. The basic component is a valve (a pentode or triode) or a transistor. Their functions were reviewed on pp. 60 — 77.

1. Coupling of the amplifier stages. Frequency characteristic

The majority of amplifiers are composed of several stages, the coupling of which must be discussed.

The input voltage is applied to the grid, which requires a constant D. C. bias for correct function. These two voltages, the signal and the bias, must be separated so as not to affect each other, particularly in order to avoid the bias producing a current flowing through the measured object. This may be achieved in two ways:

Fixed bias

A constant bias from a stable voltage source, e. g. a dry cell. This is directed through a large resistance (grid leak) to the grid. The other pole is connected to the cathode. The measured A. C. voltage is applied, through a condenser, also between the grid and the cathode. For the D. C. bias voltage the condenser represents an infinite resistance and it does not, therefore, influence the object. The measured voltage, however, is connected to a load (the grid condenser in series with the grid leak), which takes up energy from the object. This may be excessive, especially if the internal resistance of the object is too large, as in the case of single fibres, or when recording with microelectrodes. (In such cases some electrometric device must be used.) This energy is dissipated in the grid leak and in the valve (when the grid current is flowing). In addition, phase and frequency distortions occur. This may be explained as follows: The condenser and the grid leak form a voltage divider (Fig. 51). The first (upper) part of the divider — the condenser — produces a voltage drop depending on the frequency. The resistance R_g does not depend upon the frequency. Thus the amplitude and phase of grid voltage depend not only on the input voltage, but also on the latter's frequency.

Lower limiting frequency

If the harmonic input voltage is e_i, the grid voltage e_g, its frequency f, the angular frequency $\omega = 2\pi f$, the phase shift of e_g to e_i is φ, and the grid leak R_g, then

$$e_g = e_i \frac{R_g}{Z} = e_i \frac{R_g}{\left[R_g^2 + \dfrac{1}{\omega^2 C_g^2} \right]^{\frac{1}{2}}} \; ; \quad \tan \varphi = - \frac{1}{C_g R_g}$$

and since the power $P = e^2/R$, then (after squaring and dividing by R_g)

$$P_g = \frac{e_g^2}{R_g} = \frac{e_i^2 R_g}{R_g^2 + (1/\omega^2 C_g^2)} \tag{1}$$

Let us now find the frequency ω_1 for which the power at the grid falls to half its original value (corresponding to the direct coupling without the grid condenser). Without the condenser

$$P_0 = \frac{e_i^2}{R_g}$$

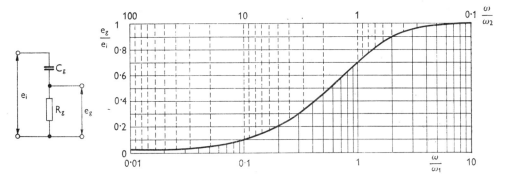

Fig. 51. Decrease in the amplitude of a low frequency signal caused by a grid condenser. Lower abscissa: relative frequency $f_r = \omega/\omega_1 = \omega RC$. Ordinate: ratio of grid to input voltage $y = e_g/e_i$ (Meaning of symbols explained in text). The same curve is valid also for the upper limiting frequency ω_2 when using the upper abscissa with the scale ω/ω_2

With the condenser

$$P_g = \frac{e_g^2}{R_g} \tag{*}$$

$$P_g = \frac{1}{2} P_0, \quad P_0 = 2P_g, \quad 2P_g = \frac{e_i^2}{R_g} \tag{**}$$

by division of (*) and (**) we obtain

$$e_i^2 = 2e_g^2$$

and by substituting in (1)

$$\frac{e_g^2}{R_g} = \frac{2e_g^2 R_g}{R_g^2 + (1/\omega_1^2 C_g^2)}$$

$$1 = \frac{2R_g^2}{R_g^2 + (1/\omega_1^2 C_g^2)}; \quad R_g^2 + \frac{1}{\omega_1^2 C_g^2} - 2R_g^2 = 0; \quad R_g^2 = \frac{1}{\omega_1^2 C_g^2}$$

$$\frac{1}{\omega_1} = R_g C_g; \quad \omega_1 = \frac{1}{R_g C_g}; \quad T_1 = \frac{1}{f_1} = \frac{2\pi}{\omega_1} = 2\pi R_g C_g$$

This frequency, for which the power falls to $\frac{1}{2}$ and the voltage to $\sqrt{(\frac{1}{2})}$ (i.e. both by 3dB) is termed the lower limiting frequency or the lower half-power frequency f_1 and is defined by the product RC, the time constant (see p. 51)

$$\tau_1 = R_g C_g, \quad \text{i. e.} \quad \tau_1 = \frac{1}{\omega_1} = \frac{T_1}{2\pi}$$

Let us now consider the phase shift (see formula p. 47) for the limiting frequency

$$\tan \varphi_1 = -\frac{1}{\omega_1 C_g R_g} = -\frac{1}{(1/\tau_1)\tau_1} = -1; \quad \varphi_1 = -\frac{\pi}{4} = -45 \text{ degrees}$$

The distortion for the limiting frequency follows from the above equations: The amplitude decreases by 3 dB and the phase is shifted by one eighth of the period. Both values change much more for lower frequencies. When amplifying harmonic voltages only the amplitude distortion is of importance. Usually 3 dB is considered as a permissible limit. When amplifying periodic non-harmonic voltages which, according to Fourier, may be considered as composed of many harmonic components with definite amplitude and phase relations, the phase shift for the limiting frequency may cause a considerable distortion of the wave form, if this frequency is one of its important components. In this case, the demands on eliminating phase shift are much greater than in the former case. This conclusion holds generally, i. e. also for the upper limiting frequency (see p. 136).

Let us repeat: The size of the time constant in the grid circuit is chosen according to the lowest frequency transmitted in such a way that:

$$T_1 = \frac{1}{f_1} = \frac{2\pi}{\omega_1} < \tau_1 = R_g C_g$$

R_g is chosen as large as the valve permits, especially its vacuum, thermal grid emission, leaks between the electrodes and the condenser leak (see p. 67 and p. 136).

Self bias (cathode bias)

The measured voltage is again applied to the grid through the condenser, not against the cathode but against the frame (earth). The grid is connected through the grid resistor R_g with the frame and not with the bias source. Bias is obtained by a voltage drop across the cathode resistor R_c, caused by cathode current. This positive cathode bias is, of course, equivalent to the negative grid bias (Fig. 52). This method has several advantages and disadvantages: Advantages: No special highly stable bias source is necessary. Considerably larger grid leaks may be used. Disadvantages: Amplification is decreased by the negative feedback: For a more positive grid voltage a larger cathode current is flowing and a larger voltage drop is formed across the cathode resistance. The cathode is thus more positive to the frame and grid, i. e. the grid more negative to the cathode, as compared to the grid-to-cathode potential without the signal. Thus the positive change in grid voltage is decreased by the corresponding change of voltage drop across the cathode resistance, which is proportional to the signal. This negative feedback decreasing the sensitivity can, however, be suppressed by preventing the cathode voltage from varying and maintaining a constant average cathode voltage. This is achieved by connecting a condenser C_c in parallel to the cathode resistor. The condenser is alternately charged and discharged and thus maintains a more or less constant cathode voltage (analogously to a filter condenser in an anode supply rectifier). A detailed analysis shows that, as for the grid condenser, the size of the cathode condenser is defined by the time constant $\tau_1 = R_c C_c$ (the cathode resistor R_c and condenser C_c). The time constant determines the limiting frequency for which the decrease in amplitude by 3 dB and the phase shift by $T_1/8$ (i. e. 45 degrees) occurs. Since, however, R_c is small (hundreds of Ω), the necessary capacitances are usually large and often even tremendous for frequencies used in physiology: tens to thousands of μF, which is difficult to obtain.

Interstage coupling

The coupling of the anode circuit to the following stage is again achieved using a condenser which is connected between the anode of the first and the grid of the second valve (Fig. 52). Mathematical considerations are somewhat more complicated, but since the anode load (R_a) is usually much smaller than

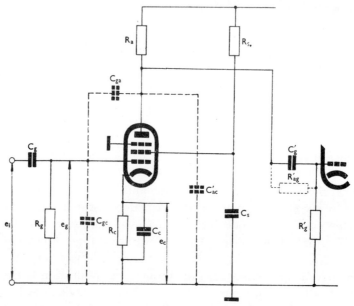

Fig. 52. Coupling between amplifier stages. For description see text.

the grid leak (R_g) it is possible, as far as transmission of the lower end of the frequency range is concerned, to consider the time constant of the grid circuit of the following stage as essential. It is important, however, that the condenser C'_g be connected between the high positive potential of the anode and the negative grid. Since every condenser also has a leakage resistance R'_{ag}, it may happen that the grid obtains an evident positive potential which considerably worsens its function, particularly when using a fixed grid bias. This usually occurs when using very large condensers of inferior quality and large grid leaks.

Upper limiting frequency

The upper limit of the transmitted bandwidth is determined by a decrease in amplification due to stray capacitances of the valve electrodes and the conductors carrying current to them. These are a) the grid-cathode (and grid-frame) capacitances, b) the anode-cathode (also anode-screen grid and anode-frame) capacitances. Both are usually connected in such a way that the grid-cathode capacitance is actually connected in parallel to the anode-cathode capacitance of the preceding stage and they thus are added. The resulting capacitance C'_{ac} is then connected in parallel to the valve. It is charged or discharged with every change in anode current. Changes in anode voltage are thus delayed with respect to the grid voltage by this capacitance (resulting in a phase shift). At the same time amplitude distortion occurs. Again the time constant τ_2 determines these effects:

$$\tau_2 = R_{eq} C'_{ac} = \frac{R_i R_a R'_g}{R_i R_a + R_i R'_g + R_a R'_g} (C_a + C'_{gc})$$

where τ_2 is the time constant, R_{eq} the equivalent resistance for the anode resistance (R_i), anode load (R_a) of the first valve and for the grid leak resistance of the following valve (R'_g) connected in parallel. C'_{ac} is the equivalent capacity for the parallel connection of the anode-earth capacitance of the first (C_a) and the grid-earth capacitance of the subsequent valve (C'_{gc}), i. e. the leak (parasitic) capacitance (C'_{ac}) of the anode circuit. The decrease by 3 dB occurs at the upper limiting frequency or upper half power frequency

$$\omega_2 = \frac{1}{\tau_2}$$

The voltage falls to 70·7% or the power to 50% with respect to the medium frequencies (for which neither this decrease due to leak capacitance nor the decrease due to the grid condenser described above plays a part). The phase shift of the output voltage from the input voltage is $+ T/8$ or $+ 45$ degrees, i. e. numerically the same as the distortion due to the input condenser C_g for the lower limiting frequency, but of opposite sign.

The above relations are important for designing the amplifier. Consider, for instance, that $C'_{ae} = C_a + C'_{gc}$ is at least 100 pF (often much more) and that it cannot be decreased to any considerable extent. If the anode resistance of the pentode is completely utilised, this being several MΩ, and if $R_a = R_i = R'_g$ then for

$$R_a = 3\text{M}\Omega, \quad R_{eq} = 1\text{ M}\Omega \quad \text{and} \quad \tau_2 = 10^{-4}\text{ sec}$$

and thus $\omega_2 = 10^4$ and $f_2 = \omega_2/2\pi = 1600$ c/sec which is often a value that is too small. Consequently (and also in order to reduce the anode supply voltage) a considerably smaller anode load is chosen — about 0·1 MΩ. Then $R_a \ll R_i$ and $R_a \ll R_g$ and thus $\tau_2 \approx R_a C'_{ac}$. The capacitances C_g and C'_{ac} can often be changed by connecting several suitable condensers with a selector switch. Thus the lower (by high-pass filters) and upper (by low-pass filters) limiting frequencies can be controlled in order to reduce noise and interference by reducing the bandwidth of the amplifier. More selective filtering can be attained with *LC* filters (Terman 1943, Smirenin 1950).

Screen grid circuit

For pentode amplifiers a further factor affecting the frequency characteristic must be considered. This is the screen grid circuit. The screen grid has a lower potential than the anode. The potential difference is usually obtained by a drop across the screen grid load due to the screen grid current. The screen grid current, however, varies with the control grid voltage, as does the cathode current considered previously. Thus negative feedback arises again, as at the cathode resistor. It can be simply suppressed by blocking the screen grid. Again $\tau_{s_1} = R_s C_s$ where $R_s = R_{si} R_{se}/(R_{si} + R_{se})$ is the equivalent resistance for the parallel connection of the screen grid resistance (R_{si}) and of the external screen grid load (R_{se}); C_s is the blocking condenser. It is necessary that

$$\tau_{s_1} \gg \frac{1}{\omega_1}$$

(ω_1 is the lower limiting frequency), otherwise phase shift and amplitude distortion occur at low frequencies.

Various types of A. C. voltage amplifiers

1) We have now described the most important amplifier circuit, the single-ended amplifier with resistance-capacitance coupling ("the *RC* amplifier"), which is very often used in electrophysiology. Triodes are usually used for the first stage of the most sensitive amplifiers. Pentodes are used for the further stages.
2) In the *LC*-and *RLC*-coupled amplifiers, an inductance is used instead of the anode resistor. It is very rarely used in electrophysiology.
3) Transformer coupled amplifiers use the primary coil of a transformer as the anode load. The grid of the following stage is connected to the secondary coil.
4) Tuned amplifiers use resonant circuits, tuned transformers, band-pass filters etc., instead of anode load (or grid leak). They transmit only a very narrow bandwidth.

The amplifiers described above are single-ended. They amplify a voltage applied between the grid and the cathode.

5) Balanced, symmetrical or push-pull amplifiers. Both valves of any stage of a push-pull amplifier are connected in exactly the same way. The amplification is only double that obtainable with one valve, but the stability of the amplifier is considerably increased. This circuit is used in nearly all EEG amplifiers and most D. C. amplifiers. For more detailed description of its operation see p. 159.

2. Impedance transformers

The voltage amplifiers described above also amplify power, which is not consumed, however, since the input impedance of the following stage is always considerably larger than the output impedance of the preceding one. For some purposes it is necessary to transform an electric signal into the mechanical movement of an ink writer. It is then necessary to amplify the power and not the voltage. The output impedance of the amplifier must be matched to the input impedance of the ink writer (see p. 56). In this case the power amplification is maximum and may attain about 25% for triodes and about 50% of the anode dissipation for pentodes. The anode dissipation is the heat produced by electrons hitting the anode. Since the work performed by one electron is equal to its kinetic energy (proportional to the accelerating anode voltage V_a) and the number of electrons is determined by the anode current I_a, anode dissipation is equal to $I_a . V_a$. Calculating and designing a power amplifier is comparatively difficult. Many insufficiently defined or variable factors must be considered and an approximate solution must be found fulfilling many conditions, some with more and some with less accuracy. Up to recent times nearly all power-amplifiers were inductance-coupled (by transformers and chokes) if the load impedance, i. e. impedance of the recording system, was low. For purely resistive anode load, capacitance coupling was used. Both approaches were difficult and awkward, especially for wide-band power amplifiers with a very low lower limiting frequency. The necessary inductances and capacitances were large and it was usually necessary to wind and test several large and expensive transformers. Only recently have direct-coupled power amplifiers been constructed, using anode loads of medium size (10^2 to 10^3 Ω). These undoubtedly will be used frequently in electrophysiological work.

Electrometer amplifiers

Power amplifiers may be regarded as impedance transformers since their input impedance is much larger than their output impedance. Hence they can also be used for matching impedances, if no amplification is required.

Often, however, demands for this are high: to attain input impedances of 10^8-10^{12} Ω and more. In other words: the input signal of 0·1 V (the maximum usual in electrophysiology) may cause a maximum grid current of $10^{-9}-10^{-13}$A. This is usually solved by using special electrometer circuits with electrometer valves. If demands are not extreme, however, satisfactory results may be obtained with conventional valves and, on the contrary, even with special valves success is not guaranteed if care is not taken to avoid many pitfalls. Consequently, only the most usual circuits will be described and more attention will be paid to how to avoid failure.

Cathode follower

The most usual electrometer circuit is a cathode follower (Fig. 53). The input voltage is applied to the triode grid, which is connected to the negative terminal of the anode voltage supply through a very large resistance R_g. The anode is connected to the positive supply terminal directly without any intervening impedance. The load resistance R_c is connected in the cathode and is much larger than the usual cathode resistance used for obtaining grid bias (see

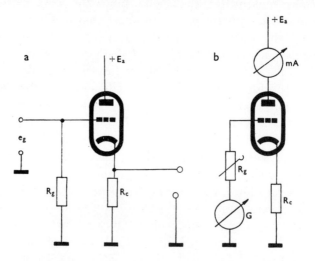

Fig. 53. Cathode follower. a) basic circuit; b) measuring grid current.

p. 135), usually 10^4 to 10^5 Ω. With a zero grid voltage a cathode current flows. This produces a drop across the cathode resistor so that the cathode has a positive potential. The grid is thus considerably negative to the cathode, since the cathode resistance is very large. The cathode current is thus very small and the cathode voltage follows the changes of grid voltage. It is in phase with the latter and has nearly the same amplitude — hence the term cathode follower. The cathode follower does not amplify the input voltage, but decreases it. Its output impedance, however, is very small. If an external load is connected in parallel to the cathode resistor, the cathode voltage decreases and the grid becomes less negative to the cathode. This increases the cathode current and consequently also the voltage drop across the cathode resistor. In other words: the effect of connecting a load in parallel to the cathode resistor is compensated for to a considerably extent by a negative feedback, so that the effective internal resistance of the cathode follower is considerably smaller than its cathode resistor. As already stated, the input voltage is always larger than the output one. For the amplification to be large (in reality for the decrease to be small) valves having large transconductance must be used.

Cathode follower input

Conditions at the amplifier input are very important. The input resistance must be as large as possible. The grid is connected not only to the measured voltage, but also to all other electrodes of the valve through the resistance of the glass envelope and by the tube-elements capacitances. In addition, different e. m. f. are connected to it (e. g. thermal emission of electrons). Let us now only consider those currents which may influence the grid,

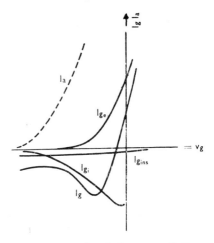

Fig. 54. Grid currents of triode (Tsarev 1953). Abscissa: grid voltage. Ordinate: current. I_a — anode current (interrupted line), I_g — total grid current, I_{g_e} — electron grid current, $I_{g_{ins}}$ — leakage grid current, I_{g_i} — positive ion grid current.

when the grid voltage $e_g = 0$ and the ways in which each of them, and thus also their sum, may be reduced to less than 10^{-12} A. The following currents are mainly concerned: the insulation leakage current between the electrodes, electron grid current, ionic grid current, thermionic grid emission (see Fig. 54).

1) Currents due to interelectrode leakages. a) Anode-to-grid leakage R_{ag}. For $E_a = 100$ V, R_{ag} must be more than 10^{14} Ω, for $E_a = 10$ V, $R_{ag} > 10^{13}$ Ω.

The current I_{ag} flows from the anode to the grid. This current direction (to the grid) will be termed negative. b) Cathode-to-grid leakage — R_{cg}. The current direction is the same, its size depends on the potential difference between the grid and the cathode. This may vary but is always smaller than E_a.

2) Electron grid current. It flows from the grid to the cathode and is thus positive. It increases exponentially with increasing grid voltage $I_{ge} = k I_k^{\alpha e_g}$ (the proportionality constant k depends on the distance of the grid from the cathode, on the grid structure, α depends mainly on the temperature of the cathode). If this current is to be maintained at the lowest possible value than a) the valve must not have an extremely high transconductance, b) R_c must be as high as possible, c) heating voltage must be decreased.

3) Ionic grid current. This flows to the grid and is thus negative. Its absolute size depends mainly on the gas pressure p and the anode current. $I_{gi} = c \cdot p \cdot I_a$ (c mainly depends on the kind of gas). It is important that this current occurs beyond a certain V_a, when the kinetic energy of the electrons suffices to ionise the gas molecules.

4) Thermionic grid emission again produces a negative grid current and depends on the material of the grid and its surface impurities, on the cathode-to-grid distance and on the temperature of the cathode. It is very effectively suppressed by underheating the valve.

Valve selection

An analysis of all kinds of grid currents gives the following results:

1) The valve must have
a) the best possible insulation of the grid and of other electrodes,
b) the best possible vacuum,
c) electrodes not soiled with getter or cathode paste and if possible, specially prepared (gilded grid, ground cathode, etc.),
d) the grid not too close to the cathode.
In addition to special electrometric valves, older types with fine electrode leads inside the envelope and the external grid terminal on the top of the valve come up to these requirements. The valve, and especially the base, must, however, be thoroughly cleaned and all screening varnish, dirt (soot) and ions (from the soldering flux) must be removed carefully.

Individual pieces must be chosen with care, since differences in the above parameters may be very great.

2) It is advantageous to underheat the valve slightly.

The circuit must be arranged in such a way that the grid currents of different directions, all decreased to a minimum, also cancel each other as far as possible. Different currents are different functions of the main variables (e_g, I_c) and parameters (E_a, R_c, p, T, and perhaps the potentials of other electrodes). Consequently, compensation cannot be obtained for the whole range of operation. It is necessary to maintain all individual grid currents at very low absolute values and to compensate for $e_g = 0$, when the same absolute error due to grid current results in the largest relative error in e_g.

It is best to proceed as follows: From a larger supply of valves, with the bakelite bases, top caps and screening varnishes removed, all those having a large anode-to-grid leak are excluded. The leak is measured in the cold valve with a voltage of 500 V and a sensitive galvanometer. All connections and the galvanometer must be carefully insulated, otherwise the surface resistance of the working table may be measured instead of the valve leak. This possibility must always be checked. The significance of removing the base and the screening varnish can be seen from the following measurement: A valve NF2 had an anode-grid leak $1\cdot5 . 10^{10}$ Ω, after removing the base and washing off the screening varnish with acetone $5 . 10^{11}$ Ω. After thorough drying in an oven at 60°C, the leak resistance rose to $5 . 10^{12}$ Ω. Thus the resistance increased 300 fold altogether.

TABLE 2

A valve NF2 in a cathode follower with cathode load 75 kΩ

E_a(V)	I_a(μA)			I_g(10^{-9} A)		Heating voltage
	$R_g = \infty$	$R_g = 50\,\text{M}\Omega$	$R_g = 164\,\Omega$	$R_g = 50\,\text{M}\Omega$	$R_g = 164\,\Omega$	
0	—	—	—	15	—	
0	—	—	—	14	—	
3	0	0·8	7·5	+13	—	
4·5	0	1·6	8·3	+11	+129	
6	0	3·2	9·2	+10	+ 60	
10·5	0·2	8·8	11·5	+ 4	+ 7	9 V
19·5	3·0	16·8	16·8	+ 8	+ 0	
30	cca55	22	22	— 0	— 0	
60	400	35	35	— 0	— 0	
80	660	47	47	— 0	— 0	
120	—	68	68	— 0	— 0	
10·5	0·1	6·1	8·3	+ 2	+ 5	
19·5	2·5	12·2	12·2	0	+ 0	6 V
30	25	16·1	16·1	— 0	— 0	
10·5	1·6	10·4	15·0	+ 4	+ 14	
19·6	17·6	20·4	20·4	+ 0	+ 0	12 V
30	100	26·4	26·4	— 0	— 0	
10·5	0·9	12·0	18·0	+ 8	+ 21	
19·5	13	22·8	22·8	+ 0	+ 0	15 V

The vacuum is checked in the remaining valves. This is done by measuring the difference ΔI_a for a high anode voltage (at least 200 V) by which the anode current I_a increases when a large resistance $10-100$ MΩ or more is inserted into the grid lead connected directly to a fixed bias of about -2V. The positive ions are attracted by the negative grid and on inserting a resistance they accumulate at the grid, charge it positively and thus increase the anode current. If the characteristics of the valves are reasonably equal, then the value $\Delta I_a/I_a = c \cdot p$ is a measure of the gas pressure in the valve as the reader can easily demonstrate himself (see p. 142). It is suitable to measure with the maximal catalogue values of I_a, which must be the same for all valves. Usually all valves with a measurable positive $\Delta I_a/I_a$ (due to the ionic current) as well as those with a negative $\Delta I_a/I_a$ (caused by excessive electron current) should be discarded. In pentodes it is often better to measure ionic current in the suppressor grid with a sufficiently large negative voltage at it.

The selected valves are connected in the cathode follower one after the other, and the total grid current is measured for different anode voltages with a very sensitive galvanometer (Fig. 53b). At the same time the anode current is also measured with a less sensitive instrument. The heating voltage is usually chosen as half to two thirds of the catalogue value. The measurements are arranged into a table for each valve as shown in Tab. 2. Valve NF2 is connected as a triode in the cathode follower with $R_c = 75$ kΩ. For circuit diagram see Fig. 53b, heating voltage 9 V.

Let us analyse the table in detail. With zero anode voltage, the initial "diode" current of the valve is measured. At a low anode voltage, only a positive electron grid current and a thermionic emission (negative) current are flowing through the valve. The ionic current is not flowing, since the kinetic energy of the electrons is not sufficient to ionise the gas. The positive grid current greatly predominates. Hence for $R_g = \infty$ (i. e. the grid disconnected) or even $R_g = 50$ MΩ the grid voltage falls considerably (grid is charged with electrons) and the anode current decreases. From 4·5 V, however, the grid current decreases and so does its effect on the anode current, because the electron grid current falls to some extent with rising anode voltage, as electrons more attracted by the anode miss the grid more easily and also because an ionic current begins to flow. Between 20 and 30 V the positive and negative grid currents are in approximate equilibrium and therefore the total grid current is zero. With higher E_a the ionic current continues to increase. This is seen indirectly from the large increase of I_a for $R_g = \infty$. E_a is chosen in such a way that $I_g = 0$ and I_a is constant for any R_g or only falls slightly for $R_g \to \infty$. Valves for which this E_a is too low are discarded. The rest are grouped according to the E_a. For a balanced amplifier (see p. 145) valves with I_a values closest to each other are selected in pairs from the above groups.

Cathode follower construction

First all components must be perfectly insulated. The leakage of the condensers (if used) must be very small, the valve envelope clean. If requirements are very strict, a drying substance (P_2O_5, silicagel) is placed into the apparatus box.

The input and output of the follower may be connected as an A. C. amplifier with capacitance coupling or a D. C. amplifier with direct coupling. In the A. C. circuit the grid leak resistor is usually also used to prevent charging of the grid with the grid current; it is not connected, however, to the frame, but to the tap of the cathode resistor. Its equivalent value then exceeds several times its nominal value and must be larger than the maximum source resistance R_i, i. e. $10^8 - 10^{11}$ Ω. In D. C. amplifiers the grid resistor is usually omitted. The input capacitance must be decreased as much as possible. A valve is chosen with the lowest capacitance of the grid to the other electrodes, i. e. a physically small valve, and this is brought as close as possible to the measured object so that the grid lead is only a few centimeters long. The grid connection is usually not screened. If screening is necessary, then the whole cathode follower and the measured object must be screened together, or the grid lead screening is connected to the cathode (see p. 156). Sometimes, however, the object must be screened from the heat produced by the valve. Insulating opaque screens (not metallic sheet) are used in such cases. All connections in the grid circuit are as short as possible and of thin wire. A longer screened cable is used to connect the follower to further stages of the amplifier. This cable also carries the supply voltages. Heating is usually D. C. from the storage battery, the anode voltage is also usually taken from the battery and is very low.

Balanced cathode follower

In the same way as the usual triode or pentode amplifier with an anode load, the cathode follower also has an A. C. output voltage, superimposed upon a D. C. voltage, produced by the flow of the mean cathode current through the cathode resistance. A coupling condenser may also be used to separate the D. C. voltage. It is, however, advantageous to use a pushpull circuit, where the input voltage is applied between the grids of symmetrically connected valves and the output voltage is taken from between their cathodes (Fig. 55a). Thus the output voltage is symmetrical with respect to the centre of the cathode resistances and a coupling condenser is no longer necessary in the output. For a balanced cathode follower, the connection may also be made as follows: cathode currents may be increased while maintaining high cathode resistances and a low anode voltage by either not earthing the anode voltage supply at all ("floating anode voltage") or by earthing a certain positive potential lying somewhere between the positive and negative supply terminals, instead of connecting the negative supply terminal to the common joint of the grid resistors with the frame.

This gives the frame, i.e. the grids, a positive bias which is compensated for by an increased voltage drop across the cathode resistor, i.e. by an increase in cathode current.

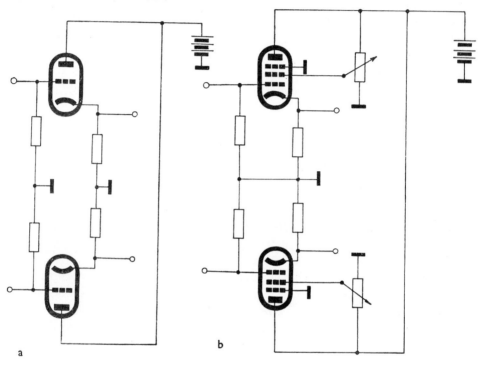

a   b

Fig. 55. Balanced cathode follower. a) basic circuit; b) equilibration of valves by changing the voltages at the screen grids.

When building a balanced follower, identity of the valves is essential. It is necessary to obtain with the same transconductance of the valves and with the same cathode resistances exactly equal cathode currents. This may be achieved by a) correct choice of valves, b) proper choice of the operating parameters (size of R_c, E_a, I_c, i. e. earthing of a certain potential, see above), c) adjustment of the valve characteristics, e. g. by using tetrodes or pentodes instead of triodes, and by changing the second grid voltage with a potentiometer in such a way that the cathode currents are exactly equalised (Fig. 55b). The same effect may be obtained by adjusting the heating of the valves. If the substrate is to be affected minimally, all that has been stated on the preceding pages concerning grid currents must be respected. If the grid currents of the two valves are equal, they cancel each other. Nevertheless it is best to maintain each at its lowest value, avoiding procedures that might also change other valve characteristics. It is better to measure each valve separately and not only the resultant current in the input circuit. New valves must not be used. They should be aged artificially; according to requirements, valves are aged for 100—1000 hours with the catalogue current. They are then permitted to rest for several days and again used for about 50 hours with such a voltage and current that will be used later. With such precautions, valve characteristics change only little and slowly. The first aging stabilises the cathode. During rest and the second aging with a considerably smaller current, the vacuum is improved.

Cathode follower analysis

In addition to these general explanations it is necessary to understand cathode followers in more detail if the finest methods such as recording from single fibres, are to be used. Hence an attempt is made here to present a quantitative analysis of the cathode follower that is somewhat more extended but also more comprehensible than the usual data in the literature (see e. g. Gray 1954, Bonch-Bruyevich 1955).

From this aspect the basis of the high input impedance and the low output impedance are most important together with an explanation of the ways in which these characteristics can further be improved and thus the frequency range widened, i. e. how to improve the fidelity of the recordings.

The gain of a cathode follower

How a strong negative feedback occurs in the cathode follower has already been explained. A quantitative examination of the follower is started by determining the relationship between U_{in} (measured input voltage) or u_{in} (small changes in U_{in} that represent the actual measured signal) and U_{gc}, i. e. the potential difference between the grid and the cathode. The U_{gc} or its small changes (u_{gc}) is the only decisive factor for the current which flows through the tube I_c or for its changes, i_c. Hence:

$$U_{gc} = U_{in} - U_c, \quad \text{and} \quad u_{gc} = u_{in} - u_c \tag{1}$$

where U_c is the whole voltage at the cathode resistor and u_c its changes.

If we change U_{gc}, i. e. by applying signal u_{gc} then the cathode current of the tube is changed with E_a and R_c remaining the same regardless of where the resistor is inserted, since the instantaneous anode current is a function of only U_a and U_{gc}. Thus instead of a follower we may consider a typical amplifier with a resistance in the anode (differences between the two appear only when considering the effect of u_{in}, while with regard to u_{gc}, no differences are present). We know already that in such an amplifier the amplified signal at the anode u_a is not the multiple of the grid voltage, the mutual conductance and the anode load, since the real anode voltage changes at the same time (and the mutual conductance is defined for a constant U_a). This causes the u_a to be smaller than the above multiple $u_{gc}g_m R_a$, the smaller the anode resistance of the tube R_i, i. e. the greater the effect of changes in anode voltage on anode current. Let us start from the defining equations

$$g_m = \frac{\Delta I_a}{\Delta U_g} \qquad (U_a = \text{const}) \tag{2}$$

$$R_i = \frac{\Delta U_a}{\Delta I_a} \qquad (U_g = \text{const}) \tag{3}$$

Here we actually have, even though the notation is not quite exact, two partial differential equations and we are seaking the total differential*) of the anode current, i. e. that change which the anode current undergoes because of a change in U_g and also in U_a. This total differential is:

$$\Delta I_a = g_m \cdot \Delta U_g + \frac{1}{R_i} \Delta U_a \tag{4}$$

The anode voltage change is then obtained from Ohm's law:

$$\Delta U_a = - R_a \Delta I_a \tag{5}$$

Hence, by rearranging equation (4), and substituting for ΔI_a, we obtain:

$$\frac{\Delta U_a}{R_i} + \frac{\Delta U_a}{R_a} = - g_m \Delta U_g \tag{6}$$

$$\Delta U_a = - \frac{g_m \Delta U_g}{(1/R_i) + (1/R_a)} = - g_m \Delta U_g R_{eq} = - g_m \Delta U_g \frac{R_i R_a}{R_i + R_a} \tag{6}$$

where R_{eq} is the equivalent resistance for a parallel connection of R_i and R_a. The negative sign in (5) and (6) expresses the fact that the voltage at the anode falls if the grid voltage rises.

If now the load resistance R_a is moved from the anode to the cathode only the sign is changed. The change in sign occurs because the cathode voltage is defined relative to the negative pole of the supply E_a i. e. in the opposite direction to ΔU_a, which is the anode voltage relative to the positive pole of E_a:

$$\Delta U_c = \frac{g_m \Delta U_{gc}}{(1/R_i) + (1/R_c)} \tag{7}$$

*) See Appendix II, p. 735.

Instead of the changes ΔU_c , ΔU_{gc} we more frequently write the voltage of the A. C. or variable signal using small letters so that finally we have:

$$u_c = u_{gc} \frac{g_m}{(1/R_i) + (1/R_c)} = u_{gc} \frac{g_m R_i R_c}{R_i + R_c} \qquad (7a)$$

According to the equation of Barkhausen (pp. 62), however,

$$g_m \cdot R_i = \mu \qquad \left(\mu = \frac{1}{D} \text{ is the amplification factor } \mu = \frac{\Delta U_a}{\Delta U_g} \right)$$

so that

$$u_c = u_{gc} \frac{\mu R_c}{R_i + R_c} \qquad (7b)$$

Up to here considerations concerning the cathode follower (amplifier with "earthed anode") are basically the same except for the notation, as for an amplifier with an earthed cathode.
Now let us take equation (1) to obtain:

$$u_c = (u_{in} - u_c) \frac{\mu R_c}{R_c + R_i} \qquad (8)$$

$$u_c = u_{in} \frac{\mu R_c}{R_i + R_c (1 + \mu)} \qquad (8a)$$

If we admit the approximation $1 + \mu \approx \mu$, which is permissible for valves having a great mutual conductance and a large anode resistance, then we have:

$$u_c \approx u_{in} \frac{1}{1 + (R_i/\mu R_c)} = u_{in} \frac{1}{1 + (1/g_m R_c)} = u_{in} \frac{g_m R_c}{1 + g_m R_c} \qquad (8b)$$

If we define the amplification of the amplifier as

$$K = \frac{u_{out}}{u_{in}} \qquad (9)$$

then we have for the follower:

$$K = \frac{u_c}{u_{in}} = \frac{\mu R_c}{R_i + R_c (1 + \mu)} =$$

$$= \frac{1}{1 + (1/\mu) + (1/g_m R_c)} \approx \frac{g_m R_c}{1 + g_m R_c} \approx 1 - \frac{1}{g_m R_c} \qquad (9a)$$

The final approximation is obtained by extending the expression (eq. (8b)) $[1 + (g_m R_c)^{-1}]^{-1}$ according to the binomial theorem and by neglecting members of higher orders. Most frequently this last form is used for followers. It is apparent that the amplification is the larger the greater the mutual conductance and the cathode resistor. (The cathode follower, however, cannot be improved by excessive increase in R_c since then the mutual conductance falls because of a fall in cathode current.)

Cathode follower output resistance

Now let us note the special characteristics of the cathode follower as an electrical appliance consuming and generating electricity. The cathode follower as a generator is the source of voltage and current for an external load which we denote Z_L. Let us first assume for simplicity's sake a purely resistive load R_L, connected to R_c in parallel. Let us find the internal resistance of the follower as a generator. By connecting R_L in parallel to R_c we actually change the size of R_c and thus also the size of u_c and i_c with the input voltage u_{in} remaining constant. The internal resistance of the generator r is defined as the ratio of the fall in the clamp voltage to the rise in current, i. e. in our case:

$$r = \frac{-\Delta u_c}{\Delta i_c} \qquad (u_{in} = \text{const}) \tag{10}$$

The simplest way is to express Δu_c and Δi_c using partial differential quotients (see Appendix II, p. 735)

$$\Delta u_c = \frac{\partial u_c}{\partial R_c} \cdot \Delta R_c \tag{11}$$

$$\Delta i_c = \frac{\partial i_c}{\partial R_c} \cdot \Delta R_c \tag{12}$$

$$r = -\frac{\partial u_c / \partial R_c}{\partial i_c / \partial R_c} \tag{13}$$

$\partial u_c / \partial R_c$ is found from equation (8a)

$$\frac{\partial u_c}{\partial R_c} = u_{in} \mu \frac{R_i + R_c (1 + \mu) - R_c (1 + \mu)}{[R_i + R_c (1 + \mu)]^2} \tag{14}$$

i_c is determined from Ohm's law $i_c = u_c / R_c$ and for u_c we insert from the same equation (8a) and differentiate.

$$\frac{\partial i_c}{\partial R_c} = u_{in}\,\mu\,\frac{-(1+\mu)}{[R_i + R_c\,(1+\mu)]^2} \tag{15}$$

$$r = +\,\frac{R_i}{1+\mu} \tag{16}$$

It is easy to calculate the electromotive force e of the generator from the calculated internal resistance of the follower as a generator. The resistance which is fed from this "generator" is R_c and this external resistance is "connected" in series to the internal resistance r of the generator. In other words:

$$e = i_c\,(R_c + r) = u_{in}\,\frac{\mu}{R_i + R_a\,(1+\mu)} \cdot \frac{R_c\,(1+\mu) + R_i}{1+\mu} = u_{in}\,\frac{\mu}{1+\mu} \tag{17}$$

Finally we calculate the output resistance of the cathode follower R_{out}, i. e. the resistance lying between its output clamps. R_c is connected in parallel to the output clamps as is also a branch in which an ideal generator with an e.m.f. e and its internal resistance r are both connected in series. Thus R_c and r are connected in parallel.

$$R_{out} = \frac{R_c \cdot r}{R_c + r} = \frac{R_c R_i/(1+\mu)}{[R_c\,(1+\mu) + R_i]/(1+\mu)} = \frac{R_c R_i}{R_c\,(1+\mu) + R_i} =$$

$$= \frac{R_c}{(R_c/R_i)\,(1+\mu) + 1} \tag{18}$$

$$R_{out} \approx \frac{R_c}{R_c(\mu/R_i) + 1} = \frac{R_c}{1 + g_m\,R_c} \tag{18a}$$

We now calculate the average R_{out} for the cathode follower with

$$R_c = 10^4\ \Omega \text{ and } g_m = 10^{-3}\text{A/V};\ R_{out} \doteq 900\Omega$$

If the output of the follower is loaded with an additional ohmic load R_L, R_{out} is evidently changed as follows:

$$\Delta R_{out} = \frac{\partial R_{out}}{\partial R_c}\,\Delta R_c \tag{19}$$

$$\Delta R_{out} = \left(\frac{R_i}{R_c\,(1+\mu) + R_i}\right)^2 \Delta R_c \tag{20}$$

because

$$\Delta R_c = R_c - \frac{R_c\, R_L}{R_c + R_L} = R_c\, \frac{1}{1 + (R_L/R_c)} \qquad (21)$$

$$\Delta R_{out} = \left(\frac{1}{(R_c/R_i)\,(1 + \mu) + 1}\right)^2 \cdot \frac{1}{1 + (R_L/R_c)} \cdot R_c \qquad (22)$$

The relative change in R_{out} is

$$\frac{\Delta R_{out}}{R_{out}} = \frac{1}{(R_c/R_i)\,(1 + \mu) + 1} \cdot \frac{1}{1 + (R_L/R_c)} \qquad (23)$$

If we again introduce the same approximation $1 + \mu \approx \mu$, then:

$$\frac{\Delta R_{out}}{R_{out}} \approx \frac{1}{g_m\, R_c + 1} \cdot \frac{1}{1 + R_L/R_c} \qquad (23a)$$

It is often, however, more important to know how the time course of the output voltage changes if the output resistance R_c is loaded with a capacitance C_c. Let us first assume an ideal amplifier without parasitic capacitances at the in- and output, which thus faithfully reproduces the voltage step entering the input.

Cathode follower output time constant

If we now connect the parallel capacitance C_c to the output terminals this capacitance will be charged with a time constant τ_c and the voltage step will be distorted at the output. Charging and discharging occurs via a resistance connected to the capacitance, i. e. via R_{out} and thus:

$$\tau_c = R_{out}\, C_c = R_c\, C_c\, \frac{1}{(R_c/R_i)\,(1 + \mu) + 1} \qquad (24)$$

$$\tau_c \approx R_c\, C_c\, \frac{1}{1 + R_c\, g_m} \qquad (24a)$$

The time constant τ_c is thus usually ten times shorter than $R_c C_c$. If we make the fairly strict demand that $\tau_c < 10^{-6}$sec and if R_c is $10^4 \Omega$ we obtain $C_c < 10\tau/R_c$, $C_c < 10^{-9}$F, $C_c < 1000\ \mu\mu$F which can easily be fulfilled even if the cathode of the follower is loaded with the capacitances of the screening etc. Nevertheless for large output (and small input) time constants the response of the cathode follower may be oscillatory, as pointed out by Flood (1951) and Nastuk and Hodgkin (1950). If the input signal also has a short time constant (i. e. is very steep) the cathode follower does not work satisfactorily in certain intervals: for a positive voltage step the voltage at the grid increases, but the voltage at the cathode rises only slowly so that grid current begins to flow until the voltage U_{gc} again attains a slightly negative value. For negative voltage step there is no danger of grid current but another complication may set in: the voltage at the grid falls to such an extent that the valve is cut off. In that case $R'_{out} = R_c$ and thus $\tau'_c = R_c R_c = \tau_c (1 + R_c g_m)$ the time constant is considerably prolonged. For details see e. g. Bonch-Bruyevich (1955).

Cathode follower input resistance

Let us now consider the input of the cathode follower, i. e. how the cathode follower as a consumer loads the measured source. (The equivalent resistance of the consumer with regards to the source is denoted as R_s). Let us consider several variants of the cathode follower, first the connection in Fig. 53a.

Assuming that no grid current flows in the valve then the measured source e_g is loaded by a resistance R_g in the same way as in an amplifier with an earthed cathode, i. e. $R_s = R_g$. Now consider the connection in Fig. 56a, where R_g is connected to the tap at R_c and let us assume that the resistance between the tap and the cathode is, in respect to the resistance between earth and tap,

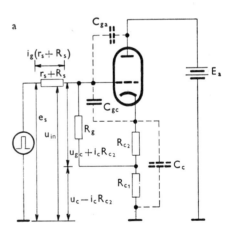

Fig. 56. a) Cathode follower with grid resistor connected to a tap on the cathode resistor.

negligible, i. e. that R_g is connected directly to the cathode. The current flowing through this resistance is $i_{gc} = u_{gc}/R_g$. For u_{gc} we insert from (1) and then for u_c from (8a)

$$i_{gc} = \frac{1}{R_g}(u_{in} - u_c) = \frac{1}{R_g}\left(u_{in} - u_{in}\frac{\mu R_c}{R_i + R_c(1+\mu)}\right)$$

$$i_{gc} = \frac{u_{in}}{R_g} \cdot \frac{R_i + R_c}{R_i + R_c \,(1 + \mu)} = \frac{u_{in}}{R_g} \cdot \frac{1 + (R_c/R_i)}{1 + (R_c/R_i)\,(1 + \mu)} \tag{25}$$

but

$$\frac{u_{in}}{i_{gc}} = R_s = R_g \frac{1 + (R_c/R_i)\,(1 + \mu)}{1 + (R_c/R_i)} = R_g \left(1 + \frac{1 + g_m R_c}{1 + (g_m R_c/\mu)} \right) \tag{26}$$

If we insert typical values then: $R_s \doteq 10\,R_g$. In basically the same way the effect of small grid currents or their changes is also reduced, even though the exact analysis is more complicated.

In electrophysiology, of course, the input of the cathode follower is usually without any resistance R_g and the e.m.f. of the source is connected in series with its internal resistance r_s and the equivalent input resistance of the follower R_s, i. e. between the grid and earth. Then $u_{in} = e_g - r_s\,i_g = R_s\,i_g$ and i_g must be sufficiently small so that $r_s\,i_g \ll e_g$. In other words, R_s should be considerably larger than r_s. If r_s is about $10^8\,\Omega$ and e_g 10^{-2}V it follows that the grid current should be considerably smaller than 10^{-10}A. The usually tolerated value is today in the range of 10^{-11} to 10^{-13}A, but undoubtedly as methods become finer demands on decreasing the grid current will increase.

Cathode follower input capacitance

Up to the present we have been considering an ideal amplifier without parasitic capacitances. In reality, however, an amplifier has capacitances, which, even though small, may cause together with large resistances long time constants and thus considerable distortion, particularly of voltage steps. Consider the source with r_s about $10^8\,\Omega$ connected to a normal pentode amplifier with an earthed cathode. Its grid (together with the connection) has a capacitance towards the cathode of about 5 $\mu\mu$F. This is increased in an amplifier by Miller's effect. In this case the anode-grid capacitance C_{ga} acts as the feedback capacitance (C_g on p. 116) and integrates an input voltage step as a phantastron. Multiplying C_{ga} by the amplifier gain we obtain the effective capacity (500 $\mu\mu$F) and time constant (50 msec) which may cause numerous difficulties. It is much larger than the output time constant of a cathode follower.

Let us hence examine what input capacitance (i. e. the capacitance between the input clamps) is found in a cathode follower. In order to make results clear and considerations simple, we must first again introduce simplifications: the output capacitance of the follower is zero, $C_c = 0$, and the grid capacitance against the anode C_{ga} is also taken as zero for the first moment.

Capacitance is defined as

$$C = \frac{Q}{U} \tag{27}$$

which for an effective capacitance between the input terminals is:

$$C_{in} = \frac{\Delta Q}{\Delta U_{in}} \tag{28}$$

ΔQ is the charge which the measured source supplied to the only present capacity, i. e. the grid capacity against the cathode, C_{gc}. Thus according to (27):

$$\Delta Q = C_{gc} \cdot \Delta U_{gc} \tag{29}$$

we insert into (28) from (29) and (9) to obtain:

$$C_{in} = \frac{\Delta U_{gc}}{\Delta U_{in}} \cdot C_{gc} = \frac{\Delta U_{in} - \Delta U_c}{\Delta U_{in}} \cdot C_{gc} = C_{gc} \left(1 - \frac{\Delta U_c}{\Delta U_{in}} \right) = C_{gc} (1 - K) \tag{30}$$

where $K \approx 1 - 1/g_m R_c$ (9a) is the amplification of the follower, so that

$$C_{in} \approx \frac{1}{g_m R_c} \cdot C_{gc} \tag{30a}$$

For $K = 0.99$, C_{in} would be only 1/100th of C_{gc}, which would improve τ_{in} in comparison to the above amplifier with an earthed cathode under otherwise equal conditions in the cathode follower to 5μsec, a value comparable to the above calculated average τ_c.

We must now return to our simplifying assumptions. In reality of course, $C_c > 0$, which means that the charge supplied by the measured source does not only charge C_{gc} but also C_c, since both capacities connected in series are connected to the input clamps. C_c is, of course, also charged by the cathode current of the tube or the output resistance of the follower. What part of its charge comes from the measured source and what part is supplied by the tube depends on several circumstances and the same circumstances also determine how C_c will affect C_{in}. If the ratio of the time constants is such that $\tau_s > \tau_{in} > \tau_c$, where τ_s is the time constant of the signal in the shape of an exponentially increasing voltage step, then the effect of C_c on the size of C_{in} is completely negligible as is easily seen without exact proof.

The second assumption was also not quite justified: $C_{ga} \neq 0$. The average values are : $C_{gc} \doteq 3\ \mu\mu F$, $C_{ga} = 1\ \mu\mu F$, $C_{ac} = 3\ \mu\mu F$. In practice these values are several times larger since the capacities of the connecting wires must be added. In order to determine how both capacitances of the anode will affect the input capacitance of the follower we must remember that the anode has a zero A. C. voltage, i. e. has a constant voltage in respect to the lower input clamp, i. e. is connected to it through a series capacitance of infinite size (realized by an anode battery with a negligible internal resistance). Thus C_{ga} is connected in parallel to the input clamps of the follower and we can write much more exactly:

$$C_{in} = C_{ga} + (1 - K)\, C_{gc} \tag{31}$$

The capacity C_{ga} thus obviously considerably worsens the input capacitance of the amplifier, often however, only apparently, since C_{ga} is increased only slightly by the capacitances of the mounting while C_{gc} is usually increased more.

C_{ac}, on the other hand, because of the above reasons, is really connected in parallel with C_{out} and because it is so small it may be neglected.

Screening of the cathode follower

The screening of the grid connection, if necessary, is not earthed but connected to the cathode. The capacity of the screening, which may be considerable, is thus not added to C_{in} but to C_{gc} and together with the latter is reduced by a factor of $(1-K)$. If the output voltage of the follower is further strongly amplified all conductors must also be screened if they are connected to the

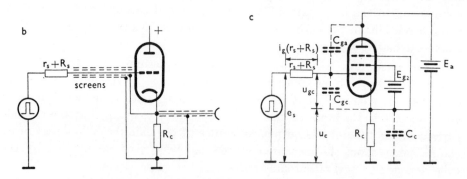

Fig. 56 b) screening of cathode follower; c) pentode cathode follower.

output clamp, i. e. those having the potential of the cathode, including the screening of the grid connection. Thus around the first screening connected to the cathode there is another screening connected to earth (Fig. 56b). The capacity between both screening envelopes is thus in parallel to C_c and can be fairly large, hundreds of $\mu\mu$F. The chamber containing the preparation or the solution containing it etc. is connected to the first internal screening envelope. The external box of the experimental set up or the Farraday cage is connected to the external screening.

Pentode cathode follower

The input capacitance of the follower can further be decreased considerably by using a pentode instead of a triode (Fig. 56c). In the pentode there is a control grid to screen grid capacitance C_{g1-2} instead of the grid to anode capacitance C_{ga} of the triode and the control grid to anode capacitance C_{ga} which is also present, is reduced to 10^{-2} to 10^{-3} $\mu\mu$F. The screening grid must of course be connected in such a way that the capacitance C_{g1-2} does not interfere, or is reduced by the same factor as C_{gc} i. e. the screening grid must have the same A. C. potential as the cathode. The D. C. potential with respect to the cathode must be highly positive. This is most easily achieved by placing a battery between the cathode and the screening grid, which again increases C_c by about 100 $\mu\mu$F, but this is bearable. C_{in}, however, is considerably decreased to:

$$C_{in} = C_{ga} + (C_{gc} + C_{g1-2})\,(1 - K) \approx (C_{gc} + C_{g1-2})\,(1 - K) \tag{32}$$

C_{in} can further be decreased only by increasing K, up to nearly 1 (i. e. $C_{in} \approx 0$), using a positive feedback, e. g. a circuit analogous to that shown in Fig. 67c, which increases i_c of the first valve and thus decreases u_{gc}.

Compensation of amplifier input capacitance

It may happen that even with such a cathode follower we still obtain an input time constant τ_{in} which is too long. This usually occurs when working with capillary microelectrodes where the microelectrode itself has in its tip not only a high resistance prolonging its time constant but also a fairly large capacity which may be considered, on first approximation, as a capacity between the grid and earth and which is thus not reduced by the cathode follower.

For more exact analysis more complex equivalent circuit diagrams are required (Amatniek 1958), since the capacity is in reality spread along the resistance of the electrode. Such a diagram may be e. g. a chain of resistances, where condensers are connected between the individual resistances, with their second poles earthed. (In reality, of course, even this diagram does not reflect the conditions exactly.) Here only an amplifier compensating for this equivalent output capacitance of the measured source and microelectrode can be of some aid, e. g. an amplifier having a negative capacity (or let us say generating one at its input) of the same size, i. e. in parallel to the capacity microelectrode-earth. What is a negative capacity? If capacity is defined as $C = Q/U$ or $C_{in} = \Delta Q/\Delta U_{in}$ then we may write formally $- C_{in} = - \Delta Q/\Delta U_{in}$, which can be realized as follows: as the charging current induced by an increase in u_{in}, flows from the measured source into C_{in}, an equal current from some generator controlled by a measured voltage u_{in} without any delay,

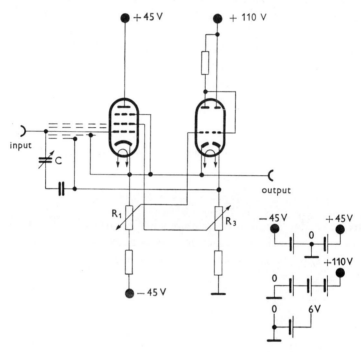

Fig. 57. Bak's amplifier with compensated input capacitance. See text.

flows into C_{in} at the same time (and in the opposite direction with respect to the current in r_s). Since this generator possesses a considerably smaller internal resistance than $R_{in} = r_s R_s/(r_s + R_s)$, a charge is supplied by it and not by the measured source. Thus the time constant of charging C_{in} is considerably shortened, i. e. the equivalent output capacity of the source or the input capacity of the amplifier is decreased. The feedback that is introduced is positive and close to 1, since it gives a voltage of the same polarity and nearly the same size as that of the measured source onto the input. The main difficulty in practical work is to attain such a small delay in the feedback branch, so that the feedback is the same for all frequencies of the required frequency band. Of the many apparatuses (valve, transistor, hybrid) of this type realized in recent years (Amatniek 1958, MacNichol 1962, Guld 1962

Moore and Gebhart 1962, Schoenfeld and Eisenberg 1962 etc.) Bak's amplifier is given as an example (1958, Fig. 57). Its first valve is a pentode cathode follower. The necessary part of the output voltage is taken from the potentiometer in the cathode of the follower to excite the second valve, a triode, the cathode of which is connected to the cathode of the first valve and the output clamp. The input signal of the second valve is thus inverted in phase. The output signal taken from its anode thus has the same phase as the first input. The third valve is again a cathode follower. Its output has two functions. It feeds the screening grid of the first valve and controls its amplification and leads the feedback voltage into the control grid of the first valve. The amplification of the amplifier is thus controlled by two potentiometers, R_1 and R_3 to (nearly) exactly 1. By mutual adjustment of those two elements the grid current of the first valve can be adjusted to zero. The size of the feedback current, i. e. the size of the negative capacity generated at the input is controlled by condenser C. This is the larger the larger C_{in} and can also be used to compensate the capacity of the microelectrode, the screening etc. The main characteristics of the amplifier are the following: a large $R_s > 10^{12}\Omega$, $K = 1$, $\tau_{in} < 3\mu sec$, frequency band $0-700$ kc, noise $20 \mu V$ in the 40 kc band. Values for the components and manipulations are given in the original paper.

Unipolar transistors make it possible to design transistor amplifiers analogous to the above type of improved cathode followers and to match even their high input impedance and low input capacitance (Oomura et al. 1967).

3. D. C. amplifiers

The causes limiting the transmitted bandwidth from below and the significance of the time constant have been discussed when describing A. C. amplifiers. If the amplifying stages are coupled together, amplitude and phase distortions occur at each stage (or even several times in each stage — at the control grid, cathode and screen grid). Thus the time constant τ_1 of the whole cascade amplifier is of necessity much shorter than any of the time constants τ_{1i} of the individual stages. Approximately,

$$\frac{1}{\tau_1} = \frac{1}{\tau_{11}} + \frac{1}{\tau_{12}} + \frac{1}{\tau_{13}} + \dots + \frac{1}{\tau_{1n}}$$

It can be shown that it is best to choose $\tau_{11} = \tau_{12} = \dots \tau_{1n}$. Then $\tau_1 = \tau_{11}/n$. For obtaining the lowest limiting frequency, n must be as small as possible, i. e. the number of capacitances must be small. These have two functions: to prevent the occurrence of negative feedback (C_c, C_s) and to separate the A. C. from the D. C. component (C_g). Fig. 58a shows a balanced amplifier without condensers. It is relatively easy to do without the first kind of capacitances. The negative feedback at the cathode is avoided as follows: If the grid voltage of one valve rises, the cathode current in this valve increases and this should result in an increase in cathode voltage (i. e. grid bias drop). This is the basis of negative cathode feedback. Since the cathode resistor R_c is common for both valves, the grid voltage in the second valve falls at the same time. Thus its cathode current also decreases, and if both valves are identical, the common

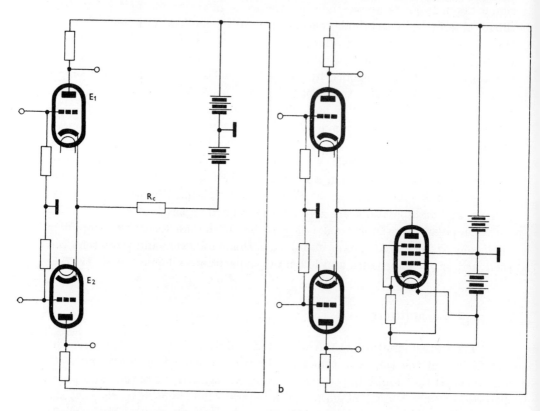

Fig. 58. D. C. amplifiers. a) increase in cathode current with a negative cathode bias; b) pentode as cathode resistance.

cathode current remains nearly constant as does the drop across the common cathode resistor. Consequently, no negative feedback occurs. The negative feedback in the screen grid circuit is eliminated in the same way, i. e. by a common resistor (Fig. 59, R_s).

Discrimination

It is evident that this circuit is much more sensitive to a difference in voltage between the grids than to a difference in voltage between the two

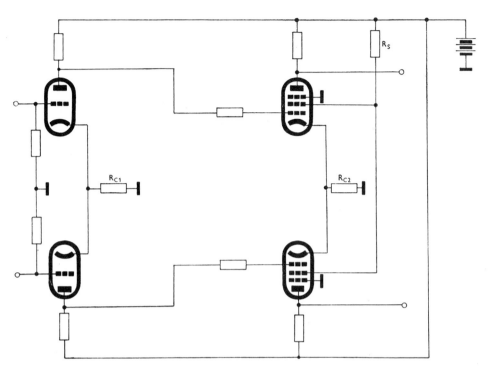

Fig. 59. Two-stage D. C. amplifier.

"connected" grids and any point on the axis of symmetry of the circuit (e. g. earth). Thus if a positive voltage is applied between the connected grids and the frame, the cathode current of both valves increases, resulting in a rise of the drop across the common cathode resistor, which limits the increase in cathode current. In this case a negative feedback does occur, which is the larger the larger the common cathode resistor. This negative feedback is important not only for the common grid-to-frame voltage changes, but also for the anode voltage variations, which are even more essential.

It is therefore best to choose a high common cathode resistor and screen grid load. It is usually necessary to compensate for the increased cathode drop by using a high negative cathode voltage supply, as described for the cathode follower, and often also a high anode voltage. This prevents the cathode current from falling below an acceptable value. It can be seen that the role of the cathode resistor is to maintain a constant current. The current will be the more constant the greater the role of the constant cathode resistor in comparison to the variable valve resistances. If a common current of 1 mA, which is relatively small, must flow through the valves, an E_a of 1000 V is necessary for $R_c = 1M\Omega$, which is excessive. Hence occasionally a current stabiliser is used (usually a pentode) instead of the high R_c. Let us recall the description of the pentode, the anode current of which does not depend on the anode voltage. In the circuit shown in Fig. 58b, a common cathode current from both amplifying valves flows through the pentode, which has more or less constant control and screen grid voltages and thus a constant anode current independent of the anode voltage, i. e. independent of the anode loads of the amplifying valves. For E_c equal to about 100 V, the current stabilisation by the pentode is about the same as that achieved with a 2 MΩ resistor, which for the same current of 4 mA would need an E_c of 8000 V. Although this connection has many advantages, it is used only rarely, since often an R_c of $10^4 - 10^5 \, \Omega$ is sufficient and also because the pentode has considerable noise, ages, complicates the power supply (heating has a high potential against the frame), etc.

Up to the present, the circuit behaviour has been described when connecting the measured voltage symmetrically between both grids. As was shown, the potential difference between the "electrical centre" of the grids and the frame is hardly important. If a certain potential difference against earth is to be measured, one grid is connected to earth, the frame is also earthed and the second grid is connected to the measured potential. The asymmetry of this connection is automatically eliminated by the negative feedback across R_c, so that the output voltage is again symmetrical. If the input signal is applied between earth and both grids connected together (i. e. in-phase in both valves), the output voltage is less by several orders of magnitude than with the same signal applied between the grids (i. e. in anti-phase in both valves). The ratio between these two output voltages is the discrimination ratio.

Interstage coupling in D. C. amplifiers

One stage of a direct coupled amplifier has been described. The lower limiting frequency of such an amplifier is evidently 0 c/sec, and hence it is termed D. C. amplifier. All that has been said for RC amplifiers holds for the upper limiting frequency. If a multistage D. C. amplifier is to be constructed, the coupling between the stages remains to be designed. It was shown that a high negative cathode voltage must be supplied to the amplifier in order to compensate for the drop across R_a. The same compensation may be obtained with a positive voltage supplied to both grids. For coupling the first and second stage, such a positive voltage is at hand: the anode voltage of the first stage (Fig. 59). Thus grids of the 2nd stage are connected directly to the anodes of the first and their positive D. C. voltage is compensated for by a large R_c. The third stage may be coupled in a similar way. This is more difficult for the 4th stage, which then requires a larger R_c and higher anode supply voltage. This complicates the power supply considerably. In a four stage amplifier it often suffices, however, to arrange the first two and the second two stages as two D. C. amplifiers and to couple those with a condenser. The time constant of this single coupling can be maintained at a high value much more easily than one composed of many separate time constants, particularly since R_g is very high and hence a relatively small C_g is sufficient.

The arrangement of a multistage D. C. amplifier used for the most delicate work is thus the following:

1) A D. C. cathode follower with battery supply. This may be disconnected. Amplification 0.8. Or a special cathode follower with amplification 1; see p. 157.
2) A D. C., battery-supplied triode preamplifier directly coupled with the cathode follower. When working without the follower, the E_c must be increased. Amplification 50.
3) The first mains-supplied D. C. pentode stage. Amplification 200.
4) The second mains-supplied D. C. pentode stage. Amplification 200.
5) The output is connected a) to the cathode ray tube. For a total amplification of 2×10^6 and a sensitivity of the screen 1 mm/V the total maximum sensitivity is 2 mm/μV, b) or to the D. C. power amplifier operating a mechanical recording system.

Valves for the D. C. amplifier must again be carefully selected in exactly similar pairs, especially for the first stage. Valves with the lowest microphony must be taken. Some firms supply paired valves, and it is worth while paying extra for these. The stability of the zero position is the most important criterion of the quality of a D. C. amplifier. This is expressed as the ratio of the equivalent imput voltage which would cause the same change in the output voltage as is produced by the amplifier instability to the time during which this shift occurs. Zero stability is measured (and amplifiers may be used) only after stabilisation, which occurs only a long time after switching on (15 min. to 3 hours). The stability of zero is most affected by variations in heating voltage in the first stages. Hence the heating supply must also be carefully regulated. This is usually attained by the use of a large by-pass storage battery (which is constantly recharged by the rectified and filtered mains heating supply).

Recently unipolar transistors (p. 76,77) made it possible to build D. C. amplifiers using circuits essentially similar to those used with valves (see p. 80), but with a much simpler interstage coupling.

Amplifying D. C. voltage with A. C. amplifiers

In some cases, especially if only a relatively low upper limiting frequency of 1 — 100 c/sec. is sufficient, but a high zero stability and high input resistance are required, the D. C. voltage can be amplified with A. C. amplifiers after first changing it to an A. C. or intermittent D. C. voltage. This may be achieved with mechanical choppers or commutators. Fig. 60 shows a mechanical mercury commutator and its connection to the A. C. amplifier. Function is reliable, but the time constant is large and the upper limiting frequency is low, since the mercury oscillations are fairly slow. More rapid switching may be obtained with spring contacts operated by an electromotor, a compressed air motor or

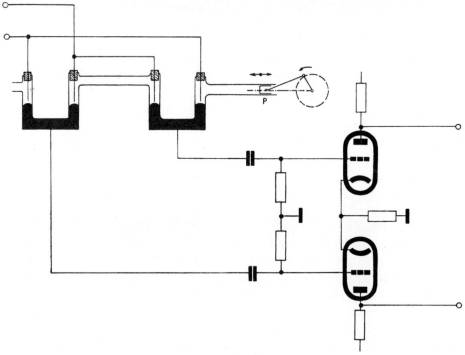

Fig. 60. Simple mercury commutator for amplifying D. C. voltage with an A. C. amplifier. *p* — a motor driven piston. By regulation of the mercury level the input in the zero position may be either short circuited or disconnected.

an electromagnetic relay. The electromotor must be at a sufficient distance from the contacts and must be screened. It is often possible to construct an amplifier perfectly tuned in a narrow frequency band (e. g. a synchronous detector) and thus to increase the signal-to-noise ratio. By appropriate connections of the mechanical contacts, several inputs can be connected one after the other to a common amplifier input, and thus a multilead recording can be obtained with a a single amplifier (Zachar 1955).

The change of D. C. voltage to A. C. modulated voltage may also be done purely electrically in a mixer. The regulated A. C. voltage is applied to one grid and the measured D. C. voltage to the other. The so-called radio-frequency amplifier is an interesting modification of this technique (Haapanen et al. 1952). It was shown in the chapter on stimulators (page 128) that complete resistive and capacitative isolation of the stimulating electrodes from earth is of advantage, and that this can be attained most easily using a radiofrequency stimulator or transformer. In some cases it may be nearly as advantageous to separate the recording electrodes from earth, although this is more difficult. Fig. 61 shows a "radiofrequency amplifier" which makes this possible.

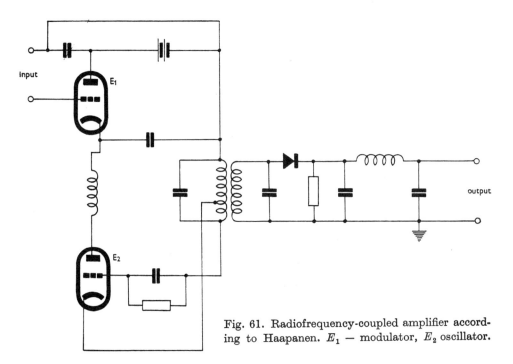

Fig. 61. Radiofrequency-coupled amplifier according to Haapanen. E_1 — modulator, E_2 oscillator.

The first valve is connected as a cathode follower acting as an impedance transformer and modulation amplifier, the cathode load of which is a Hartley H. F. oscillator. The oscillator is modulated by the follower so that the amplitude of the H. F. oscillations is proportional to the measured voltage. The heating and anode supplies are accumulators or dry cells. Neither the apparatus nor the power supply are connected to earth. An H. F. transformer is used for coupling to the further conventional stages of the amplifier. The primary coil of the transformer is the oscillator coil and the secondary coil is connected in the demodulator circuit. A germanium rectifier usually serves as the detector. In view of the fact that up to the moment of detection the signal is continuously attenuated, the adjustment is fairly delicate. Instead of this somewhat obsolete circuit a more simple and reliable circuit can be designed nowadays using unipolar transistors (p. 76). For conversion of D. C. to A. C. see also p. 190. For literature see p. 40.

4. Transistor amplifiers

A semiconductor analogue of the most common valve amplifier is the transistor amplifier with a common emitter (p. 73). Two main differences from the triode amplifier limit its use in electrophysiology: (1) The input resistance is usually lower than in triodes and is always considerably smaller than can be attained in triodes by careful selection of valves and circuits (2). Noise is still greater than in triodes although there is hope of further decreasing it. This often prevents the use of transistor amplifiers in the first stage and sometimes leads to the construction of hybrid valve and transistor amplifiers.

Recently, transistors have, however, been developed that permit the construction of all-transistor electrophysiological amplifiers for less exacting purposes. These are e. g. so called low frequency planar silicon transistors e. g. type BC 107/npn (Valvo and Siemens), BCY 50 and BCY 51 (Lorenz). Their main characteristics are the following: (1) High limiting frequency (50 Mc) so that in the usual common emitter circuit they can maintain high amplification up to about 100 kc. (2) Oxidation of the crystal surface prevents surface phenomena, particularly recombination. Hence the low leakage current (several nanoampers). (3) Only a small fall in the relative amplifying factor with a fall in the collector current, i. e. the possibility of considerably increasing the input impedance of the amplifier which decreases with rising current. (4) Better noise characteristics than in high frequency transistors. The noise of all transistors rises with falling frequency, but in low frequency planar transistors this rise is much smaller than in others. The noise figure has a minimum of about 2·5 dB for relatively small collector currents (about 0·1 mA depending on the resistance of the input voltage source).

The noise factor F is defined by any of the following relatioships:

$$F = \frac{N_{out}}{GN_R} = \frac{S_{in}/N_{in}}{S_{out}/N_{out}},$$

where N — noise power, N_R — thermal noise of the resistance connected across the input (p. 64). N_{in} — noise at input (usually we assume $N_{in} = N_R$). S — power of the signal: S_{in} — input signal, S_{out} — output signal, G — power amplification factor. When measuring and calculating F we must, of course, substitute for G the mean amplification in the f_1 to f_2 band, defined as

$$G = \frac{1}{f_2 - f_1} \cdot \int_{f_1}^{f_2} G(f) \, df,$$

or we must consider out of the whole N_{in} or N_R only that part, which belongs to a certain f_1 to f_2 range over which $G = $ const. Also N_{out} must be taken only from that range. The noise figure $NF = \log_{10}F$ (dimensionless) or $NF = lnF = \log_e F$ (in nepers) or (most often) $NF = 10 \log_{10}F$ (in decibels).

For valve noise see p. 65. Transistor noise belongs to the group termed flicker noise in valves.

In addition to the usual common emitter circuit two connections with a much larger input resistance are used particularly in input circuits. This is the common collector circuit or emitter follower which is the semiconductor analog of the cathode follower (p. 140). The zero amplification of the voltage

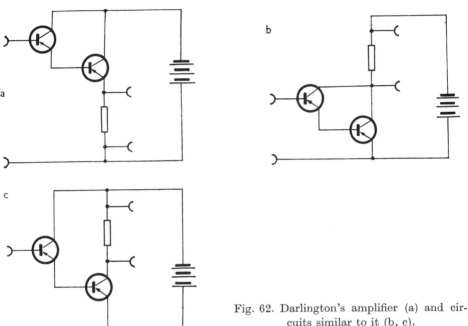

Fig. 62. Darlington's amplifier (a) and circuits similar to it (b, c).

is due to a very strong negative feed back at the emitter resistor (p. 73): The emitter current induces across the large resistor connected to the emitter a voltage drop which is proportional to the current and thus to the voltage between the base and earth. The voltage between the emitter and the base is then the difference between the two above voltages (voltage between base and earth minus voltage between emitter and earth) and remains nearly zero while the output voltage at the emitter resistor is practically the same as the input voltage between base and earth. Amplification of power occurs, however, in the circuit since the output terminals have a much smaller internal resistance than the input ones (see p. 72 for details). The input resistance may be between 1 — 10 MΩ.

Often Darlington's circuit (Fig. 62) is used. The first transistor acts as an emitter follower, the working resistance of which is the base of the second transistor connected also as an emitter follower. Both emitter-base junctions connected in series are connected in the conducting direction but a much smaller current (β_2 times) passes through the first one. The voltage at the output resistance, however as in all emitter followers, is equal to the input voltage of the whole system. The current that flows through the base of the first transistor is $i_{B1} = i_{E1} - i_{C1}$; according to definition the current amplifying factor $\beta = \Delta I_C / \Delta I_B$ and in the linear part of the characteristic $\beta_1 \approx i_{C1}/i_{B1}$; analogously for the second transistor $i_{B2} = i_{E2} - i_{C2}$,

$$\beta_2 \approx \frac{i_{C2}}{i_{B2}}, \text{ so that } i_{B1} = \frac{i_{C1}}{\beta_1} \text{ and } i_{B2} = \frac{i_{C2}}{\beta_2}, \text{ so that } i_{B2} = i_{E1},$$

Let us now introduce the current amplifying factor for the whole circuit:

$$\beta = \frac{\Delta I_{C12}}{\Delta I_{B1}}, \text{ where } \beta \approx \frac{i_{C1} + i_{C2}}{i_{B1}}$$

and finally

$$\beta \approx \frac{\beta_1 i_{B1} + \beta_2 i_{B2}}{i_{B1}} = \beta_1 + \beta_2 \frac{i_{B2}}{i_{B1}} = \beta_1 + \beta_2 \cdot \frac{i_{E1}}{i_{B1}} =$$

$$= \beta_1 + \beta_2 \frac{i_{C1} + i_{B1}}{i_{B1}} = \beta_1 + \beta_2 \beta_1 + \beta_2 \cdot$$

Since $\beta_1 \approx \beta_2 > 1$ and therefore $\beta_1 \approx \beta_2 \ll \beta_1 \beta_2$, it follows from the preceding equation $\beta \approx \beta_1 \beta_2$. If we limit ourselves only to approximate equations then the input resistance of the emitter folower (in the linear part of the characteristic) is:

$$r_{in} = \frac{e_{in}}{i_B} \text{ and } i_B = \frac{i_C}{\beta} \text{ and } i_C = i_E - i_B$$

$$r_{in} = \beta \frac{e_{in}}{i_C}; \text{ further } i_E \approx i_C \text{ and } e_{in} \approx e_{out}$$

i. e. $r_{in} \approx \beta \dfrac{e_{out}}{i_E} \approx \beta r_{out}$, where r_{out} corresponds to all resistances connected to the emitter, i. e. the series emitter resistance R_E with the parallelly connected input resistance of the following stage. With the low frequency transistors it is possible to attain β 100 to 300 using the Darlington's circuit β of the order of 10^4 and the input resistance of the order of megaohms. This, of course, applies to A. C. amplifiers with RC coupling. For D. C. amplifiers it is difficult to attain such favourable parameters since the bases must be connected to the D. C. bias via relatively small resistances and thus the input resistance is considerably decreased. Somewhat better parameters are attained in push-pull amplifiers but for these it is difficult to select transistors of exactly the same properties. Some factories supply paired transistors for slightly higher prices. For common transistor amplifiers see e. g. Popov (1964).

We are at the threshold of qualitative improvement of transistor amplifiers due to the use of unipolar transistors (p. 76). Especially the MOS transistors make it possible to use galvanic coupling between all stages of D. C. amplifiers with a low supply voltage (10—20 V). The circuitry resembles in many respects that used in valve electronics.

d) Recording devices

The recording system is the last stage of an electrophysiological recording apparatus. This always transforms electrical to mechanical energy. The movement is recorded in various ways. The recording system is characterised by its sensitivity, by the frequency characteristic and also by the constancy of the deflection. Sensitivity, however, may be expressed in various ways: in amperes or volts per unit length or unit angle of the deflection. It is more correct, however, to express sensitivity in watts or joules (as the product of watt sensitivity and of the period of mechanical oscillations) necessary for a certain angular deflection. When giving the sensitivity, the equation of motion of the recording system must also be considered. This accurately characterises the functional relationship between the sensitivity and acceleration (angular or linear). In general it may be stated that the higher the sensitivity the lower the acceleration, and thus the longer the oscillation period, but, of course, only in apparatus of the same class. In moving magnet apparatus the oscillation period increases with the square of the sensitivity, in moving coil apparatus with the square root of the sensitivity. Although today only a few types of recording instruments are used, some of the older ones, which have often been ignored unjustly, will also be mentioned. For review see Keinath (1934), Kohlrausch (1943), Trnka and Dufek (1958), Wreschner (1934):

1. Classification

Recording apparatus may be classified according to many criteria:

I. According to the basic function:

    A. Systems sensitive to current

        1) a) with a moving magnet

           b) with a moving conductor (usually coil)

        2) Dynamoelectric

a) dynamometers

b) cathode-ray tubes with magnetic deflection

3) Soft iron instruments, etc. (electromagnetic)

4) Hot-wire meter, etc.

B) Systems sensitive to voltage

1) Mechanical electrometers

2) Molecular electrometers (Lipman's capillary electrometer, Kerr's cell, piezoelectric instruments)

3) Electronic electrometers (valves, electron-ray indicators, cathode ray tubes with electrostatic deflection)

II. According to the method of reading and recording:

A. Direct reading instruments

1) Pointer instruments

2) Mirror instruments

a) with objective reading from the scale

b) with subjective reading in a telescope

3) Microscopic instruments

a) with objective reading (projected image)

b) with subjective reading (visual microscope)

4) Cathode-ray tubes

B. Instruments with photographic registration — The same instruments as in A with facilities for photographic registration

C. With mechanical recording

1) Ink writing instruments

2) Printing (point after point) instruments

3) Instruments writing with a hot needle on wax paper or with a high voltage discharge on specially prepared paper, etc.

III. According to sensitivity

IV. According to the frequency range

Instruments still of use today are described below and divided according to IV and III. Some remarks concerning criterion II will be made at the end.

1. The slowest instruments:

Rotating coil instruments (galvanometers and milliamperemeters). Sensitivity up to 10^{-11} A/mm/m, oscillation time 20—0·5 sec, internal resistance from several Ω to several kΩ. A so called aperiodic resistance must be connected in the external circuit for obtaining the most suitable damping. They are used to measure D. C. currents (in thermocouples, photocells, etc.) and for measuring D. C. voltage (e. g. in connection with a cathode follower).

Moving magnet instruments. They are often more sensitive, but more delicate. A. V. Hill (1937—1938) used them to measure temperature changes in nerves and muscles.

Quadrant electrometer. Oscillation time is the same. Sensitivity up to 10^{-4} V/mm/m depending on the connection.

2. The medium speed ($10^0 - 10^{-2}$ sec.) instruments:

Moving coil instruments. Rein's galvanometers (1940) with rotating coil, which in comparison to Moll's and to loop galvanometers are very sensitive.

Loop galvanometers are robust and simple and well damped mechanically. In all of them recording is photographic with the help of a mirror.

With a moving coil. Constructed in a way similar to dynamic loudspeaker. The coil is wound with thin wire and has a resistance up to several hundred or thousand Ω with a central tap. It is connected as a cathode load into balanced cathode follower power amplifiers. Recording may be mechanical with ink etc.

Moving magnet or soft iron instruments. Coil resistance about 10^4 Ω. They are connected as anode resistances in balanced pentode power amplifiers. Mechanical recording. Both these types are mainly used in clinical electroencephalographs. They may also be matched to the output with transformers.

Capillary electrometer. The surface tension at the boundary between mercury and diluted sulphuric acid changes in dependence upon the potential difference between the two substances. Up to about 0·1 V no current flows. Recording is photographic with a projecting microscope or subjective in a measuring microscope. Voltages below 1 mV can also be measured. This was used for ECG recording before good amplifiers were introduced.

3. Very fast instruments. Oscillation time below 0·01 sec

String galvanometer. The "coil" is reduced to one fine filament which oscillates with the current passing through it between the two poles of a strong horseshoe electromagnet. The movement is perpendicular to the lines of magnetic force. Recording is photographic with a projection microscope. Sensitivity is high. After calculating sensitivity and oscillation period, the relative sensitivity is higher than for the best moving coil galvanometers. This is due to the fact that the filament acts simultaneously as a conductor, a "former" of the coil, and the directing spring and carries no mirror.

171

Piezoelectric electrometer. The amplified voltage is put across piezoelectric plates (today ceramically treated barium or alkaline-earth titanates are used because of their high mechanical strength). A mirror is connected to the plate. This enlarges the deflection optically. Recording is photographic.

4. Inertialess instruments

Only *the cathode-ray tube* operates practically without inertia (Fig. 63). This is a valve in which electrons are concentrated into a conical beam by a series of electrodes. The vertex of the cone falls on the fluorescent screen which glows at the point where the electrons hit it. The location of this point, however, varies according to the shape of the electric or magnetic field through which the electron beam passes. In cathode-ray tubes with electrostatic deflection, the path of the electrons is controlled by two pairs of plates, deflecting the beam vertically and horizontally. The input voltage is applied across the vertically deflecting plates. The beam is deflected towards the positive plate

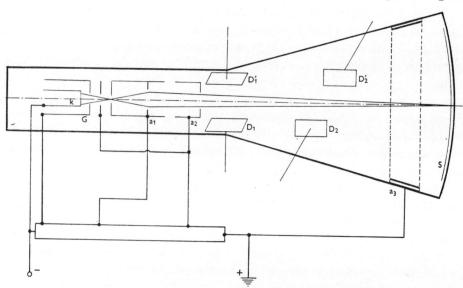

Fig. 63. Cathode-ray tube. k — cathode, g — grid, a_1 — first anode, a_2 — second anode, $D_1 D_1'$ vertically deflecting plates, $D_2 D_2'$ horizontally deflecting plates, a_3 — third anode (not essential), S — fluorescent screen.

proportionally to the potential difference between the deflecting plates. If the beam is permitted to move only in one direction, the movement of the spot is recorded on a film moving at a constant speed perpendicularly to the deflections of the beam. The variation in measured voltage is thus obtained on the film as a function of time. If only a relatively short interval is to be recorded, a constantly rising voltage, a so-called sawtooth wave, is put across the other pair of plates, usually deflecting the beam from left to right. Curves are thus plotted directly in Cartesian coordinates. The ordinate indicates the instantaneous voltage value while the abscissa shows the time when this voltage was attained. Electrical circuits for the time base may be very different. Several of the circuits described in the section on stimulators are used most frequently: thyratron time-base, blocking oscillator circuit, asymmetrical multivibrator with positive grid bias, transitron or phantastron circuits. If a periodic phenomenon with a frequency F is observed a sawtooth voltage of frequency $f = F/n$ is applied to the horizontal plates, where n is a positive whole number. For visual recordings, a frequency larger than 20 c/sec is best. For $f < 20$ c/sec and for visual observation of nonrecurrent electrical phenomena, the screen should have a long persistence of the trace. This, of course, prevents a clear record on moving film or paper but does not matter when photographing single sweeps on stationary film. Often it is necessary to have both visual control and photographic documentation. In that case a moving film recording may be made by adding a "photographic" cathode-ray tube to the "visual one". The horizontal plates of the former are, of course, without voltage. Alternatively the "visual" screen with the time base may be photographed onto a film moving slowly in a vertical direction. The recording obtained is in oblique coordinates. See Petráň (1953) for automatic transformation into rectangular coordinates.

2. Cathode-ray oscilloscope

The cathode-ray tube (see above) with the amplifier for the vertically deflecting plates, the sawtooth voltage generator for the horizontally deflecting plates and corresponding power supplies are connected together in the cathode-ray oscilloscope. The following are the main characteristics: Diameter of the screen, maximum sensitivity in volts (at the amplifier input) per 1 cm beam deflection, the type of amplifier (A. C. or D. C.), their input resistance and frequency characteristic, the frequency range of the sawtooth oscillations of the time base, possibilities for synchronisation, etc. In A. C. oscilloscopes, the sensitivity is often given in r. m. s. values (V_{eff}/cm), which may be misleading. A sensitivity of 1 V_{eff}/cm is approximately the same as that of a D. C. oscilloscope given as 3 V/cm (the transformation factor is $2 \cdot \sqrt{2}$).

Today D. C. oscilloscopes are preferred, since they are more universal and mainly because the central position of the spot on the screen is stabilised immediately, while in A. C. apparatus stabilisation takes some time, depending on the time constant, so that if an interference (e. g., the stimulus) enters the input of an amplifier having $\tau = 10$ sec, it is blocked and it lasts sometimes 30 seconds before the spot reappears on the screen. (For an antiblocking device see Baumann et al. 1956.)

Often two phenomena are to be recorded simulataneously with one cathode-ray oscilloscope. For such purposes a double beam cathode-ray oscilloscope is used, having either a cathode ray tube with two independent electron guns and deflecting systems, or containing a so called electronic switch connecting the vertical plates alternately to the two amplifier outputs in such a way that it first draws a short part of one, then a short part of the other curve. Each curve is thus composed of discrete points. It is also possible to connect the electronic switch as a special apparatus to a single beam oscilloscope. It then has a controllable switching frequency (e. g. from 0·1 c/sec to 50 kc/sec). Although the switching decreases the time resolution for higher frequencies and the curves are less exact in details, the arrangement is less expensive than the double beam apparatus, and the two curves have exactly the same time scale. In double beam cathode-ray tubes this is not the case, since both beams are independent of each other and the voltage outputs of both time-base generators may show different amplitudes and distortions. There are, of course, other ways for obtaining two records with one cathode-ray tube e. g. by splitting one electron beam into two.

Modern oscilloscopes together with their amplifiers are usually already so sensitive that for electrophysiological purposes only one or two stage battery preamplifiers (sometimes with a cathode follower input stage) are sufficient. If the oscilloscope is construced in such a way that it can stand a higher positive voltage of both input terminals against earth, then direct coupling between the preamplifier and the oscilloscope may be used. Gain is usually controlled only in the oscilloscope. The oscilloscope must be constructed in such a way that single sweeps can be started at suitable moments, i. e. when the triggering impulse enters the oscilloscope.

The function of the time-base and different methods of synchronisation

In electrophysiology the time-base is usually synchronised with the stimulator in such a way that the simulator gives a short pulse which triggers the sweep generator (one sawtooth wave is produced). The stimulus is applied only after a controllable delay, so that it arrives when the time-base is already running. The time-scale is controlled in the oscilloscope with controls marked "width" and "time -base frequency" (although the frequency is determined by the stimulator).

The synchronisation switch must be switched over to position "external synchronisation".

If a periodic phenomenon is to be observed, the frequency is adjusted as described above so that $f = F/n$, usually $f = F$. The input voltage or another voltage coupled with it (external synchronisation) is used for synchronisation or, alternatively, the synchronising voltage is taken from the amplifier for the vertically deflecting plates (internal synchronisation). If a small part of the curve is to be analysed in more detail, the voltage of the time-base is increased with "width" control to a multiple of the voltage needed for the full screen deflection of the beam. With the "horizontal" control, such a D. C. voltage is applied to the horizontally deflecting plates so that the chosen part of the curve appears on the screen.

3. Properties of mechanical recorders

Attention must also be paid to some characteristics of mechanical recording instruments. The most important are sensitivity and the oscillation time. Not taking the effect of mechanical resistances and friction (e. g. of the pen and paper) into account at first, the equation of equilibrium for an instrument having a constant current sensitivity is

$$D \cdot \alpha = kI \quad (*),$$

where D is the directive constant, which is determined by the bending strength of the spiral springs (or analogous components) maintaining the moving coil in equilibrium state, α is the angular deflection. Thus $D \cdot \alpha$ is the moment of the directive force equilibrating the moment of the second force proportional to the current I and a constant k given mainly by the shape of the magnetic field and the coil construction. The sensitivity is then $\alpha/I = k/D$, and thus inversely proportional to the directive constant.

The oscillation time $T = \text{const} \sqrt{(J/D)}$, where D is again the directive constant and J the moment of inertia, given by the mass of the movable parts and the distribution of this mass with respect to the axis of rotation. It follows that for instruments which are the same from the electrical aspect, i. e. having the same k and R_i (resistance of the coil), the sensitivity is proportional to the square of the oscillation time and inversely proportional to the moment of inertia $\alpha/I = c\,T^2/J$. Since, however, equal instruments have the same moment of inertia, $T^2 D$ is a constant for instruments of the same type or $T^2 \cdot \text{const} = \alpha/I$.

Thus, without changing the electrical characteristics, it is possible to increase the sensitivity only by decreasing D while T increases at the same time. If the sensitivity is thus raised λ times, the oscillation time increases $\sqrt{\lambda}$ times. If, on the other hand, the original oscillation time T_0 is to be shortened without changing the electrical characteristics of the instrument, its shortening to T_0/λ can only be realised by a corresponding increase of D_0 to $\lambda^2 \cdot D_0$. This means that the sensitivity is reduced to $1/\lambda^2 \cdot \alpha/I$ (where α/I is the initial sensitivity). The oscillation time can, of course, be shortened to the same extent also by decreasing J without changing D, i. e. while preserving the sensitivity. This is achieved by reducing the mass of movable parts of the instrument. It is then necessary to decrease the moment of inertia by $1/\lambda^2$ for the oscillation time to be reduced by $1/\lambda$. It is also possible to shorten T_0 to T_0/λ by simultaneously decreasing J_0 and increasing D_0.

If now the moment of inertia in mechanical recording instruments is considered, especially the fact that for a mass point $J = mr^2$ (m = mass, r = radiusvector) and for a body $J = \int_0^R mr^2 \, dr$, the following possibilities are seen: m may be decreased by substituting aluminium for copper in the moving coils. The mass of the ends with the largest r, the writing pen and its lever, must be reduced. Here every milligram is important, while decreasing the mass of the axis is of no importance. If demands on shortening the oscillation period are extreme, it is often preferable to decrease the sensitivity by shortening the recording lever and thus decreasing considerably the moment of inertia. It is also possible to increase D and to raise the sensitivity by increasing the power amplifier output to the recorder.

Just as in the microscope, where high resolution is much more desirable than magnification alone, so also in the recorder it is most important to increase discrimination power. This is given not by the absolute amplitude of the tracing but by its fidelity, particularly for the smallest details of amplitude and time.

The first important characteristic is thus the ratio of the thickness of the trace to the amplitude of the record. A line 0·1 mm thick and a recording 1 cm wide are equivalent to a line of 0·5 mm and a 5 cm wide recording. In the first case the moment of inertia may be several times smaller for the same coil.

The second important factor is the rigidity of the lever. A lever which is too long, thin and elastic may give a different deflection for rapid and for slow movements: it may oscillate mechanically. This error must be smaller than other errors and inaccuracies.

The third important group of parameters is mechanical friction. In instruments writing on paper, friction in the bearings is usually negligible in comparison to other resistances. The most important is friction of the pen against the paper. This is $F = \beta p$ where p is the pressure of the index (pen) on the paper and β the friction coefficient, dependent upon the kind of recording (ink, wax paper, etc.) and also on the quality of the surface of the index and paper. F is then the force braking the movement of the pen. The friction braking produces a moment of force $M = rF$, where r is the length of the lever. If, for instance, when the current is increasing, the pen moves from left to right then, on first approximation, its final position α_1 will be such that

$$\alpha_1 = k \frac{I_1}{D} - \frac{rF}{D}$$

If now the current begins to fall, the deflection remains constant up to the moment when

$$\alpha_1 = k\frac{I_2}{D} + \frac{rF}{D}$$

By subtracting both equations we obtain:

$$k\frac{\Delta I}{D} = \frac{2rF}{D}$$

i. e.

$$\Delta I = \frac{2rF}{k}$$

where ΔI is the error in the current measurement. If ΔI is inserted into the upper equation (*) (page 175) the error of the deflection is obtained

$$\Delta\alpha = \frac{2rF}{D}$$

$\Delta\alpha$ is the angular error, which it is better to transform to the linear error Δy. Thus $\Delta y = r\,\Delta\alpha$ and hence

$$\Delta y = \frac{2r^2F}{D}.$$

This shows how the length of the lever (r) again considerably worsens the quality of the tracing and that a thin line is preferable to a large deflection.

The problem of damping is also important. This should be such that the equilibrium position is attained as fast as possible and without overshoot (aperiodic movement). Damping differs according to construction: eddy currents, induction, shunt resistances, negative feedback of the amplifier, pneumatic or hydraulic damping. The mechanical properties and damping of the recorder are best tested with rectangular pulses.

Time mark and other markings (stimuli, injections etc.) are also performed mechanically. Either a synchronous motor moves the pen directly or the pen is fixed to the anchor of a telephone relay which receives impulses from any contact clock.

Mechanical recordings are made on continuous moving paper. The paper is either moved by an arrangement built into the apparatus (EEG, ECG etc.) or by a kymograph. The maximum utilisable speed is 50—100 mm/sec.

4. Photographic recording

For photographic recordings an optical kymograph is used. Sensitive paper is usually used in mirror instruments and string galvanometers and film in the cathode-ray oscilloscopes. This moves at a speed of 100—500 mm/sec. If more rapid movement is required, there are unnecessarily large losses when starting and stopping the apparatus. These should be limited in such a way that the motor, having a large fly wheel, runs at constant rate and movement of the film is started by a rapidly acting clutch. Often, however, a more simple device is better. This needs no time to get started and is always prepared for recording. A large drum with a turn of paper or film rotates rapidly continuously (1000—10 000 or more mm/sec at the circumference) and is protected from light by a screen box etc. but for small opening through which the recording beam is to pass. This opening is closed by a photographic shutter which is set in such a way that, on opening, nearly the whole length of film is exposed. This shutter is synchronised, e. g. with the stimulator, or is controlled manually. This arrangement can easily be improvised, e. g. from a bicycle wheel.

When recording with an optic kymograph, the image of the bulb filament is projected by a strong convex lens onto the lamp slit and the image of the slit is then projected by an achromatic convex lens via the mirror of the measuring instrument onto the kymograph slit. It is best to have the lamp as close as possible to the mirror. The sensitivity is regulated by changing the distance between the mirror of the recorder and the kymograph slit (the length of the so-called optical lever).

The time and amplitude marks are important in optic kymographs with paper recording, since the paper changes its dimensions by up to 10% during treatment and the motor is usually also not driven quite regularly. The amplitude mark is either a system of lines on the cylindrical lens in the slit of the kymograph or a system of thin wires extending across the slit. Both sharply interrupt the trace whenever the beam exactly passes the first, second and every further centimeter of the slit length. One then interpolates between these points in the recording. The time mark may be simply made by another beam and a mirror as in mechanical recordings or by illumination of the whole slit at a certain moment. This gives a system of black time coordinates across the whole width of the recording. The slit may also be covered at certain intervals. This gives white time coordinates perpendicular to the white amplitude marks. Both methods are best applied by using a wheel with cut out sectors rotated by a synchronous electromotor.

For the most accurate work, geometrical errors must be considered, arising in the recorder because of transformation of linear coil movement to rotary movement of the mirror and the tangential error inherent in every mirror registration.

The optical kymograph for recording with a cathode-ray oscilloscope has no slit, but a good photographic objective which projects the screen of the cathode-ray tube onto the plane of the film (see p. 172).

When working with optic kymographs it is sometimes not possible to place lamps, mirrors and beams of some of the recording systems into a position and direction suitable for direct registration. It is then necessary to change the direction of the beams and sometimes also to change the plane of the beam deflection. Directional changes are attained by the use of good plane mirrors having á silver- (or aluminium-) plated front surface (to prevent the formation of a double picture) or by the use of prisms (rectangular, pentagonal and others). For changing the plane of deflection without changing the direction of the beam, a so-called reversing system may be used, usually Dove's prism. When this is rotated by an angle α about its long axis, the image of the slit rotates by 2α about the beam axis. If such a prism is, for instance, placed in front of the objective of the camera, the way of recording from the oscilloscope screen may be changed without turning the camera or oscilloscope box and only rotating the prism through 45 degrees. Either the time axis

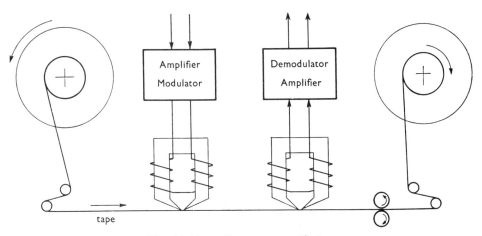

Fig. 64. Recording on magnetic tape.

coincides with the direction of the film movement (recording without the time-base), or the time axis is perpendicular to the film movement (using the internal time-base). For details see page 173 and 186.

5. Magnetic tape recording

Recording on a magnetic tape (Fig. 64) is being used more frequently since it is more reliable than photographic recording and has good frequency characteristics (much better than the ink writers), is cheap and the tape can be re-used. It is also best adapted for further treatment by computers. Magnetic recordings usually require more expensive apparatus than an ordinary tape recorder.

The magnetic tape consists of a thin plastic foil covered by ferromagnetic grains (ferric oxide) that can be magnetized in the narrow gap of the recording coil along which the tape runs. The induction in the grain layer, however, is not proportional to the induction in the coil gap because of the hysteresis of ferromagnetic substances. Even though magnetization of the tape can be linearized by D. C. premagnetization, imperfections in reproduction are not eliminated. For reproduction the tape is passed under the gap of the reproducer head in which changes in magnetic flux are induced and thus a proportional electric voltage is generated. Since, however

$$u = \frac{d\Phi}{dt} = c \cdot \frac{dB}{dt}$$

where B is induction in the tape, the voltage induced during reproduction is not proportional to the recording voltage but is its derivative. Hence if we require the reproduced voltage to be proportional to the recorded one we must record using modulation. Mostly frequency modulation is used. A high frequency voltage of constant amplitude is recorded, the deviation of which from the mean carrier frequency is proportional to the instantaneous value of the input voltage. During reproduction demodulation is performed. Impulse modulation may also be used: a change in the width of the impulse is proportional to the input voltage.

Often recordings have several traces. Several channels are used for the recorded voltages and for the stimulating or synchronizing impulses. In addition, tape recorders must have high stability of the tape transport, several speeds must be available, any part of the recording must be easily retrieved etc. Hence magnetic tape recorders are usually expensive.

6. Calibration

Amplitude calibration

Often it is sufficient to measure the amplification of the amplifiers or the sensitivity of the whole apparatus once and for all, and to provide its controls with a dial. In other cases, where there are more control elements, if exact documentation is required, it is better to use a calibrator. This is usually an accurate voltage divider with several steps fed from one dry cell (Fig. 65).

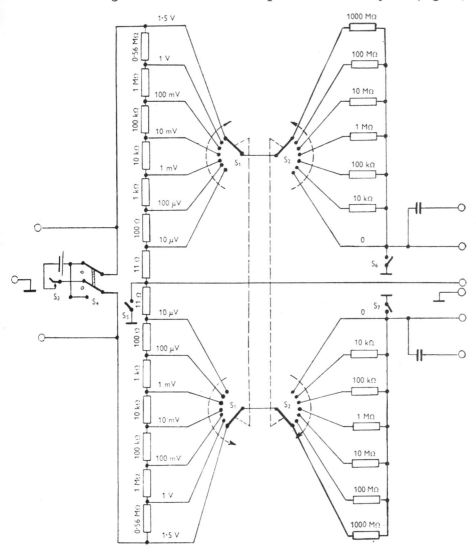

Fig. 65. Calibrator. S_1 — voltage control, S_2 — input resistance control, S_3 — key, S_4 — commutator. With the aid of plugs and switches and the commutator of input voltage, a calibrating voltage of any polarity may be directed either between both amplifier grids, between one grid and the frame, or between the connected grids and the frame, either directly or across a condenser. Thus the sensitivity, discrimination ratio, linearity, noise, etc., can be determined. As voltage source a dry cell or an external supply is used.

The calibration voltage is chosen with the selector switch and by pushing the push-button, this voltage is applied to the amplifier input. The calibrator is connected either directly to the input instead of the object or in parallel to the object (the internal impedance of the calibrator is so small that it practically short circuits the measured tissue and the recording thus contains no bioelectrical potentials but only the calibration voltage) or the calibration voltage is connected in series to the bioelectrical potential (the recording is shifted upwards or downwards). In addition the time constant can be estimated from the shape of the output.

Time mark

Time calibration is usually more important than voltage calibration and may be performed in several ways:

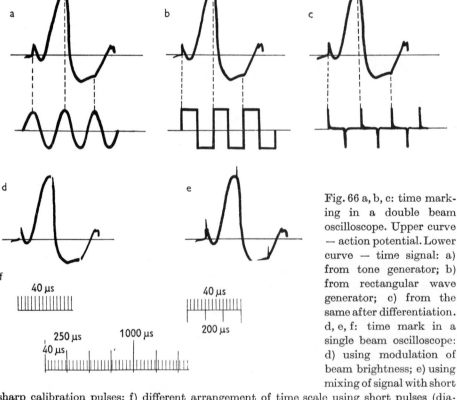

Fig. 66 a, b, c: time marking in a double beam oscilloscope. Upper curve — action potential. Lower curve — time signal: a) from tone generator; b) from rectangular wave generator; c) from the same after differentiation. d, e, f: time mark in a single beam oscilloscope: d) using modulation of beam brightness; e) using mixing of signal with short sharp calibration pulses; f) different arrangement of time scale using short pulses (diagramatic). Left: distance between lines 40 μsec. Right: 40 μsec and 200 μsec. Bottom: 40 μsec and 250 μsec. By their interference further markings (1000 μsec) are automatically formed.

1) Occasionally a periodic voltage of known frequency, e. g. 50 c/sec from the main, 1000 c/sec from e.g. a tone generator, is recorded instead of the measured voltage. The accuracy of time determination is small, maximally 10%.

2) On the second beam of a double beam cathode-ray oscilloscope a sine wave voltage from a tone generator is registered. Accuracy is small. A number of exact parallel lines must be drawn which mark the moments of the same phases of the second curve on the first curve (Fig. 66). Since the phase of a harmonic curve is difficult to determine accurately, interpolation must be used. It is, therefore, better not to use harmonic voltages but rectangular or short sharp pulses. These may easily be produced from the sine voltage of the tone generator even without valves (fig. 67a). This increases the accuracy from 5% to 0·5% of the period of the A. C. voltage used.

3) The brightness of the beam is modulated using sine wave voltage or better, short pulses applied to the modulation terminals of the oscilloscope (Fig. 66d). Evaluation is easier but accuracy is not higher.

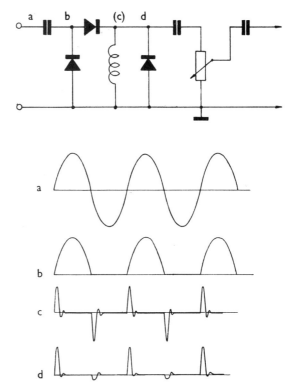

Fig. 67a. Apparatus without valves for producing time calibration peaks from sine voltage. The sine voltage is rectified and spikes are formed in the coil, which are then separated by damping parasitic oscillations with the last diode.

A

4) Short sharp pulses are injected into the vertical deflecting system. They may be applied to the amplifier input (if an asymmetrical input is used) in a push-pull amplifier directed to the free grid (or the suppressor or screen grid of the first stage) or directly to the plates. The last possibility gives the

advantage of stable amplitude. The result is a tracing as shown in Fig. 66ef. Compared to (3) there is less interference with the recording (shorter pulses are sufficient), but accuracy is only slightly increased.

5) All the above methods have a common disadvantage: The initial phase of the calibrating voltage, the delay between the first pulse of the time mark and the stimulus artifact, is accidental. Consequently the number of

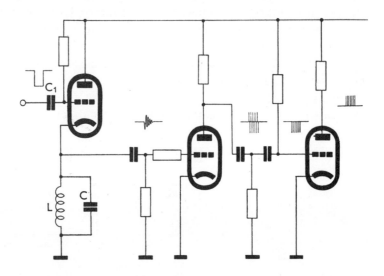

Fig. 67b. Production of synchronized calibration peaks in a resonant circuit.

whole periods of the calibrating frequency must be determined for each curve and an interpolated part at the beginning and the end must be added. This is more elaborate and less accurate than synchronisation of the calibrating frequency with the stimulus in such a way that the time mark is always started by the stimulus in the same phase, as far as possible, in the zero phase. Since the duration of a single recorded curve is usually considerably shorter than the interval between the sweeps, this is easily obtained. The circuit diagram is shown in Fig. 67b: The stimulus excites the resonant circuit *LC*, its damped oscillations are amplified, limited to a rectangular shape, differentiated and the derivative peaks are used either to modulate the beeam brightness (see 4) or are mixed with the signal (see 5). The drawback is that the damped oscillations soon fade so that only such a calibrating frequency can be used that gives 10 time marks on the recording at the most. Damping of the resonant circuit can, however, be reduced by introducing a positive feedback. Increased damping must then be introduced at the end of the sweep so that the circuit is resting at the moment when it is to start oscillating again. The whole arrangement is shown in Fig. 67c. The input exciting voltage must remain negative for the whole period during which calibrating pulses are to be formed. If there is no such voltage available (e. g. from the stimulator) then it must be produced e. g. by triggering a monostable multivibrator or transitron with a stimulus or synchronising pulse from the oscilloscope. The pulse must be slightly longer than the time during which the spot passes across the screen. Negative pulses are directed to the grid of a triode E_1 having a positive D. C. bias. An oscillation

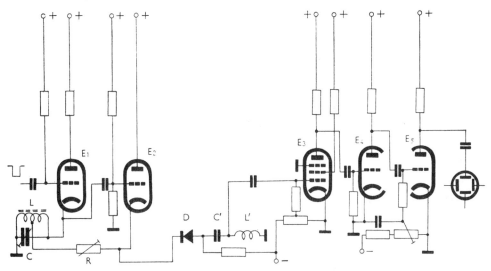

Fig. 67c. Synchronized time mark with positive and negative feedback.

circuit is connected in its cathode LC. The valve is blocked by a negative impulse to the grid. The anode current disappears and this results in damped oscillations with frequency $\omega = 1/\sqrt{(LC)}$ appearing in the circuit. The voltage from the cathode is directed to the grid of a further valve E_2, the cathode of which is connected to a tap of the coil L of the resonant circuit. During the positive period of the current passing through the coil, a positive voltage appears at the cathode. This produces a rise in current also in the second valve. This current, however, also passes through the coil. Thus there is positive feedback, the size of which is controlled by a variable resistance R in such a way that losses in the coil are compensated. When the negative pulse perishes, a cathode current begins to flow through the first valve having a positive bias. At the moment when the cathode is negative due to oscillation of the resonant circuit, a larger current passes (the grid is more positive to the cathode) and produces a voltage drop across the coil in the opposite direction: the cathode becomes more positive. Thus this is an ordinary negative feedback which very rapidly suppresses the oscillations of the resonant circuit (within $1-3$ periods). This may also be explained as follows: the first valve connected in parallel to the resonant circuit, acts as a damping resistance consuming energy accumulat-

ed in the resonant circuit. This damping resistance is infinitely large for a negative grid voltage and very small for a positive grid voltage.

As compared to the usual generators, this arrangement has one further
advantage: the frequency stability is considerable (up to 1 part per million) without any special precautions. If, however, similar absolute accuracy is required
(e. g. for measuring the conduction velocity), the apparatus must be calibrated
with a quartz crystal oscillator.

Practically, the apparatus is usually constructed with several coils and
condensers, or even several first stages which can be selected according to
the different speed of the time-base. A limiter D, differentiating circuit $C'L'$
and limiter-amplifier E_3 follow. The sharp peaks thus produced are amplified
and applied directly to the vertical plates (see 4) or to the cathode-ray tube
grid (or to the "modulation" terminal, see 3). It is also possible to have two
simultaneous time marks, one more and one less dense, one on each side of the
recording or a kind of time vernier making it possible to determine easily a
large number of scale units (Fig. 66f) even for fine division. The accuracy of
such a time marker is very high and fundamentally depends only on the sharp-
ness of the trace. The treatment of results, particularly visual measurements,
is incomparably more rapid and easier than for all other types of time markers.
The greatest advantage of this arrangement is that the time mark needs no
control and takes up no functional place in the experimental arrangement
which might be used otherwise (e. g. one beam of a double beam oscilloscope).
On the contrary it is here possible to inject the time mark on both beams:
Method 4, however, is more suitable than method 3 for a double beam oscilloscope with an electric switch. Another circuit for a synchronised time marker
is described by Saunders (1954).

e) Simultaneous recording of electrical and non-electrical quantities

It is often necessary to register both action potentials and the mechanical
reactions of living tissue or physiological (non-electrical) stimuli.

This is most simply done in mirror recordings, were a myograph or
a similar arrangement may also be a mirror instrument and the corresponding
curve is recorded with the same optic kymograph.

1. The transformation of non-electrical quantities into electrical voltage

If mechanical recording and a cathode oscilloscope are used, it is usually
necessary to transform the corresponding mechanical deflection into electrical

voltage or current. This can be done in many ways (Zhdanov 1952, Klensch 1954), the simplest of which are described below.

Electromagnetic transformation

Mechanical movement is transformed into the movement of a magnet inside a coil, or of a soft iron core increasing or decreasing the magnetic flux in coils on a permanent magnet (Fig. 68a). Examples:

a) Isometrical myograph. The muscle is stretched by a thread, the end of which is fixed to the centre of a telephone membrane (earphone). On. contraction, the movement of the membrane induces a voltage in the telephone coils. By amplifying this, a voltage sufficient for recording with an oscilloscope may be obtained.

b) Isotonic myograph. A weight, with a bar magnet below it, is fixed to the muscle. The magnet enters a flat coil. Movement induces a voltage in the coil. In both cases the voltage is not proportional to the absolute deflection, but to the rate of deflection i.e. to the deflection differentiated with respect to time.

c) Similar principles may also be used in a mechanical stimulus mark. For instance a telephone is fixed to the neurological hammer. The elastic button of the latter is attached to the membrane.

Piezoelectric transformation

Isometric myographs may be constructed in such a way that the muscle bends a piezoelectric crystal, e. g. a crystal gramophone pickup. When using an electrometric amplifier, e. g. a cathode follower, the voltage in the piezoelectric crystal itself is proportional to the deflection, provided insulation is good. If the deflection continues, the voltage disappears only slowly with a time constant of several seconds (given by the total leak).

Resistance transformation

a) Transformation of movement. A change in position is transformed into a change in resistance either by a variable wire resistor with a very fine and light sliding contact, which can be bought or improvised, or with an electrolytic potentiometer (Fig. 68c, d). A diagrammatic representation of the equipment for recording respiration in rabbit is shown in Fig. 161.

b) Transformation of tension. For this resistance wires may be used. Within the limits of Hook's law, these are prolonged slightly with increasing

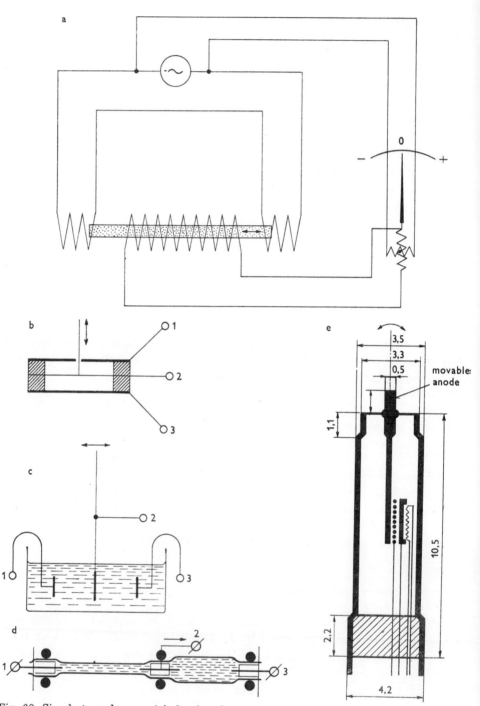

Fig. 68. Simple transducers. a) induction, b) capacitance, c, d) resistance, e) valve with movable anode (RCA 5734).

tension, while their diameter decreases. Thus their resistance increases. They are chiefly used for technical purposes (strain gauge) but may also be utilised as an isometric myograph. A bridge connection must be used, since the change in resistance is small.

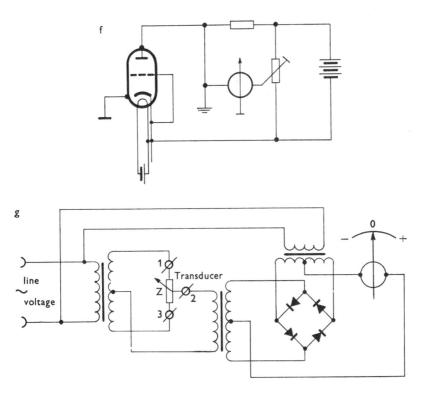

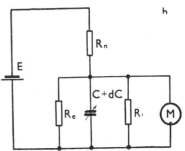

Fig. 68. f) circuit for this valve, g) a circular phase sensitive demodulator, giving DC voltage proportional to the displacement of the sensitive element of symmetrical AC fed transducers, as shown in this Fig. in a, b, c; the transducers are to be connected instead of Z to the corresponding terminals 1, 2, 3. In the case of measuring only small displacements, simpler circuits can be used. h) capacitance recorder — basic circuit.

Current transformation

Current transformation using a valve with a movable electrode (Fig. 68e, f) is somewhat similar (Klensch 1954).

Capacitance transformation

This utilises the relation between the voltage, capacitance and charge $V = Q/C$ (Fig. 68b). An air or liquid condenser is connected with the source of mechanical movement. The latter is measured using the above equation. For small deflections flat condensers with a very small inter-electrode distance are suitable (condenser microphones, capacitance recorders of vibration, etc.). For large deflections variable condensers with very small friction in the bearings or differential condensers may often be used.

Let us determine the sensitivity: By differentiating the above equation with respect to C

$$\mathrm{d}V = - C^{-2}Q\,\mathrm{d}C$$

is obtained, and by substituting for Q from the above equation

$$\mathrm{d}V = - V\frac{\mathrm{d}C}{C}$$

Hence the voltage changes are directly proportional to the voltage and the relative capacitance decrease.*)

The usual circuit is shown in Fig. 68h, where R_n is the charging resistance, R_e the leak resistance of the measuring condenser, R_i the leak of the measuring instrument, M the electrometer (electronic voltmeter, cathode follower or other electrometric instrument). The time constant of this apparatus is again $\tau = RC$, where R is the resistance of all leaks, i. e.

$$\frac{1}{R} = \frac{1}{R_n} + \frac{1}{R_e} + \frac{1}{R_i}.$$

If the time constant is not sufficiently long and R cannot be increased, C may be raised, i. e. the capacitance of the measuring condenser (or without changing the measuring condenser itself, by connecting in parallel to it a constant condenser without dielectric losses, e. g. with styroflex dielectric and the sensitivity decrement is made up for by proportionally increasing the sensitivity of the measuring instrument).

---

*) It is evident from the above equation that this principle may also be applied for measuring D. C. voltage by transforming it to A. C. voltage (see p. 163), since

$$V = -\frac{\mathrm{d}V}{\mathrm{d}C}C$$

If C is always changed by exactly the same amount, then $\mathrm{d}V$ (changes in V), as measured by an A. C. amplifier, is proportional to the voltage V. This exact periodic change in capacitance is realised by either a motor or a dynamic system (as in a loudspeaker) supplied with a regulated A. C. voltage. Such a device is termed a dynamic condenser.

Capacitance transformation may also be achieved by connecting the measuring condenser into the circuit of an A. C. bridge and by measuring changes in its capacitance with a low or high frequency. Here it is necessary for the frequency to be about ten times higher than that of the most rapid component of the movement studied (Fig. 68g).

Photoelectric transformation

Mechanical movement controls the flux of light illuminating a phototube, a photovoltaic or photoconductive cell. The voltage or current (or the drop across the anode or cathode load) is amplified and recorded in a single or double beam oscilloscope.

2. Simultaneous recording with cathode-ray oscilloscope and mirror instrument

The cathode-ray tube tracing may also be combined with one or several other recordings, e. g. mechanical phenomena using the mirror method. The procedure is as follows:

A piece of white paper is glued above or below the cathode-ray tube screen and the spot of the accessory recording system is projected onto the paper as in an optical kymograph. The lamp, however, has not a slit but a hole. The camera (optical kymograph) must be moved a little further away and again focused, so that all tracings can be projected upon the film. When recording in oblique coordinates across the film, the paper must be placed at the sides of the screen and the mirror spot must oscillate in a direction perpendicular to the movement of the film, i. e. horizontally in this case. Here, of course, it is more difficult to find corresponding points in both recordings. The advantage of this method rests in the fact that the oscillograph is not crowded with tracings for which the screen is not absolutely necessary and that, on the contrary, equipment transforming non-electric to electric functions is often saved.

E. Location of faults and interferences

Improper function of an electrophysiological apparatus may reveal itself as follows

1) the complete apparatus is not working,

2) it is working unreliably, occasionally its function changes, e. g. with vibration etc.,

3) its function is changed (e. g. the stimulator stimulates with a frequency that was not set, the registering arrangement only records excessive noise or hum),

4) the function is changed in such a way that the fault cannot be detected without special tests. For example the recording equipment registers artifacts very similar to action potentials.

a) Faults

1. Persistent faults

This group of faults is most easily repaired. The fault, however, must be sought systematically.

a) First one checks whether all components receive the necessary voltages (with voltmeter, according to the valve temperature, lighting of the valve cathodes, etc.).

b) All external connections are tested with an ohmmeter or a bulb tester for short circuits, faulty connections, correct connections, etc. The most frequent site of faulty connections is that between the conductor and the plug.

c) If the fault does not lie in a) or b), an attempt is made to localise it at a certain part of the apparatus. One tests, for instance, whether the fault is in the preamplifier or the oscilloscope by placing a short antenna into the oscilloscope input and observing whether the main A. C. frequency is of the usual size. If that is the case, the fault is in the preamplifier. When testing the stimulator the oscilloscope is used to test the shape and amplitude of pulses for each stage of the stimulator, and thus the faulty stage is found. After localising the fault, further search is continued:

d) The corresponding part is checked mechanically (i. e. visually, with a forceps) — for satisfactory joint soldering. This is the first thing done in electrotechnical factories. A drop of an alcoholic solution of an aniline dye is placed on each point as tested so as not to miss any. All suspected places are repaired (dry joints and loose connections are cleaned and resoldered etc.). If this does not help, then

e) the voltage marked at important points in the diagrams is checked. If there are no voltage values in the diagram, the following voltages must be checked: first, the voltage of the heating, anode and bias supply and the voltage at the electrodes during operation (cathode, anode, screen grid). The lighting of the valve is observed. If that is in order, the voltage at the socket contacts for the cathode, anode and screen grid are tested after disconnecting the heating voltage or after removing the valve. The results are critically evaluated. If the anode voltage with the heating on and off is high and does not change, no anode current is flowing (provided the valve is not operating in an impulse circuit with very short conducting periods). The resistance of the voltmeter must be much larger than the anode load. The causes may be the following:

A large grid bias (broken grid lead, interrupted grid leak). Screen grid without voltage (broken lead, burned load, short circuit, e. g. in blocking condenser). Large suppressor grid bias (rare). Opencircuited cathode (burned cathode resistor) or anode (in the valve). If, on the other hand, the anode voltage drops considerably on heating (i. e. the current rises considerably), this may be due to a positive grid voltage (leaky grid condenser), a short circuit between the cathode and the frame, bad vacuum, etc. Sometimes it is worth while testing not only the electrode voltages but also their currents. The voltage across electrodes connected through large resistance, especially the grid, can be measured only by using electrostatic or electronic voltmeters or a cathode-ray oscilloscope. Their voltage can, however, easily be calculated from the known or measured resistance and current. Usually the fault can be thus localised more accurately. Then

f) components at the localised site are tested. First the resistance values for resistors are checked with an ohmmeter. Then condensers are checked for short circuit and open circuit. The valves are tested by either measuring their properties in special bridges or by replacement. In sensitive instruments, only valves already aged for 50—100 hours can be used, in D. C. amplifiers both valves of the same pair must be replaced by a suitable pair. If no spare valves are available, voltage and current are tested under working conditions. The suspicious valve may also be tested in another reliable apparatus, when it is connected in the same way.

g) If the above did not give the expected result, we test

α) the amplifier using the oscilloscope and a tone or pulse generator. The voltage from the tone generator is directed through a small condenser between the grid and the frame or between both grids, depending on the construction. With the oscilloscope connected also through a condenser to the anode (anode-frame or both anodes), the amplification and distortion are measured. Insufficient amplification may be due to an interruption or short circuit along the path of the signal. Distortion is best tested with a double beam oscilloscope in such a way that the upward deflection is produced by a signal of the same and known polarity in both beams. Input voltage is recorded on one beam and output voltage on the other. The following voltages are measured: at the tone generator output, at the grid of the amplifier input, at the cathode, at the screen grid, at the anode. The grid voltage (which may be calculated from grid condenser and leak) should be somewhat smaller than the tone generator output voltage. If its positive halfwave is limited, the fault is in the bias (or a short circuit at the cathode). No A. C. voltage should appear at the cathode blocked by a condenser. If there is an A. C. voltage, the condenser is faulty (it has lost its capacitance, is interrupted). Only a fraction of the A. C. input voltage can be found between the joint of cathodes of balanced A. C. and D. C. amplifiers and the frame. If it is large, the circuit has become asymmetrical either because of an interruption of the grid, cathode or screen circuit or because of changed characteristics of one valve. The same is valid for A. C. voltage at the blocked screen grid and the joined screen grids of a balanced amplifier. The A. C. voltage at the anode (or between the anodes) must be amplified and undistorted. If amplification is small, the fault may lie in the formation of negative feedback (faulty blocking or asymmetry in the cathode or screen grid circuit), decreased voltages at the anode or screen grid, large grid bias in balanced amplifiers. Reduction of negative halfwaves at the anode is due either to grid limiting by grid current (small grid bias, see above) or to low anode voltage or high anode load. Suppression of positive halfwaves occurs if the grid bias is increased. Symmetrical limiting is due to the input signal being too large.

Time constants are tested with the pulse generator. Decreased time constants produce a larger drop in the terminal portion of the rectangular pulse top voltage (see p. 50).

The amplifier oscillates with its own frequency. This is due to feedback which sometimes may be the result of connecting the tone generator and oscilloscope to the amplifier. On excluding this possibility, the cause must be sought in interruption of a resistor, disconnected blocking condenser, internal short circuits in the valve or between other components, etc.

β) Stimulators are tested in a similar way by making sure that correct limiting and amplification occur at the proper sites.

2. Intermittent faults

The second group is more difficult to mend. We either

a) wait until the fault occurs permanently and then proceed as under 1) or

b) accelerate its occurrence and stabilise it by mechanical shaking, and then proceed as under 1) or

c) we attempt to localise the fault by gently knocking different parts of the instrument and then individual components and joints. After attaining maximum localisation, this part is tested mechanically or the suspected component is exchanged. Switch contacts and the valve sockets etc. are cleaned with tetrachlormethane and the finest polishing paper. Electrical testing only rarely gives any results here.

3. Faults distorting the function of the apparatus

This heterogenous group is repaired either as under 1) or sub 2), if it belongs here, or some phenomena of these types are repaired specifically.

a) Incorrect stimulus frequency (double the frequency set, 50 c/sec., 100 c/sec.) may occur if the frames of various parts of the apparatus are not connected together. It may be due to faulty earthing, to disconnection of inputs and outputs, to "antennas" ("receiving" and "transmitting") in different stages.

b) Large hum has the same causes as in a). For elimination of residual hum see below. Sometimes phenomena a) or b) are due to complete insulation of heating from earth and frame and may appear only after some time, when the filament-cathode resistances change. The origin of hum is often ascertainable from the frequency (or phase shift). 50 c/sec is hum from the mains. 100 c/sec is hum eihther from the anode supply, the valve heating or from the mains. In a halfwave rectifier for anode voltage a hum of 50 c/sec may also occur. In USA the mains frequency is 60 c/sec and the corresponding frequencies are 60 or 120 c/sec.

c) Large noise is either due to interruption of some resistor or bad contact of its leads, dirty contacts, "dry joint", inoperative valve. Sometimes the cause lies in a faulty anode battery, although the full voltage is still indicated by the voltmeter.

Faults belonging to this group can also be elicited by voltage changes of the regulated supplies.

4. Inconspicuous faults

The faults in this group must be systematically eliminated. They are usually not due to the electrical apparatus but lie in the experimental design, the preparation, chamber, electrodes, the mains, the electromagnetic field in the laboratory, and often it is very difficult to find their exact cause. In the description below it is assumed that the individual apparatus themselves are in order.

a) *Stimulation.* This is best tested in a nerve-muscle preparation, i. e. without electrical recording.

α) The stimulator does not stimulate. The excitability of the preparation is tested with a galvanic forceps. If that is good, the fault is searched for elsewhere. The supply of the stimulus may be interrupted or there is a short circuit between the electrodes or the leads. This short circuit is often due to an improper connection of wires: The directly earthed electrode is connected to the active output terminal and the earthed output terminal to the other, unearthed electrode.

β) The stimulator stimulates even when the amplitude of the stimulus is decreased to zero or when switched off. Either a large parasitic voltage enters the output (e. g. capacitativelly transmitted impulses from preceding stages, hum, etc.) or various external voltages - e. g. from sparking electromotors, X-ray instruments, etc. — are induced in the stimulating lead because of bad screening and a large impedance of the stimulator output.

b) *Registration*. When registering action potentials the danger of errors, interference and artifacts is even larger. Much thorough knowledge and experience are needed for awareness of these posibilities. Repeated testing of the experimental set-up and continuous checking of all variables are necessary.

b) Interferences

1. Stimulation artifacts

A stimulus artifact occurs when the stimulating pulse is transmitted to the recording apparatus. This transmission leads along several pathways and as the relative participation of these paths changes, the amplitude and shape of the stimulus artifact may change considerably even during a single experiment.

a) The resistive component. The current flowing through the tissue between the stimulating electrodes during the application of the pulse passes through a volume conductor. An electric field thus produced is characterised by two system of surfaces: tubes of electrical flux and equipotential surfaces. No potential difference is obtained at the recording electrodes only if both electrodes are in the same equipotential surface.

In addition to the connection of both pairs of electrodes across the tissue, however, they are also directly connected by external resistances, especially by different resistances of each electrode against the frame (earth). Part of the balancing current flows through them and produces a drop across them. The drop across the recording electrode resistance is then also recorded as an artifact. On the other hand, a stimulating voltage reaches the recording electrodes through the external resistances. Thus the stimulating voltage can produce excitation or at least subthreshold changes in the substrate also at the recording electrodes. This, of course, is valid not only for the recording electrodes but for all other electrodes or metal contacts, especially earthing leads. There is only one necessary, even if not sufficient, condition under which this cannot occur: if the resistance of all electrodes against each other is infinitely large (excepting the resistance between stimulating electrodes of the same pair).

From this condition follows another one: If n electrodes are connected to the tissue, at least $n - 1$ electrodes must have an infinitely large resistance against the frame (earth). Evidently it is most important that no electrode save one is ever directly earthed, for otherwise stimulation might occur below any or even every earthed electrode.

b) The rapid capacitance component. As far as the volume conductor itself is concerned, the capacitance currents may be neglected in the first approximation. Capacitance currents between the electrodes, however, are

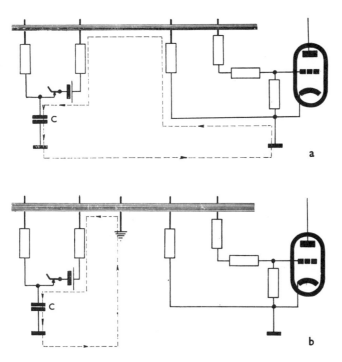

Fig. 69. a) The origin of a stimulus artifact (Schäfer, 1940). *C*-capacitance of stimulator against earth. b) removal of artifact by earthing between the stimulator and amplifier.

all the more important. As in the case of the resistive component, capacitative connection of the electrodes also produces a voltage drop across the recording electrodes and this has a time course different from that across the ohmic resistances. Since the capacitances between the electrodes are small, often only potentials corresponding to the differentiated stimulating pulse are transmitted capacitatively.

c) The slow capacitance component. Schaefer (1936) explains the slow disappearance of the artifact by charging and discharging of the common capacitance of the electrodes against earth (Fig. 69). When switching on the stimulus, the stimulator—earth capacitance is charged. The charging current flows to the other plate of the condenser through earth via stimulating electrodes — recording electrodes (interrupted line in Fig. 69). A voltage drop is thus formed across the leak resistances of the recording electrodes.

After the stimulus is over, this capacitance is discharged along the same path and an opposite voltage drop is produced. Since the longitudinal resistance of a nerve can reach a maximum value of about 10^7 Ω, the capacitance against earth maximum 1000 pF, $\tau = 10^7 . 10^{-9} = 10^{-2}$ sec, which is of the same order as the measured values.

Today, electronic stimulators with a small output resistance against earth are usually used. The prolonged tail of the stimulus artifact, however, although smaller, is still present. The whole artifact, including the tail, may be considerably reduced by directly earthing one stimulating electrode or by earthing the object (see p. 513). A number of further problems, however, arises if this is done (see above). The stimulus artifact may be nearly eliminated by completely separating the stimulating electrodes from earth (resistive and capacitative insulation — see page 128). This shows that the causes of the artifact are those described above and that the slow capacitance component (tail) of the artifact is due to current loops reaching far from the stimulating electrodes.

d) The polarization component: The current coming from an unearthed stimulating electrode divides into a larger part flowing to the second stimulating electrode and a smaller one flowing to all electrodes through their earth leaks. These currents (D. C. pulses) produce polarisation of all electrodes accordingly to the current. The polarising voltage at the recording electrodes remains there for a long time, since the equivalent capacitance which could be substituted for the polarising cell is enormous and the discharging resistances are fairly large. It is clear that, all other things being equal, the polarising component is largest for earthed electrodes. This is effectively reduced by the same means as used for the capacitative component (especially the slow one from which the polarising component can be distinguished only with difficulty).

The last traces of the stimulus artifact are usually not eliminated, but serve as the zero-point of the time axis. If they are to be eliminated completely, again a radiofrequency transformer at the stimulator output (see p. 128) or, in addition, radiofrequency recording (see p. 165) is used rather than bridge equilibration of the artifact transmission (Bishop 1927, 1928, 1929). For other techniques of artifact elimination see p. 541.

2. Artifacts due to improper leading off of bioelectric potentials

a) The single ended amplifier. This requires one recording electrode to be connected with the input valve cathode and thus also the frame of the amplifier and earth. The grid connected with the recording electrode then records the potential difference between this electrode and earth, i. e. between this electrode and all points connected to earth, even across rather large resistances. This gives rise to the long known mutual influence of several bipolar leads, since all leads, though having different recording electrodes and amplifiers, are connected through earth. Thus it is impossible to define the contribution of individual electrode to the recording, which may lead to erroneous conclusions. (For example, when recording action potentials in isolated nerves, polyphasic curves are obtained simulating different conduction velocities.) Several leads with asymmetrical amplifiers may only be used for monopolar recordings, the common earth of all amplifiers being connected to only a single common reference electrode or to an artificial average electrode analogous to the Wilson electrode used in ECG (see p. 511).

b) A symmetrical amplifier records the potential difference between the grids, which in the apparatus usually have a resistance of $0.5-5$ MΩ against earth. Nervous tissue, however, to which the grids are connected, often has only a small resistance, and thus not only the grid-to-grid resistance of the same amplifier, but also the resistance to neighbouring amplifier grids may be decreased considerably. In addition, and this is of particular importance, the grid resistance towards earth, mainly if some part of the object is earthed (indifferent electrode of another recording system, stimulating electrode, etc.) may also be decreased. From this aspect, the earthing lead used to eliminate the stimulus artifact is dangerous, since it is usually near to electrodes. This then produces a situation identical to that in an asymmetrical amplifier (see above): again, in one channel, potentials from several sites are recorded simultaneously (e. g. from the earthing plate and the recording electrode nearer to it against the other recording electrode). Mutual dependence (interaction) of the leads is thus produced. Potential of extremely small amplitude such as synaptic potentials, extrinsic potentials, local potentials, etc. are interfered with by artifacts or even simulated by them the more the smaller they are and the more complex the total layout of the experiment (a larger number of stimulating and recording systems, small interelectrode distances etc.) See e. g. Zachar (1955, 1956).

The source of such artifacts is most frequently in the recording of potentials at other points or of another kind (those having a higher amplitude, e. g. of the stimulus or action potential), which are distorted and reduced during transmission and even may simulate the phenomenon looked for.

3. Determination of artifacts

There is no universal prescription for detecting artifacts. It often happens that phenomena are found which may either be genuine or artifacts, and it may take days or even weeks to decide which is the case. It is therefore necessary:

1) To prevent artifacts by using appropriate equipment: radio frequency stimulus transmission, symmetrical amplifiers with a large discrimination factor, radiofrequency transmission of the input signal, resistive and capacitative insulation of all electrodes from earth.

2) To test systematically the character of all parts of the recorded curve, using physical and electrotechnical means:

a) To observe changes in the curve on changing the amplitude and duration of the stimulus:

α) Action potentials. Preparations with a small number (up to ten) of fibres change the shape of their action potentials in steps, and those with more fibres (the whole nerve) gradually, but the amplitude increases only up to a certain value. On further increasing the stimulus, the action potential amplitude remains either unchanged or even decreases.

β) Local potentials immediately caused by the electrical stimulus increase continuously with the stimulus as a "smooth", usually progressive function of the latter. A step-like change occurs for strong stimuli (the onset of an impulse).

γ) Local potentials due to nerve impulses (synaptic, extrinsic potentials, etc.) increase with the number of conducting fibres. They cannot be further increased after the maximum stimulus is reached.

δ) The stimulus artifact is directly proportional to the amplitude of the stimulus.

b) To determine the behaviour of all parts of the curve when interchanging the recording electrodes:

α) If the curve (wave) studied does not change its polarity, its source may either be in the amplifier itself (disturbances acting on it, etc.) or the potential is recorded from a very distant electrode somehow connected to any of the recording electrodes. Of course, this is so only if the connection is realised directly in the amplifier (e. g. in an asymmetrical amplifier).

β) The wave changes its polarity. In this case, either the actual potential difference between recording electrodes is measured, or that between a distant electrode connected with one recording electrode, if this connection is made directly in the electrode circuit. This, however, occurs only rarely.

As far as the stimulating artifact is concerned, it can be seen that it may, but need not, change its polarity on commutation of recording electrodes. This depends upon the way and point at which it enters the recording arrangement.

200

c) To determine analogous changes in the curve when changing the polarity of the stimulus. Let us first assume that the nerve impulse arises at least approximately at the same point as before.

α) The recorded wave does not change its polarity. In that case, either a real bioelectric potential is recorded (either between the recording electrodes or from another point, e. g. the earthing plate) or a stimulus artifact entering the recording arrangement not through the preparation or capacitive or resistive leakage between the leads, but directly from the stimulating apparatus.

β) The wave changes its polarity. This is a stimulus artifact passing along the usual paths to the recording electrodes.

γ) The reversal of the stimulus polarity produces changes in the amplitude and time relations of the individual parts of the curve.

The amplitude of the stimulus artifact increases (decreases) if the earthed stimulating electrode is nearer to (further away from) the recording electrodes.

The distance of the wave from the stimulus artifact decreases (increases) slightly if the impulse arises below the nearer (more distant) stimulating electrode. The distance changes considerably if the stimulus arises below one stimulating electrode and after commutation at another distant point, which is usually earthed. The conduction times must then be calculated for more accurate orientation. At the same time, the amplitude and shape of the response may also change considerably. At first it often suffices to disconnect the earthing lead; the changed time relations indicate that the point of stimulation or recording has changed.

d) Changes in the recorded waves during changes in the physical and physico-chemical state of the preparation and electrodes may be determined.

α) The resistance between the stimulating electrodes may be changed by either wetting them with a drop of Ringer solution or drying them with a piece of filter paper. This changes the current passing through the tissue (especially in stimulators with a "constant current" output) and the current density at those points at which an impulse is evoked. The stimulation threshold is thus changed. This is true only if the stimulation occurs in the region of the stimulating electrodes.

β) The resistance between the recording electrodes can be changed in an analogous manner. This changes the amplitude of all potential waves in recording of which both grids participate. Only potential changes entering the recording apparatus otherwise than through grids of the first stage do not become altered.

γ) It is also possible to change in this manner the nerve resistance between the stimulating and recording electrode pairs. If excitability changes on decreasing the longitudinal resistance, it is probable that excitation is produced not at the stimulating electrodes but elsewhere. If the wave amplitude is increased, probably this wave is due to the conductance or polarisation component of the stimulus artifact. (The capacitance, resistance or polarisation artifact component may, however, also be decreased by reduced shunting resistance.) In a similar way, the amplitude of potentials not due to impulses may be changed. Each concrete case, however, must be analysed individually. The amplitude of impulse potentials, however, hardly changes at all if actually recorded from between the recording electrodes.

e) Changes in the recording due to impairment of the normal function of the preparation:

α) Blocks due to ions, drugs, cold, etc. (see p. 334) are included here. Only the stimulus artifact remains unchanged. By suitably choosing the site of the block, an attempt is made to obtain an answer to question such as: Is the registered potential in fact recorded from the recording electrodes?

β) Interruption of transmission in the nerve due to mechanical injury (squeezing) provide the answer to the question mentioned above. The action potentials evoked by the mechanical stimulus, if a forceps of nonconductive material is used, may also provide a clue.

γ) Severing the nerve interrupts the nerve not only as a physiological, but also as an electrical conductor.

δ) More gently (e. g. without moving the nerve along the electrodes) the blocking of transmission and interruption of the nerve as a conductor may be brought about using local rapid cooling (solid CO_2, application of methyl- or ethyl-chloride). The ice, even though containing salt impurities, has a specific resistance (depending upon the rate of freezing and the temperature) of 10^5 to 10^6 Ωcm, i. e. about 1000 to 10 000 times higher than nervous tissue.

By local freezing, the nerve may also be insulated from some of the electrodes and the unipolar, apolar and other aberrant stimuli may be determined. The entry of the stimulus into the surrounding solution, interrupted insulation of the electrode leads, recording of potentials coming through the faulty insulation inside the chamber, etc., may be detected in the same way.

ε) The whole nerve is killed, as far as possible without impairing its passive electrical characteristics, using ammonia, ammonia fumes, formalin, formaldehyde fumes, heat etc.It is then necessary to maintain or renew normal moisture of the preparation. Only the stimulus and other artifacts are preserved.

ζ) A physical conductor (wick thread, band of filter paper) soaked in Ringer solution is substituted for the nerve.

η) The behaviour of the apparatus is determined without any object at the electrodes.

4. Hum and other interferences limiting the recording of low amplitude signals

Hum is an A. C. artificial voltage of a certain frequency entering the recording apparatus from outside or inside either directly in the form of an electromagnetic field or as mechanical oscillations transformed to electrical ones in the recording apparatus. Hum in the narrower sense of the term is the basic frequency of the mains (50, 60 or 40 c/sec) and its second and perhaps also higher harmonics. Low frequency hum is often due to electric motors, from which it spreads electromagnetically or mechanically. High frequency hum, especially if not continuous, is often due to bouncing of the contacts, sparking of the motor brushes, etc.

Hum is usually reduced in two ways: at the source by limiting the spread of the hum and in the recording apparatus by protecting it from the effects of hum. Protection of the recording apparatus never eliminates hum completely, but only reduces it to a certain extent, and further reduction becomes more and more difficult. Hence it is necessary to determine exactly the part played by the individual sources of hum in the formation of its final amplitude in the recording apparatus.

A. Internal hum is not affected by changing the position of the apparatus and persists on disconnecting the input. It is due to bad filtration of the anode voltage supply (50 c/sec for halfwave, 100 c/sec for fullwave rectifiers), to heating of the valves with A. C. current (100 c/sec), to the stray field of the mains transformer (50 c/sec, 100 c/sec), to mechanical oscillations of loose core-laminations of the transformer (100 c/sec).

B. External hum increases with the size of the "antenna" connected to the input and with the resistance between the electrodes (electrostatic hum) or with the number of turns, diameter and core permeability of the coil connected to the input electrodes. In both cases its size depends on the orientation of the antenna or coil. External static hum usually has a frequency of 50 c/sec, magnetic 50, 100, 150, 300 c/sec. The size of static hum depends on the intensity of the electrostatic field, i. e. for the same distance of the source of the disturbance, it is proportional to the voltage. The most important sources of static hum thus are high voltage lines, high voltage transformers, X-ray apparatus, neon lights, etc. With the source of the disturbance remaining at the same distance, the size of magnetic hum is proportional to the current and length of the conductor or the diameter and number of turns of the coil and considerably depends on the shape of the core. It is much larger for an open core than for a closed one. The largest sources of magnetic hum thus are high power electric equipment for low voltages, (apparatus for charging large accumulators or for electroplating, large motors, electric trains using A. C. current, electric furnaces).

Parts of the recording apparatus itself may also be a source of external hum if their field can affect the input electrodes.

Mechanical oscillations are a special source of external hum. If not originating within the recording apparatus, they are easily identified by testing the effect of placing the apparatus on a soft base (sand, felt, foam-rubber).

After determining the hum source and its propagation (by an electric or magnetic field) it is reduced:

1) At the source. Static sources are screened. Conductors (at a safe distance) are covered with an earthed conductive screen. Sparking contacts and brushes are connected to LC filters and the whole apparatus is again screened by earthing the metal box.

Magnetic sources may be screened only by using one or more solid, welded covers of soft iron or alloys. Since such a cover may have considerable dimensions, weight and price, it is often not used and the effect of the source is reduced by placing it further away and by suitable orientation of the recording apparatus.

Sources of mechanical vibrations are eliminated by resilient mounting of movable elements, addition of antiresonant weights, introduction of mechanical and acoustic isolation (dilating slits), pads and partitions made of foam glass or foam plastics, by placing rubber, felt or sand pads under the apparatus, etc.

2) The recording apparatus is then treated so as to decrease hum at least to a size that does not prevent the performance of the necessary electrophysiological measurements. Fundamentally the same means are used as under 1).

a) The propagation of the disturbing voltage along mains lines and earthing leads is reduced. Mains lines are supplied with LC filters and enclosed in strongwalled earthed tubing (so-called screwed tubing).

In the mains-supplied transformers, static screening is applied using one (open!) winding of copper plate between the primary and secondary coils (this only reduces the H. F. components, especially discharges and sparking brushes, X-ray interference, diathermy, etc.). The earthing connection must be as short as possible, using a thick wire (about 5 mm in diameter) leading directly to earth, where it ends at a depth of $1-2$ m in a large copper plate ($1-3$ m^2). The earth lead must never be loaded with current. This might produce a voltage drop acting as a source of static interference. The current source might cause magnetic interference and the life span of the earthing plate would also be shortened, since it would be corroded by electrolytic processes. Consequently, the earthing wire must never be connected to the neutral conductor of the mains. This, it is true, is also earthed somewhere, but since usually a large current is passing through it (and in towns also through earth) a certain, even if small, voltage drop arises across it. In view of the very low resistances of the neutral conductor and the laboratory earthing, this small difference might initiate very large currents. At our institute, a potential difference of 0·2 V was found between the laboratory earth and the neutral conductor of the mains. Thus a current of 5 A arose when connecting both these earths across a meter with a small resistance. In view of the fact that the internal resistance of the ammeter formed the larger part of the inserted resistance, the current on direct connection of the two earths would be still larger. It follows that it is always necessary to check whether such a direct short-circuit is not present in the system. It is most easily formed as follows: A mains-supplied apparatus is connected by a three pronged plug to the mains, and its box is regularly connected directly to the neutral mains conductor. If the cover is now earthed unintentionally (directly or through other parts of the apparrtus) to the special laboratory earth, the short circuit is formed. This may be avoided in several ways: most effectively by simple adaptation of the mains cable (which is usually against safety regulation). Either the earthing conductor is disconnected in the outlet (care must be taken not to exchange the mains lines) or in the mains three point cable plug the connection to the earth point is disconnected. The cover and frame of the apparatus are connected to laboratory earth permanently, so as to comply with safety regulations. The above case is a special and malicious example of an earth loop caused by earthing various parts of the circuit at several points. Such earth loops must be avoided for reasons mentioned here and sub b) and c).

b) The least sensitive parts of the apparatus are screened electrically. Care must be taken to prevent sizeable currents from flowing through the screening covers. All covers must have exactly the same potential (this is achieved by connecting them at one point). Often the experimental animal is screened together with the preamplifier, and sometimes also with the experimenter, in a metal box called Faraday's cage. If it has double walls, the two layers, separated by a distance of several cm, are connected only in one corner from which the earthing wire is led. All screening covers are connected to this corner. The inside of the box is connected to the apparatus outside by screened cables, the screening of which is also connected to the common earthing point. The frames of these apparatus are earthed at the same point. The screened cable passing through the box must be insulated from it and as short as possible. If these principles are not adhered to, interference voltages may arise, due to currents flowing in the earthing conductors or to magnetic induction.

c) Magnetic screening. Usually the voltages induced in conductors by the A. C. magnetic field of the interfering source are so small that magnetic screening is not necessary. It is sufficient to prevent conductors from forming loops with a large cross section to make them as short as possible and to lead them in the direction of the lines of force. Magnetic hum is of greater significance for sensitive apparatus working with larger currents (galvanometers, loop oscillographs, string oscillographs), particularly if they have small internal and external resistances. Sometimes the cathode-ray oscilloscope may also be affected. If magnetic screening is necessary, this is done by using a closed box of soft iron or iron alloys, in several layers if possible. Usually combined static and magnetic screening is used: the covers of the apparatus are made of solid iron plates welded at the corners and edges. Faraday's cages are made in a similar way, but here dense rivetting is sufficient. They nearly always have two layers.

d) Finally, hum may be compensated. Only primitive compensation using interfering voltages themselves is used: we test whether the interfering voltage decreases at the amplifier input when applying and earthing conductors, when changing the position of the loops when changing main plugs (turning them), when changing the phases (R, S, or T) which supply this or other apparatus, etc.

e) mechanical vibrations of the recording apparatus are eliminated as described under 1).

f) If interference becomes worse in dry weather the soil in which the earthing plates are buried is moistened.

g) In exceptional cases, when recording D. C. and very slow potentials, hum is eliminated by cutting off frequencies higher than about 30 c/sec using LC or, more rarely, RC filters.

F. Electrodes

Different types of electrodes are used in electrophysiology, depending upon the requirements for accuracy, reproducibility of results, etc.

Often simple metal electrodes are used: wires and plates of silver, platinum, nickel, stainless steel, tungsten, etc. They are simple and usually have a small resistance. They are disadvantageous in chronic experiments with long stimulation periods. Ions from the electrodes pass into the tissue under the action of current and may have a toxic effect. Other difficulties occur when recording D. C. potentials. They are due to the formation of electrochemical cells. Consequently, nonpolarisable electrodes must be used for exact measurements.

a) Nonpolarizable electrodes

Calomel electrodes

The metal is mercury covered by a layer of calomel (Hg_2Cl_2), suspended in saturated or dilute (0·9%) KCl or NaCl. If the electrode forms the anode,

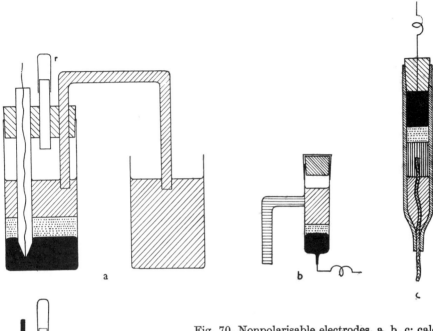

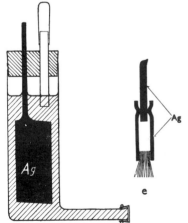

Fig. 70. Nonpolarisable electrodes. a, b, c: calomel electrodes. a) Large electrode connected via a siphon with the fluid to be measured. Bubbles in the siphon can be removed with the rubber tube r; b) small electrode with agar bridge (the agar bridge is first prepared in the side vessel and then the electrode is filled); c) wick electrode. The tube must be narrow so that cohesive forces can maintain a mercury drop in the upper end of the tube. Below the mercury is calomel and a cotton wool stopper through which the wick passes. The lower part is filled with 0.9% NaCl, d, e) Ag-AgCl electrodes; d) large; e) small wick electrode (magnified).

anions from the tissue accumulate on it, especially Cl^-. This forms Hg_2Cl_2 at the electrode and thus the latter's quality does not change. If the electrode acts as the cathode, tissue cations react with the chloride ion of the calomel and thus relase mercury. Again the electrode remains qualitatively unchanged.

Fig. 70 a, b, c shows how to prepare a calomel electrode. Distilled mercury is covered with calomel, thoroughly washed with 0·9%. NaCl. Two such electrodes are connected by a 0·9% NaCl bridge. A small D. C. current (up to 1 mA) is permitted to pass through the electrodes alternately in one and then the other direction. After this the electrodes have no voltage between them. The electrode pairs are always stored shortcircuited. When measuring, they are connected to the tissue via a liquid or agar bridge: about 2% agar is added to a NaCl or KCl solution. After swelling the solution is boiled and poured into a bent tube (avoiding bubbles). After it has cooled, the electrode is connected to the tissue through this tube containing the agar gel. The electrode may also be separated from the tissue by cellophane. This has a low resistance and takes up little space.

Silver chloride electrodes

In principle the Ag-AgCl electrode is similar to the calomel electrode. A thin layer of AgCl is formed electrochemically on the silver electrode surface. Its preparation is less difficult, it takes up less space and is easier to handle than calomel electrodes. The electrode must be protected from light, drying and friction. It is, therefore, best to store it in a special dark vessel in 0·9% NaCl and to connect this to the tissue via a liquid or agar bridge or with the aid of celophane. If there is little space, the electrode may be placed directly in the tissue or connected to it with a thin silk or cotton wick. Great care must be taken to prevent drying and mechanical injury to the AgCl layer.

Electrolytic chlorination is carried out with a very weak current density (about 0·1 to 10 A/m^2) overnight if possible, and always in the dark. Both electrodes of the same pair are connected in parallel as the anode, a piece of silver plate being the cathode. NaCl (0·9%) is the electrolyte. The lower the current density used, the lower will be the final electrode resistance. The quality of the electrodes should be tested. During electrolysis the electrode resistance rises rapidly at first and after several minutes to hours drops considerably and attains a purely ohmic character. For measuring this resistance see Vogel and Kryšpín (1956). Two types of Ag-AgCl electrodes are shown in Fig. 70 d and e.

b) Capillary microelectrodes

These are glass tubes pulled at one end to form a capillary with a tip diameter smaller than 1 μm (Fig. 71). The salt solution in the tube and the capillary tip is the actual electrode. The glass wall ensures insulation and firmness.

On inserting the tip of the microelectrode into a cell we obtain an intra-cellular lead so that the potential difference can be measured across the cell membrane against a macroelectrode in the external environment. Such microelectrodes are hence also called intracellular microelectrodes. Since electrodes can also be inserted longitudinally into the interior of the cell (p. 320) microelectrodes introduced perpendicular to the surface of nerve

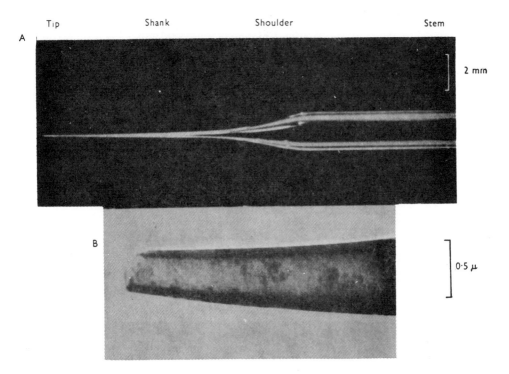

Fig. 71. A. — Photograph of a microelectrode with conventional nomenclature. B. — Electronmicrograph of the tip of a similar microelectrode (by courtesy of M. Henček).

or muscle cells are termed transverse microelectrodes. Sometimes they are named capillary microelectrodes of the Ling-Gerard type according to the authors who first used them to record an action potential (1949). Graham and Gerard (1946), however, used them earlier to measure the membrane potential.

Pulling the capillary microelectrodes

They are constructed from glass tubes either by pulling by hand over a flame or using a pulling machine. The original manual pulling (Ling and

Gerard 1949, Nastuk and Hodgkin 1950) has been superceded by defined mechanical pull. Even today, however, some workers prefer manual extension (better choice of shape and reportedly better characteristics of the electrode, e. g. tip potential). Mechanical pulling has the advantage of serial production of relatively uniform microelectrodes.

Extension can be performed by one pull but usually is performed in several steps. The number of steps depends on the properties of the parent glass tube and the method of pulling.

Today many methods for microelectrode construction are available from manual to fully automatic extension. Pull is realized either vertically or horizontally. According to the force exerted by the pull we divide such apparatus into spring, solenoid and gravitationally operated ones. In apparatus with several steps these types are sometimes combined as in the automat in Fig. 72. This is a two-stage fully automatic vertical puller where the first pull is gravitational while the second is produced by a selenoid as in the Winsburry apparatus (1956).

A glass tube 2 mm external diameter and 1·6 mm internal diameter and 8 cm long is passed through a platinum spiral (3-5 turns of 0·5 mm platinum wire). The internal diameter of the spiral is 4 mm and the individual turns of the spiral are 0.5 mm apart. The upper pipette clamp is fixed to a rack and pinion. The spiral fixed in a ceramic holder can be moved vertically by another rack and pinion. The movable pipette clamp is on a stainless steel shaft and has a plunger attached to it. The latter is pulled into the solenoid after the current is switched on. The lower shaft moves on two sets of ball bearings, each of which consists of three ball bearings 120° apart.

After mounting the glass tube, the platinum wire is heated to a bright orange colour. The current is adjusted by a powerstat (Fig. 72 A, B). Suitable heating has to be found empirically. When the glass begins to melt, it is pulled by the weight of the lower shaft and of the core. This pull decreases the tube diameter to $100-200$ μm (the shank of the microelectrode). The length of the shank is given by the distance between the ring on the lower shaft and the solenoid switch (Fig. 72 C) as measured before heating is started. The switch is operated by the ring when the shaft moves downward and causes the current to pass into the selenoid. The current and thus the final pull is controlled by an autotransformer (Fig. 72 A, B). The final pull narrows the capillary to the required diameter and breaks the parent tube into two pieces. Finally another ring on the lower shaft breaks the current in the solenoid and in the heating circuit by operating another switch (Fig. 72 D).

The movement of the lower holder is then braked and stopped. In both the upper and lower pipette clamp a finished microelectrode remains. The whole operation runs automatically after pushing the button (T) as can be seen from the diagram of the electric circuit (Fig. 72 B).

The necessary coil temperature, the weight of the lower holder, the solenoid current and the distance between the activating ring and the first solenoid switch must be found experimentally for the desired microelectrode parameters. Then the production is rapid and reproducible. Two microelectrodes can be made from one piece of standard glass within 1—2 min.

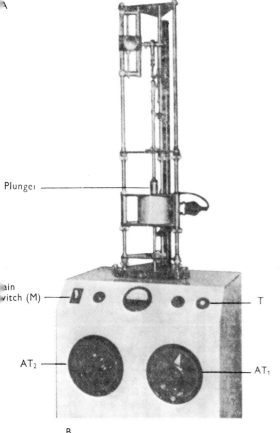

Plunger

ain
vitch (M)

AT₂

T

AT₁

The braking device is under the base; only the releasing switch is seen. B. Electrical circuit of the vertical microelectrode puller. Pulling cycle is started by the push-button T. Contacts a_1 and a_2 are closed (by means of relay A), supplying with current the heating platinum coil. The current is controlled by an autotransformer (AT_1). A ring on the falling shaft operates the solenoid switch S_1, which by means of a relay (B) closes contacts b_1 and b_2. The solenoid is thus supplied by current, controlled by means of another autotransformer (AT_2). The solenoid current is switched off by opening the switch s_2, operated by the lower ring on the movable shaft of the vertical puller. I. 1. — indicator lamp. Tr.: heating coil transformer.

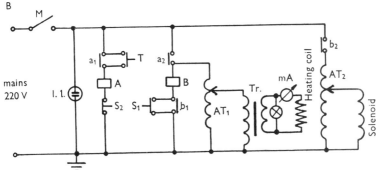

Fig. 72. Vertical microelectrode puller. A. Overall view B. Circuit diagram.

In principle construction is the same also in other apparatus, representing various modifications of the first horizontal and vertical pulling machines of Alexander and Nastuk (1953) and Winsbury (1956). Descriptions can be found in special monographs on microelectrode techniques (e. g. Kennard

C

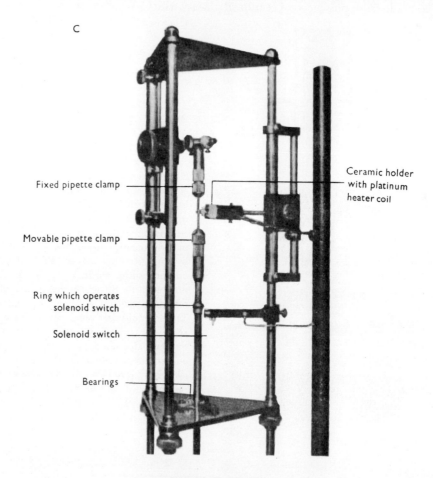

Fixed pipette clamp

Movable pipette clamp

Ring which operates solenoid switch

Solenoid switch

Bearings

Ceramic holder with platinum heater coil

Fig. 72. C. Details.

1958, Frank 1959b, Frank and Becker 1964; Kostjuk 1960, Meschersky 1960, Beránek 1965). Many are produced industrially.

The initial glass tubes are usually about 8 cm long with an external diameter of 2 mm and an internal one of 1.6 mm. They can be bought or made from large ones. Manual extension can be mechanized on a similar principle as the extension of the shank of the microelectrode from the initial tube.

The glass (Pyrex, Phoenix, Sial) has a high percentage of silica with B_2O_3 added to lower the softening temperature. The amount of alkali oxides is relatively small. Such boronsilica glass has a high volume $(3\cdot1 \cdot 10^{14}\ \Omega cm$ — Morey 1938) and surface resistance. The breakdown voltage is about 10^6 V/cm. Tubes must be thoroughly cleaned with methylalcohol followed by distilled water, before being pulled.

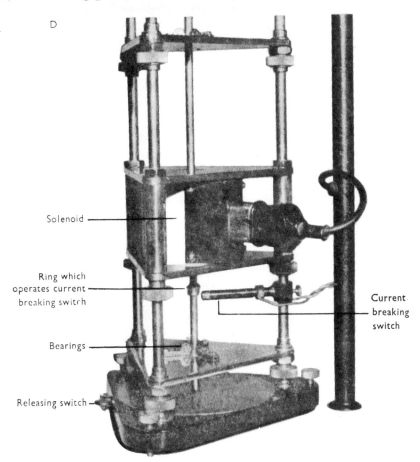

D

Solenoid

Ring which operates current breaking switch

Current breaking switch

Bearings

Releasing switch

Fig. 72. D. Details.

Filling the capillary microelectrodes

The microelectrodes are fixed to the wall of a glass beaker with their tips facing downward (Fig. 73). A glass rod is fixed to the bottom of the beaker and pushed into the rubber stopper of the vessel. The stopper contains a glass tube connected to a water suction pump. The vessel is filled with 3M—KCl so that the electrodes are completely immersed. The solution is heated to boiling point and steam is slowly sucked off. Small bubbles begin to leave the wide ends

213

of the electrodes indicating that these are being filled. When no more bubbles emerge, the electrodes are full. The whole procedure lasts 10—15 min.

Filling may be accelerated by sudden increasing or decreasing the vacuum (e. g. by allowing air into the suction line, Fig. 73).

In this procedure the tips sometimes break and it seems also expand. For very fine tips it is better to use prefilling of the microelectrode with methyl-

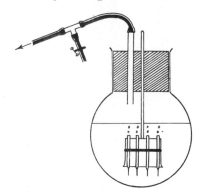

Fig. 73. Arrangement for filling capillary micro-electrodes.

alcohol (Tasaki et al. 1954). This alcohol has a low viscosity and a low boiling point so that even the finest tips can be filled rapidly. The procedure is the same as in the previous case. Methylalcohol, however, boils already at room temperature in a vacuum. Several minutes of strong boiling suffice. Then the electrode is placed in 3M—KCl. It is filled by diffusion within 2—3 days. Methylalcohol microelectrodes can also be immersed in distilled water and then be filled with solutions of substances that cannot stand alcohol. Micro-electrodes can also be filled with water directly by boiling them in distilled water.

The electrodes are filled without boiling as follows (Caldwell and Downing 1955, Kurella 1958): The electrode tip is immersed in the electrolyte. Capillary forces enable the solution to attain a certain height. The filling proceeds more slowly. To speed up this process, distilled water is poured into the wide end. An air bubble remains between both solutions. Because of the difference in vapour pressure water distils across the air bubble to the 3M—KCl. The air bubble reaches the wide part of the electrode within 1—3 days. The electrode is then easily filled with the electrolyte. (For metal filled capillary electrodes see Svaetichin 1951).

The shape of microelectrodes

The microelectrode is composed of several parts (Fig. 71). The stem is the unextended remnant of the parent tube. The first pull forms the shoulder.

The following part is the shank of the microelectrode which narrows either gradually or suddenly, due to the second pull, into the tip. The slope with which the electrode narrows towards the tip is important. It should be about 1 : 10. The greater this taper the greater the puncture opening in the cell for the same depth of insertion. The tip taper is controlled under the microscope immediately after pulling. Using water immersion objective the size of the tip is also sometimes controlled, if it is not too small and thus beyond the resolving power of the light microscope. The size of the tip, however, is more often judged from the microelectrode resistance.

Electrical characteristics of microelectrodes and their measurement

Resistance

If the tip diameter is smaller than 1 μm and the electrode is in a good conducting medium, more than 90% of the resistance arises in the tip represented by a truncated cone with bases in the tip and about 10 μm from the tip (Amatniek 1958, Frank 1959 b). It follows that for the same electrolyte the resistance is the greater the smaller the tip. This inverse relationship may be used to estimate the size of the tip. Microelectrodes with tips 0·5 μm have a resistance 10−30 MΩ (Frank and Becker 1964). The tips of the thinnest electrodes (diameter below 0·5 μ) usually cannot be distinguished under the light microscope since the wave length of the visible spectrum is about that dimension. Control under the electron microscope is very awkward and unsuitable for routine purposes.

For obtaining an idea of the electrode resistance a high impedance electronic ohmmeter is used, which is however not very suitable because of the high measuring voltage and because the flow of current through the microelectrode causes irreversible changes. Hence we must measure very rapidly.

The resistance at the start of the experiment and during its course is best measured on the screen of the oscilloscope. The principle (Fig. 74 A) rests in the fact that we connect between the input grid of the cathode follower (p. 140) and earth a known resistance R which forms with the measured electrode (R_E) a voltage divider. The following relation is valid (Fig. 74 B):

$$\frac{R_E}{R} = \frac{U_C - U_R}{U_R} \tag{1}$$

where U_C is the calibrating voltage from the calibrator and U_R the potential step due to the same voltage U_C after connecting a known R. From equation (1) we can calculate the resistance of the microelectrode. If the calibrating resistance is e. g. $R = 1\,M\Omega$ ($= 10^6\,\Omega$), the calibrating voltage 100 mV ($= 100 \cdot 10^{-3}V$) and connecting the resistance causes a deviation of 10 mV ($U_C = 10 \cdot 10^{-3}V$) then the microelectrode resistance R_E is $10^6 \cdot [(100 - 10)/10)] = 9 \cdot 10^6\,\Omega = 9\,M\Omega$. It is practical and very easy to calibrate the grid scale on the osciloscope screen directly in mV of the measured potential and also in $M\Omega$ of the measured resistance (Fig. 74 C). The procedure

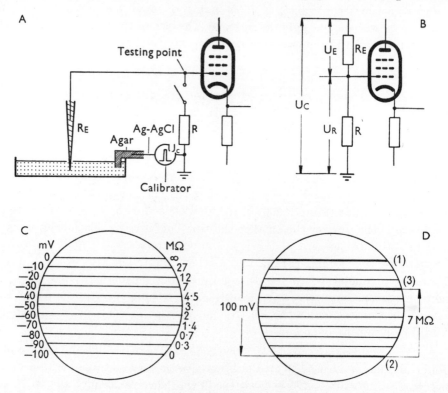

Fig. 74. A — Arrangement of test equipment for measuring microelectrode resistance. B — Equivalent electrical circuit. C — Oscilloscope screen with a grid scale calibrated in mV and in $M\Omega$. The latter is valid only if a 3 $M\Omega$ resistor is connected to the testing point. D — Measurement of resistance of a 7 $M\Omega$ microelectrode.

is as follows: The beam of the oscilloscope is set at the zero potential line which is also the line of infinite resistance (1). Amplification is adjusted so that individual 10 mV steps agree with the grid lines along the whole scale. With the calibrator the beam is deflected to the line corresponding to 100 mV (2). This is also the line of zero resistance. We connect resistance R. The deviation of the beam (3) tells us directly the value of the microelectrode resistance R_E which is read from the scale (Fig. 74 D; 7 $M\Omega$). The scale is nonlinear but by suitably choosing the fixed resistance R (in our case 3 $M\Omega$) we can adjust the range to the electrode resistance most often used. This procedure has the additional advantage that the electrode resistance can be recorded (p. 281, Fig. 101). Obviously the pulse from a stimulator may also be used as the calibrating voltage.

Capacitance of the microelectrode

This arises between the electrolyte inside the capillary and the earthed solution in which it is immersed. The glass wall of the capillary is the dielectric. The capacitance depends on the width of the microelectrode wall and the depth of immersion in the solution. Since the width of the wall decreases towards the tip with decreasing diameter the capacitance depends mainly on the depth of immersion. For an electrode with an external tip diameter twice as large as the internal one the capacitance increases by 0.4 $\mu\mu F$ per mm (Freygang 1958). Hence we take care to immerse the microelectrode as little as possible into the solution (see p. 279).

Tip potential of the microelectrode

If we want to measure the real size of the potential difference on the membrane with a microelectrode, the latter must not be itself a source of electromotive force. A large percentage of 3м—KCl filled microelectrodes have for different reasons their own potential which attains up to 70 mV and which Adrian (1956) termed the tip potential (see also p. 280). Hence we must select microelectrodes before their use and also during the experiment according to their tip potential. The tip potential does not affect the measurement of changes in membrane potential, such as the action potential, but affects the absolute value of the membrane potential, e. g. the resting potential of the cell. For measuring the latter only electrodes with a tip potential smaller than 5 mV can be used. Selection is very difficult, if we need high impedance electrodes since the tip potential increases approximately with the electrode resistance. If we use 10—15 MΩ microelectrodes about 20% have to be eliminated because of a high tip potential (Pyrex, boiling in 3м—KCl in a vacuum for 5—10 min.) Sometimes the whole set is unusable. In that case, however, the microelectrode resistance is unstable (changes with time) which is probably due to imperfect filling.

The tip potential is measured as follows (Fig. 75). The electrode is connected to the follower via a bridge forming the following half-cell: Ag wire chlorinated for 3/4 of its length (Ag—AgCl). This wire is inserted into an L tube filled with agar-Ringer solution. The other end of the tube contacts the

3м—KCl solution in the microelectrode the tip of which is immersed in Ringer solution. A completely equal half cell forms the indifferent electrode which is earthed via the calibrator. If we earth the grid of the input valve and the microelectrode has no tip potential the beam of the oscilloscope is not deviated. If the microelectrode does have a tip potential the deviation of the beam is apparent. The size of the potential in mV is determined by returning the beam

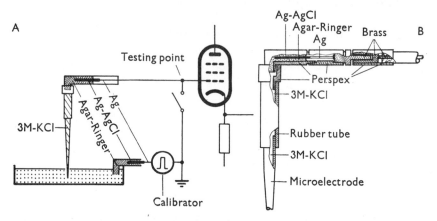

Fig. 75. A. Arrangement used for measuring the tip potential of a microelectrode. B. Method of mounting the microelectrode to the cathode follower input by means of an Agar-Ringer — Ag-AgCl bridge.

to its initial position with the calibrating voltage in 10 mV or 1 mV steps. The necessary compensating voltage gives the tip potential.

In order to be able to consider this potential as the tip potential we must make sure that it is not due to e. m. f. arising at other points of contact of the different phases of the cell. This is tested as follows: the chlorinated ends of the Ag wires are immersed into the solution. Earthing the grid should not give any deviation if both wires are equally chlorinated. If that is not the case they have to be exchanged. Then the wires are inserted into the agar tubes and the tubes are immersed into the solution. Earthing the grid should not give a deviation greater than 1—2 mV. If it is greater the agar bridges are exchanged. Only after that do we insert the agar bridge into the microelectrode and measure the tip potential. By a suitable construction of the bridge (Fig. 75 B) these procedures may be performed rapidly even during the experiment which also makes it possible to record the tip potential (Fig. 101, p. 281).

Microelectrode in the input circuit

The electrical characteristics of the microelectrode require a special adaptation of the input circuit of the amplifying system to which the microelectrode is connected. The input circuit of a usual amplifier with an anode load (Fig. 76A) is unsuitable for several reasons.

1) The input resistance R_{in} of such an amplifier ($10^5 - 10^6 \Omega$) is small in comparison to the electrode resistance R_E ($10^7 - 10^8 \Omega$) so that it would represent an impermissible load for the measured object. The measured potential would be reduced by a factor $R_{in}/(R_{in} + R_E)$.

2) Even with high input resistance the high resistance electrode would cause a considerable increase in the time constant of the input circuit which would distort the shape of rapid potential changes such as action potentials.

3) The grid current of the input valve ($10^{-6} - 10^{-7}$A) of such an amplifier would cause a considerable potential drop across the high resistance microelectrode (1 to 10 V for a 10 MΩ microelectrode), which would make measurements impossible and would also act destructively on the measured tissue.

By connecting the valve as a cathode follower, by suitably choosing valves and by further measures (p. 142), these difficulties can be reduced.

The time constant

By connecting a valve as a cathode follower the problem of the input resistance is solved since the cathode follower acts as an impedance transformer. The effective input resistance is so large that it has no effect on the time constant of the input circuit, τ_{in}, which is then given by

$$\tau_{in} = R_E \, (C_E + C_{in}) \tag{1}$$

where C_E is the capacitance of the microelectrode, R_E is the microelectrode resistance and C_{in} the input capacitance of the other elements of the input circuit excepting the microelectrode. For a good recording of the action potential we require τ_{in} to be maximally 100 μsec ($= 10^{-4}$ sec). As the resistance of the microelectrode cannot be deliberately varied since it is inherent in its principle we must reduce the total input capacitance $C = C_{in} + C_E$. When using a 10 MΩ electrode and for $\tau_{in} = 100$ μsec, C should be 10 μμF ($= 10^{-4}$ sec/10 . $10^6 \Omega = 10^{-11}$ F). C_E of the electrode is also to some extent given and represents in the optimum case about 1 μμF/mm electrode immersion. Hence if we want to attain a low capacitance we must reduce to the lowest possible value C_{in} — the capacitance of the elements of the input circuit.

In the conventional amplifier C_{in} is formed (Fig. 76A) by

1) The interelectrode capacitance of the input valve C_{ga} (grid-anode) and C_{gc} (grid-cathode) which contribute several μμF.

2) The capacitance of the input circuit against the earthed screening, C_s which may be several hundreds μμF depending on the length of the cable and on the width of the core-screen insulation.

3) Miller's capacitance, formed by the capacitance feedback from the anode to the grid. After connecting the valve as a cathode follower this last component is excluded. By shortening the input lead and connecting its screening to the cathode (hence the term cathodally screened cathode follower) the second component (C_s) is considerably reduced. The input capacity is then given

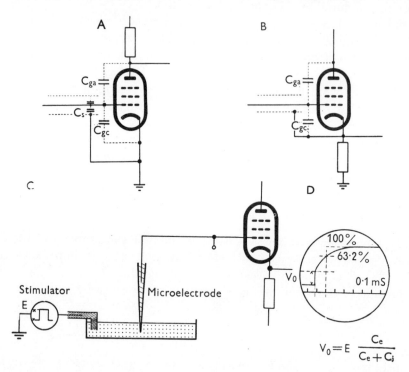

Fig. 76. Capacitance of the elements of the input circuit. A — Conventional amplifier with an anode load. B — Cathodally screened cathode follower. C — Arrangement for estimating the time constant of the input circuit. The step voltage is applied between earth and the immersion fluid. D — Response recorded at the cathode-follower output. The amplitude of the abrupt rise in potential, V_0, is a fraction of the amplitude of the applied rectangular voltage, E, depending on the capacitance across tip wall (C_e) and the input capacitance (C_i). The exponential part of the recorded voltage change has the time constant, $\tau = R_E (C_e + C_i)$; R_E is the microelectrode resistance.

by the relationship (Bonch-Bruyevich 1955, Donaldson 1958; see also p. 156 formula 31)

$$C_{in} = (1 - K)(C_s + C_{gc}) + C_{ga} \qquad (2)$$

in contrast to the case when the screening is earthed where:

$$C_{in} = C_s + (1 - K) C_{gc} + C_{ga} \qquad (3)$$

where K is the gain of the cathode follower, which is smaller than, but close to 1. From equation (2) it follows that cathodal screening considerably reduces C_s.

C_{ga} can be reduced in a simple way using a pentode cathode follower instead of a triode one. (See p. 157, Fig. 56c and formula 31).

C_{in} can be nearly completely eliminated if we use a so called "unity gain" cathode follower which is also often termed a negative capacitance cathode follower (see p. 157).

The time constant of the input circuit with the microelectrode may be simply tested as shown in Fig. 76C. The microelectrode is immersed in Ringer solution to a depth required by the experimental procedure. Between earth and the indifferent electrode a rectangular current pulse is introduced from the stimulator, the potential is recorded and the time mark is made. The leading edge is distorted because of the input capacitance. The time constant is determined from the recording as the time during which the amplitude of the pulse attains 63·2% of its final value (Fig. 76D). According to the recorded response we may also correct the shape of the recorded event (e. g. action potential) thus obtaining its actual shape (Lucas 1912, Rushton 1937).

Grid current

Connecting the valve as a cathode follower does not remove the grid current. For its elimination we must use special procedures and choose suitable input valves (p. 142). The grid current should not be larger than 10^{-11} — 10^{-12}A. In view of the critical conditions under which reduction of grid current is attained its value must occasionally be controlled (Fig. 77).

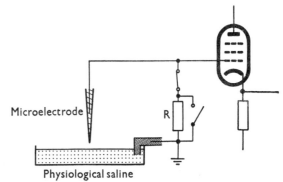

Fig. 77. Arrangement for measuring grid current.

Microelectrode

R

Physiological saline

A known resistance R is connected between the valve grid and earth and the microelectrode is withdrawn from the solution so that it is not connected in the circuit. The grid current now passes through resistance R. The beam of the oscilloscope is placed on a line of the voltage scale. After short-circuiting R we record the deviation of the beam which is proportional to the grid current. The grid current usually makes the grid more positive towards earth, (see p. 141) in contrast to the tip potential which mostly makes the grid more negative towards earth. From the size of the deviation in mV and the value of the measured resistance R we can calculate the grid current using Ohm's law. The amplification must be large since the grid current of 10^{-12}A will give for $R = 100$ MΩ a deviation of only 0·1 mV and for $R = 1000$ MΩ only 1 mV.

Special applications of microelectrodes

In addition to recording membrane potentials capillary microelectrodes are also widely used for stimulation with an electric current and for injecting substances into cells or to a limited area of their surface (e. g. the synapses), either by diffusion or, more often, using an electric current.

Microelectrophoretic injection

The microelectrode content cannot be expressed by pressure when the electrode tip is smaller than 1 μm (Falk and Gerard 1954, Grundfest et al. 1954). Ejection can be achieved by electric current if the substance is ionized in the solution contained in the microelectrode vessel. Anions may be expelled by applying a negative pulse (against earth) to the electrode, cations with currents of opposite polarity. This method is widely used for applying mediators, e. g. acetylcholine to the synaptic region (see p. 429). Another example is given on p. 619 (for review see Curtis 1964). In the same way ions are applied intracellularly. This is particularly utilized in the brain and spinal cord for changing the ionic composition inside the cell (Coombs et al. 1955a, b, Eccles et al. 1964a, b,).

Nonionized substances can be applied by pure diffusion from the tip into the medium.

Filling microelectrodes with substances used for microinjections is usually achieved by diffusion into an electrode previously filled with distilled water or methylalcohol. The water is sucked out of the stem and shank of the electrode with a micropipette (diameter 10 — 20 μm) and is exchanged for a concentrated solution of the substance to be injected. The electrode is then immersed into a solution of the same substance so that the tip is filled by diffusion. Substances that are not destroyed by boiling may be filled into the electrode in the same way as 3м—KCl.

Diffusion from the microelectrode

The rate of diffusion from the electrode is indirectly proportional to the tip size. After immersing the microelectrode into a large volume of fluid with zero concentration of the substance outflow is very rapid in the first seconds, then it becomes stabilized at a certain value which is maintained for several days. Nastuk and Hodgkin (1950) give the value of the stable outflow of KCl from a microelectrode filled with 3м—KCl and a tip of 0·5 μm as $6 . 10^{-14}$ mol/sec . This rate cannot have any significant effect on the internal K or Cl concentration in acute experiments when the microelectrode is inserted into the cell. By contrast, when using substances with a specific action, e. g. Ach and during their diffusion into a small space as e. g. the synaptic cleft, diffusion from the microelectrode may in itself have a considerable effect. In such cases we use electric blockade of diffusion with a suitably poled electric current (p. 420). On the whole application of substances by diffusion from the microelectrode is not a very suitable method since the amout of substance injected can only be controlled by varying the time interval of application. We must only consider it as a factor that may cause changes in the measured object.

Stimulating microelectrodes

Electrodes filled with 3м—KCl may also be used for applying current across the membrane either for stimulating or for measuring the electrical constants of the membrane (Fig. 78A). Examples are given on pp. 350. Since the membrane of excitable cells has a high effective resistance (about 10^5 Ω) a 10 — 100 mV potential change can be induced by a current of 10^{-7} to 10^{-6} A. For such currents a potential difference of 1 — 10 V is formed at the tip of an electrode having a 10 MΩ resistance. This difference practically exists only on the last 10 μm of the tip. The potential gradient across 0·1 μm of the tip wall thus attains 100 — 1000 kV/cm, which is close to the breakdown voltage of the glass. If the dielectric constant of the pyrex glass (ε) is about 5 the walls of the tip are compressed by a pressure of about 2 atm. This would explain the breakage of microelectrodes when a strong current is passed through them.

The passage of the electric current through the microelectrode also causes selective migration of ions across the tip as in electrophoretic ejection of ions from the electrode. If the inside is negative anions will move out and vice versa. If the mobilities of cations and anions are the same (as for K and Cl) the electric conductance of the microelectrode does not depend on a change in current polarity. If mobilities are not the same, conductance of the microelectrode depends on the direction of the current. In a microelectrode filled

223

with NaCl, conductance is lower when applying the anode to the electrode since the mobility of Na ions is lower than that of Cl ions (p. 26). The electrode rectifies. This simple rectification is no serious obstacle since the current flowing through the electrode must always be checked (p. 355).

A more disagreeable property of microelectrodes, however, is the delayed rectification which manifests itself by an increase in electrode resistance

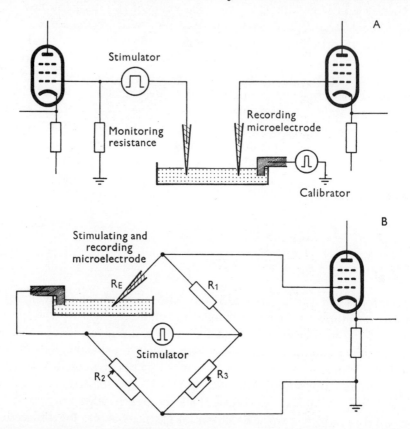

Fig. 78. A-Arrangement for stimulation (left circuit) and recording (right circuit) by means of microelectrodes. B-Bridge circuit used for simultaneous recording and stimulation through a single microelectrode.

after a certain period of current application. The amplitude of the rectangular current pulse (recorded as a voltage drop across a small resistance connected in the stimulating circuit) gradually begins to fall. The delayed rectification is much more frequent if the outflowing current is carried by cations. The current size that can be conducted by the pipette (without delayed rectification) also depends on the tip size and on the electrode resistance. For a 20 MΩ microelectrode the maximum current is 10^{-7} A. The increase of resistance is reversible after some time if the current was not too large and did not act for too long, otherwise there may be an irreversible rise in resistance up to thousands of MΩ.

Delayed rectification for outward currents is probably due to electro-osmotic events at the electrode tip (Taylor 1953), which cause a flow of the fluid from the external medium into the electrode tip. This decreases the electrolyte concentration in the tip and thus raises the resistance.

The electroosmotically transported amount of fluid (v) passing through the tip during a unit of time is

$$v = \frac{\xi \, \varepsilon \, E \, r^2}{4 \, \eta \, l} \tag{1}$$

where l is the tip length, E the potential difference at the tip, r the radius of the tip, η the viscosity coefficient, ε the dielectric constant of the tip content and ξ the electrokinetic potential causing fluid movement.

The fluid transported into the tip by electro-osmosis induces in the tip an inward pressure gradient P. It follows from Poiseuille equation and equation (1) that

$$P = \frac{2 \, \xi \, E \, \varepsilon}{\pi r^2} \tag{2}$$

It follows from equation (2) that by applying a pressure opposing P it should be possible to prevent electro-osmotic movement and thus delayed rectification. This has been confirmed experimentally (Rubio and Zubieta 1961). Applying a hydrostatic pressure of about $20 - 30$ cm of water to the wide end of the microelectrode, distortion of the rectangular pulse due to delayed rectification can be prevented.

The resistance of the microelectrode may change not only because of passage of current through it but also because of clogging by cell particles when inserting the electrode into the cell. Boistel and Fatt (1958) suggested that insoluble Ca compounds may enter the tip as the electrode passes through the membrane and block it for positive pulses. This would agree with the less frequent incidence of microelectrode blocking if the electrode is filled with K-citrate.

Combined use

The same capillary microelectrode may be used simultaneously for stimulation and recording of the membrane potential induced by this stimulation. The membrane response, however, in this method of recording and stimulation, is superimposed on the potential difference which arises as the current passes between earth and the amplifier input. This component of the recorded response can be removed by connecting the microelectrode to one shoulder of a bridge (Fig. 78B, see also p. 619 and Fig. 277). If the bridge is in equilibrium, i. e. when $R_E/R_1 = R_2/R_3$, the recording apparatus does not record any potential deviation during application of the stimulating pulse. If

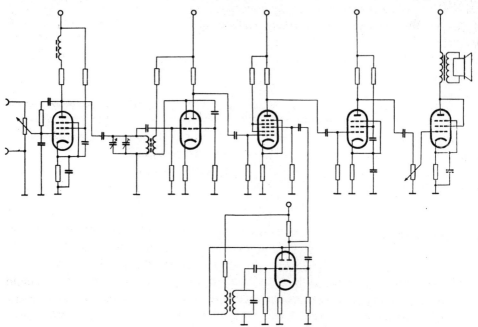

Fig. 79. Acoustic monitor of membrane impalement. For details See text.

the microelectrode is in the cell the current passing through the electrode induces a voltage step at the membrane which is recorded. This connection has been used where visual control of the cell is not possible (Araki and Otani 1955, Frank and Fuortes 1956b). If visual control of the object is possible stimulation and registration are performed by two separate microelectrodes (see p. 440). The double microelectrode (Coombs et al. 1955a) can also be used to record and stimulate a cell not accessible to the eye. One channel is used for recording, the other for stimulation. Each of these methods has its advantages and disadvantages. For details see the original papers.

Double microelectrodes can also be used for injecting substances into the cell and for studying the resulting responses (Coombs et al. 1955b Eccles et al. 1964 a, b). They can also be prepared with the same pulling machine as simple microelectrodes. The parent material is a double tube prepared by extending a pair of larger tubes placed close to each other and simultaneously heated over a flame.

Micromanipulation

Microelectrodes are introduced into cells by micromanipulators, usually similar to those used for cytological research. There is an extensive literature on this subject and many micromanipulators are commercially available (Barer and Saunders — Singer 1951, Békésy 1952, Bishop and Tharaldson 1921, Browaeys 1943, Buchthal and Persson 1936, Bush et al. 1953, Cailloux 1943, Chambers 1921, Florian 1928, Fonbrune 1932, Hansen 1938, Kopac 1929, McNeil and Gullberg 1931, Ordway 1952, Reinert 1939, Schouten 1934, Tschachotin 1912, Watanabe 1926). Many modifications have been described. Sometimes we may improvize them by utilizing e. g. the microdrive of the microscope stage or tubus, micrometers etc.

Applications of micromanipulators for microelectrode work are described in the following chapter in connection with the experiments utilizing intracellular recording and/or intracellular stimulation. Micromanipulators have been recently reviewed by Kopac (1964).

Acoustic indicator of membrane impalement.

Sometimes it is necessary to check very exactly, if the electrode is inserted into the cell. Observing with a microscope is sometimes disadvantageous and sometimes also uncertain. To employ also human senses, other than the eye, we use an acoustical signal, generated in a heterodyne oscillator, which functions as follows (Fig. 79): when the tip of the microelectrode penetrates the cell membrane, the input voltage and thus also the reactance formed by the first valve circuit is changed; this valve is in paralell to the oscillation circuit of the first oscillator (the left half of the second valve, the right half of which is connected as an amplifier). The second oscillator and amplifier (lower valve) generates oscillations of nearly the same frequency as the first one (cca 200 kc). In the third valve (pentagrid) both frequencies are mixed and the difference frequency is amplified in the last two valves and feeds the loudspeaker. The convenient tone is adjusted by the variable capacitance in the first oscillator.

c) Metal microelectrodes

Use: These are suitable for extra- or intracellular recording of action potentials or synaptic activity, but cannot be used for reliably recording DC potentials because of polarization, contact potentials and poor stability. Their main advantage is ruggedness combined with low impedance and noise. Electrical properties of metal microelectrodes are examined in the same way as those of the capillary ones and also the amplifier input circuit must comply to similar requirements. The following paragraph is limited, therefore, to the practical problems of the metal microelectrode production.

Material: Stainless steel or tungsten wire, diameter 0·15—0·3 mm. The wire is gradually heated to red and straightened by slight stretching. The surface is thoroughly cleaned with sand paper. Pieces 5—6 cm in length are cut from the wire. These are bent to an angle of 45° one cm from their end and half their length is inserted into injection needles (diameter 0·4 mm).

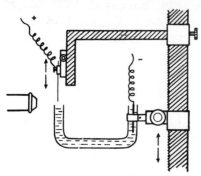

Fig. 80. Electrolytic sharpening of metal microelectrodes (see text).

The bend ensures a good electrical contact with the wall of the needle. Finally, fat is removed from the wire by immersing it in carbon tetrachloride.

Electrolytic treatment. The carrier needle is fixed into a vertically movable holder in such a way that the wire is well visible in a microscope horizontally placed (Fig. 80). When preparing steel microelectrodes, a U-shaped vessel filled with 10—20% HCl is placed below the end of the wire. The holder of this vessel can also be moved vertically. By alternately lifting and lowering the the vessel, the wire is immersed and pulled out of the acid. The positive pole (1·5—3·0 V) of a dry cell is connected to the electrode, the negative pole (silver foil) to the electrolyte. First 10—15 mm of wire are immersed into the acid and current is permitted to flow for 30—60 sec. The wire is then pulled out and carefully cleaned with a brush soaked in distilled water. Narrowing of the tip is observed under the microscope. The electrode is again immersed for a shorter period and to a lesser depth and the whole procedure is repeated until the tip is sufficiently narrow (about 2—5 μ). The length of the tip is also important. 100 μ from the tip the electrode should not be less than 15 μ nor more than 20 μ in diameter. During the final stages only the tip is immersed, for only 1—2 sec. When the desired shape is attained, the electrode is removed from the holder, washed with water and thoroughly dried in an air stream. For electrolytic polishing of tungsten microelectrodes a saturated solution of KNO_2 or $NaNO_2$ and AC current (2—5V) is used. Tungsten microelectrodes are more rigid than the steel ones and can be sharpened to tip size below 0·5—0·1 μ (Hubel 1957).

Insulation. Various substances may be used for insulation. Their choice, however, is not decisive. They vary from ordinary nail varnish to special varnishes with exact descriptions for use. The whole electrode, together with several cm of the needle, is immersed in the varnish and then slowly pulled out. The slow backward movement prevents formation of drops along the electrode. It is best to use a mechanical device pulling the electrode continuously out of the varnish (e. g. an electromotor driving a screw). In another method of insulation, the needle tip leaves the varnish first. The electrode is completely submerged. Just before pulling it out, air is blown on the surface of the varnish and thus a fine layer contributes to better insulation of the tip, which due to surface tension remains uninsulated. Depending on the type of varnish used the electrode is dried in air or in an oven, usually with its tip facing upwards. If necessary varnishing is repeated several times.

Checking insulation. Before use it is necessary to determine the size of the uninsulated tip and flaws in insulation along the electrode. This is done most simply by connecting the electrode to the negative (!) pole of a 1·5 V battery and moving it in a Petri dish filled with saline and lying on a black bottom. Where insulation is incomplete, small chains of bubbles are formed, which are easily visible under direct illumination. An electrode forming bubbles only at its tip is satisfactory. Its resistance is less than $50-200$ KΩ for a diameter of $3-6$ μ.

Metal ultramicroelectrodes (tip about 1 μ) cannot be tested in such a way, since their resistance is too large. For them it is sufficient to measure their resistance. Only those electrodes, whose shape is satisfactory under the microscope and whose resistance is between $5-15$ MΩ are used. The resistance is measured with an electrometric apparatus (a cathode-ray oscilloscope is best) as the drop across the resistance connected in series with the source and the micro-electrode.

In larger electrodes it is possible to check the insulation along the needle directly by connecting a sine wave voltage from a tone generator (0·1$-$0·5 V, 300 c/sec.) to the electrode, the other terminal being connected through an ear phone to a strip of moist filter paper. The electrode is gradually pushed through the paper and the intensity of sound in the ear phone directly reflects even very fine disturbances in insulation.

Storage of electrodes. It is best to prepare a supply of electrodes and then to use them as needed. They must be protected from mechanical damage and chemical influences. It is best to store them, tip upward, in a layer of plasticine or soft wax in large desiccators. Before use, electrodes should be examined under the microscope and their resistance tested.

G. The processing of electrophysiological data

In classical electrophysiological experiments — using stimulation and graphic recording of electrical activity — the processing of results was relatively simple but often very tedious. It was based mostly on the marking of points on the recorded curve and on measuring their voltage and time coordinates, the size and changes of which were evaluated in relation to the conditions of the experiment. Some recordings, e. g. the EEG, were often evaluated even more simply on the basis of the overall qualitative impression they gave. In this way the large amount of information contained in every recording was reduced to a few simple indicators. No other approach, however, could be used, since a more detailed analysis was humanly impossible and even today these reduced indicators may lead to important conclusions.

New approaches were suggested when the random fluctuations of the recorded values, shortly called noise, were better understood and when experiments were performed in which the noise component was a much more serious problem than in the classical experiments with isolated nerves in vitro. Particularly electrophysiological phenomena in sense organs and in isolated nerve cells or fibres as well as recordings from sites distant from the source of activity, all gave waveforms which could not be used because of noise, although the possibility of evaluation of such recordings was theoretically important.

Later demands for a more perfect information processing sprang from the fact that in randomly selected nervous elements (e. g. cells) not isolated from their surroundings and thus reacting not only to stimuli applied by the investigator but also to naturally occurring stimuli it was necessary to detect and to separate these two kinds of reactions.

Finally still later, demands were made on more rapid information processing so that the experiment could be modified according to the results obtained the previous moment. Hence the need for a feedback between the reaction of the organism and the stimulus arose, the required delay in this feedback loop being so small that the information processing had to be performed by high speed computers.

As is obvious even from this brief introduction this problem is very extensive and varied, technically difficult, intellectually demanding and also rapidly developing so that we are forced to give here a superficial outline only. The use of computers is not only complex and expensive but also dangerous for the in-experienced beginner, especially if he is working without the guidance and cooperation of other specialists.

a) Hunting the signal in noise

This is most important since without it electrophysiology could not develop and also because this is, theoretically at least, an easier problem than the others.

It is worthwhile to mention the history of this problem. The first solution was offered by physicians and physiologists: the Czech physician Škoda (1839) elaborated auscultation to a level that cannot be attained by the most modern cardiophonography, since the human ear is endowed with such a remarkable ability to suppress — with the aid of statistical methods — noise and to detect the signal that it nearly equals modern computers. Several tens of years later the Russian physiologist Wedensky (1884) again hunted signals in the noise with his ears — the action potentials of muscle and nerves. The development proper of this problem started in the forties when it was necessary to discover radar echoes in noise, the amplitude of which was several times greater (Kanevsky and Finkelshteyn 1963). On the basis of the first technical solution of this problem it was possible to apply this method to electrophysiology in the fifties.

Let us assume an ideal wide-frequency band linear voltage amplifier with high gain and a large input resistance connected to a voltage source also having a great internal resistance. In the parallel connection of both resistances noise occurs; the equation for its minimum theoretical value was given above (p. 64). The power of this noise may exceed that of the amplified signal. The signal to noise ratio is used here and is defined either as the ratio of powers or of voltages, which is the square root of the former. It has been empirically shown that there is a certain value of this ratio below which exactness of the determination of the individual phases of the signal rapidly decreases until the possibility of detecting the signal is completely lost.

Hence the problem is: is it possible to construct an apparatus which can increase the signal-to-noise ratio? According to the theory of information during the transmission through a channel the amount of information can only be decreased and never increased (in other words: the entropy of a closed system does not fall). On the other hand, we know that we can understand incomprehensible (i. e. distorted by noise) information either by suppressing all sounds which do not participate in information transmission (i. e. noise) or if we have it repeated. And this is the answer to our question.

The mathematical theory of these problems is complex and diffcult and the literature extensive.(Dern and Walsh 1963, Falkovich 1961, Helstrom 1960, Lawson and Uhlenbeck 1950, Ziel 1954, Rice 1945 etc.). In order to give a brief account of the theoretical results, we must first be acquainted with the terminology used and then we shall introduce certain limitations and simplifications. It must be realized that the noise originating in different sources

231

has different characteristics and these determine the choice of means for suppressing it. The nature of the distorted signal must also be considered.

1. Features of noise

Noise is called stationary in the narrow sense of the word, if the probability of the occurrence of a certain instantaneous voltage (or current) value is independent of time. (Stationarity in the broader sense only requires that the mean value and standart deviation of voltage, current, power etc., are independent of time.) The probability, that after a certain value of a random variable at time t_1, the value of this variable at time t_2 should fall within a definite interval of values, will depend on this interval and on the duration of the period $t_2 - t_1$, but not on time t.

We have mentioned mean values of random variables within a certain time interval. These are called time averages. The same random variable, however, may be realized in several different devices of the same type and in every such system its time course can be measured. This makes it possible to determine several time averages simultaneously. In addition other averages can be calculated, e. g. the so called ensemble averages which are the means of the random variable values attained in each device of the ensemble at the same instant; these averages are also termed expected values. If now the difference between the time average and the ensemble average has a zero limit (i. e. the time average is always approximately the same as the ensemble average) the corresponding random process is termed ergodic. It can be demonstrated

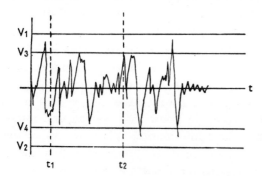

Fig. 81. Time function ("shape") of noise voltage and occurrence of different values of its amplitude. See text.

that what has been said about averages is also valid for moments of higher orders, i. e. particularly for variance or the standard deviation (for details see Loeve 1960, Khinchin 1933, 1949).

A further important characteristic of noise is the distribution of the probability that its instantaneous value will fall within certain limits. Let us consider the time course of the noise voltage (Fig. 81). We can see that (1) all values of the voltage do not occur with the same frequency: in the depicted example zero values are most abundant while values larger than V_1 and smaller than V_2 do not occur at all. If the recording, however, were shorter, e. g. between points t_1 and t_2 there would also be no values larger than V_3 or smaller than V_4, while in a longer lasting record values larger than V_1 and smaller than V_2 would probably also be encountered. If we now attempt to determine the probability that the voltage will attain a certain value, we easily find it to be always zero. Since the number of different values which may be attained by the voltage is infinite (i. e. alef one) and since we have chosen only one of these values, "$1/\infty$" = 0. Hence we must look for another expression of the incidence of different voltage values. We can take a certain interval, e. g. $\langle V_4, V_3 \rangle$ and calculate the probability that V will be within that interval. Further simplification is obtained if we first take one limit of the interval as constant, e. g. $- \infty$ and define the probability, let us say $P(V_3)$ that the value v will fall within the interval $(- \infty, V_3)$ as follows:

$$P(V_3) = P(- \infty < v \leqslant V_3)$$

and analogously

$$P(V_4) = P(- \infty < v \leqslant V_4)$$

These probabilities are termed integral probabilities.
Even without direct proof we can see that

$$P(- \infty) = 0 \quad \text{and} \quad P(+ \infty) = 1$$

and further that

$$P_{4,3} = P(V_4 < v \leqslant V_3) = P(V_3) - P(V_4) \tag{1}$$

This probability, of course, depends both on the position of the interval $\langle V_4, V_3 \rangle$ in relation to the zero level or to the average and on the magnitude of the difference $(V_3 - V_4)$. We can, however, calculate the "average" density of probability in the given interval:

$$\bar{p}(v) = \frac{P(V_3) - P(V_4)}{V_3 - V_4} \tag{2}$$

The density value depends on the width of the interval $V_3 - V_4$ to a much smaller extent than the original probability $P_{4,3}$; if now the interval $\langle V_4, V_3 \rangle$ continuously decreases, then $\bar{p}(v)$ will approach a certain limit which is called the probability density $p(v)$. (The reader may compare the above relationship with that between the average and instantaneous velocity, well known from classical mechanics.) Then we have

$$p(v) = \lim_{\Delta V \to 0} \frac{\Delta P(V)}{\Delta V} = \frac{dP(v)}{dv} \, ,$$ \hfill (3)

assuming*) that the function $P(v)$ can be differentiated. The dependence of

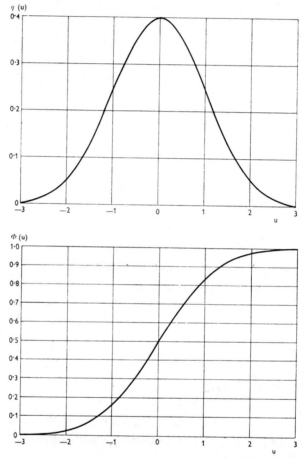

Fig. 82. Normal (or Gaussian) frequency distribution. a) Probability density p_u (ordinate) as function of normalized argument $u = \dfrac{v}{U_{eff}}$, or generally $y = p_u = \varphi(u) = \varphi \dfrac{x}{(\overline{x^2})^{\frac{1}{2}}}$

(differential Gaussian law).

b) Cummulative probability $y = \varPhi(u) = \int_{-\infty}^{u} \varphi(u)\, du$.

The heigh of the ordinate shows the area below the curve in Fig. 81a up to a given argument (abscissa), if the area below the whole curve (a) is taken as unity (integral Gaussian law).

---

*) The capital V (usually with an index) denotes definite voltage values while small v indicates voltage as a variable.

$P(v)$ on v is called the integral law of probability distribution and of $p(v)$ on v the (differential) law of probability distribution or the probability density function (Fig. 82a). Integrating this equation we obtain (Fig. 82b):

$$P_{4,3} = P(V_4 \leqslant v \leqslant V_3) = \int_{V_4}^{V_3} p(v) \; dv . \tag{4}$$

Gaussian noise

Function $p(v)$ may have different characteristics for noises of different provenience. In most cases, however, it is a function that can be constructed as the sum of an infinitely large number of voltage "quanta" of either sign and of limitless small (but constant) size. This function is called the normal or Gaussian distribution and is, as just explained, the limit of the binomial distribution for an infinite number of limitless small elements. It is:

$$p(v) = \frac{1}{\sigma \sqrt{2\pi}} e^{-(v-\bar{v})^2/2\sigma^2} \tag{5}$$

where v is the independet variable, $\bar{v}$ is its mean value defined as

$$\bar{v} = U_{D,C} = \int_{-\infty}^{+\infty} v \cdot p(v) \; dv \tag{6}$$

(in most cases it is equal to zero*) for noise), σ is its mean square deviation or standard deviation, a constant defined by:

$$\sigma = U_{eff} = \pm \left(\int_{-\infty}^{+\infty} v^2 \cdot p(v) \; dv \right)^{\frac{1}{2}} \qquad \text{**)} \tag{7}$$

---

*) For voltages this value is easily measured e. g. with an apparatus having a coil in a magnetic field of a permanent magnet, i. e. with a voltmeter with a moving coil or moving magnet. The DC component may be removed from a signal with a condenser in series.

**) From the point of view of the probability theory this value is termed σ and is called the standard or mean square deviation. If the random variable is voltage it is denoted as U_{eff} (or V_{eff} or E_{eff}) and called the effective voltage or the root mean square (r. m. s.) voltage which is related to power (i. e. to the "effect" — hence the index "eff") by the following relationship: Connected to a pure resistance it would induce in it the same average power as that fluctuating random voltage, of which it is the standard deviation. This may be written as:

$$\text{Power} = \frac{U_{eff}^2}{R} = \int_{-\infty}^{+\infty} \frac{v^2}{R} p(v) \; dv = \lim_{T \to \infty} \frac{1}{T} \int_0^T \frac{[v(t)]^2}{R} \; dt \tag{8}$$

where R is ohmic resistance, t time, T the interval of measurement. The effective voltage values are easiest to measure directly with dynamic apparatus (moving coil in the magnetic field of a stationary coil connected in series), thermocouple apparatus etc. On the other hand, the most frequently used apparatus with a rectifier do not measure effective values although they indicate them. See p. 48.

Thus if $\bar{v} = 0$, we can introduce a new variable $u = v/U_{eff}$ (9) and call it the relative noise voltage:

$$p(u) = \frac{1}{U_{eff}\sqrt{2\pi}}\, e^{-\frac{1}{2}u^2}. \tag{10}$$

If we now put $U_{eff} = 1$ we have a new function of u, which is usually written as

$$\varphi(u) = \frac{1}{\sqrt{2\pi}}\, e^{-\frac{1}{2}u^2} \tag{11}$$

and is termed the normalized probability density in the Gaussian distribution or the normalized or standardized distribution, in which the variable (noise voltage) is obviously not measured in absolute units (volts) but in volt/U_{eff}, i. e. in multiples of U_{eff}. It is evident that $\varphi(u)$ is maximum for $u = 0$, $\varphi(0) \doteq 0\cdot4$ and $p(v) = \varphi(v/\sigma)/\sigma$, $p(0) = 0\cdot4/\sigma$ (see Fig. 82). Function $\varphi(u)$ is tabulated in many textbooks together with its integral

$$\Phi(u) = \Phi(-\infty, u) = \int_{-\infty}^{u} \varphi(u)\, du = \frac{1}{\sqrt{2\pi}} \int_{-\infty}^{u} e^{-\frac{1}{2}u^2}\, du, \tag{12}$$

It can be demonstrated that every stationary random process with a Gaussian distribution of probability is ergodic, and this is a practically important conclusion.

2. Autocorrelation function of noise

Obviously the probability of the occurrence of a voltage not attaining or exceeding a certain value may be determined in two ways: (1) Absolutely — and then it is determined by the integral of the probability density (eq. 4). In the case of an ergodic process this is a constant and in the case of Gaussian distribution

$$P(-\infty < v \leqslant V) = \Phi\left(\frac{V}{U_{eff}}\right). \tag{13}$$

(2) "Historically" — taking into account the past development we may predict with a different probability the occurrence of a voltage within a definite interval for a certain next moment. Let us denote by t_1 the time when the voltage attains a certain value V_1 and by t_2 the next moment at which it has reached the value v; $t_2 - t_1$ is denoted as τ. Let us call $P_\tau(v, V)$ the probability that $v \leqslant V$. In the case of an ergodic process this probability is independent of

time t (independent variable) and evidently formally we may write:

$$P_\tau(v, V) = g(V, \tau), \tag{14}$$

where g denotes a function.

In a similar way we can express the probability that v will not differ from V by more than a small value ε:

$$P_\tau(v, V \pm \varepsilon) = h\,(V, \varepsilon, \tau) \tag{15}$$

If we consider that ε is the differential of v (if it exists) than we may write:

$$P_\tau(v, V \pm \varepsilon) = h(V, \tau)\,\varepsilon \tag{15a}$$

We can intuitively find the limits within which this probability is located:

(1) For $\tau = 0$, $\varepsilon \gg \delta_v$ (δ_v is a voltage value limitless approaching zero) and for any Gaussian and many other distributions:

$$P(v, \tau = 0, V \pm \varepsilon) = 1$$

(2) For $\tau \gg \delta_t$ (δ_t is time interval limitless approaching zero) and for $\varepsilon = 0$

$$P_\tau(v, V \pm 0) = 0$$

(3) For $\tau \to \infty$, $\varepsilon \to 0$

$$\lim_{\tau \to \infty} P_\tau(v, V \pm \varepsilon) = 2\,\varepsilon\,p(V)$$

$$\lim_{\tau \to \infty} P_\tau(v, V) = \int_{-\infty}^{V} p(v)\,\mathrm{d}v = \frac{1}{\sigma}\,\Phi\!\left(\frac{v}{\sigma}\right) \tag{16}$$

(the function Φ can only be used for the Gaussian distribution)

(4) $\quad P_{-\tau} = P_\tau \quad$ (even function) $\tag{17}$

The above question may also be posed somewhat differently: What is the probability that the random variable v attaining a value $V \pm \varepsilon$ in time t_1 will lie within the same interval $V \pm \varepsilon$ in time t_2? (where ε is in fact the voltage differential). Again we assume a process that is stationary in the broader sense and ergodic with a zero average value $\bar{v} = 0$. The validity of some conclusions is then limited to the Gaussian distribution (which does not mean that they need not be valid in other distributions). The difference $t_2 - t_1$ is again termed τ.

The probability $p(V \pm \varepsilon)$ of value $V \pm \varepsilon$ in time t_1 is

$P_{t_1}(V \pm \varepsilon) = 2\varepsilon p(V)$ and analogously the second probability is

$P_{t_2}(V \pm \varepsilon) = 2\varepsilon p(V)$ or for the Gaussian distribution

$$P_{t_1} = 2\varepsilon \cdot \frac{1}{\sigma\sqrt{2\pi}} e^{-\frac{1}{2}(V/\sigma^2)} \tag{18}$$

$$P_{t_1} = \frac{\varepsilon}{U_{eff}} \left(\frac{2}{\pi}\right)^{\frac{1}{2}} e^{-\frac{1}{2}(V/U_{eff})^2} \tag{18a}$$

and analogously for P_{t_2}

If both realizations of the process were independent of each other the known rule $p_{(1+2)} = p_1 \cdot p_2$ would be valid. Of course, as explained above, both processes are not independent except in the limit for $\tau \to \infty$. In that case we have:

$$\lim_{\tau \to \infty} P_\tau = [P_{t_1}(V \pm \varepsilon)]^2 \tag{19}$$

and for the Gaussian distribution:

$$\lim_{\tau \to \infty} P_\tau = \frac{2\varepsilon^2}{\pi U_{eff}^2} e^{-(V/U_{eff})^2} \tag{19a}$$

The distribution of quantity P_τ evidently differs from that of $p(v)$ only in the scale and again has a Gaussian shape.
The analogous limit for $\tau \to 0$ is

$$\lim_{\tau \to 0} P_\tau = 2\varepsilon \cdot p(V) \cdot 1 = \frac{\varepsilon}{U_{eff}} \left(\frac{2}{\pi}\right)^{\frac{1}{2}} e^{-\frac{1}{2}(V/U_{eff})} \tag{20}$$

The following holds for a Gaussian distribution if $0 < \tau < \infty$

$$\frac{2\varepsilon^2}{\pi U_{eff}^2} e^{-(V/U_{eff})^2} < P_\tau < \frac{\varepsilon}{U_{eff}} \left(\frac{2}{\pi}\right)^{\frac{1}{2}} e^{-\frac{1}{2}(V/U_{eff})^2}$$

Instead of the probability P_τ we can introduce the probability density $p(v) = dP_\tau(v)/dv$ remembering that according to the above definitions $dv = 2\varepsilon$.

Model of noise

The question becomes somewhat easier theoretically if we consider how the noise voltage is generated and that the Gaussian distribution is the limit of the binomial distribution. We can even form a mathematical model in the following way: We have a body of capacity C into which and out of which quantified electric current flows mediated by thermal movement of electrons (thermal noise). Due to this current the charge of the body changes by ± 1 electron for each time interval (time quantum). The constant average value of the charge (and thus of the potential) is maintained by the corresponding field strength which changes (very slightly) the probability of the entry or exit of electrons. For events lasting only several "time quanta" it is not necessary to consider in the model these forces, which, however, determine the most probable value of the charge and voltage, i. e. the so called expected or average value. Let us start at moment t_1 from the just realized value of the charge Q and ask what is the probability that Q will remain during the next moments within the limits $Q \pm n\varepsilon$, where n is a natural number and ε the charge of the electron. If the time quantum is designed by ϑ and $\tau = m\vartheta$ then, in our model $P_\tau(Q \pm n\varepsilon) =$ $= 1$ for all $\tau \leqslant n\,\vartheta$, in other words $P_\tau = 1$ as long as $m \leqslant n$. For $m > n$ will be $P_\tau < 1$ and will be a function of both, n and m. Thus the charge will at first be more or less determined by the past state of Q and will start to change on this basis in a random manner, so that the probability of value $Q \pm n\,\varepsilon$ will decrease monotonously to $P_\infty(Q \pm n\,\varepsilon) = n\,\varepsilon\,p(Q)$, where $p(Q)$ is the probability density of Q, which no longer depends on m and thus on τ. The dependence on n is also evident. In addition, of course, P_τ also depends on ϑ, the time quantum, i. e. the time between two entries or exits of an electron into or out of the body.

Now perhaps we can understand more easily all the factors determining the noise voltage in the next moments: The previous state in the conductor, the randomness represented e. g. in the simplest case by the thermal electron movement, and the regulatory effect of the generated electric field. Conditions may, however, be even more complex if the body and its surroundings make possible the generation of electric oscillations. In that case the existence, say, of a positive potential at one moment may increase the probability of the same (or nearly same) potential occurring in the next moment, but it may also on the contrary, increase the probability of a negative potential occurring later.

Another way of mathematically expressing the above considerations is to express the relationships between the voltage at two moments delayed from each other by τ, using (auto)correlation function $R(\tau)$ and the (auto)correlation coefficient

$$\varrho(\tau) = \frac{R(\tau)}{R(0)} \tag{21}$$

Definition:

$$R(\tau) = \overline{(v_t \cdot v_{t+\tau})} = \int_{-\infty}^{+\infty}\int_{-\infty}^{+\infty} v_t \cdot v_{t+\tau}\, p(v_t, v_{t+\tau})\, dv_t\, dv_{t+\tau} \tag{22}$$

i. e. $R(\tau)$ is the average value of the multiple $v_t \cdot v_{t+\tau}$ (the ensemble average).

In ergodic processes the time average can be substituted for the ensemble average. Then $v(t)$ and $v(t + \tau)$ are time functions of the voltage and T is the total time from which the average is formed. Then

$$R(\tau) = \overline{v(t) \cdot v(t + \tau)} = \lim_{T \to \infty} \frac{1}{T} \int_0^T v(t) \, v(t + \tau) \, dt \qquad (22a)$$

The limits that can be attained by $R(\tau)$ and $\varrho\,(\tau)$ are the following:

$$R(0) = R(\tau = 0) = \overline{v^2} = U_{eff}^2 \qquad \varrho(0) = 1$$

$$R(\infty) = R(\tau \to \infty) = \overline{v}^2 = U_{DC}^2 \quad \varrho(\infty) = \left(\frac{U_{DC}}{U_{eff}}\right)^2$$

in the frequent case when $U_{DC} = 0$, $R(\infty) = 0$ and $\varrho(\infty) = 0$.

Measurement of autocorrelation function

An idea about the autocorrelation function of a given noise may be obtained e. g. as follows: We take two identical photographic recordings of noise voltage from the oscilloscope screen (at least one of them on transparent paper) and exactly cover one with the other. According to the definition (for an ergodic process) we now obtain the correlation function as the average of the squares of the ordinates at individual moments. We can see that this procedure is identical to the graphic determination of the effective voltage from a single recording. Now both recordings are slightly moved in the direction of the time axis. They still nearly exactly cover each other. Now, however, two ordinates belong to every instant and for $R(\tau)$ we must take the average value of their products, i. e. add the products of the ordinates for many successive t values and then divide the sum by the number of points. The value $R(\tau)$, however, will be smaller than $R(0)$ and as τ increases it will continuously decrease since the number and size of the negative values will increase (for $R(0)$ all values are positive). As τ is prolonged $R(\tau)$ continuously falls to value U_{DC}^2, or for $U_{DC} = 0$, to zero. This autocorrelation function can be relatively easily measured (Fig. 83). The output of the amplifier is recorded on magnetic tape by the recorder head (1) and is read with two reproduce heads (2) and (3), the distance between which can be varied from 0 to l, where $l = c.\tau$ (c is the speed of the tape transport). The output of the head (2) is connected to one coil of a sensitive electrodynamic apparatus with an oscillation period $T \gg \tau_{max}$ and the output of head (3) to the second coil of the same apparatus. Since the force acting on the moving coil and thus its deflection is proportional to the product of the currents in both coils, the deflection is proportional to the average value of the product $v(t)$ and $v(t + \tau)$, i. e. $R(\tau)$. Calculations and measurements of correlation and autocorrelation functions*) are, of course, of significance not only for noise but also for detecting the deterministic and semideterministic dependences of voltages on each other or on time: e. g. when seeking to answer the question whether the voltage appearing on two pairs

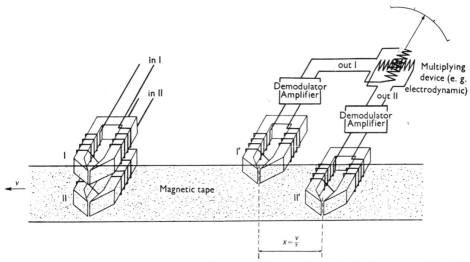

Fig. 83. Simple system for measuring autocorrelation and correlation functions. For the former both recording heads record the same event from the same amplifier and modulator and both reproducing heads are displaced each other by the lenght $x = v/\tau$, where v is the transport speed and τ the time argument of the autocorrelation function. For correlating two diferent processes head I records the first and head II the second process and reproducing heads one and two will usually not be displaced against each other.

of recording electrodes is propagated from a single or from two mutually independent sources. Using series condensers $U_{xDC} = U_{yDC} = 0$ can be attained; for a mutually dependent v_x and v_y, $R(x, y) \neq 0$, while for independent v_x and v_y, $R(x, y) = 0$. A further significance of the correlation function $R(\tau)$ is in its relation to the frequency spectrum of the signal and noise. This problem must be considered in more detail, yet we have to limit ourselves to the qualitative aspect only.

3. Frequency composition of signal and noise

Fourier demonstrated in 1822 and 1824 that any function can be expressed as the sum of trigonometric (sine and cosine) series. In periodic functions

---

*) For measuring the correlation function of two voltages $R(x, y)$ voltage from two sources is led to the multiplying device.

all periods of these series are equal to the basic period divided by k (k is a natural number). In other words the ratios of the periods are always rational numbers. In nonperiodic functions their Fourier transformations are not expressed as the sums of discrete values of the period of argument, but as integrals of trigonometric functions with a continuously variable argument. Perhaps several examples will best explain this (Fig. 84). In most cases time is the argument.

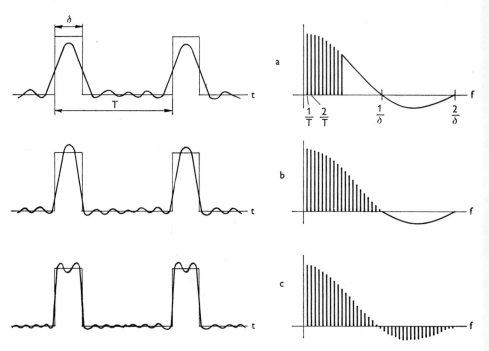

Fig. 84. Fourier transformation. Nonharmonic oscillations (impulses) and their normal harmonic components. a) not too steep impulses composed of 10 harmonic components, b) steep impulses composed of 20 harmonic components. If only the first 10 harmonic components were composed together we could get (a) instead of (b). c) very steep impulses composed of 40 harmonic components.

Since this also is the most important case in electrophysiology we shall assume (without limiting the general validity of our considerations) that voltage is a function of time. Instead of the period we shall frequently use in equations either its inverse value, the frequency, or 2π times the frequency, the circular frequency ω. The diagram in which the frequencies ω are plotted on the abscissa and the absolute values of their sine (or cosine) functions as ordinates (the phase shifts are thus ignored) is called the frequency spectrum of the given function. For periodic functions the frequency spectrum is thus composed of lines with the distance between them equal to the basic frequency ω_0, while for nonperiodic functions it is continuous (Fig. 85). For stationary ergodic noise or for a sufficiently long time interval of that noise the frequency spectrum is continuous. Its shape, of course, considerably depends on the type of source. In the case of a continuous spectrum there is no sense in defining the heigt of the ordinate as the power of one frequency since this is (for a finite total power) zero. Hence we introduce the differential power density w

$$w(\omega) = \lim_{\Delta\omega \to 0} \frac{\Delta W(\omega)}{\Delta\omega} = \frac{dW(\omega)}{d\omega} \tag{23}$$

the total noise power is then

$$W = \int_0^\infty w(\omega)\, d\omega, \tag{24}$$

or the power in the range from ω_1 to ω_2 is

$$W_{\omega_1 - \omega_2} = \int_{\omega_1}^{\omega_2} w(\omega)\, d\omega \tag{24a}$$

We can then introduce the term effective voltage in this range

$$U_{\omega_1 - \omega_2} = (r W_{\omega_1 - \omega_2})^{\frac{1}{2}} = \left(r \int_{\omega_2}^{\omega_1} w(\omega)\, d\omega \right)^{\frac{1}{2}} \tag{25}$$

where r is the internal resistance of the noise source. The actual conditions in individual noise sources of course depend considerably on the function $w(\omega)$, on its frequency dependence, i. e. on the frequency spectrum. In electrophysiology there are two most important types of this dependence:

(1) thermal noise of resistors is an example of the so called "white" noise for which $w(\omega) = \text{const}$ (26) (as long as we are only interested in dependence $w(\omega)$ on ω), or more exactly:

$$w(\omega) = \frac{2}{\pi} k \, \Theta \tag{26}$$

k — Boltzmann constant $1 \cdot 38 . 10^{-23} \text{Ws deg}^{-1}$,

Θ — absolute temperature in Kelvins

Hence

$$U^2_{\omega_1 - \omega_2} = r . \int_{\omega_1}^{\omega_2} \frac{2}{\pi} k \, \Theta \, dt = \frac{2}{\pi} k \, \Theta \, r(\omega_2 - \omega_1) \tag{27}$$

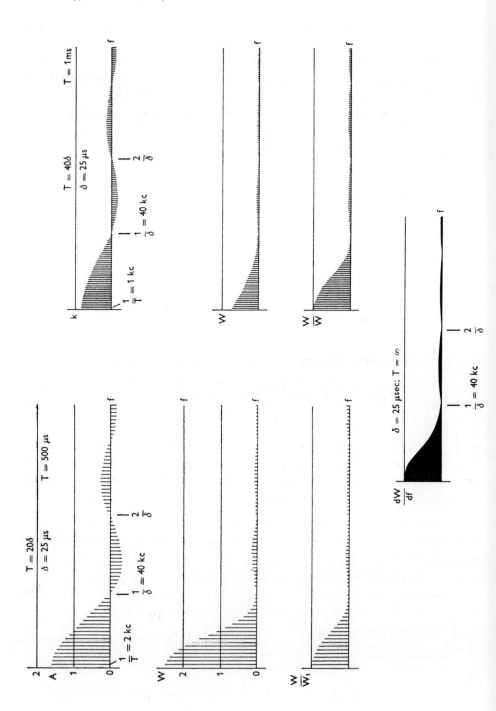

$$U_{\omega_1-\omega_2} = 2\left(\frac{1}{2\pi} k\,\Theta\,r\Delta\,\omega\right)^{\frac{1}{2}} = 2\,(\,k\,\Theta\,r\,\Delta\,f)^{\frac{1}{2}*)} \tag{27a}$$

(2) The noise of electric components. This is usually composed of several parts theoretically differing from each other. Some of them have a spectrum similar to that of thermal noise (1), others, particularly important in valves and semiconductors have

$$w(\omega) = \frac{\text{const}}{\omega + \alpha} \tag{28}$$

where α is a constant.

Then

$$W_{\omega_1-\omega_2} = \text{const.}\,[\ln\,(\omega_2 + \alpha) - \ln\,(\omega_1 + \alpha)] = \text{const}\,\log\frac{\omega_2 + \alpha}{\omega_1 + \alpha}$$

or for $\omega_1 = 0$, $\quad W_{0-\omega_2} \approx \text{const}\,\log\dfrac{\omega_2}{\alpha}$ \hfill (29)

The noise usually occurring in electrophysiological practice is

$$W_{0-\omega_2} = A\cdot\omega_2 + B\log\,(\omega_2/\alpha), \quad \text{where } A, B, \alpha \text{ are individual constants.}$$

Relationship between spectral density and autocorrelation function

There is an important relationship between the density of noise $w(\omega)$ at the circular frequency ω and the correlation function $R(\tau)$, which makes it

---

*) Although the correctness of the above relationships has been demonstrated experimentally up to a frequency of the order of 10^9 c/sec it is evident that they are incorrect: $w(\omega)$ cannot be a constant from zero to infinity as long as W is of finite size. This contradiction is due to the fact that the quantum character of the delivered power W is neglected. Hence a correction has been introduced into the equations for the quantum nature of energy (see van der Ziel 1954) but this is without significance for our purpose.

Fig. 85. Frequency spectrum. a) of periodic impulses of duration $\delta = 25$ μsec and period $T = 500$ μsec; b) of periodic impulses of duration $\delta = 25$ μsec and period $T = 1$ msec; c) of nonperiodic (single) impulse of duration $\delta = 25$ μsec. The ordinate in this case is the power density dW/df instead of power (see text). A — amplitude, W — power. Density of spectral lines of frequency spectrum is determined by the period T (the interval between lines being $1/T$), the crossing point of the envelope with f — axis by the width of the pulse δ. The number of lines in one loop is T/δ. See also Fig. 84.

possible to judge from the correlation function (see above for its calculation and measurement) the value of the noise power density and vice versa:

$$R(\tau) = \int_0^\infty w\,(\omega)\,\cos\,\omega\,\tau\,d\omega \tag{30}$$

$$w(\omega) = \frac{2}{\pi}\int_0^\infty R\,(\tau)\,\cos\,\omega\,\tau\,d\tau \quad \text{(Khinchin 1938).} \tag{31}$$

Let us assume, for simplicity's sake that white noise $w\,(\omega) = a$ in the range

from $\omega_1 = 0 \quad$ to $\omega_2 = \omega_0 \quad \left(\text{i. e.} \quad w(\omega) = \dfrac{U^2_{eff}}{r\omega_0} = a\right)$

Then $R(\tau) = \displaystyle\int_0^{\omega_0} a\cos\,\omega\,\tau\,d\omega = \dfrac{a\,\sin\,\omega_0\tau}{\tau} = \dfrac{U^2_{eff}}{r}\,\dfrac{\sin\omega_0\,\tau}{\omega_0\,\tau}$

and the other way round $\quad w\,(\omega) = \dfrac{U^2_{eff}}{r\omega_0} = \dfrac{\tau\,R(\tau)}{\sin\,\omega_0\tau}$

For a narrow frequency band of white noise between ω_1 and ω_2 $\left(\text{if } \omega_2 - \omega_1 = \right.$

$= \Delta\,\omega, \quad \dfrac{\omega_1 + \omega_2}{2} = \omega_0 \quad$ and $\quad w(\omega_0) = \dfrac{U^2_{eff}}{r\Delta\omega}\left.\right)$ we get analogously

$$R(\tau) = \int_{\omega_1}^{\omega_2} w(\omega)\,\cos\,\omega\,\tau\,d\omega = \frac{2}{\tau}\,w(\omega_0)\,\sin\,(\tfrac{1}{2}\,\Delta\,\omega\,.\,\tau)\,\cos\,(\tfrac{1}{2}\omega_0\,.\,\tau)$$

$$R(\tau) = \frac{2\,U^2_{eff}}{r\,\Delta\omega\tau}\,\sin\,(\tfrac{1}{2}\,\Delta\,\omega\,\tau)\,.\,\cos\,(\tfrac{1}{2}\,\omega_0\,\tau) \quad \text{which can be changed to}$$

$$R(\tau) = \frac{U^2_{eff}}{r}\,.\,\frac{\sin\,(\tfrac{1}{2}\,\Delta\,\omega\,\tau)}{\tfrac{1}{2}\,\Delta\,\omega\,\tau}\,.\,\cos\,(\tfrac{1}{2}\,\omega_0\tau)$$

For a very narrow band as long as

$$\alpha = \Delta\,\omega\,\tau < \frac{\pi}{12}\left(\text{i. e. } \Delta\,f < \frac{1/\tau}{24}\right) \text{ is}$$

$$\frac{\sin\,\alpha}{\alpha} \approx 1 \quad \text{and thus} \quad \lim_{\Delta\omega \to 0} R(\tau) = \frac{U^2_{eff}}{r}\,\cos\,(\tfrac{1}{2}\,\omega_0\,\tau) \tag{31}$$

This is an important result showing that noise voltage in an electric circuit with the described characteristics (e. g. a good parallel resonant circuit) has none of the characteristics usually found for noise but is pure undamped harmonic oscillation. This is easily seen when discussing the equations:

only constants (U_{eff}^2 is a function of the resistance, the temperature and $\Delta\,\omega$, and is here hence a constant, ω_0 is the resonance frequency) and the variable τ, which is really equivalent to the variable time, appear in them. According to definition $R(\tau)$ is $R\tau = \overline{v(t)\,.\,v(t+\tau)}$. If we assume that v is the harmonic voltage with a circular frequency ω_0, then

$$R(\tau) = \lim_{T\to\infty} \frac{1}{T}\int_0^T V_0 \sin \omega_0 t\,.\,V_0 \sin \omega_0\,(t+\tau)\,\mathrm{d}t$$

$$R(\tau) = \lim_{T\to\infty} V_0^2 \frac{1}{T}\int_0^T \sin \omega_0 t\,.\,(\sin \omega_0 t \cos \omega_0 \tau + \cos \omega_0 t \sin \omega_0 \tau)\,\mathrm{d}t$$

$$R(\tau) = \lim_{T\to\infty} \frac{V_0^2}{T}\left(\cos \omega_0 \tau \int_0^T \sin^2 \omega_0 t\,\mathrm{d}t + \sin \omega_0 \tau \int_0^T \cos \omega_0 t \sin \omega_0 t\,\mathrm{d}t\right)$$

$$R(\tau) = \lim_{T\to\infty} \frac{V_0^2}{T}\left[\cos \omega_0 \tau \int_0^T \tfrac{1}{2}(1-\cos 2\,\omega_0 t)\,\mathrm{d}t + \sin \omega_0 \tau \int_0^T \tfrac{1}{2}\sin 2\,\omega_0 t\,\mathrm{d}t\right]$$

$$R(\tau) = \tfrac{1}{2}\,V_0^2 \cos \omega_0 \tau$$

(the remaining two terms are equal to zero in the limit)

This demonstrates that $R(\tau)$ determined by the equation (31) may belong to harmonic oscillations. It can be shown that it must belong to them.

Conclusion: A noise voltage is formed in a resonant circuit with a high Q (narrow band) and a resonance frequency ω_0, which oscillates at that frequency.

Frequency characteristics of the signal

The Fourier theorem can, of course, also be applied to the signal which can be separated into a spectrum of harmonic components. The spectrum may be either discrete or continuous (for nonperiodic signals). The harmonic analysis of signals is best elaborated for rectangular pulses (Bubeník 1958, Meyerovich 1948) which are important particularly in radar technique and in logical circuits. If we denote the duration of the impulse δ, then δ^{-1} corresponds to a certain frequency. It is of great advantage to use this frequency as a measure in diagrams of frequency spectra of impulses (Fig. 85). All pulses have basically the same shape of the envelopes of their spectra, i. e. with the same positions of the amplitude maxima and minima. Their absolute values are determined by the repetition frequency, but the ratio of maximal (and minimal) values again is the same. The density is also a function of the repetition frequency. By limiting the band width of the transmitting channel the transmitted pulse is distorted. Its shape, however, does not depend on the number of harmonics allowed to pass through the channel but only on how many "loops" of the frequency spectrum are transmitted. (One "loop" of the spectrum is the width of the spectrum corresponding to δ^{-1}). Fig. 86 shows the shape of a recombined (transmitted) rectangular pulse dependending on the number of transmitted loops. Note that odd loops (positive) increase the amplitude of the recombined

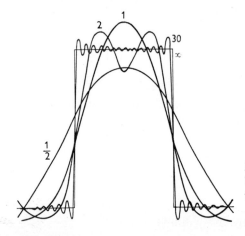

Fig. 86. Shape of rectangular pulse recombined from frequency components limited by the transmission channel to 1/2, 1, 2 and 30 loops of its frequency spectrum.

pulse while even ones (negative) improve the shape (slope) but decrease the amplitude.

The pulses that interest us (e. g. action potentials) have, of course, a completely different frequency spectrum. Their detailed analysis has not yet been published. Hence here we give (Fig. 87) the frequency spectrum of a typical action potential calculated by the graphic method (Hruška 1952, Yakovlev 1953).

4. Improvement of the signal-to-noise ratio by filters

If the noise and the signal have different frequency compositions, as is usually the case, the signal-to-noise ratio can easily be improved by limiting (cutting off) those frequency components of the signal + noise mixture, which are unnecessary for an undistorted transmission of the signal. This can be achieved in different ways, depending on the intended purpose (Kolmogorov 1941, Wiener 1949). The system which optimalizes the energetic signal-to-noise ratio in such a way that it increases the probability of the appearance of the signal in time is called an optimal filter, i. e. it is assumed that all the information is contained only in the time when the signal reaches the receiver. (This is so e. g. for radar where a target is to be found and its distance is to be determined.) Usually the filter is optimalized only for the spectral composition of the signal, i. e. the spectral composition of the noise is not considered. If the filter is optimal at a "higher" level, i. e. if also the spectral composition of the noise is considered, it is called a matched filter. Most frequently such filters are calculated for white noise, either for videoimpulses or radioimpulses (Fig. 88),

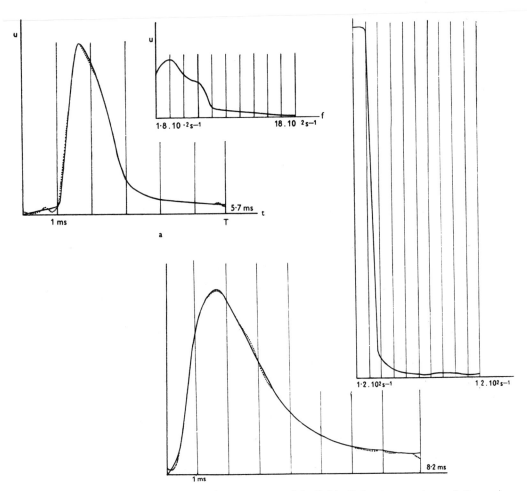

Fig. 87. Frequency spectrum (right) of action potentials (left). Only the envelope of the frequency spectrum is drawn, or the line density, since the number of spectral lines below it and thus also their amplitude depends on the repeating frequency. Dotted line: Shape of action potential calculated from its frequency spectrum (limited to 1800 c in the upper exemple (left and middle) and to 1200 c in the lower one (middle and right).

for pulses coming either in regular or irregular intervals etc. For our purposes usually only videoimpulses are important, mostly appearing in irregular intervals. In that case the optimal matched filter has for white noise a transmission band of exactly the same shape as is the spectral composition of the signal. This, of course, means that those frequencies are transmitted with the slightest decrease, the amplitude of which is greatest in Fourier's transformation, i. e. the spectral composition of the signal at the output is not the same as at the input. Such a filter distorts the time course of the pulse. This is most evident in rectangular pulses which are changed to triangular ones. The time course of "rounded" pulses is usually changed less. Filters can also be optimaliz-

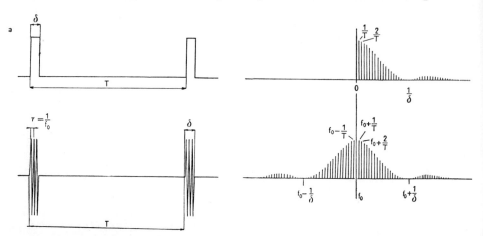

Fig. 88. a) videoimpulses and their spectrum; b) radioimpulses of the same width and repeating frequency as in a), with a carrier frequency f_o, and their spectrum.

ed according to other conditions than the maximum signal-to-noise ratio, e. g. in such a way that the probability of signal detection has greater or smaller weight than the probability of mistaking the noise for signal.

As far as the construction of such filters goes, it is difficult for frequencies which we are interested in (on filters in general see Terman 1943, Smirenin 1950, optimal filters Lezin 1963, Zadeh and Ragazzini 1952, White and Ruvin 1957) and also not very effective since an ordinary RC coupled amplifier with suitably chosen time constants (i. e. with a narrow frequency band) or with relatively simple low pass filters (see Terman 1943) gives results that are only slightly (by the order of 1 dB) worse than those attained with an optimal filter. The signal-to-noise ratio of the optimal filter for individual videoimpulses is:

$$\frac{W_S(t_0)}{W_N} = \frac{v_2^2(t_0)}{U_{effN}^2} \tag{32}$$

where W_S is the power of the signal during the pulse, W_N is the average noise power, $v_2(t_0)$ is the output signal voltage.

The use of such filters is indicated in experiments in which the effect of the experimental conditions on the statistical distribution of nervous impulses is determined, e. g. when studying unit activity of a very small amplitude (p. 615).

If we know the repetition frequency of the videoimpulses and if this frequency remains constant, the signal-to-noise ratio can be very considerably improved using otimal filters as compared with individual pulses arriving at irregular intervals (see above). In that case the spectrum is linear with the fewer lines, the more the inverse value of the pulse duration exceeds the repetition frequency. Hence it is possible to produce a filter which only transmits the frequencies of the corresponding lines, a so called comb filter. Inspite of the advantages of such a filter, its use for electrophysiological purposes is very limited. It can be constructed only for a certain case, i. e. for a constant frequency and in addition the number of elementary filters and their price are too high to make its use advantageous. We can also construct an apparatus completely equivalent to such a filter and in addition adjustable to a chosen frequency and to other parameters. See below.

If we are not only interested in the presence or absence of the impulse but also in its time course other means have to be used to suppress noise. For fairly small noise, when it is sufficient to limit the width of the transmitted frequency band, it must be suppressed in another way than in the optimal matched filter, i. e. so that the relative amplitudes of the individual components of the signal remain approximately preserved. Hence the transmitted band must be chosen so that those parts of the spectrum which decisively determine the form and size of the signal are transmitted with the same amplification (or decrement) and in the correct phase, while a certain small remnant (according to the permissible distortion) is completely cut off. The usual wide-frequency band amplifiers with RC coupling affect transmission in a similar way even though the cutting off of the upper frequencies is usually too gradual.

As far as radioimpulses are concerned, their processing is rare in electrophysiology, mainly when using wireless transmission of physiological functions. The frequency spectrum of radioimpulses is twice as broad as that of videoimpulses, since it lies symmetrically on both sides of the carrier frequency ω_0. For a series of mutually incoherent radioimpulses this spectrum is continuous and hence only individual band filters approaching the optimal ones or simple resonant circuits can be used. For coherent radioimpulses, where all impulses are modulated on an exactly periodic (usually harmonic) carrier wave and start with exactly the same phase, it is often possible to use comb filters, since the spectrum is discrete, with lines of a frequency ω_0 (middle line), $\omega_0 \pm \omega_n$, $\omega_0 \pm 2\omega_n \ldots$, where ω_0 is the carrier frequency, ω_n the repetition frequency of the radioimpulses. Such filters are usually more easily built than filters for videoimpulses themselves.

5. Statistical detection of a weak signal

Finally if noise in the frequency band of the main signal components is of the same order or larger than the signal, all the above methods for noise elimination are completely ineffective.

Now we come to the most important task, i. e. to the separation of a very weak signal from a signal—noise mixture using statistical methods. First these methods must be considered theoretically. Let the noise be ergodic and stationary, the signal periodic with a circular frequency ω_0 and a frequency spectrum between ω_1 and ω_2. Correctly, of course, $\omega_1 \equiv \omega_0$, yet it is often better to take an input filter with $\omega_1 = 0$. The system can then be used for any ω_0, since $\omega_2 \gg \omega_0$ and rather depends on the shape of the individual pulse of the signal than on ω_0 and in most electrophysiological systems ω_2 may thus be considered as a constant. Depending on the demands placed on the reproducibility of the shape of the electrophysiological impulse we usually take $10^3 < \omega_2 < 10^4 \text{ sec}^{-1}$ (or 150 cps $< f_2 < 1500$ cps). Let us assume that out of the signal and noise spectra we only receive frequencies between 0 to ω_2. The effective voltage of the noise in that range is (eq. (8) and (25)).

$$ U_{eff} = \lim_{T \to \infty} \left(\frac{1}{T} \int_0^T [v(t)]^2 \, dt \right)^{\frac{1}{2}} = \left(r \int_0^{\omega_2} w(\omega) \, d\omega \right)^{\frac{1}{2}} \tag{33} $$

The noise power in that band is then:

$$ W = \frac{U_{eff}^2}{r} = \lim_{T \to \infty} \frac{1}{rT} \int_0^T [v(t)]^2 \, dt = \int_0^{\omega_2} w(\omega) \, d\omega = $$

$$ = \frac{2}{\pi} \int_0^{\omega_2} \int_0^{\infty} R(\tau) \cos \omega \, \tau \, d\tau \, d\omega \tag{34} $$

The distribution of the Gaussian noise is given by the curve in Fig. 82a, where there is on the abscissa the scale for the normalized variable u

$$ u = \frac{v}{U_{eff}} \tag{9} $$

On the ordinate is plotted the probability density $p(u)$. The area of the whole curve is equal to 1. The area contained between the two lines u_1 and u_2 gives the probability that the random value (realization) of the variable u will lie between u_1 and u_2, thus (see eq. (13))

$$ P(u_1 < u < u_2) = \Phi(u_1, u_2) = \Phi(u_2) - \Phi(u_1) \tag{35} $$

The rest of the area gives the probability that u will be smaller than u_1 or

252

larger than u_2

$$P\,(u < u_1,\ \text{vel}\ u_2 < u) = 1 - \Phi(u_1, u_2) = 1 - \Phi(u_2) + \Phi(u_1) \qquad (30)$$

The following case is most important

$$u_1 = -u_2 \quad \text{i. e.} \quad |u_1| = |u_2| = u_0$$

Both equations are then simplified as follows:

$$P\,(|u| < u_0) = \Phi(+u_0) - \Phi(-u_0) = 1 - 2\Phi(-u_0) =$$
$$= 2\Phi(+u_0) - 1 \qquad (35a)$$

$$P\,(|u| > u_0) = 1 - \Phi(+u_0) + \Phi(-u_0) = + 2\Phi(-u_0) =$$
$$= 2\,(1 - \Phi(+u_0)) \qquad (36a)$$

If $u_{MP} = v_{MP}/U_{eff}$ contains not only noise but also the signal, then the probability $P\,(|u_{MP}| > u_0) > P(|u_{MO}| > u_0)$ for any $0 < u_0 < \infty$, where u_{MP} is the measured relative voltage at the moment when the signal is present, u_{MO} in its absence. We usually want the positive answer to the question whether the signal is present to be "practically undoubted". If the demands on the higher probability of the correctness of such an answer are increased, the probability of the error of the second type increases, i. e. we may declare the signal to be absent when it is present. Hence the probability must be chosen reasonably. For instance the probability that the absolute value of the measured voltage v_M exceeding twice U_{eff} is only due to noise is

$$P\,(|u_{MO}| > 2) = 2\,\Phi(-2) = 0.045500 \doteq 5\% \qquad (36b)$$

[v_{MO} — measured voltage not containing a signal, v_{MP} — containing signal;

$u_{MO} = v_{MO}/U_{eff}\,;\quad u_{MP} = v_{MP}/U_{eff}\,;$ see eq. (9)]

The probability that the measured voltage of this amplitude is not due to noise alone but to the sum of signal and noise could thus be obtained by subtracting the above probability from 1.

This conclusion is, however, not justified. If the signal is very large, a relative voltage larger than the given value could arise with the given small probability at the expense of noise; the choice of larger values would then give a safer assurance that we are dealing with the signal. If the signal were very small the situation would be even worse: even for zero (i. e. nonexistent) signals we would find the signal in a certain number of measurements, if we only considered the statistical characteristics of noise and not also the size of the expected signal. In other words as long as we do not know in advance the characteristics of the expected signal, we must be on the safe side using one of the following procedures:

(a) As a criterion we choose the apriori probability of noise existence sufficiently low, e. g. 10^{-6} and thus get a high value $u = 4 \cdot 89$. Thus the noise voltage $4 \cdot 9 \ U_{eff}$ has only a negligible probability of occurrence. This, of course, gives rise to the possibility that many a small signal is "lost" (the probability of an error of the II type increases).

(b) Even at the expense of considerable distortion of the signal shape we narrow the transmitted frequency band and thus also diminish U_{eff}, so that for a constant absolute value of the signal voltage the value u increases as the square root of the ratio initial band to new band. The result of such a procedure is the same as sub (a).

(c) The experiment is repeated several times and the results are compared.

Signal to noise ratio

The situation is different if we know some of the characteristics of the signal a priori. We may thus know e. g. the time of its appearance or at least the length of the period at which it is repeated. This is the most frequent case and we shall discuss it later. We may also know its size and may only want to determine the moments when it appears. This is so e. g. when studying the reactions of nervous and sensory cells to a stimulus. Finally we know at least roughly the shape of the detected signal, or its spectral composition. Each of these cases makes it possible, on the basis of an individual approach, to increase the reliability of the conclusions. In each case the reliability of distinguishing the signal in the signal—noise mixture depends on the ratio of the signal and noise powers at the output of the measuring apparatus at the time of decision. This ratio is termed ϱ. Generally it is:

$$\varrho = \frac{U_p^2}{U_{eff}^2} \tag{37}$$

where U_p is the voltage of the constant signal (rectangular pulse), lasting δ, v_M the instantaneous and U_{M-eff}^2 the effective value of the measured voltage, v_N and U_{eff} the corresponding noise voltages;

$$v_M = U_p + v_N \tag{38}$$

For different values of this ratio and for different values of the analogous ratio

$$\varrho' = \frac{U^2_{M-eff}}{U^2_{eff}} = \frac{U^2_p + U^2_{eff}}{U^2_{eff}} = \varrho + 1 \text{ *)} \tag{39}$$

we can easily find the probability P that $|v_N| > U_{eff}\sqrt{\varrho}$, i. e. that $|v_N| > U_p$ (Table 3). The smaller these probabilities, the greater the probability that $|v_M| > U_p$ belongs to the signal.

TABLE 3

ϱ	$\sqrt{\varrho}$	$\varrho' = \varrho + 1$	$\sqrt{\varrho'}$	P %
0	0	1	1	100
1	1	2	1·41	32
4	2	5	2·24	5
9	3	10	3·16	0·3
11	3·3	12	3·46	0·1

Evidently it is not possible to judge with a sufficiently large probability on the presence of the signal as long as ϱ is not larger than about 4, i. e. $U_p > 2U_{eff}$. The quantity ϱ, however, as already described generally, is the real ratio of powers at the moment of measurement and its value thus depends on the technique of measuring and hence need not always be

$$\varrho = \frac{U^2_p}{U^2_{eff}} \tag{37}$$

This value is only attained if measurements are limited to the time interval δ, in which the signal really exists (and has a finite size U_p). This is marked ϱ_p.

(1) If the signal before measurement (!) is averaged, i. e. if we realize its average value $\overline{U}_p = U_p \delta/T_0$, where T_0 is the repetition period, or in

*) From eq. (38) we get by squaring

$$v^2_M = U^2_p + 2 U_p v_N + v^2_N \tag{38a}$$

Using the equations (6), (7) and (8) and integrating we get

$$U^2_{M-eff} = \lim_{T\to\infty} \frac{1}{T}\int_0^T v^2_M \, dt = U^2_p + \lim_{T\to\infty} \frac{2U_p}{T}\int_0^T v_N \, dt + \lim_{T\to\infty}\frac{1}{T}\int_0^T v^2_N \, dt$$

Since, however for Gaussian noise

$\overline{v_N} = U_{DC} = 0$ and according to (8) $= \overline{v^2_N} = U^2_{eff}$ we finally have $U^2_{M-eff} = U^2_p + U^2_{eff}$, as mentioned in the text.

255

other words the time in which we take the integral

$$U_p = \frac{1}{T_0} \int_0^{T_0} v_p(t)\, dt \tag{40}$$

and then find the effective value of this average

$$\overline{U_p} = U_{DC} = U_p \frac{\delta}{T_0}; \tag{41}$$

then it is evident that the corresponding ϱ denoted as ϱ_{T_0} is

$$\varrho_{T_0} = \frac{U_{DC}^2}{U_{eff}^2} = \frac{U_p^2}{U_{eff}^2} \cdot \frac{\delta^2}{T_0^2} = \varrho_p \left(\frac{\delta}{T_0}\right)^2 \tag{42}$$

This method of measuring evidently completely eliminates the difference between the signal and noise and is very disadvantageous, particularly for small δ/T_0, this being the rule in electrophysiology. Yet this kind of measuring is often used (see e. g. Hill 1950), particularly for direct measurements with apparatus having a large time constant which also limits the width of the frequency band and thus decreases U_{eff}. This compensates, however, for the decrease in sensitivity only partly.

(2) If we continuously record u_n on the oscilloscope, then our judgment of the presence of the signal depends on the size of the noise, i. e. U_{eff} and on the shift of the noise "midline", i. e. on U_p and is approximately determined by the ratio

$$\varrho_p = \frac{U_p^2}{U_{eff}^2} \quad \text{(see above)} \tag{37a}$$

Synchronous accumulation of signal

(3) We can, however, if we know the length of the period in which the signal is repeated, summate several intervals δ in which alone the signal is or may be present, or divide the whole period into several (not necessarily equal) intervals and add all first intervals within n periods, all second intervals etc. up to the k-th (last) interval and only then measure these sums or calculate the arithmetic means (by dividing with n) of the voltage in each interval. The average voltage of the signal remains, as will be shown, constant, but the average effective noise voltage falls proportionally to the square root of the number of summed periods. We then obtain:

$$\varrho_{np} = \frac{U_p^2}{U_{eff}^2} \cdot n = n \varrho_p. \tag{43}$$

Only the resulting voltage is measured, or recorded. The detection of the signal in noise during measurement does not correspond to the input ϱ_p but to ϱ_{np}. The corresponding probability is found in Tab. 3 (p. 255), where, of course, instead of ϱ we look for $\varrho_{np} = n\,\varrho_p$. The mathematical processing of the measured voltage (mixture of signal + noise) is usually performed with automatic systems (see below).

(4) In a similar way to sub (3) we can also treat results distorted by procedure (1), if that procedure was for some reason necessary (e. g. because of the inertia of the measuring devices) and we can thus restore the deteriorated signal-to-noise ratio. We thus obtain:

$$\varrho_{nT_0} = n\,\varrho_{T_0} = \frac{U_p^2}{U_{eff}^2} \cdot \frac{n\delta^2}{T_0^2} = n\,\frac{\delta^2}{T_0^2}\,\varrho_p \tag{44}$$

In order to completely remove by procedure (3) the distortions due to procedure (1) it is evidently necessary that

$$n = \frac{T_0^2}{\delta^2} \tag{45}$$

Now further details concerning method (3). First the proof of the equation: The ratio of powers at the time of measurement is denoted ϱ. By procedure (3) we, of course, first collect energy and then measure. Hence it will be best to calculate ϱ_{np} not from the powers but from the total energy. Let us divide the whole long voltage recording into n sectors of T_0 duration and let us choose from them n short sectors (of length δ) in all of which a constant size signal and random noise are present. Assuming that the period T_0 is sufficiently long so that the autocorrelation function $R\,(T_0) \approx R\,(\infty)$ we may consider these n sections δ as an ensemble and form ensemble averages which, assuming ergodicity, are identical with the time averages of the original recording. Let us now calculate separately the sum of signal voltages (U_S) and from that the energy of the signal (A_S) and separately the noise energy (A_N); the energy of the signal + noise mixture is A_M

Signal:

$$U_S = n\,U_p \tag{46}$$

$$A_S = \frac{C}{2}\,(n\,U_p)^2 \tag{47}$$

Noise:

$$\left(U_N = \sum_1^n v_N = \varepsilon\right) \quad A_N = \frac{1}{r}\sum_1^n v_N^2 \cdot \delta = \frac{\delta}{r} \cdot n\,\overline{v^2} = \frac{1}{r}\,n\,U_{eff}^2 \tag{50}$$

where ε is a random quantity of the same order as individual v_N (since

257

$\lim\limits_{n \to \infty} \Sigma\, v_N/n = 0$). Then we obtain

$$\varrho_{np} = k\,\frac{A_S}{A_N} = k\,n\,\frac{U_p^2}{U_{eff}^2} = k\,n\,\varrho_p \qquad (43a)$$

where k is a constant (dimensionless) depending on the characteristics of the electric circuits. We can get some information about it if we consider that the signal voltages are added at the condenser C and that the noise power dissipates in the charging resistance r during time δ. Then $k = rC/2$. Usually we take the parameter k as equal to 1 without much thought. In that case we should obtain $\delta = rC/2$, i. e. the pulse length should be half the time constant. At that point the charging curve (exponential) is still practically linear (see p. 51). Let us further calculate

$$\varrho'_{np} = \frac{A_M}{A_N} = \frac{n^2 U_p^2 + n\,U_{eff}^2}{n\,U_{eff}^2} = n \cdot \varrho_p + 1 = \varrho_{np} + 1 \qquad (51)$$

Usually we are not so interested in the absolute values of ϱ_{np} but rather in the improvement obtained with this synchronous accumulation, i. e. in the ratio of ϱ_{np} to ϱ_p. If we had considered the size of ϱ_p more exactly we would also have found analogous coefficients. Of course, for other summation techniques (not using the condenser) or for other types of memory, the significance of this k varies and hence it is usually not, as already mentioned, taken into account.

It follow from the above equation that in one respect we are fortunate: Even if the signal-to-noise ratio is the smallest possible, this ratio can be improved in periodic signals by their repetition to a value that makes qualitative and also quantitative judgements concerning the signal possible, assuming that all noise is random, stationary in the broader sense, ergodic and has an autocorrelation coeffcient equal to zero for period T_0. (The assumption concerning the Gaussian character of noise is not necessary, but is of great advantage for calculations.) Unfortunately many authors only take the conclusion from this thesis without checking the above assumptions. It is especially important to note that the whole measured voltage has been divided into two parts, the signal and noise. In reality every measurement is distorted not only by random errors (noise) but also by systematic errors (disturbances) which usually are resistant to statistical treatment and hence must be looked for and removed, the more carefully the lower is ϱ_P and the higher hence n must be. Special attention must be paid to hum from the A. C. mains voltage, formed by the latter's basic frequency and by its harmonics (see also p. 203). If the frequency of the mains and the frequency of the signal are rationally related, it may happen that we "detect" the hum instead of the signal. (Much more dangerous than pure sinusoidal hum are short pulses arising by distortion of the mains voltage in thyratrons, ignitrons, rectifiers, motor collectors etc.). It is not only difficult to ensure complete irrationality of the relationship between the mains and signal frequency, but such an independence need not be a sufficient protection from this danger because of the finite value n and the finite length δ. In this respect it will evidently be best to make a test with artificially generated impulse disturbances derived from or synchronized by the mains.

Digital accumulation

Now let us commence the practical aspect and let us start by solving the most complex task, since logically it is the simplest one. We divide the total time of signal sampling into n periods of length T_0 using a synchronizing pulse derived from the stimulus. Each period T_0 is further divided into k equal sections of length δ, numbered with a two figure index $i\ j$, of which the first figure denotes the position of section δ in the corresponding period, thus having i values from 1 to k and the second figure j denotes that the corresponding δ belongs to a certain period and thus has values from 1 to n (i and j being thus natural numbers).

We have thus

1st period	$\delta_{11}, \delta_{21}, \delta_{31} \ldots \ldots \delta_{k1}$
2nd period	$\delta_{12}, \delta_{22}, \delta_{32} \ldots \ldots \delta_{k2}$
$\cdot \quad \cdot \quad \cdot \quad \cdot \quad \cdot \quad \cdot \quad \cdot \quad \cdot$	
nth period	$\delta_{1n}, \delta_{2n}, \delta_{3n} \ldots \ldots \delta_{kn}$

In a similar way we can denote (mean) voltage values v_{ij} measured in the different intervals δ_{ij}. These voltages are then added in columns. Each sum may be further divided by n. This gives k mean voltages $\bar{v}_i$, the signal-to-noise ratio of which is improved n times (considering energy) or $\sqrt{n}$ times (considering voltage) and these k voltages approximate the time course of the signal. The approximation is the better the shorter is δ (and thus the larger is k) but with shorter δ noise conditions become worse (the width of the frequency band increases). The most perfect systems for this operation are really single purpose digital computers with an analog-to-digital converter at the input and output (Fig. 89). The digitized input is connected to a gating circuit which connects the input voltage to the corresponding i-th channel. The gating is controlled by a clock synchronized with the stimulator, or also otherwise. At the input of the i-th channel electric mean voltage for time δ is formed. After switching over the input to the next $(i+1)$ channel the mean voltage $\bar{v}_{ij}$ in the i-th channel is stored in digital form in the machine's memory where it is added to the sum of all previous values of $\bar{v}_i$. The new sum is again stored so that it can be used when the number corresponding to $v_{i,j+1}$ arrives and may also be divided by j so that we obtain $\overset{j}{\Sigma}\bar{v}_{i,j}$ ($i=$ const), or

$$\bar{\bar{v}}_{i,j} = \frac{\overset{j}{\Sigma}\,\bar{v}_{i,j}}{j} \qquad (i = \text{const}) \qquad (52)$$

The digital form of $\bar{\bar{v}}_{i,j}$ is then changed by a digital-to-analog converter into a voltage proportional to $\bar{\bar{v}}_{i,j}$, which is recorded on the oscilloscope as a horiz-

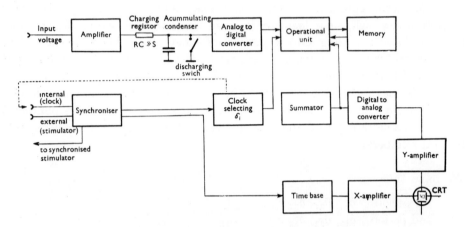

Fig. 89. Block diagram of a digital computer for statistical elimination of noise.

ontal line between abscissae δ_i, δ_{i+1} at a height proportional to $\bar{\bar{v}}_{i,j}$. In period $j+1$ the next channel is used, everything is repeated and the result is again recorded on the oscilloscope etc. Thus the oscilloscope shows a step-like approximation of the periodic signal, which improves from period to period in proportion to the increase of $\sqrt{j}$. This principle, or some similar one, is the basis for construction of some industrially produced systems, the best known being Mnemotron CAT 400 and CAT 1024.

Analog accumulation

The system is somewhat simpler if an analog computer is used. Here n cannot be chosen deliberately since analog computers are of limited exactness. The often used principle is shown in Fig. 90 (Vladimirsky 1951). The voltage corresponding to the interval δ is passed by gating circuits (controlled by multivibrators) into the i-th channel of the computer system where a condenser is charged in the i-j-th period by a charge ΔQ proportional to: $\int_{\delta_{i-1,j}}^{\delta_{i,j}} v_M dt$, then

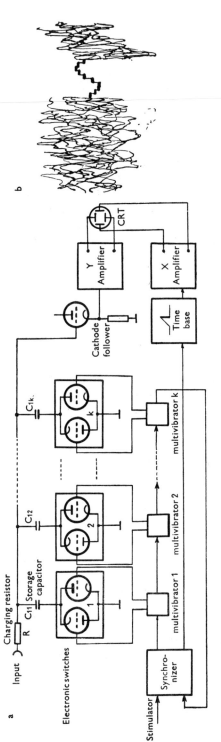

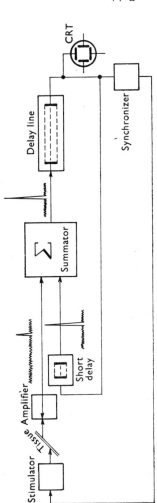

Fig. 90. Analog system for noise suppression (Vladimirsky 1951). a) Simplified diagram showing the k storage capacitors C_{11} to C_{1k} one of which is always switched on by grounding its lower pole by means of a double triode, serving, as an electronic switch. These switches are operated by a chain of monostable multivibrators in which the "impulse" is led from the first multivibrator to the second one and so on. The first multivibrator (or also all others) are synchronized by an operational unit, which gets the main synchronizing pulses from the stimulator, or, on the contrary, synchronizes the stimulator itself. b) Superimposed recording of several signal and noise periods with 10 averaged intervals in the middle (according to Vladimirsky 1951).

Fig. 91. Noise suppression by adding signals to delayed subtotals. Block diagram.

the channel is at rest up to the onset of interval $\delta_{i,j+1}$ when a further contribution $\int_{\delta_{i-1,j+1}}^{\delta_{i,j+1}} v_M dt$ starts to be formed.

In comparison to the earlier described principle, things become worse progressively as the ratio δ/T_0 rises and n increases, since thus losses through the leaks are increased. In addition the nonlinearity of the charging curve has an unfavourable effect.

From the aspect of the theory of measurements the principle shown in Fig. 91 is identical with the above one. The measured voltage (without being divided into intervals δ) arrives through a summation unit into the delay circuit through which it passes exactly during time T_0. After leaving the delay circuit it is led to the second input of the summation unit. The voltage from the just occurring 2nd period of the signal is led to the first input. Both signals are added exactly in phase, i. e. the voltage of the signal is twice as large while the voltage of noise only rises with $\sqrt{2}$ (see above). The output voltage (sum) again passes through the delay line and after leaving it, is added to the third period etc. There are 3 disadvantages of such a system:

(1) the coefficient of amplification in the delay line must be smaller than one to prevent oscillations (2) distortions occurring in the delaying device progressively increase with increasing j (because of feed back), (3) it is not easy to realize the delay line in such a way as to ensure exact and constant delay. Delay lines with distributed impedances or with concentrated impedances cannot be used because of the length of T_0. Ultrasonic delay line is technically difficult and expensive and would also be usually too long. A magnetic drum would be the best solution, there are, however, great demands on constancy of the rotation speed and on linearity (see sub 2).

The last two systems discussed have, according to their parameters, a certain optimum n for which the singal-to-noise ratio improvement is largest.

We have been describing systems that are theoretically easy to handle but expensive to realize. Many small laboratories will have to make do with less perfect but much cheaper improvizations. For them it is probably best to record many periods of the signal (and of synchronizing pulses) on a tape and to evaluate the signal later from this tape either in a CAT system or in similar single purpose machines or universal digital computers equipped with the necessary analog-to-digital converters.

In less critical cases where the signal-to-noise ratio is not too low simpler memory elements may be used, e. g.

(1) Oscilloscope screens with a long persistence (the time constant of the persistence is longer than period T_0). The signal + noise leave a trace which lasts for several further periods. Deflections due to the signal will thus be brighter since they are added, those due to the noise are more rapidly extinguished. Of course, the trace must not be too thin i. e. the amplification of the measured voltage must not be too large. Theoretically this system is

262

rather similar to those described previously, but with an amplification in the "delay line" much smaller than one. Hence also the effective n is relatively small and the improvement of the signal-to-noise ratio slight.

(2) The above system may be modified by making the amplification of the memory channel exactly equal to one by using photographic recording from a short persistence screen of the oscilloscope, instead of observation of the long-persistence screen. This, however, makes for further difficulties, since we must choose a brightness of the trace which is compatible with photometric evaluation of the recording, i. e. which is low enough so that noise deviations are exposed slightly or not at all, but which also is sufficiently high to give a clear picture of the superimposed parts of the curves, which correspond to the signal. The gradation range of the photographic material is not great enough for this purpose and hence the relationships between amplification (dimensions of recording), brightness and the number of periods (n) are rather uncertain. The method can be improved as follows:

(a) The screen is photographed by several cameras at the same time, with suitably graded diaphragms. Only the best recordings are evaluated, or parts of different frames may be combined. The processing of subsequent pictures can also be adjusted according to the first one.

(b) There are methods enabling us to obtain from a photograph a "gradation curve" that is one or two orders of magnitude longer than that obtained with photometry. These include evaluation of the relief of photographic material which rises in the darkened sites linearly with the logarithm of the light intensity also where darkening is so strong that photometrically no differences can be distinguished. See Lau et al. (1958) for details.

(3) A special screen can be used which is charged by secondary emission to a positive potential where an electron beam strikes it and the local voltage is highest where the beam strikes most frequently. Using a diffuse electron current of suitable energy this "voltage picture" can be made visible. The improvement of the signal-to-noise ratio is more or less the same as in the above methods.

Krekule (1965) critically reviews most of the methods used. Here we shall only give one more photographic method (Rumler and Švarc 1960) which, even though simple, is very effective. The beam is moved by a horizontal sawtooth voltage across the screen of an oscilloscope with a period T_0 and by a vertical sawtooth voltage with a considerably higher constant speed. The latter sawtooth voltage is synchronized, i. e. the moments of return to zero are determined by synchronizing pulses produced by a linear frequency modulator controlled by the measured voltage (mixture of noise and signal). Thus the height of the "writing" on the screen is controlled by the momentary value of the measured voltage. The screen is photographed through an optic wedge for n periods. The negative thus has a diffuse black area where darkening increases from the bottom to the top differently at different sites depending on the mean value of the voltage in the corresponding part of period T_0. The line of the same degree of darkening is then determined by contrast development, the ordinate of which is proportional to the logarithm of the mean value of the measured voltage. Lines of the same darkening can, of course, be determined much more exactly than by contrast developing (see e. g. Mütze 1961) and we can thus obtain the measured curve sufficiently exactly to evaluate it quantitatively.

We have been discussing noise reduction, correlation, autocorrelation, frequency analysis etc. This, however, by far does not exhaust the subject. In this chapter, however, we cannot go into further details. A detailed review of data analysis in electrophysiology was published by Dern and Walsh (1963). For further literature see also Rosenblith (1962) and Brazier (1961).

b) Analysis of spike distribution

In addition to mathematical analysis of electrophysiological voltages a further problem has recently arisen: mathematical and particularly statistical analysis of the time distribution of nerve and muscle impulses (spikes). For solving these problems we must apply a number of circuits used in pulse technique for pulse shaping (see stimulators) for selecting pulses according to amplitude, duration, time of appearance, coincidence with others and finally for their gating and counting in different classes, corresponding to the above classification. Logical circuits used for these purpose perform with the pulses logical operations, i. e. negation, conjuction, disjunction and are also used for gating circuits and counters which are all related to each other. Such circuits may also be used for analysing many other problems.

An example will best serve to make these problems clear: The impulse activity of several closely adjacent neural units is recorded with a single amplifier. We are interested in two units with impulses of the highest amplitude. The time course shown in Fig. 92 is obtained. This basically is a recording of two series of pulses, differing in amplitude. The spike amplitude in each one series also somewhat varies because of noise. The first task is to separate this mixture into two individual sequences of pulses of standard amplitude and, if possible, also of standard width. This can be attained e. g. in the following steps:

(1) Limiting the whole series from below. This considerably supresses noise and impulses of the smallest amplitudes and we are left with only two mixed series of the largest impulses. These must be separated. The impulses with the larger amplitude are isolated by further limiting from below and are standardiz-

ed to the same amplitude and width by e. g. a monostable multivibrator. The original series of impulses is also standardized and led to one input of the anti-coincidence system, large pulses to the other input, so that output pulses appear only when no pulse arrives at the second input. Such series of pulses can be treated in an even more complex way, e. g. using anticoincidence circuits selecting all impulses say between 8 and 10 mV and those between 10·5 and

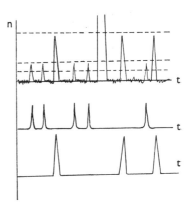

Fig. 92. Separation of a sequence of impulses of different amplitudes into several series of impulses with about the same amplitude.

13 mV from all possible impulses. This further limits noise and other impulse disturbances. Such separated and "cleaned" series of pulses are now further treated as required. E. g. we may add all pulses of one series arriving up to the moment t_1 after the stimulus, then add those up to moment t_2 etc. Gating circuits are used for this purpose which allow the arriving pulses to pass for a certain period of time. Basically these are concidence circuits which receive longer pulses from the timing system to one input and short pulses to be gated to the second input. Only those arriving at the same time as the longer gating pulse can pass. The counter counts them. In a similar way time can also be measured. Pulses, the intervals of which are to be determined switch over a bistable multivibrator which produces positive and negative pulses of the same duration as the intervals between the original pulses. A suitable pulse of this series is led to one side of the coincidence circuit to the other input of which are applied pulses produced by a clock with an exact period of 10^{-3} sec (or 10^{-6} sec). The milisecond impulses that are allowed to pass are counted with a counter and give the length of the interval.

Informations contained in time series of impulses of various provenience can be classified and processed in an analogous fashion.

Logical circuits and counters

Many circuits have been devised for these purposes. Valves, diodes, transistors, tunnel diodes, special counting and indicating valves (e. g. dekatrons) and also relays and electromechanical counters are all used. Many suitable circuits may be bought and then joined together. If we want to construct our own circuits transistors and semiconductor diodes are most suitable. Suitable connections are given e. g. in Transistor manual (1962), Petrovich and Kozyrev (1954), Mayorov (1957), Schwartz (1960), Shterk (1964). For the synthesis and

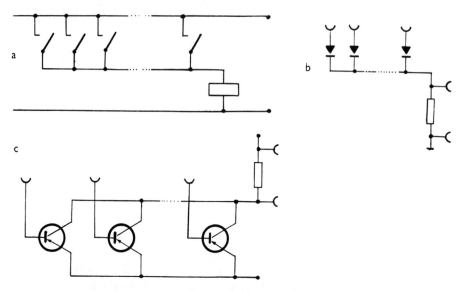

Fig. 93. Logical addition using. a) relay, b) diodes, c) transistors. Voltage (impulse) appears at the output if a signal appears at any input.

transformations of more complex systems of such circuits Boolean algebra is used (algebra of logic — see e. g. Soubies — Camy 1961, Tutenmacher 1962).

Here only a few of the simplest examples of logical circuits are given:
(1)"Or" circuits for logical addition performing the operation (Fig. 93)

$$a + b + c + \ldots = z \tag{53}$$

(a) using a relay: the contacts of the relay A, B, C etc. are connected parallel so that the relay Z is activated by any of them;
(b) using diodes: diodes A, B, C connected in parallel and in series with a common load resistance. Whenever a positive pulse of standard size arrives at any diode it also appears at the common output resistance;
(c) using transistors: the output signal is inverted in phase (it is positive for adding negative input pulses in pnp transistors);
(d) when adding with electronic valves inversion usually also occurs;
(e) similar circuits may be constructed with torroidal ferrit cores etc.

(2) "And", circuits for logical products perform the operation (Fig. 94)

$$a.\, b.\, c \ldots = z \tag{54}$$

(a) relay circuit. Contacts of A, B, C etc. are connected in series so that the relay Z is activated only if all contacts are closed.

(b) a diode circuit. Diodes are connected to the DC voltage of the source in the conducting direction, so that current passes through them, which forms a constant voltage drop at the load resistance (the series resistances of the

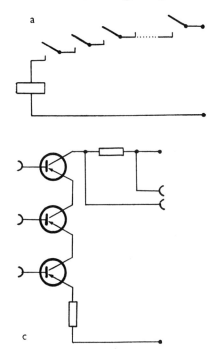

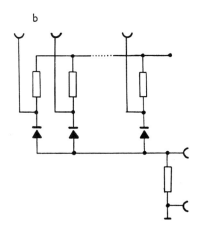

Fig. 94. Logical multiplication using. a) relay, b) diodes, c) transistors. Voltage (impulse) appears at output only if the signal appears at all inputs simultaneously.

diodes being much smaller than the load resistance). Opposite input pulses makes the diodes reverse biassed and if they are all in a non conducting state, a pulse is formed at the load resistance, i. e. an output signal.

(c) transistor circuit. Transistors are connected in series and are closed. The negative pulse must open all transistors so that a positive output pulse can appear. Thus again inversion is obtained as in 1c.

(2) Inverse circuits ("negation") perform the operation

$$z = \bar{a} \text{ (non } a).\qquad(55)$$

(a) relay. The n. c. contact a is disconnected and thus relay Z is released when relay A operates.

(b) In a transistor with an earthed emitter the signal on the load collector resistance is an inversion of the signal on the base.

(c) Similarly in a triode with earthed cathode, anode voltage is an inversion of the grid voltage.

(d) As seen in sub (1) and (2) the inverse function may be combined with the sum and product and this is usually the case when using transistors or valves.

If the output signal must not be inverted a further inversion is performed.

(4) Simple circuits fulfilling more complex tasks can be constructed, e.g. $Z = (a + b + ..) . \bar{s}$ i.e. $z = a + b + ...$ if and only if $s = 0$ (i.e. if another signal s is not present). Such more complex tasks can always be solved by using several circuits sub (1), (2), (3), e. g. the above taks by one circuit (1), one (2) and one inversion (3).

(5) Impulses are counted by counters. These are of inumerable variety and can be purchased or constructed as desired.

(a) Electromechanical counters and stepping relays, as used in telephone technique, have a limited speed of about 10/sec.

(b) Counters counting in the binary system are simplest. These are trigger circuits (bistable multivibrators) using valves or transistors connected in series and having an indicator system (glow discharge tube, electromechanical indicators) which shows which part of the corresponding couple is or is not conducting. Using such multivibrators the frequency of the input (counted) pulses is decreased 2^n times until an electromechanical counter can be connected to the output of the last multivibrator. The counter data must be multiplied by 2^n while the rest indicated by the indicators in the flip-flops can be added (i. e. numbers 0 or $1 \times 2^{n-1}$, 1 or 0 times 2^{n-2} etc. up to 0 or 1 for the first flip-flop).

Decade counters are sometimes more easy to handle since we are used to the decadic system, but their construction is less advantageous and often also less reliable. For details see the literature quoted above.

c) Application of digital and analog computers

In electrophysiology computers are beginning to be used more and more frequently but this problem cannot be discussed in detail here. Nevertheless we must indicate how to plan and direct work in such a way that we do not shut the door to using computers if necessary.

Fundamentally the use of computers can be divided as follows.

(1) Into requirements in which computer is used only to evaluate the result of the experiment, so that it need not be a part of the experimental lay out but is used only afterwards, i. e. *"off line"*. This is evidently so in most cases and we must take care to arrange the experiment in such a way that the subsequent use of the machine will be easy.

(2) It is possible to make the experimental procedure directly dependent on the results obtained and on their machine processing. Here the computer is connected directly for programming the whole experiment and takes the place, so to speak, of the investigator during the experiment. In such cases we say that the machine is connected *"on line"* or that the data processing proceeds in real time, i. e. at the time when the data are sampled. This method can be used when the investigator is sufficiently experienced and has both the intellectual and material prerequisites for getting a computer. Hence we shall not discuss it here.

Three kinds of computers must be considered: analog, digital and hybrid computers.

Analog computers deal with real continuous variable, quantities such as mechanical shifting or turnig and their functions, (sums, products, exponential, logarithmic etc. functions) and differentials and integrals where the argument may be anything but usually is time. Modern analog computers nearly always use electric voltages with time as the argument. The advantage of such computers is the simplicity with which they may perform the most complex operations, a disadvantage limited exactness (1% for the best electrical and 0·1% for the best mechanical ones). This is quite sufficient for electrophysiology; the electrophysiological data are still less exact. Analog computers are very suitable for harmonic analysis, correlation, differentiation and integration, they are not good enough for adding too small numbers or for high precision (Smolov et al. 1964).

If analog computers are found to be defficient for our purposes we use digital or hybrid ones. Since electrophysiological data are of analog character (continuously variable in time) they cannot be directly applied to the digital computer. They must be converted into a group of discrete numbers. This conversion is performed by an analog-to-digital converter which samples the variable at discrete moments sufficiently close to each other and expresses it as two numbers, usually binary, one of which defines the time, (i. e. the argument) and the other is an approximation of the function. These digits are deposited into the machine's memory and are processed in an appropriate way. The results are either used to program the experiment or are presented to the investigator or are preserved for further processing with the same or with another machine. The presentation may either be numerical — printinting of results in the decadic system or recording on a magnetic or perforated tape in binary code, or numerical results are converted by a digital-to-analog converter into analog form, again placed on a magnetic tape, or more usually are presented graphically by a recording apparatus (electromagnetic, electrooptic) or servomechanical coordinatograph etc. Rarely (e. g. when counting impulses) are the data naturally digitized and no conversion is necessary. Hybrid computers only differ from the digital ones by the fact that before conversion of analog data to digital ones some operation is performed with the input data still in analog form. For converters see Soubies-Camy (1961), Suskind (1958), Zavolokin (1962), Pélegrin (1959).

Magnetic recordings are best for recording data that are to be processed in any of the above ways. Recording on paper is much more difficult to deal with.

The use of computers in electrophysiology is described e. g. by Siler and King (1963), Moore (1963), Brazier (1961), Rosenblith (1962).

General electrophysiology of cells and tissues

A. Electric potentials of cells

Electric potential differences in living matter are related to the physico-chemical characteristics of its fundamental components — amino acids and proteins. Since it would be beyond the scope of this book to consider bioelectric phenomena at this level, however, we shall commence with the electrophysiology of the cell.

For didactic purposes the following is based on the membrane theory (cf. p. 31), according to which the inside of a cell is negative with respect to the surface because the surface structures of the cell (the cell membrane) have the properties of an ionic sieve, selectively permeable to different cations and anions. The resulting membrane potential can be measured if one electrode is inside the cell and the other in contact with its surface. Several examples will show the different ways in which this membrane potential can be demonstrated experimentally.

a) The membrane potential of large plant cells (Nitella, Chara)

Problem: Determine the potential difference between killed and intact ends of the internodal cell of *Nitella* or *Chara* and its dependence on the ionic composition of the external medium.

Principle: The large cells of fresh water algae sometimes attain a length of several centimeters and a diameter of 1 mm. The individual internodes can easily be isolated and, because of their size, are particularly suitable for studying membrane polarity (Osterhout 1936, 1958). The inside of the cell can be reached either by a penetrating microelectrode (see below) or by injuring one end of the cell and thus destroying the membrane integrity at this point. In both cases the real membrane potential will only be measured if the other

271

parts of the cell remain intact and maintain their normal polarity. The second method is, of course, more convenient than the first, provided that the resistance of the external medium between the injured and intact parts of the cell is high. The e. m. f. of the membrane (E_m), measured between the normal and depolarised parts, is usually decreased by the so called short-circuiting factor, so that the potential recorded (E_a) is only a fraction of the actual E_m as follows:

$$E_a = \frac{R_1}{R_1 + R_2} \cdot E_m$$

where R_1 is the resistance of the external fluid and R_2 the resistance of the axoplasma between the electrodes. The full potential difference ($E_a = E_m$) is measured only when

$$\frac{R_1}{R_1 + R_2} = 1$$

This ideal value can be approached by increasing R_1 and decreasing R_2. We therefore attempt to choose rather thick fibres (low R_2) and to increase R_1 by interrupting the external communication between the normal and injured ends of the internode with vaseline.

Object: The isolated internodes of *Nitella* or *Chara* at least 3 cm long and 0·3—0·5 mm in diameter. Protoplasma streaming must be well evident under the microscope.

Apparatus: A cathode follower with a galvanometer or a cathode-ray oscilloscope with a D. C. amplifier, overall sensitivity of at least 5 mV/cm. Calibrator. Compensator. A simple stimulator.

Other requirements: Wick calomel cell electrodes. A paraffin frame, mounted on a slide, with two or four chambers separated by vaseline seals of 2 mm width as shown in Fig. 95. Alcohol 20%, 0·1, 0·01, 0·001, 0·0001M-KCl. Arteficial pond water ($2·5 . 10^{-5}$ M-KCl, $5 . 10^{-5}$ M-KNO$_3$, 10^{-5} M-NaH$_2$PO$_4$, $2 . 10^{-4}$ M-NaCl, $5 . 10^{-4}$ M-CaCl$_2$, 10^{-4} M-MgSO$_4$, dissolved in distilled de-ionized water, pH 6·5 — Kishimoto 1964). Glass dissecting needles and scalpels for isolating and mounting the internodes into the chamber. An electro-cautery for thermocoagulation. A low-power microscope.

Procedure: Leaves are carefully removed from several young internodes. The cells are isolated and placed into a vessel with water for one or more hours in order to permit the injury potentials to disappear. Carefully cleaned glass instruments are used for dissection, as plant cells are very sensitive to even slight contact with metals. The functional state of the isolated cells is followed under the microscope. Only internodes showing active protoplasmic streaming (about 60 µ/sec at 20°C) are suitable for experimentation. One cell is chosen, carefully fixed to a bent glass rod and with the latter's aid pressed to the bottom of the vaseline filled slit separating the two chambers. Leakage is checked by filling one compartment with water, which must not penetrate into the other compartment along the internode. Only if the seal is watertight is the second chamber also filled with water. The slide is then set on the stage of a microscope allowing observation of the protoplasmic streaming in both parts of the cell.

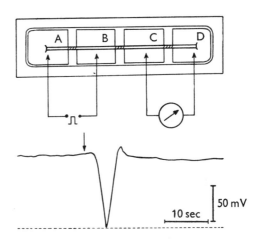

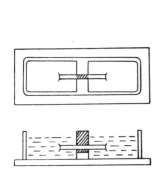

Fig. 95. A paraffin chamber on a slide for work with the isolated internode of *Nitella*. Shaded: vaseline seal.

Fig. 96. Action potential in the isolated internode of *Nitella*. Above: a four-compartment recording chamber mounted on the slide. Below: typical recording of an action potential. For details see text.

Using calomel electrodes, the potential difference between the two compartments is measured. After recording the potential difference for several minutes, the water in one chamber is replaced by 20% alcohol and measurements are continued.

In another experiment, after first determining the initial potential difference, one end of the internode is coagulated with an electrocautery and the potential difference is again measured.

In both the above experiments the cell membrane is injured irreversibly and thus is completely depolarised. The membrane potential can be changed reversibly, however, by application of solutions of KCl to one end of the internode, the other being in water. We start with $0 \cdot 0001$M-KCl applied for 5 minutes. The chamber is then filled with water for 5 minutes and only then are the more concentrated KCl solutions used ($0 \cdot 001$M, etc.).

The membrane mechanism of the action potential can also be demonstrated in the giant cells of *Nitella* and *Chara*. The chamber in this case is divided into four equally large compartments A, B, C, D (Fig. 96) through which a longer internodium passes. A, B, C are filled with water, D contains $0 \cdot 1$ M-KCl.

273

Solutions are separated from each other by vaseline. A and B are connected by silver electrodes to the output of the stimulator, C and D via calomel cell electrodes to the input of the D. C. amplifier of the cathode-ray oscilloscope the time base of which (10 sec) is triggered by the synchronizing pulse of the stimulator.

Results: The potential difference between the ends of the internode in water is usually zero (values from $+10$ to -10 mV can be explained by per-

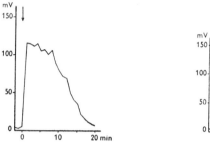

Fig. 97. a) Gradual decrease in demarcation potential produced by thermocoagulation of one end of the internode of *Nitella*. b) Depolarisation of the membrane potential of the internode of *Nitella* by different concentrations of KCl. One half of the internode is permanently washed by water, the other alternately with water (interrupted abscissa) or the test solution (full abscissa).

sistence of the injury potential at one end of the internode). After thermocoagulation or 20% alcohol, the killed end of the cell becomes 110 to 130 mV negative. Protoplasmic streaming immediately ceases in the injured part of the cell while it continues undisturbed in the other parts. The potential difference, however, gradually decreases down to zero (Fig. 97a) within several tens of minutes, evidently because the cell dies. Also the protoplasmic streaming ceases gradually.

Fig. 97b illustrates a typical experiment in which different concentrations of KCl were used. $0 \cdot 0001$M-KCl does not influence the potential difference, while $0 \cdot 001$M-KCl shows an evident depolarising effect. The change caused by potassium ions is entirely reversible and confirms the assumption that the potential difference between the inside and the outside of a cell is mainly due to different concentrations of potassium ions inside and outside the cell. The relationship may be expressed by the equation

$$E_m = 0 \cdot 058 \log \frac{[K_i]}{[K_o]}$$

(cf. p. 33).

If alcohol is added to the part of the cell which was maximally depolarized by raising the external $[K^+]$, no further increase in negativity occurs.

Fig. 96 shows a recording of the action potential. The resting potential recorded between electrodes C, D decreases towards zero during the action potential. The total duration of the action potential is about 10 sec, the rate of conduction is 5—10 cm/sec. Stimuli applied during the relative refractory phase evoke smaller action potentials the amplitude of which attains a maximum with a 60—70 sec interval between the conditioning and testing stimulus.

Conclusion: Direct measurement of injury potential (in this case identical with membrane potential) in the large cells of *Chara* or *Nitella* shows high potential differences (up to 150 mV) between the negative interior and positive surface of the cell. The electromotive force is due to selective membrane permeability and to metabolically sustained differences in concentration of ions, particularly of potassium, inside and outside the cell.

The action potential in the giant cells of freshwater algae is similar to that of nerve and muscle cells but its mechanism of origin is different, in the the first place because the concentration gradients of all participating ions are oriented from the intracellular into the extracellular fluid (pond water). Under such conditions the ascending phase of the action potential cannot be due to an increase in the permeability of the membrane for cations K^+ and Na^+ (their concentration being of course much higher in the cell than in the water) but for anions Cl^-, the equilibrium potential of which is of opposite sign. The action potential is evidently generated by the initial outflow of Cl^- immediately followed by the outflow of K^+ (Gaffey and Mullins 1958, Kishimoto 1964).

b) The membrane potential of nerves

Problem: Determine the membrane potential of myelinated nerve fibres of the frog and its dependence on the composition of the external medium.

Principle: It is possible to use the method described in the previous experiment (increasing the external resistance with vaseline) in the case of single cells or isolated nerve or muscle fibres (e. g. Tasaki 1953). For a whole bundle of nerve fibres this method would not be effective, however, since vaseline or mineral oil cannot penetrate into the capillary spaces between individual fibres of the bundle. These spaces form a considerable shunt even after the external resistance has been raised by vaseline. Stämpfli (1954), therefore, suggested a simple method to increase R_1, making it possible to approximate the true membrane potential in a bundle of nerve fibres. The R_1 is increased by threading the bundle of nerve fibres through a thin glass tube with an appropriate diameter. Isotonic sucrose flows continuously from the centre of the tube to both ends, thus separating the ends of the nerve, immersed in normal Ringer solution or the test solution. Sucrose rapidly enters the extracellular spaces of the nerve bundle and thus effectively increases R_1.

Object: Frog nerve bundles freed of connective tissue and dissected for a length of at least 5 cm.

Apparatus: A cathode follower with a galvanometer or a D. C. amplifier with a cathode oscilloscope (total amplification 5 mV/cm). Compensator and calibrator.

Other requirements: Nonpolarisable calomel electrodes. A special apparatus, as in Fig. 98 consisting of a glass T-shaped tube (diameter of the horizontal arm 0·8, 1·1 or 1·6 mm, length 14 mm) and two side tubes (diameter 2 mm, length 20 mm) joining the horizontal part of the T tube. The side tubes can be rotated, so that they can assume either a horizontal or vertical position. The vertical part of the T tube is joined by rubber tubing to a Mariotte's bottle with a supply of isotonic sucrose. The side tubes are connected to Mariotte's bottles filled with Ringer solution. One of the side tubes is further connected via a two-way stop-cock to a burette containing the test solution.

Isotonic sucrose (7·3%). The sucrose must be as pure as possible, the solution must be prepared in distilled water and its specific resistance must be greater than 100 KΩ.cm. If not, it must be cleaned in an ion exchanger. Several litres of frog Ringer solution. Solutions of KCl (0·1, 0·03, 0·01, 0·003, 0·001M) 100 ml. each, 100 ml. of 0·02M-KCl in 0·02M-CaCl$_2$. 95%O$_2$ + 5% CO$_2$.

Procedure: The frog sciatic nerve is prepared in the usual way (p. 309, Fig. 115). Connective tissue is carefully dissected away and a branch 5—6 cm long is isolated. A thread is tied to both ends of the isolated nerve bundle. The preparation is left in Ringer solution for 1—2 hours so that the injury potentials can pass off. Several preparations can thus be made from one sciatic nerve and its branches. A T-tube with an appropriate horizontal arm diameter is inserted into the apparatus. The lateral tubes are moved into a horizontal position during this phase of work (Fig. 98a). This makes it easier to pull the nerve bundle through the system of tubes. The lateral tubes are then rotated by 90° (Fig. 98b) and the threads holding the preparation are fixed with rubber tubing. In addition a cotton wick is fixed in the lateral tube extending by 1—2 cm below the lower end of the tube. The rubber tubings are fitted to the Mariotte's bottles filled with sucrose or Ringer solution and the flow of sucrose through the T tube and of Ringer solution through the lateral tubes is regulated with clamps. A two-way stop-cock is inserted on one side allowing rapid exchange of the Ringer for the test solution prepared previously in a 20 ml burette. The latency of exchange of solutions depends on the rate of flow and the volume of the rubber tubing between the cock and the preparation (with a rate of flow 2 ml./min. and a volume of the system 0·35 ml. this latency will be about 10 sec.). The test solution is usually permitted to act for 1 minute, after which the stopcock is switched back to Ringer solution. The calomel electrodes make contact with the cotton threads in the lateral tubes. In view of the fact that continuously outflowing solutions exclude diffusion from the electrodes to the preparation, calomel electrodes filled with saturated KCl and having high stability may be used.

Results: Usually only small potential differences (up to 5 mV) are registered between both ends of the nerve bundle separated by the sucrose gap.

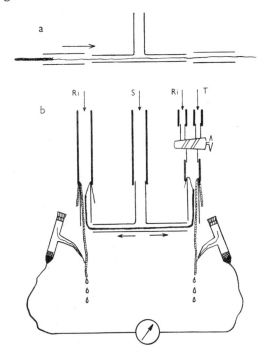

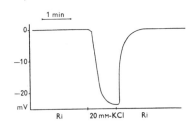

Fig. 99. Depolarisation evoked by applying 20 м-KCl to a nerve bundle of the frog tibial nerve. The sucrose gap technique.

Fig. 98. Diagram of apparatus for measuring membrane potentials of a nerve fibre bundle (Stämpfli 1954). Above: Introduction of the nerve bundle into the apparatus. Below: The apparatus during measurement. Ri — Ringer solution, S — sucrose, T — test solution.

Fig. 99 shows a recording of the change due to application of 20 mм-KCl to one end of the nerve bundle. Results are similar to those obtained with *Nitella*. Maximum potentials attain values ranging from 50 to 80 mV. When 0·02м-CaCl$_2$ is added to 0·02м-KCl no depolarisation is produced. This is a typical example of the antagonism between potassium and calcium (Woronzow 1924, Fleckenstein et al. 1951). It is possible to find for various potassium concentrations the ratio $[Ca^{++}]/[K^{+}]$ that will suppress the effect of potassium ions either completely or reduce depolarisation by 50%.

In addition to various depolarising agents an increase in membrane potential may be demonstrated with this method. Thus Ringer solution aerated with a mixture of 95% O$_2$ and 5% CO$_2$ after several minutes produces hyperpolarisation in the treated part of the nerve (Lorente de Nó 1947).

An application of the sucrose gap technique to examination of membrane properties of small areas of a smooth muscle preparation was described by Berger (1963). The arrangement consists of 2 sucrose gaps separating the short (less than 1 mm) and well delimited middle compartment (test) from both end compartments (Ringer). Artifacts due to contractions of the muscle can be avoided by passing the preparation through closely fitting holes in thin rubber membranes, separating the sucrose compartments from the adjacent solution.

Conclusion: The sucrose gap technique permits measurements of membrane potential in a bundle of nerve fibres by increasing the resistance of the extracellular fluid between the external electrodes. In this way it is possible to approach values of membrane potential recorded with microelectrodes in single nerve (Woodbury 1952) or muscle fibres.

c) The resting potential of a muscle fibre

Problem: Record the resting potential of the frog muscle fibre using intracellular microelectrode.

Principle: The potential difference between the inside and the outside of a muscle cell measured in a physiological electrolyte medium is termed the resting potential. It is due to the unequal distribution of potassium and chloride ions inside the fibre and outside it (see p. 32). If we know the concentration of potassium $[K_i]$ and chloride $[Cl_i]$, inside the cell and $[K_o]$ and $[Cl_o]$ outside the cell, we can calculate the size of the potential difference V_m using Nernst's equation. In order to be able to compare this calculated membrane potential with actual results a method must be used that measures the absolute values of the potential difference across the membrane with sufficient accuracy. Measurement of the demarcation potential (III-A-a) would give too low values because of a great short-circuiting factor due to the external environment. The sucrose gap method (III-A-b) gives absolute values but is much more difficult in the case of a muscle, than measuring transmembrane potentials with transverse capillary microelectrodes used here. The sucrose gap is used if the fibre dimensions are so small that the use of microelectrodes is very difficult or impossible, e. g. in smooth muscle (Stämpli 1952, Berger 1963).

Transverse intracellular microelectrodes make it possible to measure the potential of a single fibre in the whole muscle so that it need not be isolated. The muscle is in solution which is a further advantage of this method. The principle of measurement is described on p. 215.

After producing suitable microelectrodes (p. 208) and with the necessary recording apparatus (p. 140) the method itself is very simple. In principle the electrode is inserted into the fibre under microscope observation using a micromanipulating device. The resting potential is represented by the potential jump recorded as the microelectrode enters the cell through the membrane.

Object: The sartorius muscle of the frog.

Apparatus: A single beam oscilloscope with a D. C. amplifier and a cathode follower. Total sensitivity of the system 20 mV/cm. A calibrator with voltage steps of 10 and 1 mV. An acoustic indicator of cell impalement. A micromanipulator. A dissecting microscope (max. magnification 100 ×).

Other requirements: Arrangement for fixing the preparation. A Petri dish. Dissecting instruments. Ringer solution of the following composition:

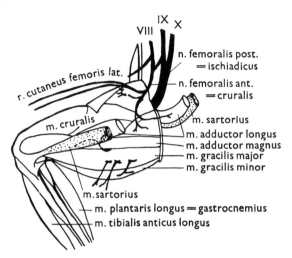

Fig. 100. Medial side of right lower limb of frog. Nerves are shown in black.

(mM/1) : Na^+ : 116·5, K^+ : 2·5, Ca^{++} : 1·8, Cl^- : 117·1, $H_2PO_4^-$: 0·4, HPO_4^{--} : 2·5; pH : 7·4. Microelectrodes with a resistance of about 10 MΩ and a tip potential smaller than 5 mV.

Procedure: The sartorius muscle is carefully prepared without the nerve (Fig. 100). When tying the fixing thread special care is taken at the pelvic end so as not to damage the muscle fibres. This is the most common mistake made. The fibre is not only injured at end but the injury spreads along it leading to a fall in the membrane potential.

The muscle is then fixed by the thread to the fixation platform in the Petri dish and is slightly tensed. The bottom of the fixation platform is bent into an arc so that the muscle is well lying on it. The internal side is facing upwards. This has less connective tissue and makes it easier to insert the microelectrode.

The microelectrode is fixed to the micromanipulator (Fig. 123, p. 324) and placed above the site of insertion. The tip is in Ringer solution the surface of which is not more than 3 mm above the site of insertion (see p. 217). The broader end of the electrode is connected to the cathode follower via an agar-Ringer bridge and a chlorinated silver wire. The indifferent electrode is in Ringer solution. It is of the same construction as the bridge joining the microelectrode with the input cable of the cathode follower, i. e. a plexiglass tube filled with agar-Ringer containing an Ag-AgCl wire.

We now measure the resistance of the microelectrode by connecting a parallel resistance to it (p. 215) and also the tip potential by following the deviation of the beam when earthing the grid of the follower (p. 217). The steps can be seen to be less than 5 mV and the electrode can be used.

Under the microscope with light falling on the muscle from below the microelectrode is brought close to the surface of the fibre (p. 354). Then using a brief rapid movement it is inserted into the fibre. For this fine movement the micromanipulator is rarely necessary. With a little practice the electrode may be inserted with a suitably oriented microdrive of the microscope carrier of the preparation. Entry through the membrane is indicated by a sudden displacement of the beam (the electrode becomes more negative). If a sound indicator is used entry into the cell is heard as a change of tone, otherwise the microelectrode must be inserted without visual control since we must observe the potential jump on the oscilloscope. After pulling out the electrode of the muscle the beam returns to its initial level. Thus the microelectrode may be inserted repeatedly into the same fibre and very similar values for the resting potential are obtained.

Such a jump is recorded on a moving film. The time base of the oscilloscope is switched off so that we see a point on the oscilloscope that writes the initial level on the film (Fig. 101A). When the electrode enters the cell there is a sudden change in the level (arrow pointing down). The new level showing the resting potential does not change with time. When the electrode is pulled out (arrow up) the beam returns to its initial level. The amplification is calibrated by increasing the calibrating voltage in 10 mV steps up to a total of 100 mV. The steps are regular and thus there is no vertical distortion of the screen.

Calibration can also be performed with the electrode in the fibre. This makes it possible to measure the membrane potential also without recording on a film. With the electrode in the fibre we compensate the resting potential with the calibrating voltage until the level of the beam returns to its initial value (Fig. 101B). The compensatory calibrating voltage is then equal to the resting potential. Evidently here we can also use greater amplification even if the beam is off the screen with the electrode in the fibre. The beam may be returned accurately by using 1 mV steps for the last stage of compensation (Fig. 101 C).

After measuring we retest the electrode resistance and the size of the tip potential.

Conclusion: We find that the resting potential E_m of the muscle fibre of the frog has a mean value of -92 mV (inside negative) at 21°C. The muscle

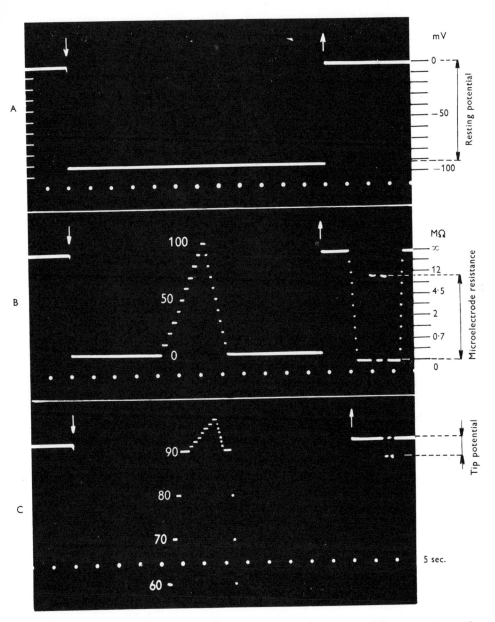

Fig. 101. Measurement of resting potential in muscle fibres by capillary microelectrodes. Arrows indicate the introduction (↓) or withdrawal (↑) of the microelectrode from the cell. For details see text.

fibre of the frog contains 139 mM-K/kg fibre water (= $[K_i]$; Adrian 1956, Hodgkin and Horowicz 1959). The K concentration in the Ringer solution is 2·5 mM/l (= $[K_o]$). Inserting these into the Nernst equation for the equilibrium

281

potential of K ions, E_K (in mV)

$$E_K = 58 \log \frac{[K_o]}{[K_i]} \tag{1}$$

we find that E_K is -101 mV. Thus the measured potential is about 10 mV lower. The explanation for this difference is not uniform. Conway (1957) ascribed this small difference to the fact that muscle fibres in vitro are not in a steady state so that they lose potassium and gain sodium. Hodgkin (1958) and Adrian (1960) assume that active sodium efflux is compensated by active potassium uptake. In that system the resting potential is given by

$$E_m = 58 \log \frac{[K_o] + \alpha[Na_o]}{[K_i] + \alpha[Na_i]} \tag{2}$$

where $\alpha = P_{Na}/P_K$ is the permeability ratio of the membrane for sodium and potassium. Equation (2) is a reduced form of the Goldman's (1943) equation (see p. 35). $\alpha[Na_i]$ may be ignored as a small quantity since $\alpha = 0.01$ and the internal concentration of sodium is also low, $[Na_i]$ being 20 mM/kg fibre water (Hodgkin and Horowicz 1959). After inserting into equation (2), with the sodium concentration in the Ringer solution being 120 mM/kg ($= [Na_i]$) we find that $E_m = -92$ mV which corresponds to the measured value.

Chloride is distributed passively on both sides of the membrane depending on the membrane potential given by equation (2). Its internal concentration may thus be evaluated from the equation:

$$E_m = 58 \log \frac{[Cl_i]}{[Cl_o]} \tag{3}$$

The chloride concentration in Ringer solution is 120 mM/l ($= [Cl_o]$) so that for $E_m = -92$ the fibre contained about $3-4$ mM-Cl/kg fibre water.

The agreement between the theoretical and measured potential is usually better for higher $[K_o]$, as follows from equation (2) where the quantity $\alpha[Na_o]$ becomes less significant. The dependence of the membrane potential on $[K_o]$ then approaches the Nernst relationship. To test this dependence the muscle would have to remain in a changed external potassium concentration for several hours to attain a new equilibrium. In isolated fibres, where the external medium can be changed within a fraction of a second (Hodgkin and Horowicz 1959, Zachar et al. 1964b) it can be tested by following changes in potential due to a sudden change in the concentration of extracellular ions.

Intracellular microelectrodes have made it possible to measure the resting potential in different types of nerve and muscle cells (for review see Hodgkin 1951, 1958, 1963, Eccles 1957, Shanes 1958).

Everywhere where the intracellular concentration of potassium and chloride is known good agreement was observed between the theory as defined by Conway (Boyle and Conway 1941, Conway 1957) and actual values.

B. Electric potentials of tissues

In metazoa, cells are organised into tissues and organs. An explanation of the electrical characteristics of tissues based on the electrical properties of individual cells is rendered very difficult by variability in cellular shape and function, by innumerable variations in their spatial organisation, and by different properties of intercellular substances. The interpretation of potentials registered in a complex volume conductor is also complex. Even here, however, at least some simple fundamental principles can be applied (cf. also p. 579).

No potential difference can be detected between two intact sites of a tissue composed of apolar cells, the membranes of which have the same surface polarity everywhere (Fig. 102).

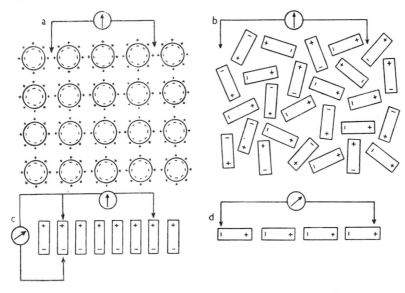

Fig. 102. Potential difference between two different points in a tissue composed a) of apolar cells, b) polar cells arranged irregularly, c) in parallel or d) in series.

In tissues composed of polar cells, the membrane of which has, for morphological or metabolic reasons, a higher (lower) potential at one site than at another, either no potential can be registered between two uninjured sites if the cells are irregularly spaced and oriented so that their potentials cancel one another (Fig. 102) or a certain polarity is present given by the vectorial addition of elementary cellular dipoles and by resistance conditions. If the anatomic structure of the tissue is very regular, the arrangement of cellular dipoles may, from an electrical point of view, approach a connection in parallel, in series or a combination of both (Fig. 102). Finally, semipermeable membranes of noncellular nature may also be a source of electrical potentials.

The demarcation potential of a tissue — the potential difference arising between an injured and a normal area of the tissue — is produced in most

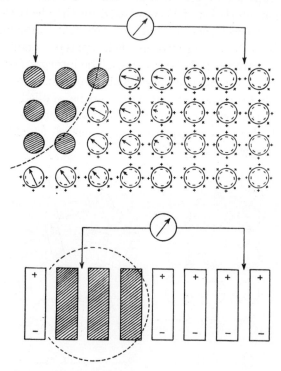

Fig. 103. Demarcation potential in tissue composed of apolar (top) and polar, regularly arranged cells (below).

cases by destruction of the inherent tissue polarity at the site of injury (Fig. 103). A certain demarcation potential, however, also occurs in tissues composed of nonpolar cells as the result of a diffusion potential arising between the normal region and the injured site, where outflowing intracellular ions accumulate. The formation of a layer of partly injured cells at the border between the destroyed and normal tissue seems even more important.

The fundamental examples of the possibilites mentioned above will be illustrated by the following experiments.

a) Polarity of frog skin

Problem: Determine the potential difference between the inner and outer surface of isolated frog skin under various conditions.

Principle: Frog skin which, according to the theory of Koefoed-Johnsen and Ussing (1956), contains in the epidermal layer bordering on the corium polar cells with different permeability of the cell membrane facing the inner and outer skin surfaces, represents a cellular membrane with parallelly-arranged dipoles. In addition to the physico-chemical causes of the membrane potential described in more detail in chapter I, so-called active transport of ions is of prime importance in the frog skin. This is a metabolically maintained ability of the cell to transfer a certain ion (ions), even against a concentration or electrical gradient, from one side of the membrane to the other. If the solutions on both sides of the skin have the same composition, the potential on the membrane or the current passing through the membrane is an expression of active ion transport only, which can be demonstrated by chemical analysis of both solutions or, even better, by the use of radioactive tracers (Ussing 1949, Kirschner 1955a).

Object: Rana temporia or esculenta, weighing about 50 g. A fresh specimen as far as possible.

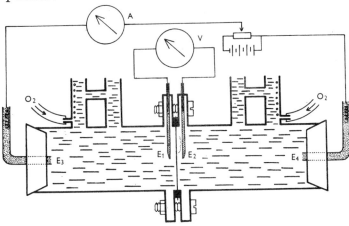

Fig. 104. Chamber for studying electroproduction in frog skin. For details see text.

Apparatus: A galvanometer with a sensitivity of $10^{-8}-10^{-9}$ A/mm/m and an internal resistance of 500—1000 Ω. A cathode follower with a galvanometer (sensitivity of 10^{-7} A/mm/m or more), or a cathode-ray oscilloscope with D. C. amplifier (overall sensitivity at least 5 mV/cm). A calibrator and compensator.

Further requirements: Nonpolarisable Ag-AgCl electrodes with Ringer-agar bridges, preparation dishes, dissecting instruments. A special chamber for fixing the frog skin (according to Ussing and Zerahn 1951 and Linderholm 1952): two tubes with an internal diameter of 3 cm and a length of 5 cm each glued to a plate of plexiglas which can be connected to the other one with screws (cf. Fig. 104). The plexiglass plates are fitted with rubber packing rings on the inside for fixation of the isolated skin. The free ends of the tubes are closed with rubber stoppers having an opening for a 5 mm glass tube with a Ringer-agar bridge of the Ag-AgCl electrodes (E_3. E_4). A capillary electrode (E_1, E_2), diameter 0·5 mm, is introduced into the inside of each tube through a vertical opening in the plexiglass plate. The electrode tip is in the centre of the tube about 1 mm from the skin surface. The apparatus has further openings for admission of air or oxygen. These may also serve for exchange of solutions.

Solutions:	Ringer	NaCl	KCl	$CaCl_2$	$NaHCO_3$
		112 mM	2·0 mM	1·0 mM	2·5 mM
	Ringer without Na	choline chloride	$KHCO_3$	$CaCl_2$	
		115 mM	2·3 mM	1·0 mM	
	Ringer without Cl	Na_2SO_4 75 mM	$KHCO_3$ 2·6 mM	$CaSO_4$ 1·0 mM	

5% CO_2 in O_2, oxygen or compressed air. Adrenaline 1/1000. Atropine. NaCN.

Procedure: The frog is killed in the usual way, the skin is carefully removed from the trunk and spread on the preparation dish filled with Ringer solution. Depending upon the size of the frog, one or two circles about 4 cm in diameter are cut from the lateral or abdominal skin. The skin is carefully mounted between the two parts of the apparatus which are then firmly screwed together. The tube in contact with the internal surface of the skin is filled with Ringer solution to the level shown in Fig. 104. The other half of the apparatus is filled in a similar way with the test solution. The solutions are aerated, thus also bringing about a slow circulation of the solutions in each half of the apparatus.

1) The dependence of the skin potential on the composition of the external solution.

The half of the apparatus in contact with the external surface of the skin is also filled with Ringer solution and the potential difference measured for 20 minutes with capillary electrodes E_1 E_2. This solution is then exchanged for 10% Ringer (obtained by mixing 9 parts of Ringer solution without Na with one part of normal Ringer solution) and potential differences are again registered for 20 minutes. Finally, this solution is replaced by Ringer without Na and recordings are taken for a further 20 minutes.

In a similar way the significance of chlorides for the membrane potential may be tested in a new preparation. After first recording potentials in normal Ringer solution, this is replaced in both halves by Ringer without Cl and again measurements are continued for a further 20—30 minutes.

2) The short circuit current of frog skin and its dependence on composition of the external solution.

If the large Ag-AgCl electrodes at the extreme ends of the apparatus are connected to a galvanometer, the electroproduction of the frog skin can be registered in units of current. The short circuiting is not complete, however, since part of the electric energy results in polarisation of the skin, which can be registered with the capillary electrodes as a certain potential difference between the two skin surfaces. In order to exclude this component and thus to obtain the actual short circuit current (Ussing and Zerahn 1951), a low variable resistance and a dry cell are inserted into the circuit of the large Ag-AgCl electrodes. With these the value of the current is adjusted so as to reduce the potential gradient across the frog skin (recorded simultaneously on the capillary electrodes) to zero. When $I . R_s + E_s = 0$ (R_s = skin resistance, E_s = skin e. m. f.), the short circuit current I reaches a maximum. If under these conditions there are identical solutions on both sides, then no concentration or electrochemical gradient exists for any ion and transport can only be maintained by active metabolic processes. The left part of the apparatus is then filled with solutions of different sodium concentrations and the short circuit current is always measured for 20—30 minutes.

3) The dependence of electrical phenomena in frog skin on metabolic processes.

A mixture of 95% O_2 and 5% CO_2 is introduced into both parts of the apparatus instead of O_2 when the short circuit current is clearly in evidence. The resulting changes in the potential or current are observed for 60 minutes. Severe interference with skin electroproduction may be obtained with metabolic inhibitors, e. g. by adding NaCN to the solution on both sides of the skin. The transport mechanism may further be influenced by some drugs. This is well demonstrated by adding adrenaline to a final concentration of 10^{-6} to the internal solution. In another experiment atropine (final concentration from 1 to 10 mM) is added to the external solution.

Results: 1) With Ringer solution on both sides, the inner surface of the skin is positive to the outer surface. With decreasing Na concentrations in the external fluid, the potential at first rises from 50—70 mV to 130—150 mV at 10 mM-Na. If the frog skin behaved like a passive membrane selectively permeable to cations, then the corresponding Na potential would be of opposite polarity (the more dilute solution positive, cf. p. 30). Only when the solution in contact with the exterior contains no sodium, does the potential drop to zero and finally reaches negative values (-3 to -5 mV — Galleotti 1904, Kirschner 1955a). If chloride ion is replaced by sulphate ion in the bathing solution on both sides of the skin, the skin potential increases considerably. This is due to low skin permeability for SO_4^{--} and resulting elimination of the

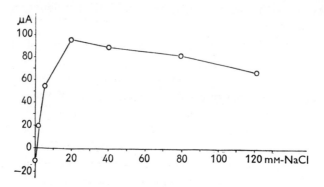

Fig. 105. Dependence of the short-circuit current of frog skin on the Na content in the external solution. Na^+ replaced by choline.

internal shunt caused by movement of chloride ions carrying charges across the skin in the opposite direction to sodium.

2) The short circuit current has similar characteristics. At a zero sodium concentration in the external solution it is negative ($10-20$ μA), then it rises rapidly and reaches a maximum of about 100 μA at a concentration of $20-40$ mM-Na (Fig. 105).

As was shown in experiments with simultaneous measurements of movements of Na^{24} and Na^{22} (Ussing 1952, Kirschner 1955a) across the frog skin in both directions, the short circuit current corresponds to net sodium transport

$$Na_{in} - Na_{out} = Na_{net}$$

(Na_{in} = influx of sodium, Na_{out} = outflux of sodium, Na_{net} = net movement of sodium). We can thus calculate the net transport of sodium directly in microequivalents from the average values of the short circuit current (I) during a certain period (t) using the equation

$$Na_{net} = \frac{I \cdot t}{F}$$

(F = Faraday constant 96 494 coulombs, I = current in microamperes, t = time in seconds). If, for instance, the average short circuit current during one hour was 200 μA, then during that time 200 . 3600/96 494 = 7·46 μequ. Na were transported.

3) Aeration of Ringer solution with 95% O_2 + 5% CO_2 suppresses the potential and the short circuit current within several tens of minutes (Ussing and Zerahn 1951). A similar effect is obtained with NaCN (Ussing 1949) but initial values are only decreased by 70—80%. Addition of adrenaline (Koefoed Johnsen et al. 1953) to the internal or atropine to the external solution (Kirschner 1955b) increases the short circuit current.

Conclusion. The microelectrode analysis of the distribution of the electric potential in different layers of the frog skin has shown (Koefoed-Johnsen and Ussing 1956, Whittembury 1964) that its main source is the layer of paralelly arranged cells of the stratum germinativum. These cells show pronounced asymmetry. The membrane of their external surface is highly permeable to sodium (the inside of the cell is positive against the external Ringer solution) while in the membrane of their internal surface permeability to potassium predominates so that a potassium equilibrium potential is formed on it (inside of cell is negative against the internal Ringer solution). The total potential difference is the sum of those two voltages. Sodium movement in the skin is realized in two phases: Na ions passively enter from the external solution into the cells of the stratum germinativum through their outer surface. From the cells they are transferred into the internal environment of the organism by an active metabolically maintained mechanism situated at the inner surface of the cells. A number of data shows that similar transport mechanisms exist in surface structures of nerve and muscle cells (sodium pump — Hodgkin 1951) and also in the cells of many other tissues (kidney, glands). The frog skin is a suitable and easily attainable object for experimental studies of active transport.

b) Positive demarcation potential of gastric mucosa

Problem: Determine the potential difference between a normal and an injured site of the gastric mucosa or serosa.

Principle: The orientation of the classical demarcation potential of nerves and muscles does not mean that an injured site must always be negative to an uninjured region. The inherent polarity of the structure at the site of a lesion is the determining factor. In muscle and nerve fibres the injury potential depends upon the polarity of the membrane (outside positive, inside negative). In structures with reversed orientation of the elementary dipoles, the demarcation potential is positive. This occurs in *Nitella* after reversing its membrane polarity with potassium (Osterhout 1936). Similar conditions also exist in tissues, e. g. in the gastric mucosa, which is negative to the serosa (Titaev 1938, Venchikov 1954, Rehm 1943).

Object: Dog weighing 5—10 kg.

Apparatus: A cathode follower with a galvanometer or a cathode-ray oscilloscope with a D. C. amplification of at least 5 mV/cm. A calibrator, 3 compensators.

Other requirements: 4 calomel electrodes. Ringer-agar bridges in glass tubes (diameter 3—4 mm) in a special holder as shown in Fig. 106a. Surgical instruments for operation. 96% alcohol, ether. Electrode stands. 10% dial.

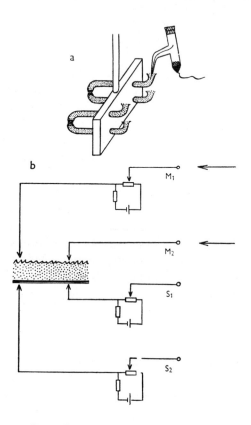

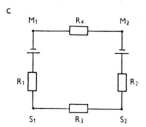

Fig. 106. Agar bridges for measuring the potential of the gastric wall (a) and diagram for connecting the electrodes with the balancing circuits (b). The input leads of the amplifier drawn opposite to M_1M_2 may be connected with any electrode pair. c) diagramatic representation of resistances and electromotive forces in the gastric wall.

Procedure: The dog is fasted for 24 hours. The abdomen is opened under dial anaesthesia in the midline, the stomach is exposed and opened by an incision 6—8 cm long between the greater and lesser curvatures, avoiding injury to the blood supply as far as possible. The lower pair of agar bridges are inserted into the incision, thus making contact with the mucosa. The upper pair is applied to the serosa just opposite to the lower pair. Such an arrangement is suitable for experiments with freely accessible serosa. If, however, we intend to work with the mucosa, the lower edge of the incision is flapped back so that the mucosa lies uppermost. The gastric wall is fixed between the pairs of bridges in that position (mucosa upward) by two sutures fastened to the holder. The Ringer agar bridges are then connected to the calomel electrodes and measurements may be started. There must initially, of course, be no potential difference between the electrodes. This is attained by balancing three of the electrodes against the fourth (Fig. 106b). The potential differences across the stomach wall (M_1S_1, M_2S_2) are recorded, followed by measurement of surface potentials (S_1S_2, M_1M_2).

After these fundamental measurements have been made, the area M_1 is treated with 96% alcohol or ether and potential M_1M_2 is recorded. Then again S_1M_1, S_2M_2 and S_1S_2 are measured. In another experiment, 96% alcohol is applied to area S_1, and again similar recordings are made.

Results: Usually only small potential differences are found between two sites of the gastric serosa S_1S_2) or mucosa (M_1M_2). The potential across the gastric wall, on the other hand, (M_1S_1, M_2S_2) usually attains several tens of mV, mucosa negative. Destruction of the mucosa below M_1 results in positivity of this area with respect to the normal mucosal surface (M_2). Thus a positive demarcation potential is formed.

On the serosa, on the other hand, the potential difference S_1S_2 remains nearly unchanged. Theoretically, the following relation holds good for the above potential differences

$$M_1M_2 = M_1S_1 - M_2S_2 + S_1S_2 ,$$

the sign of the potential always referring to the first electrode of the pair. The fact that after destruction of the mucosal surface the potential difference on the mucosa is much higher than on the serosa ($M_1M_2 \gg S_1S_2$) probably follows from the resistance and e. m. f. distribution in the circuit (cf. diagram in Fig. 106c). This corresponds to $R_3 < R_4$ and a source of the potential near to the internal surface of the stomach. Rehm (1946) could actually demonstrate, using microelectrodes, that nearly the whole potential of the gastric wall originates in secretory elements of the mucosa. This conclusion is also supported by experiments in which 96% alcohol is applied to the serosa. Usually there is no change in the potential difference between the normal and injured serosal surface and the potential difference across the gastric wall is also unaltered.

Conclusion: The finding of a positive injury potential in tissues does not contradict the classical definition of a demarcation potential. It is the result of reversed tissue polarity.

c) Cell dipoles in series

Problem: Examine the electric gradient along an onion root.

Principle: The frog skin and gastric wall belong to tissues in which cell dipoles are directed radially towards the surface, i. e. they are connected in parallel from an electrical point of view. There are, however, tissues with cellular dipoles arranged in series. A well known example is the electric organ of the electric eel, which can produce potentials of up to several hundred volts (Albe-Fessard et al. 1951, Grundfest 1957a), because several thousand units with a voltage of approximately 0·1 V are connected in series. The principle of in series connection of cell potentials can, however, be demonstrated

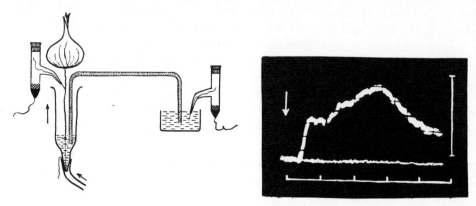

Fig. 107.a). Experimental layout for demonstrating the potential distribution along an onion root using a moving boundary. b) Oscillographic tracing. Reference electrode at the root base. An upward deflection correponds to negativity of the active electrode.

on a more accessible object — the roots of plants. The potential differences between individual sites along the root from the tip to the base are determined by the size and orientation of the potential differences between the apical and basal ends of individual cells in different areas of the root (Lund and Kenyon 1927, Marsh 1928, Rosene 1935, Scott et al. 1955).

Object: Fine young root of the onion (*Alium cepa*), water hyacinth (*Eichhornia erassipes*), or bean (*Pisum sativum*).

Apparatus: A cathode follower with a galvanometer or a cathode-ray oscilloscope with a D. C. amplification of at least 5mV/cm. A calibrator and compensator.

Other requirements: A paraffin plate for fixation of the roots. A special burette (Fig. 107). A Mariotte's bottle. Physiological saline. 20% alcohol, vaseline. Calomel electrodes and agar bridges.

Procedure: A young growing root of an onion 3—5 cm in length and not separated from the plant is placed into the ridge of a paraffine plate and fixed at its ends with vaseline. One to two hours after fixation, the potential difference between different areas of the root is measured. The tip or base is used as reference point, the other electrode being slowly moved along the root. The potential difference can also be measured using a pair of electrodes with a fixed interelectrode distance of 2 mm, that are moved together along the root. The results of both types of measurement made rapidly one after the other on the same root are compared.

After determining the distribution of the potential, a part of the root 5 mm in length is sealed off with vaseline on both sides and 20% alcohol is dropped on it. The potential difference between the apex and the base of the root is measured and the effect of alcohol is observed, especially with respect to its localisation.

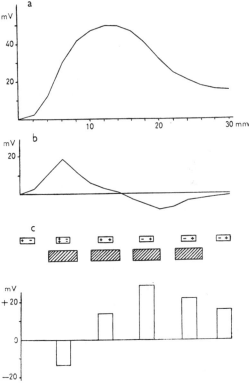

Fig. 108. Distribution of potential along an onion root a) in a mono- polar recording (reference electrode at tip-0 mm). b) in a bipolar tracing (interelectrode distance 2 mm).c) Above: diagramatic representation of cell dipoles in different parts of the root. Below: change in potential differ- ence between tip and base after ex- cluding a 4 mm section of the root (shaded rectangles) with alcohol (20%).

The potential distribution along the root can be rapidly and simply re- corded with a cathode-ray oscilloscope and a D. C. amplifier. The root is fixed with threads in a vertical position, tip downwards, in a glass burette (Fig. 107a). The burette can be rapidly filled with saline from a Marriotte's bottle. The wick of a calomel electrode is in contact with the base of the root, the other elec- trode is connected through an agar bridge with the solution in the vessel .Both electrodes (first adjusted to zero difference with the compensator) measure the potential difference between the base of the root and that area to which the level of the saline in the vessel reaches. By adjusting the rate of filling in such a way that the steadily rising level of the saline will cover the tip—base distance in 5 sec. and by adjusting the time base of the oscilloscope to this rate, we can then record the distribution curve of the potential along the root by simultaneously switching on the sweep and opening the flow of saline. Recordings are only made while the vessel is being filled. Between individual measurements, the root is washed with water and allowed to dry somewhat to prevent short circuiting of the measured potential.

Results: The tip of the root is usually positive against the middle parts and these again are negative against the base (cf. Fig. 108a). The shape of the monopolar potential distribution along the root will be the same regardless of whether the reference electrode is applied on the base or on the tip. Only the zero level will be shifted. When using a pair of electrodes with a constant interelectrode distance, the distribution of the bipolar potential differences plotted against the right border of the corresponding sectors is fundamentally the derivative of the monopolar distribution (Fig. 108b). This curve has its maximum where the first curve has the steepest slope and the minimum is found where the first curve shows constant values.

The minimum of the potential gradient at the 14th mm. and the change in polarity of the potential difference about this point is probably related to the fact that in this region of the root the cellular dipoles arranged in series change their orientation as is shown diagrammatically in Fig. 108c. This is also confirmed in experiments using alcohol. Depending on the position of the treated area and thus on the direction and size of the affected dipoles the potential difference measured between the base and the tip will be either increased or decreased (Fig. 108c). Simple short circuiting of a certain area by saline which practically abolishes the contribution of this area to the total potential difference has similar effect as alcohol (Rosene 1935).

Fig. 108c shows a typical oscillographic recording of the potential distribution along an onion root obtained by the method described. The length scale added permits direct transformation of the voltage-time function to a voltage-distance function.

Conclusion: The potential difference measured between two points of a tissue may be the result of vectorial addition of individual cellular dipoles in the interpolar area. Without detailed analysis of the electric field between the electrodes, it is not possible to draw any conclusion concerning the distribution of electromotive forces between them.

C. Electric phenomena in plants

Although plant material has already been used twice for demonstration of bioelectric phenomena (internodes of *Chara* or *Nitella*, plant roots), it is felt that a special chapter must be devoted to the electrophysiology of plants.

It was demonstrated as early as in the second half of the last century that plants may be a source of electric potentials (Buff 1854, Burdon-Sanderson

1882). Yet plant electrophysiology has remained far behind that of animals. The latter rapidly developed, particularly in connection with studies of electrical changes accompanying the function of nerves and muscles. A knowledge of electrical phenomena in plants makes it possible to demonstrate some bioelectrical processes under simple conditions and to consider electroproduction as a universal property of living matter, both in cells and tissues and in whole organisms. For the above reasons some fundamental experiments directly arising out of the last chapter will be described.

a) The bioelectric potential of photosynthesis

Problem: Determine the potential difference arising between an illuminated and a darkened area of a green leaf or between two areas with different chlorophyl content.

Principle: The intensity of metabolic processes of photosynthesis differs considerably in illuminated and darkened parts of a plant. This is indicated not only by differences in O_2 production and CO_2 consumption, but also by a characteristic potential difference. Since photosynthesis depends upon the presence of chlorophyl, a similar potential difference depending upon the intensity of illumination can be observed also between the green and unpigmented sites of variegated leaves.

Object: Elodea canadensis, Pelargonium zonale.

Apparatus: A cathode follower with a galvanometer and an optic kymograph, or a cathode-ray oscilloscope with a D. C. amplifier, total sensitivity of at least 5 mV/cm. A compensator, a calibrator.

Other requirements: Nonpolarisable wick calomel electrodes. A black dissecting plate of rubber or cork. Dissecting instruments. Pins. A light source (100 W bulb, microscope lamp). A water filter. A solution of black India-ink in water.

Procedure: An isolated *Elodea* leaf (length 10−12 mm, width 3−5 mm) is particularly suitable for demonstrating photosynthetic potentials because of its simple structure. It is composed of only two layers of cells without cuticle and has only one central rib.

1) In daylight such a leaf is fixed to the preparation plate and the wicks of calomel electrodes are placed onto two points on its surface. The electrode potential was previously balanced with a compensator. The potential difference between those two points is recorded. After some 10−20 minutes one half of the leaf is covered with a black cover through which the electrode passes, and the change in potential difference, is followed for at least 20 minutes. The cover is then removed and the course of potential changes is again observed. A similar result is obtained if a drop of water is applied at one electrode and

a drop of black India-ink covering an area of at least 5 mm in diameter at the other.

2) In a darkened room a leaf of *Elodea* is fixed onto a black plate with an opening 2 mm in diameter through which a beam of light is thrown from below, across a water filter (a 5 cm layer of water for absorbing heat radiation, distance of a 100 W bulb 50 cm). The leaf is placed in such a way that light

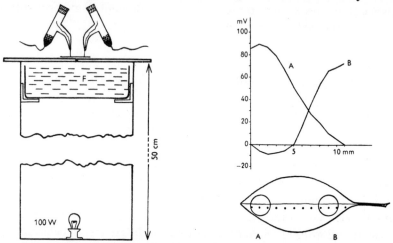

Fig. 109. a) Experimental layout for determining potential differences between illuminated and nonilluminated sites of an *Elodea* leaf. b) Distribution of potential along an *Elodea* leaf on illuminating the tip (A) or base (B). Reference electrode at base (A-11 mm) or tip (B-0 mm).

falls either onto its apical or basal portion. One electrode is permanently placed onto the dark portion of the leaf; with the other, the potential of the illuminated part and the potential gradient along the whole leaf is examined (cf. Fig. 109a). For the experiment to be successful, the leaf must first be adapted to the dark (for, at least, half an hour). The bulb must be well screened in a black tube at the top of which the dissecting plate is mounted (Fig. 109a).

3) The photosynthetic potential can simply be demonstrated in so called variegated (mosaic) leaves of some plants, e. g. *Pelargonium zonale*. One electrode is placed on an unpigmented portion of the leaf, the other on a green portion. In the dark, there will be only a small potential difference between the two electrodes (previously balanced to zero). This difference will change characteristically on suddenly exposing the plant to light.

The photosynthetic potentials can be demonstrated in all green leaves of flowers, vegetables, trees, etc. Their polarity, size and time course differ considerably in various plants.

Results: 1) The darkened portion of an *Elodea* leaf becomes 50—100 mV negative to the illuminated portions within several minutes. Within certain limits, this potential difference is proportional to the difference in the intensity of illumination. After the dark cover is removed, the original potential reappears (Fig. 110). Light and dark periods can be alternated many times, always with the same effect (Glass 1933, Uspenskaya 1951).

2) Fig. 109b shows the potential distribution around an illuminated point of the leaf. It is evident that the electrical gradient is not so steep as

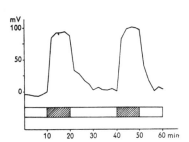

Fig. 110. Potential of photosynthesis in an *Elodea* leaf. Shaded rectangles—area of active electrode illuminated. White rectangles— the whole leaf in the dark. The illuminated site positive.

would be expected from the distribution of light and shade, but is gradual, indicating some influence of the illuminated portion on the non-illuminated parts of the leaf. The metabolic gradient, which evidently forms the basis of these potential gradients, adds to the natural polarity of the leaf (Fig. 109b). The tip of the leaf remains positive to the more central parts, even if the maximum of positivity shifts to the illuminated basal portion (Glass 1933).

3) The green parts of a *Pelargonium* leaf are the more negative with respect to the non-pigmented portions the more intensive the illumination. This potential difference, however, begins to decrease within 10—15 minutes, even though the illumination remains unchanged. Different leaves show different directions of the potential difference between the illuminated and non-illuminated portions. The majority of leaves react like pelargonium (the illuminated part is negative) but the isolated leaves of cabbage, for example, give a positive potential at the site of illumination (Klein 1898, Waller A. D. 1900, Waller J. C. 1925, 1929).

Conclusion: Potential difference between the illuminated and darkened portions of a leaf are the result of metabolic processes of photosynthesis. They depend on the intensity of illumination and on the structure and metabolism of the leaf.

b) Potential changes accompanying leaf movement in Mimosa pudica

Problem: Record the electrical phenomena due to the process mediating the spread of seismonastic movements in *Mimosa*.

Principle: In the majority of plants, reaction to external stimuli is limited to the stimulated area. There are some plants, however, which react with movements to stimulation of distant areas. This capacity may be of biological significance, e. g. in insectivorous plants (*Dionaea muscipula* — Burdon-Sanderson 1882, Di Palma et al. 1961).

The transmission of stimuli for considerable distances can be demonstrated in *Mimosa*. The path of excitation appears mechanically as closing of leaves and electrically as a wave of negativity, which may be recorded in a monopolar or bipolar recording (Bose 1926, Burr 1943).

Object: A young specimen of *Mimosa pudica*, at least 30 cm in height.

Apparatus: A cathode-ray oscilloscope with a slow time base (0·01—0·1 cycles/sec.) and a D. C. amplifier with a an input resistance of at least 2 MΩ. and total sensitivity at least 5 mV/cm. A two channel apparatus is best. A stimulator giving single electric pulses synchronised with the time base. A compensator for each channel. A calibrator.

Other requirements: An anode battery. Nonpolarisable electrodes. Physiological saline, a dropper, stands, electrocautery. A photographic camera.

Procedure: The condition of a *Mimosa* plant is tested on the previous day by observing its reactibility to mechanical stimuli. The leaves of *Mimosa* consist of a primary petiole, some 5 cm in length, divided into two or four secondary pinnae, each of which is further subdivided into 10—20 oppositely arranged pinnules. All movements are performed by specialised organs, pulvini, localised at the articulations of two leaf subdivisions (pinnules with the secondary pinna, secondary pinna with the petiole, or of the petiole with the stem). Electrodes are placed according to the diagram in Fig. 111 on carefully cleaned parts of the stem and leaves, moistened with 0·9% NaCl. The reference electrode is placed low on the stem, the further electrode (electrodes) on the petiole or pinna of the leaf, the peripheral part of which will be stimulated electrically or by heat. Great care must be taken to avoid mechanical stimulation. First the potential difference between the reference point and the other electrodes is determined. The time base is then switched on synchronously with the stimulus, and the course of potential changes is recorded. Since after stimulation the leaf displays considerable movements, good contact between the electrodes and the site of recording must be ensured using long and freely movable wicks. An attempt may be made to obtain a more precise correlation between mechanical and electrical processes by photographing the leaf at the moment when the wave of negativity commences below one of the electrodes or when it reaches its maximum.

Results: The peripheral electrode is usually 5—20 mV positive to the basal one at rest. The response is somewhat variable. With weak stimuli the spreading of the reaction may soon cease. Only a few of the nearest pinnules react by folding up, and the negativity may reach only the nearest electrode or not even that. With stronger stimuli, however, the response spreads for a considerable distance and may affect the whole plant. It is then possible to follow not only basipetal but also acropetal transmission of excitation along the leaves from the junction of individual pinnae or petioles with the stem.

Electrically, such a response appears as a wave of negativity (Fig. 111) attaining an amplitude of 50—100 mV and spreading from the site of the stim-

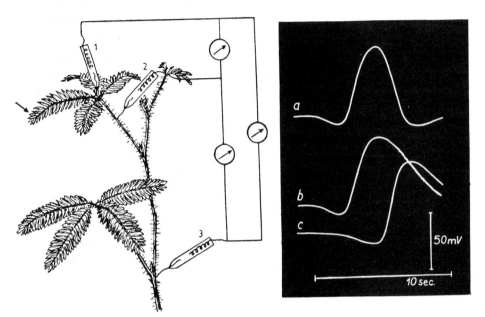

Fig. 111. Diagram of electrode distribution when recording action potentials in *Mimosa*. The arrow indicates the site of electrical or heat stimulation. Oscillographic tracing of the action potentials of *Mimosa*. a–Bipolar recording from electrodes 1—2; b, c — simultaneous monopolar recordings from electrodes 1—3, 2—3. Upward deflection — negativity of the first electrode.

ulus at a rate of about 20 mm/sec. The wave attains its maximum within 1—2 seconds. Its ebb, on the other hand, is usually slower and it is sometimes tens of seconds before the original polarity is restored. This corresponds to the slow return of the mechanical reaction, for the leaves remain folded a relatively long time. This course of the electrical reaction is observed in monopolar recordings with the reference electrode low on the stem. In bipolar recordings (cf. Fig. 111), the shape of the potential depends upon the electrode distance, and is somewhat different, since the potential wave passes below both electrodes.

As long as incomplete repolarisation persists it is not possible to produce a new reaction. The prolonged negativity is thus an expression of the refractory period in *Mimosa*.

Conclusion. The basis of the motor reactions of *Mimosa* to distant stimuli is a mechanism analogous to impulse conduction in nerve and muscle fibres. By introducing electrodes to different depths of the leaf stem Bose (1926) obtained the maximum aplitude of responses in the region of the phloem and protoxylem and expressed the hypothesis that longitudinally oriented cells are the conducting elements (120 μ in length and 10 μ in diameter). This has been confirmed more recently (Sibaoka 1962) experimentally by demonstrating with a microelectrode technique cells with a high resting potential (150—160 mV) in the region of the phloem and protoxylem. Their potential decreased to nearly zero during the electrical response. The membrane potential of cells in the other tissues is 40—60 mV and does not change during electrical surface response. It may be assumed that the excitable cells, similar to internodes of characea (see p. 271), may activate each other and thus mediate conduction of excitation along considerable distances.

D. Electric polarity of the animal organism

The characterictic differences in organisation and functional significance between apical and basal parts of the body of lower organisms have led to the formulation of a theory of axial gradients (Child 1929). A number of data were obtained concerning metabolic gradients related to gradients in physiological functions. Electrical polarity is also an expression of this axial gradient. This was observed in a number of organisms (*Hydroids* — Mathews 1903; *Amblystoma* — Burr and Hovland 1936a; chick embryo — Hermann and Gendre 1885, Burr and Hovland 1936b). According to some authors, this polarity is no chance consequence of anatomical structure, but is of fundamental significance in formative processes of embryogenesis. It was actually demonstrated that it is possible to suppress, direct or even completely reverse regeneration using an external electric field in *Obelia* (Lund 1923) and in *Dugesia tigrina* (Marsh and Beams 1952). The same was shown for the growth of nerve fibres in explanted nerve tissue from chick embryos (Marsh and Beams 1946). Electric polarity is the stronger the simpler the structure of the organism. Even then, however, it is usually not easy to determine which part of the body and which tissues are its source. All that has been said regarding polarity of whole tissues holds good in these cases, but conditions are much more complex. As an example, an experiment is described in which changes of electrical polarity are studied in the development of an organism under special conditions of registration.

Electric phenomena of early embryogenesis in the chick

Problem: Observe the development of electrical potentials in the fertilised chicken egg during the first five days of incubation.

Principle: The chick embryo up to the 5th day of incubation is positive to the albumin or the yolk (Hermann and Gendre 1885). This potential difference can by measured even through the shell. Technical difficulties caused by the large resistance of the shell and by contact potentials arising at the site of application of the electrodes are eliminated by the method described by Vorontsov and Emchenko (1947). The method utilises the free mobility of the embryo in the shell. Because it is of lower specific weight than the yolk, the embryo always floats within a few tens of seconds to the uppermost part of the egg. Electrodes situated at opposite sites of the egg (one on top, one at the bottom, cf. Fig. 112a), record the potential difference, which includes that of the embryo and the contact potentials.

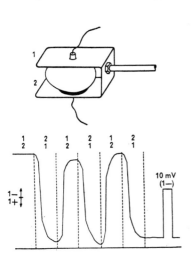 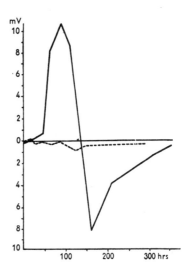

Fig. 112. Measuring the potential difference in a chick embryo. a) Above: fixation of the egg in the holder. Below: time course of potential difference between electrodes 1→2 when turnig the holder 180°. Interval between rotations 1·5 min. Upward deflection due to negativisation of electrode 1. b) Development of potential difference between the upper and lower surface of a fertilised (full line) and unfertilised (interrupted line) hen egg during incubation. Upward deflection corresponds to negativity of the upper egg surface.

If the egg, together with the electrodes, is turned 180° about its longitudinal axis, the contact potentials remain unchanged but the potential of the embryo changes its polarity with respect to the electrodes within a few tens of seconds. The embryo moves upwards and is again uppermost, but adjacent to the electrode which was underneath previously. The potential difference before and after turning the egg is double the potential difference produced by the embryo itself.

Object: Freshly-laid hens' eggs, preferably from a controlled breed.

Apparatus: A cathode follower, galvanometer with a sensitivity of 10^{-9} A. A calibrator and compensator.

Other requirements: A turnable holder for the egg with special disc Ag-AgCl electrodes (Fig. 112a). Physiological saline. A laboratory incubator at $37°C$.

Procedure: Ag-AgCl electrodes wrapped in cotton wool and stored in a dark bottle with saline are fitted to the holders and connected to the leads. Each egg is first cleaned with cotton wool soaked in warm water and then dried with a flannel cloth. It is placed into the holder in such a way that its longitudinal axis is parallel to the axis of rotation. During recording the electrodes are moistened occasionally. The deflection of the galvanometer is observed or the potential may be adjusted until it becomes stable, using a compensating circuit. The result is recorded. The egg, clamped between the electrodes, is then rotated through 180°. After a latency of several seconds, the potential begins to change, slowly at first and then more rapidly until finally (after 30—50 sec.), it comes to rest at a new value which is also recorded. Rotations are repeated several times at regular intervals (1·5 min.). Care is taken not to rotate the egg in the same direction every time. Finally, a calibration voltage of 10 mV is inserted into the electrode circuit with the egg in one of the two positions. This voltage is poled in such a way that it increases the negativity of one of the electrodes. A comparison of the direction and size of deflections when rotating the egg with deflections produced by the calibration signals makes it possible to determine the sign of the potential and its magnitude in mV. One and the same egg is first measured immediately after being laid and then at various intervals after the beginning of incubation (after 12—24 hours). If several measurements are made on one egg, the time the egg is outside the incubator must be kept to a minimum. Temporary short cooling and other procedures involved in recording potentials, however, do not impair the development of the embryo in any way.

Results: Fig. 112a shows the results of measurements made in an egg after 3 days of incubation. It also shows the way in which galvanometer deflections are changed to mV and how polarity is determined. In the example given, the average deflection, according to calibration, corresponds to 17 mV. The actual polarity of the embryo, however, as has been explained, is only half. Since a changeover from position 2/1 to position 1/2 gives a deflection corresponding to negativisation of the upper electrode (deflection in the same direction as is produced by increasing negativity of electrode 1 with a calibration signal), the upper surface of the egg is negative to the lower one. In freshly-laid eggs, on the other hand, only irregular small deflections are recorded, not exceeding 1 mV in most cases. From the second day of incubation a characteristic potential difference begins to appear in fertilised eggs, while unfertilized ones continue to be electrically inert. The potential of fertilised eggs attains a maximum on the 4th day and then decreases again (Fig. 112b), evidently as a result of the increased volume of the embryo, its decreased mobility and the altered distribution of the potential sources within the egg.

At the time when the polarity of the egg is maximal, this can be changed reversibly by warming the egg in warm water (50°C, 2 min) or by cooling it. Such procedures decrease the polarity of the egg considerably but temporarily

Conclusion: The potential differences between the chick embryo and the yolk or between different parts of the embryo produce an electrical field, which can be recorded even across the intact shell. According to changes in that field, it is possible to determine whether the egg is fertilised or not, to follow its development and reactions to changes in the environment.

IV

The electrophysiology of isolated excitable structures in vitro

The part of electrophysiology analysed in most detail to day is that of isolated excitable structures. Excitable structures are capable of sending signals — impulses — from the site of stimulation across considerable distances. Conduction occurs without decrement, since at each point energy for the impulse is obtained from local metabolic supplies.

The anatomical accessiblity of peripheral excitable structures and their relative resistance to changes in the external environment permit successful studies to be made under in vitro conditions.

Experiments in this chapter are arranged in such a way as to give, in addition to electrophysiological techniques, some orientation in the fundamentals of electrical activity in these structures and its relation to physiological functions.

A. Electric manifestations of a nerve impulse

The transmission of signals from the periphery to nerve centres, and hence to peripheral effectors, is a specific function of nerve fibres. This signalling function is realised by nerve impulses. A nerve impulse is a complex biological process which spreads from the site of stimulation at a certain rate along the nerve fibre. It manifests itself by various electrical, chemical and volume changes, changes in heat production, etc.

Changes in electrical potential at the excited site are the most conspicuous Consequently electrical signs of a nerve impulse, so called action potentials, have been studied most intensively and are the most usual indicator of the presence of a nerve impulse.

Hence it is necessary to master the technique of recording action potentials if any electrophysiological approach to the problems of nerve and muscle physiology is to be considered.

a) The action potential of peripheral nerves

Problem: Record the monophasic and biphasic action potential of the frog sciatic nerve.

Principle: The excited site is electrically negative with respect to a quiescent site. That portion of the nerve which is negative at any given moment

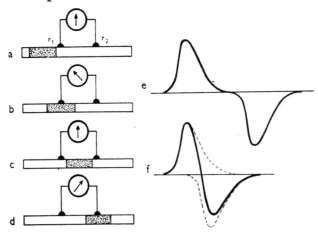

Fig. 113. Diagramatic representation of the origin of a biphasic action potential.

is of different length in various nerves, ranging from several mm to several cm, and is called the *wave length*. The negativity spreads along the whole nerve, the wave length remaining constant. The electrical potential of a site through which the nerve impulse has passed returns to resting values.

When the nerve impulse is anywhere outside the electrodes connected to a cathode-ray oscilloscope (Fig. 113a), no deviation can be recorded since there is no potential difference between electrodes r_1 and r_2. When the nerve impulse reaches electrode r_1 (Fig. 113b) the part of the nerve below it becomes negative. A potential difference arises between electrodes r_1 and r_2 and a deflection is recorded. Fig. 113c shows the moment when the impulse is between the electrodes. The tissue below both electrodes again has the same potential and consequently no deviations is recorded. When negativity reaches electrode r_2 (Fig. 113d), a situation similar to that illustrated in Fig. 113b arises. The only difference is that this time the deflection of the beam will be in the opposite direction, since the negativity is now below the second electrode.

If the potential is traced as a function of time, a curve as shown on the right is obtained (Fig. 113e). Since the onset and termination of negativity at a certain point are not instantaneous, but require a certain short time to reach a maximum, the ascending and descending parts of the curve have a certain slope. For the sake of clarity an example has been chosen which occurs only rarely in actual fact (see p. 380). Usually the wave length is longer than the distance between the electrodes. It is obvious that in such a case there will be no isopotential line between deflections corresponding to appearance of negativity under r_1 and r_2, as was the case in the previous example. The decay of negativity below one electrode will still continue while negativity already reaches the second one. Thus a curve as shown in Fig. 113f is obtained.

Similar considerations lead also to the conclusion that the amplitude of the deviation will be the smaller the smaller the distance between the two electrodes.

An action potential recorded in this way is termed *biphasic*. Negativity from the electrode closer to the point of stimulation is recorded in an upward direction and consequently negativity from the second electrode (more distant) in a downward direction*)

This second phase of the biphasic action potential is often termed positive.**)

Recording of the so called *monophasic* action potential is more suitable when studying its exact shape. In such a recording the arrival of the impulse below the second recording electrode is prevented and only its course below the first electrode is traced. This can be done in various ways. The simplest is to clamp the nerve with forceps between the electrodes or to place electrode r_2 upon the killed end of nerve. The most suitable and most effective method is depolarisation of the terminal part of the nerve using isotonic KCl solution.

The recording of a monophasic action potential is diagramatically illustrated in Fig. 114.

Object: The frog sciatic nerve with the tibial and/or fibular branches.

Apparatus: A single or double beam cathode-ray oscilloscope and an A. C. preamplifier. Overall sensitivity of the apparatus 100 µV/cm. A voltage calibrator. Time marker 1000 c/sec. A stimulator giving square wave pulses (duration 0·2 msec.) synchronised with the time base. Radiofrequency isolation unit. Screened chamber.

Other requirements: A moist chamber for the nerve preparation. Frog Ringer solution, gaseous mixture 5% CO_2 in O_2. A Petri dish with a layer of paraffin at the bottom. Dissecting instruments.

---

*) While in the technical literature upward deflection usually corresponds to positivity of the active electrode, in electrophysiology it denotes negativity in most cases. This convention is often not adhered to, however, especially in electrophysiology of the cerebral cortex, in electroretinography and in intracelullar recordings. Also in this book upward deflection does not always indicate negativity of the active electrode and the reader is adviced to check always polarity of the individual records in the text or legends. Such a usage is of advantage in both studying and writing electrophysiological papers.

**) This is not correct since this might be mistaken for the actual positive phase of the action potential, which occurs in some non-myelinated nerves of invertebrates.

Procedure: Preparation of the sciatic and the tibial nerve. The nerves of poikilotherms are particularly suitable for learning the method of recording action potentials since they are, in contrast to those of homeotherms much more easily maintained in good functional condition. For the same reasons the nerve-muscle preparation (sciatic nerve-gastrocnemius muscle) is a classical object for teaching physiology.

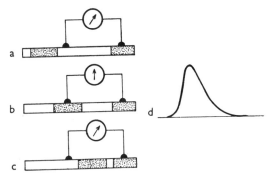

Fig. 114. Diagramatic representation of the origin of a monophasic action potential.

Fundamentally the same method is used to prepare the sciatic-tibial nerve preparation and the above mentioned nerve-muscle preparation.

The frog (*Rana esculenta* or *R. temporaria*) is decapitated with scissors. A probe is introduced into the spinal column and the spinal cord is destroyed. The frog is then fixed on its back to a board with a cork covering. The abdomen is opened and the gut is reflected so as to expose the roots of the sciatic nerve entering the spinal column. About 1 cm above that point of entry, the trunk of the frog is cut through, dividing the animal into two. The lower part is skinned, pulling as in taking off a glove. Care must be taken to prevent the skin from coming into contact with the solution in which the nerve is being prepared. The preparation thus obtained is placed into a Petri dish with about 0·5 cm of paraffin wax at the bottom. The preparation is fixed to the paraffin on its ventral side with pins, and frog Ringer is added until it is completely immersed.

The tibial nerve is cut above the talocrural joint and all side branches are cut cranially. The branches of the sciatic nerve are also removed. The fibular nerve is severed. After the tibial and sciatic nerves are thus freed of all branches they can be easily freed by raising them carefully with a glass hook. It is not necessary to remove the connective tissue in the regions of the knee and hip joints. The same is done with the other hind limb. Then the preparation is turned and the sciatic and tibial nerves are pulled to the ventral side. In such a way the left and right tibial and sciatic nerves are freed up to their entry into the spinal column. It then suffices to cut the spinal column in two and the preparation is ready. The connection between the sciatic nerve and the spinal cord is not severed. A piece of spinal column to which the sciatic

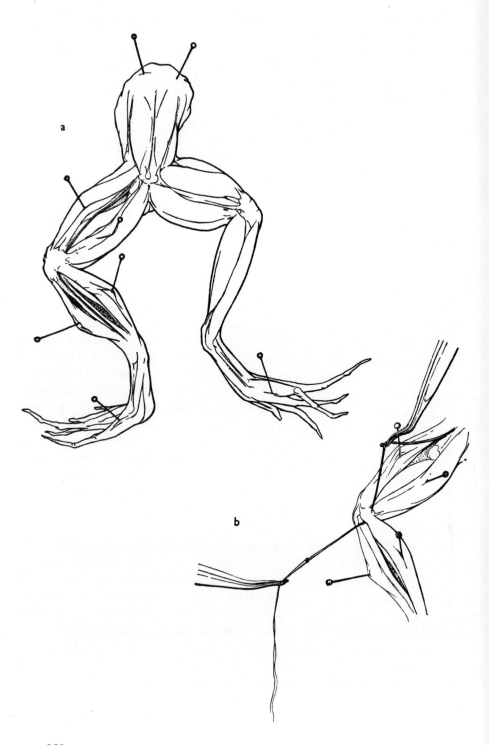

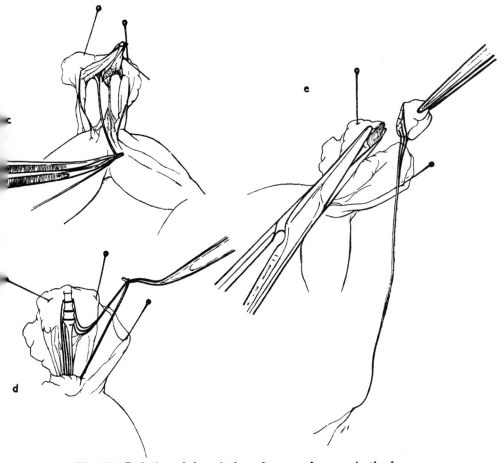

Fig. 115. Isolation of the sciatic and peroneal nerves in the frog.

nerve is attached is used to hold the preparation with forceps when manipulating with it.

The main stages of dissection are shown in Fig. 115. The nerves are placed in small Petri dishes containing frog Ringer solution.

If the construction of the moist chamber does not permit supporting the nerve along its whole length, silk loops are fixed to each end of the nerve and are used to stretch it.

The moist chamber. This is intended to serve the following purposes:

1) to fix the nerve to the stimulating and recording electrodes so that stimuli may be applied and action potentials recorded,

2) to maintain the nerve in good state, i. e. produce in vitro conditions approximating those in vivo. It would, of course, be ideal if the nerve were placed in a solution of identical temperature and composition as the extracel-

lular fluid of the animal from which it was taken. These conditions can be fulfilled ideally when using intracellular microelectrodes. Electric stimulation and recording of bioelectric potentials with external electrodes, however, require the absence of solution between the electrodes, as it is a short-circuiting factor.

Moist chambers, of which there are as many modifications as there are laboratories using them, are always a certain compromise between those two conflicting requirements.

Fig. 116 shows several diagrammatic representations of moist chambers.

The simplest is a paraffin chamber with an elevation in the centre. Several pairs of platinum wire may be placed into small grooves in this elevation. The wires are connected to insulated leads leaving the chamber. The bottom of the chamber on both sides of the elevation contains Ringer solution. The nerve is placed on the electrodes and the chamber is covered with a glass plate. Evaporation of the solution in the closed chamber is sufficient to prevent the nerve from drying (Fig. 116a).

A similar chamber may be prepared from plexiglass. Stimulating and recording electrodes on movable riders are fixed to a bridge that can be inserted into two pairs of contacts on the lateral sides of the chamber. Contacts must be made from nonoxidisable material.

Fig. 116b shows a nerve chamber suitable for recording from isolated nerve fibres. The chamber is made of plexiglass and composed of a frame carrying the stimulating (s) and recording (r) electrodes and of an exchangeable bridge with a groove below the recording electrodes. The dissected nerve fibre is placed in the groove filled with Ringer solution so that its distal part lies on the recording electrodes. The Ringer solution in the groove is then replaced by paraffin oil from 1.

Another type of chamber is shown in Fig. 116c. The nerve has no firm support but is stretched between two glass hooks (1, 2). Platinum electrodes are sealed into the glass tubes so that they are connected to their leads outside the chamber (r_1, r_2, s_1, s_2,). A fine sintered glass filter through which a mixture of 5% CO_2 in O_2 or air is bubbled into the frog Ringer solution is placed at the bottom of the chamber (f). Bubbling results in fine dispersion of the solution within the chamber, thus causing the solution to settle on the nerve and maintaining the desired gas atmosphere around the nerve.

In some experiments the method illustrated in Fig. 116d is also suitable. The nerve is suspended in the solution at a certain angle. It may be kept in place at point 3 with a glass hook. The solution is earthed. The boundaries 1 and 2 between the solution and air act as the stimulating or recording electrode. The other stimulating or recording electrode is situated on that part of the nerve which is in air. By changing the level of the solution the distance between the recording and stimulating electrodes can be varied. This arrangement is used when studying the effect of different drugs on peripheral nerves.

Fig. 116e shows a diagram of a method used very frequently. Platinum hook electrodes are sealed into glass tubes fixed to a horizontal plate. The nerve is held between two forceps and the stimulating and recording electrodes carefully placed on it. The electrode holder can be moved along a vertical steel rod. The electrodes together with the nerve can be immersed in Ringer solution when electric activity is not recorded, or raised to their original position when stimulating or recording. The arrangement is suitable for recording from isolated nerve fibres. In this case, however, the fibre is raised during the recording period into the paraffin oil instead of into the air.

Registration. The moist chamber (Fig. 116e) is placed inside a screened chamber. The stimulating and recording electrodes are connected to screened leads from the stimulator or amplifier. The nerve is fixed between forceps p_1 and p_2 and the stimulating and recording electrodes are carefully applied. For the time being the nerve is immersed in Ringer solution in a Petri dish. Stimulation frequency is adjusted to 20 c/sec. Each pulse of the stimulator is synchronised with the sweep of the oscilloscope. The nerve and electrodes are taken out of the solution and drops remaining between the electrodes are dried off. The voltage of the square wave pulse of the stimulator is gradually increased and the shape of the tracing is observed on the oscilloscope. First the stimulus artifact appears on the screen, i. e. the stimulating pulse which reaches the recording electrode by physical spread of the current. This indicates the moment of stimulation (Fig. 117a). By further increasing the voltage, the artifact grows until a slight undulation appears on the tracing. This increases on further increasing the stimulus voltage (Fig. 117b, c, d) until a full sized action potential is obtained (Fig. 117e, f). In this case a biphasic action potential is recorded. Using a sharp forceps the nerve is now clamped between the electrodes r_1 and r_2. The second phase of the potential disappears, since the nerve impulse does not reach the second electrode. A monophasic action potential is then registered (Fig. 117g).

A voltage calibrator is connected to the amplifier input and the size of the calibration pulse is photographed from the screen of the oscilloscope. Amplification must remain the same as when photographing the action potential. The time marker is then connected to the oscilloscope input (a sine wave from the audio-oscillator) and is also photographed (Fig. 117). Here again the time base must remain the same as before.

Conclusions: The action potential of a preparation: sciatic nerve — tibial nerve has an average amplitude of 2 mV. The duration of the action potential is 2·5 msec. The ascending phase lasts for 0·5—1·0 msec, the descending phase 1·0 or 1·5 msec. Apparently the action potential does not obey the "all or nothing" law, since its amplitude increases from threshold to maximum values. This, however, is due to the fact that the sciatic nerve is composed of a large number

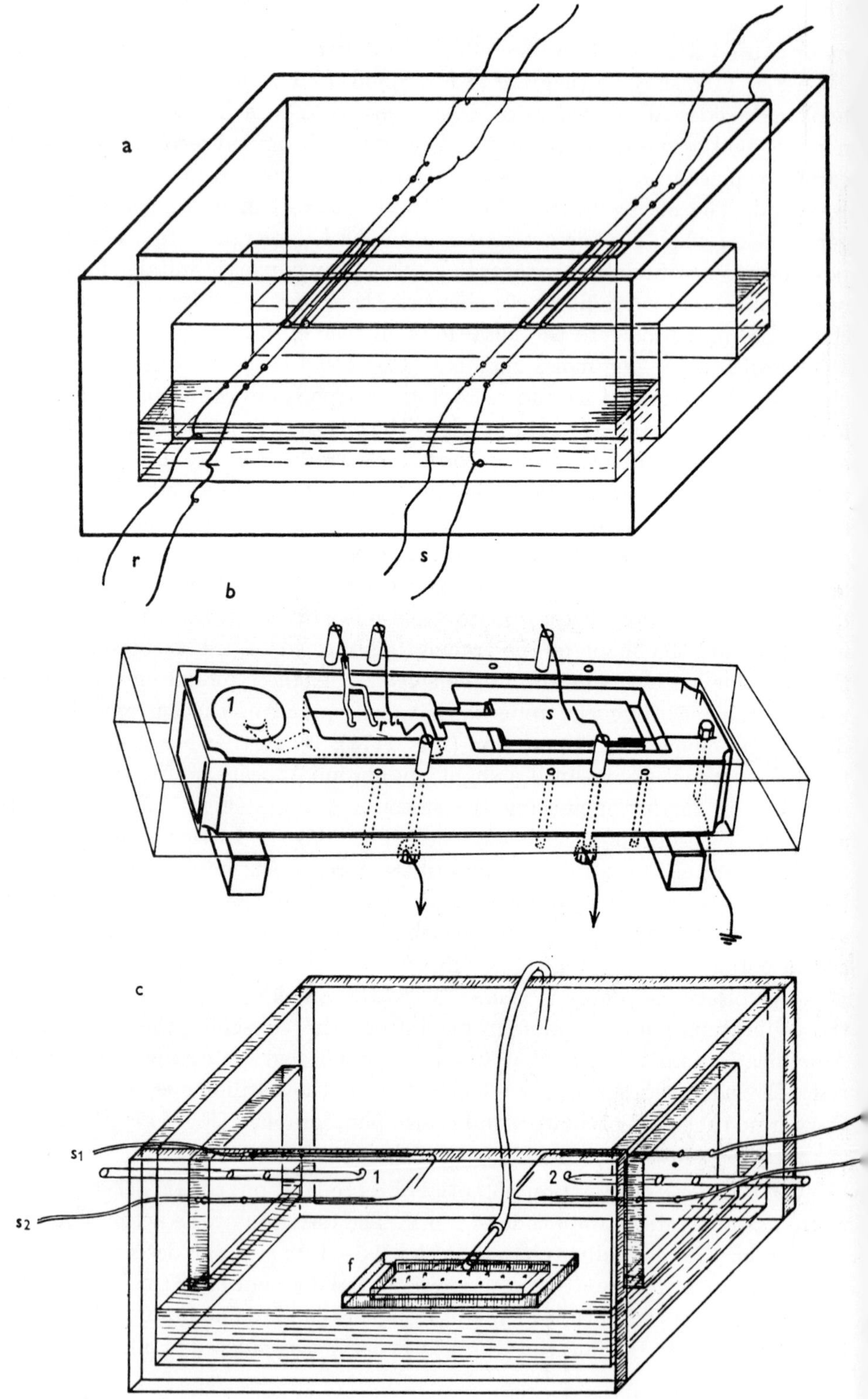

a

r   s

b

1   s

c

s₁   1   2

s₂   f

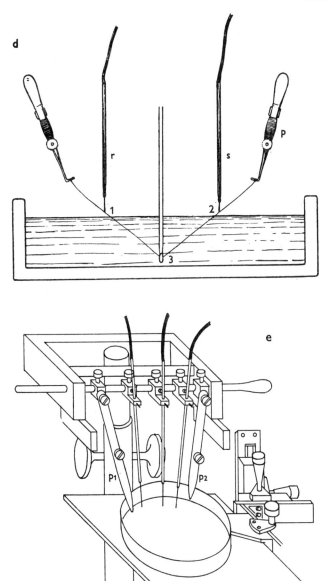

Fig. 116. Several types of moist chambers. a) nerve chamber of paraffin. b) plexiglass nerve chamber for isolated nerve fibres. c) moist chamber without a firm support for the nerve; 1, 2 — glass hooks; r_1, r_2 — recording and s_1, s_2 — stimulating electrodes, f : fine filter for aeration of Ringer solution and introduction of gases into the chamber. d) arrangement for studying the effect of drugs on the nerve; p — screw controlled forceps, r, s — electrodes, 3 — point of contact of the glass hook with the nerve; 1, 2 — electrodes formed by the Ringer-air (or oil) interface, the common indifferent electrode is connected to the saline (e. g. silver plate). e) a commonly used arrangement for fixing the nerve and mounting the electrodes. p_1, p_2 — screw controlled forceps for holding the nerve.

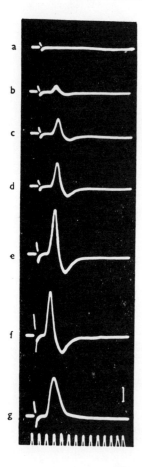

of nerve fibres. By gradually increasing the voltage of the stimulus, more and more nerve fibres are stimulated. Each fibre responds according to the "all or nothing" law but the recorded compound action potential is a summation curve of a large number of spikes with different thresholds, conduction velocities and amplitudes. Since the conduction velocity is not known, the wave length cannot be determined.

Fig. 117. Response of frog sciatic nerve to gradually increasing strength of the stimulating pulse. Time mark: 1000 c/sec. Voltage calibration: 1 mV.

b) The action potential of isolated nerve fibres. The "all or nothing" law

Problem: Record the action potential in non-myelinated nerve fibres of the crayfish (*Astacus fluviatilis*). Demonstrate the validity of the "all or nothing" law.

Principle: A nerve such as the sciatic nerve is composed of a large number of nerve fibres with different diameters, stimulation thresholds, conduction velocities and spike amplitudes. When such a nerve is stimulated, fibres with the lowest threshold are excited first.

The action potential of each individual fibre is subject to the "all or nothing" law. It is produced when a certain stimulus value is reached and does not increase on increasing the intensity of the stimulus. Other fibres however, are activated. The action potential recorded from a nerve is the sum of action potentials of its individual fibres.

The result of a gradually increasing stimulus is a gradually increasing amplitude of the action potential. When all fibres are excited, the action potential amplitude does not increase further, even on further increasing the intensity of stimulation.

Non-myelinated invertebrate nerves are suitable for demonstrating these facts experimentally, since nerve bundles composed of a small number of nerve fibres, sometimes of only two or three or even one, can be found in them. It can be demostrated that in one fibre the maximal action potential is produced by a threshold stimulus. With invertebrate nerves containing one or only a few nerve fibres it is not necessary to remove the fibres one by one in order to obtain a one fibre preparation as is the case for vertebrate nerves.

Object: Three fibre preparation from the claw of the crayfish (*Astacus fluviatilis*).

Apparatus: A single beam cathode-ray oscilloscope with a preamplifier, overall minimum sensitivity of 100 µV/cm. A calibrator. Time marker 1000 c/sec. A stimulator. A screened chamber.

Other requirements: A recording chamber. Dental drill, paraffin dish, crayfish solution (van Harreveld 1936; 12·6 g NaCl, 1·5 g $CaCl_2$, 0·4 g KCl, 0·25 g $MgCl_2$, 0·2 g $NaHCO_3$ per litre distilled water), paraffin oil, oxygen, solid paraffin. Dissecting instruments.

Procedure: **Three-fibre preparation.** The large and small nerves of the proximal joint of a crayfish (ischio-and mero-podite) are isolated and a fine nerve bundle having only three motor fibres is prepared. This separates in the carpopodite from the remaining fibres and branches multifidously at the propodite attachment.

The medial side of the large claw is used. First the hard shell is removed. This is facilitated by drilling holes into it with a dental drill. Lateral incisions are made together with incisions in the sulci of the ischio, mero- and carpopodite. A semicircular incision is drilled in the propodite at the border of its lower third. It leads from the edge of the m. abductor to the opposite edge of the claw (Fig. 118a).

The limb prepared in this fashion is cut off from the body in the coxopodite. The wound can be treated with heated paraffin to prevent bleeding.

The claw is now placed into the paraffin dissecting dish filled with crayfish saline.

It is best to start removing the shell from the ischiopodite. If the shell is not completely cut, scissors may be used to complete the incisions. The muscles attached to it are removed with a scalpel close below the inner face of the shell. The freed shell is removed up to the next incision. The same method is used for the other joints. Special attention is paid to the lateral sulcus of the meropodite and medial sulcus of the carpopodite. Here the nerve

315

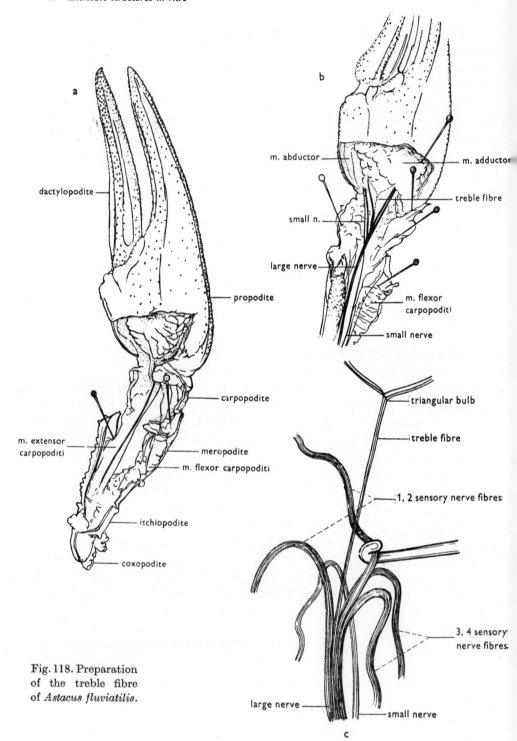

Fig. 118. Preparation of the treble fibre of *Astacus fluviatilis*.

trunk is superficially placed. The fine skin covering these areas is cut and pulled off with forceps.

The nerve trunk is then exposed. The preparation is fixed in the paraffin dish with pins and covered with saline. The site of dissection is illuminated with a focused light. Dissection is best started in the meropodite (Levin 1927, Furusawa 1929, Monnier and Dubuisson 1931, Bayliss et al. 1935, Bogue and Rosenberg 1936).

A needle is used to fix the meropodite as laterally from the sulcus as possible, to prevent injury to nerves running beneath it.

After fixation of the meropodite, the central tendon of the m. flexor carpopoditi is held with forceps. The tendon is attached to the carpopodite and can be seen in the merocarpal fold as a white rectangle. The muscle is pulled away. This is easily done, since removal of the shell has freed its attachment.

When removing the muscle care must be taken not to injure the small nerve that is more deeply situated in the muscle. After removal of the flexor carpopoditi, the small and large nerves can readily be seen along the m. extensor carpopoditi. Both nerves can be followed in the ischiopodite, and here too they can be isolated from surrounding muscle bundles.

In the carpopodite the nerves run between two muscular bundles. They are situated superficially and their course can be traced immediately upon removal of the shell. Both muscle bundles are pulled laterally (with needles) in order to expose the nerves.

1) It can be seen that in about the middle of the carpopodite the small nerve passes below the large one, reaching the opposite side of the large nerve (it approaches the edge of the abductor muscle of the claw) (Fig. 118 b).

2) The large nerve divides into several finer bundles. Two of those are directed towards the adductor of the claw and the other two run to the abductor muscle, together with a vessel, in the ridge between the abductor and adductor. The fifth and last bundle runs a nearly medial course. It is possible to see three reflecting axons (three fibres) under the dissecting microscope if good illumination is used. Each of the three axons divides near the attachment of the muscle. A triangular bulb can be seen at each bifurcation (Fig. 118b, c).

A 2—3 cm length of the fibres must be isolated. This is obtained by cutting the four sensory branches somewhat above the division of the three fibre bundle and by gradually pulling them one by one with a forceps towards the meropodite. Thus the motor fibre of the adductor muscle (three fibre bundle) is freed from the other sensory nerve fibres.

When the bundle is exposed for a length of 2—3 cm, the other nerve bundles which have been pulled away are severed and all tissue connections and side branches are removed in the exposed part. The large nerve is then isolated from the muscles in the mero- and ischiopodite. In such a way a preparation is obtained composed of the large nerve ending with a 2—3 cm long portion of the small nerve bundle.

It is advisable to leave some muscle tissue intact at both ends of the preparation, i. e. both at the end of the large nerve and the end of the three fibre bundle. This makes fixation of the preparation in the recording bridge easier.

Recording and fixing arrangement. The three fibre preparation must not remain in contact with air, since it dries rapidly. Recording can therefore only be made under paraffin oil. The arrangement shown in Fig. 116e is therefore useful. The treble fibre together with crayfish saline is sucked into a wide tube and transfered into the crayfish saline in a Petri dish. Using forceps p_1 and p_2 the ends of the treble fibre and the nerve are held. Using microscrews the preparation is slightly stretched. The recording electrodes are carefully placed on the treble fibre and the stimulating electrodes on the nerve. A layer of paraffin oil is carefully placed on the crayfish saline. Using a microscrew the whole electrode holder is now carefully raised into the paraffin oil. The solution forms a very fine film surrounding the preparation. This ensures a normal ionic environment around the nerve and also makes it possible to stimulate and record. The paraffin must be pure and it is advisable to bubble oxygen through it before the experiment is started.

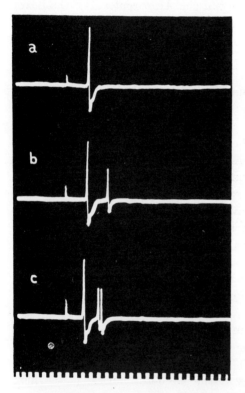

Fig. 119. Response of the treble fibre of *Astacus fluviatilis* to gradually increasing stimulus strength. Time mark: 1 msec.

Registration. The procedure is similar to that in the previous experiment. The voltage of the stimulating square wave pulse is gradually increased. First the stimulus artifact appears on the screen. This gradually increases with increasing stimulus. At a certain value of the stimulating voltage, the action potential of one fibre suddenly appears. On further increasing the stimulus intensity its amplitude does not change but a new action potential appears, when the threshold of another fibre has been attained. Then the stimulating voltage is further increased until the action potential of the third fibre appears. It can be seen that the amplitude of the previous action potentials has not increased. After the threshold of the third fibre has been reached, further increase of stimulus voltage will not result in the appearance of further action potentials (three fibres), nor will the amplitude of the action potentials change (Fig. 119).

When decreasing the intensity of stimulation, the action potentials of the individual fibres disappear in the reverse order. The time mark (sine-wave 1000 c/sec. from the audio oscillator) and the amplitude calibration pulse are photographed without changing the time base and sensitivity of the apparatus.

After registration is over, the electrodes and the preparation are immersed in crayfish saline. The experiment may be repeated several times. The fibre is capable of functioning for several hours.

Conclusion: The action potential of an isolated fibre arises at a certain threshold value of the stimulus. This occurs suddenly and the amplitude does not change on further increasing the intensity of the stimulus. The "all or none" law holds good for one fibre. The distance of the action potential from the stimulus artifact indicates a conduction velocity that is much smaller than in vertebrate nerves.

c) The action potential of a giant axon. Intracellular recording with microelectrodes

Problem: Using a capillary microelectrode of the Ling-Gerard type record the action potential from the giant axon in the nerve cord of the crayfish.

Principle: It is not possible to determine the absolute value of the amplitude of action potentials with extracellular electrodes described in the previous experiment since the extracellular fluid shortcircuits the recording electrodes. It is possible to do so if the conductivity of the interelectrode sector is artificially reduced, e. g. using the sucrose gap method (Stämpfli 1952) described in III-A-b. Another way of obtaining absolute values for the action potential is to record it between one electrode inside the axon (intracellular electrode) and another in the extracellular fluid. Intracellular electrodes were originally introduced along the longitudinal axis of the fibre (Curtis and Cole 1940, 1942, Hodgkin and Huxley 1939, 1945). This made it possible for the first time to determine the absolute values of the resting and action potentials. The use of such intracellular electrodes (Fig. 120A) is limited, however, to axons of relatively large dimensions (giant axons of cephalopods). Transverse capillary microelectrodes are more versatile (Ling and Gerard 1949). A capillary filled with a concentrated electrolyte solution with an external diameter less than 0·5 μ (Fig. 120B) inserted into an axon is equivalent to a longitudinally introduced microelectrode. The membrane is not functionally disturbed by such a micro-puncture. The leak along the walls of the capillaries is slight as if the surface structures were firmly pressed to the wall. Microelectrodes of the Ling-Gerard

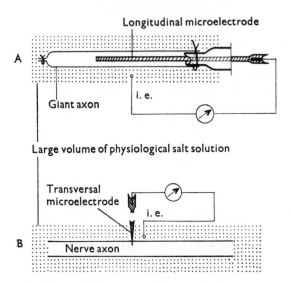

Fig. 120. Two types of capillary microelectrodes. A — Hodgkin-Huxley type introduced longitud-inally into the axon for approx-imately 30 mm (capillary diameter about 50 μ). They are filled with isotonic KCl or extracellular fluid. B — Ling-Gerard type introduced transversely into the axon (tip dia-meter less than 0·5 μ; i. e. — indifferent electrode.

type have permitted exploration of those parts of the neurons that are inac-cessible to direct observation (e. g. somata of the motoneurons in the spinal cord). Their applications are, however, also limited by the dimensions of the cells. It is already very difficult to record from muscle or nerve cells the dia-meter of which is smaller than 20 μ, which is the upper limit of the diameter of vertebrate axons. Hence for intracellular recording from axons invertebrates still remain an advantageous object since they have many nonmyelinated axons of the order of hundreds of microns.

The crayfish (*Astacus fluviatilis*) is chosen here as the experimental object. It is obtainable both near the sea and also far from it. Two large nerve axons run in its nerve cord on both sides and these can easily be made accessible for the insertion of a microelectrode of the Ling-Gerard type. The lateral giant axon is interrupted at certain distances by septa which do not prevent conduction in either direction but can decrease the safety factor of conduction. They send branches to other axons (Furshpan and Potter 1959 a, b, Kao 1960). The medial giant axon has no septa.

Object: The medial and lateral giant axon in the isolated nerve cord of the crayfish (*Astacus fluviatilis*).

Apparatus: A single beam oscilloscope with a D. C. amplifier and a

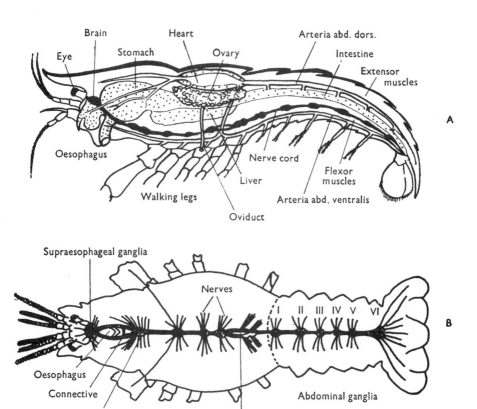

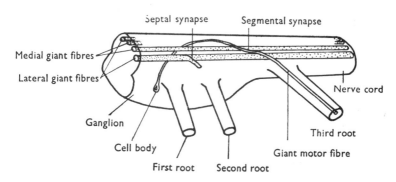

Fig. 121. A — Topography of the nerve cord and of some organs in the female crayfish. B — Dorsal view of the nerve cord. C — Location of lateral and medial giant axons in the 3rd abdominal ganglion of the cord. Axons are situated at the dorsal surface. The motor giant axon is also shown. The soma of this axon lies in the ventral part of the ganglion. The synapse with the lateral giant axon and its point of departure from the 3rd root is also seen.

cathode-follower. Total sensitivity of the system 20 mV/cm. Calibrator. Time mark. Stimulator. Indicator of impalement. Micromanipulator. Microelectrodes with resistance of about 10 MΩ.

Other requirements: Recording chamber. Paraffin dish. Van Harreveld solution (see experiment IV-A-b). Surgical instruments for rough work (heavy scissors for cutting the carapace, anatomical pincette) and for fine preparation (Wecher's scissors and pincette). Condensor for dark field illumination.

Procedure: Isolation of the nerve cord. The ganglionic chain in the crayfish is situated ventrally (Fig. 121). It can be prepared from both the ventral and the dorsal side. The ventral approach is better if the giant axons only are to be used, as is the case in this experiment, where we need not stimulate the roots.

The abdominal part of the cord is approached after first removing the head (head ganglion) and the legs. Part of the thoracic carapace, covering the gills and the gills themselves are removed. The thorax is cut into a ventral and dorsal part. The latter which contains the internal organs (see Fig. 121A) is removed and we are left with the ventral part where the nerve cord is seen to run below the fine skeletal wall and to leave the thoracic carapace channel in the lower part of the thorax. We start to prepare the abdominal nerve cord from the tail region. The ventral part of the abdomen is cut fairly superficially on both sides, one segment after the other. The dorsal part of the abdomen is pulled back and the nerve cord is continuously followed on the ventral side, i. e. on the soft covering of the abdomen. After complete removal of the dorsal part of the abdomen the preparation is placed in a Petri dish. The tail is fixed. The nerve cord is then prepared starting from the point where it leaves the carapace canal. Here it is separated from the rest of the thoracic nerve cord. The ventral part of the remaining thorax is also removed. Then we continue to isolate the abdominal cord from the surrounding tissues and the ventral abdominal artery caudally down to the last ganglion. The ganglionic roots are cut off as far as possible from the cord. They will serve for fixing the preparation. The isolated cord is transferred to the recording chamber.

Fixation of the preparation in the recording chamber: Threads are fixed to the end of the cord and are used to tie it in a canal made of plexiglass. This canal is closed on both sides by paraffin blocks (Fig. 122). Incisions are made with a scalpel in the blocks and the thread is pressed into them. The threads are then firmly fixed in these incisions by pressing on the latter (Fig. 122b). The nerve cord lies at the bottom of the canal with its dorsal surface upwards since the giant axons run dorsally in the cord (Fig. 121C). In order to prevent rotation of the cord the roots of some ganglia are pressed into the paraffin part of the lateral wall of the canal (Fig. 122a). During this procedure the recording chamber and the canal are in a Petri dish filled with van Harreveld solution. The recording chamber is then placed on the stage of the microscope (Fig. 123). A dark field condensor is used, and this approaches closely the

322

bottom of the chamber. In the dark field both pairs of lateral and medial giant axons appear as dark strings on a lighter background. The diameter of the lateral axons is greater (about 100 μ) than that of the medial ones (about 50 μ). The preparation of the cord is completed on the microscope stage by carefully removing the connective tissue envelope which covers the axons and prevents introduction of the microelectrode.

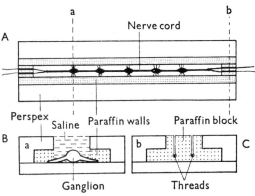

Fig. 122. Fixation of nerve cord in recording chamber. A — overall view. The channel is closed with paraffin on both sides. The free ends of the thread holding the cord are pressed into the paraffin as seen in section (C) in position b. B — section in position a. The fixing of roots into the paraffin which fills the lower part of the channel.

Stimulation: The stimulating electrode for extracellular stimulation is a tube 2 mm in diameter with a conical end and a tip of about 50 μ. A broken microelectrode may be used if its sharp edges are made smooth over a flame. A chlorinated silver wire is inside the tube. The tip touches the axon in this kind of stimulation. The tube is filled with Ringer solution. The reference electrode is an Ag-AgCl electrode in contact with the solution surrounding the preparation.

Registration: The microelectrode is placed into the holder of the micromanipulator and under control of the preparation microscope (magnification 30 ×) it is made to approach the preparation. The input of the cathode-follower is connected to the microelectrode via a chlorinated silver wire. Before introducing the microelectrode in to the fibre the tip potential of the electrode and its resistance are measured (experiment III-A-c and page 217). If the potential is 5 mV and the resistance about 10 MΩ the microelectrode may be inserted into the fibre. The beam of the oscilloscope is in the middle of the upper half of the screen and amplification is set so that a 100 mV change fits comfortably onto the screen. Using the coarse vertical movement of the micromanipulator the microelectrode is moved under control of the microscope (magnif. 100 ×) to the surface of the giant fibre. The beam of the oscilloscope remains at its original level. The tone in the acoustic indicator remains unchanged. Then by using the vertical microdrive the electrode is lowered until we hear a sudden change in tone (e. g. by one octave) which corresponds to a fall in the height of the beam on the screen by the value of the membrane potential. In Fig. 124 the membrane potential, as judged from the change in beam position, is −80 mV (inside negative) which corresponds to the values of the resting potential found in the crayfish *Cambarus* (80·75 ± 5·67 mV, Kao 1960). After withdrawing the electrode the beam returns to its initial position on the screen (Fig. 124D). If the time base is switched off so that a point appears on the screen, the insertion into the fibre can be recorded better on a moving film.

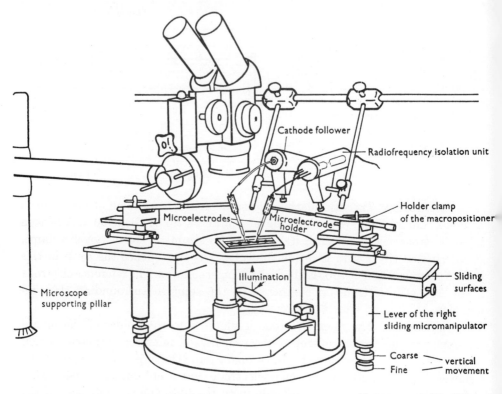

Fig. 123. Experimental set-up using intracellular capillary microelectrodes. The preparation is illuminated from below using a mirror. The dark field condensor is just below the preparation and is not shown nor is the light source and the lense visible. The preparation is observed through the microscope from above. The recording electrode is manipulated with the left, the stimulating one with the right sliding micromanipulator. The recording chamber is on the object plate of the microscope.

We then again introduce the microelectrode into the fibre and stimulate the nerve cord with a stimulating capillary with a pulse of 0·5 msec duration. When the stimulus attains the threshold value an action potential appears according to the "all or none" rule (Fig. 124A-C). The spike amplitude exceeds that of the resting potential (Fig. 124C, D). The part of the action potential above the initial potential is called the overshoot. The electrode is withdrawn from the fibre by turning the microscrew in the reverse direction. Amplification is calibrated by connecting 10 mV voltage steps to the input. Finally the time mark is recorded.

In the same way the action potential from the medial giant axon is recorded.

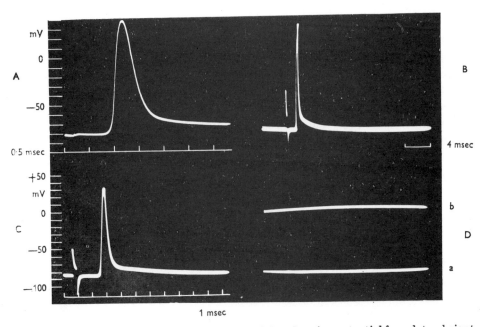

Fig. 124 Intracellular recording of resting potential and action potential from lateral giant axon of the nerve cord of *Astacus fluviatilis*. A — C — action potentials recorded at three different speeds. Action potential in C was registered at lower gain. D — a-potential level after inserting microelectrode, b-after withdrawing the microelectrode. The same amplification as in C.

Conclusion: While the resting potential of the lateral giant axon in the cord of the crayfish is —80 mV (inside negative) the action potential attains absolute values of 120 mV. The *overshoot* is thus 40 mV. The existence of the overshoot is in evident disagreement with the concept of Bernstein (1902) on the nature of the action potential. This fact was first demonstrated by Hodgkin and Huxley (1939) and Curtis and Cole (1940) and since then has been repeatedly confirmed for different excitable structures (for review see Hodgkin 1951, 1958, 1963). The explanation of this fact led to the formulation of the ionic hypothesis of the action potential (Hodgkin et al. 1949; Hodgkin and Huxley 1952).

The experiment also shows that intracellular recording makes it possible to register from a bundle of fibres an action potential complying to the "all or none" rule even with nonselective stimulation. This fact has made transverse electrodes of the Ling-Gerard type a powerful tool for exploring unit activity in the CNS where anatomical isolation of units is practically impossible.

Repeated insertions of the microelectrodes have shown that the dimensions of the electrode tip do not injure the membrane of the fibre since the same action and resting potentials can be recorded repeatedly.

The membrane and action potentials of the medial and lateral giant axons do not differ much from one another. The time to peak amplitudes of the spikes is about 0·3 msec, the descending phase is longer, lasting about 1 msec. In the lateral giant axon there is a slow return to the resting potential (about 15 msec) while in the medial one return is much more rapid. During after-depolarization recordings from the lateral axon show sometimes small variations in potential due probably to activity in the other lateral axon. There is indirect evidence of connections between the two lateral axons which show a kind of ladder structure (Furshpan and Potter 1959a).

B. Propagation of nervous impulses

It is generally agreed that the cause of spreading of nerve impulses along a nerve fibre is the ability of the excited zone to stimulate the adjoining regions. The active point is depolarised and negative. Thus a potential difference arises between this point and the neighbouring region. This potential difference produces local currents (hence the term: *local circuit theory*), which flow from the intact surface to the depolarised point and thus progressively depolarise the adjacent inactive region.

The potential difference caused by an action potential is several times larger than that necessary for the production of a nervous impulse in a neighbouring point. There is thus a large safety factor, permitting conduction to occur even under adverse conditions, e. g. during the relative refractory period after a nervous impulse.

Examples given in this section are chosen to show the technique of measuring the rate of conduction as well as some classical experiments which led to the formulation of the local circuit theory.

a) Measurement of conduction velocity of a nervous impulse

Problem: Determine the conduction velocity of a nervous impulse in A-fibres of the frog sciatic nerve.

Principle: The sciatic nerve of the frog is a mixed nerve. It contains fibres having diameters from 3—29 μ. The corresponding differences in conduction velocities are not evident at small distances from the point of stimulation. The recorded action potential is smooth, although it represents the sum of potentials from hundreds of fibres with different conduction velocities. At greater distances from the point of stimulation, the different conduction velocities manifest themselves as small humps on the descending limb of the action potential. At very great distances the action potential of the most rapidly conducting fibres is separated from that of more slowly conducting ones.

Hence recordings must be made at sufficient distances from the stimulating electrode in order to differentiate between groups of fibres.

The conduction velocity can be calculated in two ways. Both require a determination of the path travelled by the nerve impulse and the time necessary for it.

The first method records the action potential at a certain distance from the stimulating electrode. Time is measured from the stimulus artifact to the onset of the ascending phase of the action potential. This method is less exact, since the point of stimulation cannot be determined acccurately enough because of the possibility of stimulating with virtual cathodes (p. 343).

The second method uses registration of action potentials from two points that are separated by a given distance. This avoids the difficulty of determining accurately the point of origin of the action potential, but is more elaborate. A dual-beam oscilloscope with two differential amplifiers is needed. A single beam oscilloscope may also be used by alternating the connection of the electrodes to the amplifier input.

Object: A sciatic nerve — peroneal nerve preparation from the largest frogs available. Length more than 13 cm.

Apparatus: A cathode-ray oscilloscope with an A. C. preamplifier. Overall sensitivity of the apparatus 100 μV/cm. A time marker 1 msec. A square wave stimulator with radiofrequency output.

Other requirements: A moist chamber for the nerve with 5 platinum electrodes. Otherwise as in experiment IVAa.

Procedure: The nerve is fixed in the moist chamber using silk loops tied to glass rods and slightly stretched (Fig. 116c). Glass hooks are placed under the nerve at several points. The stimulating electrodes are placed close to the origin of the sciatic nerve from the spinal cord. The first recording electrode (r_1) is situated at the beginning of the peroneal nerve, the second (r_2) is close to the end of the preparation. The reference electrode r_i is at the end of the nerve (monophasic recording).

First r_1 against r_i is connected to the amplifier input. The intensity of the stimulus is increased and the amplitude of the action potential is observed. An action potential first corresponding to A fibres, group α is formed. Gradually group β appears, and finally γ fibres are also visible. After photographing the recording from the first electrode, electrode r_2 is connected to the amplifier input and, using the same intensity of stimulation, the action potential from a more distant part of the nerve is recorded (Fig. 125).

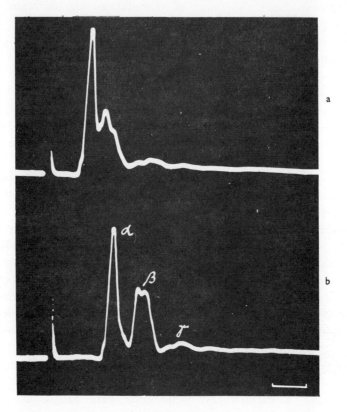

Fig. 125. Conduction velocity in frog sciatic and peroneal nerve A fibres. Time mark: 2 msec.

The time signal and amplitude calibration are photographed and the distance between electrodes r_1 and r_2 is measured with calipers. This gives all the data necessary for calculating the rate of conduction of different groups of A-fibres.

The time from the onset of the action potential of A α fibres at the first electrode r_1 to the onset of the action potential at electrode r_2 is either measured in enlarged positive pictures, care being taken to have the same enlargement, or from negative films projected onto paper on which the time scale has been marked previously. In A β and A γ fibres the conduction velocity is estimated from the time interval between the corresponding peaks.

The same sweep, starting from the same point of the screen must be used when recording the time mark and the action potentials in order to avoid error due to a non-linear course of the time base.

When enlarging, systematic errors must also be avoided. These may be due to thickness of the enlarged tracing when measurements are not made from a standard point, e. g. always from the upper contour of the trace.

With these precautions, the time difference between action potentials of groups α, β and γ are determined. The mean conduction velocity is calculated by the equation:

$$\bar{v} = \frac{s}{t}$$

where $\bar{v} =$ the mean conduction velocity in m/sec., $s =$ the length of the path in m (the distance from the first recording electrode to the second one), $t = $ = time between the appearance of the action potential at r_1 and r_2 in seconds.

Results from an experiment similar to that shown in Fig. 125 are given in Table 4. The distance $r_1 - r_2 = 0.055$ m.

TABLE 4

	t	$\bar{v}$
α-fibres	0.0013 sec	42 m/sec
β-fibres	0.0022 sec	25 m/sec
γ-fibres	0.0034 sec	16 m/sec

Conclusion: The conduction velocity of the most rapidly conducting group of fibres (A) was determined in the frog's peroneal nerve. It has 4 components α, β, γ, δ. The conduction velocity was determined for the first three. With greater amplification, the conduction velocity in the B and C fibre groups can also be measured. These have a smaller spike-amplitude and a lower conduction velocity (B-4.2 m/sec; C-0.4 m/sec).

b) Relation between fibre size and conduction velocity

Problem: Compare the conduction velocities of two nerve fibres of different diameters.

Principle: Within the same phylogenetic species it is possible to demonstrate a relationship between fibre diameter and conduction velocity of a nervous impulse. The original concept, that conduction velocity is proportional to the square of fibre diameter (Blair, Erlanger 1933) has been modified. Today it is believed that the relationship is linear (Gasser and Grundfest

1939) in mammalian nerves. Multiplying the diameter of the myelinated nerve fibre (in μ) by 6 gives the approximate conduction velocity in m/sec.

This cannot be applied to fibres from different species. During phylogeny the rate of conduction is increased by increasing fibre diameter (e. g. giant fibres) in invertebrates, whereas in vertebrates, this come about rather by myelinisation and development of the mechanism of saltatory conduction. In different species, the product of fibre diameter times birefringence of the fibre sheath is approximately constant (Taylor 1942).

It is difficult to determine the relationship between the conduction velocity and the fibre diameter in mixed nerves. One of the methods used is to correlate the maximum conduction velocity in different mixed nerves with the corresponding maximum fibre diameter as determined histologically (Hursh 1939a, b).

A preparation with few fibres is better for demonstrating these relations. The simplest way is to compare the rate of conduction in two fibres with differing diameter. The rate of conduction of each fibre is determined in the usual way and its diameter is then measured under the microscope.

Object: A motor double fibre, an inhibitory and excitatory nerve, innervating the abductor of the claw of the crayfish (*Astacus fluviatilis*).

Apparatus: A cathode-ray oscilloscope with an A. C. preamplifier, total sensitivity 100 μV/cm. Stimulator with radiofrequency output (pulse duration 0·5 msec.). A time marker 1000 c/sec.

Other requirements: Two pairs of recording electrodes. Otherwise as in experiment IV. Ab.

Procedure:

a) Isolation of the double fibre. The preparation consists of a small and large nerve trunk and a 1—3 cm portion of the inhibitory and excitatory fibre innervating the abductor muscle of the crayfish claw. In the proximal portion the inhibitory fibre is contained in the large and the excitatory fibre in the small nerve trunk. Both fibres separate from the other fibres in the distal part of the limb. Enveloped in the same sheath they form in the propodite the so called double fibre (Fig. 126).

Opening of the shell and isolation of the nerve trunks is basically performed in the same way as described for the treble fibre (page 315). Differences in the preparation are described below (Eckert and Zacharová 1954).

When incising the shell in the propodite four longitudinal incisions are made in addition to the semicircular one at the boundary of the lower third of the claw. These are joined in the upper parts in such a way that three rectangular fields are formed. Then the shell is easily removed. It is important to make a special hole on the dorsal side of the propodite in the ridge between the abductor and adductor muscles. The incision does not reach the lower

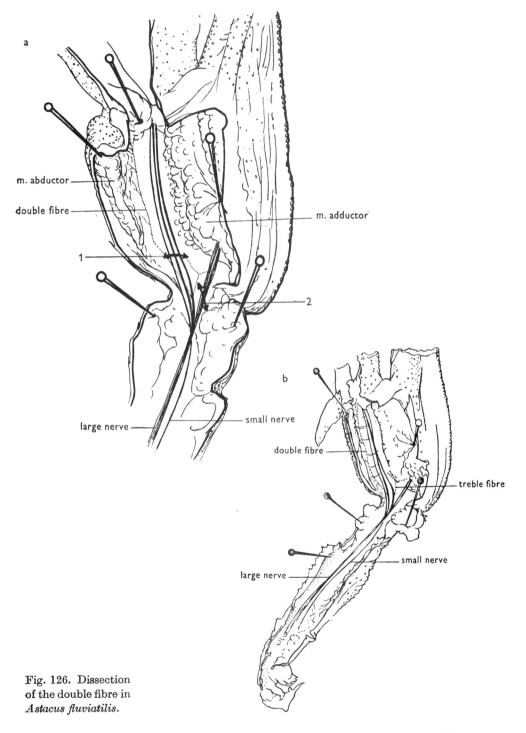

a

m. abductor

double fibre

m. adductor

1

2

large nerve

small nerve

b

double fibre

treble fibre

large nerve

small nerve

Fig. 126. Dissection
of the double fibre in
Astacus fluviatilis.

edge of the propodite so that on deflecting the shell it remains connected to the rest of the shell.

When preparing the trunks it is necessary to avoid injuring the small nerve trunk in the lower parts of the mero-, ischio- and carpopodite. Special care must be taken in the carpopodite in which the small nerve passes below the large one and comes to lie on the opposite side. This point must remain free from injury.

The double fibre is isolated under the microscope. The propodite must be fixed. The adductor muscles is pulled away with a firm needle which is also used to fix the propodite to the board. The other side (edge) of the adductor muscle is also pushed onto the board and a second needle is passed through the internal side of the cut stump of the dactylopodite and fixed as far laterally as possible to the bottom of the preparation dish.

After cutting the connective tissue between the two muscles these are separated from each other. It is best to leave the small nerve and the two sensory branches of the large nerve together with the vessel on the side of the adductor muscle and to separate the abductor. Then it is possible to pull away carefully the freed edge of the abductor muscle and to look for the double fibre. Using good concentric illumination and correctly turning the dish the axons reflect the light and are seen as two bright fibres running parallel to each other and enveloped in a connective tissue sheat. The double fibre sends off branches while passing along the abductor muscle. These are cut. In the lower third the double fibre changes its direction and runs obliquely down ward towards the sensory fibres. Here it passes below a vessel. After separating the double fibre from the muscle along its whole length the sensory fibre is severed just where it passes below the vessel. Usually the double fibre is only prepared up to this point but sometimes, especially if a longer sector is required, it may further be isolated and separated from the vessel (arrow 1 — Fig. 126).

Finally the nerve trunks are separated from the muscle in the ischio-, mero- and carpopodite up to the free part of the double fibre. Here special care must be taken in the carpopodite. The double fibre runs deeply submerged in the muscles and separates. The inhibitory fibre joins the large nerve, the excitatory one the small nerve. A binocular lens is always used for dissecting. The sensory fibres are severed and separated from the muscle (arrow 2 — Fig. 126).

Then the most distal portion of the fibre is exposed and separated where it finally branches. A piece of muscle in which the fibre terminates is also taken so that the preparation may be held in forceps by this muscle.

b) The preparation is fixed as in experiment IV. Ab. First the large nerve which contains the inhibitory nerve is placed on the stimulating electrodes. On stimulation the action potential of the inhibitory fibre is recorded at the recording electrodes (Fib. 127b). Stimulation of the small nerve, on the other hand, evokes an action potential in the excitatory fibre (Fig. 127c). When both are stimulated at the same time the action potential of both fibres is obtained (Fig. 127a). The conduction velocity in both fibres can be estimated from the distance between the stimulus artifact and the action potential. More accurate results are obtained when two pairs of recording electrodes, which can be connected alternately to the amplifier input, are placed on the double fibre. First the action potential from electrodes r_1, and then that from the more distant electrodes r_2, is recorded. Finally the distance between the first electrodes of both pairs of recording electrodes is measured.

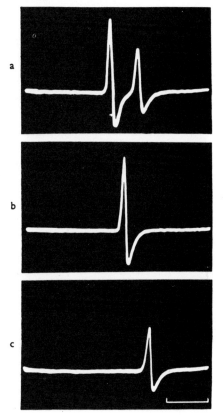

Fig. 127. Action potential of the excitatory and inhibitory nerve fibre of *Astacus fluviatilis*. a — double fibre, b — inhibitory fibre, c — excitatory fibre. Time mark: 10 msec.

Conclusion: The inhibitory fibre has a diameter of 18 μ and the excitatory fibre one of 14 μ (Eckert and Zacharová 1957). It is therefore to be expected that the rate of conduction in the inhibitory fibre is greater. The experiment recorded in Fig. 127 confirms this assumption. Fig. 127b shows a recording from the double fibre when stimulating the large nerve, which contains the inhibitory axon, and the action potential recorded, consequently, belongs to that of the inhibitory fibre. Fig. 127c shows the action potential of the excitatory axon evoked by stimulation òf the small nerve, which contains this fibre. It can be seen that the inhibitory axon, having a larger diameter, conducts impulses at a greater rate than the excitatory fibre. This is shown in Fig. 127a where both nerve trunks were stimulated simultaneously and action potentials from both fibres were evoked. The action potential can be identified not only from the conduction velocity, but also by its amplitude. The spike-amplitude of the inhibitory axon is greater than that of the excitatory one. This is in full agreement with the rule that the conduction velocity and spike amplitude are directly and the threshold to electric stimulation indirectly proportional to the fibre diameter.

c) Extrinsic potentials

Problem: Record the extrinsic potentials and changes in excitability of a peripheral nerve beyond a cold block.

Principle: This experiment tests the hypothesis that local currents are important for the propagation of nerve impulses. The excited point causes increased excitability in adjacent regions of the nerve. The same is true for the nervous impulse arriving at a nerve block. Changes in excitability are produced beyond the latter even if the impulse itself cannot pass it. This effect has been long known in the literature as *Wedensky facilitation* across a block (1903). The method for ascertaining such changes in excitability is shown in Fig. 128a. A block is induced at site B, e. g. by clamping the nerve. One pair of stimulating electrodes is above the block (s_1), the other beyond the block (s_2). At a certain distance from them the recording electrodes are placed at C. When a block is formed at B no potential is recorded from C on stimulating at s_1. At s_2 subthreshold stimulation is applied and again no action potential is recorded at C. If, however, s_2 is stimulated at the moment when the nerve impulse elicited at s_1 reaches block B, then it may be expected,

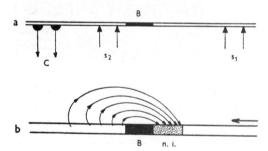

Fig. 128. Determination of excitability changes beyond a block. a) — electrode arrangement. s_1, s_2 — stimulating electrodes, C — recording electrodes, B — blocked nerve segment, n. i. — nerve impulse. b) — flow of local currents beyond the block after the arrival of the nerve impulse in the blocked segment.

according to the local circuit theory, that part of the local currents will flow from a site beyond the block towards the stretch of the nerve in front of the block depolarised by the arriving impulse (Fig. 128b). In such a way local currents produce increased excitability beyond the block. A subthreshold stimulus at s_2 will become effective now, and an action potential is recorded at C.

This experiment was already performed by Wedensky, who concluded that the nervous impulse increases the stationary excitation in the blocked site for a certain short period, and produces corresponding *periparabiotic* excitability changes (Wedensky 1903).

Hodgkin in 1937 recorded electrical changes occurring beyond the block after arrival of the nervous impulse and corresponding to the excitability

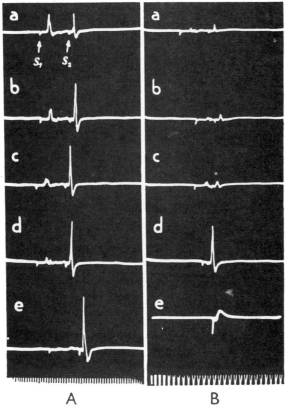

Fig. 129. Changes of excitability in the vicinity of a cold block. A) — increase of excitability in the vinicity of a developing cold block. Time mark: 0·5 msec. B) — changes in excitability produced by a nerve impulse beyond the cold block. Time mark: 1 msec.

increase predicted by the local circuit theory. This potential was termed the "extrinsic" potential by him. It differs from the action potential in having a lower amplitude and a spatial decrement (a decrease to $1/e$ of its value in two mm).

Apparatus: Cathode-ray oscilloscope with a preamplifier, total amplification at least 100 μV/cm. Two pairs of stimulating and one pair of recording platinum electrodes. A stimulator with two outputs. The time interval between the first and second output pulse can be changed continuously. Time marker 1000 c/sec. A voltage calibrator.

Other requirements: A moist chamber (e. g. Fig. 116c). An arrangement for producing cold block. This is an L shaped glass tube (diameter 5 cm) closed at both ends with rubber stoppers. The tube is filled to ¼ its volume with ether through an opening in the upper stopper. Ether vapours are sucked off through another opening in the same stopper using a water vacuum pump (exhauster). A temperature of −10 to −15°C can easily be attained. A silver wire (diameter 0·5 cm) passes through the lower stopper. This serves to lead off heat from the nerve to the tube. The area of the wire touching the nerve can be changed at will by soldering silver plates of different dimensions to the end of the wire. The wire is introduced into the chamber through a longitudinal fissure in its side, thus permitting movement of the cooling apparatus and placing of the cooled wire below any part of the nerve. The temperature acting on the nerve can be changed by changing the rate of exhaustion or by switching off the pump. Nerve blocks below −6°C, i. e. supercooled (Boyd and Ets 1934) are usually irreversible. It is advisable to use reversible blocks of about −1°C. This may be simply achieved (Hodgkin 1937a) by encasing the whole wire in paraffin and leaving a slit along its whole length, thus maintaining contact with the air. A layer of ice begins to form at the cooler end of such a slit. This gradually progresses along the wire and reaches the nerve, and thus supercooling does not occur.

Object: Sciatic nerve − peroneal nerve of the frog (*Rana esculenta* or *Rana temporaria*).

Procedure: The nerve is stretched as described previously (p. 310) in a moist chamber. Two pairs of stimulating electrodes, recording electrodes and the silver cooling wire are placed on the nerve as shown in Fig. 128a.

The nerve is stimulated with a maximal stimulus at s_1. A subthreshold stimulus is applied to point s_2 after a certain time interval greater than the duration of the refractory period. A maximum action potential of A fibres is obtained in response to stimulus s_1. No action potential appears after stimulation of s_2. Then cooling with the silver wire is started. Gradually a cold block develops in more and more nervous fibres. This is evident from the fact that the action potential produced at s_1 and recorded beyond the block becomes smaller and smaller. Since, on the other hand, the excitability near the block increases (in Wedensky's terminology: near the parabiotic focus is a periparabiotic increase in excitability), the subthreshold stimulus at s_2 becomes a suprathreshold one, and consequently a submaximal action potential is evoked also in response to stimulus s_2 (Fig. 129Aa). On further cooling, the action potential, evoked at s_1 continues to decrease until it finally disappears altogether, while that elicited from s_2 continues to rise.

After warming the nerve, the experiment is repeated as follows. When the block at B is nearly complete, the stimulus intensity at s_2 is lowered so as to make the stimulus subthreshold or just threshold once again (cf. Fig. 129Ba). When now the interval between stimuli is reduced so that the impulse from s_1 reaches block B just at the instant when the stimulus from s_2 begins to act, an action potential appears on the recording electrodes in response to stimulus s_2. This indicates that the impulse from s_1, although already blocked, produces an increase in excitability beyond the block (osc. d, Fig. 129B).

If the stimulating electrodes s_2 are disconnected from the stimulator and connected to the amplifier input instead of electrodes C, the extrinsic potential is recorded beyond the block.

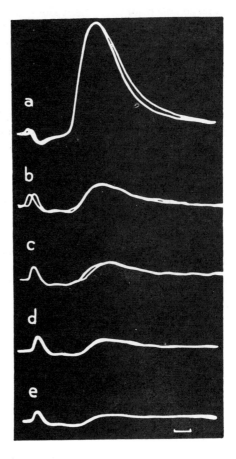

Fig. 130. Spatial decrement of extrinsic potentials. Time mark: 1 msec.

The next experiment is arranged in that way. One recording electrode is proximal to the block and four beyond the block. One after the other, these five electrodes are connected, against the distal recording electrode, to the amplifier input so that the spatial decrement of extrinsic potentials can be observed. Fig. 130 shows a recording from such an experiment.

Conclusion: The first experiment showed that a periparabiotic excitability increase occurs around the parabiotic focus (using the terminology of Wedensky). The mechanism of this excitability change can be explained in a similar way as the excitability increase ahead of a nerve impulse, i. e. by assuming a flow of current from the periparabiotic areas to the parabiotic region. The only difference is due to the fact that in this case stationary depolarisation (Wedensky's stationary excitation) is developing much more slowly.

Further experiments confirmed the assumption deduced from the local circuit theory. After arrival of the impulse at the block, an excitability increase is found beyond the block accompanied by an electric potential having a large potential decrement along the nerve. The extrinsic potential is an electrotonic effect of the action potential. A nearly identical spatial distribution was obtained by Hodgkin (1937a, b) when he applied a short catelectrotonic pulse to the blocked site. The extrinsic potential precedes the impulse wave when the latter is normally propagated along a nerve and represents the foot of the ascending phase of the spike potential. This phase is not active. The active phase follows it only after a certain period, as is also shown by simultaneous recording of the action potential and of impedance changes during a nervous impulse (Cole and Curtis 1939).

d) The law of independent conduction

Problem: Demonstrate that α and β spikes of the A group of nerve fibres in the frog sciatic nerve are conducted by separate fibres.

Principle: This demonstration is based on differences in threshold, and also utilises the refractory period (p. 372). If α and β spikes are conducted by separate nerve fibres, the test stimulus maximal for α and β fibres following a stimulus maximal only for α fibres, should produce an action potential only in β fibres, since it falls into the refractory period of the previously stimulated α fibres.

Object: Frog sciatic nerve — peroneal nerve preparation, at least 12 cm in length (*Rana esculenta* or *Rana temporaria*). Large frogs must be chosen so that separation of the α and β spikes caused by different conduction velocities in α and β fibres will be apparent.

Apparatus: A cathode-ray oscilloscope with a preamplifier. Total sensitivity 100 μ/cm. A time calibrator. A stimulator giving paired square wave pulses with independently adjustable delay and amplitudes. A radiofrequency output unit. A moist chamber and other requirements as in experiment IVAa.

Procedure: The nerve is fixed in the moist chamber in the usual way. The stimulating electrodes are placed close to the exit of the sciatic nerve from the spinal column, and the recording electrodes on the end of the peroneal nerve. Since the distal recording electrode is situated at the point where the silk loop is attached, a monophasic recording is made. A sufficiently large interval between stimulating pulses is chosen; the second stimulus is applied only after the end of the refractory period following the first stimulus. The voltage of the first stimulating pulse (s_1) is adjusted so that only α fibres are stimulated, that of the second (s_2) is maximal also for β fibres. A recording as shown in Fig. 131a and b is obtained.

With the voltage of both pulses remaining unchanged the delay between them is shortened. The action potential of the α fibres in response to stimulus s_2 gradually decreases (Fig. 131d) until it disappears completely (Fig. 131e). Only the response of β fibres to the second stimulus s_2 remains. The stimuli may be permitted to merge, but the β spike evoked by s_2 does not change.

Conclusion: This experiment, first performed by Erlanger and Gasser (1937) demonstrated that the β spike is conducted by other fibres than the α spike. According to the conduction velocity it may be judged that the β fibres have a smaller diameter (cf. p. 329).

The low of independent conduction can be very clearly demonstrated when using a nerve with reduced number of nerve fibres (Tasaki 1953). For this purpose the double-fibre (page 330) or treble-fibre of the crayfish is very suitable.

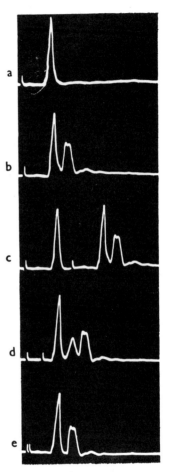

Fig. 131. Demonstration of independent conduction in α and β frog nerve fibres. The time interval between the shock is 5·5 msec. in c.

e) Interaction between nerve fibres

Problem: Determine excitability changes in one fibre when an impulse passes in another closely adjacent fibre.

Principle: Under physiological condition a nervous impulse cannot pass from one nerve fibre to another. This, however, does not mean that the impulse does not affect the excitability of adjacent fibres (Katz and Schmitt

1939, 1940). This effect can also be deduced from the local circuit theory. Imagine two fibres lying next to each other as in Fig. 132. The shaded area indicates the nervous impulse moving from left to right (fibre I). Arrows indicate the direction of current flowing ahead of and behind the impulse. Any point of the active fibre will be successively a source, sink and again source of current, a small fraction of which will flow through the inactive fibre. With

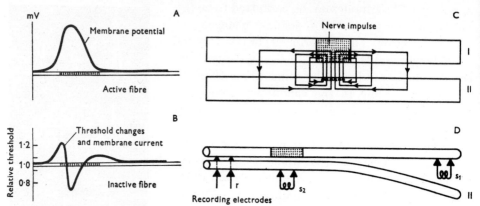

Fig. 132. Interaction between two nerve fibres. An action potential in fibre I (A) changes the threshold in fibre II as shown in B. The excitability changes follow the same time course as the second derivative of the action potential. C — course of lines of current flow in fibre II when the impulse is in fibre I (dotted area). D — diagramatic representation of experiment designed to determine the interaction between two nerve fibres.

respect to the inactive fibre, the current flow is successively anodal, cathodal and anodal. Accordingly (see p. 345), thresholds are at first increased, then decreased and finally again increased. The course of excitability changes in the inactive fibre is thus opposite to that in the active fibre.

To demonstrate this process, two fibres lying next to each other are necessary. It must be possible to stimulate one fibre separately as is shown diagramatically in Fig. 132. An impulse is evoked in fibre I with stimulating electrodes s_1. Excitability changes in fibre II are determined with stimulating electrodes s_2 during the passage of the impulse through fibre I. Action potentials from both fibres are recorded with the same recording electrodes.

Apparatus: A cathode-ray oscilloscope with a preamplifier. Total sensitivity about 100 μV/cm. A stimulator with two independent outputs, continous adjustment of pulse delay, independent regulation of the amplitude of both pulses.

Other requirements: as in experiment IVAb.

Object: An isolated double fibre (inhibitory and excitatory axons) of the claw abductor muscle of the crayfish (*Astacus fluviatilis*) — with the large and small nerve trunks (cf. p. 331).

Procedure: The object is prepared as described previously (p. 330). The nerve trunk containing the excitatory axon (I) is placed on electrodes s_1. Electrodes s_2 and r are placed on the isolated double fibre. It was shown previously that the spike amplitude of the excitatory axon is lower than that of the inhibitory one. This helps to identify the fibres. In addition the threshold is lower in the inhibitory axon (II). The fibre I is given a suprathreshold stimulus with electrodes s_1. The action potential of this fibre appears on the recording electrodes r. If a suprathreshold stimulus is applied through electrodes s_2 the action potential of both fibres is obtained on electrodes r (Fig. 133a). The larger spike is from fibre II, the smaller one from fibre I. The intensity of stimulation at s_2 is then gradually decreased. The spike of fibre I disappears first (fig. 133b), since it has a higher threshold. The spike of fibre II then also falls out (Fig. 133c). The intensity of the stimulus at s_2 is kept just

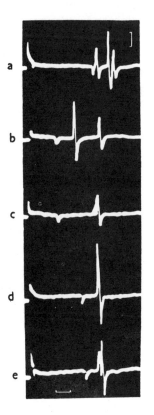

Fig. 133. Demonstration of interaction between inhibitory and excitatory axons in the double fibre of the crayfish. Voltage calibration: 2 mV. Time mark: 5 msec.

below threshold while a suprathreshold stimulus is applied at s_1. Now the interval between the stimulating pulses s_1 and s_2 is changed. If it is such that the cathode s_2 begins to stimulate with subthreshold intensity at the moment when negativity of the nervous impulse occurs in the adjacent region of the I fibre, the spike of the inhibitory fibre also appears at the recording electrodes (Fig. 133d,e). For other delays between the pulses s_1 and s_2 the spike of the inhibitory fibre disappears again.

The second experiment is performed as follows: The intensity of the stimulus s_2 is kept just at threshold value for fibre II. At s_1 the intensity of the stimulus is suprathreshold for fibre I as in the preceding experiment. If now the delay between the two pulses is changed the following results are obtained. If cathode s_2 stimulates much earlier than the I impulse wave arrives, both action potentials are registered. If the s_2 stimulus immediately precedes the arrival of the I impulse into the stimulated area the action potential from fibre II disappears. This lasts until the beginning of stimulation at s_2 coincides with the moment the I impulse arrives at the point, next to which fibre II is stimulated. At that moment the action potential of fibre II again appears. It is sufficient, however, to increase the interval between the impulses only slightly for the action potential of the inhibitory fibre to disappear again. If the interval is further increased a point is reached after which the action potential again appears.

Conclusion: The first experiment shows that the excitability of a point in fibre II increases during the passage of a spike through a corresponding point of fibre I. The second experiment indicates that for a certain period before and after the passage of nerve impulse in fibre I, a decrease in excitability occurs in fibre II. These experiments thus confirm the deduction made from the local circuit theory. In such a way the exact time course of excitability changes in fibre II could be mapped out when a nerve impulse passes in the adjacent fibre I. Fig. 132 shows such a map using a similar preparation. It is evident that the course of changes is the same as was demonstrated qualitatively in the experiment presented here.

The threshold changes shown in this experiment are small and so do not contradict the principle of isolated conduction. Impulses cannot jump from one fibre to another under natural conditions. This may be achieved artificially by increasing the excitability of the fibre e. g. by decalcification. Local currents entering the adjacent inactive fibre become suprathreshold. This was shown by Kwasow and Naumenko (1936) in the nerve-muscle preparation and Arvanitaki (1942) in the nerves of *Lolligo*. She called this artificial synapse *ephapsis*.

C. The initiation of nerve impulses

Nerve impulses may be elicited by different chemical, mechanical and electrical stimuli. In the organism either stimulation of a receptor organ or excitatory activity of preceding neurones constitutes physiological stimuli. During propagation of the nerve impulse along a nerve fibre the impulse itself is the stimulus as is postulated by the local circuit theory. Fundamentally this is an electrical stimulus formed successively along the nerve fibre by the nervous tissue itself.

Practically, electric stimuli are the most suitable artificial stimuli. They can be characterised well and produced simply and repeatedly even in a short period of time, since their action is completely reversible. For those reasons they are very frequently, or even exclusively, used when analysing nerve impulses.

Natural or artificial stimuli result in changes in the nerve membrane that make the latter more permeable to Na ions (cf. chapter I).

Experiments to be described here are intended to give practical examples of conditions under which an electric stimulus evokes excitation of a nerve. Changes occurring in nervous tissue under the influence of an electric current and the effect of Na ions on the action potential of a nerve will be demonstrated.

a) Electrotonus and nerve excitability

Problem: Demonstrate changes in excitability occurring in the frog nerve as the result of cat- and anelectrotonus. Record electrotonic potentials.

Principle: It is instructive to consider a nerve fibre as a core-conductor (Hermann 1879) when studying the effect of an electric current on a nerve. Such a conductor consists of a core (the interior of the axon) that has a small resistance or high conductivity and is separated from the extracellular space

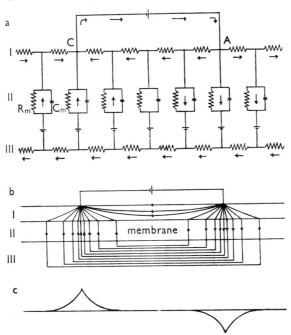

Fig. 134. a) — diagram of the electrical equivalent of the nerve membrane. Arrows indicate current flow after switching on the current in the model. On the right — anode, on the left — cathode. I. — equivalent of extracellular space, III. — intracellular space, II. — membrane. b) — current distribution in the membrane. c) — current density below the cathode and anode.

(also having a high conductivity) by a membrane composed of high resistance and capacitance units (Fig. 134a).

If a current is applied to a nerve through extracellular electrodes, it will flow as shown by arrows in Fig. 134b. Part of the current flows through the extracellular space and part across the membrane into the axon, through the conductive interior and again across the membrane to the cathode.

The density of the current flowing across the membrane is greatest at the site of application of the current, i. e. at the point of electrode contact. It decreases in an aproximately exponential manner on both sides (Fig. 134c). When the current flowing from the axoplasma across the membrane to the outside attains a critical level, a propagated response is produced. Changes occuring in the nerve at the cathode when applying subthreshold currents result in an increase in excitability. The spatial distribution of increased excitability corresponds to that of the current density, i. e. it decreases exponentially with increasing distance. At the anode opposite changes occur — excitability decreases.

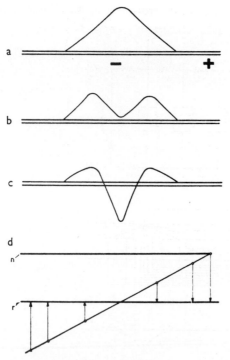

Fig. 135. Diagram of the development of Werigo's cathodal depression. a) — excitability changes immediately on switching on the cathode. b, c) — after prolonged cathode action on the nerve. d) — diagramatic representation of the anode effect on the nerve in different states of excitability according to Werigo. n — normal excitability, r — line of reversal of anodal effect. Arrows indicate the direction of excitability changes due to the anode acting at different initial states of the nerve.

In general these changes produced by a flow of current in the nerve are termed *electrotonus*. At the point where the current enters the axon anelectrotonic changes occur, whereas at the point where it leaves the axon catelectrotonic changes are encountered.

The above changes in excitability form the basis of Pflüger's rule (1859). Werigo (1883, 1901) found that the cathodic increase in excitability represents only a temporally limited part of the effect of a cathodic current on the nerve. If the cathode is permitted to act on the nerve for a longer period, the increase in excitability is not so evident as in the beginning. Finally, immediately below the cathode, a decrease in excitability, so-called cathodic depression, is produced. The zone of increased excitability shifts towards the sides as is shown in Fig. 135.

If during this cathodic depression the anode acts on the nerve, it has the opposite effect from that predicted by Pflüger's rule. It restores the decreased excitability — or in other words, raises it to normal. According to Werigo (1901) the anode effect depends upon the state of the nerve's excitability. When excitability is decreased by strong cathodic current below r (Fig. 135d) the anode returns it to normal value, i. e. it increases excitability. If excitability is increased by weaker cathodic action, the anode again returns it to value r — i. e. in this case it decreases excitability..

Woronzow (1924, 1925) discovered that the anode acts in this way (increases excitability) also if excitability is decreased by the action of a KCl solution (Woronzov's phenomenon).

Changes in membrane potential produced by an external current are termed electrotonic potentials (see Exp. IVCb).

In order to determine qualitative cat- and anelectrotonic excitability changes in a mixed nerve it is sufficient to observe the effect of electrotonus on the height of the spike amplitude after submaximal stimulation with electrodes near the cathode or anode of the polarising external current.

Object: Sciatic peroneal nerve preparation of the frog.

Apparatus: A cathode-ray osciloscope with an A. C. preamplifier, total sensitivity of at least $100\ \mu V/cm$. A stimulator giving rectangular pulses 0·5 msec in duration. A source of D. C. current (4·5 V battery with variable resistor). A microampermeter. A nerve chamber. Nonpolarisable electrodes (Fig. 70).

Other requirements- As for experiment IV-A-a. 0·15м-KCl solution.

Procedure:

1) *Changes in nerve excitability near the cathode and anode.*

The nerve is fixed in the nerve chamber by silk loops on both ends of the preparation. Electrodes are distributed as shown in Fig. 136. Electrodes p_1 and p_2 are 30 mm apart. The nerve is severed in the middle of the interpolar distance. The first stimulating electrode s_1 is situated 3 mm from electrode p_1. The remaining electrodes are 3 mm apart each from the other.

Using the pole-changing switch (k_1), electrodes p_1 and p_2 are connected so that p_1 is the cathode of the applied current. Using the second switch (k_2),

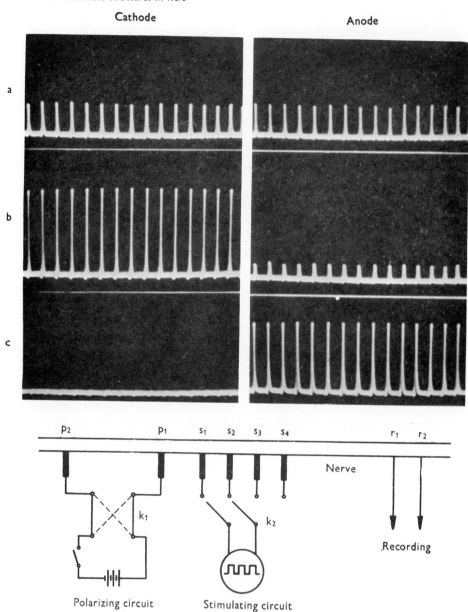

Fig. 136. Effect of cathode and anode on the excitability threshold of the nerve. a) — before the make, b) — 2 sec. after the make, c) — 3 sec. after break of the D. C. current. Below is the arrangement of electrodes for qualitative determination of excitability changes in the vicinity of the cathode and anode of the applied current. p_1, p_2 — polarizing electrodes, s_1, s_2, s_3, s_4 — stimulating test electrodes, r_1, r_2, — recording electrodes, k_1 — commutator, k_2 — selector switch.

stimulating electrodes s_1 and s_2 are connected to the stimulator. The nerve is stimulated with a train of submaximal stimuli and the result is observed (Fig. 136). The current is allowed to flow in the circuit $p_1 - p_2$, the electrode p_1 being the cathode. The amplitude of the spikes has increased considerably while the intensity of stimulation in s_1-s_2 remained the same. After switching off the polarising current, the spike amplitude decreases until it nearly disap-

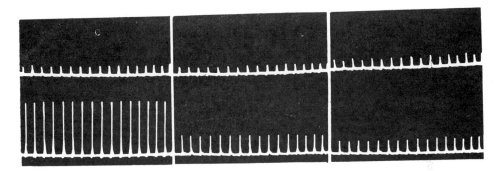

Fig. 137. Excitability changes produced by the cathode at various distances from the point of its application. Upper oscillograms: activity before the cathode is applied. Lower oscillograms: at distances 3, 6 and 9 mm from the point at which the cathode is applied.

pears (Fig. 136; *post-cathodal depression*). The experiment is repeated with p_1 as the anode. Spike amplitude decreases greatly (Fig. 136), and it increases again when the current is switched off. Using the same current intensity the experiment is repeated for different distances between the stimulating electrodes and electrode p_1 (Fig. 137).

2) Werigo's cathodic depression and the phenomenon of Woronzov.

A freshly prepared nerve is fixed as in the previous experiment. It is stimulated with a frequency of 100 c/sec through electrodes $s_1 - s_2$ using a stimulus maximal for A fibres. The intensity of stimulation is then halved (submaximal stimulus), and a polarising current of 7 μA is allowed to pass. The amplitude of the spikes increases at first. As the polarising current is permitted to continue, this increase gradually becomes less and less evident until spikes are as low or lower as at the beginning — Werigo's cathodic depression sets in. Finally the action potentials disappear.

The polarising current is switched off. The amplitude of the spikes begins to increase only very slowly and when the anodic current is switched on action potentials increase.

Woronzov's experiment is performed on another nerve. This is placed on the electrodes in a nerve chamber as shown diagramatically in Fig. 138. A stimulus maximal for A fibres is applied to the nerve through electrodes

$s_1 - s_2$ situated in the middle between p_1 and p_2 electrodes. A vessel containing 0·15M-KCl solution is placed between electrodes p_1 and r_1. At first neither anodal (7 μA) nor cathodal polarisation causes any changes. After some minutes, however, the cathode already produces a block in a large number of nerve fibres (Fig. 138b) and the anode restores spike amplitude to normal (Fig. 138c). Later, until complete block occurs, these changes aré even more pronounced.

Conclusion: The above experiments demonstrate two of three fundamental theses of Pflüger's rule (1859):

1) The cathode of an external current decreases and the anode increases the stimulation threshold in a nerve.

2) After the break of current, excitability is decreased at the cathode and increased at the anode.

3) Both effects — the excitability increase at the cathode and decrease at the anode, are maximal after the make of the current. If the excitability of the nerve is tested at various intervals after the make of the polarising current, the curve of excitability changes during a long lasting polarisation may be drawn. Decay of the initial excitability changes during the prolonged action of the same current is an expression of accomodation (Nernst 1908), i. e. of the ability of the nerve to act against the tendency of an external current to excite or to inhibit the nerve (p. 361).

When the applied current exceeds a definite threshold value a nervous impulse is initiated at the cathode. This can be demonstrated by compression of the nerve between the stimulating electrodes. If the stimulating electrode proximal to the recording ones is the anode, no response is obtained. It happens frequently, however, that the action potential does not disappear after this procedure. Consequently, especially in the older literature, statements are found that an impulse may arise also at the anode after the closure of the current and at the cathode after the opening of current. This phenomenon was explained by Bernstein (cf. Biederman 1895) by the formation of a so called *virtual cathode*. This is especially easily formed during transcutaneous stimulation of nerves in situ, i. e. lying in a volume conductor.

A virtual cathode, however, may also arise when working with isolated or exposed nerves, though much less often. The formation of such virtual cathodes is explained by irregularities in the cross-section of the interpolar segment of the nerve (Lorente de Nó 1947). If this segment has a uniform diameter, the applied current will only flow across the membrane near the cathode and anode. If, however, irregularities are present the current may also leave the axon at the narrowest stretch of the nerve between the electrodes as if a cathode were present at that point (hence the term virtual cathode).

When switching on the current an impulse may arise even if the point below the cathode is anaesthetised. With currents of sufficient strength such virtual cathodes may occur close to the anode if a branch is cut or a small drop of Ringer solution left nearby.

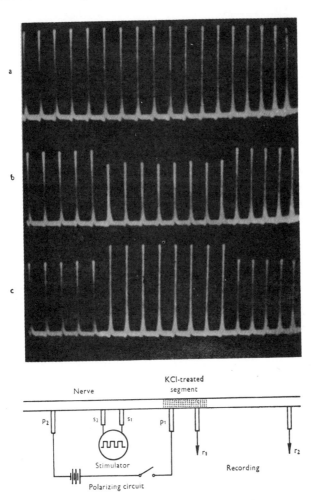

Fig. 138. Woronzov's effect. a — control responses to maximal stimuli for A-fibres; b — 2 min. 8 sec. after application of 0·15 M-KCl solution (decrease due to the cathode); c — increase due to the anode. Stimulation frequency 100 c/sec. Diagram of the arrangement of electrodes

The majority of cases described in the literature of impulses appearing after the closure of current at the anode, are due to the existence of virtual cathodes. It is, however, possible to initiate the impulse also by the action of the anode when according to Lorente de Nó (1947), the following conditions are fulfilled 1) the current is very weak, 2) the nerve is in a nearly complete rhytmical state so that its threshold is very low.

b) Electrotonic potentials. The measurement of electrical constants

Problem: Record the electrotonic potentials evoked by rectangular current pulses. Measure, using these potentials, the following electric constants of the membrane: specific membrane resistance, space constant, time constant, and membrane capacity.

Principle: The potential change occuring in a nerve or muscle in response to the action of an electric current is termed the electronic potential. Depending on the polarity of the external electrode through which current is applied to the nerve or muscle fibre the potential is either anelectrotonic or cateelectrotonic. An external anode is equivalent to an intracellular cathode; in both cases the current flows from outside into the inside. With an external cathode or intracellular anode current flow is in the opposite direction. Anelectrotonus increases the potential difference across the membrane. The membrane potential has a higher absolute value in anelectrotonus — we speak of hyperpolarization. Catelectrotonus decreases the potential difference. It has a depolarizing effect.

The electrotonic potential is a complex function of the current which evokes it. Its value depends on both the time and the distance from the stimulating electrode. This is related to the cable structure of the nerve or muscle fibre (IV-C-a).

The situation is simpler if current is applied between one internal and one external electrode in such a way that it flows equally across a large area of the membrane. This can be achieved with longitudinal electrodes (IV-A-c). The dependence of the potential on the distance is thus eliminated. The current crossing the membrane is then equivalent to the current passing through a circuit composed of a condenser with capacity C_m and a parallel resistance R_m (Fig. 134a). For current I passing throug this circuit the following is valid:

$$I = C_m \frac{dV_m}{dt} + \frac{V_m}{R_m} \tag{1}$$

where V_m is the potential difference on the condenser and dV_m/dt the rate of change of this difference. In other words the current has two components, a capacitive ($C_m \, dV_m/dt$) and an ohmic one (V_m/R_m).

If the current, resistance and capacitance are constant then the potential V_m depends on time t (measured from the moment of current application) according to the relationship derived from equation (1)

$$V_m = I R_m (1 - e^{-t/R_m C_m}) \tag{2}$$

where $e = 2 \cdot 7183$ is the base of natural logarithms.

If the current is allowed to act for a very long time the expression in brackets is very close to unity and equation (2) is reduced to Ohm's law

$$V_m = I R_m \tag{3}$$

Under these conditions the amplitude of the electrotonic potential V_m evoked by a constant current depends on the resistance R_m of the membrane. This resistance can thus be measured if we know the current passing across the membrane and if we record the size of the potential difference induced by this current. If we know the area of the membrane through which the current passes (A/cm^2) we obtain the value for the specific membrane resistance ($\Omega \cdot cm^2$). If we want to determine the membrane capacity we must record the initial changes of membrane potential induced by constant current application as follows from equation (2).

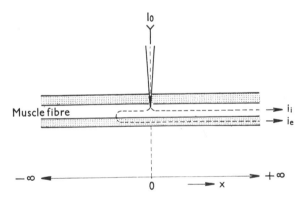

Fig. 139. Diagram of current (I_o) application with the intracellular microelectrode. i_i — current flowing through the inside of the fibre, i_e — current flowing through the extracellular fluid, x — distance from point source of current.

At time $t = 0$ the current suddenly attains value I. As this is achieved by a rectangular current pulse the method used to determine the electric constants of excitable membranes is termed square wave analysis.

Such a relatively simple way of determining R_m and C_m could be used for the giant axons with longitudinal intracellular microelectrodes (Hodgkin et al. 1952). When applying current to one point only (Fig. 139) with either an extra- or intracellular electrode the mathematical relationships are more complex since V_m varies both with time and with the distance from the source of the current.

Here differential equations define the current and the potential and can be written on the basis of certain assumptions (Hodgkin, Rushton 1946) defining the cable structure of the nerve axon or muscle fibre. These equations can be solved under certain mathematical conditions which are met by the following experimental conditions:

1) The extrapolar and intrapolar sections (distance from end of axon to electrode which applies the current or distance between anode and cathode) are sufficiently long so that they can be considered as infinite mathematically.

2) The electrodes are very thin so that their width may be considered mathematically as zero.

3) A rectangular current is applied (square wave analysis).

In square wave analysis with point application of intracellular current and with intracellular recording of the electrotonic potential the amplitude of this potential is not measured directly at the site of application but at a certain distance x from that site. Equation (4) holds for a stable state of the electrotonic potential (theoretically at time $t = \infty$ from the onset of the current but practically several hundred miliseconds) and defines the relationship between the amplitude of the electrotonic potential V_x at distance x, and the amplitude V_0 at the site of injection of the current

$$V_x = V_0\, \mathrm{e}^{-x/\lambda} \tag{4}$$

where λ is the space constant defined by equation (5)

$$\lambda = \sqrt{[r_m/(r_e + r_i)]} \tag{5}$$

where r_i and r_e are resistances per unit length (Ω cm^{-1}) of the axo- or sarcoplasma and of the extracelullar fluid respectively

r_m is the resistance multiplied by unit length of the membrane (Ω cm)

x is the distance between the current and recording electrodes (cm)

For the electrotonic potential (V_0) at the site of current application and in a steady state ($x = 0$, $t = \infty$)

$$V_0 = \tfrac{1}{2}\, r_i\, \lambda\, I_0 \tag{6}$$

where I_0 is the amplitude of the rectangular current pulse (A).

From equations (4) and (6) the basic practical equation of square wave analysis for the intracellular injection of current and for intracellular recording can be derived.

$$\frac{V_x}{I_0} = \tfrac{1}{2} r_i \lambda e^{-x/\lambda} \tag{7}$$

If r_e as a small number is ignored equation (7) is changed into (8) after substituting for λ

$$\frac{V_x}{I_0} = \tfrac{1}{2} \sqrt{(r_m r_i)} \, e^{-x/\sqrt{(r_m/r_i)}} \tag{8}$$

The equation contains two unknown — r_m and r_i. In order to find them the electrotonic potential must be recorded at least at two distances (x_1, x_2) from the microelectrode which applies the current (I_0) to the fibre and thus two values V_x (V_{x1}, V_{x2}) are obtained. This gives us two equations with two unknown and values r_m and r_i can be found using elementary calculus. Once we know r_m and r_i the total membrane resistance $(\Omega \cdot cm^2)$ can be calculated:

$$R_m = 2\pi a \, r_m \tag{9}$$

where a is the radius of the fibre in cm.

The specific resistance R_i $(\Omega \cdot cm)$ can be found from

$$R_i = \pi a^2 \, r_i \tag{10}$$

and the capacity C_m $(\mu F/cm^2)$ from

$$C_m = \frac{\tau_m}{R_m} \tag{11}$$

where τ_m (msec) is the time constant which can be evaluated in several ways (Hodgkin and Rushton 1946). Here the most frequently used method is given:

$$\tau_m = 2 \frac{\Delta t}{\Delta x} \lambda \tag{12}$$

where Δx (mm) is the distance between the two insertions of the recording electrode and Δt (msec) the time difference between the increase in amplitude of the electrotonic potential to half its final value below the first and the second recording electrode. The expression $\Delta x/\Delta t$ has a velocity dimension; τ_m is thus determined by this method on the basis of the spreading rate of electrotonus.

Object: The sartorius muscle of the frog (*Rana esculenta* or *R. temporaria*).

Apparatus: A double beam cathode-ray oscilloscope with a preamplifier and a cathode follower. The total amplification of the channel for recording the electrotonic potential is 2 mV/cm. A voltage calibrator with 1 mV steps.

A stimulator giving rectangular current pulse (pulse duration: 50 msec). A pre-amplifier for recording the current pulse (total amplification 10-20 mV/cm). An indicator of electrode insertion. Time mark 100 c/sec. An isolation unit for the stimulator.

Other requirements: Surgical instruments. Capillary microelectrodes filled with 3M-KCl. Frog Ringer solution of the following composition: 120 mM NaCl, 2·5 mM KCl, 1·8 mM $CaCl_2$, 0·85 mM NaH_2PO_4, 1·1 mM Na_2HPO_4 (pH = 7·2). Recording chamber for holding the preparation. Micromanipulator. Binocular microscope. Ocular micrometer. Objective micrometer.

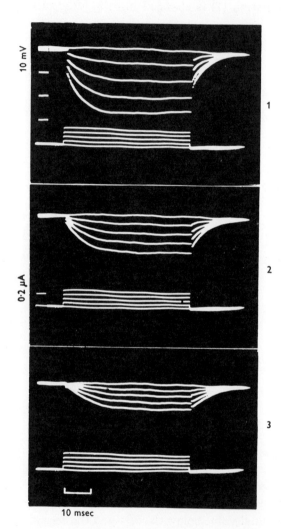

Fig. 140. Electrotonic potentials of a muscle fibre of sartorius muscle of *Rana temporaria*. Upper beam: anelectrotonic potentials, lower beam: rectangular current pulses. The oscillograms were obtained at the following distance from the current electrode: 1—72 µ, 2—890 µ, 3—1720 µ (in the sequence 1, 3, 2). In each position of the recording electrode responses to 4 current intensities were registered. The membrane potential was −80mV.

Procedure: The sartorius muscle (see page 279) is prepared without the nerve and fastened in a Petri dish with its internal surface upwards. On this side the muscle contains less connective tissue and insertion of the glass electrode into superficial muscle cells is easier. The preparation is illuminated from below and observed under a binocular microscope (total magnification 40 ×). Fluid is maintained 3 mm above the muscle in order to decrease the capacitance to

354

earth across the glass wall of the microelectrode. The microelectrodes are inserted into the same muscle cell with a micromanipulator usually under a greater magnification (100 ×). First the recording electrode is inserted into a cell on the surface and this is seen as a sudden fall in potential of about 90 mV. The microelectrode is left in place, but the cathode follower input is connected to the stimulating electrode which is inserted into the same fibre as close as possible to the recording microelectrode (about 50 μ). The membrane potential is the same or only a few mV smaller than that recorded with the recording electrode. Then the cathode follower is again connected to the recording electrode and the output of the stimulator giving rectangular inward current pulses (the tip is negative) is connected to the polarizing electrode. If polarity were reversed the fibre could be depolarized. When both electrodes are placed in the same fibre the anelectrotonic potential is recorded (see Fig. 140). After recording electrotonic potentials to current pulses of different intensities in this position the recording electrode is withdrawn and inserted at a different site in the fibre. The distance between the electrodes is measured with the ocular micrometer. Electrotonic potentials are recorded thus at least at 3 distances from the polarizing electrode (about 50, 800 and 1500 μ). Do not forget to calibrate both oscilloscope channels. During recording the shape of the current pulse is carefully noted. It happens very frequently that the pulse shape is distorted by an increase in the resistance of the microelectrode during current application (see page 224). In such cases it is better to change the electrode. Sometimes it is sufficient to pull the electrode out of the fibre and to pass a pulse of inverse polarity through it. These difficulties can be considerably reduced if the stimulating electrode is filled with K-citrate (Fatt and Ginsborg 1958).

Results: The membrane constants of the muscle cell can be found from the recorded electrotonic potentials, the corresponding currents and the calibrations. Fig. 141 shows how this is done.

The individual points on the diagram (141A) represent the ratio of the amplitude of the electrotonic potential at the end of a 50 msec pulse (ordinate) to the amplitude of the current pulse (abscissa). They were obtained from recordings similar to those shown in Fig. 140 at three different distances from the polarizing electrode. The lower curve (a) is from a distance of 70 μ, the middle one (b) from a distance of 696 μ and the top one (c) from a distance of 1390 μ from the polarizing electrode. These voltampere characteristics can also be recorded directly on the oscilloscope (Henček and Zachar 1965).

The membrane of the muscle cell behaves according to Ohm's law only with low current intensities, at higher intensities it shows rectifications. Since the theory for measuring membrane conductance by "square wave analysis" assumes the validity of Ohm's law for the studied system, the ratio is evaluated from the linear part of this relationship. The mean value for a given distance is obtained graphically from the slope of the line which connects experimental points in the interval $(-0.15; 0)$ μA.

Thus mean values of V_x/I_0 for the three above distances from the polarizing electrode are obtained.

There is a linear relationship between $\log V_x/I_0$ and the distance at which this value has been measured (Fig. 141B).

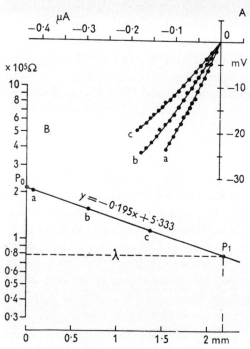

Fig. 141. A — Relation between amplitude of electrotonic potential in mV (ordinate) and intensity of current pulse in μA (abscissa) in Ringer solution. Individual points were obtained from recordings similar to that in Fig. 140 at three distances (a, b, c) from the current microelectrode. B — Dependence of the electrotonic potential to applied current ratio (V/I_0) — ordinate (log) — on the distance from the current electrode — abscissa (in mm). Points a, b, c are slopes (V/I_0) from diagram A. The straight line constructed according to the least squares method cuts the ordinate at P_0 ($= 1/2\sqrt{r_m r_i}$). The point P_1 has coordinates ($P_0/e; \lambda$).

The experimental points lie on a straight line as might be expected theoretically from equation (8) which appears as follows after being changed into the logarithmic form:

$$\log \frac{V}{I_0} = -\frac{\log e}{\lambda} x + \log \tfrac{1}{2} \sqrt{(r_m r_i)} \tag{8.1}$$

This is the equation of a straight line with the slope:

$$b = -\frac{0.4343}{\lambda}$$

where $y = \log V/I_0$; $c = \log \tfrac{1}{2} \sqrt{(r_m r_i)} = \text{const.}$

The line in the given case has been calculated by the method of least squares which gives the equation of the straight line. From it all necessary data for calculating the membrane resistance R_m can easily be obtained.

For $x = 0$ equation (8·1) is reduced to

$$\frac{V_0}{I_0} = \tfrac{1}{2}\sqrt{(r_m r_i)} \tag{8.2}$$

In this case $\tfrac{1}{2}\sqrt{(r_m r_i)} = 2 \cdot 2 \times 10^5$ and the slope $b = -0 \cdot 195$ and thus $\lambda = 2 \cdot 2$ mm while

$$\lambda = \sqrt{(r_m/r_i)} \tag{8.3}$$

After substituting the values obtained for the space constant λ and the input resistance R_0 ($= V_0/I_0$) into equations (8·2) and (8·3) values for r_m and r_i are obtained (Tab. 5). R_m can be calculated from equation (9) if the radius a of the fibre is known. In the whole muscle it is difficult to determine the fibre diameter since the fibre is partly covered by other fibres and its section need not be circular. Here the fibre radius can be calculated from equation (10) if, however, the internal specific resistance R_i of the fibre is known. This resistance is constant. For the frog sartorius muscle R_i is 250 Ω . cm (Katz 1948). From this value and $r_i = 1 \cdot 9 \cdot 10^5$ Ω . mm^{-1} the radius $a = 63$ μ was calculated. This is a reasonable figure and corresponds to values found in isolated fibres (Hodgkin and Horowicz 1959).

TABLE 5

Electrical constants of a frog muscle fibre

$1/2\sqrt{(r_m\,r_i)}$ (Ω)	λ (mm)	r_m (Ωmm)	r_i (Ω/mm)	a^* (μ)	R_m (Ωcm^2)
220000	2·2	960000	190000	63	3830

* Radius calculated from Eq. 10, assuming $R_i = 250$ Ω cm (Katz 1948).

The capacity of the membrane C_m can be calculated from the measured R_m if we know the time constant τ_m. The electrotonic potential attained half its amplitude below the second electrode $20 \cdot 10^{-4}$ sec later than below the first one. If the distance between the first and second position of the recording electrode was $626 \cdot 10^{-4}$ cm, then $\Delta t/\Delta x = 20/626 = 0 \cdot 05$ sec . cm^{-1}. Using equation (12) it is seen than $\tau_m = 2 \cdot 2, 2 \cdot 10^{-1}$. $5 \cdot 10^{-2}$ sec $= 22 \cdot 10^{-3}$ sec $= 22$ msec. According to equation (11) the capacity of the membrane can then be calculated. $C_m = 22 \cdot 10^{-3}/3830 = 5 \cdot 7 \cdot 10^{-6}$ sec Ω^{-1} . cm$^{-2} = 5 \cdot 7$ μF . cm^{-2}.

Conclusion: The recorded anelectrotonic potentials (polarity according to the accepted convention) have a wave form that can be predicted from the equivalent circuit of the membrane. Their amplitude increases linearly with the current only for very small currents. The amplitude of catelectrotonic potentials ($> 10 \, \text{mV}$), however, commences to deviate from Ohm's law (rectification).

Since the validity of Ohm's law is one of the basic assumptions of the" square wave analysis" the latter must be carried out with very small currents. Catelectrotonic potentials are less suitable for analysis since their shape begins to be obscurred by subthreshold active responses (IV-C-e) when half the threshold is reached. In addition in muscles spikes may be accompanied by a subsequent contraction with the possibility of fibre injury.

In the fashion outlined above electrical constants of nerve axons may also be obtained. Special relationships hold for the spherical structures of neuron somata (Eccles 1964a). In nerve axons conditions for analysis are more favourable than in muscle since their length is usually several times greater than the space constant λ. If the extrapolar sector is short and comparable to the space constant the relationships mentioned in the introduction do not hold.

The electrical constants in such a case are calculated from relationships that take a correction for the finite length of the fibre into consideration (Weidmann 1952). The corrections, of course, differ according to whether the ends of the short conductor are enclosed within the membrane or represent a short circuit between the extra and intracellular environments. The first example is typical for short muscle fibres which are limited by muscle membrane at their points of attachment to the tendon. In this case the following holds instead of equations (4 and 6):

$$V_x = V_0 \frac{\cosh (L - x/\lambda)}{\cosh(L/\lambda)} \tag{4.1}$$

$$\frac{V_0}{I_0} = r_i \cosh \left(\frac{L}{\lambda} \right) \tag{6.1}$$

where L is the length of the fibre and the stimulating electrode is in its centre.

Another source of errors is due to the fact that the conductor is not so thin as required by the condition of analysis. In big fibres a potential difference may arise in the transverse plane when current is applied. Such anomalous results have been found empirically by Fatt and Katz (1953a) in giant muscle fibres of crustaceans as a deviation from the exponential course in the neighbourhood of the stimulating electrode. Falk and Fatt (1964) have analysed this phenomenon theoretically.

The physical significance of the electrical constants has not been fully explained. G_m ($= 1/R_m$) is thought to be an expression of ion conductance of the membrane. In the sartorius muscle this value is in good agreement with the G_m found from resting flows of K^+ and Cl^- ions. The membrane capacity C_m as determined by square wave analysis is too high for a dielectric membrane, e. g. 20 $\mu F \cdot cm^{-2}$ in the crayfish (Fatt and Katz 1953a, Henček and Zachar

1965). These difficulties are overcome to a considerable degree by the model of the muscle fibre with two time constants (Falk and Fatt 1964).

c) Strength-duration curve. Chronaxie

Problem: Determine the relation between the strength of the stimulating current and its minimum duration necessary to excite. Measure the chronaxie of the frog's sciatic nerve.

Principle: The threshold intensity of a stimulating current depends on its duration. This relation is characteristic for various excitable tissues and is expressed by the Hoorweg-Weiss' curve (Fig. 142), which can be described by a purely empirical equation:

$$i = i_r \left(1 + \frac{ch}{t} \right)$$

in which i is the threshold current, i_r the rheobasic current and t is the duration of the current pulse. The constant ch represents the time during which a current of twice rheobasic strength must flow in order to excite, as it follows from the equation, when $i = 2i_r$. This constant termed as chronaxie by Lapicque (1926) was extensively used in the past as a measure of excitability of different tissues.

The strength-duration curve (Fig. 142) can be constructed by plotting the time needed for initiation of excitation at different stimulus strength. The

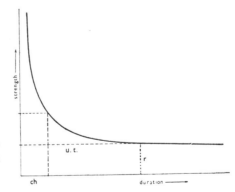

Fig. 142. Strength-duration curve. r — rheobase, ch — chronaxie, u. t. — utilisation time.

chronaxie can be estimated from the strength-duration curve or directly by measuring the rheobasic strength. Rheobase is a threshold intenzity of sufficiently long lasting stimuli.

Object: The sciatic-peroneal nerve preparation of *Rana temporaria* or R. *esculenta*.

Apparatus: A double beam cathode-ray oscilloscope with a preamplifier, total sensitivity 100 μV/cm. A stimulator giving rectangular pulses of variable duration and amplitude with the possibility of doubling the amplitude. A time and amplitude marker.

Other requirements: A nerve chamber with one pair of platinum (recording) and one pair of Ag-AgCl (stimulating) electrodes.

Procedure: The preparation is placed into the chamber on the recording and stimulating electrodes in the usual fahion. The threshold for a pulse of 100 msec duration is determined. The pulses are then progressively shortened and for each time interval the threshold intensity is determined. The appearance of slight activity on the oscilloscope screen indicates the threshold. The threshold is that stimulus intensity which just fails to evoke activity. The intensity and duration of the pulse are found by photographing it from the oscilloscope together with amplitude and time markings for each point of the curve. If the duration of the pulse is graduated on the stimulator, photographing is not necessary, and strength values may be expressed in relative figures. The stimulator, however, must be calibrated in volts, amperes or arbitrary units.

Chronaxie is determined on the same nerve. A pulse of 100 msec duration is applied and the threshold determined. Using the voltage or current doubler the pulse amplitude is increased to twice the threshold value. A large action potential appears on the screen. The duration of the pulse is then shortened and the value at which the action potential just disappears is found. This time interval is the chronaxie.

Conclusions: The paired values obtained and plotted onto paper give the strength-duration curve. The chronaxie may be determined graphically from it. On comparing it with the value determined directly good agreement is found.

The chronaxie determined in the above experiment is that of the most excitable nerve fibres.

Before the development of electronic stimulators, chronaxie was determined with simple chronaximeters in which the duration of the stimulating pulse was determined by the duration of a condenser discharge. Chronaxie was then calculated from the equation:

$$ch = \frac{1}{e} RC = 0{\cdot}37\ RC$$

where ch = chronaxie (msec),
  C = capacity (μF),
  R = resistance of the discharge circuit (k Ω).

If the tissue and electrode resistances are to be neglected, a constant current stimulator (p. 43) must be used. A high internal resistance is chosen, so that

360

to obtain chronaxie, capacity C must be multiplied by a whole number, thus simplifying calculation.

d) Accomodation

Problem: Determine the accommodation constant λ in frog nerve.

Principle: In addition to a certain strength and duration, an electrical stimulus must also have a minimum rate of rise in order to produce an impulse.

Evidently, this phenomenon is related to the fact that the size of excitability produced at the cathode is not maintained but declines to a lower level (Exp. IV-C-a). There are a number of theories attempting to explain this phenomenon (Monnier 1930, Schriever 1931, Rashevsky 1933, 1948, Hill 1936, Polissar 1954, Hodgkin 1951, Tasaki 1959).

The measurement of the accomodation constant λ, is based on Hill's theory of excitation. When a current pulse is applied to a nerve two processes occur, which take place simultaneously, but with different time constants. A state of excitation (V) is build up with a time constant k, accompanied by an opposing process, which manifests itself by an exponential increase of the excitation threshold (U) with a time constant λ. If the increase in the former process (V) is slower than in the latter (U) excitation does not occur (Fig. 143, V_1). It does occur, however, if the increase in the state of excitation is faster than the threshold rise (Fig. 143, V_2).

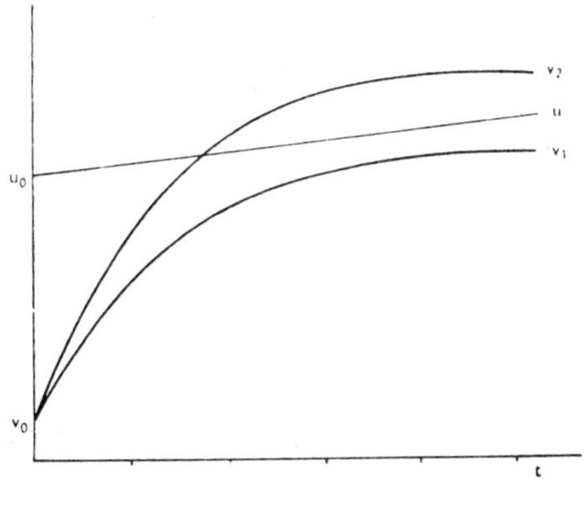

Fig. 143. Origin of excitation in a nerve according to Hill's theory. u — rise in excitability threshold. v_1 — development of a state of excitation for subthreshold stimuli, v_2 — for supraliminal stimuli.

The time constant λ can thus be regarded as an expression of the accomodative ability of excitable tissue. The accomodation constant λ should not be confused with the space constant λ of 'square wave analysis' (Exp. IV-C-b).

In order to determine accomodation constant λ, a curve of the threshold increase U must be ploted by determining the threshold values for stimuli with various rates of rise (Solandt 1936).

In practice measurements are made according to the principle shown in Fig. 144. Resistance R_1 and R_2, a capacitance C and a switch are connected to the square wave stimulator output. The slowly increasing voltage across the condenser C is used for stimulation (for discussion of similar RC circuits see p. 51). The capacitance C is variable and with its aid the rate of increase of the stimulus can be regulated. At a certain value of R_1, the threshold is determined without the capacitance C (E_0) and then with the capacitance C, i. e. with a definite time constant (E).

The values (E/E_0) are then plotted against the time constant RC of the stimulus rise (Fig. 145). The slope of the curve is determined and its reciprocal gives the value of λ.

Apparatus: A cathode-ray oscilloscope with a preamplifier, total sensitivity 100 μV/cm. An amplitude and time marker. A stimulator giving rectangular pulses. A circuit for obtaining exponentially rising pulses which can be connected to the output.

Other requirements: as in Exp. IV-A-a.

Object: The sciatic-peroneal nerve preparation.

Procedure: The nerve preparation is fixed in the nerve chamber in the usual way (p. 312c). A pair of platinum recording electrodes and a pair of non-polarisable Ag-AgCl stimulating electrodes are placed on the nerve. The thres-

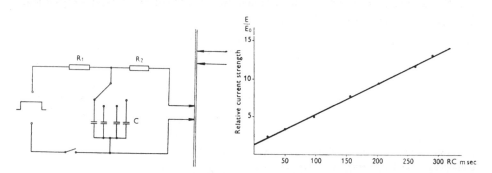

Fig. 144. Electrical circuit in the output of a rectangular pulse stimulator for determining accommodation. R_1, R_2 — resistances, C — variable capacitors, r — recording electrodes.

Fig. 145. Accommodation curve. E_0 — threshold voltage of a rectangular pulse, E — threshold voltage of a pulse with RC circuit inserted. RC — time constant of pulse rise.

hold is determined with a given value of the variable capacitance C starting with small values of C. The threshold current or voltage E is read from the amplitude regulator at the stimulator output. The threshold E_0 is then determined without the condenser. The time constant of the rate of rise may be obtained by recording and photographing the pulse on the screen of the oscilloscope and the time constant R/C is then calculated from the pulse shape. The time constant can also be computed directly from the R and C values of the circuit. The whole procedure is repeated several times with different C values. The values obtained are plotted onto a graph with (E/E_0) in arbitrary units of the output amplitude control on the ordinate and the time constant RC in msec on the abscissa (Fig. 145). A curve is drawn through those points. The reciprocal value of the slope of the curve is λ (in msec).

Conclusion: The determined value of λ for the frog nerve in this experiment is 21 msec. The time constant k can be calculated from the chronaxie determined in the preceding experiment, since a simple relation exists between k and the chronaxie ($k = 1\cdot44\ ch$). The time constant k is then $0\cdot25$ msec, i. e. nearly two orders smaller than the accomodation constant λ. The accomodation constant λ varies a great deal in different excitable tissues and is smaller in motor than in sensory nerve fibres. The accomodation of motor nerve fibres is thus greater than in the afferents.

Hodgkin (1951) attributed the process of accomodation to an increased potassium conductance of the neuronal membrane during prolonged stimulus action.

Accommodation can also be measured with other methods. Values thus obtained can be converted to λ by simple calculation (Schaefer 1940).

e) The local response of a nerve

Problem: Record the local response in the giant axon of the crayfish at the site of subthreshold stimulation.

Principle: The propagated action potential appears only when the intensity of the stimulus under the cathode attains the threshold value. A subthreshold stimulus evokes two kinds of changes at the site of stimulation. Weak subthreshold stimuli evoke electrotonic potentials (IV-C-b) which are of equal shape and amplitude but of different polarity under both the cathode and anode. Stimuli approaching the threshold value induce a local subthreshold potential under the cathode. These potentials already represent an active response of the nerve but differ significantly from the action potential since their amplitude is stimulus dependent and they are propagated with a decrement. Hence they can only be recorded near the stimulating cathode. Since their amplitude is 5—10 times lower than that of the action potential it is practically impossible to record them from the whole nerve where the amplitude of the action potential itself is considerably lowered by the short-circuiting effect of the inactive tissue and of the interstitial fluid. Hence they were first recorded from the isolated fibre (Hodgkin 1938) where the stimulating cathode served at the same time as one recording electrode. This electrode was connected to earth and located in the middle between the anode and the second recording electrode. Those performing experiment IV-A-b can record the subthreshold response in the same preparation and in the same way using extracellular electrodes.

In the present experiment a larger axon must be chosen so that it can be stimulated with an intracellular microelectrode and the subthreshold response can be recorded in its close vicinity with a second intracellular microelectrode.

Object: The medial or lateral giant axon in the nerve cord of the crayfish (*Astacus fluviatilis*). For preparation see Exp. IV-A-c.

Apparatus: The same as in experiment IV-C-b.

Other requirements: The same as in experiment IV-C-b and IV-A-c. An ocular micrometer.

Procedure: The nerve cord is prepared and fixed in the recording chamber as described in experiment IV-A-c. This procedure is more difficult because two intracellular electrodes are used. First the microelectrode serving as the recording electrode is inserted. Then the input of the cathode follower is connected to the stem (p. 209) of the second microelectrode which will serve for stimulation of the axon and this is introduced into the fibre with a second micromanipulator at a distance of about 50 μ from the recording electrode. The input of the cathode follower is then connected to the recording electrode again. If the recording electrode has not been pushed out of the fibre during the insertion of the stimulating electrode the position of the beam will correspond to the resting potential of the fibre. Now the output of the stimulator can be connected to the stimulating electrode. The stimulator is set to give rectangular pulses lasting about 25 msec and the intensity of the stimulus is increased. If both electrodes are in the same fibre we record the electrotonic potential. The stimulus intensity is gradually raised and we record for equal stimulus intensities the responses to the depolarizing pulses (Fig. 146a — outward current flow, positive pulse) and to the hyperpolarizing pulses (inward current flow; the stimulating intracellular electrode serves as cathode). As we approach the threshold, the response to the depolarizing pulse starts to deviate from the electrotonic potential: a local response appears on its ascending phase, the amplitude of which rises with the stimulus intensity. The response to the hyperpolarizing pulse retains its shape of an electrotonic potential (Fig. 146b). When the local potential attains the critical level a spike is generated, the latent period of which is shortened as the stimulus intensity increases (Fig. 146a). On subtracting the response to the hyperpolarizing pulse from that to depolarization of the equivalent intensity we obtain the local subthreshold response in its pure form uncomplicated by the electrotonic potential. Here it is better to use at least 3 times greater amplification.

In the same way we record the responses to both current directions also for shorter pulses. Then we set the right subthreshold value of the stimulus, pull the recording electrode out of the fibre and insert it at distances of 100, 300, 500 and 1000 μ from the stimulating electrode so that the spatial spreading of the local response can be studied.

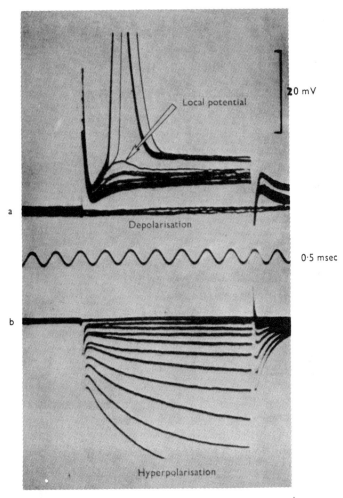

Fig. 146. Response of the lateral giant axon of the nerve cord of *Astacus fluviatilis* to application of rectangular current pulses. a: depolarization of fibre by intracellular anode. b: hyperpolarization with intracellular cathode. The distance between the stimulating and recording electrodes was 50 μ.

Conclusion: It follows from the experiment that subthreshold shocks induce two different effects. A weak cathodic stimulus or an anodic one of any strength evoke an electrotonic potential. When the cathodic stimulus approaches the threshold value another negative wave is generated — the local response — which is gradual, local and disappears several mm from the stimulating electrode (propagation with a decrement). The action potential appears only when the subthreshold local potential has attained a certain critical amplitude — the so called threshold (or critical) depolarization. Threshold depolarization is practically independent of the duration of the stimulus although for short stimuli a greater stimulus intensity is required to attain this critical level. That the threshold remains constant with stimuli of different duration can be also demonstrated when using uniform stimulation of the membrane with longitudinal electrodes. Under these conditions the subthreshold response is not complicated by the spatial factor (Tasaki 1956, 1959).

The subthreshold response is considered to be a sign of the commencing regenerative process which does not turn into the fully developed spike, however. According to the sodium hypothesis (IV-C-f) the subthreshold response is ascribed to a small increase in sodium conductance resulting in a small decrease in impedance (Tasaki 1959). The subthreshold response is accompanied by changes in excitability which had been described before the potential changes were discovered (Katz 1937, 1939).

f) The effect of sodium ions on the action potential. The sodium hypothesis

Problem: Demonstrate the effect of lack of sodium ions on the action potential of the giant axon in the nerve cord of the crayfish.

Principle: The rise of the action potential above the value of the resting potential (see Exp. IV-A-c) is explained, according to the sodium hypothesis (Hodgkin and Katz 1949), by the relatively high permeability of the membrane to sodium ions during excitation in comparison to the resting state (see p. 35). As a result of this sodium ions move into the axon interior along the concentration gradient if the concentration of sodium ions in the external medium of the axon $[Na]_o$, is greater than in the axoplasm, $[Na]_i$. The amplitude of the overshoot of the action potential should thus approach (at a temperature of 21°C) the equilibrium potential given by Nernst's equation:

$$V_{Na} = 58 \log \frac{[Na]_o}{[Na]_i} \tag{1}$$

which, e. g. for 10 times higher $[Na]_o$ than $[Na]_i$ would represent 58 mV. Sodium concentration gradients of such size have been found in many excitable structures (Hodgkin 1951).

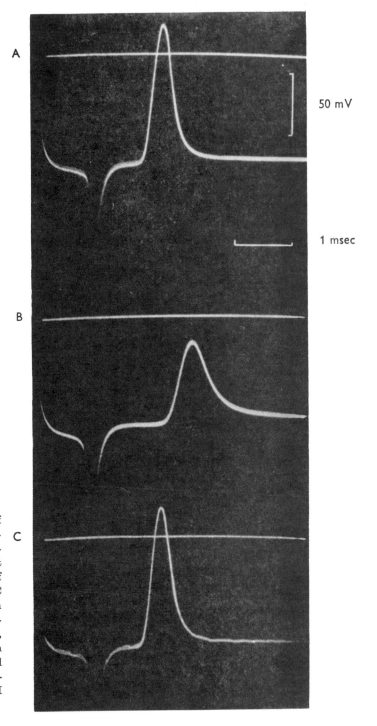

Fig. 147. The effect of sodium ions on the amplitude of the action potential in the lateral giant axon of the nerve cord of *Astacus fluviatilis*. A, C — action potentials in van Harreveld's solution($[Na_o]$ = 208 mM-Na), B — action potential in solution with a reduced outside sodium concentration ($[Na_o]$ = 26 mM - Na).

It follows from this hypothesis that the amplitude of the action potential should be dependent on $[Na]_o$ and should vary linearly with log $[Na]_o$.

Object: The medial and lateral giant axons in the nerve cord of the crayfish (*Astacus fluviatilis*).

Apparatus: As in experiment IV-A-c.

Other requirements: Van Harreveld's solution and solutions in which 1/2, 3/4, 7/8 and 15/16 of the NaCl are replaced by dextrose (20 mM dextrose is the osmotic equivalent of 10 mM NaCl). Perfusion arrangement making possible the exchange of solutions in the canal in which the nerve cord is situated (see Exp. IV-A-c).

Procedure: This is similar to Exp. IV-A-c. Before inserting the microelectrode we fix the holder of a glass capillary (diameter 1 mm) to the microscope stage. Through the capillary the test solutions will be applied. A holder for a suction capillary is also fixed to the stage. This capillary serves also as a regulator of the level of solutions in the canal. Hence it is necessary for its tip to be gradually movable vertically. The rate of inflow and outflow is regulated in such a way that the preparation does not move. We also check that the preparation does not move by exchanging the normal solutions. This test exchange is performed from the second reservoir of van Harreveld's solution. The above tests are important since the microelectrode must remain in the axon during the exchange.

Then the microelectrode is fixed, its resistance and its tip potential is measured as in Exp. IV-A-c and then it is inserted into the fibre. An action potential is evoked with a stimulus about twice the threshold value. Several action potentials are recorded and the flow of van Harreveld's solution is exchanged for the test solution containing less sodium. The amplitude of the action potential rapidly begins to fall to a certain level where it remains, indicating that an equilibrium state has been attained (Fig. 147). The less sodium in the perfusion fluid the lower the amplitude of the action potential in the steady state conditions. After recording the action potentials the test solution is again replaced by van Harrenveld's solution. The spike amplitude returns to the initial value. In this way solutions with different sodium concentrations are tested.

Finally the amplitude and time calibrations are recorded.

Conclusion: The reduction of sodium in the external medium decreases both the amplitude and the maximum slope of the action potential. The resting potential changes only slightly. The change in the active membrane potential should be linearly proportional to the logarithm of the sodium ion concentration in the test solution as follows from equation (1) if the active membrane behaved in the physico-chemical sense as an ideal sodium electrode. The dependence on increased $[Na]_o$ can unfortunately not be tested without changing the osmolarity of the solution. From the graph it is evident (Fig.

148) that agreement with the theoretical straight line is very good as was shown in the axons of lobsters (*Homarus americanus*) and of other types of crayfish, *Orconectes virilis* and *Procambarus clarkii* (Dalton 1957, 1959). In the giant axons of the squid the agreement is good for smaller changes in $[Na]_o$ but the deviation from linearity at greater concentration of $[Na]_o$ is considerable (Hodgkin and Katz 1949).

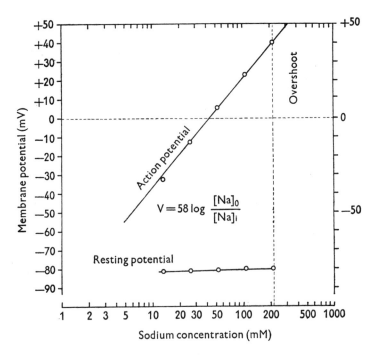

Fig. 148. Dependence of the action potential amplitude on the external Na concentration (log scale). Individual points denote the amplitude changes attained in the steady state conditions. The slope of the line is 58 mV/tenfold change in Na concentration.

The experiment shows that Na ions are indispensable for the occurrence of an action potential also in the axons studied. Since the first demonstration on the giant axon of *Loligo* (Hodgkin and Katz 1949) this fact has been demonstrated in a number of excitable nervous and muscle tissues (for review see Hodgkin 1951, 1958, 1963) and this together with other facts has led to the generalization that the action potential is formed by a sodium current. It has been ascertained that part of the ionic current in experiments with the voltage clamp method is formed by sodium current (Hodgkin et al. 1952) and the inside of axons has been found to be enriched with sodium after activity (Keynes 1949, Rothenberg 1950, Grundfest and Nachmansohn 1950). An exception is the muscle of crustaceans where the primary inward current is probably carried by Ca^{++} ions (Fatt and Ginsborg 1958) and the plant cell *Chara* (Gaffey and Mullins 1958) where Cl ions leave the vacuolar sap (see also p. 275).

D. Recovery processes following a nerve impulse

During the passage of a nerve impulse through a certain stretch of the nerve fibre, the interior of the axon gains a certain amount of sodium and loses a corresponding amount of potassium. The quantities concerned are so small that the distribution of ions on both sides of the membrane is changed only insignificantly and further nerve impulses may pass through the axon. After prolonged activity it is nevertheless necessary to restore the original ionic composition. This is achieved by active transport of ions against concentration gradients which occurs in the resting axon and requires metabolic energy. This is the recovery cycle of the nerve fibre in general. In the narrower sense the recovery cycle of the nerve impulse signifies changes occurring in the membrane immediately after the termination of the action potential and lasting tens of msec. During this period membrane permeability becomes normalised and the membrane becomes again capable of generating a new impulse. Electrophysiologically this period is characterised by after-potentials and by changes in excitability as absolute refractory period, relative refractory period, supernormal period, subnormal period and return to normal excitability.

From the physiological point of view, excitable tissues can be characterised by their ability to give a maximum number of responses per unit time. This maximal frequency is given by excitability changes following a nerve impulse.

Experiments described in the following section are concerned with this recovery cycle in the narrower sense of the word.

a) After-potentials

Problem: Record after-potentials in the sciatic nerve of the frog following a single stimulus and a train of stimuli; demonstrate the effect of veratrine on the negative after-potential.

Principle: In addition to the spike potential, a sequence of small changes in membrane potential may be recorded during the passage of a nerve impulse if high amplification and a slow sweep are used. They can be recorded only in monophasic leads. The *negative after-potential* is represented by slow decrease of the spike negativity towards the original level. With a slow sweep its beginning on the descending limb of the spike potential is well in evidence. This potential is followed by a *positive after-potential.*

370

After-potentials are a very sensitive component of a nervous response. They change during repetitive stimulation and when the external medium of the nerve is altered. The negative after-potential changes vary considerably under the influence of veratrine.

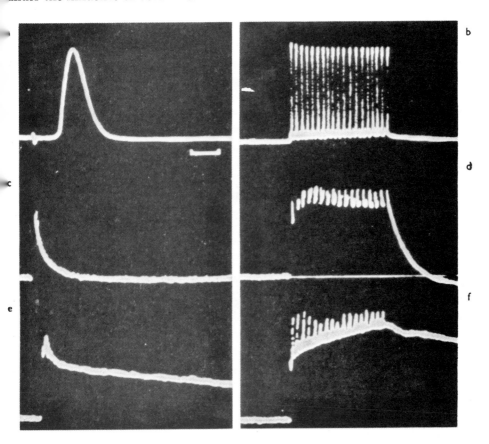

Fig. 149. After potentials in the frog sciatic nerve. a — monophasic action potential. c — the same on fivefold amplification and slower time base. b-d — after-potentials on repetitive stimulation. e — after-potential in veratrine-treated nerve to one stimulus. f — same as e), with repetitive stimulation. Time scale: for a) 1 msec, for b-f) 40 msec.

After-potentials are accompanied by corresponding changes in excitability. During the negative after-potential an increase, and during the positive one a decrease in excitability can be observed.

Object: The sciatic-tibial or sciatic-fibular nerve preparation of *Rana esculenta* or *Rana temporaria*.

Apparatus: A cathode-ray oscilloscope with a preamplifier (D. C. or with a long time constant); total sensitivity 100 μV/cm. A stimulator giving at its output either a single stimulus or a train of electric shocks (10—20) with a variable frequency. Time marker; voltage calibrator.

Other requirements: A vessel for veratrine application to the nerve. Veratrine (1 : 50 000).

Procedure: The preparation (sciatic-tibial nerve) is fixed in a nerve chamber (cf. experiment IVAa.) and placed on one pair of stimulating and one of recording electrodes. The nerve is crushed betwen the recording electrodes. It is then stimulated with a single electric pulse and the monophasic maximal action potential is recorded (Fig. 149a). The sweep is slowed down so that the ascending and descending limbs of the spike potential fuse together and the amplification is increased fivefold (Fig. 149c). The crest of the spike is now beyond the screen of the oscilloscope. A slowly falling negative after-potential changing to prolonged positivity can be seen on the descending part of the spike potential.

The nerve is then stimulated with a volley of 20 pulses with a frequency of 140/sec. and the responses are recorded (Fig. 149b, d).

Subsequently the nerve is immersed in veratrine (1 : 50 000) and the response of the nerve to a single stimulus (149e) and a train of stimuli is observed at 15 min. intervals (Fig. 149f).

Conclusion: During tetanic stimulation negative after potentials of the individual responses at first summate, but the effect of the positive after-potentials predominates. After interruption of stimulation, a large positive afterpotential is observed following the last response.

In 1938 Gasser showed that during long lasting and frequent stimulation, two components can be seen on the positive after potential. Sometimes these are separated by a negative hump. In Gasser's terminology the first component P_1 is connected with the spike process, the second component P_2 with the negative afterpotential.

As follows from the recordings 149e, f, veratrine enormously enhances and greatly prolongs the negative afterpotential. Since veratrine increases oxygen consumption and heat production in the nerve, afterpotentials are considered to be connected to metabolic recovery processes in the nerve after the passage of a nerve impulse. This is indicated by their sensitivity to changes in the compositon of the external medium of the nerve and also by changes in excitability after a nerve impulse.

b) The absolute and relative refractory period. The supernormal and subnormal period

Problem: Determine the absolute and relative refractory period, and the supernormal and subnormal period of A-fibres in the sciatic nerve of the frog.

Principle: The nerve is stimulated with a stimulus maximal for A fibres. This stimulus is termed conditioning stimulus. After a variable time interval, he nerve is excited with a test stimulus of variable intensity. If the time nterval between those stimuli is sufficiently long, the same amplitude of ction potentials is obtained, provided the intensity of both stimuli is the ame. The amplitude of the action potential produced by the test stimulus is

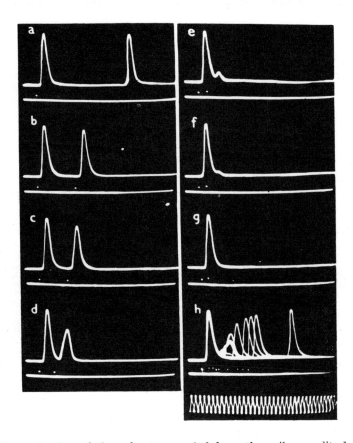

Fig. 150. Determination of the refractory period from the spike amplitude (a-g). The lower beam shows the intervals between stimulating pulses. Time mark: 1000 c/sec. Oscillogram h — the spike amplitudes at different intervals following the conditioning stimulus correspond to the course of the decrease in excitability during the relative refractory period.

decreased when the interval between stimuli is shorter than the relative refractory period. If the test stimulus falls into the absolute refractory period, no impulse is produced by its action.

The relative refractory period is due to a decrease in excitability of the nerve fibres. If the curve of changes in excitability during the relative refractory period is to be plotted quantitatively, the nerve must be stimulated at different intervals after the conditioning stimulus with a test stimulus the intensity of which must be adjusted so that the same action potential is obtained. The relative refractory period is followed by further, less pronounced excitability changes. Excitability is increased at first, so that a less intense stimulus, as compared with the conditioning stimulus, is required to produce the same spike amplitude. This supernormal period is followed by a period of subnormal excitability.

Apparatus: A double beam oscilloscope with one preamplifier, total amplification 100 μV/cm. A pair of recording and a pair of stimulating platinum electrodes. A stimulator giving paired rectangular pulses (variable delay between pulses and independently variable amplitude of both pulses). Time marker 1000 c/sec. A voltage calibrator.

Other requirements: As in experiment IVAa.

Procedure: The nerve is fixed in the moist chamber using silk loops tied to both ends. The latter are fixed to glass hooks and the nerve is slightly stretched. Stimulating electrodes (*s*) are placed at one end of the nerve. At the other end recording electrodes (*r*) are applied. The stimulator output is connected simultaneously to the (*s*) electrodes and to the input of the second channel of the oscilloscope so that the amplitude of the pulses can be observed.

First the nerve is stimulated with the conditioning stimulus alone, so that a maximal action potential for A fibres is just obtained. The sensitivity of the calibrating beam of the oscilloscope is increased so that the amplitude of the stimulus can be easily observed. The time interval between the pulses is then adjusted to about 20 msec. and a test stimulus of the same intensity is applied (Fig. 150a). It can be seen that the amplitude of both action potentials is the same. Maintaining the same pulse strength, the delay of the test pulse is gradually decreased. Starting from a certain delay, the amplitude of the action potential produced by the test stimulus begins to decrease. This is one limit of the relative refractory period. Finally the action potential disappears. This is the end of the absolute and the beginning of the relative refractory period (Fig. 150b, c, d, e, f, g).

The experiment is repeated, but this time the strength of the test stimulus is increased while shortening the time delay until the original amplitude of the action potential is obtained (Fig. 151).

The increase in amplitude of the test stimulus is calculated from the tracing for each delay used as percentage of the amplitude of the conditioning stimulus. The existence of the supernormal period is demonstrated as follows: With a 200 msec interval between the pulses a threshold intensity of the test pulse is chosen. With unchanged pulse strength the delay is decreased. At a certain delay the threshold stimulus becomes a suprathreshold one.

The percentage decrease in the excitation threshold is determined for the supernormal period as in the case of the relative refractory period.

The subnormal period can be demonstrated in a similar way.

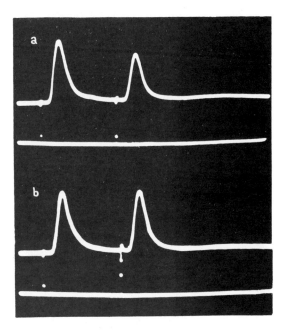

Fig. 151. A study of the refractory period according to changes in excitability. a) test stimulus of equal intensity as the conditioning stimulus in the relative refractory period. b) necessary increase in test stimulus for overcoming the decrease in excitability during the same time interval.

Fig. 152. The course of excitability changes after a nerve impulse. Ordinate: excitability in terms of resting threshold set at 100%. Abscissa: intervals between conditioning an test stimuli on a relative time scale.

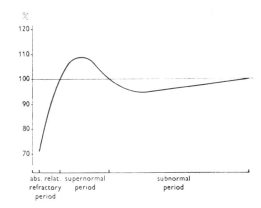

If the obtained values are plotted on a graph a curve as shown in Fig. 152 is obtained.

Conclusion: The curve shows that for 1·5 msec. after the stimulus the frog nerve is absolutely refractory since no impulse can be produced in it even when using very strong test stimuli. After 1·5 msec the nerve is relatively refractory since an impulse can be produced only by applying more intense test stimuli. The relative refractory period ends after 4 msec when the threshold has returned to normal. The supernormal period, with increased excitability, then begins. This phase lasts 15—40 msec; the excitability threshold then returns to normal a second time, and is followed by a period of decreased excitability, the subnormal period. Such a course of changes can be observed in the majority of excitable tissues. The supernormal period is very dependent on the composition of the external medium (Adrian 1920, Lorente de Nó 1947).

c) The maximal rhythm of a nerve fibre. The lability of Wedensky

Problem: Determine the maximum stimulus frequency to which the isolated inhibitory and excitatory nerve fibres of *Astacus fluviatilis* can respond.

Principle: The maximum number of impulses which a nerve is capable of transmitting in unit time in agreement with the rhythm of stimulation is termed the maximal rhythm of the nerve. The maximum rhythm is a measure of the *"lability"* of excitable tissue (Wedensky 1884). It is mainly, but not solely, determined by the absolute refractory period, since it also includes cummulative changes occuring during longer lasting prosesses of the recovery cycle of the nerve impulse.

Maximal rhythm differs for different tissues.

If the stimulus frequency is greater than the maximal rhythm, the nerve does not respond to every stimulus, but transforms the rhythm of responses to a new one. At very high frequencies it only reacts with one response at the beginning. An activity similar to that due to a direct current or spontaneous activity may be started.

In general, determination of the maximal rhythm of a nerve requires application of maximal stimuli with variable frequencies to the nerve.

Transformation phenomena are very clearly demostrated when using a constant frequency of maximal stimuli close to the maximal rhythm while simultaneously cooling, thus decreasing the "lability" of the nerve (in the sense used by Wedensky).

Object: The double fibre with the small and large nerve of *Astacus fluviatilis* as in experiment IVBb.

Apparatus: A cathode-ray oscilloscope with an A. C. preamplifier, total sensitivity $100\,\mu V/cm$. A stimulator giving rectangular pulses with a variable frequency and amplitude (an audio oscillator is also sufficient). Time marker.

Other requirements: A fixing bridge (Fig. 116). Cooling arrangement (page 336).

Procedure: The preparation is placed in the fixing arrangement. The recording electrodes are placed on the double fibre, the stimulating electrodes once on the large and the second time on the small nerve.

The stimulator is connected to the stimulating electrodes. The threshold is determined and then the inhibitory fibre (the small nerve) is stimulated with a short train of stimuli of twice the threshold strength. The frequency is gradually increased (Fig. 153). It can be seen that the inhibitory axon is capable of transmitting 100, 150, 200, 300 and 350 c/sec without transformation. On applying 400 c/sec transformation commences and is very regular, since every third action potential is dropped. There is always a longer time interval of about 4·3 msec between the second and third action potentials, while the first two action potentials are separated by about 2·8 msec. The resultant frequency is about 290 c/sec. When applying 350 c/sec it can also be seen that the one but last action potential dropped out of the series of responses traced in Fig. 153e. This indicates that the inhibitory fibre is not capable

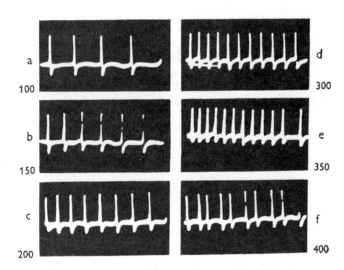

Fig. 153. Determination of the maximal rhythm of the inhibitory fibre of *Astacus fluviatilis*
Stimulation frequency in cps.

of responding with the maximal rhythm for prolonged periods. The maximal rhythm of the excitatory fibre is found in a similar way. It is 400 c/sec.

Conclusion: The maximal rhythm of the inhibitory fibre is 350 c/sec., and that of the excitatory fibre 400 c/sec. The transformation of the response to the 400 c/sec. stimulus may be explained on the basis of excitability changes during the recovery cycle following the nerve impulse. The second impulse of the train falls into the relative refractory period of the preceding impulse. This results in a prolongation of the latent period of the second response. Hence the time interval between the first two spikes is longer than in the preceding oscillogram, when a frequency of 350 c/sec was applied. The third stimulus falls into the absolute refractory period of the second impulse and consequently no action potential is produced. Changes in the spike amplitude corresponding to changes in rhythm of responses may also be observed. At higher frequencies of 300 to 350 c/sec, the action potential is about $1/_7$ smaller than at a frequency of 100 c/sec. The dependence of the amplitude on the time interval between individual spikes is evident also when stimulating with a frequency of 400 /sec. The amplitude of a spike potential produced after a longer interval is greater than that of one occurring after a shorter time interval.

The same experiment can be performed in a frog sciatic nerve, but a large number of nerve fibres of different lability makes impossible the exact determination of the maximal rhythm.

E. Electrophysiology of the isolated skeletal muscle

In comparison to the isolated nerve, the isolated muscle is relatively much less used for studying fundamental questions of excitability by extracellular techniques. The main reason is the contraction accompanying the membrane excitation, which may produce artifacts during recording.

In intracellular recording, this drawback is, however, compensated for by a large size of muscle fibres in comparison with nerve fibres. When resting potential (Exp. III-A-c) and subthreshold potential changes (Exp. IV-C-b) are to be followed, the muscle is an object of choice. In some circumstances the contraction can be suppressed leaving the action potential mechanism intact (Exp. IV-E-b). The isolated muscle is then the most suitable object for mastering the techniques of intracellular recording and stimulation.

A muscle may be activated either *directly*, i. e. by directly stimulating muscle fibres with an artificial stimulus, or *indirectly* by a physiological stimulus, a nerve impulse.

In this section only direct stimulation of a muscle will be considered. Indirect stimulation is described in connection with problems of neuromuscular transmission (p. 405). Chapter V deals with problems of the recording of action potentials from muscle fibres in situ.

All the characteristics of excitable tissues as described for nerve fibres can be demonstrated in muscle. In addition, it is possible to study the mechanism of activation of the contractile process by the muscle action potential.

a) The action potential of a skeletal muscle

Problem: Record the biphasic and monophasic action potential of the frog skeletal muscle evoked by direct electrical stimulation, and determine the latent period of the mechanical response.

Principle: The method of stimulation and recording is the same as for nerves. In order to eliminate indirect stimulation of the muscle through nerve endings, the site of stimulation on the muscle must not contain nerve endings and plates. The muscle is stretched so that only isometric contractions with minimum movement occur.

Both the muscle action potential at the point of stimulation and the muscle contraction must be recorded simultaneously in order to determine the true latent period of the mechanical response.

Object: The sartorius muscle of *Rana esculenta* or *R. temporaria*. This muscle is particularly suitable since its end-plates are concentrated at one or only a few points, and certain parts of the muscle are free of nervous elements. It is best to determine the area containing the end plates first (see. p. 406). Usually the pelvic part has no nervous elements.

Apparatus: A cathode-ray oscilloscope with an A. C. preamplifier, total sensitivity 100 μV/cm. A double beam oscilloscope with one D. C. channel or at least a long time constant (4 sec) is required for recording the "mechanogram" (record of muscle contraction). A stimulator giving rectangular pulses. A radiofrequency output unit for reducing the stimulus artifact. A time marker. A calibrator. A mechanoelectrical transducer (e. g. transducer RCA 5734).

Other requirements: AgCl electrodes with cotton wool wicks for stimulation and recording. A multiple electrode consisting of several platinum wires 3—5 mm apart each from the other for stimulation of the muscle all over its length.

Procedure: Biphasic and monophasic action potentials: The muscle is stretched in the recording arrangement as described on p. 406. The stimulating electrodes are placed 3 mm apart on the pelvic end of the muscle. The first recording electrode is situated about 5 mm peripherally from the stimulating electrodes, the other is placed successively 3, 5, 7 and 9 mm from the first. The action potential is recorded for each position of the recording electrodes (Fig. 154a, b, c, d). The further removed the second recording electrode the more distant are both phases of the biphasic action potential. For the last position they are separated by a clear-cut isopotential line.

The monophasic recording is simply made by placing the second recording electrode onto the knee tendon of the sartorius muscle.

The amplitude and duration of the individual parts of the muscle spike are measured on the oscillogram.

The mechanical latent period of the mechanical response. The muscle is mounted in a recording arrangement with the pelvic tendon fixed and the knee tendon connected to a transducer working in an isometric regime. The stimulating multiple electrode is slightly pressed onto the upper surface of the muscle over its entire length. The muscle is in Ringer solution and is stimulated with maximal single shocks against an indifferent electrode in the solution. The multiple cathode ensures activation of the contractile machinery of the muscle simultaneously from the membrane. The output of the transducer is connected to the osciloscope D. C. amplifier of a low gain (30 mV/cm). The isometric twitch is registered on one beam, the stimulus on the second beam.

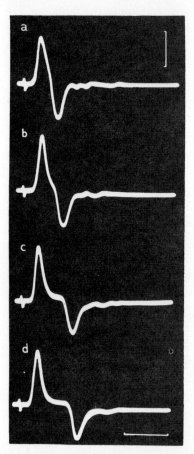

Fig. 154. Action potential of the frog sartorius muscle produced by direct stimulation. The interelectrode distance of recording electrodes (a) 3, (b) 5, (c) 7 and (d) 9 mm. Time mark: 10 msec. Calibration: 10 mV.

Conclusion: The shape and duration of a muscle action potential are similar to that of a nerve. The ascending phase lasts 1 msec, the total duration of the monophasic spike is 2 msec. The rate of conduction calculated indirectly (p. 326) is very slow: 2 m/sec. The short spike duration and the small conduction velocity indicate a small wave length of the impulse. This also explains the fact that both phases of a biphasic action potential can be separated by increasing the interelectrode distance to only several mm.

The time interval between the onset of the stimulus and the twitch is defined as the mechanical latency of the twitch. At 0°C it lasts about 20 msec. During this period gradual activation of the myofibrils occurs, accompanied by manifold changes in physical and physico-chemical properties of the muscle (Hill 1949, 1951a, b, Sandow 1955, 1964, Huxley 1957b). If the gain of the amplifier is increased several times, the *latency relaxation* of the muscle can be recorded before the onset of the tension rise (Sandow 1944, 1945, 1948).

b) Membrane excitation and development of tension. Excitation — contraction coupling

Problem: Record the muscle action potential with an intracellular microelectrode and simultaneously also the isometric muscle contraction. Demonstrate the dissociation of the action potential from the contraction.

Principle: The muscle action potential is the first link in a chain of events which starts off contraction of the muscle. These events which are the basis of excitation-contraction (E-C) coupling (Sandow 1952) occur during a very short period of time between the onset of the action potential and of the rise intension. If these processes are impaired at any link, uncoupling of the excitation-contraction link occurs. Consequently the muscular contraction disappears even though the action potential is present. There are many agents that suppress or completely inhibit contraction, mostly, however, because they eliminate the action potential. In such cases we cannot speak of a dissociation of the E-C coupling nor of dissociation between the action potential and contraction. Thus in order to decide whether contractions have disappeared as a result of impairment of the E-C coupling or because of a membrane effect, both the action potential and muscle contraction must be recorded simultaneously. Recording such functions from the whole muscle, however, only gives us a rough picture. Ideally a single muscle fibre should be used since both processes are studied in the same functional unit.

With the use of intracellular microelectrodes it is, however, possible to obtain usable semiquantitative data from a whole muscle. Two variations of the experimental procedure are possible. In both cases contractions are recorded from the whole muscle and the action potential from a single fibre with an intracellular microelectrode. In one case, however, stimulation is extracellular while in the other the fibre from which the action potential is recorded is also stimulated. In the latter case greater demands on the tension recording are made. The twitch of the stimulated fibre is considerably weakened in this case and distorted by parallel elastic components, i. e. by the nonstimulated fibres. If the preparation has a relatively small number of fibres and if recording and stimulation are both intracellular it may be used for solving a number of problems related to E-C coupling. In view of the parallel connection of the elastic components it is better to record muscle tension (*isometric* recording) than muscle length at a constant tension (*isotonic* recording).

A suitable preparation is used here to demonstrate the dissociating effect of hypertonic solutions on the muscle (Hodgkin and Horowicz 1957). In this medium the muscle does not contract, but the action potential is preserved. Such a muscle is very suitable for gaining experience with the technique of recording action potentials intracellularly and for beginners it may be recommended as an initial object.

Recording from muscles in normal ionic and osmotic media with intracellular microelectrodes is, however, much more difficult than from nerve axons (exp. IV-A-c). The twitch, which is initiated by the action potential causes the electrodes to be pushed out of the fibre, breaks them and injures the fibre unless flexibility of the microelectrodes is ensured. Hence isometric recording of contractions is of advantage.

Object: M. extensor longus dig. IV of the frog. This is one of 5 muscles situated on the dorsal side of the 4th finger of the hind limb. It is a thin muscle which together with the m. tarsalis anticus terminates in a tendon on one end and joins the m. extensor brevis medius digit IV. internus on the other (Fig. 155). The tendon of this extensor is attached to the basal phalange. The muscle runs close to the main vessel and nerve trunk in this region. In view of the complicated situation it is isolated very carefully, preferably with pieces of surrounding muscles and connective tissue. After removal from the body it is cleaned from these remnants under the dissecting microscope. The muscle has an eliptic cross section with a maximum diameter of 500 μ (depending on the size of the frog) and is about 2 cm long.

Apparatus: As in Exp. IV-C-b. A mechanoelectrical transducer (RCA 5734). A double-beam oscilloscope. One channel for recording of the membrane and action potentials, the other for tension recording. Overal sensitivity of the latter channel is about 10 mV/cm.

Other requirements: As in Exp. IV-C-b. Instruments for macro- and microdissection. A silver wire ($\varnothing$ 50 — 80 μ). Ringer solution of normal composition and with 2·5 × the normal concentration of NaCl. Recording chamber. Transducer holder enabling fine movement of the transducer along the longitudinal axis of the muscle and perpendicular to the base of the recording chamber. Since the total tension of the object does not exceed the permissible loading of the transducer, no reduction is necessary. A sewing needle is fixed to the anode with a small piece of plexiglass. The eye of the needle is cut so that a miniature hook is formed.

Procedure: A loop of silver wire is attached to the upper tendon of the extensor muscle and by it the muscle is hung on the miniature hook of the needle of the transducer. The lower tendon is fixed (Fig. 156; see also Hodgkin and Horowicz 1959; Zachar and Zacharová 1966). The muscle is extended to 125% of its length in the body. In order to facilitate insertion of the microelectrodes the muscle is slightly pressed to a small pedestal situated nearer to the fixed end. As in Exp. IV-C-b the stimulating electrode is first introduced into the fibre under visual control using the microscope. The electrode hangs on a piece of rubber tubing (Zacharová and Zachar 1965). This enables it to follow movements of the fibre but does not prevent its insertion into the

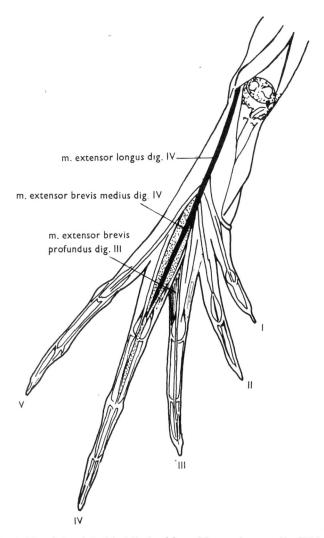

m. extensor longus dig. IV

m. extensor brevis medius dig. IV

m. extensor brevis profundus dig. III

Fig. 155. Spinal side of the right hind limb of frog. M. ext. longus dig. IV is shown in black The m. extensor brevis medius dig. IV internus is stippled.

fibre. The recording electrode is fixed in the same way and inserted into the same fibre. A negative pulse (inward current) is first applied to the stimulating electrode. This evokes a hyperpolarizing electrotonic potential, if both electrodes are in the same fibre. Before changing the polarity of the current an amplification corresponding to 50 mg/cm is set in the channel for recording tension. The transducer is calibrated before the experiment by following the deviation of the beam of the oscilloscope due to different weights (10—500 mg) hung vertically on the needle of the transducer.

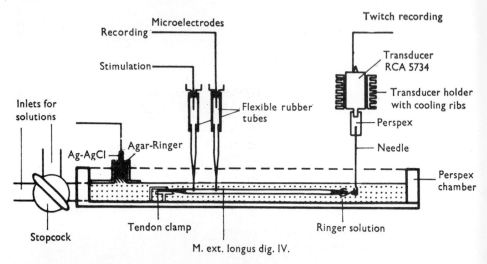

Fig. 156. Simultaneous recording of membrane changes and contractions in a single muscle fibre without its isolation from the remaining fibres.

Then we start to apply a positive (depolarizing) pulse to the stimulating intracellular microelectrode. The increasing amplitude of catelectrotonic potentials, the appearance of a local response and action potential are followed. Together with the spike we also record on the second beam the twitch which does not increase with increasing stimulation ("all or none" response). Usually, however, only one twitch is recorded since the stimulating microelectrode is pushed out of the fibre. This is most frequently the case even with microelectrodes of the type described here (Fig. 157A).

After every twitch the resistance of the microelectrode is tested to see whether the electrodes are broken. Damage to the fibre is indicated by a fall in the membrane potential. If it is considerably decreased, we choose another fibre (with an isolated fibre, of course, the preparation is lost).

After recording the action potential and twitch in a normal medium, hypertonic Ringer solution is introduced (2·5 times the normal NaCl concentration). Two to 3 min after such an exchange the microelectrode is again inserted and an action potential unaccompanied by contraction is recorded (Fig. 157B).

The effect is reversible. In normal Ringer solution the action potential is again followed by a contraction.

Conclusion: It follows from the simultaneous recording of spikes and contractions that the onset of the twitch is delayed by 3 msec after that of the action potential. During that period processes occur that lead to activation of contraction. Signs of these processes are various changes in optical, elastic, thermic and other characteristics of the muscle (reviews: Sandow 1955, Huxley 1957b,

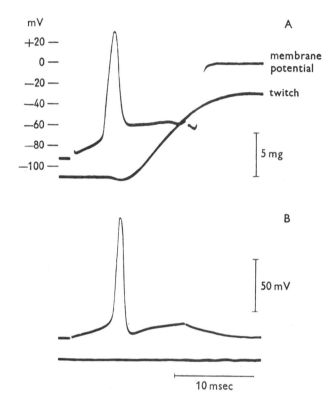

Fig. 157. Recording of action potential with an intracellular microelectrode (upper beam) and of contraction with a transducer (RCA 5734) (lower beam) from a single fibre of the m. ext. longus dig. IV. Rectangular current pulses applied via the intracellular micro-electrode were used for stimulation. Contraction of the stimulated fibre was registered from the whole muscle. A : in Ringer solution, B : in Ringer with 2·5 times the normal Na ion concentration (hypertonic solution). No contractions present in B.

Reichel 1960). During that time excitation-contraction coupling also occurs (Sandow 1952). This coupling proceeds only at certain sites of the muscle membrane as shown by Huxley (1957a, b) by local activation of the membrane above certain sections of the sarcomere. The structual basis for transmission of the stimulus from the membrane to the sarcomeres is represented by the system of transverse tubules of the sarcoplasmatic reticulum (Veratti 1902,

Bennet 1960, Porter and Palade 1957, Franzini-Armstrong 1964). E-C coupling occurs during the early stages of the latent period before the appearance of the activation heat (Hill 1950). Interference with the processes underlying the E-C coupling at that time leads to a true E-C dissociation. The action potential and contraction uncoupling caused by hypertonic solutions, howeever, is not due to an E-C dissociation since the activation heat in this case remains unchanged (Hill 1958). Contractions disappear because they are slowed down in an hypertonic solution as follows from a study of the mechanical properties of muscle in such a medium (Howarth 1958). The muscle cannot extend the series compliance during the short active state period.

Real dissociation of E-C coupling is caused by a decrease in Ca^{++} concentration in the external environment (Frank 1960) as can be shown in the giant muscle fibre of the crayfish with simultaneous recording of depolarization and contractures induced by increasing the external K^+ concentration (Zacharová and Zachar 1963, 1967; Zachar and Zacharová 1965).

Technically the simultaneous recording of contractions and action potentials is difficult since it is not easy to maintain the electrode in the muscle during the twitch. Relatively good results are obtained with broken microelectrode tips fixed to very fine wires connected to the input of the cathode-follower. This technique has been used by Woodbury and Brady (1956) for the intracellular recording of action potentials in the beating heart.

V

Electrophysiology of peripheral excitable structures in situ

Electrophysiology of peripheral excitable structures (muscle, nerve) in situ is based mainly on the recording of action potentials. These may be produced by normal reflex activity or by electrical stimulation.

Fundamentally two methods are used to record action potentials under such conditions: recording from organs made accessible surgically or without surgical preparation. Surgical exposure may be acute or may consist in implanting electrodes for a chronic experiment.

Electrophysiological recording from excitable structures made accessible surgically does not differ from that from the same structures in vitro. Usually no special modification of these methods is necessary for the latter, and thus the only new problem involved is a surgical one.

When recording from structures in the normal organism a new factor must be considered: registration from a nerve or muscle situated in a volume conductor. The fundamentals of the distribution of electric current in a volume conductor must be known if recordings are to be correctly interpreted.

Many electrophysiological methods of recording in situ have become independent clinical methods: electrocardiography, electromyography etc. The literature concerned with these methods is extensive and will not be considered here.

Experiments given in this section are examples of recording action potentials from nerves and muscles in situ.

a) Recording of an impulse in a volume conductor

Problem: Record the action potential from the frog sciatic nerve in a volume conductor in situ and in a model experiment.

Principle: When a nerve conductor is situated in an electrically insulating medium (a narrow film of extracellular fluid or Ringer solution surrounding the nerve, otherwise surrounded by air, paraffin oil etc.) a recording instrument will register changes in electrical potential only at the point of electrode contact. If the nerve conductor is in an electrically conducting medium the manifestations of the nerve impulse are more complex.

The external current field of the nerve impulse acts on the volume lead for as long as the impulse passes through the nerve located in the volume conductor.

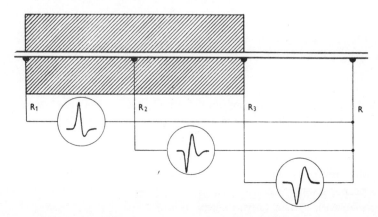

Fig. 158. Diagramatic representation of nerve impulse recording in a volume conductor. R_1, R_2, R_3 — recording electrodes in the volume conductor (shaded area). R — indifferent recording electrode. The circles contain diagrams of action potentials recorded with the corresponding electrodes.

The nerve impulse forms in the nerve a mobile, double dipole connected by negative poles. Negativity is represented by depolarisation of the membrane during the nerve impulse, the positive poles by segments proximal and distal to that point. The current flows first from the interior of the axon across the membrane to the exterior and then in the opposite direction. Finally it again flows towards the exterior (see Fig. 132). The external current field in a volume conductor is not limited to a narrow film of the conducting medium, as is the case in an insulating medium, but spreads a long way into the conducting surroundings.

The current flow from the axon to the exterior acts as a source, and in the opposite direction as a sink. From that aspect a nervous impulse is characterised by the sequence: source — sink — source.

Depending upon what part of this sequence passes the volume lead, corresponding electric changes will be recorded.

When the nerve impulse enters that part of the nerve which lies in the volume conductor (Fig. 158) the recording electrode (R_1) situated here will record a negative wave and later a positive one. At R_2 a triphasic positive-negative-positive wave is recorded. At the point where the nerve impulse leaves the volume conductor a positive followed by a negative wave is recorded.

Action potentials may be recorded in a volume conductor from a nerve in situ, or even better, in an artificial volume conductor which can easily be made by placing the nerve on filter paper soaked in Ringer solution. Such a model, which is also very instructive for understanding more complex three-

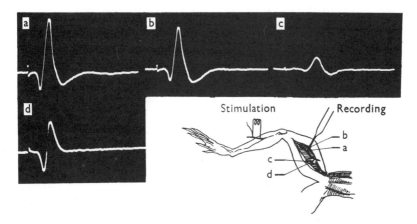

Fig. 159. Recording of action potential of the frog sciatic nerve with a volume lead in situ at the site marked on the drawing. Letters a-d in the diagram denote the corresponding oscillogram. Stimulating electrodes are on the peroneal nerve. The indifferent earthed electrode is on the opposite limb.

dimensional volume conductors, such as structures in the brain (see p. 579) may be used to map out the electric field of a travelling nerve impulse.

The principle of such mapping is as follows. The nerve is placed on filter paper soaked in Ringer solution. One recording electrode is at the edge of the paper at a sufficient distance to prevent the external current field of the impulse from acting on it. The other electrode records responses at different distances from the axis of the nerve.

The values of the potential at some instant after the application of the stimulus are used to reconstruct graphically the equipotential contours and lines of current determining the sources and sinks of current flow. For the theory and practical procedure of such mapping see the papers by Lorente de Nó (1947), Lloyd (1952), Kostyuk (1956, 1960), Tasaki (1959) and page 579.

Object: The sciatic and peroneal nerve of the frog *Rana esculenta* or *R. temporaria*.

Apparatus: A cathode-ray oscilloscope with a preamplifier. Total sensitivity 100 μV/cm. A voltage calibrator. Time marker 1000 c/sec.

Other requirements: A glass capillary semi-microelectrode with an opening $20-30 \mu$ in diameter, filled with frog physiological saline and connected to the input via an AgCl coated silver wire immersed into the broader end of the tube. Filter paper 30×30 cm.

Procedure: The preparation (lower half of the trunk with both hind limbs of a frog) is skinned and fixed in a Petri dish having a cork bottom. The left limb is earthed. The peroneal nerve of the right limb is exposed, cut distally and placed on the stimulating electrodes. The semi-microelectrode is used to record action potentials from the points marked by letters a-d in Fig. 159.

The responses recorded near the entry of the nerve impulse into the volume conductor are diphasic. Those recorded from more distant parts are triphasic and their amplitude decreases with increasing distance from the nerve.

The nerve is then freed from surrounding tissues for a length of about 25 mm and placed on filter paper. The nerve lies in an horizontal ridge that permits the surface of the nerve to be at the same level as the surface of the paper. The paper is soaked in frog saline. One end of the nerve is in a small bath containing paraffin oil. The stimulating electrodes are located here. Then the whole preparation is covered with paraffin oil to prevent drying of the paper and thereby a change in its conductivity.

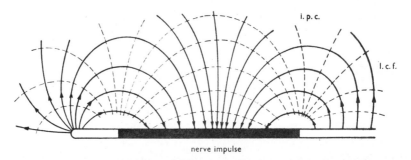

i. p. c.

l. c. f.

nerve impulse

Fig. 160. Diagramatic representation of the distribution of the external current field in the nerve at the moment when the impulse wave is at the marked point, i. p. c. — isopotential contours, l. c. f. — lines of current flow, (diagramatically, according to Lorente de Nó (1947) and Kostyuk 1960).

As before, the action potential registered when the nerve impulse enters the volume conductor is diphasic, then triphasic and the amplitude decreases with increasing distance from the nerve.

Several hundred action potentials must be recorded for mapping the external current field. Then, at a certain time interval from the beginning of the stimulus, which represents a certain position of the nerve impulse in the nerve conductor, the corresponding potential amplitude is measured. Points of equal amplitude are connected thus forming isopotential contours. The *lines of current flow* and the *isopotential lines* intersect one another at right angles. Zero isopotential lines separate zones of opposite polarity and determine the direction of the lines of current flow and thus the sinks and sources of the external current field of the nervous impulse at the chosen time.

The field diagram in Fig. 160 is characterised by the sequence: source — sink — source. In a similar way it is possible to map the external current field from the same oscillograms for any position of the nerve impulse in the nerve conductor.

Conclusion: The volume distribution of the external current field of the nerve impulse is obtained by rotating the two dimensional current field drawn in this way through 360° about the axis formed by the nerve. The field strength will decrease more rapidly, however, with the distance.

When recording with volume lead, it is important to keep in mind that this lead records changes in the voltage drop caused by membrane current, while the monophasic lead in an insulating environment records changes in membrane potential. This explains the fact that recording with á volume lead depends upon the duration of the registered process. A potential change having a certain amplitude and a shorter space extent produces a greater current density than a change due to the same amplitude but of greater space extent. Hence it is difficult to record after—potentials (their wave length is several m and they have a small amplitude), in a volume conductor.

Recording in a volume conductor is important when registering and interpreting electric activity in the brain (see chapter IX K).

b) Impulse activity of somatic and vegetative nerves

Problem: Record impulse activity from a somatic and autonomic nerve in situ in the rabbit.

a) from the nerve leading to the gastrocnemius muscle when the latter is stretched,

b) from the cervical sympathetic trunk,

c) from the cervical sympathetic trunk together with respiration,

d) from the cervical sympathetic trunk together with discharges from the phrenic nerve.

Principle: Electrophysiological recording from whole nerve in the organism is simple, if recordings are not made in a volume conductor. The point of the nerve from which impulses are to be recorded is made accessible surgically and an electrode is placed on it. This must not touch the surrounding tissues.

If electric stimulation of such a nerve is used the method is the same as for nerves in vitro. In situ this method, however, is only rarely used.

Usually impulse activity due to reflex activity of the organism is studied.

Electrical activity of a mixed peripheral nerve evoked by reflex activity gives a very complex picture. It is similar to spontaneous activity produced in a nerve in vitro, e. g. by decalcification. The physiological stimulus does not activate all fibres at the same time, so that the recording is characterised by asynchronous activity of nerve fibres. Such tracing may be used to determine the presence or absence of stimulation. They are not suitable for quantitative evaluation.

For such purposes it is best to use a nerve bundle with a small number of fibres, or even better a single nerve fibre isolated for a certain length. This is difficult and time consuming but permits quantitative evaluation.

If the direction of the impulses is to be determined the nerve must be severed on the corresponding side of the recording electrode since the nerve trunk contains both afferent and efferent fibres.

Sources of error during recording.

1) Artifacts due to movement occur if the nerve shifts along the recording electrodes during movement of the animal. Such movement along the metal electrodes may also produce mechanical stimulation of the nerve and thus an interfering impulse acitivity. This artifact is avoided if the nerve hangs freely on the electrodes and if the part of the nerve near the recording electrodes is supported by glass hooks.

2) Drying of the nerve and disturbances of circulation. Drying of the nerve may either produce automatic activity or, on the contrary, decrease of excitability. The same holds true for impairment of the circulation. This may arise as the result of incorrect preparation. Drying is prevented by having the animal in a thermostat with adequate temperature and moisture, or by covering the nerve with a layer of paraffin oil at an appropriate temperature.

3) Electric activity of other organs: Most frequently this is the heart, particularly when working in its vicinity. If assymetrical amplifiers are used, this is prevented by a treble electrode. The grid electrode is in the middle, the other two are earthed. Symmetrical amplifiers should have a high discrimination coefficient (see page 160).

4) Finally external interference may be more important when registering in situ than when working with isolated preparations.

Object: A rabbit under urethane anaesthesia with the required nerves exposed.

Apparatus: A double beam cathode-ray oscilloscope with A. C. preampli-fiers. Total sensitivity 100 μV/cm. Time marker 1000 and 50 c/sec. Respiratory movements are recorded oscillographically as shown in Fig. 161. Movements of the thorax are recorded in the same way as when using a kymograph. The movements of drum 2, however, are transformed to electric potentials by a liquid rheostat — 3 — from which they are lead to the input of the D. C. amplifier of the oscilloscope. A power amplifier with a loudspeaker.

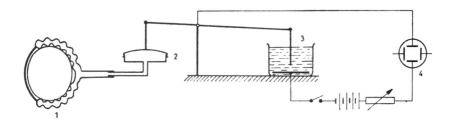

Fig. 161. Diagram of the oscilloscopic recording of respiratory movements from the thorax. 1 — pneumograph, 2 — Marey's drum, 3 — liquid rheostat, 4 — oscilloscope.

Other requirements: Surgical instruments. Adrenaline. Nicotine.

Procedure:

1) *Impulse activity in the nerve leading from the gastrocnemius muscle, when the latter is stretched.*

The gastrocnemius muscle of a rabbit under urethane anaesthesia is exposed, together with the nerve leading to one of its heads. The nerve is cut centrally and attached to a stand by a silk thread. The recording electrodes are placed on the nerve in such a way that they do not touch the rest of the preparation. On connecting them to the oscilloscope with attachements of the muscle intact, a continuous irregular discharge is observed on the screen This discharge is centripetal and is produced in muscle proprioceptors, since it considerably decreases on cutting the muscle tendon.

If a load is put on the muscle the discharge increases (Fig. 162). The figure shows that the response from the whole nerve is completely asyn-chronous and that a more detailed analysis of the discharge is not possible. The same experiment may be easier performed also in vitro with the frog sciatic-gastrocnemius muscle preparation.

2) *Impulse activity from the cervical sympathetic trunk.*

The cervical sympathetic trunk is exposed and cut distally. A silk loop is tied to its central end and the nerve trunk is placed on recording electrodes. A discharge is shown in Fig. 163a. The figure shows that activity from this nerve is not so asynchronous as in the preceding experiment, when a somatic nerve was used.

The amplitudes of the recorded waves vary, but as a rule the majority do not vary much with respect to duration and shape of the action potential.

This is due to the fact that groups of nerve fibres become active synchronously. Activity is sometimes so well synchronised that it may be considered as a unit discharge.

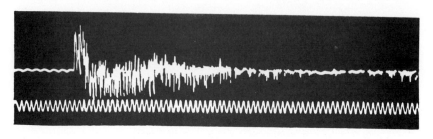

Fig. 162. Discharge from the gastrocnemius nerve of the rabbit after stretching the tendon. Time mark : 25 c/sec.

Activity is exclusively due to centrifugal fibres since activity of centripetal fibres was excluded by cutting the nerve distally. Centrifugal fibres may by either pre- or postaganglionic. The participation of both kinds of fibres may be determined using Langley's nicotine method. The superior cervical ganglion is treated with 0·5% nicotine and the discharge is recorded (Fig. 163b). It can be seen that activity decreases to only a small extent. This indicates that the spontaneous discharge in the sympathetic nerve is mainly preganglionic in origin.

If 0·5 mg adrenaline is given intravenously this natural discharge disappears (Fig. 163c) for 5—10 minutes and then returns to normal levels only slowly. It has been shown that this natural sympathetic activity has a vasoregulatory effect (Adrian et al. 1932).

3) *Simultaneous recording of impulse activity from the cervical sympathetic trunk and of respiration.*

The impulse discharge in vegetative nerves is grouped in a more or less regular rhythm. In rabbits this grouping often coincides with respiratory movements.

This effect is present even with the vagus intact but is more evident when it is cut. Fig. 163d shows this grouping in a recording with very slowly moving film. Fig. 163e is a recording of respiration.

4) *Discharges from the sympathetic and phrenic nerves.*

A double beam oscilloscope with two A. C. amplifiers is used. The electrodes from the cervical sympathetic nerve are connected to the input of one preamplifier. Electrodes placed on the distally cut branch of the phrenic nerve in the neck are connected to the other one. The vagi are cut. Fig. 163f, g shows tracings from such an experiment. The upper beam is a recording from the sympathetic trunk, the lower one from the phrenic nerve. The discharge from the phrenic nerve (Fig. 163g) starts before the actual sympathetic discharge and ends before the sympathetic discharge begins to decrease. It can be seen that this occurs at different frequencies of respiration.

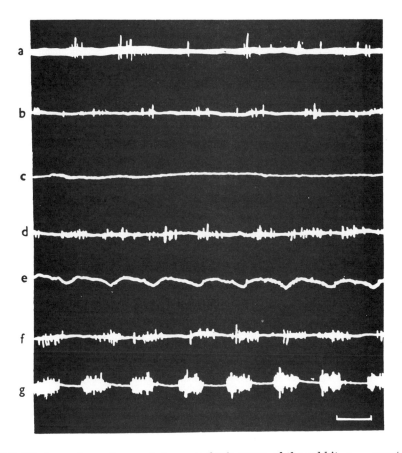

Fig. 163. Discharge from the cervical sympathetic nerve of the rabbit. a — spontaneous discharge. b — after treating the superior cervical ganglion with 0·5% nicotine. c — after i. v. adrenaline. d — discharge in the sympathetic nerve after cutting the vagi. e — respiration. f — discharge in sympathetic trunk. g — discharge in phrenic nerve. Time mark: 2 sec.

Acoustic monitoring of nerve activity may be of advantage. The output of the preamplifier or of the oscilloscope amplifier is connected to the power amplifier with a loud-speaker. The coincidence between respiratory movements and the acoustically displayed nerve discharges can be easily followed in this way.

It follows that in the absence of sensory impulses from the vagi the sympathetic centres in the rabbit are stimulated directly by every period of activity in the respiratory centre.

Conclusion: Recording of electric phenomena due to impulse activity from whole nerves gives only a very rough idea of the activity of these nerves. Vegetative nerves differ from somatic ones in that respect since during reflex activity the discharge is less asynchronous and has a smaller number of relatively larger potential waves. As shown above, this enables one to compare the function of these nerves with the activity of other systems in the organism.

In general whole nerves are not suitable for exact analytical work, for which smaller units must be used.

c) Impulse activity of muscle in situ. Electromyogram

Problem: Record the electrical phenomena accompanying muscular activity using surface and concentric electrodes:
a) during shivering evoked by cold,
b) during metrazol convulsions

Principle: If recording electrodes are applied to a muscle, a very complex recording of electrical phenomena due to muscular activity is obtained when an asynchronous volley of nerve impulses enters the nerve, such as during reflex stimulation. Impulse activity obtained by this surface electromyography is then difficult to analyse.

It is, therefore, better, also in the case of muscle, to work with smaller units. This may be achieved in two ways, either by using an isolated muscle fibre, or with the aid of electrodes having a very small interelectrode distance. The first method is possible either by means of intracellular microelectrodes in vivo (Beránek 1959, Rieker et al. 1963) or with external leads in vitro. Both are not easy to perform. For in situ work the method of Adrian and Bronk (1929) is more feasible. The principle of this method is as follows.

When recording impulse activity from a muscle with two electrodes with a certain distance between them (Fig. 164A), the electrodes will record the algebraic sum of potential changes from several fibres, in addition to those in contact with the electrodes. The potential gradient in the fibre at a distance d produces a considerable potential difference between a and b. If the electrodes are very close to each other (a', b', Fig. 164B) they can record the same potential gradient only from distance d'.

The small interelectrode distance is attained in a way suggested and used by Adrian. A very fine wire (about 100 μ) is used, covered with an insulating layer. This is inserted into an injection needle (Fig. 264). The needle forms one electrode and the wire in the centre, insulated from the needle, the second electrode. The needle also serves to pierce the muscle and to give stability to the electrode. In addition the needle serves as the screen for the grid lead. This electrode is termed *Adrian's concentric needle electrode.*

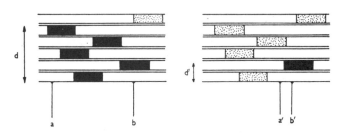

Fig. 164. Recording of muscle activity with needle electrodes, with a large (a) and a small (b) interelectrode distance.

In addition to this concentric needle, a bipolar needle electrode may also be used. This differs from the above only by the fact that two fine wires are inserted into the needle.

Object: rat.

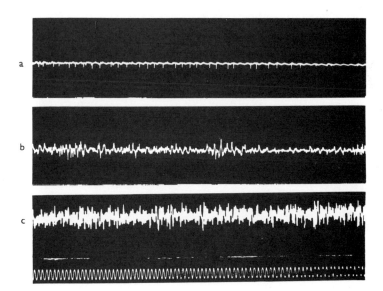

Fig. 165. Electromyogram during shivering due to cold recorded with Adrian's concentric electrode. Time mark: 25 c/sec.

Apparatus: A cathode-ray oscilloscope with an A. C. preamplifier, total sensitivity 100 μV/cm. An amplitude and time marker.

Other requirements: Plate silver electrodes (3 × 3 mm). Adrian's concentric needle. An ice bath. Arrangement for mechanical recording of muscular activity.

Procedure:

1) *EMG during shivering produced by cold.*

The rat is anaesthetised with ether. Plate electrodes covered with ECG paste are placed above the previously depilated region of the gastrocnemius muscle, using collodium. The rat is tied on a board on which a polyethylene bag containing ground ice is placed. At first continuous muscle activity is not observed. Only occasionally does a discharge of irregular activity appear. As the animal begins to lose heat, shivering sets in as a compensating mechanism and gradually more and more intense muscular activity can be seen.

The recording shows a very complex, intense, asynchronous activity.

If Adrian's concentric needle is inserted into the muscle the picture becomes much simpler. During weak shivering, provided the electrode is in a suitable position, unit muscular activity of low frequency may be recored (Fig. 165a). This increases with increasing shivering. In addition, the electrode now records activity from more than one muscular unit (Fig. 165b, c).

2) *EMG during cardiazol convulsions.*

A rat under ether anaesthesia is tied to a special stand (Servít 1958). One hind limb is free, the other fixed. The free limb is connected to a liquid rheostat via a mechanical arrangement. The rheostat will record the muscular contraction of the limb. The fixed hind limb is depilated as above and either surface electrodes are attached to it or Adrian's concentric needle is inserted into the muscle. Metrazol (pentamethylentetrazol) is then given intraperitoneally (80 mg/kg).

During the first convulsion, the muscle contractions and the surface electromyogram are recorded. During the second seizure the electromyogram is registered using the concentric needle (Fig. 166):

The muscle recording shows typical tonic-clonic convulsions and the electromyogram intense muscular activity of the gastrocnemius muscle during the tonic phase and intermittent activity during clonus. Acoustical display (see p. 227) of EMG makes it possible to observe the behaviour of the animal and to follow simultaneously its muscle activity.

Conclusions: Electromyography has developed into an independent clinical discipline used as an aid for diagnosing neuromuscular disorders. It is often supplemented by EMG recording after percutaneous stimulation of nerves. It makes it possible to distinguish between fasciculation (spontaneous contraction of muscle bundles) and fibrillation (spontaneous contractions of individual fibres), (Denny-Brown and Pennybacker 1938).

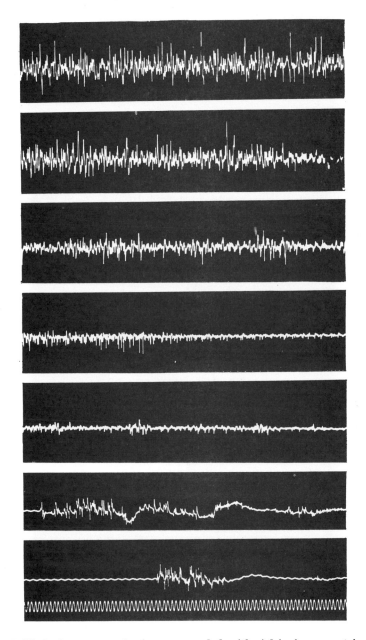

Fig. 166. EMG during metrazol seizures recorded with Adrian's concentric electrode. Time mark: 25 c/sec.

It is also possible to distinguish between neurogenic atrophies due to disorders of nerve conductors and those due to impairment of the anterior horn cells according to the degree of synchronisation (Buchthal and Clemensen 1943, Buchthal and Madsen 1950).

The amplitude may be used for determining disturbances in neuro-muscular transmission (Hodes 1948, Buchthal and Honcke 1944).

Multilead electromyography gives a broader picture of peripheral changes due to disturbances in the CNS, such as disturbances of reciprocal innervation (Hoefer 1942).

d) Recording of nerve and muscle action potentials through the intact skin in man

Problem: Record the action potential from the ulnar nerve and the abductor digiti V muscle in response to stimulation of the ulnar nerve.

Principle: Transcutaneous stimulation and recording of nerve and muscle action potentials through the intact skin is principally the same as in experiments in vitro. The following two circumstances account for the differences in technique. First, the tested neuromuscular object is in a volume conductor and second, stimulation and recording occur through two layers with different electric conductivity (skin and subcutaneous tissue). The layers between the current source and the excitable tissue are considerably inhomogeneous and this leads to inhomogenity in the spread of the electric current. This makes for considerable uncertainty of stimulation and recording particularly as far as the determination of the site of stimulation and recording is concerned. Conditions for the occurrence of a virtual cathode (see p. 348) are formed. As a result of this, the nerve potential is very much reduced and its detection requires a considerable gain of the amplifying system; sometimes special techniques for detection of the signal from noise must be applied. With such amplifications the stimulation artifact introduces considerable difficulties (p. 196) particularly if we consider also the shunting effect below the stimulating electrodes and the resultant necessity to use more intense stimuli.

Hence the experiment in the first place requires considerable gain of the amplifying system and adjustment of the stimulator output for reduction of the stimulation artifact. The other technical arrangements aim at placing the stimulating and recording electrodes as close to the nerve stem in situ as possible, e. g. by searching for the anatomically most suitable emergence of the nerves below the skin, by scarification and moistening of the skin below the electrode etc.

The transcutaneous recording of evoked muscle potentials is not as difficult as the recording of evoked nerve action potentials as follows already from a comparison of the geometry and of the anatomical localization of muscle and nerve. The amplitude of the transcutaneously recorded muscle action potential is higher by two orders of magnitude than that of the nerve action potential. This explains why it is far more difficult to record the latter during muscle movement and its associated electrical activity. This source of artifacts can be eliminated only by ensuring complete muscular relaxation at the site of recording of the nerve action potential.

The technical problems and difficulties are similar to those encountered when recording evoked potentials from the brain with scalp electrodes. Experience gained in this experiment can hence be used for recording evoked potentials in human EEG. On the other hand, if the amplitudes of the evoked action potentials from the peripheral nerve are near the noise level the summation techniques developed for detection of evoked potentials in scalp EEG can be applied (Dawson 1951, 1954).

Object: man.

Apparatus: A cathode-ray-oscilloscope with A. C. differential amplifier. Total sensitivity $10-20$ μV/cm. A stimulator giving rectangular current pulses lasting $0.1-0.5$ msec. A radiofrequency isolation unit. Time mark 1000 c/sec. Amplitude calibration 10 and 1000 μV.

Other requirements: Stimulating and recording Ag-AgCl electrodes, an earthing Ag-AgCl plate. Rubber bands for fixing the electrodes. Electrocardiographic paste. Physiological saline. Needle for skin scarification.

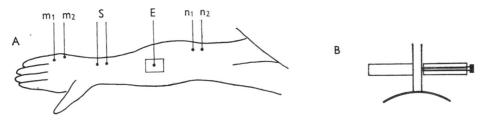

Fig. 167. A — Arrangement of stimulating (S) and recording electrodes on the volar side of the forearm and hand. E — earthing plate, n_1; n_2 — electrodes for recording the nerve action potential, m_1, m_2 — electrodes for recording the muscle action potential. B : a section through the recording electrode.

Procedure: A circular ($\varnothing$ 2 cm) earthing electrode is fixed to the volar side of the forearm with the rubber bands. The electrode is from chlorinated silver foil covered by gauze and moistened with physiological saline. Before applying the electrode fat is removed from the skin surface and ECG paste is applied. The electrode is earthed. With an auxillary stimulating electrode we search for the closest localization of the ulnar nerve on the wrist and elbow as judged from the hand movement. The auxillary stimulating electrode is a silver rod ($\varnothing$ 2 mm) inserted into a non-conducting holder. The silver is chlorinated and covered with gauze in a similar way to the earthing plate and the other electrodes. The sites from which the same current intensity induces the greatest motor effects are marked with a dermograph. Those who are performing such an experiment for the first time are advised to look for these sites on their own arm so that they become acquainted with the sensory sensations during electric stimulation. They are not painful. Then the stimulating and recording electrodes are placed definitely onto those sites. The localization of the stimulating (S) and two pairs of recording electrodes is shown schematically in Fig. 167A (n_1, n_2 for nerve, m_1, m_2 for muscle action potential). The shape of individual electrodes is adapted as required. For recording, saddle-shaped electrodes (Fig. 167B) are most suitable. These are silver foil strips 1 mm thick, 2—3 mm long and 5 mm wide, bent into an arc. Their middle part is connected to a silver tube the other end of which is joined with the lead to the amplifier input. The electrodes are mounted on an insulating plate. The distance between the electrodes is 2—3 cm. The plate with the electrodes is fixed to the site of recording with a rubber band. If the plate carries several electrodes it is advantageous to have the possibility of moving the silver tube perpendicular to the plate so as to ensure uniform pressure on the skin at the site of recording. The same electrodes may also be used for stimulation. It, however, we want to compare the muscle and nerve action potentials elicited by the same stimulus such an arrangement is not suitable. During stimulation of the ulnar nerve in the wrist the more distally situated anode might cause a block of conduction in a certain portion of the nerve fibres and thus decrease the amplitude of the muscle action potential. Therefore disc shaped electrode ($\varnothing$ 5 mm) are preferable. The cathode is placed on the marked site on the wrist and the anode laterally or medially to it. The stimulating electrodes are not earthed but are connected to the output of the radiofrequency isolation unit.

Electrodes and the skin are treated as for placing the earthing plate (E). Care is taken to avoid interelectrode resistance (measured with D. C. current) larger than 10 KΩ.

The placing of the stimulating electrodes on the wrist and the recording electrodes on the elbow ensures that the pulse activates hand muscles only, so that this source of interference with nervous activity is thus eliminated. It is necessary, however, to eliminate postural reflex activity in the upper arm. Relaxation must be achieved by suitably placing the arm on a support.

Now the experiment proper can be performed. First recording electrodes n_1 and n_2 are connected. The stimulus intensity is gradually increased (frequency of stimuli 1 c/sec, pulse duration 0·5 msec). After attaining the threshold the amplitude of the response increases in an S-like fashion. The maximum is reached when the threshold value is approximately doubled. With further increases in the stimulus intensity the response amplitude no longer changes. The response is triphasic (registration in a volume conductor) as is seen in Fig. 168b, obtained by Dawson's technique of superimposing the recorded responses for the same stimulus intensity and frequency (1 c/sec).

After recording the amplitude and time calibration the input of the amplifier is switched over to the muscle recording electrodes (m_1, m_2). For the

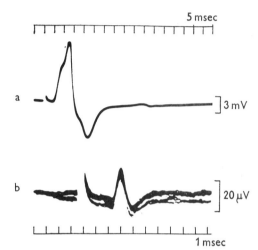

Fig. 168. Recordings of muscle action potential from m. abd. dig. V in man (a) and of nerve action potential from ulnar nerve (b) after stimulating the latter. Recordings are made by superimposition of 20 responses to the same stimulus.

same intensity of stimulation the muscle action potential is recorded. Its amplitude is a 100 times greater than that of the nerve action potential (Fig. 168a). The stimulus intensity is then decreased until the muscle spike nearly disappears. With this intensity the nerve spike recorded from n_1 and n_2 is still present and is about 25% of the initial amplitude evoked by a maximal stimulus.

Conclusion: The nerve spike evoked by a maximal stimulus is a mixed response consisting of the action potentials transmitted antidromically along nerve fibres and action potentials running orthodromically along sensory fibres. The sensory fibres have a lower excitation threshold as indicated by the presence of a nerve spike when the muscle spike had already disappeared. Since the amplitude of the nerve spike was 25% of its initial value at that moment, more than 1/4 of the fibres of the ulnar nerve have a lower threshold than motor fibres. These fibres also have a greater rate of conduction and their stimulation elicits evoked potentials in the brain (Dawson 1956). Thus in man sensory fibres from the fingers are more readily excited than motor fibres. In view of the general relationship of the above parameters (threshold of stimulation, rate of conduction etc.) to the fibre diameter (p. 334). it is probable that the diameter of finger afferent fibres is greater than that of motor fibres. Larger fibres are more sensitive to oxygen lack and pressure than smaller ones. This is in agreement with experiments using ischaemisation of the upper limb (Magladery et al. 1950). The voluntary action is nearly normal even after 17 min of inflation of the rubber cuff to 200 mm Hg while the perception of sensory stimuli excepting deep pain is considerably decreased. At that time the amplitude of the nerve spike is also reduced evidently because of a block of coarser sensory fibres.

The method of recording action potentials across the intact skin was first introduced by Dawson and Scott in 1949. Since then it has become a clinical method (Magladery and MacDougal 1950, Gilliatt and Sears 1958, Bannister and Sears 1962, Mayer 1963).

VI

Electrophysiology of peripheral synaptic junctions

Peripheral synaptic junctions are those sites where the nerve impulse is transferred from the nerve endings to the effector organ (muscle, gland) or to another neurone situated outside the central nervous system (autonomic ganglia).

The fundamental characteristics of synaptic transmission have been studied chiefly on peripheral synapses, due to their easy accessibility with external leads. Most of our knowledge about synaptic transmission originates in these structures. This holds true even after the intracellular microelectrodes enabled the central synapses to be directly investigated. It was actually revealed, that peripheral synaptic junctions show the same fundamental properties as synapses in the central nervous system including praesynaptic and postsynaptic inhibition.

This is in agreement with the results of the ultrastructural research of synaptic junctions, showing a remarkable uniformity of the functionally most significant synaptic structural units.

Anatomically they contain the basis of a divergent and a convergent mechanism. The divergent mechanism ensures only diffusion of excitation into a larger area of the effector. Convergence, on the other hand, is the basis of elementary integrating activity.

Electrophysiologically, transmission across synapses is characterised by synaptic potentials (Eccles 1943).

In this chapter experiments are described which demonstrate the fundamental characteristics of peripheral synaptic transmission in the most commonly studied peripheral synapses, in the end-plate and in the superior cervical ganglia. Examples of both chemical and electrical transmission are given.

A. Neuromuscular transmission in skeletal muscle

The anatomical basis of neuromuscular transmission is the end-plate. Before its termination the nerve fibre is no longer myelinised. There is no continuity between the nerve ending and the muscle fibre but only very close contact. The muscle tissue immediately below the nerve endings has special structural features. It appears as a palisade of small rods. Developmentally it is of mixed origin, the rods originating from Schwann's sheath. This specialised junction is called the motor end-plate.

Very thorough electron-microscopic pictures of neuromuscular junctions are already available (Palade and Palay 1954, Robertson 1956, 1960, Reger 1958, Anderson-Cedergren 1959, Birks, Huxley and Katz 1960; for light microscopy see Couteaux 1944, 1960).

a) Electrophysiological localization of the end-plates in skeletal muscle

Problem: Localize the end-plate region of the sartorius muscle of the frog using the latent period.

Principle: In addition to microscopic examination, two electrophysiological methods may be used to localize the end-plate region (Eccles et al. 1941). One method is based on the measurement of the spike latency. If a nerve leadding to a muscle is stimulated, the distance between the stimulus artefact and

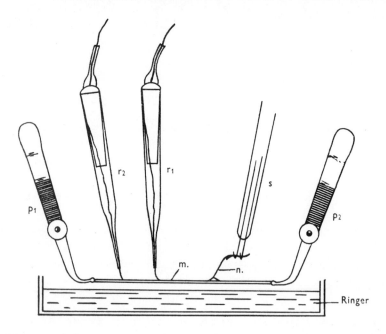

Fig. 169. Diagram of the arrangement for recording action potentials from indirectly-stimulated frog sartorius muscle. p_1, p_2 — screw-controlled forceps, s — stimulating electrodes, r_1, r_2 — recording electrodes, n — nerve, m — muscle.

the muscle action potential will be the shorter the closer the recording electrode to the point at which the impulse in the muscle fibre originates, i. e. the nearer to the end-plate region. If the recording electrode is above the end-plate region a certain minimal latent period is obtained. This includes the time required for conduction of the impulse along the nerve from the point of stimulation and the synaptic delay at the site of transmission (Eccles et al. 1941, Kuffler 1948) The time necessary for conducting the impulse along the muscle fibre in this case is not added. In the majority of muscles the end-plate region is not localized at the same point in all muscle fibres. For that reason when recording from the whole muscle, muscles like the frog sartorius, are suitable in which the majority of fibres are innervated in two or several discrete end-plate zones.

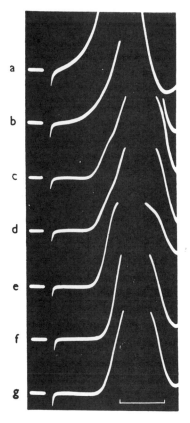

Fig. 170. Localization of end-plates according to the latent period of the muscle action potential. Indifferent electrode on knee tendon. a — action potential above synaptic junction, b-g — at 2 mm distances each from the other. Time mark: 3 msec.

The number of such zones can be determined by this method as those points having a minimal latent period of muscular action potentials.

The second method which can be used for the curarised muscle is based on evaluation of the amplitude of the end-plate potential and will be described in the chapter on end-plate potentials.

Object: The frog (*Rana temporaria* or *R. esculenta*) sartorius muscle with the nerve leading to it.

Apparatus: A cathode-ray oscilloscope and an A. C. preamplifier. The over-all sensitivity of the recording system about 100 μV/cm. A square wave stimulator and a time marker.

Other requirements: Fixing arrangement for the nerve-muscle preparation. A microdrive for one recording electrode.

Dissecting instruments, frog's Ringer solution, a pair of stimulating platinum electrodes and a pair of Ag-AgCl nonpolarizable electrodes.

Procedure: The preparation is fixed as shown in Fig. 169. Both the pelvic and the knee ends of the sartorius muscle are held by a screw-controlled

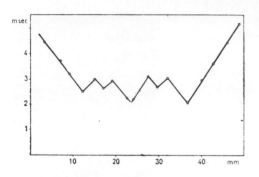

Fig. 171. Graphical evaluation of the end-plate localization in frog sartorius muscle. Abscissa: distances at which action potentials were recorded. Ordinate: latent period of muscle response in msec.

forceps. The nerve is placed on the stimulating electrodes so that it is freely pendant. This prevents it from shifting during muscular contractions and thus producing artifacts in the latent period. The cotton wick of one recording electrode is placed on the tendon, the other is placed on the muscle. The electrode is held in a holder that has a microdrive permitting movement along the muscle fibre.

The stimulating electrodes are then connected to the stimulator giving rectangular pulses. The time base of the oscilloscope is synchronised with this pulse. The nerve is stimulated with a suprathreshold stimulus for the muscle. Then the "active" recording electrode is moved by 1 mm and again the muscle action potential is recorded. In such a way the recording electrode traces the whole muscle (Fig. 170).

Conclusion: If the distances from which the action potentials were recorded are plotted along the abscissa and the corresponding latent periods on the ordinate, a curve with three minima is obtained (Fig. 171). Minimal latent periods occurred at distances 12, 23 and 37 mm from the knee tendon. At those points the end-plate zones of the muscle are situated. The curve rises equally on both sides of these minima. This means that no end-plates are present in those muscular areas and the increase in latent period is due to conduction of the impulse along the muscle fibre.

The rate of conduction of the muscle impulse can be calculated from the slope of the curve.

b) End-plate potential (EPP)

Problem: Record the end-plate potentials in the neuromuscular junction of the sartorius muscle of the frog under normal conditions and after treatment with curare. Determine the spatial distribution and Q_{10} of the individual components of the EPP.

Principle: The synaptic potential in the neuromuscular junction is termed end-plate potential. This differs from a nerve impulse in several ways. It does not comply with the "all or none" law, but depends upon stimulus strength, it arises at the point of transmission and does not spread into the surroundings and its amplitude decreases exponentially with increasing distance from the neuromuscular synapse. It is capable of summation and is of longer duration than a nervous or muscular impulse. When its amplitude attains a certain level a muscular impulse is produced and this spreads along the whole muscle fibre (in vertebrates).

Consequently the end-plate potential must be recorded directly from the synaptic junction and, in order to determine its shape, stimuli which are subthreshold for the production of a muscular impulse must be used.

The first condition is fulfilled by locating end-plates as described in the previous experiment. The second condition can be complied with in several ways. Three methods for the normal and curarised muscle are given below.

When recording from the whole normal muscle it is not possible to obtain an EPP uncomplicated by muscle spike, e. g. by making the nerve volley very small. Two stimuli are necessary. The second nerve volley is applied when the end-plate is in the refractory period after the first stimulus. During the refractory period the muscle spike either does not occur at all (absolute refractory period) or occurs with some delay on the EPP, and the latter thus becomes evident. The potential added in this way by the second nerve impulse is then obtained by subtraction (Eccles and O'Connor 1939).

An easier method is to separate the action potential from the EPP using graded curarisation. Subparalytic curarisation delays and eventually decreases the muscle spike, while the EPP becomes progressively smaller without altering its latency or time course.

The third method consists of recording changes in action potential due to small shifts of the recording electrode from the focus of the end-plate zone. Here separation of the two potentials is obtained, since the muscle action potential arives later owing to the time necessary for conduction. The EPP however, decreases exponentially with distance, so that the shift of the recording electrode must be such that the muscle spike is delayed and yet the EPP does not disappear on increasing the distance (Eccles and O'Connor 1939, Eccles and Kuffler 1941).

The EPP has the highest amplitude in the end-plate zone. It decreases exponentially with increasing distance. This fact may be utilised for the spatial mapping of end-plate zones and the points with a maximum amplitude of EPP thus obtained can then be correlated with those having a minimum latent period of the muscle potential in the same muscle before curarisation.

The Q_{10} of the different EPP components permits some insight into the processes occurring during the different phases of EPP.

Object: frog sartorius muscle with its nerve.

Apparatus: as in experiment VI-A-a. A D. C. preamplifier. A square-wave stimulator with two pulses at the output.

Other requirements: d-tubocurarine chloride in Ringer solution in increasing concentrations from 1 μmol. Otherwise as in experiment VI-A-a.

Procedure: The muscle with its nerve is fixed as in the preceding experiment. First the end-plate zone is found using the method of minimal latent periods. With the electrode on the end-plate the dissociation of the EPP from the spike is tried by an early second nerve volley.

The muscle is then immersed for about half an hour in Ringer solution containing 1 μM of d-tubocurarine. The solution is then sucked off and the response to single nerve volleys is recorded. Care is taken to keep the recording electrode in the end-plate zone. In such a fashion the action potential from the end-plate zone is recorded after treatment of the muscle with different concentrations of d-tubocurarine (Fig. 172 a, b).

If these tracings are projected and thus enlarged several-fold and redrawn in the same time-response coordinates, the changes in slope of the ascending limbs may be determined during progressive curarisation.

As in the preceding experiment for determining the minimal latent period, the recording electrode is shifted along the whole muscle in one mm intervals and the EPP is recorded. It can be seen that the amplitude varies at different points. There are two or more amplitude maxima along the muscle. Beyond those maxima no EPP is observed in the direction towards the tendon, i. e. in those areas having no end-plates. The exponential decrement of EPP in this zone is shown in Fig. 172c-e.

The Q_{10} of the end-plate potential is then determined. The EPP is recorded immediately after the muscle was equilibrated for some minutes in Ringer solutions of two different temperatures t_1 and t_2. The *temperature coefficient*, Q_{10}, of the rate of rise and fall of the EPP and of the other characteristics of EPP is then calculated according to the relation

$$Q_{10} = \left(\frac{q_1}{q_2}\right)^{10/(t_1 - t_2)}$$

where q_1 and q_2 are the values of a given parameter at temperature t_1 and t_2.

Conclusion: The end-plate potential evoked by a single nerve impulse in curarised muscle exhibits a characteristic time course, a relatively rapid rise to a summit and a slower exponential decay. The rate of decay of the EPP is very little affected by temperature changes ($Q_{10} = 1 \cdot 3$), while the ascending phase is prolonged nearly three-times when decreasing the temperature by 10°C. It is evident that different processes are responsible for the build up and decay of EPP. It is assumed that the decay is mainly due to passive dissipation of the electrotonic charge along and across the muscle membrane. On the other hand, the onset of the EPP is the main period during which the mediator acts.

EPP exhibits 1—3 amplitude maxima along the muscle length which indicates that there are 1—3 end-plate zones in this muscle. Evaluation of the decrease in EPP amplitude shows that the EPP decays by half in 0·9 mm (fall to 1/e in 1·25 mm).

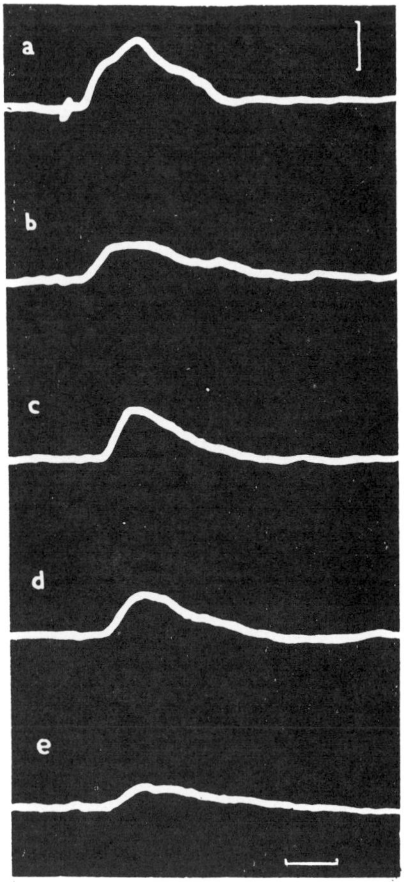

Fig. 172. End-plate potentials. a — during partial curarisation (3 μM d-tubocurarine). b — during complete curarisation (5 μM d-tubocurarine). c-e — spatial distribution of e. p. p. in curarised muscle. c — 7 mm, d — 6 mm, e — 5 mm from the pelvic end of the frog sartorius muscle. Time mark: 5 msec; voltage calibration: 0·2 mV.

The characteristic exponential decrement of the EPP along the muscle and the accompanying slowing down of its temporal course is similar to the spread of an electric charge along a leaky capacitative cable (Cremer 1909, Rushton 1934).

The extent of the spread of the EPP along the muscle fibre depends on the resistance of the fibre, chiefly on the transverse impedance of its surface membrane. Since the surface membrane has a large distributed capacity (Cole and Curtis 1939), the electrotonic potential does not rise suddenly but gradually, depending upon the rate at which local membrane capacitances are charged. Hence the larger the distance from the end-plate focus the slower the rise in the recorded potential (Eccles et al. 1941). It may be objected that the spatial decrement of the EPP is due to diffuse distribution of motor endings rather than to spread of extrinsic current from the depolarisation focus. In such a case, however, the time course of EPP would be unalterable and would not have the characteristic slowing related to electrotonic spread.

Sometimes it is possible to observe, particularly when working with homeotherms, hyperpolarisation preceding the EPP at some distances from the point of depolarisation. This is due to a shunt across excess fluid or tissue as was demonstrated by Bishop (1937) and Bishop and Gilson (1929).

The conclusions derived from extracellular recording of EPP were confirmed and extended by intracellular recording techniques (Fatt and Katz 1951, 1952a), which also made it possible to investigate the ionic mechanisms of the EPP (Del Castillo and Katz 1956, Fatt 1959).

An example of an intracellular recording from the postsynaptic membrane of a neuromuscular junction is given in Exp. VI-A-d.

c) Repetitive stimulation of the end-plate. Recruitment. Wedensky inhibition

Problem: Record the EPP set up by two and by a train of nerve volleys of different frequency. Demonstrate facilitation and Wedensky block in frog neuromuscular junction.

Principle: Depending upon the stimulation frequency, it is possible to restore transmission in a blocked neuromuscular junction or block transmission in a normal muscle.

The fact that a neuromuscular block can be overcome by repeated stimulation is termed *facilitation* or *recruitment* (Adrian and Lucas 1912, Bremer 1927). The EPP acts on the muscle fibre as a subthreshold catelectrotonic potential and can add its excitatory effect to that of the subthreshold current spreading in front of an action potential (Eccles et al. 1941). It follows that two EPPs produced by two successive nervous impulses will summate as do successive catelectrotonic potentials (Schaefer and Haas 1939). When the EPP attains a threshold amplitude, an action potential in the muscle fibre is produced. If the neuromuscular synapse is in a critical state of curare block, neuromuscular transmission may be renewed by summation of successive EPPs.

Facilitation in the neuromuscular junction of frog sartorius muscle can be demonstrated by using a critical curare block of neuromuscular transmission to a single nerve volley. The nerve must then be stimulated by two stimuli varying the interval between them in order to determine the optimal interval for facilitation.

The following is *Wedensky inhibition:* If the stimulation frequency from the nerve exceeds a certain value, muscular tension decreases until the muscle relaxes and muscular electric activity disappears. This is not due to fatigue, since as soon as the stimulation frequency is decreased the muscle

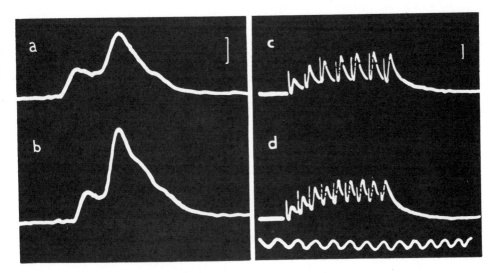

Fig. 173. Summation of EPPs in the frog sartorius muscle. a-b — summation of two EPPs for different intervals between stimuli. c-d — summation in response to a train of impulses. Time mark: 100 c/sec (for c, d). Voltage calibration: 2 mV.

immediately responds normally or even supranormally. Wedensky attributed this phenomenon to differences in lability of the nerve, muscle and neuro-muscular end plate (Wedensky 1892).

Demonstration of this fact is achieved as for facilitation, but the normal end-plate is stimulated with repetitive nerve volleys at different high frequencies.

Procedure: The sartorius muscle with the supplying nerve is prepared as in the preceding experiment. The muscle is immersed in 6 µM-d-tubocurarine chloride for 20—30 min until there is complete block of neuromuscular transmission.

1) Summation of two EPPs.

The nerve is stimulated by two maximal stimuli with different time intervals. First longer intervals (50 msec.) are used. The EPPs are separated from each other. The interval is then gradually shortened (Fig. 173a, b). The oscillograms show that, commencing with a certain interval, the potentials begin to summate. With the shortest time intervals the EPP to the second stimulus has a higher amplitude than with longer intervals. This is called *potentiation*.

If the experiment is repeated under weaker curarisation, which just produces a block, the summated potentials reach a critical level for the occurrence of muscle spikes, the block is overcome and "neuromuscular facilitation" or "recruitment" occurs.

2) Summation in response to repetitive nerve volleys.

Transmission is completely blocked with d-tubocurarine and the nerve is stimulated with a volley of maximal stimuli at 100 c/sec. The time base is slowed down so that all stimuli are visible on the oscilloscope screen.

It can be seen from the tracing that the first end-plate potentials which summate are larger and larger (Fig. 173c, d). Curarisation, however, is fairly strong, so that the critical level for spike formation is not attained. Then, however, both the amplitude and the total summated level decrease. Finally both stabilise at a certain level. After discontinuing stimulation the potential slowly returns to zero line.

3) Wedensky inhibition.

The nerve of a normal nerve-muscle preparation is stimulated with repetitive nerve volleys at frequencies from 30/sec. upwards. The action potential is recorded from the end-plate free region of the sartorius muscle.

The muscle stimulated indirectly responds faithfully to frequencies up to 80 per sec. At higher frequencies the spike from the muscle decreases progressively with time and transformation of the rhythm occurs. If the frequency of stimulation is reduced, the action potentials rapidly reach the original height again.

Conclusion: Potentiation and the subsequent decline may be ascribed to corresponding changes in acetylcholine release (Eccles 1948, Eccles and Macfarlane 1949). Mobilisation of preformed acetylcholine is probably responsible for the potentation. Partial exhaustion of acetylcholine reserves may be reflected in a subsequent decrease of potential. The plateau might indicate that both formation and secretion of acetylcholine have become stabilised.

Mammalian muscles do not show this potentiation phase. In them a decrease in a successive EPPs can be seen right from the beginning (Liley and North 1953). This might indicate that mammalian nerve endings contain much lower acetylcholine reserves.

d) Excitatory and inhibitory junctional potentials

Problem: Record the junctional potentials in response to stimulation of the excitatory and inhibitory nerve in the neuromuscular synapse of the crayfish using intracellular microelectrodes.

Principle: Synaptic potentials evoked by an excitatory nerve in the myoneural junction of *Crustacea* are termed excitatory junctional poten-

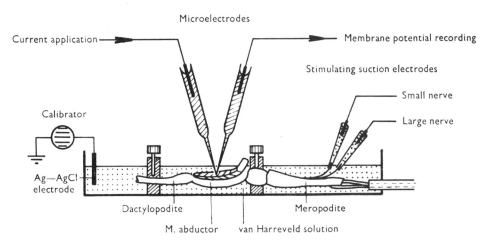

Fig. 174. Recording of junctional potentials from the m. abductor dactylopoditi of the crayfish. One microelectrode is used for current application (when recording i. j. ps). Short parts of the large and small nerve are sucked into separate stimulating electrodes.

tials (e. j. p.) a term analogous to that used for electric activity in the slow muscle system of the frog (Kuffler and Gerard 1947). They are similar to EPP of vertebrate muscles and are also an expression of the electrogenic action of the mediator. On summation of these potentials, however, no conducted spike arises. The muscle fibres of Crustacea are multiply innervated and hence the synaptic potentials have nearly the same size along the whole length of the fibre (Fatt and Katz 1953b). They are, therefore, very suitable for learning the technique of intracellular recording of synaptic potentials, as a search for the synaptic area is not necessary. It suffices to stimulate the muscle nerve and to insert a microelectrode into the fibre. This insertion is also made easier by the diameter of the crustacean fibres which is 2—3 times larger than in vertebrates.

Using a similar procedure it is possible to demonstrate inhibitory junctional potentials (i. j. p.) in the same preparation. These arise on stimulation of the inhibitory nerve. The situation here, however, is somewhat more complex because the inhibitory potential in this synapse becomes manifest only after

415

a change in the membrane potential. For this purpose we must introduce a further microelectrode into the fibre for current application (Fig. 174). If current flows outwards the muscle fibre membrane is depolarized. The nervous inhibitory effect appears then as a hyperpolarizing junction potential (Fatt and Katz 1953a).

Object: M. abductor dactylopoditi of the crayfish (*Astacus fluviatilis*) with the excitatory and inhibitory nerves.

Apparatus: As in exp. IV-C-b.

Other requirements: Two capillary suction electrodes. Microelectrodes filled with 3 M-KCl and 2 M potassium citrate. Arrangements for holding the claw firmly. Perfusion and suction arrangements. Van Harreweld's solution (p. 315).

Procedure: The claws are removed as described previously (p. 315) at the site of natural autotomy. The m.abductor is made accessible by opening the carapace in the propodite from the medial side and by removing the adductor muscle (Fig. 126). During removal of this large muscle the Ringer solution in the preparative dish is constantly renewed to prevent injury of nerve fibres by crushing of muscle fibres. The inhibitory and excitatory nerves run separate courses in the so called large and small nerve which is exposed in the meropodite (Fig. 118).

The preparation is firmly held in a Petri dish filled with van Harreveld's solution (Fig. 174). The large and small nerves are sucked into separate capillary suction electrodes which contain platinized wire. The indifferent electrode is in the solution.

The KCl microelectrode is fitted into the holder of the micromanipulator. After its resistance is measured, it is introduced into the fibre. The membrane potential is —80 mV. Then the small nerve is stimulated with a short (1 msec) rectangular pulse. In response to this a potential change is registered with the microelectrode the shape of which is similar to the EPP of vertebrates (Fig. 175A). The recording electrode is withdrawn from the fibre and reinserted at different distances so that the whole fibre is tested. Everywhere e. j. ps are recorded. Their amplitudes do not vary much.

The nerve is then stimulated with a volley of stimuli spaced at 50 msec intervals. The e. j. ps to successive stimuli are summated (Fig. 175 B-D).

Stimulation of the large nerve containing the inhibitory axon does not lead to any changes in membrane potential. Using a second micromanipulator a second microelectrode filled with 2M potassium citrate is introduced into the same fibre about 100 μ from the recording electrode. According to Boistel and Fatt (1958) electrodes filled with citrate salts (sometimes neutralized with citric acid) are more suitable for applying depolarizing currents than electrodes filled with 3M-KCl. Probably the citrate ions released within the muscle fibre decompose insoluble Ca compounds which otherwise would migrate to the tip of the microelectrode and block it when it becomes positive. The insertion of the current electrode is controlled with the aid of the cathode follower of the recording electrode. After insertion of the current electrode the cathode follower is connected to the recording microelectrode and the output of the DC current source to the citrate electrode. The positive pole is applied to the broad end of the microelectrode (outward current). After switching on the current the membrane is depolarized. We choose current intensities that lead to a membrane depolarization of about 10 mV. Now the inhibitory nerve is stimulated. In this case the nervous effect becomes manifest. The inhibitory junctional potential (i. j. p.) is of a hyperpolarizing character (Fig. 175E).

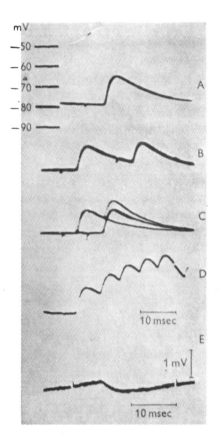

Fig. 175. A — D — intracellular recording of the excitatory junctional potentials (e. j. ps) from a muscle fibre of the crayfish in response to stimulation of the excitatory nerve. B, C — e. j. ps evoked by two stimuli, D — by stimulating the excitatory nerve with a train of stimuli. E — intracellular recording of inhibitory junctional potentials (i. j. ps) to stimulation of the inhibitory nerve. The membrane potential was changed by applying current via the second microelectrode inserted into the same fibre. Hyperpolarizing i. j. p. during depolarization of the fibre.

If the current polarity is reversed the membrane potential increases (hyperpolarization). Stimulation of the inhibitory nerve under the same conditions evokes a depolarizing inhibitory junction potential in the muscle membrane.

Conclusion: It follows from the results that excitatory and inhibitirory neuromuscular junctions are numerous and are distributed along the whole muscle fibre (Fatt and Katz 1953b). A second explanation for this spatial distribution of the e. j. ps over the whole muscle would be the existence of electrotonic currents. The space constant ($\lambda = 2$ mm), however, is too small to ensure this extensive spreading of the e. j. ps. The constancy of the time course of e. j. ps also contradicts such an explanation. In addition Dudel and Kuffler (1961a, b, c) using extracellular microelectrodes demonstrated that numerous individual junction areas are scattered over the surface of each fibre. Using extracellular capillary microelectrodes they recorded sharply localized potentials of individual junctional areas. Intracellular recordings summate or integrate several such junction potentials. Since e. j. ps in individual junctional areas are "quantal" (Dudel and Kuffler 1961a, b, c) it is possible that the variations in e. j. ps recorded along the length of the muscle fibre with intracellular leads reflect the low probability of quantum release in individual areas.

The distribution of i. j. ps may be explained in a similar way. The absence of potential changes in response to the nervous inhibitory effect indicates that the inhibitory reversal potential is identical with or very close to the resting potential as is also shown by the polarity of the potential change during variation of the membrane potential. It can be shown that the reversal potential is the same as or is close to the equilibrium potentials of Cl ions. The inhibitory nervous effect in the inhibitory junction of the crayfish increases mainly the chloride conductance of the membrane (Boistel and Fatt 1958). Gamaaminobutyric acid acts in the same way simulating the effect of the inhibitory mediator in this junction (Boistel and Fatt 1958; Zachar et al. 1964b, c). This acid is present in inhibitory nervous axons only (Kravitz et al. 1963a, b).

e) Excitation of the end-plate by acetylcholine. Microelectrophoretic injection

Problem: Demonstrate the excitatory effect of acetylcholine on the region of the end-plate in the sartorius muscle of the frog.

Principle: The neuromuscular synapse in amphibians is a chemical one, i. e. it is cholinergic with acetylcholine as the mediator. The process of this mediation may be divided into several stages according to modern opinion (for review see e. g. Eccles 1963, McLennan 1963):

(1) Synthesis of acetylcholine (Ach) by the cholinacetylase enzyme system at nerve endings.

(2) Release of Ach from presynaptic terminals by the approaching nerve impulse.

(3) Diffusion of Ach through the synaptic cleft to the receptors of the postsynaptic membrane.

(4) Combination of Ach with the receptor which leads to changes in the conductance of the postsynaptic membrane.

(5) Destruction of Ach by acetylcholinesterase (AchE) which occurs simultaneously with (4).

(6) Generation of the end-plate potential and spikes as a result of conductance changes.

The synaptic cleft opens into the extracellular space. It is hence understandable that Ach appears in the perfusion fluid after stimulation of the nerve (Dale and Feldberg 1934, Dale et al. 1936, Dale 1937). On the other hand it may be expected that an "injection" of Ach into the synaptic cleft from a nerve ending could be simulated by injection of Ach from the external source. Buchthal and Lindhard (1937, 1939) and Kuffler (1943) had already shown in isolated fibres that Ach locally applied in the region of the synapse simulates the nervous action. The time parameters of the effect of external Ach, however, were much slower than when stimulating the nerve. Only the introduction of the iontophoretic application of Ach with microelectrodes (Nastuk 1953, 1957, 1959, Del Castillo and Katz 1955, 1956, 1957) made it possible to record the electrogenic effects of both kinds of Ach "injections" (natural and artificial) on the same time scale (Krnjević and Miledi 1958).

In principle this method involves (see also Expt. on p. 619) expulsion of Ach ions out of the microelectrode filled with Ach, using an anodic pulse when the electrode is close to the synaptic region. The amount of externally applied Ach can be regulated by altering the duration and intensity of the pulse. The potential change induced at the synapse by the microelectrophoretic injection or by nerve stimulation is studied with a normal microelectrode filled with 3M-KCl (Fig. 176). The Ach-microelectrode can also be placed

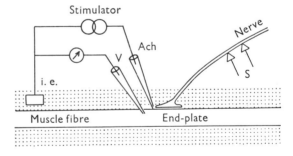

Fig. 176. Microelectrophoretic application of acetylcholine (Ach) to the region of the end plate. The resulting potential change is recorded with an intracellular microelectrode (V). S — stimulating electrodes.

intracellularly and the effect of Ach on the inside of the membrane can thus be studied.

Object: The frog's sartorius muscle and nerve.

Apparatus: As in exp. IV-A-c. A stimulator giving rectangular pulses (for microinjection). A source of D. C. current.

Other requirements: As in expt. IV-A-c. Frog Ringer solution (for composition see p. 354). Frog Ringer solution with 3 times the normal Ca concentration. Acetylcholine chloride (2—4 µM). Capillary suction electrode.

Procedure: The sartorius muscle of the frog with its nerve supply is prepared in the usual way (p. 278) and firmly secured as in expt. III-A-c. The nerve branch to the sartorius is sucked into the capillary suction electrode which will serve as the stimulating electrode against the indifferent electrode in the solution.

Normal frog Ringer solution is exchanged for Ringer solution with a higher Ca content which increases the threshold at which twitches occur. The Ach microelectrode is placed into one holder of the micromanipulator and connected to the source of D. C. current in such a way as to prevent diffusion of Ach from the microelectrode (negative pole in the electrode, voltage of several tenths of volts). The microelectrode is brought close to a branch of the nerve above the muscle fibre and the tip of the electrode follows the branch to its termination (as long as we can see it). Occasionally the braking current is switched off and twitches of the muscle fibre are observed. If the resistance of the electrode is high (several hundred M Ω) diffusion of Ach itself is insufficient to evoke a reaction and it is necessary to apply a current pulse of opposite polarity for a short time (inside of electrode is positive). For lower resistance of the Ach-electrode (about 30 M Ω) the switching off of the current alone is sufficient to induce twitches. By varying the braking or injecting current and by moving the electrode above the nerve ending in the fibre we roughly determine the site most sensitive to Ach. At that point the KCl microelectrode is inserted. The membrane jump is about 90 mV (inside negative). A short "injection" current (about 10 msec) is applied into the *Ach electrode*. Its intensity is lower than that used when looking for the myoneural junction. The depolarizing potential (Fig. 177Aa) is recorded with the KCl electrode. This potential is similar to the EPP evoked by direct stimulation. It is termed the *Ach-potential*. If the intensity of the injection pulse is raised sufficiently an action potential appears on the Ach potential (Fig. 177Ab, B). The resulting twitch, however, usually pushes the recording microelectrode out of the fibre. After making sure that it is not broken we move both the recording and Ach-microelectrodes to find the site where the amplitude of the Ach-potential is greatest and the rate of increase of the ascending part of the Ach-potential highest. These indicators vary extraordinarily steeply with the distance between the injecting microelectrode and the synapse. Several responses to subthreshold injections of Ach are recorded. Then the Ach microelectrode is inserted into the fibre. The application of the same or of a larger injection current pulse does not evoke an Ach potential. In response to a rectangular injection current pulse we only obtain an electrotonic depolarizing potential (the current flows out of the fibre).

We stimulate the nerve branch with the capillary suction electrode using a weak short pulse and record the EPP uncomplicated by any spike activity in order to compare it with the Ach-potential.

By moving the Ach and KCl electrodes to a non-innervated part of the muscle fibre we make sure that the microelectrophoresis of Ach does not cause depolarization of the fibre.

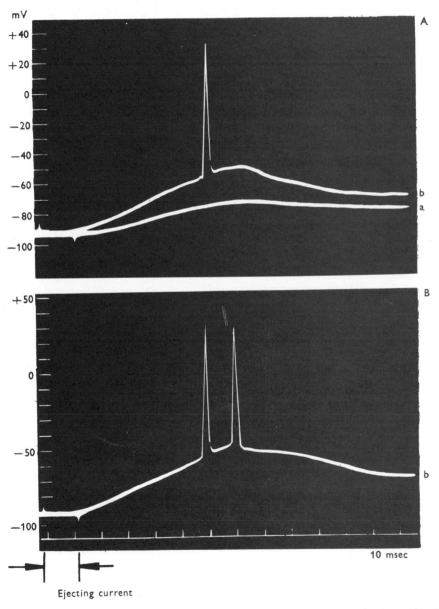

Fig. 177. Ach potentials induced by ejecting Ach from the microelectrode situated extracellularly above the end plate using a short positive rectangular current pulse applied to the microelectrode. a — subthreshold, b — superthreshold Ach-potential.

Conclusion: It follows from the experiment that Ach has an electrogenic effect limited to the region of the myoneural junction if applied from outside the membrane. The ineffectiveness of the intracellular application of Ach indicates that Ach receptors are localized on the outer surface of the post-synaptic membrane. The Ach potential is similar to the EPP in shape and time parameters. If the tip of the microelectrode is carefully placed close to the synapse (several μ) and Ach is released from the pipette by a very short current pulse the course of the Ach potential and of the EPP are nearly identical (Krnjević and Miledi 1958). Complete identity is too much to expect in view of the different geometric arrangement of the natural and artificial Ach sources. The latter first activates receptors at the periphery of the synaptic cleft. The duration of the ascending phase (T) and the amplitude of the Ach potential (A) vary with the distance of the Ach source from the receptor according to the following relationships (Del Castillo and Katz 1955):

$$T = \frac{r^2}{6D} \tag{1}$$

$$[\text{Ach}]_{max} = \frac{Q \cdot e^{-1.5}}{8r^3 \, (\pi/6)^{1.5}} \tag{2}$$

where [Ach] is the concentration of Ach in the receptor. The amplitude A of the Ach potential is proportional to this concentration. Q is the amount of released Ach, r the distance between the tip of the injecting microelectrode and the receptor, D the diffusion coefficient (about 8×10^{-6} cm^2 sec^{-1}). From equations (1) and (2) it can be seen that the amplitude A of the Ach potential falls as the third power of the distance between the electrode tip and the receptor while the duration of the ascending phase is prolonged with the square root of this distance. This accounts for the great changes in these parameters for small movements of the injecting microelectrode. With the aid of these equations and recordings of the Ach potential we can calculate the amount of injected Ach that can evoke an EPP. This amount is very close to the amount of Ach released by 1 nerve impulse (1.5×10^{-5}g/impulse in the rat diaphragm — Krnjević and Miledi 1958). This good agreement is a serious argument in favour of the Ach mediation of neuromuscular transmission. Externally injected Ach also induces the same conductance changes of the postsynaptic membrane as injection of Ach by a presynaptic impulse (Del Castillo and Katz 1956).

f) Miniature end-plate potentials

Problem: Record the miniature EPPs in the neuromuscular junction of the frog.

Principle: Using an intracellular microelectrode inserted into the endplate region it is possible to record with sufficient amplification (1 mV/cm) spontaneous discharges the shape of which is similar to the EPPs. Their amplitude, however, is only 1% of that of the EPP. Because of their similarity to EPPs Fatt and Katz (1952b) who discovered them, called them miniature end-plate potentials. There is much evidence to show that miniature EPPs are due to quantal release of Ach from presynaptic nerve endings. One such Ach quantum contains several thousand molecules of Ach. Ach quanta are relatively stable in size in the same muscle fibre and hence the amplitude of the miniature EPP is uniform (for review see Del Castillo and Katz 1956, Katz 1958, 1962, Eccles 1964a). Synaptic vesicles are considered as the histological correlate of Ach quanta. These are parcels of Ach molecules visible in the presynaptic endings of chemical synapses (de Robertis 1958).

Since the size of the vesicles is independent of the size of the postsynaptic muscle fibre the amplitude of the miniature EPP is indirectly proportional to the muscle fibre diameter (Katz and Thessleff 1957) as follows from the cable characteristics of the muscle fibre and the corresponding equations (experiment IV-C-b). Hence more clear-cut min. EPPs can be recorded from thin muscle fibres.

The amplitude of the min. EPP falls similarly to that of the EPP with the distance between the microelectrode and the end-plate and hence the recording microelectrode has to be inserted close to the neuromuscular synapse. In a nerve-muscle preparation the end-plate is found in one of the ways described in experiment VI-A-a. Usually, however, it is sufficient to follow a nerve branch which remains on the muscle fibre after cutting off the nerve proper and then to look for the region of the largest amplitudes of these potentials by repeatedly inserting the electrode. Electrophoretic application of Ach as described in exp. VI-A-e can also be used to localize the end-plate.

If we want to avoid a search for the neuromuscular synapse min. EPPs can easily be recorced from muscle fibres with multiple innervation, e. g. those of the crayfish (Exp. VI-A-d). In this case all we have to do is to insert the microelectrode into any part of the fibre and to use great amplification in the recording system. Since the space constant of the membrane is longer than the distance between individual neuromuscular synapses on the fibre there is no attenuation with distance. Such fibres are very large, however, and the amplitude of the min. EPPs is relatively very small (Dudel and Kuffler 1961a) so that a high gain of the amplifying channel is necessary for their recording.

For demonstrating min. EPPs we have chosen the m. ext. dig. IV. of the frog where it is possible to record min. EPPs that are 5 times larger than those obtained in the classical object (m. sartorius) where their amplitude is about 0·1 mV (Fatt and Katz 1952b).

Object: M.ext.long.dig.IV of the frog.

Apparatus: As in experiment VI-A-c. Total amplification of the recording system 1 mV/cm.

Other requirements: Normal Ringer solution (p. 354) and Ringer solution with twice the concentration of K ions (addition of KCl). Hypertonic Ringer

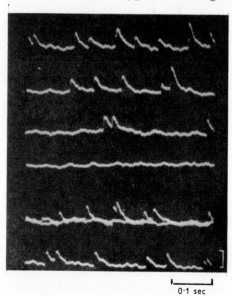

0·1 sec

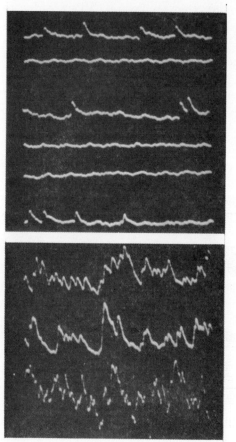

Fig. 178. Miniature EPPs from the m. ext. longus dig. IV of the frog. A, C — in Ringer solution, B — in hypertonic solution (NaCl added to Ringer).

solution (addition of 120 mM saccharose), Ringer solution with physostigmine (10^{-6}M), or with d-tubocurarine (5 . 10^{-7} and 10^{-6}M). A suction device.

Procedure: The muscle is prepared as in exp. IV-E-b. Under microscopic control we follow the nerve branch along a muscle fibre and at the site of contact with the fibre we insert the recording electrode. A membrane potential of about −80 mV appears. The gain is raised from 10 mV/cm to 1 mV/cm. At this amplification, if we are in the endplate region, potentials are seen at irregular intervals on the oscilloscope screen. These are the min. EPPs (Fig. 178). The microelectrode is then withdrawn from the fibre and inserted at another site until the best place is found. At first it is well to use Ringer with prostigmine since in such a solution the amplitude of the min. EPP is several times greater (2—3 mV). Not until we have gained some experience with the localization of end-plates and the discharge characteristics of the min. EPPs do we use normal Ringer solution.

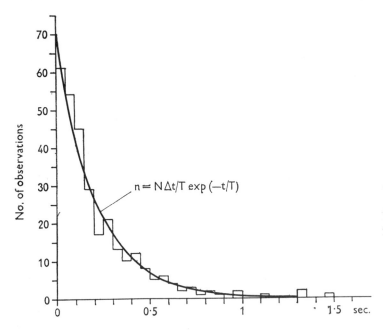

Fig. 179. Distribution of time intervals between 300 miniature EPPs, covering a total period of 63 sec (the mean interval, $T = 210$ msec). Full line is distribution expected for random sequence [n = $N \Delta t/T$ exp $(-t/T)$]. From m. ext. longus dig. IV of the frog in Ringer solution.

This is then exchanged for a solution with twice the K ion concentration. The microelectrode is in the fibre. After the exchange of solutions the frequency of the min. EPP is approximately doubled, but their size remains unchanged. After exchanging this solution for a hypertonic (increase of osmotic pressure by 50%) solution a still greater increase (30—50 times) of the frequency of spontaneous potentials is obtained (Fig. 178B). In normal Ringer solution this effect disappears again (Fig. 178C).

We now test the effect of d-tubocurarine. A concentration of $5 . 10^{-7}$M decreases the amplitude of the min. EPPs by half and they disappear completely if a concentration greater than 10^{-6}M is used.

Evaluation: A graph of the distribution of the intervals between individual spontaneous discharges is made. Intervals are grouped in 50 msec classes (Fig. 179 abscissa). The shape of the histogram is in good agreement with the theoretical curve which represents the expected Poisson distribution characteristic for a random sequence of potentials according to the equation (Fatt and Katz 1952b):

$$n = N \frac{\Delta t}{T} e^{-t/T}$$

where n = the number of potentials observed in the individual classes,
N = the number of observations (300 for Fig. 179),
T = the mean interval (210 msec in our case),
Δt = the size of the histogram class (50 msec),
t = the observed interval between individual potentials.

Conclusion: Miniature end-plate potentials (min. EPPs) are similar in shape to end-plate potentials evoked by a nervous impulse. Their amplitude, however, is smaller by two orders of magnitude than that of the EPP. Min. EPPs show the same spatial decrement as EPPs. Curare decreases and anticholinesterase increases the amplitude of both the min. EPPs and EPPs indicating that their mechanism of generation is similar. Differences are only quantitative, in the amount of Ach released from nerve terminals. Spontaneous release of Ach which induces the min. EPP proceeds in equal quanta as indicated by the very constant amplitude of these discharges. In contrast to this stable amplitude the frequency of the min. EPPs considerably varies under the influence of different agents which affect the mechanism of Ach release from the nerve terminals, e. g. by their depolarization in an excess of K^+ ions. This depolarization may be regarded as a model of what occurs during the action potential in the nerve ending. It causes a synchronous discharge of a considerable amount of Ach from the nerve terminals leading to the formation of the EPP. This is well shown in experiments on the effect of Ca ion lack (Fatt and Katz 1952b). If Ca ions are removed from the medium the action potential induces an EPP, the amplitude of which is reduced down to that of the min. EPP. Under this condition there is also a fluctuation of responses (EPPs) to successive stimuli. The probability of occurrence of such individual responses is met by Poisson distribution as is the occurrence of spontaneous min. EPPs shown in Fig. 179. This indicates that there is no interaction between discharges. Miniature discharges thus occur randomly and the probability of their occurrence in any given time interval remains independent of the preceding discharge.

Min. EPPs have been recorded from many chemical synapses (for review see Fatt 1959, Del Castillo and Katz 1956, Eccles 1964a) even where there is no cholinergic mediation. They occur in the noncholinergic junctions of the crayfish (Dudel and Kuffler 1961a, b, c) and in adrenergic neuromuscular synapses (Burnstock and Holman 1961, 1962). They have even been described

426

in the CNS (review: Eccles 1964a), where they still have to be distinguished from random synaptic bombardment.

As there is a correlation between the occurrence of min. EPPs and the presence of synaptic vesicles in presynaptic terminals it is probable that the quantal release of the mediator is a general phenomenon. The presence of synaptic vesicles, however, does not necessarily result at rest in the spontaneous release of quanta of the mediator and thus in electrophysiological manifestations of this release — miniature end-plate or synaptic potentials, since the membrane potential of some presynaptic terminals can be below the threshold level of the background liberation of mediator in the resting state.

B. Synaptic transmission in a sympathetic ganglion

Sympathetic ganglia represent a very convenient agglomeration of synaptic junctions in the periphery. The ratio of the total synaptic area to the mass of the ganglion is high in comparison with the end-plate, where the synaptic area forms only a minute part of the nerve-muscle preparation. Furthermore it is possible to perfuse ganglia with arteficial solutions, which enables quantitative measurements of the mediator output to be carried out during rest and activity. Ever since the discovery of the cholinergic nature of transmission (Kibjakov 1933), the sympathetic ganglia represent a favorite object for those interested in physiology and pharmacology of chemical mediation. Most of our knowledge about acetylcholine metabolism stems from investigations carried out on these structures. A comprehensive picture of cellular mechanisms of acetylcholine manufacture was given by Birks and MacIntosh (1957, 1961). Electrophysiologically, the synaptic transmission is characterised by synaptic potentials, which can be registered easily by extracellular leads. Intracellular registration is more difficult to perform as compared with the end-plates, owing to the resistance offered to the microelectrode tip by connective tissue (R. M. Eccles 1955, 1963).

a) Action potentials from the superior cervical ganglion

Problem: Record the action potential from the isolated superior cervical ganglion of the rabbit.

Principle: In order to determine the responses which a preganglionic nerve volley evokes in the presynaptic and postsynaptic elements of the sympathetic ganglion, various lead positions are used as shown diagramaticaly in Fig. 180.

In analysing the responses obtained with these lead combinations it is necessary to keep in mind that the cross section of the preparation is not uniform. The cross-section of the ganglion is about ten times larger than that of the nerve trunks. Hence both resistance per unit length and the density of longitudinal current produced by potentials in preganglionic terminals are much lower than in the nerve trunks. Consequently, the recorded potential is attenua-

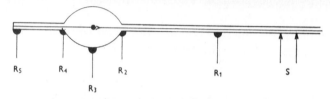

Fig. 180. The usual arrangement of recording electrodes when recording from the superior cervical ganglion.

ted and the discrimination of recording is lessened. This means that electrode R_3 will record preganglionic potentials as if it were at R_2 rather than potentials of the terminals in the ganglion.

In this experiment only action potentials recorded with R_1R_2 and R_3R_5 leads are described. Electrodes in position R_1R_2 record action potentials from the preganglionic nerve fibres. With leads R_3R_5, potentials generated in the activated ganglion cells are recorded rather than those of postganglionic nerve fibres.

For interpretation of records with other leads and in other ganglia, the reader is referrred to the papers by Bishop (1936), Bronk (1939), Eccles (1935a, b, 1936, 1937), Llyod (1937, 1939a, b), R. M. Eccles (1952ab).

Object: The isolated superior cervical ganglion of the rabbit.

Preparation: The rabbit is anaesthetised with dial. The preganglionic trunk is exposed in the neck and cut several cm from the ganglion. The postganglionic trunk is separated from the internal carotid artery, the surrounding connective tissue is removed and the trunk is cut where it enters the skull base. Care is taken to preserve the blood supply to the ganglion. Finally the ganglion is rapidly removed after tying silk loops to both its ends and immersed in Krebs solution at room temperature. In the solution the sheaths of the ganglion are carefully removed so that ions can diffuse into the ganglion and the nerve.

Apparatus: A cathode-ray oscilloscope with a preamplifier, total sensitivity 100 µV/cm. D. C. amplification or a capacity coupled amplifier with a large time constant (about 8 sec).

Other requirements: A recording chamber made of perspex with openings for the electrodes, for admitting gases and for exchanging solutions.

428

Procedure: The ganglion is fixed by the loops to the chamber and the stimulating and recording electrodes are placed on it. The chamber is filled with Krebs solution aerated with 95% O_2, 5% CO_2. The chamber is put into a Magnus bath warmed to 37°C. When recording, the ganglion must be taken out of the solution. This is achieved by running the solution into another vessel in the same bath connected to the chamber by a rubber tube.

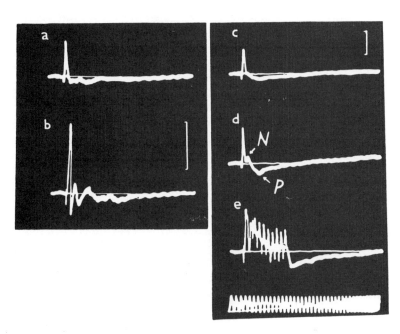

Fig. 181. Excised superior cervical ganglion of the rabbit. a — R_1, R_2 leads for a lower, b — for a higher stimulus strenght applied to the preganglionic trunk. c-d — recordings (R_3, R_5 leads) for gradually increasing stimulating pulse strength. e — response to a train of impulses. N — slow negative, P — slow positive potential. Time mark: 100 c/sec. Voltage calibration: 0·5 mV.

Leads R_3R_5: As in the ganglion in vivo, a relatively sharp and synchronous spike is produced by a weak stimulus. This is folloved by slow negative (N) and slow positive (P) waves (Fig. 181c, d). With stronger stimuli, the initial spike which represents a discharge of ganglion cells splits into components S_a and S_b (R. M. Eccles 1952a). The S_a spike is compounded of the S_1 and S_2 waves in the cat.

When stimulating rhythmically (Fig. 181e) with a volley of impulses, it can be seen that on cessation of the stimulation the N and P waves are much larger than after a single preganglionic volley (Bronk et. al. 1938, Eccles 1944).

Lead R_1R_2*:* Fig. 181a, b shows a tracing from these leads. A weak stimulus produces a synchronous spike with a conduction velocity of 20—25 m/sec (Fig. 181a). On applying stronger stimuli more irregular activity appears after the first spike (Fig. 181b). The large preganglionic spike forms the S_a component in lead R_3R_5, and the small one the S_b component.

Conclusion: The action potential recorded from the superior cervical ganglion (lead R_3R_5) consists of a spike that is produced by impulses arising in ganglion cells. In addition there is a large negative and positive after-potential which is several times larger than any recorded from a peripheral nerve.

The interpretation of the complex sequence of waves, which the pre-ganglionic volley evokes in sympathetic ganglia of both turtles and rabbits (Laporte and Lorente de Nó 1950, R. M. Eccles 1952a, b), was enabled by the results of simultaneous electrophysiological and pharmacological investigations of synaptic transmission in the ganglia (R. M. Eccles 1952a, b, R. M. Eccles and Libet 1961, Libet 1962a, b). Some examples of this approach are given in the next experiment.

b) Synaptic potentials in the superior cervical ganglion

Problem: Record the synaptic potentials and the late negative wave in the isolated superior cervical ganglion of the rabbit and determine the effect of d-tubocurarine and prostigmine.

Principle: As in the neuromuscular junction, it is possible to prevent the initiation of spikes in the postsynaptic neurone with relatively low concentrations of d-tubocurarine. This substance acts as a competitive inhibitor of the depolarising effect of acetylcholine. Under such conditions, the action potential disappears in the ganglion cells and only a prolonged synaptic potential termed postsynaptic (Eccles 1943) remains, since this occurs on the postsynaptic parts of the synapse.

The summation of postsynaptic potentials may be obtained if the time interval between the stimuli is suitably chosen. If the blockade with d-tubocurarine is not very deep, the critical level for spike initiation in the postsynaptic neurone may be attained by summation. If recordings are made using a very slow time base it is found that the postsynaptic potential (N) is followed by a prolonged late negative wave (LN). The LN wave is absent at the neuromuscular junction. Obviously the transmission in the sympathetic ganglion is much more complicated.

Object: The isolated superior cervical ganglion of the rabbit. Preparation as in the preceding experiment.

Apparatus: A D. C. oscilloscope with a D. C. preamplifier, total sensitivity 100 μV/cm. A stimulator giving one pulse or a volley with a variable number of pulses. The pulse or the beginning of the volley is synchronised with the sweep of the time base. Time marker 1000 c/sec and 10 c/sec.

Other requirements: As in the preceding experiment. Krebs solution with different concentrations of d-tubocurarine.

Procedure:

1) Synaptic potential.

The electrode arrangement $R_3 - R_5$ is used. With them the action potential is recorded (Fig. 182a). The solution in the chamber is then exchanged for

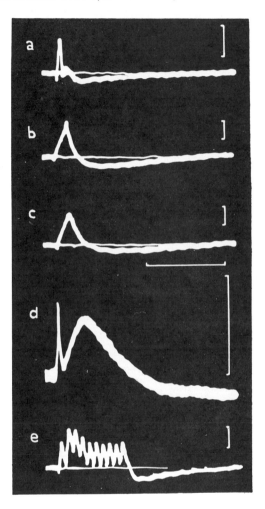

Fig. 182. The effect of d-tubocurarine on transmission in the superior cervical ganglion of the rabbit. R_3, R_5 leads. a — action potential before curarisation. b — partial curarisation ($1 \cdot 5 \cdot 10^{-5}$M). c — complete curarisation ($2 \cdot 85 \cdot 10^{-5}$M). d — synaptic potential and late negative wave. e — response to repetitive preganglionic nerve volleys (60 c/sec). Voltage calibration for a — 2 mV, for b and c — 0·25 mV, for d — 0.2 mV, for e — 0.25 mV. Time mark for a, b, c 500 msec, for d 2 sec.

Krebs solution with $1 \cdot 5 \cdot 10^{-5}$M d-tubocurarine. The preparation is left in this solution for 20 minutes and the response to the same stimulus applied to the preganglionic trunk is recorded (Fig. 182b). A synaptic potential is produced. This still has a spike on its peak. Curarisation is not complete. The preparation is then placed in a more concentrated d-tubocurarine solution ($3 \cdot 10^{-5}$M). This produces complete curarisation and a pure synaptic potential is recorded (Fig. 182c).

If the sweep speed of the oscilloscope is slowed down it is seen that the synaptic potential is followed by a late negative wave (Fig. 182d). This is usually preceded by a small positivity.

2) Response to repeated stimulation.

If the preganglionic trunk is stimulated with a train of impulses summation of synaptic potentials is obtained (Fig. 182e). This results in the block being overcome, with a discharge of impulses arising in the ganglion cells if a concentration that just blocks the spikes is used. This is facilitation similar to that in neuromuscular transmission.

3) The effect of anticholinesterases on synaptic processes.

The action potential in a synaptic ganglion is recorded with leads $R_3 R_5$. After blocking with d-tubocurarine ($2 \cdot 8 \cdot 10^{-5}$M), a typical synaptic potential with waves P and LN is recorded. After increasing the curarine concentration to $8 \cdot 10^{-5}$M, the N wave decreases but the P wave increases. If $3 \cdot 10^{-6}$M prostigmine are added, waves N and LN are decreased and wave P is considerably increased.

Conclusion: It can be seen from the tracings that the postsynaptic potential in the superior cervical ganglion has characteristics similar to EPP. Theoretically there are as yet no proofs that acetylcholine has a depolarising effect by shortcircuiting the membrane as is the case for the end-plates.

The experiment with different curare doses showed differences in the effect of deeper curarisation on the N wave on the one hand (decrease), and the P and LN waves on the other hand (increase).

Anticholinesterase also has different effects on the P and LN waves. The P wave is increased and the LN wave decreased. The key observation which enabled the interpretation of the individual waves was performed by R. M. Eccles and Libet (1961). They showed that botulinum toxin exerts a uniform depressant action on each wave. Since this toxin acts by depressing the output of Ach from the nerve terminals (Burgen et al. 1948), it follows, that the N, LN and P waves are initiated by cholinergic terminals. The shape and the time relations of these waves can be accounted for by a different structural pattern of praeganglionic nerve terminals (Fig. 183). The N wave is produced by the depolarising action of Ach released from the presynaptic fibre and is responsible for generation of the postsynaptic spike. The LN wave is caused by the diffusion of Ach to a more remote place, where the LN wave is initiated. The generation of P — wave is supposed to be produced by disynaptic action. The synapsis on the chromaffin cell is cholinergic (the effect of botulinum toxin) and that of the chromaffin cell on the ganglion cell adrenergic. Adrenalin exerts its depressing action in sympathetic ganglia possibly by a hyperpolarising action (Marrazzi 1939, Büllbring 1944, Lundberg 1952), which would explain the polarity of the P-wave. This explanation is supported by the blocking

432

effect of dibenamin (Libet 1962). Curarising agents (dTc) increase the P and LN-waves only apparently by depressing the N-wave (in a similar manner as in the end-plate). The increase of the P-wave during anticholinesterase action is due to its depressing action on the LN-wave (competitive inhibition with Ach at LN-sites). For the effect of other drugs, the original papers or the review of Eccles (1964b) should be consulted.

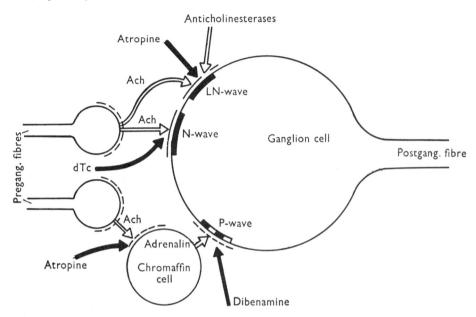

Fig. 183. The generation of N, P and LN waves in the superior cervical ganglion of a cat (diagrammatically after R. M. Eccles and Libet 1961; Eccles 1964a). The inhibitory nervous or pharmacological action is indicated by black arrows, the excitatory action by white arrows. The depressant action of botulinum toxin on the terminals of pregang-lionic fibres is not shown.

c) Occlusion and facilitation in the superior cervical ganglion

Problem: Demonstrate occlusion and spatial facilitation in the isolated superior cervical ganglion of the rabbit.

Principle: These phenomena, observable in the sympathetic ganglion, follow from the principle of convergence. If the preganglionic nerve is divided into two parts, both bundles may be stimulated separately or together. From the amplitude of the action potentials from the postganglionic trunk it may be determined that the resultant amplitude is not a simple arithmetic sum of amplitudes obtained when separately stimulating the nerve bundles. It can be larger than their sum (spatial facilitation is produced) or smaller (occlusion occurs).

Object: The isolated superior cervical ganglion of the rabbit.

Apparatus: A cathode-ray oscilloscope with an A. C. amplifier, total sensitivity 100 μV/cm. A stimulator with two independent outputs giving pulses with changeable amplitudes.

Procedure: The ganglion is prepared as before. The preganglionic trunk, however, is carefully divided along its longitudinal axis into two approximate halves of about 15 mm length. The preparation is then fixed in the chamber so that each trunk lies on a separate stimulating electrode. The recording electrodes are situated on the postganglionic trunk.

Facilitation: The response to maximal stimulation for S_a fibres of one and the other preganglionic bundle is recorded (Fig. 184a, c). Then the stimulus is decreased to submaximal until the amplitude is half the maxi-

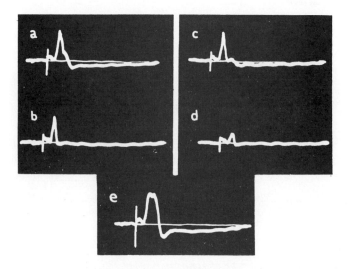

Fig. 184. Spatial facilitation in the superior cervical ganglion of the rabbit. R_4, R_5 leads. a — response to a maximal stimulus in one preganglionic bundle; c — response to a maximal stimulus in the other one; b, d — responses to submaximal stimuli in corresponding preganglionic bundles; e — response to simultaneous submaximal stimulation of both bundles.

mum value (Fig. 184b, d). The preganglionic trunks are then stimulated together (Fig. 184e). If now the action potentials for the 1st and 2nd stimulus are subtracted from the common tracing, a certain negativity remains. This indicates that the potential due to simultaneous stimulation of both bundles is greater than the sum of those potentials. This is spatial facilitation.

Occlusion: Again both nerves are stimulated maximally. If both trunks are stimulated simultaneously with a maximal stimulus it can be shown that the recorded potential is smaller than the sum of the amplitudes of the action potentials obtained on separate stimulation. This is occlusion.

Conclusion: Fig. 185 best explains occlusion and facilitation. Fig. 185a shows the mechanism of occlusion. Each nerve bundle (n_1 or n_2) "supplies" a

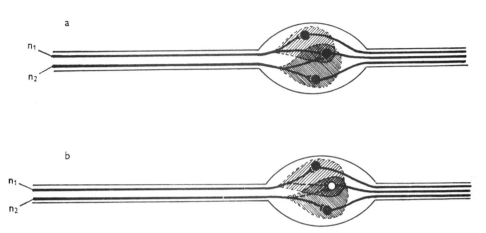

Fig. 185. Diagramatic representation of the mechanism of occlusion (a) and spatial facilitation (b).

certain number of ganglion cells and these become active on maximal stimulation. Since some fibres of trunk-n_1 converge onto ganglion cells innervated also from trunk-n_2, the resultant effect of simultaneous stimulation of both bundles must be less than the sum of the amplitudes obtained when stimulating each bundle separately. Ganglion cells supplied from both trunks do not partake in the addition.

Spatial facilitation is shown in Fig. 185b. This was produced by weak stimuli. Each preganglionic fibre on submaximal stimulation produces activation of several ganglion cells, some of which are stimulated only subliminally. The region of cells stimulated is termed the discharge zone, the region of subliminally stimulated cells is termed the subliminal fringe. Some cells of the subliminal fringe are supplied from both n_1 and n_2 bundle. If the latter are stimulated separately, these cells are not activated, but if they are stimulated together a discharge is produced in the cells. Hence the resultant amplitude of the postganglionic trunk is higher than the sum of amplitudes when stimulating separately.

C. Electrically transmitting junctions

It is a characteristic sign of such junctions that transmission is achieved by the depolarizing effect of the electric current which is generated by nerve impulses in the presynaptic axon. This reminds one of impulse transmission in earlier models of the synapses which were called *ephapses,* false synapses (Hering 1882, Kwassow and Naumenko 1936, Arvanitaki 1942). Impulses were transmitted from one axon to another at the altered site of their contact, by electrical transmission. In view of this common transmission mechanism the naturally occurring junctions with electric transmission are sometimes also called ephapses (Grundfest 1959). In model ephapses transmission is not polarized, it is capable of propagating an impulse in either direction. Naturally occurring ephapses are of great variability from the simplest septal nonpolarized junctions and electric bridges between neurones to electrically transmitting synapses with one-way transmission. Mixed types with electrical and chemical transmission in the same synapse have also been described (review Eccles 1964b, Grundfest 1959).

a) Septal synapses of giant axons

Problem: Demonstrate the bidirectional transmission of impulses and of electrotonic potentials in the septal synapses of the lateral giant axon in the nerve cord of the crayfish (*Astacus fluviatilis*).

Principle: The lateral giant axons of the nerve cord of the crayfish are composed of a series of segments seperated by membranes which lie at an angle of 45^0 to the axis of the segments (Fig. 186, see also Fig. 121). The membranes at the junctions of these segments are termed septa and hence the name of this type of axon. Individual segments arise by the fusion of small fibres derived from several cells (Young 1939). In some species individual segments fuse. In *Annelidae* and *Crustaceae,* however, they remain separated by membranes (septa). In contrast to the earth-worm the septal surfaces in the crayfish are covered by Schwann cells and connective tissue. This layer is about 2 μ thick. The width of the septum and the greater interseptal distance (in comparison to worms) make the crayfish septate axons a suitable object for studying septal transmission of excitation (Kao 1960, Watanabe and Grundfest 1961).

The septa represent nonpolarized junctions, i. e. they transmit an impulse in both directions. In contrast to the interseptal parts of the axon, however, they have a lower safety factor of transmission and thus block of transmission first occurs in those places. The postseptal potential formed during

436

a block behind the septum is equivalent to an artificial extrinsic potential (IV-B-c). In a damaged axon the presence of the septum is seen in the recording of the action potential as an inflection on the rising phase. The resistance of the septal membrane causes the attenuation of the electrotonic potential recorded postseptally.

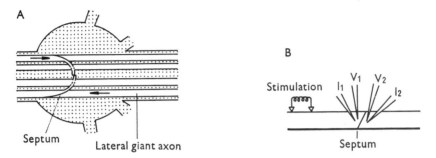

Fig. 186. A: dorsal aspect of the third abdominal ganglion of the nerve cord of the crayfish (diagrammatically). B : electrode arrangement when studying the septal transmission in the giant lateral axon. I_1 and I_2 : microelectrodes for current application, V_1 and V_2 : voltage-measuring electrodes. The stimulating electrodes are situated rostrally to the septum.

Fig. 186B is a diagrammatic representation of the experimental lay-out necessary to demonstrate these effects. V-microelectrodes are for recording potential changes against the indifferent electrode outside the fibre, I-electrodes are for current application.

Object: The nerve cord of a crayfish.

Preparation: The nerve cord is isolated as in exp. IV-A-c from the middle of the thorax down to the 6th abdominal ganglion.

Apparatus: The amplifier and stimulator as in exp. IV-C-b.

Other requirements: As in exp. IV-A-c. Four micromanipulators, a source of D. C. current (a 4·5 V battery with a potentiometer).

Procedure: The nerve cord is held in the recording chamber as in exp. IV-A-c and also to the bottom of the canal by two pieces of soft paraffin. Thus 3 compartments are formed, fairly well isolated from each other. The 2nd abdominal ganglion which we shall work with, is in the middle compartment. Fine silver wires placed one into each compartment will be used as the stimulating electrodes against the earthed middle compartment.

We find the septum which is localized above the ganglion. The site where the commisural branch leaves the base of the segment helps us to find the septum (Fig. 186). A dark field condensor and a 100 fold magnification of the microscope is used.

The first electrode (V_1) is inserted rostral to the septum. The external stimulating electrodes are also located rostral to the septum. The resting potential has a value of −80 mV. We stimulate with a short (0·5 msec) rectangular pulse (using an isolation unit). The action potential has a smooth rising phase without any inflection (Fig. 187A) and lasts for 200—500 µsec at half the amplitude. The descending phase is less steep and some after-depolarization can be seen. With the second micromanipulator the second microelectrode (V_2) is inserted caudally to the septum (at about 50 µ from the preseptal one). The same cathode follower is used (if we do not have a separate amplifying system for each channel). We then stimulate. The action potential differs from the previous one by having an inflection on the rising phase (Fig. 187B). The inflection point becomes more apparent, as the stimulation frequency is increased. At a critical frequency the transmission is blocked and only a septal potential is recorded postseptally (Zachar, unpublished). Another way of demonstrating the block of septal transmission is as follows.

The third microelectrode (I_2) is inserted postseptally with the third micromanipulator (again the same cathode-follower is used, its input cable being connected to the third electrode). After insertion into the fibre an inward hyperpolarizing current from a D. C. current source is applied to the electrode. We use the external electrodes to stimulate and record postseptally with V_2. As the hyperpolarization increases, the spike appears later and later on the septal potential until only the septal potential is recorded. We note at which level of hyperpolarization the transmission block occurs and then connect the input of the cathode follower to the preseptal electrode V_1. As the action potential is present here, the block neccessarily occurred in the septum. We note that the hyperpolarization is smaller at V_1 than at V_2. This means that the septum forms a barrier for D. C. current.

If the stimulation frequency is just critical (Fig. 187C) a slight hyperpolarization at I_2 causes the block of almost all incoming preseptal impulses (Fig. 187D).

Now the cord is stimulated from the other end. The D. C. current is applied at the same site I_2 (which is now on the side of the stimulating electro-

---

Fig. 187. Septal transmission in the giant lateral axon in the nerve cord of *Astacus fluviatilis*. A — action potential recorded with microelectrode V_1 (preseptally), stimulating electrodes are rostrally (see Fig. 186), B — postseptal action potential recorded at V_2; delaying of postseptal action potential from the septal potential by increasing the stimulation frequency. C — postseptal action potentials elicited with stimulation frequency just below the blocking magnitude. D — delaying of postseptal action potentials and blocking of transmission across the septum by hyperpolarising the fibre at I_2. The same stimulation frequency as in C.

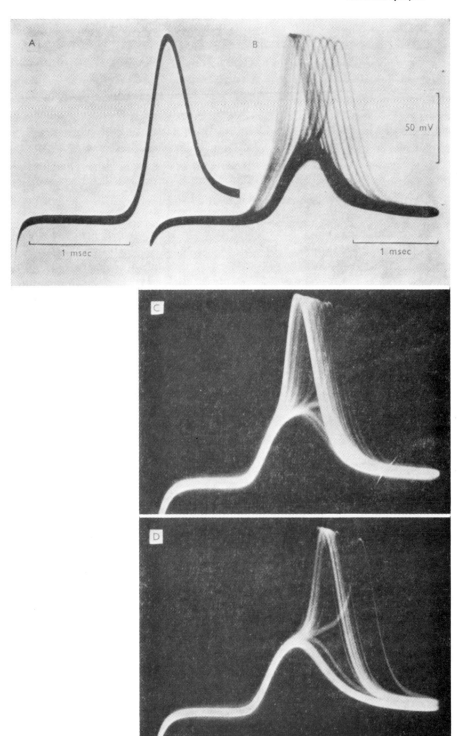

des). In this case, however, (for an equal hyperpolarizing current) action potentials are recorded both with electrodes V_1 and V_2.

We insert a fourth electrode (I_1) or transfer the electrode I_2 to the other side of the septum so that it will act as electrode I_1. The whole experiment is now repeated from the beginning. If the external electrodes are on the side of the septum opposite to V_1 and I_1, an action potential without inflection is recorded in V_2 and one with inflection in V_1. Thus the septum is not a valve for the transmission of impulses. If I_1 is hyperpolarized then the action potential at V_1 disappears after a certain hyperpolarization is attained and only the septal potential remains. The action potential is present in V_2. Hyperpolarization is greater in V_1 than in V_2. When stimulating the cord from the other end and plarizing at I_1 spikes are recorded in both V_1 and V_2.

Now the experiment is arranged in such a way that rectangular pulses lasting 50 msec are applied through electrode I_1 and the amplitude of the electrotonic potentials in response to current flowing in both inward and outward directions is recorded at electrodes V_1 and V_2. The same procedure is repeated with the current applied to I_2. In both cases the electrotonic potential is lower postseptally.

Conclusion: The experiment showed that the septum in the lateral giant axon of the crayfish represents a nonpolarized junction since it permits impulses to pass in either direction. The safety factor of the junction is lower than in the interseptal segment of the axon as shown by the block of transmission when the postseptal part (with respect to the direction of the impulse) is hyperpolarized. The blocking effect of an inward current injected through the microelectrode into the postseptal segment is evidently due to its compensating effect on the current generated by the spike in the preseptal segment which flows in outward direction through the postseptal sector.

The attenuation factor, as follows from a comparison of the pre- and postseptal voltampere characteristics is about 2·5 and this may be nearly fully ascribed to the resistance of the septal membranes (Watanabe and Grundfest 1961).

Thus in the propagation of an impulse along the septate axon each septum acts as a resistance which is added to the conductivity of the core. In this way electrotonic transmission of depolarization responsible for impulse propagation is weakened. Practically the situation in septal synapses differs only slightly from transmission along continuous fibres.

b) Electrical synapses with one-way transmission

Problem: Demonstrate the rectification characteristics of the giant motor synapse in the crayfish.

Principle: The giant lateral septate axon in the nerve cord of the crayfish is the presynaptic fibre of this junction. The giant motor fibre leaving the cord in the third root (Fig. 121) is the postsynaptic fibre. The synapse proper is at the point of contact of these axons (Hardy 1894, Johnson 1924). The lay-out of the experiment (Fig. 188) is principally the same as in the study of septal junctions (VI-C-a). The stimulating extracellular electrodes serve to induce the

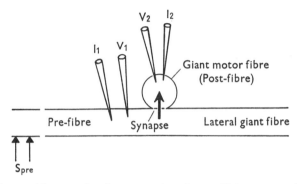

Fig. 188. Experimental lay-out for demonstrating the rectifying properties of the electric synapses between the giant lateral and the giant motor axons. The arrow shows the direction in which the impulses are conducted through the synapse. V_1, V_2 — recording electrodes, I_1, I_2 — microelectrodes for current application. S_{pre} — external stimulating electrodes on the presynaptic axon. S_{post} — not shown on the postsynaptic axon.

impulse in the presynaptic (S_{pre}) or postsynaptic (S_{post}) axon and thus the irreciprocity of transmission across the synapse (Furshpan and Potter 1959a) can be demonstrated. Presynaptic and postsynaptic potential changes are recorded with microelectrodes V_1 and V_2 respectively. Microelectrodes I_1 and I_2 are used to apply rectangular current pulses to the pre- or postsynaptic fibre and electrotonic potentials are recorded with microelectrodes V_1 and V_2 (Furshpan and Potter 1959a, b).

Object: Nerve cord of crayfish.

Apparatus: As in the preceding experiment.

Other requirements: As in exp. VI-C-a. Capillary suction electrodes.

Procedure: Preparation. Since the third root must be stimulated, the cord is exposed not by the ventral (as previously) but by the dorsal approach. After removal of the thoracic and abdominal exoskeleton the first task is to reach the attachments of the third roots to the flexor muscles. These are made accessible by separating the musculature along the midline. The third roots are cut as far away from the cord as possible. The nerve cord is held as in the previous experiment, but no paraffin partitions are used. Stimulation is performed with capillary suction electrodes. The third root of the 2nd abdominal ganglion is sucked into a glass tube of about 0·5 mm diameter together with the solution. The negative pressure is produced with a syringe attached to the other end of the glass tube via a rubber tubing. The glass tube contains a chlorinated silver wire. The second capillary with a smaller opening is placed under microscopic control above the lateral axon in the region of the lower abdominal ganglia. The slight negative pressure attained by the syringe causes partial sealing of the capillary to the fibre and reduces the shunting of the stimulating current. The stimulating current is applied between the inside of the capillary and the bath.

The microelectrode V_1 is introduced and the presynaptic axon (S_{pre}) is stimulated. This is usually not difficult since the lateral axon is well visible in the dark field. The tip is localized about 100 μ from the border of the giant motor axon. The same cathode-follower (as in exp. VI-C-a) is used to guide insertion of the microelectrode V_2 into the giant motor axon above the point where it crosses the lateral giant axon. This is more difficult since these axons are not so well visible as the lateral ones. It is of some aid to vary the illumination using the microscope mirror. Since the axon is situated superficially we attempt to introduce the microelectrode without direct visual control. A further difficulty rests in the fact that penetration of the microelectrode through the superficial structures of the axon is not easy. After recording the intracellular potential drop stimulation of the third root (S_{post}) is started. If the electrode is in the motor giant axon an action potential appears. It is to be expected that the motor axons of many preparations of the nerve cord will, mainly at first, respond with graded responses to stimulation of the third root since they are more readily injured (because of their surface localization). These preparations are not discarded since the experience of Furshpan and Potter (1959a) has shown that the synaptic junction under these conditions works normally. Conduction across the synapse is always tested by stimulating the dorsal surface of the cord (S_{pre}). In good preparations an action potential is recorded with electrode V_2 the ascending phase of which has a well dicernible inflection due to an ephaptic potential similar to that seen in septal synapses (VI-C-a).

We now stimulate with the electrodes S_{post} and record with microelectrode V_1. No action potential occurs. If the sensitivity is increased several times, a potential may be recorded, the amplitude of which is about 50 times smaller than that of the potential recorded at V_2. Thus the synapse conducts in one direction only.

In order to demonstrate rectification current microelectrodes I_1 and I_2 must be inserted. A rectangular current pulse (lasting 50 msec) is applied to electrode I_1 and several (5—10) electrotonic potentials are recorded with electrode V_1 in response to current pulses of different intensity. The current intensity is such that it generates electrotonic potentials of max. 20—30 mV when applied in the outward direction and 50—60 mV in the inward direction (hyperpolarization). Then we record, when stimulating with the same electrode I_1 and within the same range of current intensities, postsynaptic electrotonic potentials with electrode V_2. It is at once obvious that electrotonic potentials in response to depolarizing pulses are conducted much better than those in response to current of opposite polarity.

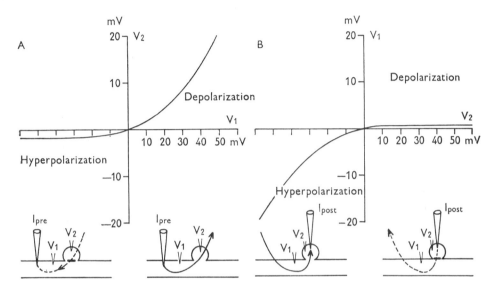

Fig. 189. Transfer characteristics of the electric one-way synapse between the lateral and motor giant axons of the crayfish. A — dependence of amplitude of electrotonic potentials (in steady state) in the postsynaptic fibre (V_2) on the amplitude of the electrotonic potential in the presynaptic fibre (V_1) during current application to the presynaptic fibre (I_{pre}). B — as in A but current applied postsynaptically (I_{post}). The direction of the applied current is indicated.

The experiment is repeated while the current is applied via electrode I_2 (postsynaptically) and electrotonic potentials are recorded first at V_2 and then at V_1. In this arrangement it is obvious already from the amplitude of the electrotonic potential on the monitor that electrotonic potentials are transmitted much more readily from the postsynaptic to the presynaptic axon in response to hyperpolarizing pulses than if depolarizing currents are used, i. e. in the opposite way as compared with the preceding experiment.

Using the recordings of electrotonic potentials and current values we first construct, as in exp. IV-C-b, graphs of the voltampere characteristics $I_1:V_1$, $I_1:V_2$, $I_2:V_2$, $I_2:V_1$ for both current directions. From these curves the conduction characteristics of the synapse can be obtained by substituting for V_1 and V_2 values corresponding to the same current intensity applied presynaptically — I_1 (Fig. 189A) or postsynaptically — I_2

(Fig. 189B). This time-consuming procedure can be speeded up by either direct recording of the volt-ampere characteristics (Furshpan and Poter 1959a, Henček and Zachar 1965) or by direct recording of the conduction characteristics ($V_1 : V_2$ — Furshpan and Potter 1959a), but this requires two recording channels.

Conclusion: The experiment showed that transmission through the giant motor synapse has a very short latent period (0·1 msec) which is in contrast with the long latency (1—2 msec) of structurally similar but chemically transmitting giant synapses of the squid stellate ganglion (Bullock and Hagiwara 1957, Hagiwara and Tasaki 1958). The junctional or ephaptic potential has the typical EPSP (excitatory postsynaptic potential) form. In these characteristics the synapse is similar to septal synapses (IV-C-a) but differs from them in that it is polarized, unidirectional both for transmission of the action potential and the passage of current. The orthodromic direction is thus from the lateral giant axon to the motor giant axon. The results obtained with application of current and a study of its transmission across the synapse by electrotonic potentials can be explained by a single mechanism both quantitatively and qualitatively. The synapse permits current to pass orthodromically but represents a resistance in the antidromic direction, i. e. it acts as an electrical rectifier. The *"synaptic rectifier"* is oriented in such a way that local currents related to the action potential in the presynaptic fibre stimulate the postsynaptic fibre but not the other way round. (For electron microscopy see Robertson 1955, 1961.)

VII

The electrophysiology
of sensory receptors

Electrophysiological methods for studying the function of sensory receptors may be divided into two large groups: Recording of nervous impulses from nerve fibres leaving the receptors, and recording of electrical manifestations of processes in the receptor occurring during application of a specific stimulus. Such recordings are difficult in an anatomically complex receptor organ such as the eye. Recently progress has been made in this direction by the use of ultramicroelectrodes. It has thus become possible to analyse the activity of individual components of such complex receptor organs or of some suitably shaped simple receptors.

Electrophysiology has been used for studying many receptors. Results have been generalised to a certain extent and several monographs and reviews have been devoted to this subject (Adrian 1928, Granit 1955, Gray 1959a, b).

Some fundamental rules of receptor activity are demonstrated below for some receptors. The application of the two methical approaches to other receptors can be studied in the monographs mentioned above. Experiments in this chapter were chosen so as to show basic principles of the methodical approach to the investigation of sensory receptors.

A. Models of sensory organs

Characteristic discharges of nervous impulses in nerve fibres from specifically stimulated receptors can be demonstrated in the isolated nerve fibre to which a weak constant current is applied.

Nonmyelinated crustacean nerves are particularly suitable for this purpose. These respond to a D. C. current by repetitive discharge.

In such a manner the fundamental characteristics of a discharge in nerves in response to receptor stimulation may be demonstrated:

a) dependence of discharge frequency on the stimulus strength,
b) adaptation,
c) "on" and "off" effect,
d) spontaneous receptor activity,
e) initiation of receptor potentials.

a) The effect of D. C. current on the isolated nerve fibre

Problem: Determine the dependence of the discharge frequency on the intensity of a weak constant current pulse in the isolated inhibitory nerve in the claw of the crayfish (*Astacus fluviatilis*).

Principle: The response to constant current varies considerably in different axons (Fessard 1936, Arvanitaki 1938, Hodgkin 1948). Hodgkin studied several kinds of axons in *Carcinus meanas* and divided them into 3 groups according to the kind of response.

His first group of axons is most interesting. They have no significant supernormal phase in the recovery cycle and are capable of responding within a large frequency range.

Zacharová (1957) found that the inhibitory nerve of *Astacus fluviatilis* belongs to this group.

Responses of this group of fibres to constant current are similar to those of sensory receptors.

In principle, then, the isolated inhibitory fibre of *Astacus fluviatilis* is stimulated with constant currents of different strengths and the resulting activity is recorded.

Object: The isolated double fibre (inhibitory and exitatory nerve fibre) leading to the abductor muscle of the claw of *Astacus fluviatilis*. The nerve is prepared as in experiment IVBb.

Apparatus: A cathode-ray oscilloscope with an A. C. amplifier; total sensitivity 100 μV/cm. A source of D. C. current. A microampermeter. A time marker.

Other requirements: As in experiment IVAb. Non-polarisable Ag-AgCl electrodes.

Procedure: The preparation is fixed to the recording arrangement as in experiment IV-A-b. The double-fibre to which the recording electrodes are applied is under paraffin oil. The nerve trunk with the nonpolarisable electrodes is above the paraffin oil. The distance between the nonpolarisable electrodes is 2 cm.

That strength of the stimulating pulse is chosen at which one action potential appears. This current strength is taken as unity. The number of spikes increases with the strength of the constant current (Fig. 190).

At a certain current strength (Fig. 190c) the rheobase of the excitatory fibre is reached and its spike is registered (smaller spike — see experiment IVBb).

Conclusion: The recordings show that the discharge frequency depends on the strength of the stimulus. If the stimulus continues to act, the discharge frequency decreases. This is accomodation of the nerve.

There is a general similarity between this recording and the response from a receptor (Adrian 1932, Matthews 1931a, b, Hartline and Graham 1932).

A graphical evaluation of the relationship between current strength and repetition interval, on the one hand, and the response time from the beginning of the stimulus, on the other hand, gives curves as shown in Fig. 191.

These curves are very similar, indicating that both are determined by the same process.

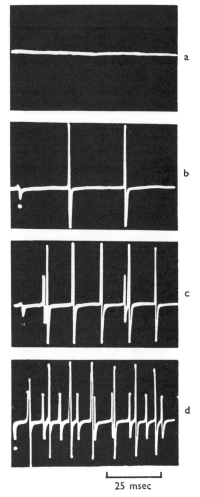

Fig. 190. The effect of a weak D. C. current on the inhibitory fibre of *Astacus fluviatilis*. Dots indicate the current make.

25 msec

This model experiment led Hodgkin to assume that for the frequency of a repetitive discharge, the response time of a sensory organ to a constant stimulus may be as important as its refractory period.

Several explanations have been offered for the repetition interval between pulses in a volley produced by a constant current. The first was given by Adrian 1928. This is based on the recovery cycle of nervous impulses. It is

447

assumed that a given stimulus strength produces a constant degree of excitation in the axon and that the second response appears only when the nerve has sufficiently recovered from the effect of the preceding impulse. Thus a weak stimulus will excite late in the refractory period and the repetition interval will be great. Strong stimuli excite sooner at the beginning of the refractory

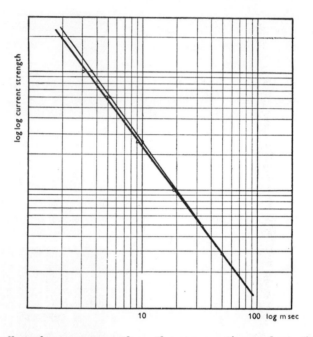

Fig. 191. The effect of current strength on the response time and repetition interval in a single axon of *Astacus fluviatilis*. Filled circles: relation between current strength and response time (interval between make of current and the first spike). Open circles: relation between current strength and repetition interval (interval between first and second spike). Abscissa: interval in msec. Ordinate: log . $\log_{10}$ of current strength, rheobasic strength taken as unity

period and the interval is short. The upper frequency limit depends on the absolute refractory period.

This hypothesis requires an extremely long refractory period for receptors in order to explain the long intervals between natural impulses.

This is partly overcome by Hodgkin's hypothesis as supported by experiments of the same kind as reported here (Hodgkin 1948).

With the recording electrode close to the polarizing cathode and using a D. C. amplifier, it can be demonstrated that the discharge frequency is related to the magnitude of depolarisation of the fibre in a similar fashion as the sensory discharge in a receptor depends on the amplitude of the receptor potential.

B. Impulse activity in sensory nerve fibres

Afferent impulses from muscles. Stretch responses

Problem: Record the response from the nerve bundle, leading to the m. extensor brevis profundus dig. III. in the frog, to stretching of this muscle.

Principle: In muscle, the muscle spindles and Golgi tendon organs form the proprioceptive sense organs. The muscle spindle system of the frog consists of a bundle of intrafusal muscle fibres with a series of sensory organs distributed along it. The intrafusal fibres of the sensory bundle (about 8) are running the entire length of the muscle from tendon to tendon. The sensory region is the place where the branches of a sensory axon contact with the intrafusal muscle fibres. The sensory branches at first intertwin between and coil around the individual intrafusal fibres and then, after losing their myelinisation, they break up into fine varicose endings ('flower-spray' sensory endings). The sensory region is encased in a capsule filled probably with lymph. The striation of intrafusal muscle fibres appears to be lost for a short distance within the sensory region.

The efferent inervation of the extrafusal muscle fibres is accomplished by two motor systems. The 'large' motor axon system innervates the extrafusal twitch fibres while the 'small' motor axon system exerts its influence on the tonic muscle fibres.

Both systems participate also in the motor innervation of the intrafusal muscle fibres. The intrafusal end-plates occur between the sensory regions and are of both the twitch and tonic type. A motor axon often branches to the intra- and extrafusal muscle fibres.

A detailed anatomy and light-microscopy of the frog's muscle spindle system in the m. ext. longus dig. IV was given by E. G. Gray (1957). For electron-microscopy see Robertson (1957) and Gray (1959a, b).

The mammalian muscle spindle differs in many respects from that of a frog (see Barker 1948, 1962a). The spindle sensory innervation in mammals is double with the IA afferent fibres arising from the primary, or the annulospiral endings of the nuclear bag and the II fibres arising as 'flower' spray (secondary) endings in the myotube. The efferent control of the spindle activity in mammals is accomplished only by 'small' motor axon system which appears to be exclusively intrafusal in mammals. The mammalian extrafusal muscle fibres are innervated by 'large' motor axon system (γ — motoneurons), as the tonic fibres in mammals, with the exception of the extrinsic eye muscles (Tiegs 1953), are all intrafusal muscle fibres.

The muscle spindles both of the frog and of mammals are brought into the action by lengthening of the muscle in accord with their anatomical arrangement (in parallel with the extrafusal muscle fibres).

449

The Golgi tendon organ is, on the other hand, susceptible to tension changes as it is located in series with the muscle. The impulses arrising in tendon organs are conveyed by I B afferent nerve fibres.

This anatomical organisation makes it possible to identify fibres and to state wheter they belong to the spindles or tendon organs (Matthews 1933).

The transformation of the mechanical energy into the nerve impulses can be studied quantitatively only if single units are available. In principle this may be achieved in two ways: isolation of one end-organ or recording from single nerve fibre.

The first method may be used with some species of animals. In frogs it is possible to find small muscles with one spindle and one sensory nerve ending. In mammals this is practically impossible. They have many spindles and their muscles contain also many other nerve endings (tendon-organs, small nerve fibres, etc. — Matthews 1933).

The second method is very tedious and a large sample must be used so that the receptors of nerve fibre activity may be identified.

Electrophysiologically this is not difficult since only the standard amplifying apparatus and oscillograph and recording electrodes placed on an exposed segment of the isolated nerve fibre are required. The problem that is difficult and tedious is preparing the fibre.

Here the first method as used by Matthews (1931a, b) has been chosen.

Object: The small muscle on the superior lateral aspect of the middle toe of the frog is used (m. ext. brev. prof. dig. III; Fig. 155). This muscle extends the toe and produces adduction. It is supplied by the peroneal nerve. Only the lateral branch is exposed. On the average two out of three such preparations give unit discharges. The other either does not respond, or more than one unit is recorded.

The muscle is then freed from the rest of the body and only the proximall end remains attached to the bone.

Apparatus: A cathode-ray oscilloscope with an A. C. preamplifier, total sensitivity 100 μV/cm. Time marker. A camera for the oscilloscope with continuous movement of the film. A set of weights.

Other requirements: A binocular dissecting microscope. A preparation dish, dissecting instruments, threads.

Procedure: The preparation is fixed in a moist chamber. One tendon is fixed. A loop of thread is tied to the other tendon. A metal hook (weights will be hung on it) is fixed to the other end of the thread outside the chamber. The nerve is placed on platinum recording electrodes in such a way that it hangs freely between its entry into the muscle and the first recording electrode. This prevents movement artifacts when applying stretch to the muscle.

The part of the chamber containing the muscle is filled with frog Ringer solution and paraffin oil so that the muscle is completely submerged with the nerve in paraffin oil.

If a 2 g load is suddenly put on the muscle, a frequent discharge is obtained. This gradually decreases as shown in Fig. 192d, e.

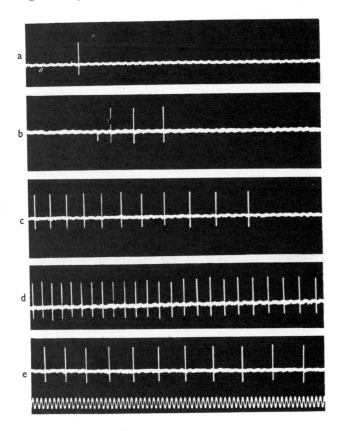

Fig 192. Discharge in sensory nerve fibre after proprioceptive stimulation The muscle is loaded with increasing weights at 5 minute intervals. The last recording is continuation of the preceeding one. M. extensor brevis profundus digiti III. Time mark: 25 c/sec.

If the experiment is repeated with a smaller load, a slow random discharge continues for several minutes. With heavier loads the discharge stops much sooner. With very large loads it ceases after several seconds; probably the end organs have become permanently injured. After several such large loads, no response is obtained.

If the muscle is not loaded supramaximally the experiment can be repeated very exactly as long as at least 5-minute intervals are observed between stretchings.

The dependence of the frequency of discharges on the size of the stimulus.

The response of a sensory nerve to stimuli of different strengths is recorded. At 5-minute intervals the muscle is loaded with increasing weights. A linear relationship is obtained if the discharge frequency within the first two seconds after application of the stimulus is plotted against the logarithm of the load.

This relationship is a concrete example of Fechner's rule which states that a geometric increase in stimulus strength results in an arithmetic increase in sensation.

Adaptation.

The number of impulses per second not only depends on the stimulus strength but also on the time during which the stimulus acts. At first the discharge frequency is high, then slows until a certain constant frequency is attained. This is called adaptation.

Adaptation is not the same in all sense organs. The muscle spindle belongs to the group of slowly adapting sense organs as do other stretch receptors (e. g. receptors in the lungs), cold receptors, nociceptive receptors in the cornea, pressure receptors of the carotid sinus. Rapidly adapting receptors include cutaneous touch receptors, especially those connected to fibres with much encapsulated Paccini bodies.

Conclusion: The above experiments demonstrate, in addition to the fundamental characteristics of receptors, the fact that the sensory end apparatus complies to the "all or none" rule, as has been known since the first experiments with isolated sensory fibres were performed (Adrian 1932). Stronger stimulation does not produce a larger action potential. Sense organs signalise changes in stimulus strength by changing the frequency of discharges. Weaker stimulation produces a slower discharge of impulses, a stronger one a more frequent discharge.

Stimulus strength is also signalised in another way — by so called recruitment of end organs. This is activation of additional sensory end organs, with production of discharges in further sensory fibres leaving them.

As far as the muscle spindle function in the frog is concerned, the experiment has shown, that the receptor is operated by the increase of the muscle fibre length. During contraction shortening we would then expect the sensory discharge to cease. In normal animal this stoppage of sensory firing during contraction is avoided by automatic activation of intrafusal muscle fibres. It was shown, indeed, that the sensory discharge from muscle spindles increases when stimulating motor axons (Eyzaguirre 1957). It should be remembered that the innervation of extrafusal and intrafusal fibres is common (Katz 1949,

Eyzaguirre 1957, Gray 1957), so that during reflex excitation both the extra-
and intrafusal fibers are activated.

The independent efferent innervation of the extra- and intrafusal muscle
fibres in the mammalian muscle spindle (Leksell 1945, Kuffler et al. 1951,
Kuffler and Hunt 1952) furnishes a finer regulation but requires a more complex
nervous organisation (γ-loop, Leksell 1945). The participation of the muscle-
spindle in the stretch reflex system can be elucidated only when working with
a preparation in which the loop is left intact. For this purpose only a negligible
tiny strand of the dorsal root is cut to sample spindle activity, leaving the remainder of the dorsal root and the ventral root intact (see Granit 1955, 1962,
Barker 1962b).

C. Receptor potentials

A receptor potential is a potential change which depends upon the size
or the rate of increase of the stimulus and is produced in the receptor. The
discharge frequency in sensory axons leaving the receptor is dependent upon
the size of the receptor potential.

This potential is thus an intermediary link in the process of transforming external energy into a nervous impulse.

The receptor potential is similar to local processes in nerves or synapses,
i. e. to a local or a synaptic potential. It is localised at the point at which it
arises, it spreads electrotonically with a decrement for a small distance only,
its amplitude is gradual, depending on the stimulus strength. The nervous
impulse arises in the sensory axon when the receptor potential attains a certain
amplitude. Finally the receptor potential is a depolarising potential.

This potential was found in different receptors, e. g. the muscle spindle
(Katz 1950), the eye of *Limulus* and *Dytiscus* (Hartline and Graham 1932,
Bernhard 1942, Granit 1947), the cones (Svaetichin 1954), the Paccini body
and the lobster stretch receptors (Gray and Sato 1953a, b, Kuffler and Eyzaguirre 1955, Alvarez-Buylla and Ramirez de Arellano 1953). Gray (1959a) uses
the term 'receptor potential' for potential changes generated in single receptors
and the term 'generator potential' for any graded potential change in a sensory
receptor or in a complex sense organ.

Other local changes in receptors are also known, e. g. the electroretinogram of vertebrates, microphonic potentials, electro-olfactogram. Further
research will show whether they also belong into the category of receptor
potentials.

The existence of receptor potentials does not exclude the possibility that
some end organs may be activated by a chemical mechanism without receptor
potentials as an intermediate link.

The experiments described below are intended to show some fundamental characteristics of receptor potentials, recorded both with external (muscle spindle) and internal leads (stretch receptor organ). They also give some examples of other local changes in complex receptors, the basis of which is not yet clear (ERG, EOG).

a) The receptor potential of a muscle spindle. Extracellular recording

Problem: Record the receptor potential from a muscle spindle of the frog for different stimulus strengths.

Principle: When a muscle is stretched mechanical energy is transformed into nervous impulses through local potential changes which can be recorded from the sensory axon at a point near the muscle spindle. This potential change has all the characteristics of a receptor potential: it increases with stimulus strength, it is local and spreads with a decrement. When a certain level is attained, impulses are produced in the sensory nerve.

Using local anaesthetics it is possible to separate these two processes, i. e. local depolarization and impulse discharge, and thus to study the shape of this receptor potential undisturbed by spike activity.

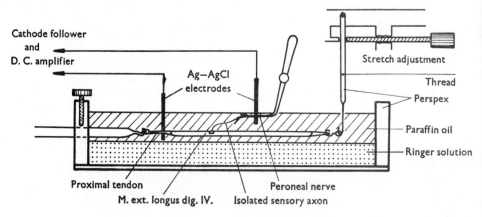

Fig. 193. Experimental set-up for recording the receptor potential from a muscle spindle of the frog. The stretch is applied either by hand (thread) or by means of an electromagnet.

Object: The M. extensor longus digiti IV of the frog with its nerve branch (Fig. 155). This muscle is supplied by a branch from the peroneal nerve. The nerve divides into several bundles before entering the muscle. The most proximal part often contains one motor and two sensory axons.

The muscle and the nerve are isolated, one of the proximal axons is selected and the remaining nerve branches are cut.

Apparatus: A double beam cathode-ray oscilloscope with a D. C. amplifier and cathode follower. Total sensitivity of 100 μV/cm. A time marker. A mechanical stretcher of the muscle (Fig. 193).

Other requirements: Ag-Cl coated silver wires as recording electrodes with

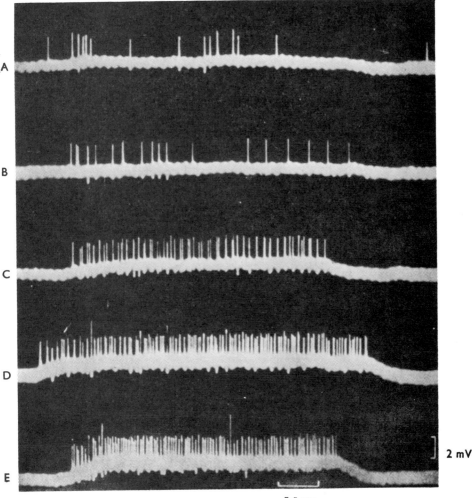

2 mV

0·1 sec.

Fig. 194. Potential changes in the sensory axon from a muscle spindle of the frog's m. ext. longus digiti IV, subjected to a transient stretch of progressively increasing size (A—E). Both the discharge frequency and the depolarization (receptor potential) increase with increasing intensity of stretching.

a potential difference between them as low as possible (<1 mV). Paraffin oil. Frog Ringer solution with 0·3% procain.

Procedure: The sensory axon together with its sheaths is about 30—60 μ in diameter and several mm long (Katz 1950). The resistance of this preparation is very high and may increase the time constant of the input circuit and thus slow down and attenuate the recorded response. This difficulty can be overcome when using a cathode follower in the input circuit.

In this preparation terminal structures remain enclosed in the spindle capsule and thus a change occurring in them is only recorded after electrotonic spreading along fine terminal branches, i. e. with a decrement, so that actual values of registered potentials are not obtained.

Every potential change in the muscle lying between the electrodes may be a serious source of artifacts. These errors may be tested by shortening the axon or by squeezing it where it enters the muscle. The electrodes are situated as shown in Fig. 193. First a so called resting discharge, when the muscle is not stretched and is under physiological tension, is registered. The muscle is then subjected to stretches of increasing intensity and the corresponding responses are recorded. The same procedure is repeated using monophasic recording after the sensory axon has been squeezed below the recording electrode.

The recordings show that during stretching a volley of impulses is produced. The frequency of this discharge is the higher the greater the stimulus strength — the greater stretching. In addition it can be seen that the level shifts upwards during a discharge of impulses. This is the receptor potential. With stronger stimuli the discharge frequency rises as does the level of depolarization (Fig. 194).

The effect of local anaesthetics.

The preparation is then soaked for 10 minutes in a 0·3% buffered procain-Ringer solution and the experiment is repeated. The impulse discharge from the muscle spindle disappears but the receptor potential remains unchanged.

If now the nerve is crushed just before it enters the muscle, no response is obtained from the muscle spindle. This indicates that the recorded depolarization is not an artifact.

Conclusion: The experiment showed that the origin of impulses in the muscle spindle is preceded and accompanied by a depolarizing potential. These spindle potentials may be explained as a local link between the stimulus and the nerve response. This explanation is supported by the fact that there is a correlation between the amplitude of this depolarization — receptor potential — and the impulse frequency in the sensory nerve (Katz 1950). A similar depolarization may be found also in other mechanoreceptors: Paccini's bodies (Gray and Sato 1953a, b) or the stretch receptors of the crayfish (Eyzaguirre and Kuffler 1955a, b).

The intimate mechanism of the conversion of mechanical energy into the receptor potential of the muscle spindle has not yet been fully elucidated. On isolated muscle spindles of the frog Ottoson (1964, 1965) demonstrated the dependence of the receptor potential on the external sodium and calcium concentration as is also the case in other mechanoreceptors: Pacinian corpuscle (Gray and Sato 1953 a, b, 1955) and crayfish stretch receptor organs (Edwards et al. 1963).

b) The receptor potential of a crayfish stretch receptor cell. Intracellular recording

Problem: Record the receptor potential from the stretch receptor cell in the crayfish using intracellular microelectrodes.

Principle: Stretch receptors in the abdominal segments of the crayfish (Alexandrowicz 1951, 1952, Florey and Florey 1955) are physiologically similar

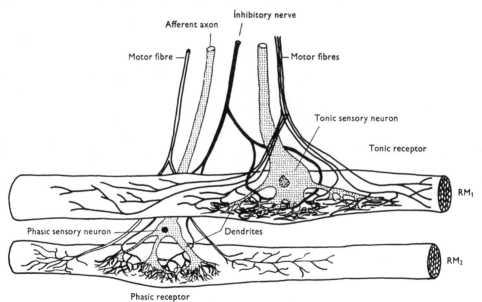

Fig. 195. Stretch receptor organ of crayfish. RM_1 and RM_2 are the muscle bundles of the stretch organ (modified after Autrum 1959).

to muscle spindles in vertebrates. They do not, however, indicate the functional conditions in one muscle but the relative position of a certain segment, i. e. of muscle groups. Hence their number is much smaller than that of muscle spindles.

Two stretch receptor organs are found in one abdominal segment and these are situated symmetrically on both sides of the body. Each stretch receptor organ is composed of two muscle bundles (Fig. 195), a tonic RM_1 and a phasic RM_2. Dendrites from a peripherally located sensory nerve cell run to each of these bundles. The dendrites are activated during stretching of the corresponding muscle bundle. A receptor potential is generated in them and this can be recorded with an intracellular microelectrode from the cell body whence the potential spreads electrotonically in view of the short length of the dendrites.

The division of muscle bundles into tonic and phasic ones follows from the character of the nerve cell response to stretch and also from the character of the contraction of the muscle bundles in response to stimulation of the corresponding efferent nerves. Tonic receptors have a small adaptive capacity. They can react to stretching with a discharge lasting even for several hours (Wiersma et al. 1953). Phasic receptors adapt quickly (within one minute) even to extreme stretching.

The muscle bundles are innervated by separate efferent nerve fibres. Muscle bundle RM_2 responds to nerve stimulation with a rapid twitch, RM_1 with a slow contraction. Receptor organs also have inhibitory innervation which in the crayfish consists of a nerve fibre common for both receptor organs. The inhibitory fibre suppresses the discharge of the sensory neurone (Kuffler and Eyzaguirre 1955).

The receptor potential can be demonstrated in this receptor in several ways depending on the type of activation, either by passive stretching of the muscle bundles or by their contraction due to stimulation of efferent nerves. We shall use the isolated receptor organ and activation by stretching as the most simple method for demonstrating the generation of a receptor potential and of a sensory discharge recorded with a microelectrode inserted into the body of the sensory cell.

Object: The stretch receptor organ from the second abdominal segment of the crayfish.

Topography: The muscle bundles of the receptor organ are situated dorsally from the most medial fibres of the m. superficialis medialis when facing the second abdominal segment from the ventral side. Even though they are localized just below the dorsal carapace of the crayfish they are approached from the ventral side to avoid injuring them.

After decapitation of the crayfish (cutting off the head ganglion) the limbs are removed and the thorax is separated from the abdomen. The abdomen is placed on its back in a dish made from bee's wax and retained with steel needles. The ventral cartilagenous cover of the abdomen and the flexor muscles are removed. The intestine helps us to get oriented since it lies at the margin separating the flexors from the extensors. After its removal the characteristic arrangement of the extensor musculature is evident (Fig. 196, left). The deep extensors (the anatomical nomenclature denotes the dorsal surface as the most superficial one) are closest to us. The mm. profundi mediales form arcs around the axis of symmetry and meet near the tail end. They are removed together with the mm. profundi laterales which form several bundles and pass across the whole segment. Below them the superficial musculature is

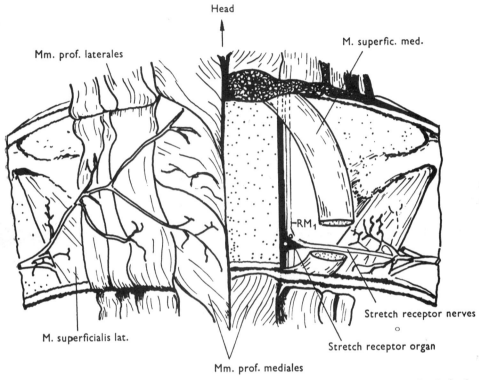

Fig. 196. Semidiagrammatic drawing of position of stretch receptor organ in 2nd abdominal segment of the crayfish. The rostral part above. The medial superficial extensor muscle is cut to show the position of the organ.

found: m. superficialis lat. part of which is visible also without removal of the deep extensors and m. superficialis medialis. Behind this muscle at its medial edge we see the desired muscle bundles: the medial bundle RM_2 and the lateral one RM_1 (nomenclature of the discoverer: Alexandrowicz 1951). The fibres of both bundles are thinner than in the other muscles and also have smaller sarcomeres. The bundles become coarser posteriorly (nearer the tail end). Here the sensory cell of the corresponding bundle is situated and innervation is realized. Under illumination from above no cell contours can be seen and nerve fibres are discernible only with difficulty because of the large amount of connective tissue. The whole bundle of entering nerve fibres, however, is well

459

discernible and is cut as far as possible away from the coarse part of the muscle fibre in order to avoid injuring the sensory cell and to make manipulation of the preparation easier.

Then we follow the muscle bundles to their points of attachement where we free them from surrounding tissues in such a way that enough tissue remains for attaching them to the retaining pincettes of the recording device.

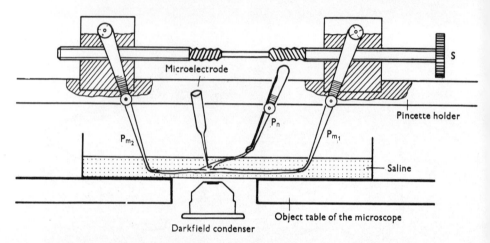

Fig. 197. Experimental arrangement for recording from isolated stretch receptor organ. Pm_1, Pm_2 — pincettes for holding and stretching the receptor muscle. Pn — pincette for holding the stretch receptor nerve bundle. S — screw of the stretcher.

The preparation is transferred in van Harreveld's solution to a Petri dish where we work with translumination.

Apparatus: As in exp. IV-A-c.

Other requirements: Van Harreveld's solution (p. 315) of normal composition and with chymotrypsin (0·5 mg/ml). Macro- and microdissection instruments. Microelectrodes filled with 3M-KCl and with a resistance of about 10 MΩ. Arrangement for holding and stretching the preparation. Dark field condensor.

Procedure: The ends of the muscle bundle of the tonic receptor RM_1 are caught in the retaining pincettes (Fig. 197), the ends of which are bent and finely ground. The second receptor muscle is either cut away or is left hanging freely and is used later for studying activity of the phasic muscle. The pincettes are fixed on riders that move towards each other when turning the screw (S) since their threads are cut in opposite directions. This maintains the sensory cell in the same position even when the muscle is stretched. The pitch of the threads is steep so that the desired change in length is quickly attained. The rest of the nerve trunk is held in a further pincette (Pn) which can be moved by a micromanipulator. In this way a suitable position of the sensory cell is set for inserting the microelectrode fixed in the other micromanipulator. Fig. 197 shows the

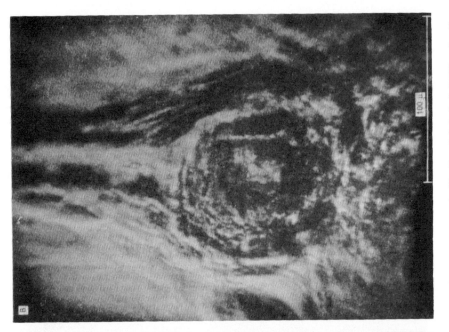

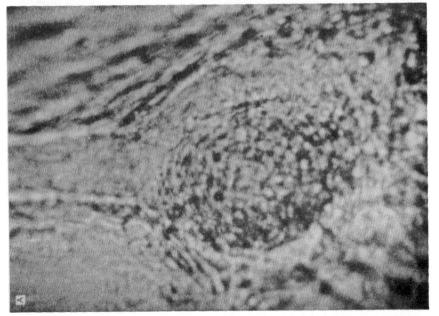

Fig. 198. Photomicrograph of a living, unstained stretch receptor cell of *Astacus fluviatilis*. Partial dark field illumination of the same cell in B.

overall arrangement of the experiment. The retaining apparatus is attached to the object table and the preparation is submerged in van Harreveld's solution in a Petri dish. The receptor preparation is moved towards the bottom of the dish by means of the vertically moving pincette holder so that it is as close as possible to the dark field condenser. The solution is removed so that it is not more than 3 mm above the preparation. The indifferent agar Ag-AgCl electrode is submerged in the Petri dish.

The outlines of the sensory cell are well visible in the dark field (Fig. 198) and if the magnification is sufficient the dendrite contours are also discernible. The tip of the microelectrode is directed towards the centre of the cell which appears dark and has an illuminated rim due to the surrounding tissue envelope.

This envelope causes considerable difficulties when introducing the micro-electrode into the cell. The entry of the electrode tip into the cell is indicated by the appearance of a membrane potential (70—80 mV). Once the microelectrode is within the cell it usually remains there reliably. It does not leave the cell even during small lateral movements of the microelectrode. It is as if the cell were attached to it. In the beginning the impalement will cause difficulties. Before experience is gained the connective tissue may be softened with enzymes. Fifteen to 20 min after application of chymotrypsin or trypsin it is possible to insert the microelectrode easily. The enzyme concentration used has no effect on cell activity but makes for other difficulties due to "softening" of the tissue in the whole preparation (tearing of muscle fibres from the tendon and tearing of muscle bundles during stretching).

When inserting the microelectrode into a cell, the receptor muscle of which is completely relaxed, a resting potential of 70—80 mV without spike activity and without fluctuations is found.

By turning the screw we stretch the muscle bundle continuously and observe the membrane potential. The amount of stretch is increased until evident depolarization of the receptor occurs. The recorded depolarization is the receptor potential.

On further increasing stretch the amplitude of the receptor potential gradually rises until for a depolarization of about 10 mV the first action potential appears. Then only a small increase in stretch suffices to get the cell to respond with repeated discharges (Fig. 199). The frequency of discharges and the level of depolarization increase with increasing stretch until for a certain value of the receptor potential no discharge occurs in the cell. This phenomenon has been termed "overstretch" by Wiersma et al. (1953).

The discharge frequency is maintained for the whole time of stretching. After the first spike it rapidly attains a constant value which is maintained unaltered and only falls slightly at the end of the stretch period. The receptor shows very slow adaptation.

Fig. 199. Slow receptor cell in second abdominal segment. The effect of continuous stretch (A—D) and continuous relaxation (E—H) progressing over 30 sec. presented in small sections. I — continuous record (20 times slower speed) from the same cell to show the receptor potential. r. p. — resting potential of the relaxed cell; in I indicated by an interrupted line.

463

The slowly adapting stretch receptors generally have a low threshold for spike generation (Eyzaguirre and Kuffler 1955a, b). It may hence happen that in some slow cells we do not succeed in obtaining a subthreshold receptor potential undistorted by spikes. In such cases or if we want to study the isolated receptor potential with larger stretches we use 0·1% novocaine which, as in the preceding experiment (Exp. VII-C-a) blocks spike activity without affecting the receptor potential.

These difficulties with subthreshold potentials are eliminated when working with phasic receptors where spikes are generated at higher (15—20 mV) threshold amplitude of the receptor potential. The procedure for working with phasic receptors is identical with the above. Instead of muscle bundle RM_1, muscle bundle RM_2 is gripped in the stretching apparatus and the corresponding sensory cell is impaled with the microelectrode. By applying several 2—3 second-stretches the stretch value giving an evident subthreshold receptor potential is first found. The level of the potential, in contrast to tonic receptor, does not remain stable but visibly decreases after several seconds. In the same way the suprathreshold receptor potential declines. This is correlated with the frequency of spike discharges which is high at first but soon (after several seconds) slows down and then disappears. A small increase in stretch is sufficient for the discharge to reappear. Soon, however, it disappears again as before. This characterizes a phasic, rapidly adapting, stretch receptor.

Conclusion: The experiment showed that the generation and frequency of action potentials in the sensory cell depends on the amplitude of the receptor potential. This arises at the site of contact between dendrites of the sensory cell and fibres of the muscle bundle during stretching of muscle fibres. Moving the dendrites alone does not cause depolarization as can be shown by moving the sensory cell fixed to the microelectrode. The receptor potential spreads from the primary region along the dendrites electrotonically to the body of the nerve cell and to the axon, where it initiates (near to the cell body) an action potential as was demonstrated by Edwards and Ottoson (1958). They recorded spikes with extracellular electrodes simultaneously from the soma and the axon near to the cell body and found that the cell body was invaded by antidromic propagation about 0·3 msec after the initiation of the spike in the axon. Their findings suggest that the threshold of the cell body is probably higher than that of the axon. This also appears to be true of the spinal motoneuron, where the conducted impulses are believed to originate in the initial axon segment (see Exp. VIII B a).

Since the stretch receptor cell is not covered by synaptic knobs and glial cells as is the cell body of the motoneurone, the higher threshold of the somato-dendritic membrane must be due to the difference in the intrinsic properties of this membrane compared to the axonic membrane. Possibly the same is true also for the motoneuron membrane (Eccles 1964a). Conduction along the dendrites does not obey the "all or none" rule as shown by the fact that the receptor potential, i. e. the secondary electrotonic potential in the soma does not disappear when conduction is blocked with novocaine.

In summary the course of events leading to impulse generation in the receptor is as follows: deformation of the dendrite terminals by stretch → receptor potential → electrotonic spread to the cell soma and axon → axon impulse → simultaneously: (1) orthodromic propagation along the axon, (2) antidromic propagation back into the soma (somato-dendritic impulse).

The action potential is thus generated at a place other than the receptor potential, as in other mechanoreceptors (review: Gray 1959a, b).

Also in other respects the stretch receptor of the crayfish has similar basic mechanisms as other mechanoreceptors (Eyzaguirre and Kuffler 1955a). The possibility of recording receptor potentials with microelectrodes has made this receptor a suitable object of study for the sequence of events which change external energy into a nerve impulse. The finding of an inhibitory nervous control of the receptor mechanism was of equal importance (Kuffler and Eyzaguirre 1955).

From the point of view of general neurophysiology the sensory nerve cell of the receptor is also suitable for studying the conduction of excitation in all three main parts of the neurone: in the dendrites, soma and axons under much simpler conditions than in the CNS of vertebrates or in the ganglia of invertebrates.

c) The electroretinogram

Problem: Record the electric response of the frog's eye, adapted to light and darkness, to continuous stimulation.

Principle: The production of impulses in this receptor is accompanied by (a) changes in steady potential, (b) the electroretinogram and (c) a discharge of impulses in the optic nerve.

The potential change arising between the cornea and the back of the eye bulb on illuminating the eye is termed the electroretinogram. Different ways of recording electrical activity from the eye are shown diagramatically in Fig. 200 (Granit and Therman 1938). Lead I is termed the *standard lead* in which electroretinograms are usually recorded.

The formation of the electroretinogram (ERG) depends upon the favourable orientation of some retinal structures capable of conducting current in one direction. Such structures are the layers of receptors and bipolar cells. Both these structures are thought to give rise to the electroretinogram.

The shape of the ERG depends upon the kind of optic receptor (a retina containing only rods or cones or a mixed one) and on previous adaptation to light or darkness.

Four components are described in the retinogram. An initial negative *a*-wave, a positive *b*-wave followed by a slow *c*-wave and a positive *d*-wave which appears on cessation of light stimulation (Fig. 202). With the exception of the *c*-wave the rods and cones of the eye show the same components, although their relative size may vary from one animal to another.

Object: Large frogs (*Rana esculenta* or *R. temporaria*). Recording the ERG from the frog's eye is of advantage for two reasons:

a) the retina has about the same number of rods and cones,
b) excised bulbs can be used. These are easy to prepare.

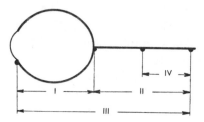

Fig. 200. Four ways of recording electrical activity from the eye (I, II, III, IV). I is the standard lead.

Frogs are adapted by keeping them for 1 or more hours in a completely dark room. Under dark red light the bulb is then excised and placed into an arrangement for recording the ERG.

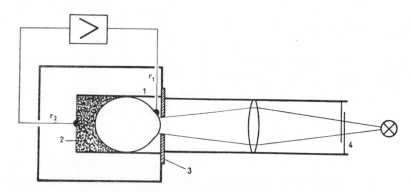

Fig. 201. Arrangement for ERG recording in vitro. 1 — glass tube, 2 — cotton wool soaked in frog Ringer solution, 3 — iris diaphragm, 4 — shutter, r_1, r_2 — recording Ag–AgCl wick electrodes.

In curarised frogs the eye may be left in situ. The cornea, lens and corpus vitreum are removed and the different electrode is applied directly to the exposed retina.

Adaptation to light is carried out during the experiment by exposure to light for a long period.

Apparatus: A cathode-ray oscilloscope with a D. C. amplifier. An A. C. amplifier is satisfactory if the registration of c-wave is not required. Total sensitivity 100 μV/cm. A photostimulator. A time marker.

Other requirements: An arrangement for fixing the bulb and applying the recording electrode (Fig. 201). Nonpolarisable wick Ag-AgCl electrodes.

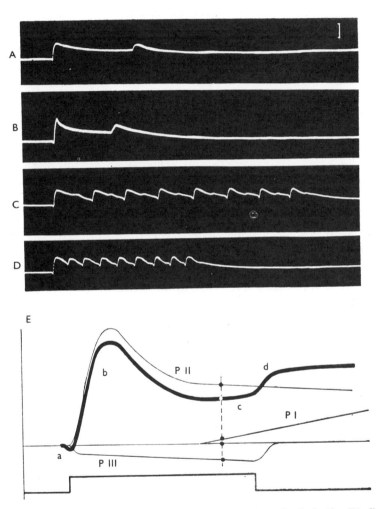

Fig. 202. Electroretinogram of excised frog's eye adapted to the dark (A—B). C — repetitive stimulation at 0·5/sec. and D — 1/sec. Voltage scale: 1 mV. E — diagrammatic representation of various components of ERG.

Procedure: The bulb is placed into a short glass tube. The space behind the bulb is filled with cotton wool soaked in frog Ringer solution. The tube is placed into a small dark chamber so that the front surface of the cornea is close to the iris diaphragm. The recording electrodes are placed as shown in fig. 201. One is on the anterior, the other on the posterior border of the bulb where it touches the cotton wool. Care must be taken to prevent light from acting directly on them in order to avoid photochemical electrode artifacts. The recording electrodes are connected to the input of a D. C. amplifier. The shutter is opened admitting a light beam lasting 2 seconds. The intensity of the beam is gradually increased and the bulb is stimulated at 3 min. intervals. Fig. 202A, B shows the response to such stimulation. The shutter is then left open and light is permitted to fall on the bulb for 10 min. The shutter is again closed for 3 sec. and a beam of the same intensity is admitted for 2 sec. The response is recorded. The retinal response to dark includes all the components: a small negative a-wave, followed by a rapidly increasing b-wave, which falls relatively quickly and gradually changes into the c-wave (not shown in Fig. 202). After switching off the light a positive off-effect or d-wave appears. The response of a retina adapted to light differs from the above by a decreased b-wave, and a more rapid increase and a higher amplitude of the off-effect which also has a shorter latent period. In addition, the c-component is absent in an eye adapted to light.

Conclusion: The ERG is the final result of several processes occurring in the retina. It has been possible to separate it into several components (PI, PII, PIII). The figure 202 E (diagram below) shows that the cornea-negative a-wave is due to the early appearance of wave PIII. This is interrupted by the cornea-positive PII component which forms wave b. Wave c is formed by component PI.

The ERG may be separated into its components by different methods (anaesthesia, asphyxia, adaptation to light, chemical means — Granit 1933, 1955, Noell 1951, 1952a, b).

d) The electro-olfactogram

Problem: Record the electro-olfactogram in the frog.

Principle: The electric potential change occurring in the olfactory receptor in response to specific stimulation is termed the electro-olfactogram (Ottoson 1956). Most probably this is an electrophysiological effect of reactions between substances in the olfactory epithelium and particles of the stimulating agent.

The electro-olfactogram is a negative monophasic potential which can be recorded with an electrode placed on the olfactory part of the nasal mucosa.

Object: The olfactory epithelium of the frog is very suitable since it is easily accessible and since recordings from both the olfactory epithelium and the olfactory bulb can be made simultaneously.

Fig. 203 is a diagramatic representation of the anatomy of parts of the olfactory organ. In comparison to the same organs in higher vertebrates the nasal cavities of the frog are very simple. The external nasal aperture is connected through the vestibulum to the cavum principale nasi covered with sensory epithelium. The areas containing the sensory epithelium have a slightly yellow tinge and thus differ from the respiratory tract which is pink in colour.

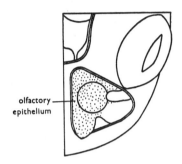

olfactory
epithelium

Fig. 203. Diagramatic representation of the anatomy of the frog's olfactory organ.

It is not difficult to reach the cavum principale in the frog. The animal is decapitated and the olfactory epithelium is exposed by removing the dorsal wall of the cavity of the olfactory organ.

Apparatus: A cathode-ray oscilloscope with a D. C. amplifier, total sensitivity 1 mV/cm. An amplitude marker.

Other requirements: a) Agar Ag-AgCl electrodes. The electrode situated on the sensory epithelium has a tube tip diameter of 0·1—0·2 mm.

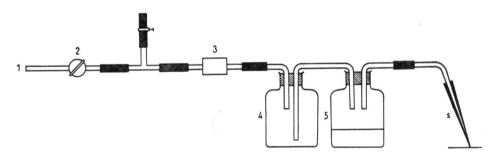

Fig. 204. Arrangement for olfactory stimulation of the olfactory epithelium. 1 — air pump, 2 — cock, 3—4 — filter with active charcoal, 5 — test vessel, s — stimulating pipette.

b) An arrangement for olfactory stimulation as shown in Fig. 204. The principal equipment is an air pump. At every turn the pump expells about 50 ml. of air into the air line. The major part of the air leaves through the cock (2), the rest passes through activated charcoal (3, 4) and enters a 150 ml.

test vessel (5) containing 10—15 ml. of an aqueous solution of a volatile substance. The odorised air is blown onto the nasal mucosa through a glass pipette (s) connected to the outflow of the test vessel. The tip of the pipette is at a distance of 5—8 mm from the epithelial surface. The volume of air expelled for each blow depends on the tip diameter of the pipette.

c) n-butanol in concentrations of 0·001M, 0·005M, 0·025M, 0·05M, 0·25M is used as the odorising substance. 15 ml. of each concentration are placed into the 150 ml. vessels.

Procedure: The preparation is placed on cotton wool soaked in frog Ringer solution. One agar Ag-AgCl electrode is applied to the cotton wool,

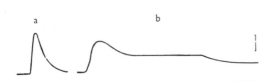

Fig. 205. Electro-olfactogram from the olfactory epithelium of the frog. a) EOG with 0.001 M butanol. b) EOG on continuous stimulation (0.005 butanol). Time mark: 3 sec; voltage calibration : 1 mV.

the other to the centre of the eminentia olfactoria. The stimulating pipette is situated 5 mm above the point at which the recording electrode is applied. After connecting the electrode to the D. C. amplifier and preparing the camera the odorised air is blown onto the olfactory epithelium. The electric response to this is shown in fig. 205a. If the electrode and pipette are moved by 1—2 mm to different points of the mucosal surface, the spatial amplitude distribution of the EOG can be obtained.

If a 0·5% cocaine solution is dropped onto the mucosa the amplitude is partly decreased but does not disappear.

This insensitivity to cocaine also applies to receptor potentials of other organs (Katz 1950, Gray and Sato 1953a, b, Eyzaguirre and Kuffler 1955a, b etc.). This supports the assumption that the EOG originates in olfactory sensory cells.

Distilled water, ether and chloroform, on the other hand, rapidly eliminate the EOG.

The stimulating arrangement is changed for continuous stimulation. The air current has a speed of 1 ml/sec and lasts for 15 sec. The recordings in Fig. 205b show that the potential in response to continuous stimulation rapidly attains a maximum, returns to a certain level and remains at the plateau. This decrease to a certain plateau indicates adaptation of the olfactory receptors.

Olfactory receptors may consequently be termed slowly adapting end organs.

VIII

Electrophysiology of the spinal cord

The basic functional unit of the central nervous system is the reflex. The fundamental knowledge concerning reflex activity of the spinal cord was obtained by classical methods of stimulating the afferent part of a reflex arc and recording the reaction of the effector, usually the muscle. A classical summary of this work is given by Sherrington in his "Integrative Action of the Nervous System" (1906) and later by the Oxford school (Creed, Denny-Brown, Eccles, Liddell and Sherrington "Reflex Activity of the Spinal Cord", 1932).

With classical methods it was only possible to draw indirect conclusions concerning processes in the cord itself (the central component of spinal reflexes). Refinement of electrophysiological methods has made it possible to study this component directly.

Two periods are evident in the development of electrophysiological research into spinal activity. During the first period previous myographic analysis of reflex activity was defined more exactly using action potential recordings from efferent and afferent pathways of the reflex arc (review: Eccles 1953, Kostyuk 1959). Transmission of nervous impulses from afferent to efferent neurones was studied by electrotonic recording of synaptic potentials generated within the spinal cord from the ventral and dorsal spinal roots. The second period (since 1952) was made possible by the development of a new electrophysiological technique (Brock, Coombs and Eccles 1952, Woodbury and Patton 1952, Frank and Fuortes 1955). Introduction of intracellular micro-electrodes enabled direct observation of processes taking place during transmission of impulses from afferent fibre terminals to motoneurones. The most significant contribution of this period is the discovery of hyperpolarization of the subsynaptic membrane during direct spinal inhibition (Brock et al. 1952).

A. Extracellular recording

Some expriments that are made possible by the anatomical structure of the spinal cord with separate input and output fibres are summarized under this heading. By extracellular recording we mean registration of an action potential with electrodes applied to the anterior or posterior spinal roots. It thus does not include focal registration in the white or grey matter with extracellular microelectrodes or macroelectrodes when the action potential is recorded in a volume conductor (pp. 579).

When studying the input-output relationships in the cord the input signal may be varied by stimulating different types of afferent fibres. According to Lloyd's (1943) nomenclature afferent fibres are divided into several groups that differ in their diameter: group I ($21-12\ \mu$), group II ($12-6\ \mu$), group III ($6-2\ \mu$) and IV (unmyelinated). The input signals sent into the spinal cord through the posterior root may be classified into these groups already on the basis of anatomical relationships. Group I afferents run in muscle nerves only and are not present e. g. in cutaneous nerves. Stimulation of these two types of nerves gives different input-output characteristics (VIII-A-a).

A study of the input-output relationships contributed in this way not only to a better understanding of central excitatory and inhibitory processes but also founded functional anatomy of the spinal cord and of its afferent and efferent pathways.

Extracellular recording from the anterior and posterior roots also makes it possible to record local processes at spinal synapses. Synaptic potentials reach the root by electrotonic transmission in view of the small distance between the site of their origin and the site of their registration.

Electrophysiological research into functions of the spinal cord requires not only a good knowledge of stimulation and registration techniques which in principal are the same as in peripheral excitable structures in vitro but also a good acquaintance with the anatomy of the spinal cord and somatic nerves.

The classical object of electrophysiological research on the cord is the cat. Most work has been performed in the lumbosacral cord because anatomical conditions are favourable there (length of roots) and because its location is also better in relation to disturbing sources (movements of thorax, electrical activity of heart). The essential anatomical data necessary for performing the experiments are presented in Figs. 206-208. For more detailed data see special studies: the anatomy of the cat (Reighard and Jenkins 1923), innervation (Sherrington 1890, 1894, Romanes 1951, Jefferson 1954), localization of motor nuclei in the cord (Romanes 1951, Rexed 1952, Balthazar 1952).

A

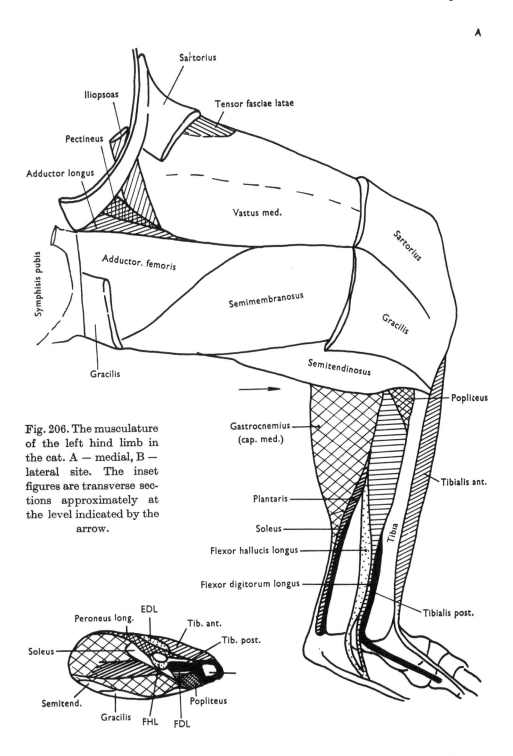

Fig. 206. The musculature of the left hind limb in the cat. A — medial, B — lateral site. The inset figures are transverse sections approximately at the level indicated by the arrow.

473

B

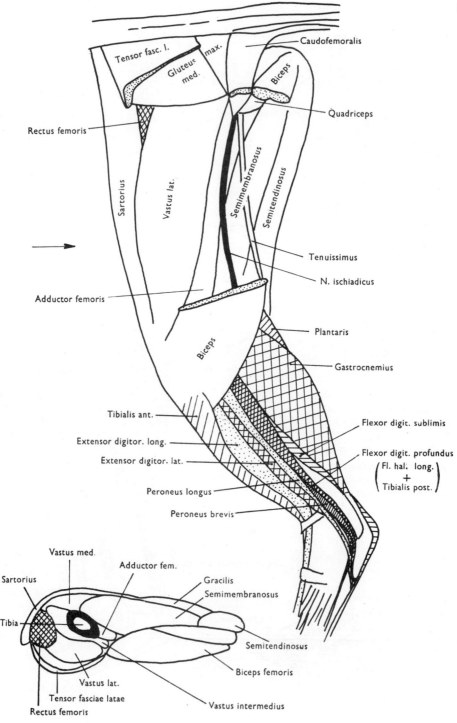

Tensor fasc. l.

Gluteus med.

max.

Caudofemoralis

Biceps

Quadriceps

Rectus femoris

Sartorius

Vastus lat.

Semimembranosus

Semitendinosus

Tenuissimus

N. ischiadicus

Adductor femoris

Biceps

Plantaris

Gastrocnemius

Tibialis ant.

Extensor digitor. long.

Extensor digitor. lat.

Peroneus longus

Peroneus brevis

Flexor digit. sublimis

Flexor digit. profundus
(Fl. hall. long.
+
Tibialis post.)

Vastus med.

Adductor fem.

Sartorius

Gracilis
Semimembranosus

Tibia

Semitendinosus

Biceps femoris

Vastus lat.

Tensor fasciae latae

Rectus femoris

Vastus intermedius

474

a) Electrophysiological manifestations of a monosynaptic and polysynaptic reflex arc

Problem: Record the action potential from the ventral roots of the cat in response to afferent stimulation of a cutaneous nerve and a nerve from a muscle.

Principle: Cajal (1909) simplified the anatomical complexities of the spinal reflex system to two fundamental types as shown in Fig. 209. Fig. 209a illustrates the circumscribed reflex mechanisms. The afferent fibre terminates directly on the motoneurone in a certain narrow segment. This reflex arc is termed monosynaptic. Fig. 209b shows a disynaptic reflex as the most simple example of a polysynaptic reflex. An interneurone is inserted between the afferent terminals and the motoneurone. This causes diffusion of impulse activity from the afferent neurone to a large number of efferent fibres along the spinal cord.

The functional picture of Cajal's diagram may be obtained by stimulating the posterior roots with an electric pulse and recording from the ventral root of the same segment. Thus a recording of a segmental reflex discharge is obtained.

All afferent influxes reach the cord via the dorsal root. A rough separation of this influx may be obtained if instead of the dorsal root the muscle and cutaneous nerve are stimulated separately (Lloyd 1943).

Object: The decerebrate cat with the lumbal cord exposed and the ventral roots cut peripherally (7th or 8th post-thoracic).

Preparation: The cat is decerebrated during brief ether anaesthesia. A trephine opening is made above one hemisphere and is then rapidly widened with rongeurs, until the whole dorsal surface of the hemispheres is exposed. Bleeding from the bones is stopped with bone wax or plasticine. The brain is lifted with a spatula introduced along the tentorium and rapidly removed in parts. A small amount of brain tissue is left in place at the base to prevent injury to the circulus Willisi. The cavity thus formed is filled with cotton wool soaked in warm Ringer solution. Then both carotid arteries, which were compressed during decerebration, are released. The animal breathes spontaneously and no respiratory pump is necessary.

The skin is incised in the lumbal region, the fasciae are loosened and the paravertebral musculature pulled aside so that a small enclosure is formed over the cord. The spines of the vertebrae are removed with plyers and then the dorsal and lateral parts of the vertebrae forming the osteous covering of the spinal cord are removed. The dura mater is then opened by cutting it longitudinally. Its edges are fixed with silk threads to the edges of the laterally displaced paravertebral muscles.

The cord is divided between the 2nd and 3rd lumbal level. All ventral and dorsal roots are cut on one side. Recordings will be made from the other.

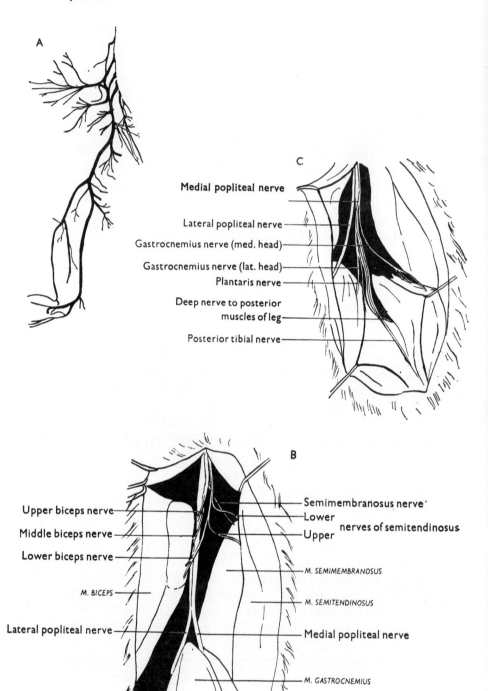

A

C

Medial popliteal nerve

Lateral popliteal nerve
Gastrocnemius nerve (med. head)
Gastrocnemius nerve (lat. head)
Plantaris nerve
Deep nerve to posterior
muscles of leg
Posterior tibial nerve

B

Upper biceps nerve
Middle biceps nerve
Lower biceps nerve

M. BICEPS

Lateral popliteal nerve

Semimembranosus nerve
Lower
Upper   nerves of semitendinosus

M. SEMIMEMBRANOSUS

M. SEMITENDINOSUS

Medial popliteal nerve

M. GASTROCNEMIUS

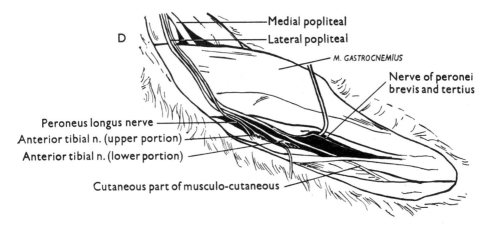

Fig. 207. Innervation of hind limb muscles of the cat. A — a diagram of the prefixed type of lumbosacral plexus and the peripheral nerves. B — hamstrings nerves. C — medial popliteal nerve. D — lateral popliteal nerve (modified after Romanes 1951 and Jefferson 1954).

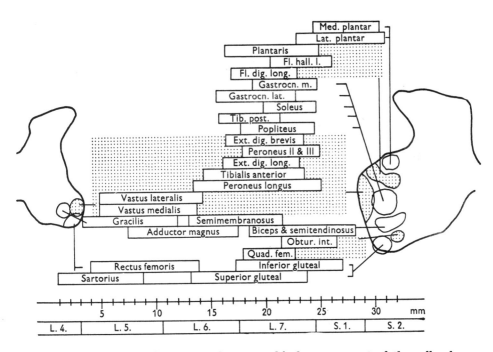

Fig. 208. The longitudinal extent and topographical arrangement of the cell columns and of the individual muscles they supply (modified after Romanes 1951, 1964). Lumbosacral spinal cord of the cat.

477

This permits turning of the spinal cord and thus makes it possible to identify the roots. Now the ventral roots (7th and 8th posthoracic) of the ipsilateral side are found and are carefully cut peripherally so that at least 20 mm of them are preserved. The remaining ventral roots are cut nearer the cord.

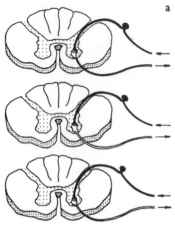

Then the nerve leading to the gastrocnemius muscle and the sural nerve are exposed.

Apparatus: A cathode-ray oscilloscope with an A. C. preamplifier (100 µV/cm). A simple stimulator giving rectangular pulses with variable amplitude.

Other requirements: Platinum electrodes: 2 pairs of stimulting and one pair of recording electrodes, paraffin oil.

Procedure:

1) Segmental reflex discharge.

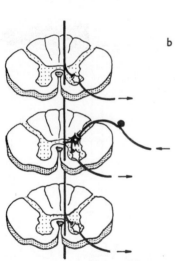

Fig. 209. Diagramatic representation of the basic types of connections in the spinal cord, according to Cajal. a — circumscribed reflex mechanism. The afferent neurone terminates directly on the motoneurone. b — diffuse reflex mechanism. One interneurone is inserted between the afferent neurone and the motoneurone.

The stimulating electrodes are placed on the dorsal root S_1. The recording electrodes are on the ventral root of the same segment at a suficient distance from the cord. Recordings are monophasic so that the more distal recording electrode is on the cut end of the ventral root. The spinal cord and the region of stimulation are covered with paraffin oil, filling the through formed by the retracted muscle and skin edges. The preparation is allowed to recover from the effects of anaesthesia so that reflex discharges may be recorded.

Fig. 210a shows a response obtained in these conditions. The tracing is composed of a sharp spike at the beginning and of diffuse activity with lower and different conduction velocities.

2) Response of the ventral roots to stimulation of the muscular and cutaneous nerves.

The stimulating electrodes are placed on the sural nerve and on the nerve leading to the gastrocnemius muscle. The former is a cutaneous nerve, the latter contains fibres with a maximal diameter for mixed nerves.

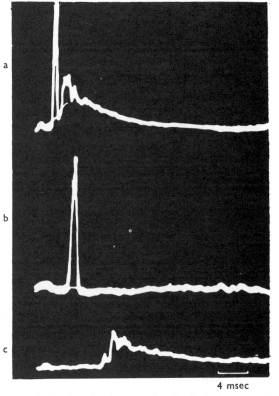

Fig. 210. Action potentials from the ventral root S_1. a — produced by stimulating the dorsal root of the same segment. b — action potentials recorded from the ventral root evoked by stimulating the nerve leading to one head of the gastrocnemius muscle. c — in response to stimulation of the cutaneous nerve. Time mark: 4 msec.

4 msec

A response as shown in Fig. 210b is obtained on stimulating the nerve leading to the mucle, and that shown in Fig. 210c on stimulating the cutaneous nerve.

Conclusion: The response to stimulation of the nerve from the muscle is a typical monosynaptic response. It can be calculated that the interval between the stimulus artifact and the onset of the action potential includes, in addition to the time necessary for conduction in the nerve fibres, only a single synaptic delay.

Transmission through several synapses must be expected in the response to stimulation of the cutaneous nerve. This is also indicated by its irregularity lasting for more than 10 msec.

b) Facilitation and inhibition in the motoneurone

Problem: Demonstrate facilitation and inhibition of response in a monosynaptic path of the spinal cord.

Principle: Facilitation in a monosynaptic reflex arc (Lloyd 1946) is due to convergence of presynaptic neurones onto the body of the motoneurone.

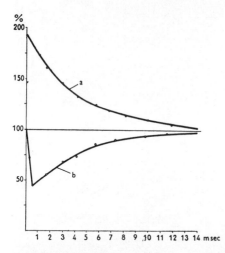

Fig. 211. Facilitation and inhibition in a monosynaptic reflex arc. a — facilitation curve. Abscissa: time interval between conditioning stimulus and submaximal test stimulus. Nerves to both heads of the gastrocnemius muscle. Ordinate: Percentage increase in test stimulus. Response to isolated test stimulus was taken as 100%. b — inhibition curve in a monosynaptic reflex arc. Abscissa: time interval between maximal conditioning volley to nerve from gastrocnemius muscle · and testing volley applied to the nerve leading from the antagonist (anterior tibial muscle). Ordinate: Percentage decrease in amplitude of the response to the test stimulus.

Two afferent nerves from two synergistic muscles are to be stimulated at different stimulus intervals. The amplitude of the conditioning volley is maximal; that of the testing one submaximal. The increase in the amplitude of the latter is the expression of the magnitude of facilitation.

Direct inhibition (Lloyd 1941) exists because of the presence of direct inhibitory synapses on the body of the motoneurone. According to contemporary opinion, direct inhibition occurs in a disynaptic reflex arc. An inhibitory neurone is inserted between the afferent and efferent neurone (Eccles 1957).

The existence of direct inhibition can be demonstrated in the same way as facilitation, but by stimulating afferent nerves of two antagonistic muscles (e. g. the anterior tibial and gastrocnemius muscles).

Object: The spinal cat prepared as in the preceding experiment. The nervous branches leading to both heads of the gastrocnemius muscle and to the anterior tibial muscles are exposed. Recording are made from the ventral roots L_7 or S_1.

Apparatus: As in the preceding experiment. A stimulator with two independent outputs is, however, needed in which the time interval between the two stimuli can be changed continuously.

Procedure:

1) Demonstration of facilitation.

The recording electrodes are placed on L_7 or S_1. One pair of stimulating electrodes is situated on one nerve leading to the gastrocnemius muscle, the other pair on the other nerve leading to the same muscle. The first stimulus is of maximal strength, the second submaximal. First a time interval of about 20 msec is used. Then it is gradually shortened. It can be seen that the amplitude begins to increase starting from a certain time interval until a maximum value is obtained when both impulses coincide.

2) Demonstration of direct inhibition.

The recording electrodes remain in the same position. One pair of electrodes is left on the nerve leading to the gastrocnemius muscle, the other pair lies on the nerve to the anterior tibial muscle. A maximal conditioning stimulus is applied to one of these muscles. The test stimulus is applied to the antagonist. From a certain limit, the amplitude of the response to the test stimulus decreases with the time interval between the stimuli.

Conclusion: Curves as shown in Fig. 211 are obtained if the increase (facilitation) —a— or decrease (inhibition) —b— of the amplitude of the test stimuli is plotted against time. Inhibition, save for an initial increment phase, takes the same course as facilitation.

A similar time course and configuration is seen for postsynaptic excitatory and inhibitory potentials, which may be recorded intracellularly from motoneurone cells.

c) Relation between afferent influx and efferent efflux in the spinal monosynaptic arc

Problem: Demonstrate the extent of the subliminal fringe and of the discharge zone in the motoneurone pool of the gastrocnemius muscle.

Principle: The existence of facilitation in the motoneurone indicates that a *subliminal fringe* is present in the motoneurone pool (Lloyd 1943). This means that in addition to motoneurones discharged by a presynaptic volley, a number of motoneurones are activated *only subliminally*. If the extent of this area is to be estimated, first the *discharge zone,* i. e. the number of activated motoneurones must be known. This may be estimated from a correlation between the amplitude of the presynaptic spike and that of the postsynaptic one. The curve obtained in this way represents the limit between the discharge zone and the subliminal fringe. The other limit of the subliminal fringe is given by the facilitation curve (see the preceding experiment).

In principle this experiment requires recording from the ventral and dorsal roots and stimulation of two afferent nerves from two synergistic muscles. The conditioning volley applied to one of these nerves is maximal. The strength of the testing one is an independant variable.

Object: The spinal cat with the ventral and dorsal roots of L_7 and S_1 exposed together with the nerves leading to the gastrocnemius muscle.

Apparatus: A double beam cathode-ray oscilloscope with A. C. amplifier, total sensitivity 100 μV/cm. A stimulator with two independent outputs.

Other requirements: As in the preceding experiment.

Procedure: The stimulating electrodes are placed on the nerve leading to one head of the gastrocnemius muscle. One pair of recording electrodes is situated on the dorsal and one pair on the ventral root.

The stimulus strength is gradually increased and the response from both recording electrodes is traced. The first postsynaptic discharge appears only when the presynaptic spike attains a certain value. On further increasing the stimulus strength, the postsynaptic response increases rapidly with increase in the presynaptic spike until the latter spike reaches 50% of its maximum value. The amplitude of the postsynaptic spike increases only slightly with further increases in the presynaptic response (Fig. 212).

The experiment is then repeated, but in addition a constant submaximal stimulus s_2 is applied to the afferent nerve from the second head of the gastrocnemius muscle. Both nerves are stimulated simultaneously and again responses from ventral and dorsal roots are recorded. This procedure is repeated several times using different stimulus strengths on electrodes s_1 with the stimulus intensity on s_2 remaining unchanged.

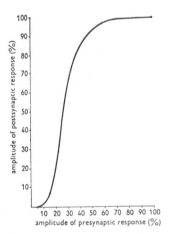

Fig. 212. Relationship between the size of the presynaptic influx and the postsynaptic efflux. Abscissa: amplitude of the presynaptic spike in percent. Response to maximal stimulation of afferent nerve from gastrocnemius muscle taken as 100 %. Ordinate: amplitude of postsynaptic response in percent. Maximal amplitude to orthodromic stimulation taken as 100%.

Both experiments are expressed in graphic form. If the size of the presynaptic influx is expressed as the percentage of the maximal amplitude in response to stimulus s_1 on the abscissa, and the size of the postsynaptic response plotted on the ordinate (also as the percentage of the maximum postsynaptic response), a curve as shown in Fig. 212 is obtained. This shows the development of the discharge zone.

In the second experiment the size of the spike-amplitude recorded from the ventral root in response to separate stimulation of the two nerves is subtracted from the amplitude obtained on simultaneous stimulation, and thus the value for the subliminal fringe is obtained for each value of the presynaptic pulse at s_1. The curve obtained shows the limit of the subliminal fringe.

Finally both curves are plotted into the same coordinates.

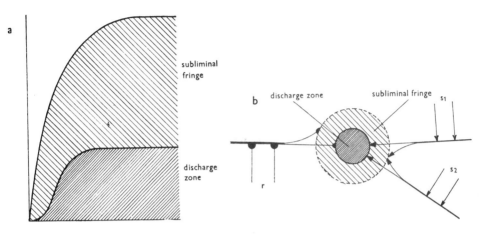

Fig. 213a. Relationship between subliminal fringe and discharge zone. b — diagramatic representation of determination of the discharge zone and of the subliminal fringe (for detail see text).

Conclusion: The diagram in Fig. 213 shows that the subliminal fringe is large in comparison to the discharge zone of the monosynaptic reflex arc studied. This diagram, of course, shows synchronous afferent excitation and the two-neurone-arc discharges. As excitation is realised in the motoneurone pool as the result of a diffuse influx of impulses, the changes shown in the figure as a function of the intensity will occur with time as an additional parameter. At first they will move to the right along the abscissa as the impulse activity increases, and then to the left as this influx decreases.

d) Recovery cycle in the motoneurone

Problem: Determine changes in excitability of the motoneurone following transmission of the nerve impulse.

Principle: Since the motoneurone pool has a large subliminal fringe, it is not possible to determine the course in excitability changes following a maximal orthodromic volley. Consequently, antidromic excitation is used to determine such changes. This causes excitation simultaneously in all motor cells, that are to be tested orthodromically.

In principle, this is application of a maximal antidromic stimulus to a ventral root (the conditioning stimulus) and a submaximal orthodromic stimulus to the corresponding dorsal spinal root — the test stimulus (Lloyd 1951).

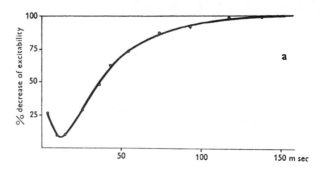

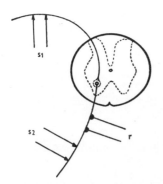

Fig. 214. a — Changes in excitability in the motoneurone pool following maximal antidromic activation. Abscissa: time interval between conditioning and test stimuli. Ordinate: percentage decrease of the spike amplitude due to orthodromic stimulation. b — Arrangement of electrodes for determining the recovery cycle in the motoneurone. s_1, s_2 — stimulating electrodes, r — recording electrodes.

Apparatus: A cathode-ray oscilloscope with an A. C. amplifier, total sensitivity 100 μV/cm. A stimulator with two outputs and a variable interval between the stimulating pulses.

Other requirements: As in the preceding experiments.

Procedure: The arrangement of the experiment is shown in Fig. 214b. The stimulating electrodes s_2 for antidromic stimulation are on the ventral root, and the stimulating electrodes s_1 for orthodromic stimulation are on the dorsal root. The recording electrodes *r* are situated on the ventral root.

First, the maximum antidromic stimulus is determined from the height of the action potential amplitude in the ventral roots by stimulating at s_2. The maximal spike value for orthodromic stimulation is then determined.

The interval between the conditioning and test stimulus is then changed and the size of the action potential in the ventral roots is observed.

If the amplitude of the action potentials as the percentage of that of the maximal orthodromic spike is plotted on the ordinate and the interval between the stimuli on the abscissa, changes in excitability in the motoneurone pool are obtained (Fig. 214a).

Conclusion: Fig. 214a shows that the motoneurone remains subnormal for 120 msec. after the first response. This subnormality is one factor determining the discharge frequency during reflex or voluntary contractions (Adrian and Bronk 1929).

e) Synaptic potentials in a monosynaptic reflex arc

Problem: Record synaptic potentials from the ventral spinal roots in the cat.

Principle: As in peripheral synaptic junctions, transmission in a monosynaptic reflex arc is characterised electrophysiologically by a synaptic potential, in this case referred to as excitatory postsynaptic potential. This is more difficult to record than at the periphery, since the synaptic region is situated in the spinal grey matter and, therefore, can be approached experimentally only with difficulty. Introduction of intracellular microelectrodes has made direct tracing of postsynaptic potentials possible.

Before this method was introduced, these potentials were recorded with an electrode placed directly at the point where the ventral root leaves the cord. Electrotonic spreading of these potentials was utilised. They spread from the synaptic area along the ventral roots. Since this synaptic region is close to the point where the ventral roots leave the cord, the potentials can still be recorded, although with a decrement (Eccles 1946, Eccles and Malcolm 1946).

In order to record a pure EPSP (excitatory postsynaptic potentials) the occurrence of spike discharges must be avoided. Many more ways may be used to achieve this than in peripheral synapses. The following means are suitable

1) block using curare,

2) block with deep barbiturate anaesthesia,

3) a very small presynaptic influx. This utilises the large subliminal fringe of the motoneurone pool.

4) inhibition produced by stimulating antagonist nerves.

Interneurone discharges may interfere when recording EPSP from the spinal cord. These may be prevented by using a monosynaptic discharge — stimulating afferent fibres from muscular nerves.

Object: A decerebrate cat, spinal preparation. The ventral roots L_7 and S_1 and nerves to both heads of the gastrocnemius muscle.

Apparatus: A cathode-ray oscilloscope with a D. C. amplifier or an A. C. amplifier with a long time constant. A stimulator with the following possibilities: 1 pulse, 2 pulses with variable intervals, 2 independent outputs with variable intervals between pulses.

Other requirements: d-tubocurarine, nembutal.

Procedure:

1) Synaptic potential.

The stimulating electrodes are placed on one branch of the nerve leading to the gastrocnemius muscle. The recording electrodes are on the ventral root.

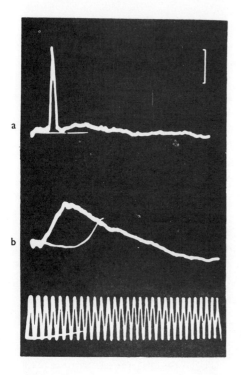

The nerve is stimulated and the response from the ventral roots is recorded (Fig. 215a). Under light anaesthesia a monosynaptic reflex discharge is obtained. Then 70 mg/kg of nembutal are give intravenously. From time

Fig. 215. a — action potential from the ventral roots evoked by stimulating of the afferent nerve from the gastrocnemius muscle. b — postsynaptic potential. Time mark: 1000c/sec. Voltage calibration: 1 mV for a), 0.1 mV for b).

to time the nerve is stimulated and changes in the recordings are observed. The spike gradually disappears and in its stead a long negative potential finally appears (Fig. 215b). When 90 mg/kg are given the potential decreases and the descending phase of the potential is steeper. This is a typical synaptic potential. It arises about 0·7 msec after the beginning of the small spike, which is recorded by the electrode by direct electric spread from the dorsal root.

2) Spatial decrement of synaptic potentials.

If the recording electrodes are moved further away from the spinal cord, and if the synaptic potentials are recorded at different distances from it using a fast sweep, it is found that the rate of rise in the synaptic potential decreases together with its amplitude. If the amplitudes of the synaptic potentials are plotted in a logarithmic scale on the ordinate against the distance in mm from the cord, an approximately straight line is obtained, indicating an exponential relationship for the spatial decrement (Fig. 216).

3) Summation of synaptic potentials.

If two successive stimuli are applied to the afferent nerve, summation of synaptic potentials may be obtained.

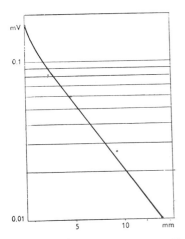

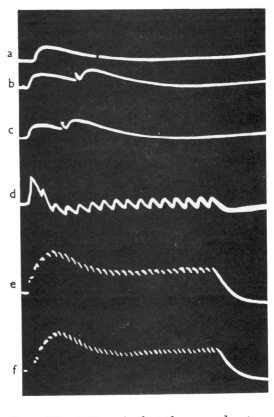

Fig. 216. Logarithm of the amplitude (ordinate) of the synaptic potential plotted against the distance (abscissa) from the spinal cord.

Fig. 217. Summation of synaptic potentials in response to two stimuli (b, c) and a train of stimuli (d, e, f) (after Eccles 1946).

It can be seen from the recordings (Fig. 217a—c), that the second potential is the same or slightly smaller than the first one. The sum of the synaptic potentials exceeds the height of the single potential if the interval is shorter than 16 msec. The absence of the second synaptic potential with intervals shorter than 1—2 msec is due to the fact that the second stimulus falls into the refractory period of the afferent nerve.

If the afferent nerve is stimulated with a volley of several tens of impulses with different frequencies, recordings as shown in Fig. 217d—f are obtained. The after-positivity observed for one synaptic potential prevents the formation of a plateau of synaptic potentials. There is summation of the first few responses, and then this after-positivity causes a rapid fall to a low plateau.

On cessation of stimulation, the synaptic potential decreases with a slope very similar to that of a single response.

Conclusion: Synaptic potentials are produced when a synaptic area is bombarded with afferent impulses. They reach the ventral root electrotonically. The fibres of these roots lead very selectively since, if synaptic transmission is blocked, the adjacent roots have much smaller potentials than the ventral roots corresponding to those dorsal ones, and the contralateral roots show no potentials at all. It may, therefore, be concluded that the central site of catelectrotonus is in the motoneurone bodies belonging to the corresponding ventral roots.

These potentials have characteristics similar to peripheral synaptic potentials; they can be summated, spread with a decrement and their time course is fundamentally the same. They represent the average response of the whole motoneurone pool.

f) Dorsal root potentials. Presynaptic inhibition in the spinal cord

Problem: Record the dorsal root potential and the slow positive potential (P-wave) in response to stimulation of a cutaneous afferent nerve.

Principle: After entering the spinal cord the afferent fibre forms synaptic contacts with the soma or dendrites of the motoneuron or interneuron in the cord and also sends out collaterals to other specialized interneurons, the axons of which terminate on terminals of the same or of another afferent fibre (Fig. 218), forming an *axo-axonic synapse.* Thus the terminals of afferent nerve fibres in the cord form both the presynaptic part of the axo-somatic or axo-dendritic synapse and also the postsynaptic part of the axo-axonic synapse. The latter is also a chemical synapse as indicated by pharmacological evidence and by electronmicroscopic findings of synaptic vesicles in the presynaptic part of this synapse (Gray 1962, 1963). The presynaptic impulse causes in the postsynaptic part of the axo-axonic synapse (i. e. in the terminal of the afferent fibre) depolarization, termed *primary afferent depolarization,* PAD (Eccles 1964 a, b). The amplitude of the action potential entering this depolarized terminal is decreased. This, of course, diminishes its postsynaptic excitatory effect in the axo-somatic or axo-dendritic synapse. The EPSP may fall below the critical depolarization value so that the impulse does not pass across the axo-somatic synapse. This is the basis of the mechanism of presynaptic inhibition (Frank and Fuortes 1957, Eccles, Eccles and Magni 1961, Eccles, Magni and Willis 1962, Eccles, Schmidt and Willis 1962).

The postsynaptic part of the axo-axonic synapse is too small so that microelectrodes cannot be used to study transmission in this synapse in the same way as in the axo-somatic synapse (see VIII-B-a). The characteristics of the axo-axonic synapse can thus be studied only indirectly (Fig. 218).

Using the method shown in Fig. 218A Frank and Fuortes (1957) and Frank (1959a) were the first to give convincing evidence of presynaptic inhibition. The excitatory postsynaptic potential (EPSP) recorded with a microelectrode in the motoneuron (postsynaptic part of the axo-somatic synapse) decreases if the conditioning stimulus is properly timed. This in itself is not a sufficient proof of presynaptic inhibition. It must be shown that membrane conductance is not changed by the conditioning stimulus, and this can be inferred from the absence of potential changes at the normal resting potential and when the membrane potential is altered by the applied current.

The "postsynaptic potential", PAD, which appears in the axo-axonic synapse may be recorded with a microelectrode inserted into the afferent fibre in the dorsal region of the cord where the fibres are fairly coarse (Fig. 218B). PAD is recorded with a considerable decrement because of the decremental electrotonic spread of PAD to the site of registration (Eccles et al. 1962).

PAD phenomena may also be recorded with extracellular electrodes applied to the peripherally cut dorsal root filament or to the dorsal roots. The central electrode is 1—2 mm from the entry of the filament into the cord, the other on the cut end (Fig. 218C). PAD induced by the conditioning stimulation spreads electrotonically and is recorded as a negative dorsal root potential (DRP). In a similar way EPSP can be recorded from the ventral roots as the ventral root potential (Exp. VIII-A-e). DRP was recorded already by Barron and Matthews (1938) who also postulated that this dorsal root potential was produced by the same potential generator that gave the P wave of the cord dorsum (Gasser and Graham 1933). The method of recording the P wave is also shown in Fig. 218C. The electrode is applied to the cord medial to the entry of a dorsal root. The earthed reference electrode is applied to the back musculature.

The time course of PAD may also be obtained by observing the excitability of the postsynaptic part of the axo-axonic synapse at different intervals after the conditioning stimulation (Fig. 219D), which technique has been developed by Wall (1958). After the conditioning stimulation the threshold at the site of application of the extracellular microelectrode (filled with 4M-NaCl, resistance about 1 MΩ) is decreased because of PAD. This appears as an increase in the amplitude of the submaximal action potential recorded antidromically in the nerve, the terminals of which are the postsynaptic part of the axo-axonic synapse. The same stimulus evidently activates now a larger population of nerve terminals.

The antidromic discharge in the dorsal roots can also be elicited by synaptic activation of the axo-axonic synapse if the PAD attains the critical level

A

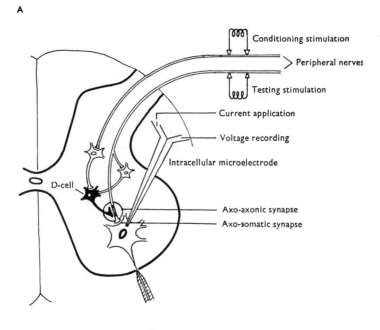

B

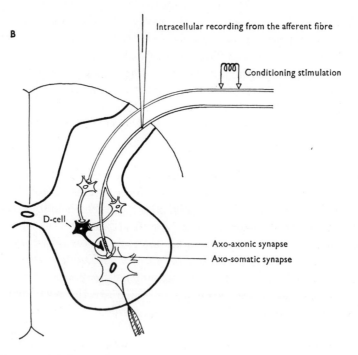

Fig. 218. Diagrams of the experimental arrangement for studying the effects and/or

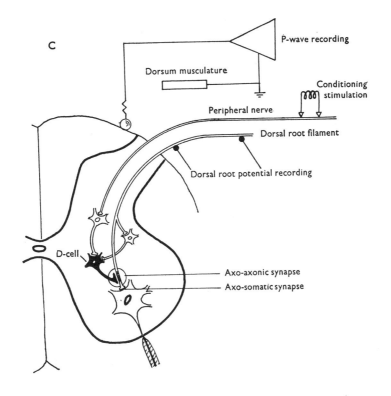

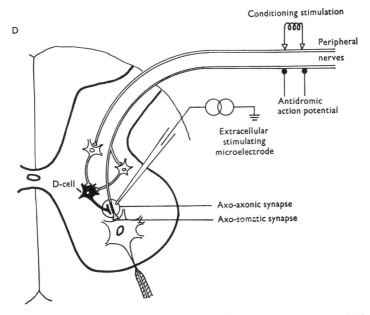

manifestations of presynaptic inhibition in the spinal cord. For further description see text.

for the occurrence of an impulse. This gives rise to a dorsal root reflex. Elicitation of these reflexes is facilitated by hypothermia. Dorsal root reflexes occur more easily on stimulating cutaneous than muscle nerves (Toennies 1938, 1939).

Object: A decerebrate cat, spinal preparation. Cutaneous nerves: PT (tibialis posterior), SU (suralis), SP (peroneus superficialis).

Apparatus: Cathode-ray oscilloscope with a D. C. amplifier. Overall sensitivity 1 mV/cm. A stimulator giving a train of 3—6 impulses with a frequency of about 250/sec. Time marker 100/sec.

Other requirements: Three pairs of platinum electrodes for stimulation and recording. Ball tipped platinum electrode. Earthing silver plate.

Procedure: Dorsal root potentials. A fine filament is dissected from dorsal root L7 and cut as far peripherally as possible. One recording platinum electrode is placed on the filament as close as possible to its entry into the cord (1—2mm), the electrode must not touch the surface of the spinal cord however. The second recording electrode is placed close to the cut end of the filament. A cutaneous nerve, e. g. PT, is placed on the stimulating platinum electrodes. A further pair of recording electrodes is applied to the dorsal or ventral roots. The cutaneous nerve is first stimulated with one supramaximal stimulus. DRP of relatively low amplitude is recorded from the filament. After stimulation with a train of impulses the response is well discernible (Fig. 219A). A typical polysynaptic discharge is observed on the ventral roots.

Slow positive potential (P wave) of the cord dorsum. Medial to the entry of dorsal root L7 we apply the ball tip of the platinum electrode onto the spinal cord surface. The indifferent earthing electrode is a silver plate (3 × 2 cm) fixed to the exposed muscles. On stimulating the cutaneous nerve we record a response with polarity opposite to that obtained when recording DRP. This is the P wave of the cord surface (Fig. 219B). The ventral roots have, of course, the same polysynaptic discharge as before.

Conclusion: DRP and P waves are an expression of the same process — — depolarization of the primary afferent terminals. Their opposite polarity is due to recording in a volume conductor, this is also indicated by the biphasic shape of the P wave, the positive phase of which is preceded by a low amplitude negative wave. The P wave can, of course, be recorded from different depths of the cord thus making mapping of its source possible (Eccles 1964a, b). The technique of Wall (Fig. 218D) can also be used to obtain an approximate idea of the source generating PAD.

The inhibitory effect of PAD initiated in the given case by terminals of a cutaneous nerve (PT) may be demonstrated by observing the reflex discharge on the ventral roots using suitably timed stimulation of another cutaneous nerve, e. g. SP. PAD is formed on the afferent terminals of the latter. Presynaptic inhibition exerted by a cutaneous nerve affects most effectively

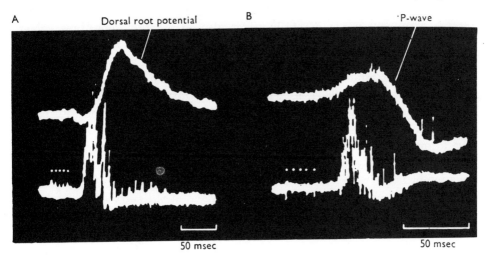

A — Dorsal root potential   B — ·P-wave

50 msec   50 msec

Fig. 219. Manifestation of praesynaptic depolarization of afferent fibres in dorsal roots and on the cord dorsum. A — Dorsal root potential. B — Slow positive potential (P-wave), (upper beam). Lower beam: discharge from ventral roots evoked by cutaneous nerve stimulation (a train of five impulses). Recording conditions are given in Fig. 218 C.

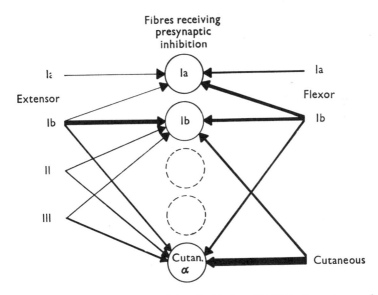

Fibres receiving presynaptic inhibition

Ia   Ia   Ia

Extensor   Flexor

Ib   Ib   Ib

II

III

Cutan. α Cutaneous

Fig. 220. Diagram showing the amount of depolarization (PAD) in three types of afferent fibres (Ia, Ib and α — cutaneous; circles in the middle) produced by various types of afferent fibres. The thicknesses of the arrows are proportional to the amount of depolarization (modified after Eccles 1964b).

another cutaneous nerve, it acts however, also on other afferent fibres, though to a lesser extent (Fig. 220). A similar situation also prevails for Ib fibres where the effect of Ib fibres on Ib fibres is again best expressed as shown in Fig. 220, summarizing the numerous studies of Eccles' group on presynaptic inhibition (for review see Eccles 1964a, b). The same fibre forms a negative feed-back to its own terminals by sending out collaterals to the interneuron (Fig. 218), which forms synaptic contacts with a D-cell (Eccles et al. 1962a). This last neurone forms an axo-axonic synapse on synaptic knobs. The effectiveness of this negative feed back can be tested by studying DRP elicited by stimulation of the afferent nerve with a frequent stimulus (200—300 /sec) for several seconds. Both DRP and presynaptic inhibition are largest at the start and then decline exponentially, in spite of continuing stimulation, to a certain plateau, the height of which is not the same for all fibres (Eccles et al. 1963a, b).

Synaptic transmission at axo-axonic synapses is mediated by a different agent than postsynaptic inhibition as indicated by the different pharmacologic al properties of the participating synapses (reviews: Eccles 1964a, Schmidt 1963, 1964). The simple method of DRP recording still offers valuable service when studying the pharmacological characteristics of presynaptic inhibition.

Presynaptic inhibition occurs immediately at the input into the central nervous system where it acts as a negative feed-back on the inflow of sensory information into the central nervous system (Eccles 1963). At this level it is also much more effective than postsynaptic inhibition, which predominates in the brain. Higher parts of the central nervous system, however, control presynaptic inhibition in the spinal cord or exert their influence on spinal functions by means of presynaptic inhibition (Andersen et al. 1962, Carpenter et al. 1962a, b).

Phylogenetically presynaptic inhibition has been described in the crayfish (Dudel and Kuffler 1961a, b, c), in Mauthner cells of the fish brain (Furakawa et al. 1963a, b) and in the giant ganglion cells of *Aplysia* and *Helix* (Tauc 1960). Presynaptic inhibition in the crayfish represents a suitable object, since axo-axonic synapses are situated at the periphery and are thus much more accessible to a study of conductance changes in the postsynaptic membrane (Dudel 1965).

g) Post-tetanic potentiation

Problem: Demonstrate potentiation in the monosynaptic reflex arc after tetanic stimulation.

Principle: In addition to changes in excitability lasting for several miliseconds, changes lasting for minutes may also be observed in synaptic structures. Such a change is the phenomenon of post-tetanic potentiation

(Lloyd 1949). This is an increased postsynaptic response lasting for several minutes following intense repetitive synaptic stimulation.

Post-tetanic potentiation occurs in peripheral synaptic junctions, but has been most often studied in the monosynaptic reflex arc (Lloyd 1949, Eccles and Rall 1951; for review see Hughes 1958).

In principle, changes in amplitude height of the action potential from ventral roots, evoked by orthodromic activation of the monosynaptic reflex arc lasting several seconds are studied.

Object: The spinal cat with exposed ventral root L_7 or S_1 and the nerve leading to the gastrocnemius muscle.

Apparatus: A cathode-ray oscilloscope with an A. C. amplifier, overall sensitivity 100 μV/cm. A stimulator giving pulses synchronised with the time base at 3 sec intervals and a generator of rectangular pulses with a variable frequency from 1—500 c/sec. Two pairs of stimulating and one pair of recording electrodes.

Other requirements: as in experiment VII A a.

Procedure: The recording electrodes are placed on the ventral root further away from the spinal cord, in order to avoid electrotonic spreading of postsynaptic potentials. Both pairs of stimulating electrodes are placed on the nerve to the gastrocnemius muscle with the electrodes for tetanic stimulation situated more peripherally. First the action potential maximal for *Ia* fibres from the ventral root is recorded. Then the stimulator for repetitive stimulation is switched on with a frequency of 500 c/sec. On cessation of the tetanic stimulation, the action potential from the ventral roots evoked by single pulses is recorded at regular 3 sec intervals for 6 minutes. Figure 221 shows that the amplitude increases rapidly to a maximum. It remains thus for some time and then decreases gradually and very slowly.

Fig. 221. Post-tetanic potentiation. Abscissa: time in minutes after tetanic stimulation of the afferent nerve from the muscle. Duration of tetanic stimulation 50 sec. Ordinate: increase in postsynaptic response as compared with normal values.

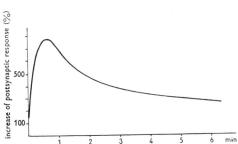

The experiment may be repeated several times in succession and it may thus be determined how different durations or frequencies of tetanic stimulation affect posttetanic potentiation.

It can be shown that the longer the duration of stimulation the longer post-tetanic potentiation. During potentiation two phases may be observed. The first is shorter, the second prolonged.

Conclusion: It can been demonstrated that postsynaptic elements do not change during the period of potentiation, and that changes which occur are present only in those fibres stimulated tetanically. It is generally accepted that post-tetanic potentiation is related to some specific change in presynaptic fibres. The basis of this change has not been fully elucidated. Several hypotheses have been put forward. It has been attributed to the effect of post-tetanic positive after-potentials, morphological changes in synaptic knobs and to release of larger amounts of mediator (Eccles 1964a).

Potentiation is related in some way to restoration of synaptic activity which has become ineffective because of prolonged disuse (Eccles and McIntyre 1951).

Post-tetanic potentiation and restoration of action impaired by disuse or by repetitive stimulation are examples of *plastic changes* which occur in the central nervous system.

B. Intracellular recording

The relatively large dimensions of the somata of neurons in the grey matter of the central nervous system make it possible to introduce capillary microelectrodes into nerve cells without injuring them. Since the somatic membrane is the place where postsynaptic activation occurs, the intracellular microelectrode in the neuronal soma has enabled us to study not only neuron activity at the unit level but also to observe directly the postsynaptic part of the neuro-neuronal synapse and thus to study the mechanisms of excitatory and inhibitory transmission at central synapses.

In principle the technique of intracellular recording and stimulation from somata of central neurons is the same as for peripheral excitable structures. As in those structures it is possible to record activity from a single cell without the need to isolate it anatomically. In contrast to recordings from peripheral structures, however, cells of the central nervous system are not accessible to direct abservations with the microscope so that in the CNS microelectrodes have to be inserted into the cell "blindly".

This fact causes the greatest problem when using microelectrodes in the CNS, i. e. identification of the structure generating the potential. Identification at the present is still mainly physiological, based on a knowledge of the structure and reflex activity of the part of the CNS being studied. The marking techniques (electrocoagulation, deposition of stains or substances giving a colour reaction) are suitable only for the localization of gross microelectrodes, since their resolving power (maximum 500 μ) is insufficient for identifying individual cells or their parts. Up to the present only Rayport (1957) has succeeded in reaching the cellular level by the microelectrophoretic (Exp. VI-A-e) injection of iron into the cell and its subsequent detection with Hess' reaction.

In addition to the visual inaccessibility further difficulties in the application of the microelectrode technique to neurons of the CNS are due to the fact that it increases some critical elements of this method to a disagreable extent. In the first place this is due to a rise in the capacity across the microelectrode wall when impaling a cell situated deeper in the nervous tissue. This results in distortion and attenuation of the recorded signal (p. 217).

Visual inaccessibility also complicates direct stimulation of the cell. Successive introduction of two electrodes into the same cell is practically impossible and hence when activity evoked by direct cell stimulation has to be recorded, we must use a special procedure (see page 223).

The performance of the following experiment is intended as an introduction to the specialities of intracellular recording in the central nervous system using as an example the most frequently studied cell of the CNS, the spinal motoneuron of the cat in vivo. The in vivo registration with microelectrodes demands a further factor: complete immobilization of the animal so as to avoid withdrawing the microelectrode out of the cell during movements.

Orthodromic and antidromic activation of the motoneuron

Problem: Record the action potential in the spinal motoneuron of the cat in response to an afferent volley from physiological extensors of the hind limb using intracellular microelectrodes.

Principle: The microelectrode is inserted into the motoneuron situated in the ventral horn from the dorsal side. Appearance of the membrane potential serves as indication that we have impaled a cell. Whether this is a motoneuron is determined by the presence of a spike in response to stimulation of the ventral roots which reaches the soma of the motoneuron by antidromic conduction (antidromic spike). Closer identification of the motoneuron (motor nucleus) can be achieved by stimulation of extensor muscle nerves. Hence it is advisable to have several nerves supplying the physiological extensors prepared for stimulation. In response to stimulation of the nerve which synapses directly with the impaled motoneuron an excitatory postsynaptic potential (EPSP) or a spike potential (orthodromic spike) is recorded.

The above criteria are necessary and usually also sufficient for identifying the motoneuron as a structure generating spikes. Sometimes, however, further criteria must be used for identifying the electrogenic structure.

If the microelectrode is fine enough axons, dendrites or primary sensory fibres (Frank and Fuortes 1955) may also be impaled.

497

The action potentials of axons differ from somato-dendritic spikes in shape and duration. They last for a short time only, are not accompanied by slow potential changes and have no inflection point on the ascending phase.

The primary sensory fibres produce intracellular spikes with characteristics similar to those of motor axons from which they differ by not responding to antidromic activation. They differ in the somatic responses also by their shorter latent period (shorter by the synaptic delay) and by high lability of Wedensky. They can follow frequencies exceeding 500 c/sec. Finally these fibres respond with a regular rhythm to maintained natural stimulation (e. g. touch, stretching of the muscle) unless they are cut off from the peripheral receptor.

The interneuron, from the aspect of intracellular technique, is defined as the postsynaptic unit which does not send axons to the ventral roots (Frank and Fuortes 1955). This, of course, does not apply to Renshaw's interneurons (Renshaw 1946, Eccles et al. 1954) which are also fired by antidromic stimulation via the recurrent axon collaterals. These, however, can be distinguished by their characteristic high frequency discharge to a single antidromic stimulus.

At the present it is not possible to distinguish reliably between activity from dendrites and motoneuron somata. The activity of both, however, differs from that of the other parts of the neuron by the presence of synaptic potentials in response to subthreshold stimulation, by the inflection point on the ascending phase of the spike and by the long lasting afterpositivity which is particularly noticeable when using higher gain.

Object: The spinal cat.

Apparatus: D. C. amplifier, cathode follower, double beam oscilloscope. Total amplification of the recording channel 10 mV/cm. A calibrator giving 10 mV voltage steps. Time mark 1000 c/sec. Acoustic indicator of impalement. Stimulator giving rectangular output pulses (0·5 msec duration). Radiofrequency isolation unit. For more systematic work an A. C. amplifier with a total gain of 100 μV/cm is necessary for extracellular recording of the root discharges on the second beam of the oscilloscope.

For routine work it is of great advantage to use 2 double beam oscilloscopes with two cameras. On one screen the time mark is registered on the first beam and the membrane and action potential on the second one using a D. C. amplifier. The output of this amplifier is connected via a condensor to one channel of the second cathode-ray oscilloscope. Here the intracellular spike without membrane potential is recorded. The second beam is used for extracellular recording from the roots with an A. C. coupled high gain amplifier.

Other requirements: Several pairs (5—6) of platinum electrodes for extracellular stimulation and recording. Capillary microelectrodes filled with 3M-KCl.

A micromanipulation device for insertion of microelectrodes into the spinal cord. Arrangement for good fixation of the animal and cord. Respiratory pump. Surgical instruments, anaesthetics and other requirements as in exp. VIII-A-a.

Procedure: Preparation of the cat and surgical exposure of the spinal cord and nerves in the hind limb is the same as in exp. VIII-A-a. The fixation of the animal, however, must be much more rigid than for extracellular recording.

Fixation of animal: Fig. 222A-C is a photograph of a cat held in a special fixation device. The cat is held by two clamps in the thoracic (1) and sacral region (2). The thoracic clamp is placed somewhat rostrally from the laminectomy, the sacral clamp holds the sacrum. The animal hangs in the clamps in such a way that the abdomen and thorax do not touch the metal base in order to decrease disturbances due to respiratory movements. An electric heating pad is situated between the metal plate and the cat's body. It is covered with a PVC cover and is used to warm the animal. The head is held in a stereotaxic holder (3) (see p. 586). The upper limbs are placed on the support. The section of the spine between the clamps is fixed at the level of the microelectrode insertion by horizontal supports that press against each other on the transverse processes (4). The horizontal holders form 10 cm from the point of contact an arc to which the loose skin is sown. The bath formed in this way with the cord at the bottom is filled with paraffin oil maintained at body temperature.

A similar bath is formed for the stimulated nerve. It is made above the fossa poplitea by sewing the skin to two semieliptic metal holders (5) fixed to the base plate of the fixation device.

Fixation of external electrodes: The electrodes for stimulating and recording from nerves and roots are made from platinum wire. The wires are melted into glass capillaries thus insulating connections to the leads against fluid. The glass tube also serves as an electrode carrier. It is inserted into a spring holder mounted on a ball joint (6) which makes it possible to set the electrodes as desired.

The ball joints can be placed on a rod connected either with the frame holding the main clamps or with the stand fixed in the base plate.

Manipulation of the microelectrode: Rough placing of the microelectrode above the site of the intended insertion may be achieved with three screw drives enabling movement in 3 planes perpendicular to one another (7,8, 9). The drives are placed on the frame on which the cat hangs and are sufficiently massive so that they do not vibrate. A mechanical microdrive is fitted to the vertical carriage of the coarse drive and is controlled from the anterior part of the frame (10) by a Bowden cabel (11). The extent of the movement is indicated by an indicator (12) divided into micron units. Another indicator (divided into 10 μ units) indicates gross movements perpendicular to the long axis of the animal (13).

Fig. 222. A — Photograph of the apparatus employed in recording intracellularly from nerve cells in cat spinal cord. B — View of a cat fixed in fixing and recording system as shown in A and described in text. C — Detail of spinal bath with cord, with extracellular electrodes at the roots and with a microelectrode.

The rider of the microdrive holds the microelectrode (14) which is connected with the cathode follower (15) in the usual way. The contact with the broad end of the microelectrode is formed by an Ag-AgCl agar bridge. The cathode follower is fixed to the frame.

The experiment proper: At the selected site the pia mater and arachnoidea is removed under microscopic control from several square milimeters of spinal cord surface in order to facilitate insertion of the microelectrode. Then one of the cut ventral roots (6th, 7th or 8th post-thoracic) is placed on the stimulating electrodes in the cord bath. The corresponding dorsal root is placed on the recording electrodes (A. C. amplifier). Several nerves supplying a physiological extensor are placed on the electrodes in the nerve bath. They are cut distally. Using a switch the output of the stimulator is successively connected with each nerve and the presence of activity in the ventral and dorsal roots is checked.

The microelectrode is fitted into place over the site of insertion and then slowly inserted into the spinal cord using the vertical microdrive.

As the microelectrode passes through the cord we observe sudden potential jumps (electrode negative) which, however, rapidly return to the initial level. They are caused by passage of the electrode through nerve fibres and small nerve cells. In some cells the membrane potential is maintained for a while but soon declines and an injury discharge appears characterized by an initially high frequency, which rapidly decreases indicating death of the unit.

The density of motoneurons in the anterior horn is not large (Fig. 223). It may be expected that 3—5 motoneurons are encountered for one vertical insertion. The proximity of a motoneuron is signalled by a focal potential which the microelectrode records in a volume conductor from a motoneuron activated antidromically. When a focal potential appears we move the microelectrode in small (several μ), but sudden steps (rectangular movement) with intervals between movements until we succeed in inserting it into a motoneuron. This is indicated by the appearance of the membrane potential of about -70 mV (inside negative) and of the antidromic spike. The antidromic spike is characterized by an inflection point on the ascending phase of the action potential and by a short latent period. The total amplitude of the spike is about 100 mV with an overshoot of 25 mV.

The antidromic spike indicates that we are in the motoneuron. We now attempt to identify the motoneuron more closely by orthodromic activation. One after another we stimulate the nerves on the stimulating electrodes in the nerve bath and according to the nerve which activates the impaled motoneuron we determine the corresponding motor nucleus of the cord. Usually a certain motoneuron is monosynaptically activated only from one synergic group. In response to less intense stimuli which nevertheless induce a discharge in the dorsal roots we record a synaptic potential (Fig. 224), the amplitude of which

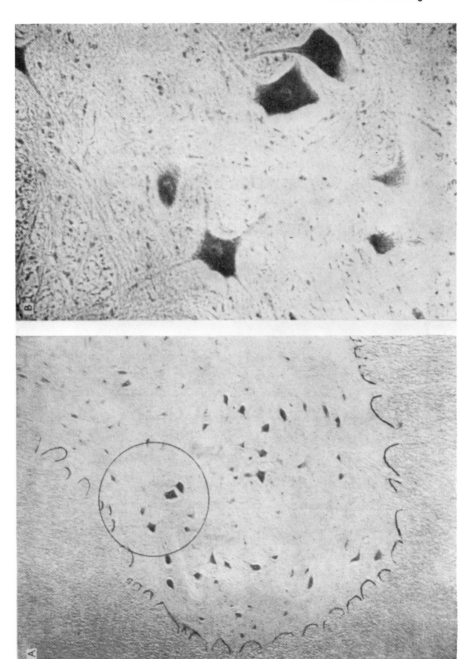

Fig. 223. Section of cat's spinal cord at C 7. The encircled area in A is shown at greater magnification in B.

increases until at a certain level of depolarization an orthodromic spike is initiated (*spatial facilitation*). Spikes also occur on successive summation of two synaptic potentials (*temporal facilitation*). Often, however, even the maximum stimulus in group I afferent fibres does not induce an impulse in the motoneuron.

The orthodromic spike has the same amplitude as the antidromic one and is characterized by a slow afterpositivity (Fig. 224A). Rapid recording reveals an inflection on the ascending phase of the action potential.

Conclusion: If the number of activated synapses on the motoneuron is small, a presynaptic volley evokes only a synaptic potential (EPSP — excitatory postsynaptic potential). Activation of a larger number of synapses by increasing afferent influx increases the synaptic potential until a critical level of depolarization is attained which leads to the appearance of an action poten-

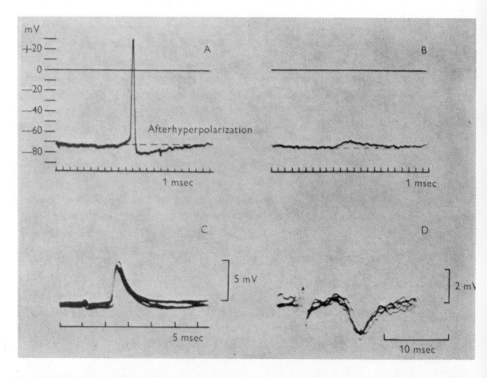

Fig. 224. Intracellularly recorded potentials of a cat motoneurone. A: Spike potential evoked by single orthodromic nerve volley in a cat gastrocnemius motoneurone. B: Excitatory postsynaptic potential (EPSP) in the same motoneurone evoked by single nerve volley of reduced size as compared with that used in A. C: EPSP recorded intracellulary in a medial gastrocnemius motoneurone by an afferent volley from medial gastrocnemius nerve. D: Intracellular record of inhibitory postsynaptic potential (IPSP) of a gastrocnemius motoneurone to a train of four impulses from the anterior tibial nerve.

Records C and D are formed by the superposition of about ten faint traces.

tial. The critical level of depolarization and thus the occurrence of the spike may be attained equally well by temporal facilitation or by artificial depolarization of the membrane (Coombs et al. 1957b).

The orthodromic spike recorded in the motoneuron is a complex waveform consisting of 3 components (Fig. 225): the synaptic potential, IS and SD spikes. The synaptic potential which arises in the excitatory synapses on

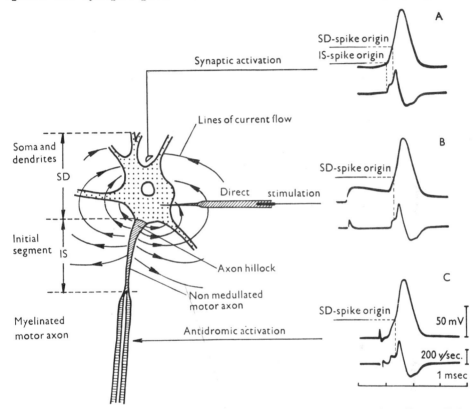

Fig. 225. The composition of the motoneuronal spike potential. A—C — Intracellular records of spikes evoked by monosynaptic activation (A), by applying a depolarizing current to motoneurone through microelectrode (B), by antidromic stimulation (C). Lower traces are electrically differentiated records (diagrammatically after Fatt 1957b, Coombs, Curtis and Eccles 1957b).

the motoneuron body first induces an IS-spike in the initial segment of the motoneuron. The initial segment, from which the spike derives its name, includes the non-myelinated part of the axon and the axon-hillock of the motoneuron. This part of the membrane is electrically more excitable than the non-synaptic membrane of the motoneuron. In the recording of the orthodromic spike the development of the IS-spike from the synaptic potential appears as the first inflection point (at a depolarization of about 10 mV). The synaptic potential spreads to the initial segment electrotonically. The distribution of local currents is shown in Fig. 225. When the amplitude of the IS-spike attains the threshold value for spike generation in the nonsynaptic membrane of the motoneuron, it initiates an SD-spike in the somato-dendritic membrane. Hence the name of this spike. The lines of current run in the opposite direction to those in Fig. 225. The critical level of the depolarization at which SD-spikes are generated is indicated in fast recordings by the second inflection point, which is particularly obvious in recordings that are electrically differentiated (Fatt 1957a, b).

The level at which the SD-spike is initiated is the same as that of the inflection point of the antidromic spike (Fig. 225C). The IS-spike in this case is not initiated by the synaptic potential but by local currents of the nerve action potential which enters the initial segment of the motoneuron antidromically.

Experiments with direct stimulation of the motoneuron (Coombs et al. 1957a, b, Fuortes et al. 1957, Araki and Terzuolo 1962) also confirm that the stimulation threshold for the somatodendritic membrane is 2—3 times higher than that of the initial segment membrane. This higher stimulation threshold is due not only to the mosaic structure of the somatodendritic membrane with electrically inexcitable (Grundfest 1957a, b, c, 1959, 1960a, b, 1961a, b) postsynaptic sites which shortcircuit the depolarizing effect of the spike surrounded by the electrically excitable membrane matrix but also to specific intrinsic characteristics of the somatodendritic nonsynaptic membrane (Eccles 1957, 1964a).

The SD-spike, being the same for antidromic and orthodromic stimulation is followed by a short (2—6 msec) depolarization which changes into *after-hyperpolarization* analogous to the positive after-potential in the extracellular recording. It is not seen after an isolated IS-spike. Afterhyperpolarization is due to the net diffusion flux of K^+ ions out of the cell. The increased permeability for K^+ ions during afterhyperpolarization and during the descending phase of the spike are two separate events (Eccles 1957). Afterhyperpolarization is a good diagnostic aid for differentiating somatodendritic spikes from spikes in other parts of the neuron (Frank 1959a, b).

In a similar way to the excitatory postsynaptic potentials (EPSP) inhibitory postsynaptic potentials (IPSP) may also be recorded from the motoneuron (Fig. 224 D). This can be done immediately after recording EPSP and the orthodromic and antidromic spikes. The microelectrode remains in the motoneuron, but stimulation is switched to the antagonistic nerve (in our case from the nerve supplying m. gastrocnemius to that innervating m. tib. ant. The response is a hyperpolarizing synaptic potential (Brock et al. 1952, for review see Eccles 1959, 1964a).

Electrophysiology
of the cerebral cortex

In higher mammals the cerebral cortex is by far the largest and most accessible part of the central nervous system. The most important analytical and synthetic processes, forming the basis of the higher nervous activity (Pavlov 1926), occur here. It is consequently not surprising that in electrophysiological research on the brain, electrical characteristics of the cerebral cortex play an important part. The majority of electrophysiological experiments on the cerebral cortex can be realised using relatively simple means, and this organ will, therefore, be used as a suitable object for many experiments in this chapter.

A. Electroencephalography and electrocorticography in general

Registration of potentials arising in the brain and especially in the cerebral cortex, using electrodes placed on the cortical surface or introduced into the cortex, is called electrocorticography — ECoG. Recording of potentials from the surface of the head through the skin and bone or by means of subcutaneously situated needles is called electroencephalography — EEG. In man and higher mammals potentials recorded in the latter way are also due to activity of the cerebral cortex, their amplitude, however, is considerably decreased. This is not due to the series resistance of dura mater, bone and skin, which need not be taken into account because of the high input impedance of the EEG amplifiers, but is mainly caused by parallel tissue resistance, especially of the galea, short circuiting the electrodes and distorting the distribution of equipotential lines. Scalp acts as a spatial averager of brain activity (De Lucchi et al. 1962) which attenuates strictly localized potentials as much as 5000 times while the amplitude of waves sychronous over large cortical areas is reduced much less (1 : 2). A cortical potential can be recognized in the human scalp EEG only when it involves a brain surface of at least 2·5 × 2·5 cm (Cooper et al. 1965). In electroencephalography there is an increased possibility of interference by electrical potentials from lower parts of the brain, especially when using occipital and temporal leads. This is best demonstrated in experiments after hemispherectomy, when it is possible to record considerable EEG activity from the surface of the head above the extirpated hemisphere (Cobb and Sears 1956).

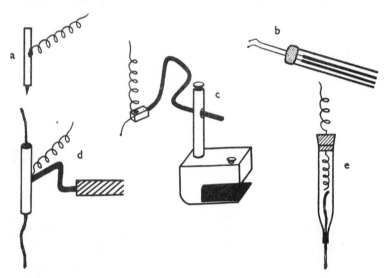

Fig. 226. Electrodes used for recording ECoG in acute experiments. a — Steel needle electrode for fixing into the skull bone. b — Ball tipped silver wire electrodes mounted in an insulating holder. c — Silver wire electrode on flexible lead bar with holder permitting direct fixation to bone. d — Wick Ag-AgCl tube electrode on a flexible lead rod with an insulating holder. e — Wick Ag-AgCl electrode vessel.

In both acute and chronic animal experiments electrocorticography is preferable, because, among other things, muscle potentials of the large subcutaneous muscles of the scalp or adjacent temporal muscles (cat, dog) are easily eliminated.

The ECoG can be picked up either from exposed parts of the brain after extensive craniotomy or using needle electrodes introduced through the bone to the surface of the dura mater (epidural registration) or the cortex. Other methods — e. g. recording from the cortical surface exposed by only small trephine holes can also be used. Suitable recording electrodes are used.

a) When recording from the exposed cortical surface we usually use electrodes with wick contacts (non-polarisable Ag-AgCl or $Hg-Hg_2Cl_2$ electrodes — Fig. 70b, c, e and 226d, e) or ball tipped silver or platinum wires (Fig. 226b, c) fixed to steel springs. The contact must be as reliable as possible but must not cause mechanical stimulation or injury of the brain. In larger animals it is of advantage to fix the electrode holder directly to the bones of the skull with a screw or by sealing wax or phosphate cement.

b) In corticography in the closed skull, needle electrodes fixed directly into bone are advantageous. Stainless steel needles have a stop, limiting the insertion of the tip to the thickness of the skull bones of the species used (Fig. 226a). The needle held with a special holder is hammered or screwed into place.

If more secure fixation is required silver steel or plastic screws, with one or more silver, platinum or tungsten electrodes (Fig. 227a) are used. The screws are fixed into prepared trephine holes that have a diameter a little smaller than that of the electrodes. In soft bone the electrode will thread its own way. In hard bone it is better to use a tap beforehand. If the bone surface is sufficiently dry neither needles nor screws need be insulated.

Finally, it is possible to place freely into small trephine openings platinum, silver, tungsten or steel foil, or wires of suitable size, that are connected to appropriate sockets and fixed to the surrounding dried and roughened bone with a drop of dental cement or acrylate. An example of such a fixation is shown in Fig. 227c. Electrodes are simply and effectively connected to their leads by using the following technique: (Fig. 227b). Injection needle tubing is cut into pieces 8—10 mm in length. The silver wire, the end of which forms the electrode proper, is inserted into a needle of appropriate diameter. By compressing its lower part a reliable contact is formed between the two metals. From the point of compression to the recording facet the electrode is insulated with lacquer. Such electrodes are then placed into small trephine openings in the dried skull bone and fixed with a 3 mm high layer of acrylate or phosphate cement, the connection of which to the bone is ensured by anchoring screws (Fig. 227e). The ends of the needles protruding 5—6 mm above the acrylate layer are about 2 mm apart and serve as contact plugs. Sockets are made from stronger injection needle tubing that just fit over the fixed electrodes. The exposed end of the input cable is inserted into the upper third of the socket needle and connected with it by compression. The point of connection is then covered with a piece of polyethylene tubing. The reliability of the contact may be increased by slightly bending that part of the inner needle which is introduced into the outer socket. In order to prevent disconnections during movements of the animal the cable is tied with a strong piece of thread to a ring fixed in the skull bone (Fig. 227d).

A more detailed description of electrodes and their fixation is given in the individual experiments. For review of literature see Delgado (1955, 1964), Batrak (1958) and Kogan (1949, 1952).

Arrangement of electrodes is a more complicated problem. Potentials recorded from the surface of the cerebral cortex are the sum of potentials arising in ganglion cells, dendrites and axones. Their amplitude depends on the distance of the electrode from active area, on the resistance of the tissue and on the localisation of the reference electrode. Here too, as elsewhere, two fundamental methods of recording are used — monopolar and bipolar.

a) Monopolar recording. Potentials from individual electrodes are recorded against a common reference (indifferent, non-active) electrode. It is assumed that in this way activity of a region below the active electrode is recorded. This assumption, however, is only justified if the amplitude of potentials below the active electrodes is much higher than that below the reference electrode.

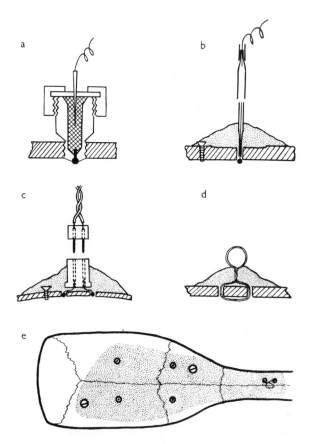

Fig. 227. Electrodes used for ECoG recording in chronic experiments. a — A plastic screw with a silver tip. Central hole is filled with mercury and closed with a rubber sheet. The input cable is connected to a steel needle which is simply inserted into mercury through the rubber (Steiner and Sulman 1961). b — A silver wire electrode with plug and socket made from injection needle tubing. For details see text. c — Silver wire electrodes connected to a subminiature socket. d — A steel wire fixation ring carrying the main weight of the animal's movements. e — Location of the electrodes, anchoring screws and of the fixation ring in the skull bones of the rat.

It is therefore important to choose the correct position of the reference electrode in electrically inactive regions. In electroencephalography the following sites are most frequently used for the reference electrode: the ear lobe, skin of the forehead, the nape of the neck or even the neck. In electrocorticography the cranial bones, ear lobes, injured parts of the brain, etc., are used.

In distinction to monopolar stimulation, where an increase of the area of the indifferent electrode and resulting decreased density of stimulating current in the indifferent electrode is desirable, in monopolar recording an increase in the area of the reference electrode only slightly improves its "indifference" (potentials of opposed polarity from different asynchronous foci below the electrode may cancel each other).

Earthing of the indifferent electrode also in no way changes conditions of recording and only helps to remove electrostatic interference. Care must be taken to earth the object only via the reference electrode, as otherwise the latter will receive potentials from other earthed parts of the body. When using push-pull preamplifiers the indifferent electrode should not be earthed.

The so called "average" electrode is a special solution of the problem of the indifferent electrode (Offner 1950, Goldman 1950). This is the EEG alternative to Wilson's ECG electrode. The average electrode is a point to which all electrodes are connected through $0.5—2.0$ MΩ resistors (R). Any pair of electrodes is thus short circuited by a resistance of $2R$. The average electrode in a system of n electrodes has a resistance of

$$R_a = \frac{R}{n}$$

Potentials arising below only one electrode will also appear at the average electrode but with reversed polarity and n times smaller if n electrodes are connected. The average electrode is useful especially when recording asynchronous activity and when using a large number of electrodes ($12—16$), as is the case with modern multi-lead EEG apparatus.

b) Bipolar recording. Potential difference between two electrodes relatively close to each other are recorded. The closer are the electrodes, the smaller is the area from which potentials are recorded and the better limited sources of po-potentials can be uncovered. The differential character of recording, on the other hand, may leave even potentials of high amplitude unrevealed. Interpretation of results obtained with bipolar registration is usually more difficult from the point of view of electrical fields than with monopolar recordings. The relation between monopolar and bipolar registration is llustrated by experiment IX B. The advantages and limitations of the monopolar and bipolar techniques are discussed by Malliani et al. (1965) and Meshchersky (1965).

Various apparatus may be used to register EEG and ECoG. Today, direct writing apparatuses predominate. Cathode-ray oscilloscopes are used for special purposes, especially for recording of short-lasting potentials or if an exact picture of the shape of cortical potentials is required. The different types of apparatus, their characteristics and use, have been described in more detail in the second part of this book. Here only a few instructions as to the correct procedure necessary for the majority of apparatus used are given:

1) switch on the apparatus, wait the necessary interval for the valves to warm up,

2) control batteries, power supplies and operational voltages of the amplifiers,

3) calibrate and set amplification in all channels to the desired value,

4) choose filters of low and high frequency adequate to the phenomenon registered,

5) connect electrodes to the individual channels and measure their resistance.

Only after having carried out the above can actual registration be started.

B. Spontaneous EEG and ECoG in animals

Problem: Record spontaneous electrical activity of the cerebral cortex of the dog, rabbit and rat using the monopolar and bipolar methods.

Principle: The cerebral cortex is source of rhythmical potentials with an amplitude of tens or hundreds of μV and a characteristic frequency that can be recorded with suitably placed electrodes using the bipolar or monopolar method of recording. The shape of the EEG (ECoG) depends to a certain extent on the recording techniques used, but mainly on the functional state of the animal.

Object: rat 150—200 g, rabbit 2—3 kg, dog 5—10 kg.

Apparatus: Ink writing multichannel EEG apparatus of conventional type or a cathode-ray oscilloscope with A. C. amplifier, overall amplification at least 50 μV/cm.

Further requirements: Screw and needle electrodes of suitable size and shape, surgical instruments, a series of trephines and a screw tap, electrode holder, a fixation board with a stereotaxic head holder for the rat, an animal board for the rabbit, a solution of collodion, dish shaped silver electrodes for sticking to the skin, an electric fan, dental cement or acrylate, procaine penicillin.

Procedure: The rat is under light ether anaesthesia. After infiltration of subcutaneous tissue with novocaine the scalp is cut along the midline from the level of the eyes up to 1 cm. behind the level of the external auditory meatus. The skin is reflected towards the sides as far as possible and the exposed bone is cleaned of muscle and fascia so that it appears dry and the bone sutures are clearly visible. These serve as important reference points further on. The animal is then tied to the board with the head holder. Firm fixation of the head is a necessary condition for a good recording. A holder ensuring fixation of the head in three points (the maxilla and both auditory meati) similar to those used in stereotaxic apparatus (cf. page 586) is best. According to a diagram prepared beforehand three needle electrodes (diameter 1 mm, 12 mm length, 1·5—2 mm length of the tip, shoulder 0·3 mm — Fig. 226a) are then placed above each hemisphere. They are either pinned or hammered into the bone using a special needle holder. A further, stronger needle electrode is hammered to a depth of about 3 mm into the frontal bone above the front edge of the olfactory bulbs (7—9 mm rostrally to the coronal suture). Flexible leads 20—30 cm long (thin high frequency cable is best) are fixed to the electrode input panel in a stand above the animal's head. The panel is connected to the EEG apparatus by a shielded cable. After the electrodes are fixed, anaesthesia is interrupted. After a further 10—20 minutes, during which the fixation of the head is corrected, if necessary, registration is started. Depending on the number of channels of the EEG apparatus the following connections are tested simultaneously or one after the other:

1—2, 2—3, 4—5, 5—6 — longitudinal bipolar leads
4—1, 5—2, 6—3   — transversal bipolar leads
0—1, 0—2, 0—3, 0—4, 0—5, 0—6 — monopolar leads

The following leads are then recorded, simultaneously if possible:

0—1, 0—2, 1—2, 2—1

With each pair of electrodes activity is recorded for several minutes at a speed of 1·5 or 3 cm/sec.

In the rabbit, implanted electrodes and chronic scalping are used. The electrodes are put into place under short lasting barbiturate anaesthesia (i. v. evipan or pentothal). After a midline incision of the scalp and removal of the galea trephine holes (2·4 mm diameter) are made into the skull bones reaching the dura. A 3 mm tap with a ground off tip is used to thread the bone into which a silver or steel screw is finally placed. The whole procedure is carried out under semisterile conditions with sterile instruments and the use of formalin (37%), "Ajatin", "Zephiran" or alcohol (70%) for sterilising the skin, screws etc. The skin is then adapted to the dimensions of the exposed skull and sutured to the parieto-temporal bone edge where it easily heals. The leads from the electrodes are insulated with polyethylene tubing and are connected to a common socket carried on a harness fixed to the neck or back of the animal. In order to prevent a direct pull on the electrodes, the leads are fixed with stainless steel wire to the edge of the occipital bone perforated with a sharp needle. Both the anchoring wire and the leads are covered with acrylate. Procain penicillin (100,000 I. U.) is given 12 hrs. before the operation and immediately after it. The animal is placed in a separate box with smooth walls.

Registration itself is carried out in the freely moving animal only 1—3 days after fixation of the electrode. The animal is placed into a circular container with a diameter of about 50 cm and a height of 50 cm. Above the centre of the container a cable connecting the socket on the neck of the animal with the input panel of the EEG is elastically fixed on a pulley, a steel spring or a rubber band. To reduce the movement artefacts individual wires of the cable must be tightly wound together. Tension on the cable must be uniform during movement of the animal. Sometimes it is of advantage to connect the animal with the pulley via a special string that bears the main weight of the animal's movements and prevents the cable from being pulled. A suitable choice of the fixation string also prevents undesirable rotation of the animal to one side only. Since cable twisting may be a serious problem in long lasting experiments, special swivel-devices are used providing reliable electrical contacts during rotation of the animal (Sutton and Miller 1963, Simpson and Valenstein 1963). Registration is fundamentally the same as in the rat. The distribution of the electrodes and their connections are shown in Fig. 228.

In the dog electroencephalography may be attempted. It is best to use an animal used to work with conditioned reflexes in a Pavlov stand. On the day before the experiment the scalp is clipped and the skin is shaven with a suitable instrument and then washed with soap and water. After thorough drying, the electrodes are stuck to the head. Fat at the sites of application of the electrodes is first removed from the skin by rubbing with cotton wool soaked in ether. Then a drop of EEG paste is applied to these points and is thoroughly rubbed into the skin. Rubbing is easily done with a smooth rotating axle of a small electro-motor. Only then is a dish-shaped electrode, also filled with EEG paste, applied. While holding the electrode with a glass rod placed into a small hollow on the upper surface, this and its nearest surroundings are covered with a few drops of collodion from a dropper. The assistant must hold the head of the dog firmly to prevent any movement. Drying of the collodion is speeded up using a current of warm air from an electric fan. The skin must be perfectly dry and without fat for the collodion to hold in place firmly. The connecting wires from the electrodes are again fixed to a small socket on the collar. The dog is placed into a stand for experiments with conditioned reflexes and as soon as it settles down registration may commence. This is preferably bipolar. E. g.:

1—2, 2—3, 4—5, 11—12, 12—13, 14—15 — longitudinally
11—1, 12—2, 13—3, 14—4, 15—5 — transversally

Results: Fig. 228 shows a characteristic record of EEG activity in the rabbit. There are no fundamental differences between the EEG in different

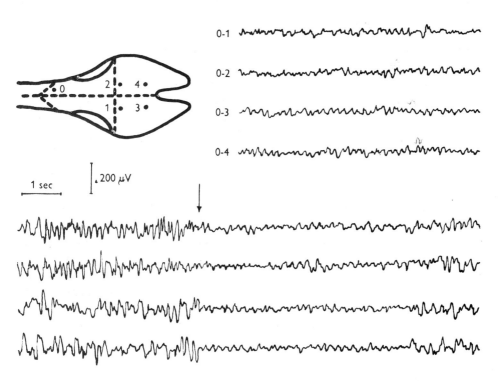

Fig. 228. Characteristic ECoG activity in an unanaesthetised unrestrained rabbit. Above: waking type of EEG activity. Below: slow sleep activity and arousal on whistle (arrow).

animals used, however, and individual leads also do not differ substantially from each other in the same animal. The record in the rabbit has the highest amplitude. This is due to the relatively large electrodes placed directly onto the dural surface. The amplitude of the EEG in the dog is smallest. This again is due to the large short circuit decrease in potential when recording across the skin. Several basic types of activity corresponding to different levels of wakefulness can be distinguished in the EEG of the unanaesthetized animal.

1) Active waking state. The animal reacts promptly to stimuli, is oriented, moves around. Low voltage (less than 50 μV) rapid (about 20—30/sec) activity predominates in the EEG.

2) Relaxation. The animal is at rest, the eyes are open, the head is held upright. The heart and respiratory rates are slightly slowed down. Regular low activity of 5—8/sec appears in the EEG.

3) "Slow sleep". The head gradually drops and the typical sleep posture is assumed. The eyes are closed. In the EEG high voltage spindles ($100-200\,\mu V$, about 12/sec) appear at the onset of this stage in the frontoparietal cortex. They are called spindles because in a series of 10—20 such waves the amplitude first rises, attains a maximum and then falls again. Later slow (2—4/sec) irregular waves of high voltage (about $200\,\mu V$) predominate and this phase is termed according to them.

4) "Rapid sleep". This phase is seen as a sudden loss of tonus of the neck muscles (further dropping of the head), movement of the eye bulbs, vibrissae, ears and sometimes the limbs and tail. In the EEG rapid low voltage activity similar to the waking state paradoxically appears (hence "rapid" or "paradoxical" sleep and since the source of this activity is the pontine reticular formation it is also termed rhombencephalic sleep). In man dreams most frequently occur in this phase.

The different stages follow each other and form thus the characteristic wakefulness-sleep cycle. Rapid sleep always follows only after slow sleep and usually changes into waking activity. If the animal is in quiet surroundings slow sleep again appears after a time and the whole cycle is repeated with a period of about 25 min in the rabbit and 12 min in the rat (Roldán at al. 1963). The electrophysiological manifestations of the different stages of the sleep cycle in subcortical structures are described in detail by Jouvet (1962) Especially the hippocampogram displays a highly regular theta activity (5—7/sec) during the rapid sleep (see also chapter X D).

Sleep activity is most frequently found in our experimental conditions in the rabbit and dog. This corresponds to the very natural conditions during registration. On the contrary EEG of the restrained rat displays permanent alertness due to the continuing attempts of the animal to escape.

If various external stimuli (clapping, touching, changes in illumination, olfactory stimuli etc.) act on an animal displaying the slow sleep type of EEG activity, low voltage fast activity suddenly appears in the records from all regions. This, again, is most apparent in the fronto-parietal leads. In the rabbit a regular 4—6/sec. activity (theta rhythm, see also p. 605) occurs sometimes in the limbic cortex during arousal. The change from sleep activity to waking activity is sudden and very obvious. The term "arousal reaction" (waking reaction, blockade, activation of EEG) is therefore used. The blockade of alpha rhythm in the EEG in man when opening the eyes or during psychic activity (Berger 1930) is also an arousal reaction. The EEG arousal may be accompanied by motor reactions corresponding to behavioral arousal of the animal. At other times, however, the animal remains motionless. As a slow, high EEG can be explained by synchronisation of potentials of individual cortical neurones, the transition to rapid, low voltage activity is related to desynchronisation of electrical activity of individual cortical elements (Adrian and Matthews 1934).

A high frequency due to muscle movements (especially the temporal muscles) often occurs in the dog EEG. Sometimes it is not easy to differentiate

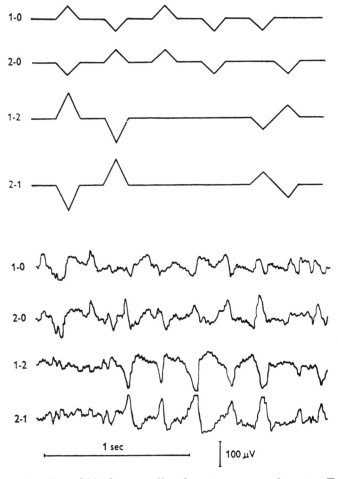

Fig. 229. Reconstruction of bipolar recording from two monopolar ones. Top: diagrammatically. Bottom: example from rabbit's ECoG.

between such muscle artefacts and high frequencies originating in the brain tissue. More abrupt movements of the head are further accompanied by so called movement artefacts in the record. These are irregular, high voltage changes, usually of long duration, blocking the amplifier because of the excessive input signal. The cause of such artefacts is usually a change in contact

517

fo poorly fixed electrodes, large changes in electrode resistance, skin potentials, capacitive influences on the moving cables, etc. If such artefacts (that can usually be identified without difficulty) appear, the electrode on which they occur must first be localised. Movement artefacts are most reliably prevented by reducing electrode resistance, by better fixation of the electrode, and by decreasing the animals movements. A high electrode resistance is usually noticed even without movement artefacts by a frequency of 50 cycles/sec occurring in the record (the lead is "noisy").

Records from the various electrodes usually differ from one another. Some characteristic waves, however, may be found simultaneously in several leads from the same or even both hemispheres (especially during sleep activity). In general it is true that the bipolar longitudinal record shows higher amplitudes than the bipolar transversal one, the distance between the electrodes being the same. This is so because connection of symmetrical points gives a smaller potential difference since the potentials are to a certain extent synchronous. In addition to these physiological conditions, purely physical factors influence the shape of the recording. Thus, for instance, a comparison of records 1—2 and 2—1 (Fig. 229) shows that both tracings are identical but of opposite polarity (mirror images), of course assuming an equal function of amplifiers in both channels. It is important to bear in mind that the polarity of a record is determined by the way the electrodes are connected to the left and right half of the amplifier input. It is therefore necessary to pay special attention to this when using the multi-lead method.

A comparison of records 1—0, 2—0 and 1—2 (Fig. 229) shows that the latter is a curve of the difference between the first two. This also determines the relation between mono- and bipolar records, assuming, of course, that the 0 electrode contributes equally to records 1—0 and 2—0. The amplitude of the bipolar record is the lower the smaller the distance between the two electrodes, i. e. the greater the similarity between both monopolar records.

Conclusion: The ECoG and EEG of unanaesthetised animals is an expression of basic states of cerebral activity and is fundamentally similar in different mammalian species.

C. Drug effects on EEG

Problem: Observe changes in EEG activity during increasing barbiturate or ether anesthesia. Demonstrate the dissociation between EEG and behavior after administration of physostigmine and atropine.

Principle: During anaesthesia characteristic changes occur in the EEG. These correspond to different depth of anaesthesia as defined by reflexes.

According to initial changes anaesthetics may roughly be divided into two groups:

1) Barbiturates. Their effect is characterised by synchronisation and increased amplitude of cortical potentials (e. g. Forbes et al. 1956).

2) Ether. This, on the contrary, causes desynchronisation of the EEG and a decrease in its amplitude (e. g. Schlag and Brand 1958). These differences indicate that anaesthesia is not the result of a simple inhibition of brain metabolism but is due to complex relations between various parts of the brain, differently affected by the anaesthetic. The effect on the cortex is not primary, rather does the EEG reflect changes in excitability of brain stem formations, especially of the reticular activating system (French et al. 1953, Bovet and Longo 1956, cf. also p. 609).

During the deeper stages of anaesthesia initial differences disappear and inhibition of the metabolism of all groups of neurones in the brain sets in. As a result of this, EEG activity decreases and gradually disappears altogether. These phases of anaesthesia, in particular, can be characterised exactly and quantitatively, e. g. by integration of EEG amplitude, and can thus be used to control the depth of anesthesia.

The correlation between slow synchronous activity and inhibition is, however, not absolute. As shown by Bradley and Elkes (1957) EEG activity very similar to slow sleep can be induced by the i. p. injection of atropine in doses of about 6—15 mg/kg. The animal shows no signs of sleep. It can move normally and reacts to stimuli. Injection of 1 mg/kg physostigmine, on the other hand, induces considerable desynchronization corresponding to the waking state even though the behaviour of the animal indicates relaxation.

Object: Rabbit weighing 2—3 kg.

Apparatus: An ink-writing multichannel EEG apparatus or a cathode-ray oscilloscope with a pre-amplifier, total A. C. amplification of 50 μV/cm. Integrator of EEG activity (Fig. 230).

Other requirements: Fixation board for rabbit, silver screw electrodes for leading off the ECoG, a mask for ether administration, a polyethylene cannula for infusion of anesthetic, apparatus for automatic injection. Ether for anesthesia, Pentothal, atropine sulphate, physostigmine salicylate.

Procedure: The rabbit with chronically implanted electrodes as in the previous experiment is tied to the board. When using ether a gauze mask is fixed to its snout. After the animal has quietened down the actual experiment is started. First a normal record is made for 5—10 minutes characterised by alternating waking and sleep activity. Then ether is dropped onto the mask, rapidly at first (10—20 drops/min), later more slowly. The assistant is ready to prevent artefacts due to obvious restlessness of the animal during the excitatory phase of anaesthesia. The depth of anaesthesia is controlled by reflexes, especially by the presence of corneal and pupillary reflexes. During the deeper phases of anaesthesia ether is added very carefully and the reflexes are controlled continuously together with respiration and heart activity, which can suitably be recorded in one of the EEG channels using needle electrodes located in the skin on the sides of the thorax. After EEG activity

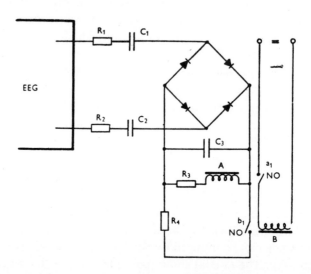

Fig. 230. A simple integrator of EEG activity. Condenser C_3 is continuously charged by the rectified output voltage of an EEG channel, and discharged across a large resistance R_3 and a sensitive relay A. After a certain voltage is reached at condensor C_3, relay A makes contact. This activates relay B, through the contacts of which C_3 is rapidly discharged through a small resistance R_4. The arrangement for automatic injection is connected to further contacts of relay B.

has been suppressed completely, ether administration is discontinued and arousal from anaesthesia is observed. Usually, of course, a certain rudimentary activity is preserved when using high amplification even during the deepest stages of ether anaesthesia.

When using barbiturate anaesthesia the anaesthetic is given i. p. (Nembutal, Pentothal 45 mg/kg with a little heparin) after the control recording has been made. The development of anaesthesia during absorption of the anaesthetic from the peritoneal cavity is then studied. When surgical anaesthesia has been produced a polyethylene cannula is introduced into an ear vein for further infusion of the anaesthetic i. v. The cannula is connected to a syringe, the plunger of which is pushed by a threaded block moved in guide bars by a screw (driven via an adjustable gear by an electric motor or by a stepping relay). One of the channels of the EEG is connected to an integrator —

fundametally a storage condenser charged by the rectified output current. After reaching a certain voltage this condenser is discharged through a thyratron or a sensitive relay. The thyratron discharge either directly activates the stepping relay or connects a circuit that for a certain period of time (1—2 sec) switches on the motor for automatic injection. Fig. 230 shows a diagram of the whole circuit and gives a more detailed description of the apparatus. Other

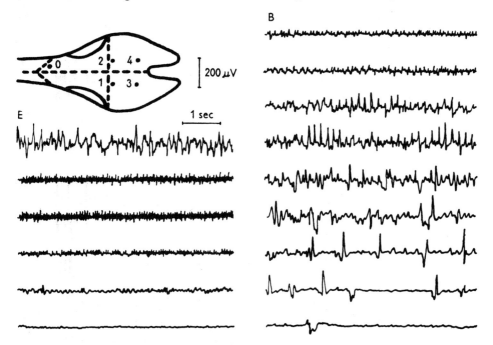

Fig. 231. ECoG activity during gradually increasing ether (E) or barbiturate (B) anaesthesia. Recording from electrodes 1—3. E: first row — control, further rows in 2 min intervals while continuously adding ether. B: first row — control, further rows in 3 min intervals after i. p. injection of 35 mg/kg Pentothal.

solutions of the same problem are given in the work of Bickford (1950, 1951), Verzeano (1951), Forbes et al. (1956), Degelman (1956).

Automatic regulation of the depth of anaesthesia is carried out as follows: when EEG activity, after having reached the maximum depth of anaesthesia, again reaches maximum amplitude, the amplifier connected to the integrator is regulated in such a way that the latter gives approximately 50 discharges per minute. Then the integrator is connected to the apparatus for automatic injection and the gear is so adjusted that the number of discharges (and thus the depth of anaesthesia) remains roughly the same (i. e. it neither increases, which would indicate a decrease in the depth of anaesthesia, nor does it decrease, indicating a deepening of anaesthesia). It is important to note that during light anaesthesia a decrement of ECoG amplitude and a corresponding reduction of integrator discharges may, on the contrary, indicate desynchronisation and arousal.

Large individual differences exist between animals and the control of anaesthesia must be adjusted to them. It is relatively easy to regulate deep anaesthesia. Maintenance of a medium level of anaesthesia, on the other hand, requires exact adjustment of the rate of infusion to a certain type of EEG activity and occasional re-regulation.

In another experiment the rabbit is given 0·5 mg/kg physostigmine. When its effect is fully in evidence (after 10—15 min) we inject 10 mg/kg atropine and observe EEG changes for a further 20—30 min. Since a special sign of the physostigmine effect is regular theta activity in the hippocampus it is useful to work with animals with electrodes implanted in the hippocampal region (see chapter X D).

Results: The development of ether anaesthesia is shown in Fig. 231. Immediately after the first drops of ether are given a rapid, low-voltage activity, typical for the aroused animal, replaces the EEG sleep pattern. This type of activity is preserved up to the stage of deep anaesthesia even though it may be combined with slower, more or less regular waves. Only during deepest anaesthesia, when severe respiratory disturbances appear does EEG activity disappear completely. After application of ether is discontinued all changes recur in the opposite sequence.

Barbiturate anaesthesia develops quite differently. After the initial i. p. dose of Pentothal (45 mg/kg) there is a short phase of waking activity due to the injection. Then sleep activity rapidly increases changing into high voltage discharges, synchronous in the whole cortex — so called spindles. These are separated by intervals of lower asynchronous activity. In the beginning it is still possible to wake the animal using sensory stimuli (cf. arousal reaction p. 516). Later on, however, peripheral stimuli lose their generalised effect and only cause primary responses in the corresponding cortical areas (cf. p. 525).

This phase is also rapidly displaced by a new one: irregular, synchronous, high voltage activity predominates. The frequency of this activity becomes less and less until finally individual high voltage waves or short series of such waves are separated by intervals of electrical silence. During most profound anaesthesia electrical activity disappears completely and only an uninterrupted iso-electric line is recorded. Respiration during this phase of anaesthesia is considerably slowed down but is still regular and deep. In long lasting experiments it is important to prevent a larger decrease in body temperature of the animal, which might also influence the EEG picture.

Thus after an i. p. injection of Pentothal EEG activity reaches a stage characteristic for a certain degree of deep anaesthesia and then, as the concentration of the anaesthetic decreases in the blood it returns to a stage typical for moderate or slight anaesthesia. It is evident from the above that it is not easy do determine the different phases of anaesthesia with the help of EEG. It is even more difficult to find a quantitative expression for the depth of anaesthesia. Forbes et al. (1956) for instance, suggested a simple indicator: in one channel the amplitude of the highest waves seen on the record during

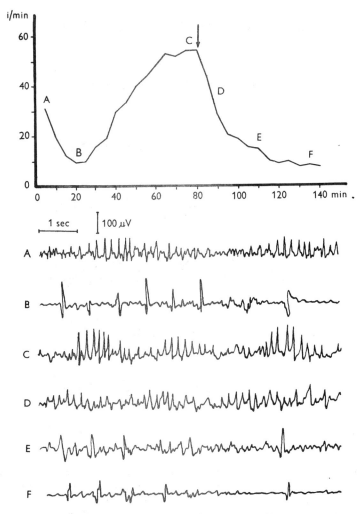

Fig. 232. Control of the depth of barbiturate anaesthesia by feedback between the EEG integrator and the injection system. Top: recording of integrator discharges. Bottom: examples of ECoG curves at corresponding stages of anaesthesia. A-B — deepening of anaesthesia after i. p. injection. B-C — gradual recovery of the animal. Arrow — connection of the automatic injection to the integrator. C-D-E-F — stabilisation of the system at a new anaesthesia level maintained by a strong feedback. Calibration: 1 sec, 100 μV.

increasing or decreasing anaesthesia is determined. At regular intervals all waves exceeding one third of this value are then counted on 10—20 sec strips of the tracing. The average number of such waves/sec. during a given interval then serves as an index characterising the depth of anaesthesia. More exact results are obtained with an integrator. Even in that case, however, the result depends on many circumstances so that integrator data are also best related

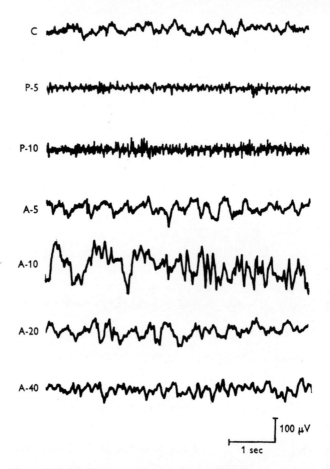

Fig. 233. The effect of physostigmine and atropine on ECoG activity in an unanesthetized rabbit. Bipolar recording from the frontoparietal regions. C — control recording; P-5, P-10 — 5 and 10 min. after 0·5 mg/kg physostigmine salicylate; A-6, A-10, A-20, A-40 — 5, 10, 20 and 40 min. following 10 mg/kg atropine sulphate. Calibrations: 1 sec, 200 μV.

to a certain characteristic level of anaesthesia (e. g. to the maximum rate of the integrator discharges in the same channel). Absolute comparison between individual leads of different animals is possible only approximately.

Fig. 232 shows a characteristic EEG record together with integrator data

524

during anaesthesia produced with Nembutal before and after introduction of the automatic control of drug injection.

Injection of 0·5 mg/kg physostigmine evokes in the rabbit a pronounced cortical desynchronization with a latent period of 3—6 min. This lasts without interruption for about 20 min and then slowly changes into the normal sleep cycle which is fully restaured after 40—60 min. Larger doses of physostigmine induce shivering, salivation and signs of motor unrest. If atropine is applied at the height of the physostigmine effect cortical desynchronization disappears within several minutes and in its stead slow irregular activity of high amplitude very similar to slow sleep is seen (Fig. 233). This activity persists even when the animal reacts to external stimuli and performs rather complex behavioral acts.

Conclusion: During anaesthesia characteristic changes occur in the electrical activity of the cerebral cortex, making it possible to identify the type and depth of anaesthesia and to regulate it. On the other hand the effect of other substances is characterized by dissociation of EEG phenomena and behaviour which shows that no definite conclusions can be drawn from the EEG picture alone.

D. Primary cortical responses

Problem: Register electrical changes in the cerebral cortex due to brief acoustic, optic or somaesthetic stimuli. Determine the extent of the cortical projection of these afferent systems.

Principle: Receptors are usually a source of complex patterns of impulse activity in paths connecting them to higher centres. Electrical signs of a large number of asynchronous action potentials coming continuously, e. g. from the eye that is watching its surroundings, and entering the optic areas of the cortex, cancel each other and disappear during registration from the surface of the brain in the electrical activity of other neurones. A different situation arises if receptors are stimulated with a relatively strong and short lasting stimulus eliciting a synchronous volley of impulses in a large number of fibres of the afferent system. Their nearly simultaneous arrival into different relays of the specific pathway and their consequent entry into a limited area of the cortex results in synchronous changes of membrane polarity in a large group of cortical ganglion cells and dendrites. Characteristic electrical potentials thus generated are very clear cut because of the regular radial structure of cortical elements and can easily be distinguished on the baseline of spontaneous EEG activity. Their localisation makes possible exact electrophysiological limitation of the extent of cortical projection of the analyser studied. In view of the fact that responses are the more apparent the lower the amplitude of sponta-

neous EEG activity, it is of advantage to use relatively deep barbiturate anaesthesia in order to demonstrate them, as during such anaesthesia spontaneous EEG is suppressed much sooner than primary responses.

Object: Cat, 2—3 kg, rat, 200 g.

Apparatus: Ink-writing multichannel EEG apparatus. A cathode-ray
oscilloscope with overall A. C. amplification of at least 50 μV/cm, linear from
1 to 5000 cycles/sec. A two channel apparatus is best. A power amplifier and
a loudspeaker. A stimulator giving single square wave pulses or condenser
discharges, paired stimuli for determining the refractory phase and rhythmic
stimuli from 0·5—200 c/sec. Isolation output unit for the stimulator. A photostimulator (stroboscope) giving light flashes of high intensity singly or rhythmically. A photographic camera for recording from the screen of the cathode-ray
oscilloscope.

Further requirements: Surgical instruments. A set of trephines, a dental
drill. Animal boards for rat and cat. 10% Dial solution. Needle electrodes for
corticography in the rat (cf. p. 508). Electrodes for corticography from the
exposed brain surface (silver wire ball-tipped electrodes mounted on springs
in an electrode carrier fixed to the bone, or wick Ag-AgCl or Hg-Hg$_2$Cl$_2$ electrodes in separate carriers).

Procedure:

1) Localisation of primary responses in the rat.

The rat under dial anaesthesia (40 mg/kg) is operated as in experiment
IX B. In addition the whole lateral surface of the temporal bone is exposed.
The temporal muscle is either cut off or deflected and fixed together with the
skin to the head holder. The sciatic nerve is exposed in both thighs for a length
of about 2 cm. The rat is tied to the animal board and the head is fixed to the
holder as described in experiment IX B. If an auditory stimulus is to be used
the screws in the external meati must have a central opening. Near the animal
an electrically shielded loudspeaker is installed. The lamp of the stroboscope
is aimed at the head of the rat from above and in front. One of the exposed
sciatic nerves is placed on a pair of electrodes connected through an isolation
transformer or a radiofrequency unit to the stimulator. A row of needle
electrodes, at a distance of 2 to 2·5 mm from each other, is placed above the
expected projection area according to the diagram in Fig. 234. A stronger
needle electrode pierces the frontal bones above the anterior edge of the olfactory bulbs to a depth of about 3 mm.

The arrangement shown in Fig. 234 can, for instance, be used to localise
the projection area of the right sciatic nerve. Depending on the number· of
available channels of the EEG apparatus the following combinations of electrodes are connected to the individual amplifiers either simultaneously or con-
secutively:

Monopolar: 0—1, 0—2, 0—3, 0—4, 0—5
0—6, 0—7, 0—3, 0—8, 0—9

Bipolar: 1—2, 2—3, 2—4, 4—5
6—7, 7—3, 3—8, 8—9

The stimulus mark is recorded in another channel.

A control recording lasting for 30—60 sec. is first made with each combination of the electrodes. Then stimulation is applied for 1 min. using

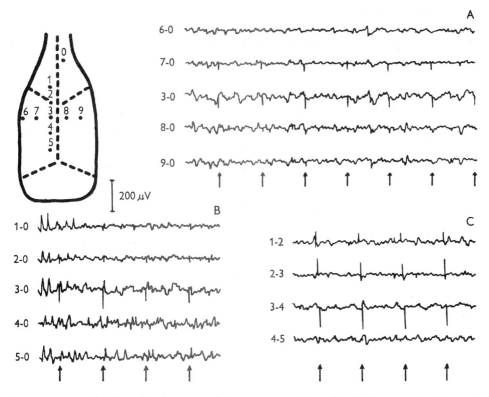

Fig. 234. Localisation of primary projection area of the left sciatic nerve in the rat in monopolar (A, B) and bipolar (C) ECoG recordings. Arrows indicate electrical stimulation of the sciatic nerve with single stimuli repeated at 1 sec intervals. The downward deflection in the monopolar recording corresponds to positivity of the active electrode.

the various stimuli at regular 2—3 sec. intervals. For acoustic stimulation pulses from the stimulator are fed into a power amplifier, the output of which is coupled with the loudspeaker. The intensity of stimulation of the sciatic nerve is just sufficient to produce a slight twitch of the limb, visual and auditory stimuli may be maximal. When all stimuli have been tested with a certain arrangement of the electrodes a new combination is chosen according to the above table and the whole procedure is repeated. Depending on the results obtained in these recordings a further recording is taken if necessary with a different connection of the electrodes or a different combination of simultaneously recorded leads. Thus, for instance, for a better determination of the somaesthetic area so called triangulation is used — connection of 3 electrodes forming the tops of a triangle to three leads corresponding to its sides, e. g.:

2—3, 3—7, 7—2

3—7, 7—4, 4—3, etc.

If large responses are obtained from peripheral electrodes further electrodes are introduced located towards the periphery of the expected projection focus.

In order to decrease the stimulation artefact during stimulation of the sciatic nerve the animal is grounded. A clamp fixed to the cut skin of the thigh above the stimulating electrodes is best. Other methods for decreasing the stimulation artefact (described on p. 544) may also be applied here.

2) Localisation of primary responses in cats using a cathode-ray oscilloscope.

The skin of the head of a cat under deep Dial anaesthesia (40—50 mg/kg i. p.) is cut in the midline for a length of 6—9 cm. The skin is reflected towards the sides as far as possible and fixed with haemostats. The temporal muscle is dissected away from the skull, retracted as far as possible and fixed. The cranial bones are thoroughly cleaned. An 8—10 mm opening is made with a trephine fixed to an electric drill about 1 cm laterally from the midline and 1 cm caudally to the coronal suture. This opening is widened with bone pincers anteriorly, dorsally and laterally. Only 2 mm from the midline a strip of bone is left in situ so as not to injure the sinus and thus cause haemorrhage that is difficult to stop. Haemorrhage from the cranial bones is controlled with bone wax. The dura mater is lifted with the help of a needle having a fine hook at its tip and a Y shaped cut is made with fine scissors. The cut dura is folded back over the borders of the opening. The exposed cortex is covered with cotton wool or gauze soaked in warm Ringer solution until the experiment is commenced.

After the surgery the cat is tied to the animal board and the head is fixed in a stereotaxic head holder by the nasal bones, palate and zygomata. The auditory meati remain open. If no stereotaxic head holder for the cat is available, good fixation of the animal suffices. Movement of the head against the electrodes is prevented by fixing the electrode carrier directly to the skull bones using two short screws. The electrodes must than be ball-tipped spring mounted silver wires. If the electrodes are in special stands independent of the head holder, Ag-AgCl or $Hg-Hg_2Cl_2$ electrodes with a 3—4 cm wick, following the movements of the cortex, are of advantage. Inspite of the great advantages of a multi-channel ink-writing EEG apparatus for routine work, the recording of rapid components of primary responses is distorted mainly because of the considerable inertia of the mechanical recording system. If an exact picture of potential changes in the cerebral cortex is required a cathode-ray oscilloscope must be used for recording. In that case either an independent pre-amplifier together with the amplifier of the cathode-ray oscilloscope is used, or the output of the EEG amplifier is directly connected to the Y plates of the cathode-ray tube. In that case only the time base and cathode-ray tube of the cathode-ray oscilloscope are used. When using a double beam C. R. tube or an electronic switch primary responses from two cortical areas may be recorded simultaneously with the ink-writer recording.

It is important to achieve exact synchronisation of the time base and the stimulus. Any of the methods described on p. 174 is therefore used in such a way as to ensure that the sweep starts 1—2 msec before the stimulus. The stimulation artefact is important for calculating the latent period. The duration of the sweep is chosen from 500 to 50 msec depending on which phase of the primary response is being studied. Pictures are taken with an open shutter during the single passage of the beam using an ordinary (p. 172 or special (p. 178) camera. If synchronisation of the time base and the stimulus is perfect it is possible to photograph several responses one over the other with the shutter permanently open (Dawson 1947, Collins 1964). Thus one picture shows both variations in the individual responses and their most typical shape. When photographing in the above way the brightness of the spot is decreased so that only very weak curves are drawn on the photographic material. At the sites where the curves most frequently superimpose the tracing is then most distinct. It must be stressed, however, that the superimposed composite line corresponds to modal waveform rather than to average waveform. Some information on the waveform variability is also available: bright and narrow portions of the summation line indicate greater reliability of the corresponding components.

A more complex averaging technique is based on the use of brightness modulation of the beam (see also p. 629). The time base of the recording oscilloscope is triggered by the synchronizing pulse of the stimulator but the amplifier is disconnected from the vertical plates. The amplified EEG activity is led to the Z input of the oscilloscope and thus used for modulation of the beam intensity around a certain mean value. A positive input voltage increases the brightness of the trace, a negative one decreases it. Individual recordings then represent narrow rows consisting of irregularly alternating light and dark sectors. By introducing a suitable vertical sweep or with vertical movement of the film it is possible to get the rows closely one below each other. Since they are all synchronized by the stimulus application they give a good notion which components of the response appear in a definite time relationship to the stimulus. A quantitative evaluation can be made by photometering 20—1000 rows of the recording. The slit of the photometer is vertical to the rows, its length corresponds to the number of evaluated sweeps and its width to 1—3 msec of the recording. The galvanometer connected to the photocell records changes in light flow when the film is moved in the direction of the rows. Even though the nonlinear relationship between the amplitude of the modulating signal and the brightness of the trace and also the nonlinearities of the photographic process and the photoelectric conversion do not permit us to consider the result as the statistical mean of individual responses, the resultant curve nevertheless gives us valid information about the polarity, latency and duration of the individual components of the reaction and even about their relative amplitudes. Another advantage is that the recording contains all information about the gradual development of the response and hence makes it possible to analyse the statistical variability of the individual components with time. A detailed description of this method and of its application is given by Kozhevnikov and Meshchersky (1963). Chapter II G surveys methods of averaging evoked potentials using analog and digital computers.

Oscillographic registration is carried out in similar leads as when using ink recording, but monopolar registration is preferable. When determining the extent of the auditory projection area, either the exposed skull bone, the metal electrode holder fixed to the skull, or the clamp on the cut skin, is used as a reference electrode. The active electrode is moved in a certain direction — e.g. in the parasagittal plane — from the border of the expected projection area towards its centre. At 2 mm intervals several responses to an auditory stimulus (short clicks) are recorded. The sites from which recordings were taken are marked with a drop of india ink and drawn into the diagram of the exposed brain surface prepared beforehand. After having thus gone through the whole row a new parallel one is started 2 mm more laterally or medially. The whole lateral surface of the hemisphere is thus mapped out. Poor responses may be increased by local application of 0·1—1% strychnine nitrate (see page 547).

Evaluation of the recording further requires time and amplitude calibration. In a two channel apparatus one beam can be used for recording the sine wave signal of 50, 100 or 1000 cycles/sec or a special time mark (cf. p. 182) together with the observed phenomenon. In a one channel apparatus the time signal is photographed now and again on the assumption that in the interval between these recordings the function of the time base did not change.

Amplitude calibration is performed by introducing a known potential into the amplifier input. A square pulse is best, making it possible to characterise also the time constant of the apparatus.

3) Primary responses to repeated stimuli.

After ascertaining the localisation of the maximum response, e. g. to stimulation of the sciatic nerve in the rat or to sound stimuli in the cat, the ability of the cerebral cortex to react to repeated stimuli of the same intensity is tested in the lead giving the best response. If a stimulator equipped for measuring the refractory period is available, we use two slightly supramaximal stimuli and change the interval between them from 200 to 5 msec. The paired pulses are applied at 2—4 sec intervals. After several responses the interval between both pulses is shortened. The relative refractory period is examined with particular thoroughness. If a stimulator giving repetitive stimuli in the range from 1 to 100 cycles/sec is at hand, we use short-lasting (about 1 sec) trains of stimuli of varying frequency adjusted beforehand. Here, of course, we determine not only the refractory period (the response to the first two stimuli) but also the ability of the cerebral cortex to respond to stimuli rapidly succeeding each other. It is therefore important, especially when using higher frequencies, to use longer intervals between short trains of stimuli (20—60 sec or more) in order to prevent exhaustion of cortical neurones.

A certain variability of primary responses in spite of constant characteristics of the stimulus depends, among other things, on the phase of the spontaneous ECoG at which the afferent signal arrives. As spontaneous activity is controlled by the non-specific afferent system (Morison and Dempsey 1942, Li et al. 1956b) whose fibres terminate in all cortical layers, interaction between specific and non-specific impulses may occur in the cortex. In order to achieve synchronisation of specific impulses with normal or pathological cortical activity an electronic trigger circuit is used (Shipton 1949, Hewlett 1951) with which some modern stimulators and stroboscopes are equipped. The stimulus is released when the output voltage of the EEG channel registering the activity of the corresponding projection area attains a certain value. Either increasing negativity or positivity may be used and also the interval between the releasing phase of EEG activity and the stimulus itself can be controlled.

Results:

1) The monopolar and bipolar recording of a typical experiment in a rat is shown in Fig. 234. It can be seen that responses to stimulation of the sciatic nerve by single electrical discharges have a characteristic distribution. The sites of maximum positive waves correspond to the maximum projection, provided, of course, that the indifferent electrode does not influence the shape of the recording. There is a certain overlap of the optic and somaesthetic projection on the periphery even though the areas of their maximum response

are well defined and fundamentally correspond to the projection areas as determined in the degeneration experiments of Krieg (1946a).

It is somewhat more difficult to determine the limits of a focus in bipolar recording. The latter's advantage, however, is a more acurate localisation of activity in smaller cortical areas. The principle of localisation is based mainly on the so called phase reversal. The characteristic focal activity will appear in tracing from a row of electrodes passing through the focus (Fig. 235A)

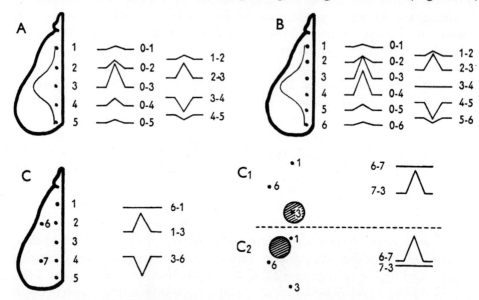

Fig. 235. Principle of localising focal activity. Above: localisation in monopolar (A) and bipolar (B) recording. The curve in the diagram of the brain shows the spatial distribution of focal activity along the line joining the electrodes. Below: Triangulation. C — basic triangle, C_1 C_2 — two possibilities that correspond to the results obtained with the basic triangle leads. Shaded area indicates the focus.

as waves whose phase has been shifted by 180°. Let it be assumed that the source of these characteristic waves — e. g. of primary response in the somaesthetic area — is situated directly below electrode 3, while electrodes 2 and 4 are almost outside the focus. Then, obviously, recorded activity will be highest in leads 2—3 and 3—4 while in leads 1—2 and 3—5 it will be either absent altogether or very low. As electrode 3 is connected once to the right and once to the left half of the EEG amplifier input, the same potential change at this electrode will cause the opposite movement of the pen in lead 2—3 and in lead 3—4. Synchronous potentials of the same shape but of opposite phase (mirror images) occurring in two tracings may thus be regarded under such conditions of recording as an expression of electrical activity below the electrode common to both leads.

Further diagrams illustrate other possibilities of localising foci with regard to a row of electrodes. In Fig. 235B the focus is between electrodes 3—4, so that focal potentials influence both equally. For that reason signs of synchronous activity will be smaller in lead 3—4 than in leads 2—3 and 4—5, which record po.ential differences between inactive areas 2,5 and electrodes 3, 4 situated in the area of the focus. The phase reversal then appears between leads 2—3 and 4—5. Practically, however, this theoretical possibility occurs only rarely. Usually the focus is nearer to one of the electrodes, and the recording shows a typical phase reversal. The amplitude of both waves shifted by 180° is not quite the same, however (see Fig. 234C, leads 2—3, 3—4).

Another frequently used method of bipolar localisation is the so called triangulation. In this method we connect any three electrodes to form a triangle in such a way that its sides correspond to the individual leads (Fig. 235C) When connecting we usually go in one direction (6—1, 1—3, 3—6). The leads in which the typical phase reversal appears contain the electrode nearest to the focus (Fig. 235 C_1). Theoretically, of course, there is also the possibility (Fig. 235 C_2), that a focus of opposite polarity is situated between the other two electrodes in such a way that its activity affects both symmetrically and thus is not apparent when recording from them. The use of a further electrode or electrodes (Fig. 235C) helps to distinguish between these two possibilities. A new triangle will then decide.

A bipolar recording from the same electrodes as in Fig. 234B is shown in Fig. 234C. If we apply the principles of localisation explained above we can easily find the focus of maximum responses to stimulation of the sciatic nerve.

All the above considerations are, of course, only valid on the assumption that the electrical phenomenon studied is limited to a certain exactly defined area of the hemisphere and that its time course is the same in all parts of that area. This theoretical assumption, of course, holds good only exceptionally as primary responses and other characteristic potentials spread immediately from the projection area into the surrounding regions (Adey et al. 1954, Lilly and Cherry 1954). It is therefore necessary to define not a stationary but a moving electric field. This task can be fulfilled only with help of special techniques (toposcopy — Goldman et al. 1948, Walter and Shipton 1951, Lilly 1954, Petsche and Marko 1954, Livanov and Ananyev 1955, Ananyev 1956 for potentials moving along the cerebral surface) or by reconstructing the three dimensional movement of an electric field from individual tracings recorded from various depths (Howland et al. 1955). It must therefore be kept in mind that use of phase reversal for bipolar localisation of a focus of electrical activity is possible only if the following conditions are satisfied:

a) a characteristic, constant and well reproducible activity,

b) perfectly simultaneous activity in the compared leads,

c) the assumption of a stationary electric field.

Reproducibility is especially important as this makes it possible to localise phenomena whose time course is exactly determined even when using a single-channel apparatus. Thus e. g. the primary cortical response to stimulation of the sciatic nerve has an exactly defined latency that practically does not change during the experiment, provided depth of anaesthesia, blood flow and other physiological parameters remain unchanged. We can therefore compare electric responses in the same intervals from the stimulus in different regions and thus evaluate the phase reversal in recordings taken at different periods. This method is especially suitable when using a cathode-ray oscilloscope. If there is a two channel apparatus available it is even better to connect the lead giving the most characteristic response permanently to one

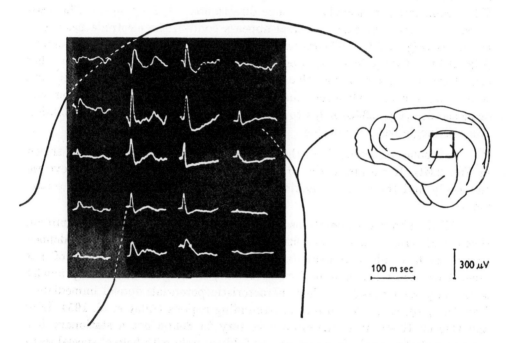

100 m sec   300 μV

Fig. 236. Topography of primary responses to auditory stimuli in the cerebral cortex of the cat. Upward deflection indicates positivity of the active electrode.

channel and the other leads successively to the other channel, so that we can compare them simultaneously with the first one.

2) Fig. 236 shows characteristic oscillographic recordings of primary responses from the auditory area of the cat's cerebral cortex to short acoustic stimuli. The following are easily distinguishable: the latent period, the positive phase of the primary response, the negative phase of the primary response.

Since the sweep starts at the same moment when the stimulus is applied, no stimulus artefact is visible. The positive phase is constant and changes little. The negative phase, on the contrary, is variable. The latent period corresponds to the physical transmission of sound, stimulation of the organ of Corti and conduction of impulses along nerve fibres and nervous pathways across a number of synapses into the cerebral cortex. The positivity of the cortical surface is an expression of the negativity arising at synapses of deeper cortical layers (4—5) where fibres of the specific afferent pathway terminate. The change of positivity to negativity is an expression of the fact that the area of negativity has spread to the cortical surface due to ortho- or antidromic activation of the superficially situated parts of the dendrites. Responses from all areas have a similar shape, as might be expected from the uniform structure of the cerebral cortex.

Oscillographic recording can be used for monopolar mapping of projection areas. Fig. 236 shows the result of such mapping of the auditory area in the cat. Fig. 237 shows the extent of the main analysers in the most important laboratory animals. Data from electrophysiological work from recent years and our own experimental material was used for its construction (Bremer and Dow 1939, Adrian 1940, Woolsey and Walzl 1942, Marshall et al. 1943, Walzl and Mountcastle 1949, Kempinsky 1951, Artemev 1951, Mickle and Ades 1952, Patton and Amassian 1954, Chernigovsky 1956, Woolsey 1960, Berman 1961a, b, Thompson et al. 1963a, b, Celesia 1963 — cat; O'Leary and Bishop 1938, Woolsey and Wang 1945, Talbot et al. 1946, Woolsey 1947, Thompson et al. 1950, Adey et al. 1954, Cazard and Buser 1963a, b — rabbit; Woolsey and Le Messurier 1948, Le Messurier 1948, Benjamin and Pfaffmann 1955, Libouban and Oswaldo-Cruz 1958, Kimura 1962, Libouban-Letouze 1964 — rat).

Detailed mapping of responses to certain stimuli in extensive areas of the cerebral cortex shows characteristic differences not only in amplitudes but also in latent periods of primary responses in different parts of a cortical projection area.

It is possible to construct isolatent lines connecting sites at which the onset or peak of positive waves (or other characteristic parts of the response) appears at the same time after stimulation. The isolatent lines concentrate around one or more foci in which the positive wave appears first and then continues to spread irregularly into more distant areas (Mickle and Ades 1953, Lilly and Cherry 1954). According to the type of response we distinguish primary cortical projection areas (I) with the shortest latent period and maximum positivity and secondary projection areas (II or III). These are parts of the cortex immediately adjacent to area I through which they receive corresponding afferent impulses (e. g. auditory area II in the cat — Ades 1943, Bremer et al. 1954). The latent period is much longer than in projection

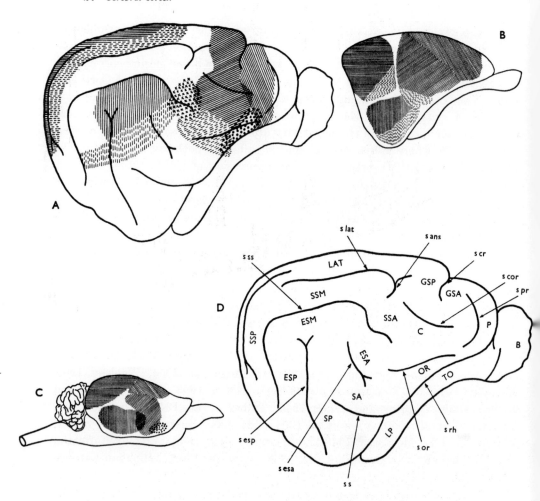

Fig. 237. Map of the projection areas of the cerebral cortex of the cat (A), rabbit (B) and rat (C). Horizontal shading — optic area, vertical shading — acoustic area, oblique shading (from top right to left bottom) — somatic area, oblique shading (from top left to bottom right) — motor area, crosses — vestibular area, dotted — gustatory area. Interrupted shading indicates corresponding secondary and tertiary projection areas.

(D) The main anatomical structures in the cat's hemisphere. B — bulbus olfactorius, C — g. coronarius ESA — g. ectosylvius anterior, ESM — g. ectosylvius medius, ESP — g. ectosylvius posterior, GSA — g. sigmoideus anterior, GSP — g. sigmoideus posterior, LAT — g. lateralis, LP — lobus pyriformis, OR — g. orbitalis, P — g. proreus, SA — g. sylvianus anterior, SP — g. sylvianus posterior, SSA — g. suprasylvius anterior, SSM — g. suprasylvius medius, SSP — g. suprasylvius posterior, TO — tractus olfactorius; s ans — s. ansatus, s cor— s. coronarius, s cr — s. cruciatus, s esa — s. ectosylvius anterior, s esp — s. ectosylvius posterior, s lat — s. lateralis, s or — s. orbitalis, s pr — s. praesylvius, s rh — s. rhinalis, s s — s. sylvius, s ss — s. suprasylvius.

536

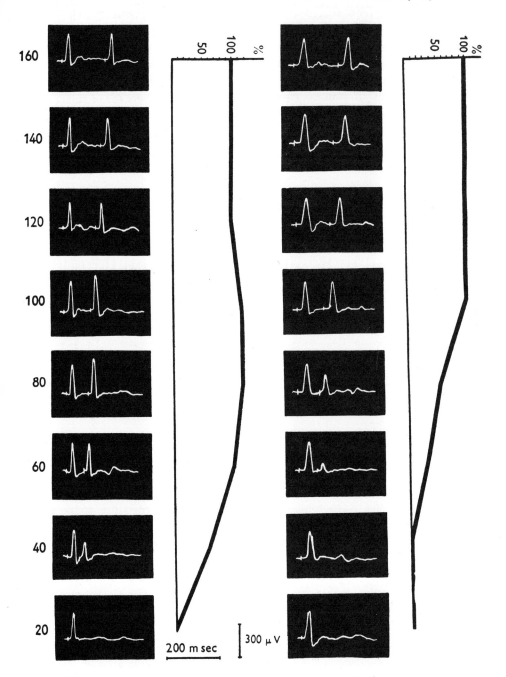

Fig. 238.The refractory period of the primary cortical response to acoustic clicks in the cat. Left: light anaesthesia. Right: deep barbiturate anaesthesia. Figures denote the delay of the second stimulus in msec. Upward deflection denotes positivity of the active electrode.

area I (20 msec instead of 6 msec in the cat) and their activity follows all changes occurring in area I. At other times we find responses with a short latent period even in those areas (e. g. auditory area III, the activity of which does not depend on the functional state of area I — Tunturi 1945, Mickle and Ades 1952). In such cases direct afferent connections with subcortical centres of the specific pathway evidently exist.

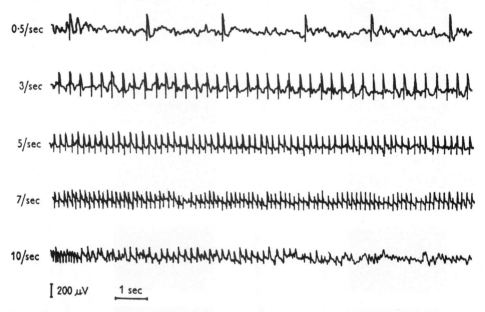

Fig. 239. Cortical responses to repetitive stimulation of the sciatic nerve in rat (Dial anaesthesia). Monopolar recording. Downward deflections indicate positivity of the active electrode.

3) The absolute refractory period of the primary response to acoustic stimuli in the cat is 30—60 msec, the relative refractory period lasts 60—100 msec according to the depth of anaesthesia. With longer intervals between stimuli it is possible in some experiments to register a transitory increase in the second response above the control level. Such excitability curves were described in the optic cortex by Marshall (1949), Gastaut et al. (1951), Chang (1952), in the somatosensory cortex by Heinbecker and Bartley (1940), Marshall (1941), Jarcho (1949), in the auditory cortex by Chang (1950), Rosenzweig and Rosenblith (1953). Fig. 238 shows a typical recording of the refractory period to paired acoustic stimuli in the cat. The amplitude of the response to the second stimulus as a percentage of that of the response to the first stimulus for different intervals between both stimuli, is illustrated by the curves in Fig. 238. They differ characteristically according to the depth of anaesthesia and the overall state of the animal.

538

A train of stimuli of a given frequency produces a series of cortical responses (Roytbak 1956, Smirnov 1953). If the frequency of stimuli is low, the cortex responds typically to every stimulus. With increasing frequency of stimulation the positive phase begins to predominate; the negative phase disappears occasionally at first and permanently later on. On further increasing the frequency of stimulation the response assumes a rudimentary character, the amplitude of the positive phase decreases and finally the projection area begins to respond to every second or third stimulus only. In some cases it responds irregularly or ceases to respond altogether. The ability to follow the rhythm of stimulation (so called lability — Wedensky 1903) changes during stimulation. It can often be observed that the projection area responds to every one of the first few impulses, then only to alternate ones and finally without any regularity. The complete picture may develop during a train of stimuli with a frequency of 10/sec lasting for several sec. (Fig. 239).

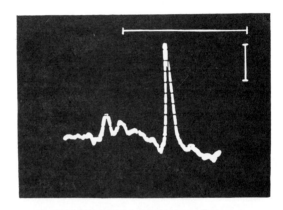

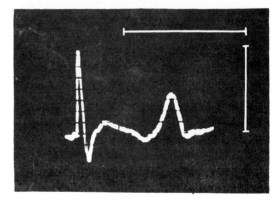

Fig. 240. Secondary response to acoustic stimulus in cat (Dial anaesthesia). Top: recording from g. sigmoideus posterior. Bottom: recording from g. ectosylvius anterior. Monopolar recording. Upward deflection indicates positivity of the active electrode. Calibration: 100 msec, 200 μV.

Secondary response

In deep barbiturate anaesthesia (Dial 50 mg/kg), when nearly complete suppression of spontaneous EEG activity has been achieved, sensory stimuli, particularly stimulation of the sciatic nerve, produce in addition to the primary response a further electrical response with an amplitude several times higher and with a latency of 50—100 msec (Derbyshire et al. 1936, Forbes et al. 1949, Forbes and Morison 1939, Purpura 1955). In shape the secondary response is similar to the primary one but both its positive and negative components are prolonged to about 50 msec (Fig. 240). Stronger stimuli are usually needed to produce the secondary response.

In contrast to the primary response the secondary one is not limited to the specific projection area. In the cat it can be recorded from the whole surface of the exposed hemisphere even though its shape is not the same in the different areas. The secondary response is thus diffusely distributed regardless of the stimulus producing it — be it an acoustic or optic stimulus or stimulation of the sciatic nerve. Simultaneous stimulation of 2 nerves (e. g. both sciatic nerves) results in considerable facilitation of the secondary response which is then often several times larger than the response produced by stimulation of each nerve separately.

The long latent period indicates the complex polysynaptic nature of the secondary response. This is also confirmed by experiments in which the refractory period of the secondary response is determined. The absolute refractory period lasts 0·2—0·5 sec, the relative 0·7—1·5 sec. The absolute refractory period of the primary response is usually less than 30—60 msec. With repeated stimuli the refractory period of the secondary response rapidly increases, individual responses fall out or disappear altogether by the time when a stimulus frequency of about 1 cycle/sec is reached.

The mechanism of the secondary response has not, as yet, been elucidated completely. It would appear that this phenomenon is closely related to the synchronous barbiturate spindles, to the recruiting response and to the activity of the reticular system of the brain stem (cf. p. 609).

The term "secondary responses" is often also used for other extraprimary potentials appearing after stimulation of afferent systems outside the corresponding cortical projection areas in unanaesthetised animals or under chloralose anaesthesia — in the cat especially in associative cortical areas (g. suprasylvius, g. lat. ant.). Again these are mostly bilateral surface positive waves which do not change after extirpation of the corresponding primary projection area. Their latency is always longer than that of primary responses. It is, however, shorter than that of secondary responses in deep barbiturate anaesthesia (Buser and Borenstein 1956 ,1959, Albe-Fessard and Rougeul 1955, 1956).

Conclusion: Monopolar or bipolar recordings of primary cortical responses make it possible to map the projection areas of individual analysers or of their components and to determine their excitability.

E. Responses of the cerebral cortex to direct electrical stimulation

Problem: Record the direct cortical response to a short electrical stimulus and the spreading of this response over the cortex.

Principle: Stimulation of the exposed cortical surface with a short electric stimulus (1 msec or less) produces a characteristic biphasic response spreading a certain distance from the stimulated point (Adrian 1936, Burns 1950, 1951, Chang 1951a, Clare and Bishop 1954, Roytbak 1955, Ochs 1956, 1962, Purpura and Grundfest 1956, Goldring et al. 1961, Li and Chou 1962, Holubář 1964b, Suzuki and Ochs 1964). In unanesthetised animals (encéphale isolé preparation) a weak stimulus produces a brief surface negative wave (Adrian's "surface response"), spreading at a rate of 2 m/sec to a distance of 4—8 mm. A stronger stimulus evokes a surface positive wave of longer duration which spreads in all directions at a rate of 20 cm/sec without any evident decrement (Adrian's "deep response"). In animals under barbiturate anaesthesia a surface negative wave predominates. The detailed nature of this response has not yet been elucidated. Participation of apical dendrites of pyramidal cells, Cajal cells of the molecular layer and short associative cortico-cortical connections have been postulated.

From the technical point of view the main problem is the reduction of the stimulus artefact to such an extent as to make possible recording of the response from close to the stimulating electrodes.

The following conditions must be achieved:

1) perfect isolation of the stimulating circuit from earth (isolating transformer, radiofrequency output or a transistorized stimulator with floating battery — cf. p. 128) in order to minimise the capacitance of the stimulating electrodes to ground;
2) the use of differential amplifiers with a high discrimination of in-phase signals (cf. p. 160);
3) bipolar stimulation ensuring minimum current spread;
4) the use of non-polarisable stimulating and recording electrodes in order to limit poststimulation changes in the circuit;
5) the use of very short pulses (about 0·1 msec) to prevent interference with the reaction;
6) geometry of the stimulating and recording electrodes. The monopolar recording electrode is placed roughly in the line midperpendicular to the stimulating electrodes. For the stimulus used this line (Fig. 241) corresponds to the zero potential level. The reference electrode is placed approximately on the same line but, of course, at a considerable distance from the site of stimulation.

In addition, various methods of electrical compensation of the stimulus artefact may be used. These will be mentioned later on.

Apparatus: D. C. or A. C. (time constant at least 0·5 sec) preamplifier with cathode-ray oscilloscope, total amplification at least 50 μV/cm. A two channel apparatus is preferable. A stimulator giving single square wave pulses or condenser discharges synchronised with the time base of the oscilloscope. An isolation unit (transformer, radiofrequency output). A time marker or a tone generator.

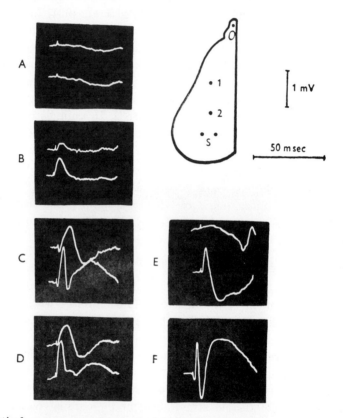

Fig. 241. Cortical response to direct electrical stimulation in rats (Dial anaesthesia). Upper tracing 0-1, lower tracing 0-2. ABCD — responses at the more proximal and more distal electrodes, with gradually increasing stimulus strength. E — 5 min following application of a filter paper with 5% novocaine between electrodes 1 and 2. F — 5 min after applying filter paper with 1% strychnine between 1 and 2. Recording from the more distal electrode. Calibration: 50 msec for A-E, 100 msec for F. Negative deflection upward.

Other requirements: Surgical instruments, a 5 mm trephine, ball-tipped or wick electrodes for corticography with electrode carriers. An animal board for the rat with a stereotaxic head holder.

542

Procedure: The rat is deeply anaesthetized with Dial (40 mg/kg). Curarisation is unsuitable since, according to Purpura and Grundfest (1956), d-tubocurarine depresses direct cortical responses. If anaesthesia is to be avoided succinylcholine is used for immobilisation.

A trephine opening 5 mm in diameter is made over the parietooccipital region of one hemisphere. The cerebral cortex is widely exposed by rongeuring away the parietal bone. The dura is cautiously cut and the dural flaps are reflected. In young animals both stimulation and recording can be performed transdurally, however.

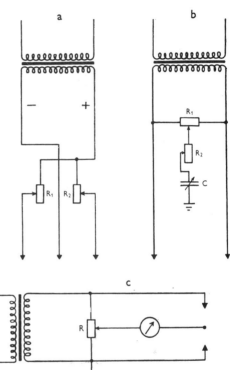

Fig. 242. Methods for electrical compensation of the stimulus artifact. For details see text.

The electrodes consist of plexiglass bars with thin steel springs (3 × 30 mm) on which flexible, ball-tipped (0·5 mm) silver wires (0·2 mm in diameter, 2·5 cm in length) are mounted. The spring ensures a constant pressure on the cortical surface, the thin flexible wire permits limited movement of the electrodes. Wick Ag-AgCl electrodes serve equally well for recording. Silver wire low-impedance electrodes are preferable for stimulation. As reference electrode a needle in the frontal bones is used. The electrodes are situated as shown in Fig. 241. The stimulating electrodes are 0·5—1 mm apart; the recording electrodes are placed 3—5 and 6—10 mm from the former.

After the electrodes are connected, the sweep is synchronized with the stimulus onset and regulation of the stimulus artefact is begun. With stimuli of low intensity and with a maximum brightness of the spot the amplitude and direction of the stimulus artefact are observed while moving the recording electrode along a line parallel to the stimulating electrodes. Usually a point is found, in the vicinity of which the stimulus artefact changes its polarity and thus reaches its minimum amplitude. Finer regulation of the stimulus artefact may be obtained by the use of one of the following methods.

a) The use of tripolar electrodes (Bishop and Clare 1953). The stimulating cathode is situated in the middle (cf. Fig. 242a). The positivity of both anodes is controlled by two variable resistors 1 MΩ permitting gradual changes in the shape of the electric field around the electrodes. This system may, of course, only be used if all three electrodes are firmly fixed in position. The most suitable position for the recording electrode is approximately found and the stimulus artefact is then adjusted with the help of the variable resistance in the anodes.

b) The use of balancing reactance. Part of the stimulating voltage from the potentiometer R_1 (10 kΩ), connected in parallel to the stimulator output, is earthed through a variable R_2C network (0·0001—0·1 μF, 1—2 MΩ potentiometer — Fig. 242b). By suitable adjustment of the output potentiometer and the earthing reactance a signal is directed to earth, the polarity, amplitude and duration of which compensate for the stimulus artefact. A simpler application of the same method of compensation is the use of so called Wagner's earth. The middle point of the output potentiometer is simply earthed. With this method of compensation the advantages of the radiofrequency output are lost.

c) Bridge arrangement of the stimulating and recording electrodes (Phillips 1956a, b). One of the stimulating electrodes may be earthed (the indifferent one in monopolar stimulation). The middle point of the potentiometer connecting both stimulating electrodes is used as reference (cf. Fig. 242c). A potential is found by trial and error corresponding to that of the recording electrode. The bridge is thus equilibrated. Between stimuli the stimulating electrodes are short-circuited by the low impedance output of the stimulator. The potential difference between these 2 electrodes and the active electrode is then registered. Another bridge arrangement of the stimulating and recording electrode which is especially suitable for working with microelectrodes is described on page 619.

After the stimulus artefact has been satisfactorily minimised, the intensity of the stimulus is gradually increased and the response is registered simultaneously from the proximal and distal electrode. Thus the shape of responses to threshold and to maximum stimuli is determined together with the extent of their spread in the cortex. The rate of spread of the response may be calculated from the distance of the recording electrodes and the relative delay of the different components of the response in the distal lead.

The physiological nature of the response may be tested by its sensitivity to various procedures. Asphyxia is most suitable. This is obtained easily by clamping the trachea or holding the thorax in the maximal expiratory position for 1—3 minutes. The response can also be affected by topical application of drugs on strips of filter paper 1—2 mm by 4—8 mm, applied between the stimulating and recording electrodes or between the proximal and distal recording electrodes. A 5% solution of Novocaine is used to suppress spreading of the response. A 1% strychnine nitrate solution increases the response.

For elucidation of the spreading mechanism subpial incisions are made, cutting the cortex between the stimulating and recording electrodes to a depth of 1—3 mm.

It is further possible to determine the refractory period of the response, its changes during frequent stimuli, interaction of two pairs of stimulating electrodes etc.

At the end of the experiment the animal is sacrificed and the response recorded from the dead brain with stimuli of different intensities. Such an experiment gives especially valuable information concerning the character of the electric field arising in the volume conductor of the brain around the stimulating electrodes.

Results: The direct cortical response is a biphasic negative-positive wave. Its positive component decreases with increasing anaesthesia. A weak stimulus produces a response only at the proximal electrode, a strong one also at the distal electrode. The spreading rate of the negative maximum is about 1—2 m/sec. (Fig. 241). Anoxia suppresses first the positive and later also the negative component of the direct cortical response, so that after 1—1·5 min only the stimulation artefact can be obtained, as in the dead animal. After recovery of normal ventilation these parts of the response reappear in the opposite order.

A strip of filter paper saturated with 5% novocaine suppresses spreading of the response from the stimulating to the recording electrodes within a few minutes (Fig. 241E). Care must, however, be taken to prevent this change occurring as a result of shortcircuiting of the stimulating electrodes by excess fluid. Records obtained after removal of the filter paper and drying of the cortical surface are most reliable. The effect of novocaine persists for a short time following application and is rapidly reversible. A similar but irreversible effect is obtained with transcortical incisions. Depending on their depth they result either in a decrease or complete suppression of the direct response at the more distant electrode.

A strip of filter paper saturated with 1% strychine nitrate produces a considerable increase of the direct response at both the proximal and distal electrodes. If the intensity of the stimulus is gradually decreased a value is found at which the negative potential is evidently divided into two components (Chang 1951b). The shape and latency of the first corresponds to the original negative wave of the direct response. The second wave, however, is higher, arises suddenly at a certain stimulus intensity and evidently corresponds to synaptic activation of deeper cortical layers (Fig. 241F).

Conclusion: The cortical response to direct electric stimulation represents a reaction in which probably various cortical elements participate. It can be used, therefore, to study different aspects of cerebral cortical activity.

F. Mapping of nervous pathways in the central nervous system using neuronography

Problem: Determine the projection of a cortical area of one hemisphere via neurones of the corpus callosum onto the contralateral hemisphere.

Principle: Electric stimuli of sufficient intensity cause a synchronous discharge of neurones situated in the area in which the electric field of the stimulus attains suprathreshold intensity. If these neurones give rise to a group of long fibres, it is possible to register along this path an electric wave corresponding to the synchronous volley. In particular, activation of the terminal synaptic area of the pathway manifests itself by very conspicuous electric signs (e. g. cortical primary responses). Areas showing an electric response synchronous with the stimulus must therefore be connected via nervous pathways with the stimulated point.

Local application of strychnine can be used instead of an electric stimulus. This is applied either using filter paper on the surface of nervous tissue or in the form of a paste applied through a needle into deeper regions. Strychnine applied topically to the grey matter of the brain produces spontaneous electric discharges, so called "strychnine spikes" (Gozzano 1936, Dusser de Barenne and McCulloch 1936a, b, 1939a, b, 1939, Dusser de Barenne et al. 1941, Frankenhaeuser 1951), predominantly negative waves of high amplitude (1—3 mV) generated by the synchronous activity of the poisoned neurones. A corresponding volley of impulses along axons of discharging neurones produces synchronous electric potentials even in remote areas. Strychnine spikes are limited to the distribution of neurones whose synaptic area is affected by strychnine. Only low voltage variations are registered from localities to which tracts from the strychninised area have to pass through one or more synapses. In distinction to electric stimulation

a) strychnine does not act on nerve fibres but exclusively on synapses (for that reason application to white matter remains without effect),
b) strychnine spikes spread only orthodromically. They can never be recorded from nerve fibres sending afferent impulses to the synapses affected by strychnine.

These characteristics of strychnine spikes make it possible to use topical strychnine application for tracing centrifugal projection of a group of neurones. The disadvantage of this procedure lies on the fact that not all neuronal systems produce convulsive waves following local application of strychnine (Frankenhaeuser 1951), so that negative results obtained with strychnine neuronography do not exclude the possibility of nervous connection existing.

Local application of strychnine can also be used in another way. Low concentrations of strychnine, inadequate to cause spontaneous discharges, dinstinctly increase evoked potentials. If these are small, so that they are difficult to identify in the basic activity, local strychninisation of the point of recording results in their accentuation. At a site not connected with the area of stimulation, application of strychnine will, however, be ineffective.

Object: Cat, 2—3 kg weight.

Apparatus: A dual beam cathode-ray oscilloscope with D. C. or A. C. amplifiers (total time constant of at least 0·5 sec. and a maximum amplification of at least 50 µV/cm); or an ink-writing multichannel electroencephalograph. A stimulator giving single square wave pulses or condenser discharges synchronised with the sweep of the oscilloscope. An isolation unit to the stimulator output. A time marker.

Other requirements: Surgical instruments, a set of trephines, a dental electric drill. Wick Ag-AgCl recording electrodes and bipolar ball-tipped silver wire electrodes for stimulation, with electrode-carriers. An animal board with head holder for fixing the animal. 10% Dial, strychnine nitrate (crystalline and 1% solution).

Procedure: A cat in Dial anaesthesia is operated as in experiment IX D2. Wide craniotomy, however, is carried out over both hemispheres. Only a narrow strip of bone (3—5 mm) above the sagittal sinus is left in place to prevent accidental injury. The dura is cut and the dural flaps are reflected.

1) The stimulating electrodes are placed on the gyrus suprasylvius of one hemisphere and spaced 1 mm apart. Stability of contact is ensured with a fine spring pressing the electrodes to the cerebral surface. The recording wick Ag-AgCl electrode is placed on the corresponding area of the contralateral hemisphere, the reference electrode on the cut skin. The stimulus artefact is decreased by means described in experiments with direct cortical responses (cf. p. 541), its regulation, however, is not too difficult. Stimuli are applied at 1—2 sec intervals, responses are observed or photographed. If only rough evaluation of polarity and amplitude is required, recording responses with a conventional ink-writing electroencephalograph suffices. First the site of the maximum response is approximately located. One recording electrode is permanently left at this point and the latter's response is recorded in one channel. The other recording electrode is moved systematically in the parasagittal and frontal planes in 2 mm steps and responses from various points of the grid are compared with the simultaneously recorded maximum response. Each position of the electrode is marked with a drop of india ink. After mapping

one projection area the stimulating electrodes are moved 1·5—2 cm and a further projection area is mapped.

2) The extent of the projection of the stimulated site onto the contralateral hemisphere can also be tested in the following way. The stimulating and recording electrodes are placed on symmetrical sites in the anterior part of the gyrus suprasylvius. The maximal response to slightly supramaximal stimuli is found and then the stimulating electrodes are moved in 2 mm steps in the parasagittal plane occipitally or frontally until the response on the fixed recording electrode disappears. Then the stimulating electrodes are again placed on the original site and the area of the recording electrode is treated with a 1% solution of strychnine using a filter paper 2 × 2 mm.

The area surrounding the electrode is first carefully dried and excess fluid is removed from the paper in order to decrease strychnine diffusion as far as possible. Then the change in the callosal response in the strychninized area is studied. As soon as there is a clear-cut increase in the negative phase of the callosal potential the paper is removed. The potentials continue to increase under the influence of strychnine for several more minutes and then reach a constant level which is maintained for 20—40 minutes. When stabilisation of the increased response is attained the stimulating electrodes are again moved along the suprasylvian gyrus in the above manner and the responses are recorded. In a similar way the extent of various projections onto a certain point of the cerebral cortex can be determined after its treatment with strychnine by stimulating other cortical or subcortical areas. It must be emphasised that the success of this method depends largely on the conditions for stimulation in different stimulated sites being completely identical. The distance between the stimulating electrodes must remain constant.

3) The experiment is continued by again applying strychnine onto the area treated. A filter paper soaked in saturated strychnine solution or a small crystal of strychnine may be used. Great care is taken to prevent diffusion of strychnine into the surrounding areas. The paper (or crystal) is removed as soon as spontaneous strychnine spikes appear, usually within 2—5 min. Then one electrode is placed directly on the strychnine treated cortical point. The second electrode (or with multichannel EEG a row of electrodes) is placed symmetrically onto the opposite hemisphere (which had been stimulated earlier) and is gradually moved as in the previous experiments. Tracings are either photographed onto a continuously moving film with a stationary spot or on single frame pictures with sweep triggered by discharges in the strychnine treated area. Recording of the high and relatively slow strychnine spikes with an ink-writing multichannel EEG is especially suitable. After mapping the projection of strychnine spikes onto the opposite hemisphere the area treated with strychnine is rinsed with warm saline and may be covered with a bit of cotton wool soaked in Ringer solution. The

slow disappearance of the strychnine spikes (30—60 min) is observed in the EEG recording. The wick of the electrode in contact with the strychnine-treated area is exchanged since it might be soaked with concentrated strychnine solution and thus cause strychnine potentials in normal areas to which it is applied later. If strychnine is used repeatedly it is applied to a region

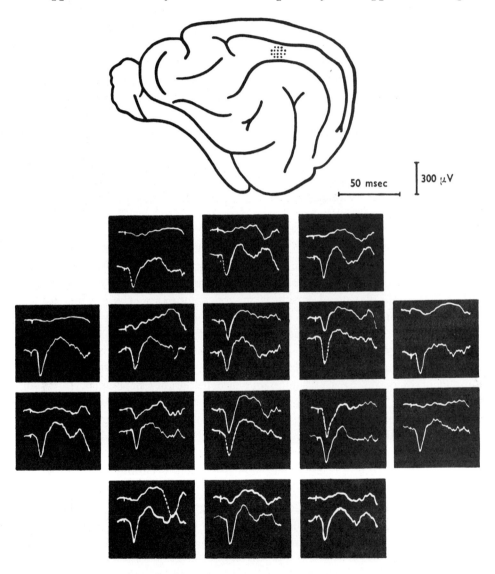

50 msec | 300 μV

Fig. 243. Distribution of callosal responses to stimulation of a symmetrical point in the g. suprasylvius in the cat. Lower tracing — recording from the point of maximal response. Upper tracing — response from points indicated in the upper diagram. Positive deflection downward.

of the hemisphere remote as far as possible from the previously treated area.

Results: 1) The electric response to stimulation of a symmetrical area of the contralateral hemisphere is similar to the primary response to sensory stimuli (Curtis 1940, Chang 1953, Bremer 1958, Grafstein 1959, 1963). First a positive wave appears lasting for 10—15 msec and having a latency of 3—4 msec. This is replaced by a longer-lasting negative wave. The ratio of the amplitudes of the two phases changes according to the area stimulated and also depends on the depth of anaesthesia. The strongest response (up to several times higher than elsewhere) in the cat is obtained in the gyrus suprasylvius and the weakest in the sensorimotor cortex. These differences are probably due to the relative density of callosal neurones in different areas. The area from which maximum responses to stimulation of a certain fixed contralateral point may be obtained is very small — hardly more than 4 to 6 mm². For a further 2—3 mm it is surrounded by a zone in which the amplitude of the response rapidly decreases (Fig. 243). Potentials in this zone may, of course, be explained in part by physical spread of the current from the area of maximum projection.

2) Results with moving the stimulating electrodes are similar. The fixed recording electrode gives a maximum response on stimulation of a symmetrical site. If now the stimulating electrode is moved in the parasagittal or other direction the response rapidly diminishes until it disappears altogether if the stimulating electrode is moved 4 mm from the site giving the maximum response.

The effect of local application of 1% strychnine is seen as a nearly instantaneous increase in electrical response in the intoxicated area (Fig. 244A). The positive component is hardly changed while the amplitude of the negative component increases up to 10 fold and is considerably prolonged. If during this initial stage of the strychnine effect the filter paper is removed, strychnine potentials will be produced only by a volley of impulses coming into the area treated with strychnine. Prolonged application of strychnine produces spontaneous strychnine spikes occurring without evident relationship to external stimuli. Electric stimuli, of course, then also trigger strychnine spikes, provided they do not coincide with the refractory period of a spontaneous discharge.

If now the stimulating electrodes are again moved the extent of the cortex activating the strychnine treated area on stimulation is large (Fig. 244B). This is due to the fact that even a weak afferent influx, hardly discernible in recording from the normal cortex, produces a maximum all-or-none response in the intoxicated area. For this reason the transfer of the stimulating electrodes into an area not connected by nervous fibres with the strychnine treated region results in a nearly instantaneous disappearance of the response. If previously the spike occurred after every stimulus, now it appears only occasionally and on further moving the electrodes it disappears altogether. The same technique may be used for mapping other projections and establishing fine differences within them, as was shown for the auditory cortical area by Tunturi (1950), Hind (1953) and others.

In experiments 1) and 2) the extent of the projection depends, among other things, on the strength of the stimulus and the spacing of the stimulating

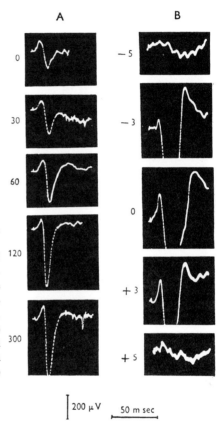

Fig. 244. The effect of strychnine on the shape and distribution of callosal potentials. A — Gradual increase in the negative component of the callosal response, after applying 1% strychnine. Figures indicate time in seconds following strychnine application. B — Maximal responses of the same point after strychninisation. The stimulating electrodes are moved along the contralateral g. suprasylvius. 0 — symmetrical point. Other figures denote distances in mm from 0. Upward deflection — positivity of the active electrode.

electrodes. With increasing intensity of the stimulus and spacing of the stimulating electrodes the number of stimulated neurones increases and so does the size of the callosal projection area. It is therefore preferable, especially in experiments with strychnine, to use weaker (liminal) stimuli. Transsection of the corpus callosum is a reliable way of demonstrating that the recorded potentials are of callosal origin.

3) Spontaneous strychnine discharges are characterised by a negative wave having an amplitude of 1—2 mV and lasting for about 100 msec followed by a less distinct positivity. They appear in irregular intervals (0·2—2/sec), with higher concentrations of strychnine they may form trains with a fre-

quency of 5/sec. Single spikes running at 1−2 sec intervals are most suitable for neuronography. In addition synchronous spikes in the contralateral hemisphere are also registered. Their amplitude is highest in the symmetrical area (Fig. 245). The distribution of potentials is roughly the same as in experiments with electric stimulation — the area of the maximum response is limited to several mm, then there is a rapid decline in amplitude. Callosal con-

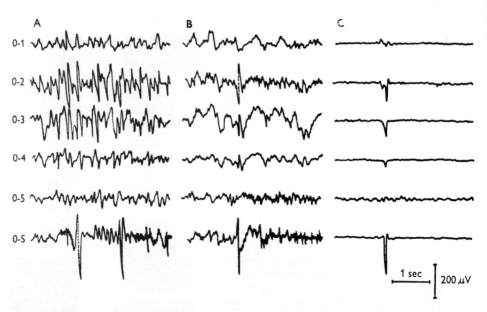

Fig. 245. EEG recording of projection into the contralateral g. suprasylvius of spontaneous strychnine spikes, evoked by application of strychnine crystals on the g. suprasylvius (S). 1-5 — parasagittal row of electrodes in contralateral g. suprasylvius. Interelectrode distance 3 mm. A, B, C — deepening anaesthesia.

nections are always reciprocal. The electric current stimulates both ganglionic cells of homolateral callosal neurons (orthodromic conduction of impulses) and axons terminals of the contralateral callosal neurons (antidromic conduction of impulses). On the other hand a strychnine discharge in the cerebral cortex is only connected with orthodromic activity.

In the case of callosal responses this circumstance is of no particular importance since the antidromic component only plays a small role when using threshold stimuli (Grafstein 1959). In addition the pure antidromic responses do not basically differ from the sum of orthodromic and antidromic potentials, evidently because they reach not only the body of the corresponding neurone but also spread into the recurrent collaterals branching in their neighbourhood. Via these collaterals synaptic activation of the same region as in orthodromic responses occurs (Clare et al. 1961).

With prolonged strychnine application secondary epileptic foci may appear in the nonspecific thalamic structures, which may influence the propagation of the strychnine spikes to the contralateral hemisphere. Under such conditions checking of extracallosal routes of propagation becomes necessary.

Strychnine neuronography can be succesfully used for determining various cortico-subcortical or cortico-cortical connections. If the response is localised subcortically the stereotaxic method is used and the sites of maximum response are subsequently identified histologically (p. 555).

Conclusion: Electrical or strychnine neuronography is a valuable means for tracing direct fibre connections between various regions of the brain. The advantage of strychnine neuronography is that the direction of the tract is also determined, absence of the strychnine spikes in a given region, however, does not exclude direct pathways leading to it from the strychninized focus. Increasing evoked potentials by local application of strychnine to the projection area permits detection of even very weak afferent activity.

G. Depth recording in the cerebral cortex

Problem: Record the response to stimulation of the sciatic nerve from different cortical layers.

Principle: In earlier experiments different methods of recording electric potentials from the cerebral cortical surface were described. Such a record, of course, is an expression not only of the activity of superficial cortical layers but also of the electric fields created by activity of deep cortical elements. Thus the positive phase of the primary response, signalising the arrival of impulses into a specific projection area corresponds in part to action potentials reaching the cortex via afferent pathways (in a similar way as positive action potentials from an electrode placed on the injured end of a nerve fibre are recorded — cf. p. 388), in part to synaptic potentials produced by these impulses on cell bodies and dendrites of pyramidal cells and interneurones (for review of the literature see Eccles 1951, Roytbak 1955, Albe-Fessard 1957, Amassian et al. 1964, Holubář 1964a). Surface positivity corresponds to movement of impulses from deep cortical layers towards the surface and the negative wave to activation of the more superficial levels.

Precise data are obtained by experiments with microelectrodes introduced to different depths into the cortex. By comparing recordings from different depths with a simultaneous recording from the surface, the depth of primary negativity and its spread in the cortex can be determined relatively accurately. When interpreting electric recordings thus obtained it is assumed that the excited part of the neurone is depolarised and becomes negative with respect to other non-excited areas. This potential difference gives rise to a current flowing from the less negative (positive) pole of the neurone to the more negative one. The distant indifferent electrode applied to the point where current density is already insignificant will be negative relative to the source of this current and positive relative to the sink of the same current. Such positivity of one area produced by negativity of an adjacent area is due to differences between different parts of the same neurone (between dendrites and the cell body). The circuit is closed through the inside of the neurone. A brief account of the volume conductor theory is given in chapter IX K.

Object: Rat 150—200 g weight.

Apparatus: Two channel cathode-ray oscilloscope with a D. C. or A. C. amplifier (time constant at least 0·5 sec, input resistance 5—10 MΩ, total amplification at least 50 μV/cm. A stimulator giving single rectangular pulses or condenser discharges synchronized with the time base of the oscilloscope. An isolation unit for the stimulator. A time marker.

Other requirements: Surgical instruments, trephine (diameter 5 mm), an animal board with a stereotaxic head holder for fixing the animal. Micromanipulation equipment with vertical microdrive of at most 0·5 mm per revolution and an electrode holder. Insulated steel semimicroelectrodes with a tip diameter of 20—40 microns. Electrodes are best prepared from entomological needles or stainless steel wire (0·2 mm) by electrolytic treatment and consequent insulation with a varnish. For details see p. 227. Wick Ag-AgCl electrodes for corticography, bipolar silver wire electrodes for stimulating the sciatic nerve. Electrode carriers. 10% Dial. A binocular dissecting microscope.

Procedure: Under Dial anaesthesia the skull of a rat is exposed and thoroughly cleaned. A trephine hole is made over each hemisphere (diameter 5 mm) situated so that one third is in front of the coronal suture. Trephining is carried out carefully in order to avoid injuring the dura mater as far as possible. The sciatic nerve is exposed in both thighs for a length of 2 cm and placed onto the stimulating electrodes. The nerves are prevented from drying by paraffin oil filled in a skin cup formed around the exposed nerve.

The experiment is commenced by determining the area of maximum primary response to stimulation of the sciatic nerve as in experiment IX D 2 — by gradual mapping of the exposed cortical surface moving the wick of the Ag-AgCl electrode. The reference electrode, a strong needle, is fixed in the frontal bone above the level of olfactory bulbs. At the point where the maximum surface positive response is obtained the dura is pierced with a sharp hook and the microelectrode is introduced. This is controlled visually under the dissecting microscope. The dura (and in the cat and rabitt also the arachnoidea and pia) must be cut in order to prevent excessive pressure on the brain surface during penetration through the strong connective tissue. Indentation of the cortex makes precise determination of depth impossible. The actual depth to which the electrode has been introduced can be estimated from the movement of the micrometer drive while watching for indentation of the cortex at the site of entry with the microscope. Since even after penetration of the dura the surface is depressed to a certain extent it is best to introduce the electrode deeper by several tens of microns and then to raise it under microscope control until the cortical surface is again at the initial level.

When the tip of the electrode just touches the cortical surface the primary response is registered both with the microelectrode and the wick electrode immediately adjacent to the point of insertion. The shape of the two responses is usually the same but the amplitude of the microelectrode record is sometimes lower particularly if the ratio of the electrode resistance to the input impedance of the amplifier is too high (cf. p. 130). In order to make comparison of deep recordings and surface response easier, amplification in the two channels is adjusted so as to obtain the same amplitude on both tracings. Then electrode is moved down 0·1 mm at a time until a depth of 2·5 mm is reached. Several responses are recorded simultaneously from both leads at every level. The same procedure may be applied when raising the electrode. If histological verification of the electrode position is required, a direct current is allowed to pass through the microelectrode (5—20 µA, 5—30 sec, microelectrode +) after the recording has been made. Iron atoms are thus deposited in the tissue and these can easily be detected histologically using potassium ferrocyanide, a 1% solution of which in 10% formalin is injected into the carotid artery after termination of the experiment. The localisation of the electrodes is shown by blue spots easily discernible in histological preparations stained according to Nissl or with haematoxylin-eosin.

Results: Fig. 246 shows responses from different depths of the rat's somatosensory cortex as compared with those from the surface. It can be seen that the amplitude and duration of the positive phase of the response decreases with depth while the negative phase is prolonged until it completely replaces the surface positivity. At a depth of 1 mm this negative wave becomes the direct mirror image (phase shift by 180°) of surface positivity, but its amplitude may be higher. The negative phase of the deep response is usually followed by a slight positivity. A more detailed analysis of changes in the primary response at different depths is attained by relating the polarity and amplitude of the deep response to the amplitude and polarity of the surface response at the moment when the latter attains its positive or negative peaks. If the amplitude of the deep response is expressed in % of the coincident positive or negative maximum of the surface response and if these values are plotted against the depth of registration, characteristic curves are obtained (Li et al. 1956a, b). It follows from these curves that the positive and negative components of the surface response have a different depth distribution. Changes of positivity to negativity occurring at a depth of about 1 mm demonstrate where the primary active area of the surface positive wave commences. The maximum negativity in deep recordings indicates the centre of this area. In the time interval corresponding to maximum surface negativity, on the other hand, negativity in the deep recordings decreases more slowly and is replaced by positivity only at a depth of 1·2 mm. This indicates that the upper layers of the cortex are the primary active area for this component (Li et al. 1956a).

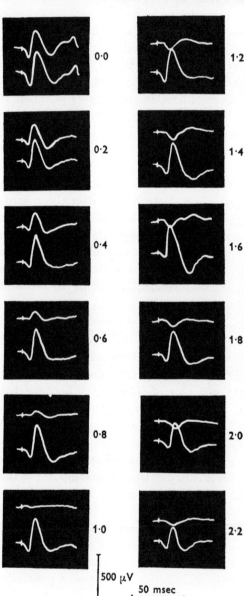

On the basis of deep recordings an attempt may be made to reconstruct the process of spreading of the primary cortical response. When the surface positive wave commences, a focus of negativity is formed in deeper cortical layers which are depolarised by the arrival of the afferent volley from specific thalamic nuclei. The negativity immediately begins to spread towards the surface at a rate that can be estimated from the interval separating maximum negativity or the zero potential level at different depths. This rate is usually less than 0·1 m/sec and is probably due to activation of the dendritic membrane at successively more superficial levels by polysynaptic conduction of the deep response through the cortical interneuronal chains. The original focus of deep negativity disappears first, negativity of surface layers persists for another few tens of msec.

Fig. 246. Deep registration of primary responses on stimulating the sciatic nerve of the anaesthetised rat. Upper tracing — microelectrode recording. Lower tracing — surface recording. Figures indicate the depth of the tip below the cortical surface in mm. Positivity upward.

Real situation may, of course, be more complex. In addition to radial dipoles also tangential dipoles may play a role in the electrogenesis of the primary response. The source region need not arise only passively as a result of depolarization of another part of the neurone but may also be an expression of hyperpolarization of the corresponding region by the IPSP mechanism. Several dipoles arising simultaneously at different depths of the cortex may participate in the primary response. These dipoles may differ considerably in polarity, intensity and duration. The change in the shape of the primary response as the electrode is inserted deeper into the cerebral cortex would then correspond to a changed participation of different generators in the resultant curve (Holubář 1964a). In order to verify such a possibility a more detailed analysis of the voltage field (using bipolar microelectrodes if necessary) has to be carried out or measures have to be taken to enhance or suppress some of the expected generators (anaesthesia, anoxia, surgery etc.).

Inspite of some difficulties in the interpretation of results deep recording is an important source of information concerning the character of the electric field of other characteristic potentials, e. g. callosal responses (p. 552), secondary responses (p. 539), antidromic responses to stimulation of the pyramidal pathway (p. 557), recruiting responses (p. 610), epileptiform activity (p. 645), strychnine spikes (p. 356) etc.

Conclusion: Simultaneous registration of electrical potentials from the surface and deep layers of the cerebral cortex permits precise characterisation of the distribution and movement of electric dipoles arising in the cortex during spontaneous or evoked activity and thus contributes to a better understanding of the functional anatomy of the cortex.

H. Antidromic and orthodromic stimulation of pyramidal pathway

Problem: Determine the extent of the motor cortex in the rabbit by mapping responses of Betz pyramidal cells to antidromic stimulation of the medullary pyramids. Trace the pyramidal discharge produced by stimulating the motor cortex in fibers of the pyramidal tract at the level of the medulla.

Principle: Motor areas of the cortex may be mapped by stimulating the cerebral cortex and observing peripheral motor responses or electrical activity of nerve paths. The opposite approach, however, is also possible. Since the pyramidal tract is composed of axons arising directly from giant pyramidal cells of the fourth cortical layer (Betz cells) it is possible to produce by stimulation of these fibers an antidromic volley that reaches the cortex and causes depolarisation of the cell bodies and perhaps dendrites of the pyramidal neurones. The technique of stimulation and recording is similar to that in experiment IX C. The situation, however, is complicated by the fact that the pure antidromic response is distorted by activity of afferent fibres terminating in the same cortical area which may be stimulated in the medullary region together with fibres of the pyramidal tract (particularly fibres of the lemniscus medialis — Landau 1956). Another factor making interpretation of results more difficult are recurrent collaterals of pyramidal cell axons which may affect further cortical cells orthodromically (Chang 1955b). A thorough analysis of tracings, however, makes it possible to distinguish between true antidromic responses and potentials of other origin, and thus makes it possible to obtain information not only on the localisation and extent of cortical areas contributing to the pyramidal tract but also on antidromic activity in cortical neurones and especially in the dendrites.

With the opposite arrangement of stimulating and recording electrodes — stimulation of the cortex and recording in the medulla — characteristic responses can be recorded in the pyramidal tract. These consist of a sharp wave with a latency of 1—2 msec, followed by one or more successively smaller waves at about 2 msec intervals. It has been shown by Patton and Amassian (1954) that the first wave (D) corresponds to direct stimulation of pyramidal cells or their axons in the cortex while the other waves (I) are an expression of synaptically mediated activation of pyramidal cells. The intensity of cortical stimulation remaining constant, the amplitude and area of the D and I deflections of the evoked pyramidal response may further be analysed from the aspect of excitability changes in pyramidal cells (D waves) and cortical interneurones (I waves).

Object: rabbit, 2—3 kg.

Apparatus: A cathode-ray oscilloscope with an A. C. amplifier (time constant at least 0·3 sec, total amplification at least 50 μV/cm). A stimulator giving either single or paired rectangular pulses or condenser discharges, synchronised with the time base of the oscilloscope. An isolation unit (transformer, radiofrequency output). Time marker.

Other requirements: Surgical instruments, dental drill, animal board and head holder or a stereotaxic instrument for rabbit, permitting ventral and dorsal approach to the head. Bipolar needle electrodes for stimulating the pyramids (spacing of the tips 0·5 mm). Spring-mounted, ball-tipped silver or platinum electrodes (diameter 0·25 mm) for corticography. Electrode carriers. 1% Nembutal or Pentothal. 1% chloralose in 10% urethane.

Procedure:

1) The rabbit, anaesthetised with i. v. Pentothal, is fixed to the animal board on its back. After low tracheotomy and insertion of a tracheal cannula the pharynx is exposed and pulled sidewards. Deep muscles are carefully removed and the base of the occiput is exposed over about 10 × 10 mm.

An opening of 4—6 mm diameter is made with the dental drill between the foramen magnum and the caudal border of the pons. The dura is not cut for the time being and the borders of the cut skin are sutured. Then the animal is released and again fixed prone on the board. The skin of the head is cut sagittally and trephine holes are made above the fronto-parietal area of the hemisphere (roughly above coronal suture). After an interval of 30—60 min the animal is fixed in the head holder of the stereotaxic apparatus in such a way that the skull base is easily approachable from below. Light anaesthesia is maintained by repeated injections of Pentothal or by gradually replacing Pentothal by a solution of 1% chloralose in 10% urethane. The latter in amounts of 4—5 ml/hr is recommended by Chang (1955b) as especially suitable in this experiment.

The trephine opening at the base of the occipital bone is again exposed and the dura cut in the mid-line. The bipolar stimulating needle electrodes are placed onto the medullary pyramid 3—5 mm below the border of the pons. Electrodes are applied with the help of the vertical microdrive of the electrode carrier under visual control (dentist's mirror). Recordings are taken directly from cerebral surface after cutting the dura, the arrangement being similar to that in experiments with primary cortical responses. In order to ensure a good functional state of the cortex the trephine opening may be covered with warm paraffin oil, previously shaken with Ringer solution. The recording electrodes are placed onto the cortical surface at 2 mm intervals. At each point several responses to stimulation of the pyramids are recorded (stimuli in at least 1 sec intervals). The reference electrode is a metal clamp fixed to the cut skin above frontal bones or a needle fixed into those bones.

After finding the area of maximum response the following procedures may be used for analysing its individual components:

a) determination of its refractory period. This is done as in experiments with primary responses (IX D 3),

b) three-minute anoxia (introduction of N_2 into the tracheal cannula),

c) local application of 1% strychnine to the area of the recording electrode (as described in experiment IX F 2),

d) micro-electrode recording of cortical responses from different depths, (as in experiment IX G).

2) The second part of the experiment is done on the other hemisphere if possible. Two silver ball-tipped stimulating electrodes are placed at a distance of 2—3 mm from each other onto the surface of the motor cortical area. Registration is made in the homolateral medulla with the electrodes used for stimulation in the first part of the experiment. Stimuli are applied at minimum intervals of 2 sec. The values for threshold and maximal stimuli are determined in the cortex and by moving both the cortical and medullary electrodes the

optimum responses are found. For more detailed analysis of their components anoxia is applied for one minute by introducing N_2 into the tracheal cannula while the cortex is being stimulated rhythmically with submaximal stimuli (submaximal, in order to make it possible to observe not only a decrease but also an increase in excitability as judged from changes in the response). The development of the medullary response is studied during anoxia and subsequent restitution.

Results: 1) The extent of antidromic projection in the rat, rabbit and cat according to the work of Porter (1955), Woolsey and Chang (1948), Landau (1956), Lance and Manning (1954), Woolsey (1958), Porter and Sanderson (1964) roughly corresponds to the motor regions, shown in Fig. 237. Responses, however, are recorded not only from the motor cortical area but also from somaesthetic area I. The typical responses in the centre of the projection area are nearly monophasic surface positive waves with an amplitude of about 200 µV and a duration of about 5 msec (Fig. 247A). These are an expression of the arrival of antidromic volleys in the cortex and of depolarisation of cell bodies of pyramidal neurones. They are characterised by a very short latency, not exceeding 2 msec, thus excluding the possibility of considering them as orthodromic responses caused by stimulation of the medial lemniscus. The positive wave changes smoothly into a low negative potential lasting for about 10 msec and corresponding to antidromic propagation of depolarisation from the cell body to the dendrites (Chang 1955a). In the rabbit under chloralose-urethane anaesthesia it is further possible to record a surface negative wave of a high amplitude and changing shape, with a latency of 15—20 msec. This, according to Chang (1955b), is an expression of activation of cortical neurones via collaterals of pyramidal axons at the cortical level. According to Landau (1956) and Porter (1955) it is the result of orthodromic stimulation of fibres of the medial lemniscus.

The initial positive wave has a very short refractory period as compared with the primary response (2 msec instead of 6—20 msec). During anoxia the positive wave is preserved for a longer period than the negative one and also reappears sooner (Fig. 247A). The antidromic responses are more resistant to anoxia than the primary responses. The local application of strychnine in no way changes the antidromic response or its early phase but usually increases late waves if they are present (Fig. 247B). This indicates that they are postsynaptic in nature. The micro-electrode recordings from different depths of the cortex show a phase reversal of the antidromic response at a depth of 1—1·5 mm. The latency of the positive component is the same in all cortical layers, that of the peak of the negative component is the longer the closer it is to the surface. This may indicate conduction of depolarisation in the dendrites from the cell body towards their apical branching at a rate of about 0·5 m/sec (Chang 1955a).

2) Fig. 248 shows the characteristic response registered from the medullary pyramids when stimulating the motor cortex. A first wave (D) is well in evidence as are also subsequent, usually less distinct wave (I) or waves (I_1, I_2). The threshold for D waves is lower than for I waves. During anoxia the amplitude of both D and I waves at first increases for a short period (if a submaximal stimulus is used) and then first the I, and later also the D waves decrease considerably or

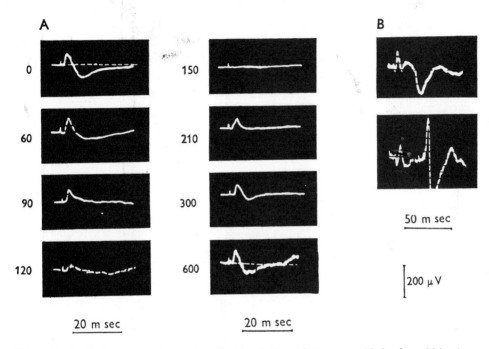

A

0

60

90

120

150

210

300

600

20 m sec

20 m sec

B

50 m sec

200 μ V

Fig. 247. Cortical response to antidromic stimulation of the pyramids in the rabbit. A — Effect of anoxia. Figures denote time in sec from start of anoxia. B — Late components of the antidromic response (above) and the effect of strychnine (below). Positivity upward.

disappear altogether. Restitution occurs in the opposite order. This result is in agreement with the assumption that I waves are connected with activity of cortical interneurones, which are much more sensitive to anoxia than pyramidal cells. The greater resistance of D waves might, to a certain extent, be due to the direct action of the stimulating current on axons of pyramidal cells in the cortex.

Even without stimulation of the cerebral cortex, spontaneous potentials occasionaly appear in the medullary electrodes. These indicate synchronous discharges of pyramidal cells. Simultaneous recordings from the cerebral cortex and the pyramids show that these discharges coincide with the appearance of sleep spindles (p. 516) in the motor area (Whitlock et al. 1953, Brookhart and Zanchetti 1956, Arduini et al. 1963).

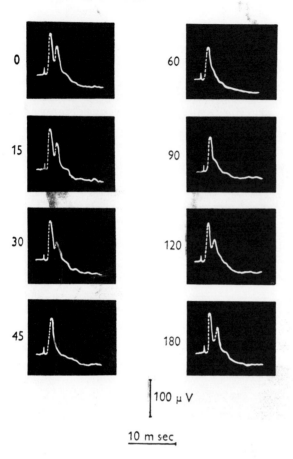

Conclusion: Antidromic stimulation of the pyramidal tract makes it possible to define the extent of motor areas and to demonstrate some aspects of antidromic conduction in the cortex. Pyramidal tract responses to stimulation of the cortical motor area, on the other hand, make it possible to study changes in excitability of pyramidal cells and cortical interneurones.

Fig. 248. Responses to stimulation of the motor cortex in the rabbit, recorded from the region of the bulbar pyramids, and the effect of anoxia. Figures denote time in sec from the start of anoxia.

100 μ V

10 m sec

I. Steady potentials and impedance of the cerebral cortex

Problem: Determine the polarity and impedance of the cerebral cortex in rats and register its changes during temporary cortical ischaemia.

Principle: The cortical surface of mammals is 15—25 mV positive to other electrically indifferent areas. This steady potential may be considered as a baseline on which phasic changes of EEG activity occur. Cortical polarity is the result of the regular structure of the cerebral cortex and probably depends, in the first place, upon the potential difference between dendrites and bodies or axons of pyramidal cells. The cortical DC potential is very stable. Small variations of cortical polarity (of the order of 1 mV) accompany the arousal reaction (negativization of cortical surface) while onset of sleep is signalized by a surface positive slow potential shift (Arduini 1958, Caspers 1959, 1962, Wurtz 1966). More pronounced changes appear only following severe interference with the metabolism of nervous tissue (anoxia, asphyxia, some intoxications — Leão 1947, Van Harreveld and Stamm 1953b, Goldring et al. 1953, Bureš 1957a, b, Bureš and Burešová 1957), or in the course of some processes simultaneously affecting extensive groups of cortical neurones (epileptic seizure, spreading depression — cf. p. 568). Even considerable changes of short duration are reversible (depolarisation lasting for tens of minutes). The D. C. potentials of the brain were recently reviewed by O'Leary and Goldring (1964).

From the technical point of view it is of advantage to use simultaneously several methods for determining cortical polarity. The choice of the reference point is particularly important. It is best to determine cortical polarity as the demarcation potential between a normal and damaged area of the cortex. Recording is carried out according to directions described in the previous experiments concerned with registration of steady potentials (Chapter III).

Another parameter characterising elementary functions of the cortex is the cortical impedance, measured between two cortical areas (Leão and Martins-Ferreira 1953, Aladzhalova 1954, Freygang and Landau 1955, Van Harreveld and Ochs 1956, 1957, Adey et al. 1962, 1966, Ranck 1963, 1964, Van Harreveld 1966). The real component of cortical impedance (resistance) is especially important. This can easily be determined using a low frequency A. C. bridge and non-polarisable silver disc electrodes coated with silver chloride. Of course the current must be such as not to stimulate the cerebral tissue itself, and the resistance of the electrodes must be considerably less than that of the cerebral tissue between them. Finally electrodes must be placed in such a way that the majority of the current flows through the cortex, for otherwise actual cortical impedance would not be measured. Determination of resistance of cerebral tissue corresponds to measuring the conductivity of the extracellular space of the cerebral cortex.

Object: Rat 150—200 g.

Apparatus: Double beam cathode-ray oscilloscope with D. C. amplification of at least 1 mV/cm or a cathode-ray oscilloscope with an A. C. amplifier and a chopper circuit permitting registration of steady potentials after their transformation into short pulses (see also page 163). A recording milivoltmeter with a cathode follower input can also be used. An A. C. bridge for determining resistance and capacitance with a measuring frequency of 500—1000 cycles/sec. Calibrator. Balancing potentiometers.

Other requirements: Surgical instruments, 2·8 mm trephine, non-polarisable Ag-AgCl screw electrodes (diameter 3 mm), $Hg-Hg_2Cl_2$ wick electrodes with stands, an animal board for the rat with a stereotaxic head holder. A rubber band (10 × 200 × 0·5 mm) with a clamp, a respiration pump with a rubber tube fitting over the rat's snout. Dial 10%.

Procedure: A rat under Dial anaesthesia (40 mg/kg) is tied to the animal board on its back and the trachea is exposed in such a way that between it and the other tissues of the neck there is a slit about 1 cm long. Through the latter a rubber band is passed (1 cm in width). The animal is then placed on its abdomen, the scalp is cut and the skull bones are cleaned.

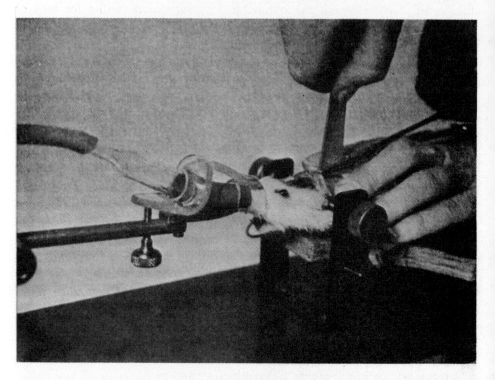

Fig. 249. Arrangement for experiments with transient ischemia of the brain. The snout of the rat is fixed to the tube leading from the respiratory pump. A rubber band passes through a bracket on the neck of the animal.

When registering cortical polarity two trephine holes 5 mm in diameter are made above the frontal and occipital regions of one hemisphere and one hole of the same diameter above the occipital region of the other hemisphere. The rat's head is then fixed in the stereotaxic holder and a rubber tube 1 cm in diameter connected to the respiration pump is slipped onto the snout (Fig. 249). Both ends of the rubber band running below the trachea are passed round the neck and through a special bracket (Fig. 249).

An earthed calomel cell electrode is placed into the anterior trephine opening (R) and further calomel cell electrodes, connected through compensators to the inputs of the D. C. amplifiers or to mechanical switch, are placed into the openings A, B. Before placing them onto the cerebral surface the elec-

trodes are shortcircuited by connecting their wicks, and their potential differences to the reference electrode are adjusted to zero with compensators.
After applying the electrodes to the brain, calibration is carried out by introducing a known D. C. potential (5—10 mV) in series between the reference electrode
and earth.

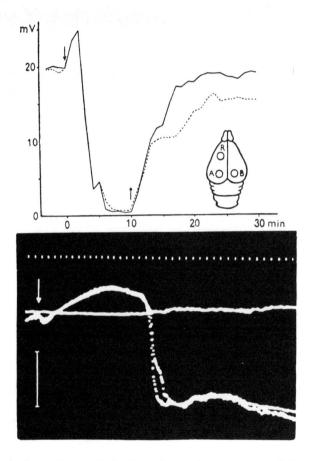

Fig. 250. Terminal anoxic depolarisation. Top: redrawn course of depolarisation and
repolarisation. The arrow indicates the beginning and end of 10 minute ischaemia. Diagram: distribution of electrodes. Bottom: oscillogram of rapid depolarisation. Arrow —
beginning of ischaemia. The horizontal trace shows the potential level from which deflections RA and RB must be measured. Calibration: 10 mV. Time mark: 10 sec.

After briefly recording potential differences AR and BR, the brain
tissue in the anterior trephine opening R is thermocoagulated by applying
a heated metal rod (diameter 4 mm) for several seconds to the cortical surface
and the cortical demarcation potential produced between the injured area R
and areas A and B is recorded. Sometimes a slow negative potential appears

in area B. This is due to a wave of spreading depression (cf. IX J) produced by thermocoagulation. Otherwise, however, the potential differences AR and BR remain unchanged. After ten minutes of recording the rubber band is tightened by a pull of about 5 kg and fixed with a haemostat. Simultaneously artificial respiration is started (30—60/min). An open system is sufficient. The changes of cortical polarity are recorded in leads AR and BR. The rubber band is released after 10—20 minutes of ischaemia and recording is continued for another 15—30 minutes.

When measuring impedance only two trephine openings (diameter 2·8 mm) are made over one hemisphere in another rat. The distance between the centres of the openings should be 6 mm. Silver screws coated with AgCl on their lower surfaces with a diameter of 3 mm are screwed into those openings. They are connected through flexible insulated wires with the measuring bridge. In order to avoid polarisation of the electrodes a low frequency alternating current (preferably 500—1000 cycles/sec) is used for measuring impedance. Stimulation of the cerebral cortex by the current is prevented by checking the potential difference on the cortical electrodes with a cathode-ray oscilloscope connected in parallel to them. The voltage at the electrodes must not exceed 20 mV. After recording the original resistance of cerebral tissue for several minutes the experiment is continued in the same way as when determining terminal anoxic depolarisation. Individual determinations of resistance are made at 15—20 sec intervals. Later, particularly during resuscitation, intervals of 1 minute suffice.

Results: The interruption of blood supply to the head, as applied in the above experiment, does not result in complete circulatory isolation of the head but does produce complete arrest of blood flow in the carotids and considerably decreases the supply of blood from the spinal and vertebral arteries. The effectiveness of circulatory arrest may be demonstrated by comparing ischaemic changes after strangling and decapitation. No significant differences between the effects of these two procedures can be observed as far as cortical polarity and impedance are concerned. The chief advantage of the method used here is the likelihood of recovery, with normal cardiac function during the period of brain ischemia (Grenell 1953). A typical experiment with terminal ischaemic depolarisation is shown in Fig. 250. The polarity of the cerebral cortex which initially has a value of about 20 mV (Bureš 1957b) is preserved for 1—3 min after interruption of the blood supply, but then it begins to fall to zero, at first rapidly and then more slowly. Depolarisation appears earlier in the hemisphere in which an area has been coagulated. If normal cardiac activity is maintained during brain ischaemia, resuscitation may be successful even after complete disappearence of cortical polarity. Revival is indicated by the recovery of cortical polarity. The rate of repolarisation is inversely related to the duration of ischaemia.

Cortical impedance is measured by adjusting the resistance and capacitance boxes so as to balance the bridge. From the values of resistance (R) and capacitance (C) the impedance (Z) can be computed using equation:

$$Z = \frac{R}{\sqrt{(1 + \omega^2 R^2 C^2)}}$$

As the balancing capacitance (C) does not usually exceed $0.02\,\mu\text{F}$, Z differs only slightly from R. For R values of the order of 1000 Ω and measuring frequency 500—1000 c/sec, resistance values read directly from the bridge at balance may be used as a measure of tissue impedance.

The resistance of the cerebral cortex begins to rise from the moment blood flow is interrupted. This rise is most rapid when the slope of the depolarisation curve is steepest (Fig. 251). Within 10—20 minutes, depending apparently on the completeness of the circulatory isolation of the head, the resistance of the cerebral cortex increases from the original 800—900 Ω to 1400—1600 Ω.

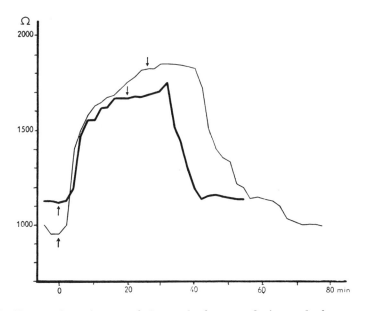

Fig. 251. Changes in resistance of the cerebral cortex during and after transient brain ischemia (20 min — heavy line, 25 min — thin line). Arrows indicate the beginning and end of ischaemia.

Then it continues to rise slowly until, after 3—5 hours a maximum of 3000 to 4000 Ω is reached. These changes are reversible up to a certain increase in resistance (and thus a certain duration of ischaemia). If, however, the resistance increases by more than 100%, there is little hope of complete recovery of normal nervous functions.

Intracellular electrolytes, enclosed within a membrane of high resistance, do not significantly contribute to the conductivity measured with a low frequency current applied via extracellular electrodes. The increase in resistance is, therefore, produced primarily by changes in conductivity of extracellular fluid. According to Van Harreveld and Ochs (1956) it is possible to calculate directly how much the relative volume of the extracellular space must decrease in order to achieve a corresponding decrease in conductivity, assuming a stable composition of the extracellular fluid. It has been shown histologically that the initial sudden increase in cortical resistance is accompanied by swelling of nerve cells and dendrites (Van Harreveld 1957). This is evidence for extensive movement of water and electrolytes between intra- and extracellular spaces during ischaemia. A corresponding increase in permeability of cellular membranes helps to explain the depolarisation of the inherent potential difference between the surface and deep layers of the cerebral cortex.

Conclusions: Reflex activity as well as phasic EEG potentials disappear within several tens of seconds following brain ischaemia. In the subsequent period, during which successful resuscitation is still possible, cortical steady potentials and resistance of cerebral tissue are the main indicators of changes occurring in the brain.

J. Spreading depression

Problem: Produce spreading depression in the cerebral cortex of the rat with different stimuli, and record its manifestations. Show the antagonistic actions of K^+ and Ca^{++} ions in producing spreading depression.

Principle: Electrical, thermal, mechanical or chemical stimuli, acting on the surface of the cerebral cortex, produce after a short latent period a local decrease in amplitude of the ECoG. This does not remain confined to the area of the stimulus but spreads over the cortical surface in all directions at a rate of about $2-5$ mm/min (Leão 1944). The advancing front of the wave of depression is accompanied by a slow negative potential of about 8 mV in amplitude followed, after $1-2$ min, by a less pronounced but longer lasting positive wave. The wave of steady potential can also be recorded from deeper cortical layers but disappears completely in the white matter. The depression of spontaneous EEG activity lasts $5-6$ min at a given cortical locus. Coincident with the reduction in EEG amplitude there is a decrease of primary responses in cortical projection areas and in excitability of the motor areas. The wave of depression is also accompanied by increased cortical blood flow, decreased cortical pO_2 and an increase in cortical impedance. For review of literature see Marshall (1959), Ochs (1962), Brinley (1963).

Recording of electrical manifestations of spreading depression is achieved with the usual techniques (cf. IX B, IX I) which need not be further explained. The characteristic wave of increased blood flow accompanying depression can be demonstrated by thermoelectric registration of blood flow changes in the cerebral cortex. Using a conventional EEG apparatus with a chopper in the input, D. C. potentials are recorded from cooled thermocouples applied to the cortical surface (Burešová 1957). The temperature of the thermocouple depends upon the equilibrium between the constant withdrawal of heat from the metal of the thermocouples by a constant temperature cooling bath (0°C) and the changing supply of heat to the thermocouples from blood circulating in the surrounding tissues. The difference in temperature between two thermocouples of exactly similar construction placed on symmetrical areas of the brain reflects differences in blood flow through these areas.

Spreading depression can be used as a means of studying some problems of cortical metabolism, in particular electrolyte metabolism. The antagonism between K^+ and bivalent cations, well known in other excitable tissues, can also be demonstrated in the cerebral cortex. The effect of K^+ is suppressed by adding Ca^{++} or Mg^{++} to KCl solutions of a concentration producing depression. Local application of Ca^{++} or Mg^{++} onto the cortical surface prevents the invasion of spreading depression into the treated area.

Object: Rats weighing 150—200 g.

Apparatus: An ink-writing EEG apparatus with at least 4 channels and slow rate of paper transport (1—2 mm/sec). A simple mechanical chopper driven by a shielded electric motor for the channels in which steady potentials are to be recorded. A recording milivoltmeter with cathode follower input can be used for the slow potential recording.

Further requirements: Needle electrodes for electrocorticography (cf. p. 509). Nonpolarizable wick calomel cell electrodes filled with Ringer solution. Glass capillary electrodes (200 µ) filled with saline. An animal board with the stereotaxic headholder for the rat. A stereotaxic apparatus for the rat with vertical microdrive. Surgical instruments, dental drill. Cooled thermocouples (Ag-constantan) for recording changes in blood flow in symmetrical areas of the brain (Fig. 252). An anode battery with a 1 MΩ potentiometer and nonpolarisable wick Ag-AgCl electrodes for polarising the brain. Different solutions of KCl and KCl + $CaCl_2$ or KCl + $MgCl_2$, see below.

Procedure: The skull bone of a rat under dial anaesthesia is exposed in the usual way. The animal is tied to the board and fixed to the stereotaxic head holder. Trepanation and fixation of the needle electrodes is carried out according to the kind of experiment.

1) Initiation of spreading depression with different stimuli.

Trephine holes are distributed according to Fig. 253: a large opening (diameter 4—5 mm) above the occipital pole of the hemisphere, two small openings (2.5 mm) rostrally from the first. Along these openings a row of 4—5 needle electrodes is placed. Above the other hemisphere openings and electrodes may be placed symmetrically. The needle electrodes are connected in a bipolar fashion (Fig. 253). Steady potentials are recorded from points A'', A' (or B'', B') against the reference electrode situated above the non-stimulated hemisphere. The reference electrode may be earthed. Care must, however, be taken

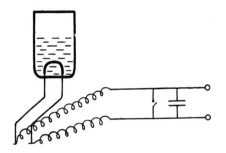

Fig. 252. Cooled thermocouples: a vessel with crushed ice and a 1 mm silver wire. Constantan leads to EEG amplifier. Mechanical switch in the input.

not to earth the animal at any other point (head holder, board etc.). If the needle electrodes are earthed through the input resistors of the EEG amplifiers a condenser must be introduced into every lead interrupting the galvanic connection of the electrode with earth. Steady potentials would otherwise be measured against all other earthed points and would thus be unpredictably distorted.

The EEG is recorded in 3 or 4 leads, the steady potential in another two. The steady potential is first adjusted to zero by compensators in series with the active electrodes (cf. p. 564). Since it is the amplitude that changes most in the EEG recording, the speed of paper movement may be limited to 1—2 mm/sec. After 10—15 min of control recording, during which we make sure that the level of EEG activity is stable and that the steady potential difference between the hemispheres does not vary by more than 1 mV, spreading depression is produced.

a) Faradic stimulation with bipolar Ag-AgCl electrodes (distance between the tips 1—2 mm) may be used to produce spreading depression. A direct current, however, gives more constant results (Leão 1944, Leão and Morison 1945, Whieldon and Van Harreveld 1950, Van Harreveld and Stamm 1953a, Marshall 1959). The current from an anode battery is regulated by a potentiometer and measured with an ammeter in series. The intensities used vary from 0·2—3·0 mA, the duration of the stimulus is limited to 5—10 sec. A wick Ag-AgCl electrode is applied to the large trephine opening, the other is fixed to the cut skin in the neck. During polarisation recording of the steady potential with the calomel electrodes is interrupted.

b) Spreading depression can be produced easily by mechanical stimuli, e. g. tapping the cortical surface with blunt pincers (Leão 1944). A more exact determination of threshold, however, is made by measuring the energy of a body falling onto a circumscribed cortical area from a known height. By gradually increasing the path of free fall and thus the kinetic energy of the falling body, the threshold for producing spreading depression can be determined exactly (Zachar and Zacharová 1961).

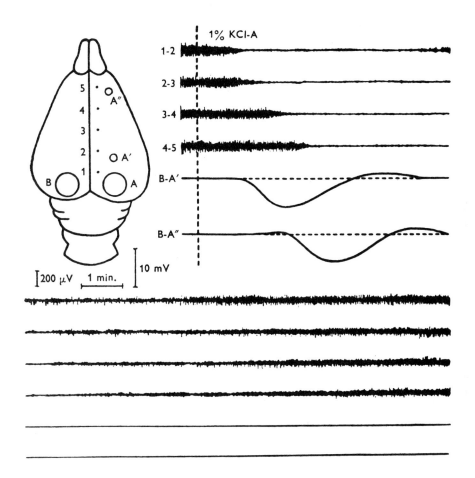

Fig. 253. Cortical spreading depression in an anesthetized rat. The interrupted vertical line indicates application of the stimulus. For details see text.

c) Spreading depression may be produced by topical application of a number of substances onto the cortical surface — KCl, glutamine, asparagine, NH_4Cl, veratrine, NaCN, 2,4—dinitrophenol, NaF, strophanthine etc. They are applied using squares of filter paper 2 × 2 to 3 × 3 mm in size soaked in the solution desired. The concentration just producing depression is determined. The papers are applied to an area exposed by a large trephine opening. If spreading depression develops, the filter paper is removed and the cortical surface is washed with Ringer solution.

2) Recording of a wave of steady potential from different cortical layers.

The procedure is the same as in paragraph 1) but only two trephine openings (5 mm in diameter) are made over the right hemisphere (cf. Fig. 254), one

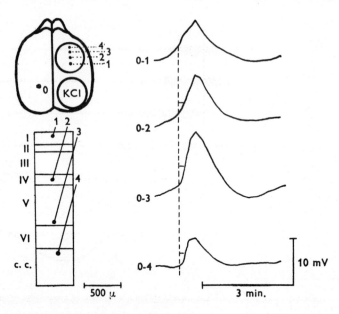

Fig. 254. Deep registration of the slow potential change accompanying cortical spreading depression in the rat. Position of the multiple electrode system indicated in the inset diagram.

above the frontoparietal region (for introduction of deep electrodes) and the other above the occipital region (for eliciting spreading depression). A small trephine opening is made above the contralateral hemisphere for application of the indifferent electrode. Glass capillary electrodes of 200 μ external diameter filled with physiological saline and connected by a salt bridge with calomel electrodes are used for recording slow potentials from deep layers of the cortex. Four such capillaries are arranged in parallel in one plane and their ends are ground to a line containing an angle of 28—34° with the longitudinal axis of the capillary (Monachov et al. 1962a). The distance between the centers of electrodes is 500—700 μ. Under the same angle the whole system is introduced into the brain with the stereotaxic apparatus in a parasagittal plane in such a way that the electrode tips are approximately perpendicular to the brain surface and facing the approaching slow potential wave. If the resistance of the capillary electrodes is not too high they may be directly connected to the inputs of four channels of the EEG apparatus, which are periodically shortcircuited by electrically driven switches (1—2/sec). Spreading depression is evoked by application of 1% KCl onto the occipital trephine opening and amplitudes of the slow potential wave in simultaneous deep and surface recordings are compared. Calibration of the microelectrode leads is carried out by introducing a known D. C. potential between the reference electrode and the EEG input. Intervals of 30—40 min should be left between individual waves of the depression.

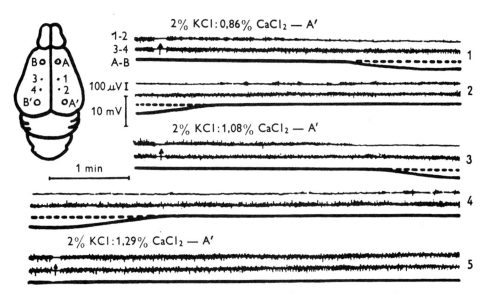

Fig. 255. Blockade of spreading depression by adding CaCl₂ to an effective concentration of KCl. For details see text.

3) The effect of bivalent cations on the production and spread of spreading depression.

Trephine openings and electrodes are arranged as shown in Fig. 225. A filter paper soaked in an effective KCl solution (e. g. 2% KCl) or in a mixture of KCl with MgCl₂ or CaCl₂ is applied to the posterior trephine opening. After depression has been produced in one hemisphere with KCl solution, a KCl solution containing a certain concentration of the bivalent cation is applied to the other hemisphere. If depression occurs KCl solution of the same concentration but containing more bivalent cations is again applied to the first hemisphere, which has meanwhile recovered from the previous application. Intervals of at least 40—60 minutes are left between experiments on the same hemisphere. The experiment is continued until a certain ratio R^{++}/K^+ is no longer capable of producing spreading depression. This ratio is determined for both hemispheres.

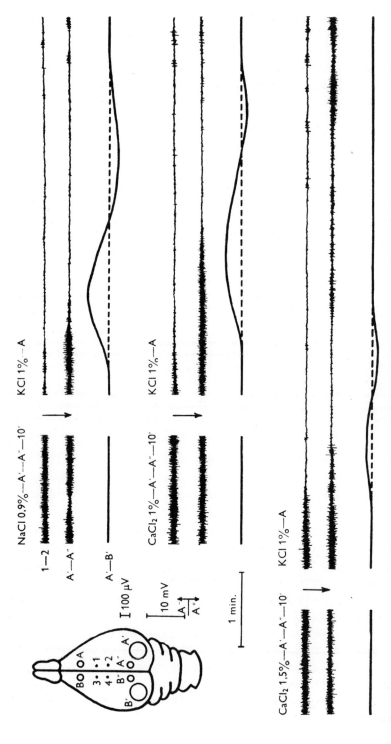

Fig. 256. Blockade of spreading depression in cortical areas treated with CaCl₂.

In another experiment (see Fig. 256) it can be demonstrated that waves of spreading depression do not enter a cortical area locally treated for 10 minutes with sufficiently strong solutions of $CaCl_2$ or $MgCl_2$. First a wave of spreading depression produced from the region A by 1% KCl is recorded in the area A'A''. Then a solution of $CaCl_2$ ($MgCl_2$) is applied to these areas. 10 minutes later the protecting solution is removed and a new wave of depression

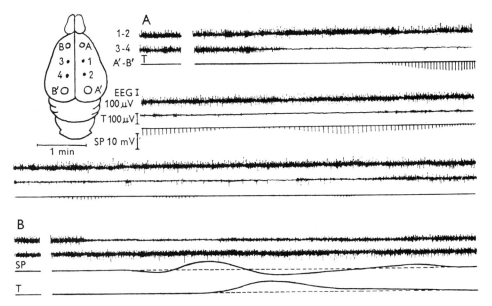

Fig. 257. Recording of vasomotor changes accompanying spreading depression. A — Potential difference between cooled thermocouples, registered by transforming D. C. potential to rectangular pulses. B — Comparison of the tracing of a slow potential wave with vasomotor changes from the same point on the cerebral cortex. Both curves redrawn.

is produced from the distant trepanation A. The change in amplitude of EEG and of the slow potential is observed.

4) Changes in blood flow during spreading depression.

Trephine openings are made as shown in Fig. 257. Spreading depression is produced from the small opening A (B), steady potential waves and changes in blood flow are recorded from the large openings A'(B'). In addition EEG activity is registered with needle electrodes. Accurate recording of blood flow largely depends upon the correct placement of thermocouples. Their round silver facets (2 mm in diameter) must be previously adapted to the curvature of cerebral surface by adjusting slope and distance between them. Only then are the suspended thermocouples, together with the cooling vessel, carefully lowered until they lightly touch the surface of the hemispheres without exerting pressure. Contact is ensured by a thin film of fluid formed between the thermocouples and the cerebral surface. Measurement will be successful only if the surrounding temperature is perfectly stable and if air currents are reduced to a minimum. Very careful exclusion of factors setting up temperature gradients (light bulbs, opening and closing of doors etc.) is therefore recommended. The direction of the steady potential recording while increasing the temperature of one thermocouple is checked by approaching it with a warm object. After control recording lasting for 10—15 minutes, during which time the difference between the cooled thermocouples does not change by more than $0.2—0.5^{\circ}C$ (10—20 μV), spreading depression is produced by applying 2% KCl to the trephine opening A (B).

When depression has passed, the cooled thermocouples are removed and replaced by calomel electrodes. A new wave of spreading depression, produced in the same way, is recorded. The two recordings are synchronized by the onset of the ECoG depression, and thus the time relations between changes in blood flow and steady potentials are determined.

Results:

1) Fig. 253 shows spreading EEG depression and the wave of negative potential after stimulating area A with a direct current (either cathode or anode may be used but cathodal stimulation is more effective) or with a mechanical stimulation (e. g. a 4 g metal rod falling onto an area of about 10 mm^2 from a height of 5—10 mm).

Chemical threshold stimuli usually produce spreading depression after a much longer latent period (up to 10 min) during which time a local reaction develops at the site of application of the solution.

Spreading depression resembles the "all or none" type reactions. Once it leaves the limits of the focus it continues to spread regardless of the former's continuing existence until it extends over the whole cerebral cortex. The rate of spread depends, among other things, upon the temperature of the brain. It can be calculated from the time lag between different phases of EEG depression or of the slow potential wave in two distant leads. The steepest part of the negative wave, corresponding to the moment when negativity has attained 50% of the maximum, is most suitable for calculation. The velocity of spreading EEG depression is calculated from the time interval between these points in the slow potential recordings from the two cortical areas and from the spatial distance between them.

The refractory period, during which a new wave of spreading depression cannot be produced from the same site and by the same stimulus, lasts for about 4 min. Since EEG activity does not recover within such a short period, supraliminal chemical stimuli applied to the cerebral surface (e. g. 25% KCl) produce a whole series of slow potential waves recurring at intervals of about 6 min. The amplitude of the slow potential waves following rapidly one after the other gradually decreases, the EEG remains permanently suppressed. Following a series of slow potential waves, convulsive potentials may appear on the recording instead of decreased EEG activity.

2) The steady potential recorded with capillary electrodes from different depths of the cerebral cortex has the same shape and duration as in the surface recording (Fig. 254). Its amplitude remains unchanged or even increases down to a depth of 1·5 mm. When the white matter or subcortex is entered by the electrode the slow potential decreases and finally disappears altogether. By comparing recordings from different depths with simultaneously made surface recordings it can be ascertained that depression spreads also vertically in the cortex. The negative maximum occurs later in deeper layers in proportion to depth. The rate of radial and tangential spread is roughly the same. (Leão (1951).

3) It is possible in all experiments to find a certain minimum ratio $[Ca^+]$: $[K^+]$ which prevents an otherwise effective KCl concentration from producing spreading EEG depression (Fig. 255). The mean values of this ratio are best calculated using probit analysis of the relation between the value $[Ca^{++}]:[K^+]$ and the percentage of experiments in which spreading depression was produced with this ratio.

The concentration ratio $[R^{++}] : [K^+]$ varies for different bivalent cations. Mg^{++} is most effective, Ba^{++} least so. The blocking ratio also depends on the blocked concentration of K^+.

Finally the spread of a wave of depression into a cortical area protected by Mg^{++} or Ca^{++} can be prevented (Fig. 256). Application of $CaCl_2$ or $MgCl_2$ decreases EEG activity and primary responses at the site of application. This decrease, however, is of a different character than that evoked by spreading depression. Protection manifests itself by a considerable decrease of the steady potential wave entering the treated area and by maintenance of spontaneous and evoked activity in that area although other cortical regions are depressed. The protective effect is most evident during the later phases of the action of $MgCl_2$, when its inhibitory effect on the EEG and primary responses has already disappeared but when the protective effect against spreading depression is still fully preserved. The relationship between the concentration of Ca (Mg) and the amplitude of the slow potential wave entering the treated area 10 minutes after application of the bivalent cation is shown in Fig. 258.

4) Temperature changes recorded with cooled thermocouples during spreading depression are shown in Fig. 257. An evident increase in the temperature of the hemisphere affected with depression can be seen on a relatively stable base line. A comparison of the temporal course of the steady potential wave and thermoelectric potential shows that the temperature increase sets in later than cortical negativity and attains a maximum when the cerebral cortex is already completely repolarised. Often several gradually decreasing waves can be seen in the temperature recording. These follow the first wave of increas-

ed temperature and indicate an undulating return of blood flow to normal. On the basis of relations between the steady potential and the change of blood flow it may be concluded that increased blood flow is not a cause but rather an after-effect of spreading depression. The vasomotor change described here is, however, characterised only qualitatively and without further elaboration of technique more precise quantitative evaluation is impossible.

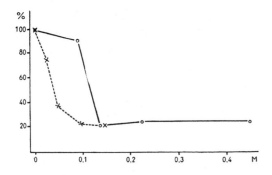

Fig. 258. Decrease in the amplitude of negative potential of spreading depression in cortical areas treated for 10 min. with different concentrationss of $MgCl_2$ (dotted line) or $CaCl_2$ (full line). Ordinate: amplitude of negative potential in % of normal. Abscissa: molar concentration of the applied solution of the $CaCl_2$ or $MgCl_2$.

Conclusion: Spreading depression can be evoked in the cerebral cortex of many laboratory animals including cats and monkeys but is most readily elicited in the lissencephalic cortices of rodents — rats, guinea pigs and rabbits. It was demonstrated also in several non-neocortical structures — in hippocampus, thalamus and caudate of mammals and in striatum of birds and reptiles (Marshall 1959, see also p. 634). From the anatomical point of view spreading depression does not depend on a particular cytoarchitectonic arrangement of neurons. Since myelinated axons and probably also glial cells do not actively participate in the spreading depression process, white matter bands and glial scars prevent spreading. Spreading depression is a humorally mediated process. When enough neurons in the area of stimulation are depolarized, the local increase of potassium concentration in the extracellular space may cause the depolarization of adjacent neurons and the reaction starts spreading as an autoregenerative process of depolarization (Grafstein 1956, Brinley at al. 1960, Křivánek and Bureš 1960). Although spreading depression can be evoked only by arteficial stimuli and does not play any role in normal brain functions, a study of this phenomenon makes it possible to analyse the relationship between electrical events and brain metabolism (Bureš 1956, Křivánek 1962). Spreading depression may further serve as a method enabling the temporary and reversible suppression of normal cortical function in behavioral studies (Bureš and Burešová 1960) and in research into the functional organization of brain (Bureš and Burešová 1962).

K. Theoretical basis of recording electrical potentials in a volume conductor

The complex structure of the central nervous system with millions of active elements makes it exceptionally difficult to interpret electrical records of brain activity. Potentials recorded with macroelectrodes are the vector sum of elementary electrical fields of a large statistical assembly of neurons, the individual elements of which may show contrary changes. For that reason the relation between macro- and microelectrode recording will be solved only by detailed analysis of the behaviour of a sufficiently large and representative statistical sample of neurones from a given area. The problem cannot be discussed in detail in this short chapter. Only some fundamental working principles will be presented in order to characterise more exactly the source of electric activity examined by macroelectrodes.

1) Membrane potentials of neurones are the main source of electric potentials recorded from structures of the brain. Since in different parts of an inactive neurone the membrane potential is evidently not the same (the membrane of the cell body, dendrites, axons) a potential difference can arise between its parts. This is perhaps the basis of the resting electrical polarity of nervous structures (cf. steady potential of cerebral cortex, p. 562, Sutin and Campbell 1955). During synaptic or antidromic stimulation there is always an asymmetrical change in membrane potential in different parts of the neurone and this again results in external potential differences.

It must be emphasised, that potential differences always arise only within one single neurone. The electric current flows in one direction through the extracellular space and in the opposite direction through the interior of the cell. The assumption that a potential difference is produced between active and inactive cells has no physical justification whatsover. If an inactive and an excited cell with a homogenously distributed membrane potential are placed in a conducting medium, no potential difference can be recorded between them. Such cells would be isopotential on their external surface and thus no current or corresponding external field would be produced. This is evident from the diagram in Fig. 259.

2) The asymmetric distribution of the membrane potential along the neurone (permanent or temporary) results in external electrical asymmetry of the nerve cell, which in a simplified form appears as an electric dipole. The external current passes from the site with the larger membrane potential to sites of lower potential. The external positivity of a certain part of the neurone does not necessarily indicate hyperpolarisation of the corresponding part of the membrane. The membrane potential at the source may be normal or even decreased, but of course is less than at the sink.

The source will be positive when recording against a distant reference electrode, the sink negative. If the reference electrode approaches the source it will gradually acquire the latter's potential. This results in a decrease in positivity recorded directly from the area of the source. Placing the reference electrode near the sink, on the contrary, decreases the negativity recorded directly from the site of depolarisation. The voltage is the product of the integrated value of the current density component and specific resistance of the medium.

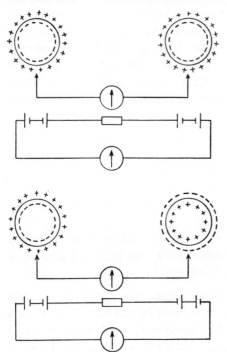

From the integrated components of the current density along the line joining two electrodes. the magnitude of the potential measured can be calculated by applying Ohm's law. The specific resistance of the medium is assumed to be constant. If the sink is spatially smaller than the source, the density of the current in the area of the sink is greater than in the area of the source. This manifests itself by more pronounced but circumscribed negativity.

Fig. 259. Potential difference between two cells with symmetrically distributed potential. Top: both cells in resting conditions. Bottom: one cell excited. In both cases the electrical diagram of the circuit is shown.

3) In a volume conductor the current arising between two areas of a neurone with different membrane potentials spreads in far reaching lines of current. In the distant parts of the field the direction of the lines of current flow is nearly reversed in relation to the intrapolar sector. Model experiments studying configuration of the electrical field of the nerve action potential in a flat conductive medium (Lorente de Nó 1947, Kostyuk 1960) and theoretical treatment of this problem (Offner 1954) show that under such conditions the monophasic action potential changes to a triphasic one (cf. experiment on p. 387). This model experiment is of fundamental significance also for the interpretation of electrical manifestations of the movement of excitation along neurones in the brain. Assuming that the fundamental manifestation of excitation is negativity, the positive deviation may be explained by the fact that the electrode is localised in the area of the current source or by an IR drop, produced by backcurving of the lines of current flow. The change of positive into

negative potential indicates that excitation has spread to the area of the electrode. A positive wave may thus be an expression of excitation approaching the electrode, a negative wave an expression of excitation reaching the area of the electrode.

4) The shape of the physiological dipole and its position in the volume conductor of the brain, may be determined either directly, by exploring the

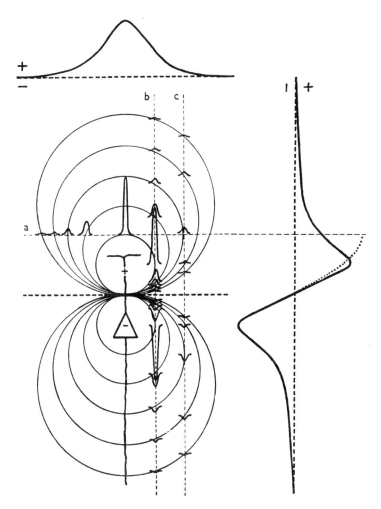

Fig. 260. The field arising around an electric dipole, and characteristic equipotential lines. The thin, interrupted lines represent sections through this field perpendicular to the axis of the dipole (a) or parallel to it (b, c). The amplitude and polarity of deflections at the corresponding points of the field are illustrated. Top: spatial distribution of the dipole potential in plane *a*. On the side: spatial distribution of the dipole potentials in plane *b*, assuming a medium of indefinite extent (full line) and the actual voltage plot found in the cerebral cortex (dotted line). For details see text.

electrical field with microelectrodes, or indirectly, by projecting the dipole onto the surface of the volume conductor.

The first method was described in detail in the experiment with depth recording of primary responses (p. 553). Introducing a microelectrode vertically into deep layers of the cortex at the site of maximum surface response makes it possible to compare activity in different cortical layers with the simultaneously recorded surface potential. Initial surface positivity of the primary response is replaced by an initial negativity at a certain depth. This indicates that the microelectrode has reached the other side of the corresponding dipole. On further advancing the electrode down the amplitude of negativity gradually decreases. Fig. 260 shows the characteristic shape of the electrical field through which the microelectrode passes when introduced parallel to the axis of the dipole in a medium of indefinite extent. The exposed cortical surface (horizontal dashed line a) introduces an asymmetry of the medium, which together with the asymmetry of the dipole alters the ideal symmetrical curve to the voltage plots actually obtained (dotted line). Evoked potentials characterized by a surface sink (recruiting waves) should give a reverse of this potential distribution (signs $+$ and $-$ exchange).

Characteristic voltage distribution may also be found on the surface of a spherical homogeneous conductor containing a radial dipole. It is evident that its projection will be the more pronounced the greater is the dipole moment and the closer it is to the surface of the conductor. It may thus be stated that the site of maximum response corresponds to the area of localisation of the dipole. Using more exact analysis the depth of the dipole can be calculated from surface distribution of the potential. According to Shaw and Roth (1955) the potential (V_p) on the surface of a homogeneous spherical conductor with radius R which contains a dipole whose axis lies on a radius is defined as:

$$V_p = \frac{M^2}{R^2} \left(\frac{1 - f^2}{f(1 - f^2 - 2f \cos\ \varphi)} \right) - \frac{1}{f}$$

where M is the dipole moment, f the distance of the centre of the dipole from that of the sphere, φ the angle between the radius for point P and the radius of the dipole. Fig. 261 shows the theoretical distribution of the potential for different depths of the dipole. The depth of the dipole can be determined from the angular distance between the site of the maximum response and the site of phase reversal.

5) In a volume conductor the potential recorded at a given point is the resultant of activity of close and distant elements, the electric fields of which often reach to a considerable distance. It is difficult to distinguish between such distant and close activity using conventional techniques. The simple solution is to use concentric needle electrodes. The outer sheath forms

then a low resistance ring which averages all adjacent voltage gradients. For sinks and sources of current lying outside the ring the inner needle has a potential approaching that of the outer needle (Fig. 262 A). On the contrary sinks or sources of current located inside the ring give rise to a clearcut potential difference between the inner and outer needle (Fig. 262 B). It is obvious that best results can be expected when the dimensions of the generator are smaller than the diameter of the outer needle. This condition is well fulfilled e. g. with muscle fibers in electromyography (Nakao et al. 1965).

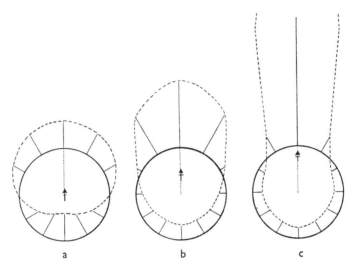

a   b   c

Fig. 261. The surface potentials on a sphere containing a radial dipole at a depth 1r (a), 0·6r (b) and 0·2r (c) below the surface. The potentials are drawn in radial coordinates. Zero corresponds to the surface of the sphere, plus values are plotted outside and minus values inside. The angle denotes the angular distance of a recording point from the radius containing the dipole.

More exact information about the distribution of sources and sinks in a small volume of tissue can be obtained using methods based on recording of current instead of potential (Howland et al. 1955). Let us consider a square area A, B, C, D (Fig. 262C) in a section plane or on the surface of nerve tissue. The parts of neurons contained in it are either isopotential in relation to those lying outside of sources or sinks of current. If the currents passing in or out through the individual sides of the square are added algebraically, the total contribution of the given area in an outward direction is given. Currents arising outside this square area are excluded automatically since if they enter through a side of the given square they must leave it by another one (Kirchhoff's law). Thus they cancel each other.

Currents flowing through individual sides of the square are determined from potential differences between monopolar leads from electrodes localised in the centre of the square and against the mid-points of its sides at distance $l/2$, according to the equation:

$$i_1 = \frac{l}{\varrho} \frac{E_0 - E_1}{l} = \frac{E_0 - E_1}{\varrho}$$

where ϱ is the specific resistance of the medium.

The current out of the whole square is

$$i_0 = \frac{1}{\varrho}(4E_0 - E_1 - E_2 - E_3 - E_4)$$

The so called Laplacian is a practical realisation of this principle (Perl and Casby 1954). Five electrodes arranged as shown in the diagram in Fig. 262C are placed on the cortical surface. The potential recorded between E_0 and the average from electrodes $E_1\ E_2\ E_3\ E_4$ is proportional to the flow i_0 and makes it possible to achieve a more exact localisation of primary projections than the usual recording of primary responses.

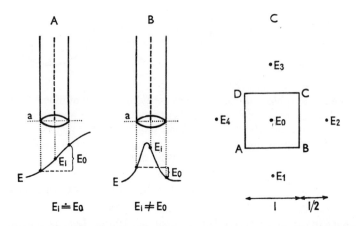

Fig. 262. Distinguishing between the local and remote generators of electrical activity. A, B — potential difference between the outer sheath (E_0) and the inner wire (E_i) of a concentric needle electrode, characterizing distant (A) and local (B) generators. E — potential gradient along the line a. C — Determination of sinks and sources on the basis of the average current flowing across the sides of a hypothetical square ABCD. For details see text.

If the distribution of electrical sources in a cross section of the volume conductor is to be analysed in detail another procedure must be adopted (Howland et al. 1955). The whole plane of the section is mapped out by inserting a microelectrode into different depths at regular steps. A series of recordings is made from each position of the electrode. This method can only be used if variability in the responses is minimal. The exact position of the electrodes is checked histologically and marked on a large photograph of the section. Using cross interpolation of the potential values the irregular net of the actual electrode positions is transformed to a regular square lattice which serves for calculating the elementary currents. In such a way the planar distribution of sources and sinks of currents at different phases of the realisation of a certain reflex act may be determined. If a three-dimensional current distribution is required an elementary cube would have to be used. This could be determined from potentials from further points distributed in such a way that they would form a regularly spaced cubical lattice.

6) Most above principles can be easily demonstrated using simple model experiments. Physiological saline in a large glass vessel is used as a conducting medium. Two pieces of polyethylene tubing filled with NaCl-agar and connected through gross Ag-AgCl electrodes to 6·0 V battery or to a square wave generator are immersed into the middle of the vessel. Their distance, orientation, size and shape can be changed according to the requirement of the model (e. g. by making more holes into the polyethylene tubing). The resulting potential field can be systematically explored with a small Ag-AgCl recording electrode which can be moved in all directions while the referent electrode is placed into the far end of the vessel. A pair of exploratory electrodes can be used for bipolar exploration of the field. The potential differences are amplified with a D. C. amplifier and recorded with a cathode-ray oscilloscope. The above basic arrangement can be further modified by inserting additional current electrodes and by dividing the medium into layers with different conductances (Byzov 1960, 1966).

Mapping the potential contours of several simple model dipoles (symmetric dipoles, asymmetric dipoles, dipoles at different depths below the surface of the medium etc.) can be recommended as the best way to practical understanding the volume conductor theory.

X

Electrophysiology
of subcortical structures

A. Stereotaxic method

The use of the stereotaxic method described by Clarke and Horsley in 1906 is a necessary prerequisite for electrophysiological and anatomical research into subcortical centers. The principle is simple: an electrode (metal needle, glass capillary) or other instrument is introduced mechanically without direct visual control into a certain nerve structure the position of which is given with reference to a three-dimensional system of coordinates determined by external landmarkers on the skull or by other reference points. Stereotaxic apparatus of various constructions for different animals have been described by Horsley and Clarke (1908), Clarke and Henderson (1920), Ranson (1934), Harrison (1938), Clark (1939), Krieg (1946b), Stellar and Krause (1954), Carpenter and Whittier (1952), Hume and Ganong (1956), Cort (1957). A number of modified stereotaxic apparatuses for man has also been described (e. g. Hayne and Meyers 1950, Bailey and Stein 1951, Spiegel et al. 1947 and others). For review of literature see Szentágothai (1957), Pavlov (1958), Meshchersky (1961) and Delgado (1964).

The usual type of stereotaxic apparatus consists of two fundamental parts: a head-holder for fixing the head of the animal in an exactly defined position and an arrangement permitting measurable movements of the electrode carriers in rectilinear or radial coordinates (Fig. 263).

The first usually consists of two adjustable screws or plugs that fit into the external auditory meati and a bar (or bars) holding the upper jaw. Fixation determines the basal reference plane permitting orientation of the system carrying the electrodes with respect to the skull and brain.

In principle this is a system of three planes perpendicular to one another. The basal plane is chosen so as to be roughly parallel with the base of the brain. It passes through both external acoustic meati and the anterior reference point differing according to the animal and apparatus used. In the majority of laboratory animals the lower surface of the upper jaw just behind the canine or incisor teeth (Cort 1957) or the lowest point on the orbital margin (Jasper and Ajmone-Marsan 1952 b) is chosen as the anterior point determining the basal plane. Some authors, however, also use other basal planes. Thus, e. g. Whittier and Mettler (1949) in the macacus use a plane passing through the external auditory meati and the upper orbital margin. In stereotaxic atlases for the rat (Krieg 1946b) and rabbit (Sawyer et al. 1954) the basal plane is defined as perpendicular to the sagittal plane and passing through the bregma and a point 1 mm (rat) or 1·5 mm (rabbit) above lambda.

Fig. 263. Universal stereotaxic apparatus permitting simultaneous introduction of a number of electrodes at different angles and basal approach to the brain. Adaptable for work with the cat, rabbit and rat (Šterc and Dvořáček 1964).

The sagittal plane passes through the head medially in the region of the sagittal suture, the frontal plane is given by the line joining both auditory meati, through which it passes perpendicularly to the basal plane. Since in the majority of apparatuses the basal plane is identical with the horizontal one, the other two planes take a vertical course. In a number of stereotaxic apparatus there is an arrangement permitting inclination of the fixed head in a rostrocaudal direction and thus also inclination of the basal plane to the horizontal.

The electrode carrier assembly makes it possible to achieve precisely measurable movement of the electrodes in rectilinear (most frequently used today) or radial coordinates. The vertical movement is sometimes equipped with a micrometric drive or an hydraulic system (Li et al. 1956a). This permits exact introduction of the electrode into the depth of the brain. Movement in the horizontal plane is achieved either by moving the electrode carrier mounted on a special frame permitting antero-posterior and mediolateral movements and their measurement on a scale, or, if the electrode carrier is fixed, by moving the head holder fitted to a lathe slide (Fig. 263). Both these methods may be combined in various ways. It is also possible exactly and mea- surably to incline the electrode carrier in the sagittal or frontal plane by at least 45°. This permits approach to those parts of the brain that are not directly ac- cessible from the dorsal surface of the skull (e. g. mesencephalon in the cat).

The electrode itself is either firmly connected to the carrier or, if it is to be implanted, in such a way that it can easily be released. Baumgarten (1957) for instance, recommends fixing the electrode to a platinum wire with a drop of paraffin. Electric heating of the platinum loop at a suitable moment melts the paraffin if rapid release of the stereotaxically introduced electrode is requi- red (Fig. 264h).

An animal-borne electrode holder frame oriented with respect to the stereotaxic planes was described in the classical work of Hess (1932) and is used at present mainly for work with rabbits (Monnier and Laue 1953, Monnier and Gangloff 1961). Insertion of electrodes into the brain is guided by auxillary rods perpendicular to the frame.

Electrodes applied stereotaxically are used for three fundamental pur- poses: for recording activity of deep structures, for stimulating them and for producing deep lesions. Similar technique is used to implant cannulas serving for introduction of drugs (in crystalline or liquid form), for perfusion of small brain areas (chemitrodes, push - pull cannulas — Delgado 1962) or for localized cooling of brain (Dondey et al. 1962).

In all cases the use of the stereotaxic instrument may be acute, i. e. the anaesthetised or curarised animal remains in the head holder of the stereo- taxic instrument throughout the experiment, or it may serve as a preparation for a chronic experiment. In the latter case after implanting the electrodes or producing the lesion in an anaesthetised animal there follows a long term period of observation of behaviour or recording electric activity from chronically implanted electrodes. In chronic experiments the whole procedure must be aseptic. The fur is removed from the head of the animal by first clipping and then applying a depilatory (barium sulphide). The head is fixed in the stereotaxic head holder by bars introduced into the external meati and by bars or clamps for the upper jaw. Care is taken to place the head symmetrically and sufficiently firmly in a position corresponding to the basal plane. Then the scalp is painted with iodine and washed with 70% alcohol. Surgical instruments are sterilised in the usual manner. Trepanation and the introduction of the electrodes are usually carried out in two stages.

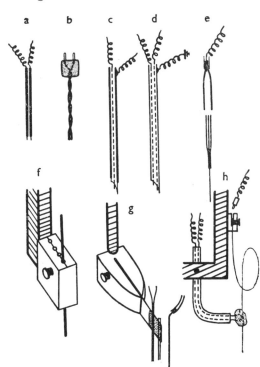

Fig. 264. Different types of electrodes and electrode carriers used with the stereotaxic apparatus. a) Bipolar needle electrodes. b) Bipolar electrodes made from insulated twisted silver wire. c) Concentric electrodes. d) Bipolar electrodes with a needle sheath. e) Electrodes for implantation with the connecting plug and socket arrangement (for details see page 509). f) Different types of electrode carriers. g) An electrode-cannula assembly used for deep microinjections and slow potential recording (see p. 635). Arrangement for rapid release of the stereotaxically introduced floating electrode according to Baumgarten (1957).

In the first stage, after cutting the skin and cleaning the skull bones, the inclination of the basal plane is adjusted to bregma and lambda if necessary. Coordinates of the zero point are then established using a heavy marking needle. The sites of trepanations are marked on the skull surface with reference to the zero point. If the point of intersection of the 3 reference planes is used as zero its coordinates must be determined before fixing the skull in the head holder for the marking needle and each of the electrodes. The use of an auxillary zero point on a horizontal bar exactly 70 or 100 mm above the interaural line makes it possible to obtain zero coordinates without removing the animal from the head holder of the apparatus. Trephine openings are drilled with a dental drill or a hand trephine at the sites marked. The openings must be of adequate size, small if the electrode is to be introduced chronically, larger if the localisation of the electrode is going to be monitored during the experiment by physiological responses to stimulations.

In the second phase the actual stimulating, recording or coagulating electrodes (coaxial, concentric, bipolar etc. — Fig. 264) washed with ajatin, zephiran or another disinfectant are fitted into the carrier. The horizontal coordinates of the trepanation are again tested (to make sure that the head was not moved during trepanation), the dura mater is pierced with a sharp needle and the electrode is introduced. In chronic experiments the upper bone layer surrounding the trephine opening is previously roughened with a dental drill or a scalpel. Several stainless steel screws are fixed into the bone in order to improve the attachment of the electrode to the scalp. The bone is thoroughly dried and the electrode is introduced. It is fixed to the roughened bone and screws with phosphate cement or acrylate. Only after the latter becomes solid is the electrode released from the electrode carrier.

Determination of stereotaxic coordinates

The position of a nervous structure is determined using a system of three coordinates indicating its distance from the point of intersection of the three zero planes or from a certain point on the skull (bregma, lambda). Stereotaxic maps showing the position of brain structures in such systems of coordinates have been worked out for the monkey (Clarke and Henderson 1920, Atlas and Ingram 1937, Olszewski 1952, Eidelberg and Saldias 1960), the cat (Clarke and Henderson 1911, Winkler and Potter 1914, Ingram, Hannett and Ranson 1932, Hess 1932, Gerard et al. 1936, Jiménez-Castellanos 1949, Snider and Niemer 1961, Jasper and Ajmone-Marsan 1952b, Reinoso-Suárez 1961, Verhaart 1964), the rabbit (Sawyer et al. 1954, Monnier and Gangloff 1961), the rat (Krieg 1946b, de Groot 1959a, b, König and Klippel 1963, Albe-Fessard et al. 1966), the guinea pig (Blobel et al. 1960, Luparello et al. 1964, Tindal 1965), the dog (Lim et al. 1960), the golden hamster (Smith and Bodemer 1963) and the chicken (van Tienhoven and Juhasz 1962). Stereotaxic coordinates for the brains of developing animals were determined by Bernardis and Skelton (1965) in the rat and by Volokhov and Shilyagina (1965, 1966) in the rabbit.

There are, however, sometimes considerable discrepancies between the data from such atlases and actual findings. These are due to different shapes of the skull in the used representatives of a certain species. If we intend to study a certain structure it is of advantage to use the atlas only as a first approximation and to determine its more precise localisation using animals of the laboratory breed and the stereotaxic apparatus actually to be employed. The procedure is as follows: the anaesthetised animal is killed by perfusing the carotid arteries with saline followed by 10% formalin. The whole head is then placed into 4% formalin for several days or weeks. Then it is fixed in the head holder of the stereotaxic apparatus in the usual position. The soft parts are removed and the bone is carefully cleaned. Using a strong marking needle fixed in the electrode carrier the coordinates of the zero reference point are obtained. Then an extensive craniotomy is made uncovering the whole of the dorsal cerebral surface down to the first spinal segment. This is done very carefully since the head must not be moved. The dura is carefully removed. A piece of razor blade is then fitted into the electrode carrier in such a way that it is parallel to the frontal plane and frontal sections are made at 0·25—0·5 mm intervals at the sites of the expected localisation of the required center, using the mediolateral movement of the electrode carrier for cutting. This is done in rostrocaudal sequence and the cut nervous tissue is removed. Sections are made until the required structure appears. The marking needle is then again fixed to the electrode carrier and the coordinates of the required point are determined.

If the structure cannot be identified macroscopically serial sections must be examined histologically. This method is only succesful if the inclination of the coronal sections corresponds to the frontal plane of the brain in the animal fixed in the stereotaxic apparatus. This is made easier if a block of cerebral tissue is cut out of the skull fixed in formalin in the way described above (Rocha-Miranda et al. 1965).

Fixation in formalin results in a certain decrease in cerebral volume. This may change the spatial relations between different cerebral structures and the skull. Still greater deformations occur during embedding in paraffin (the freezing technique is preferable), cutting and preparation of serial sections. It is therefore of advantage to verify the coordinates obtained from heads fixed in formalin, by lesions or iron deposits (see page 555) placed in normal brains. Small electrolytic lesions (cf. below) are produced in several points within and around the structure looked for. Then the animal is sacrificed, the brain fixed and prepared for histological examination. The position of the required structure is determined in serial sections with respect to the known stereotaxic coordinates of the lesions. The stereotaxic coordinates from the atlas or the formalin treated brain are corrected according to the results obtained.

Completely homogeneous material (animal of the same breed, age, weight, proportions of the skull etc.) is necessary if the stereotaxic technique is to be used successfully. Even so it is necessary to treat the data obtained from a fairly large number of animals statistically and to determine average values and their variation.

Stereotaxic lesions

Many methods are used to produce nervous lesions stereotaxically: mechanical destruction of tissues, injections of poisons or chemical substances, freezing, cauterisation, heating of implanted metal wires within an electro-magnetic field of shortwave diathermy, implantation of beta or gamma em-miters (gold or radon seeds) etc. A recent development in production of stereo-taxic lesions are trackless focal alterations of predetermined size produced in any desired location in brain by irradiation with focussed ultrasound (Lele 1962). The method of electrolytic lesions, however, still remains the most valuable and is most frequently used.

Either single or concentric bipolar electrodes are introduced into the structure to be destroyed. The electrodes are well insulated with glass or varnish and have a smooth, conical, carefully cleaned tip. Platinum, silver, steel or constantan are used as material. Before use the insulation of the electrode is tested. The electrode is connected to the —pole of a 1·5 V battery and immersed in saline. The + pole is connected to a plate electrode placed in the solution and the formation of bubbles is observed. When using the mono-polar technique the needle electrode has a positive potential against the large indifferent electrode placed on the cut skin, on the back, in the rectum or on the exposed muscles of the nape of the neck. Anodal lesions are preferable to cathodal since during electrolysis less gas is formed at the anode. Thus mechanical tissue injury is less and the shape of the lesion is more regular. An anode battery or rectifier is used as current source (voltage about 500 V). The intensity of the current is regulated by a 5 MΩ potentiometer connected in series and continuously measured with a miliammeter. The size of the lesion is roughly proportional to the quantity of electricity that has passed through the electrode. Usually a current of 2—5 mA is used for 15—30 se-conds. The resulting lesions are spherical or oval in shape with a diameter of 2—6 mm depending on the intensity and duration of the current. When using bipolar concentric electrodes the outer part is used as the anode and the core as the cathode. With the same intensity and duration of current lesions are slightly larger and more irregular than with the monopolar tech-nique. High frequency current (0·5—1·0 megacycles per sec) may be used instead of D. C. current but the insulation of the electrodes is then critical. Imperfect insulation results in irregular lesions along the electrode track due to current leaks.

The location of a lesion can usually be controlled only anatomically after the termination of the experiment. In some cases in which a successful lesion causes the disappearance of a characteristic function, the effectiveness of the stereotaxic intervention can be monitored during the experiment by observing changes in that function. Fig. 265 shows primary cortical response to acoustic stimuli during electrolytic destruction of the lemniscus medialis at the level of the inferior colliculi in the rat. Only after the second lesion has been placed does the response disappear. If must, of course, be remembered that the disappearance of a response immediately after electrolysis does not necessarily imply complete destruction of the corresponding nervous structure. In a number of neurones at the periphery of the lesion injury is reversible and their function may be recovered within a variable period.

In addition to electric current, nervous tissue can be stimulated or destroyed by implanting some slowly diffusing substances (e. g. powdered NaCN, alumina cream, a suspension of penicillin, strychnine paste etc.).

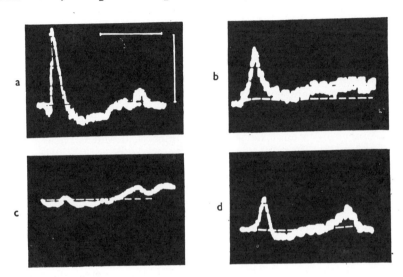

Fig. 265. Cortical response of the anaesthetised rat to acoustic stimuli before and after bilateral lesion in the region of the inferior colliculi. a — Control tracing. b — After bilateral lesion. c — After enlarging the lesion laterally. d — Two hours after placing lesions. Calibration: 200 µV, time: 50 msec.

The use of the stereotaxic apparatus for introducing recording or stimulating electrodes in acute and chronic experiments will be described in detail in further experiments.

B. Primary responses in the subcortical centres of afferent systems

Problem: Record in the rat responses to stimulation of the sciatic nerve from centres of the somatosensory pathway (posteroventral thalamic nucleus, cortical somatosensory area), to acoustic stimuli from auditory centres (corpus geniculatum mediale, colliculus inferior, cortical auditory area) and to visual stimuli from centres of the optic pathway (corpus geniculatum laterale, colliculus superior, visual cortex).

Principle: A synchronous volley of impulses produced by a short stimulus spreads along nervous paths into the cerebral cortex with a slightly increasing temporal dispersion. It elicits characteristic potentials in individual synaptic relays of the specific path. These usually are diphasic waves, an initial positivity followed by a negative phase. The response lasts for 10—30 msec and reaches its maximum amplitude in the corresponding centre. If microelectrodes with a tip diameter of less than 30—40 μ are used, unit activity can usually also be recorded together with the slow synaptic potentials.

When simultaneous recordings are made from the subcortex and the corresponding cortical projection area, the interval necessary for conduction of the impulse from the subcortical relay into the cortex can be well determined. The following means can be used to check the location of the recording electrode in the nuclei of the path studied:

1) The difference in latency between the subcortical and cortical response is small (it corresponds roughly to the time of conduction and the expected number of synapses),

2) stimulation with the recording electrodes elicits a similar cortical response with an appropriate latency,

3) application of strychnine to the area of the subcortical electrode results in strychnine spikes spreading to the next higher centre.

Technically the most difficult task is to find the corresponding subcortical centres. Usually it is necessary to use the stereotaxic apparatus and to refer to stereotaxic atlases for coordinates of the centre. It is therefore of advantage to use animals for whom such atlases are available — especially the cat, rabbit or rat. Yet even if such a map is strictly followed, the optimum localisation of the electrode must always be found experimentally by probing the expected area until the maximum response is obtained. In any case, it is of course necessary to verify the localisation of the electrode histologically at the end of the experiment.

Object: Rats, weighing about 200 g.

Apparatus: A two channel A. C. or D. C. amplifier (time constant at least 0·5 sec, input resistance 5—10 MΩ) and a cathode-ray oscilloscope for recording. Total sensitivity at least 50 μV/cm. A stimulator giving single square wave pulses or condenser discharges synchronised with the time base of the cathode-ray oscilloscope. An isolation unit (isolating transformer, RF output). Time marker.

Other requirements: Surgical instruments, a set of trephines, a dental drill. Stereotaxic apparatus for the rat. Steel needle electrodes (0·2—0·3 mm wire gradually tapering to a tip diameter of 30—40 μ, well insulated except for 50—100 μ of the tip). Wick Ag-AgCl electrodes and steel needle electrodes fitted for corticography (cf. p. 508). Bipolar hooked platinum electrodes for stimulating the sciatic nerve. A stroboscopic apparatus giving light flashes of of high intensity. 10% Dial, 10% formaline with a 1% ferrocyanide solution.

Procedure: The dorsal surface of the skull of a rat under Dial anaesthesia is exposed by cutting the skin in the midline. The parietal, occipital and temporal skull bones are scraped clear and dried, after dissecting the temporal muscle. The animal is fixed in the stereotaxic apparatus in such a way that bregma is 1 mm higher than the lambda. The external auditory meati are left free and fixation bars are applied to the rostral part of the temporal bone since sound stimuli are to be applied. Then a sharp strong needle with which the co-ordinates of the zero reference point (bregma) have first been determined is used to mark the centres of the trephine openings on the skull from which it will be attempted to reach the individual subcortical structures. Table 6 gives their coordinates in mm from the bregma assuming that the bregma — lambda distance is 8 mm.

TABLE 6

Structure	Coordinates			Diameter of trepanation
	sagittal	lateral	depth	
Coll. inf.	8	2	4—5	4
C. g. med.	5	2·5	6	3
Coll. sup.	5·5	1·5	4	4
C. g. lat.	4	3·5	5—6	4
Posterovent. nucl. thal.	2·5	2·5	5	4

Only one opening is made over each hemisphere. The surface of the bone surrounding the trephine opening is roughened with the dental drill to ensure a good contact for the phosphate cement applied in further stages of the ex-periment. After trephining the stereotaxic coordinates of the zero point are again checked.

The area of maximum cortical response to the corresponding stimulus is found first (cf. p. 526). Wick Ag-AgCl electrodes are used for mapping in areas exposed by trepanation, needle electrodes in those covered by bone. The in-different electrode is fixed in the frontal bone above the level of the olfactory bulbs. The cortical response is recorded during the further course of the experi-ment in one of the two channels. The subcortical needle electrode is fixed to the electrode carrier and the corresponding stereotaxic coordinates for the zero point are again determined.

The electrode is then inserted through the previously exposed cortical area to a point 1·5 mm above the expected centre. Then the electrode is further advanced gradually at 0·25 to 0·5 mm intervals and one minute after each step several (10—20) responses to stimuli, the effect of which is checked simultaneously by the cortical primary responses, are recorded with this electrode (Fig. 266). In such a way the electrode is advanced 1·5—2·0 mm deeper than would correspond to the depth of the centre according to the stereotaxic atlas. If no potential is found throughout the path which should be related to the

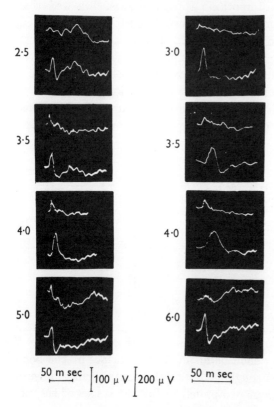

Fig. 266. Recording of primary responses to acoustic stimuli from the inferior colliculus (upper tracing) and the cerebral cortex (lower tracing) during progressive introduction of the deep electrode (AP8, L2). Figures denote the depth coordinate of the tip, according to the stereotaxic atlas. At a depth of 3·5 and 4·0 mm, tracings with a more rapid time base were made. Positivity of the active electrode upward. Calibration: 100 μV for the deep and 200 μV for the surface electrode.

cortical primary response, the electrode is pulled out and shifted 1 mm rostrally (or caudally, medially, laterally) and again introduced. Large pial vessels are of course avoided (special care to avoid injuring the sinus is necessary when trephining above the inferior colliculus).

If a response is obtained in a certain position or positions of the electrode, we attempt to define the extent of the responsive area and to find the maximum response by systematically exploring the surrounding area. It must be borne in mind, however, that the introduction of several relatively large electrodes into a small volume of brain tissue may in itself adversely affect function. It is therefore necessary to limit the number of exploratory insertions to a minimum.

As a test for the specifity of the potential produced by a certain stimulus other receptors or nerves are stimulated. The electrode is fixed at the site of the maximum response if further subcortical electrodes are to be introduced into the same animal stereotaxically. A drop of phosphate cement is placed on the bone at a point near to the trephine opening. The electrode is fixed with another drop and is then released from the stereotaxic electrode carrier after the cement has hardened. Thus several electrodes may be introduced into different cerebral regions in the same rat. The localisation of the electrodes may be verified further by using the deep electrode for stimulation and observing the cortical response. For monopolar stimulation a large indifferent electrode (an Ag-AgCl plate wrapped in cotton wool soaked in saline) is applied to the cut skin and muscles of the nape of the neck. Bipolar stimulation is, of course, better since it permits more exact localisation of the focus.

Results: A number of papers, mainly using the cat, have been devoted to detailed study of subcortical responses in specific relays of sensory pathways. The somatosensory pathway has been studied by Therman (1941), Marshall (1941), Mountcastle and Henneman (1949), Berry et al. (1950), Hunt and O'Leary (1952), Morin (1952), Morin (1953), Cress and Harwood (1953), Schricker and O'Leary (1953), Harwood and Cress (1954), Cohen and Grundfest (1954) and others, the auditory projection by Kemp et al. (1937), Kemp and Robinson (1937), Coppée (1939), Ades (1944), Tunturi (1946), Ades and Brookhart (1950), Thurlow et al. (1951), Rose and Galambos (1952), Galambos et al. (1953), Desmedt (1962) and others. Papers concerned with optic projection are by Bishop and O'Leary (1942a, b), Bishop and Clare (1951), Bishop (1953), Bishop and McLeod (1954), Vastola (1955), Lennox (1956), Cohn (1956) and others. Responses in subcortical relays of sensory pathways in rat were thoroughly described by Libouban-Letouze (1964).

In spite of the large variability in responses due to different orientation of the active elements with regard to the electrode and the small dimensions of the centres studied the majority of monopolar responses obtained with electrodes of larger diameter has the character of a positive—negative sequence. The initial positivity recorded in the area of maximum response corresponds to the afferent activity, the negativity to synaptic potential and to the postsynaptic volley. The response, usually, lasts about 15—30 msec. The latent period is always shorter by 2—3 msec than that of the primary cortical response for a response at the thalamic level (c. g. lat., c. g. med., posteroventr. nucleus) and even shorter for lower centres (Fig. 267). Colliculus superior forms an exception: collicular visual responses have a somewhat longer latency than the cortical ones probably because of slower conduction rate in the retinocollicular nerve fibres. If electrodes with a smaller tip diameter are used (less than 30 μ) unit spikes can be recorded in addition to the slow potentials. They are usually observed during the negative deviation of the synaptic potential, as an expression of the postsynaptic discharge (Fig. 267c).

597

The shape of the response depends also on the synchronisation of the afferent signal. Direct stimulation of the optic nerve after enucleation elicits in the corpus geniculatum laterale of the cat a short (4 msec) response with clear cut presynaptic and postsynaptic spikes (Bishop and McLeod 1954). The lateral geniculate with its "one to one" synapses is particularly suitable for responses of this type. When using an intensive flash in rats, a prolonged potential is obtained from this structure. This is similar to potentials from other subcortical structures and is due to increased temporal dispersion of impulses arriving from the retina.

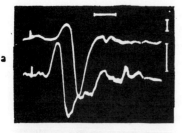

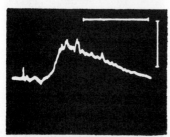

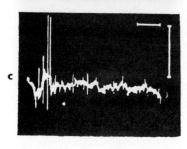

As far as lateralisation of subcortical responses is concerned there is a pronounced contralateral response to stimulation of a sciatic nerve or one eye. This is not the case with unilateral acoustic stimuli which elicit nearly equal responses in both the homo- and contralateral centers.

Fig. 267. Examples of specific responses in subcortical centres. a — Colliculus superior, optic cortex — response to a visual stimulus. b — Thalamus — slow potential with superimposed unit activity in a bipolar recording. Stimulation of the sciatic nerve. c — Colliculus inferior — acoustic stimulus. Microelectrode recording. Calibration: 100 μV, time: 20 msec.

At the end of the experiment the position of the electrodes is marked either by an electrolytic lesion or by depositing a small amount of iron from the steel electrode into the surrounding tissue (30—50 μA, 10 sec, electrode positive). The iron ions are then detected by fixing the tissue in 10% formaline with 1% ferrocyanide. The iron appears as small blue spots that are clearly visible in serial sections stained with conventional technique.

Conclusion: A study of electrical responses at various levels of a specific pathway makes it possible to follow the spread of a synchronous afferent volley in the brain.

C. Electric activity of the cerebellum

Problem: Record the spontaneous electrocerebellogram and responses of the cerebellar cortex to afferent stimulation.

Principle: Spontaneous electrical activity of the cerebellum of unanaesthetised mammals recorded with surface macroelectrodes manifests itself as very fast low voltage potentials (Adrian 1935, Dow 1938: 150—300/sec, 10—50 μV). Recording of this type of bioelectric potentials makes great demands on the amplifier which must not only have low noise (less than 5 μV) but also high amplification, linear up to 2—5 kilocycles/sec. A cathode-ray oscilloscope must be used as the recording instrument.

An electrocerebellogram registered with surface macroelectrodes by the usual ink-writing EEG apparatus does not, of course, show any of this activity because of the inertia of the pen. The recording does not differ fundamentally from the ECoG but is usually of lower amplitude (Swank and Brendler 1951, Cooke and Snider 1954).

The cerebellum also reacts characteristically to stimulation of peripheral receptors and nerves or of the cerebral cortex and other central structures (Dow 1939, 1942, Dow and Anderson 1942, Grundfest and Campbell 1942, Adrian 1943, Snider and Stowell 1944, Bremer and Bonnet 1951, Snider and Eldred 1951, Bremer 1952, Szabo and Albe-Fessard 1954, Combs 1954, Bremer and Gernandt 1954, Albe-Fessard and Szabo 1955, Jansen 1957, Morin et al. 1957, Koella 1959, Levy et al. 1961, Fadiga and Pupilli 1964). Depending upon the type of anaesthesia used, the shape of cerebellar responses is fairly irregular and their projection unstable. They always show an initial positive wave if monopolar recording is used.

Object: Cat weighing 3—4 kg.

Apparatus: A cathode-ray oscilloscope with an A. C. amplifier (time constant from 0·01 to 0·5 sec, overall amplification 20—50 μV/cm linear to 5000 c/sec, peak to peak noise maximally 5 μV). A multichannel ink-writing EEG apparatus of conventional design. A stimulator giving single square wave pulses or condenser discharges synchronised with the time base of the oscilloscope. An output isolation unit for the stimulator. A time marker.

Other requirements: Surgical instruments, trephines. Dental drill. Stereotaxic apparatus for the cat. Bipolar steel or silver stimulating electrodes. Ball-tipped platinum or silver recording electrodes mounted on springs or wick Ag-AgCl electrodes. Electrode carriers. A loudspeaker. A photostimulator. 10% Pentothal. D-tubocurarine or other relaxant. Ether. 1% novocaine.

Procedure: The trachea of the cat under ether anaesthesia is exposed low in the neck, tracheotomy is performed and a tracheal canulla inserted. D-tubocurarine is then given i. p. and as soon as it starts to take effect, arteficial respiration is begun (30 breaths/min). The scalp is incised in the midline after infiltration with 1% novocaine. Using a dental drill openings are trephined above the right hemisphere and in the occipital bone in the midline (diameter 8—10 mm). The openings are enlarged by rongeuring away the bone until nearly the whole area of the right hemisphere and the major part of the dorsal and dorsocaudal surface of the cerebellum in the region of the vermis and the hemispheres is exposed. Bleeding is continuously stopped with bone wax.

a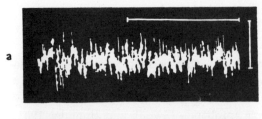

b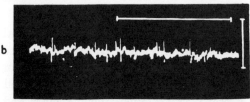

Fig. 268. Surface electrocerebellogram in the curarised cat before (a) and 10 min. after (b) an i. p. injection of 50 mg/kg Pentothal. Calibration: 200 μV, time: 100 msec.

Then, also under local novocaine anaesthesia, the sciatic and saphenous nerves in the left thigh and the radial nerve in the right forearm are exposed. After this we wait for 30—60 minutes. Small additional doses of curare are given if necessary.

Electrical potentials are recorded with electrodes fixed in electrode carriers of the stereotaxic apparatus. A bipolar recording of spontaneous activity is made with a distance of 0·5 to 1·5 cm between the electrodes. For monopolar recordings of evoked potentials a metal clamp fixed to the cut skin is used as the reference electrode.

Fast spontaneous activity is recorded with the shortest time constant available (0·01 sec) and with the H. F. filter completely switched off so that the upper part of the frequency range of the amplifier can be fully utilised. The sweep velocity or the rate of film movement must be adjusted to the frequencies recorded. For control purposes the recording from the cerebellum is compared with tracing from the surface of the cerebral cortex made under identical conditions with the same electrodes and same setting of the amplifier. The noise level is checked either at the end of the experiment by recording with the same electrodes, amplification and filtres several minutes after the death of the animal or during the experiment by substituting an equivalent resistance for the electrode and the animal.

In addition an ink recording may be made with an amplifier of the EEG apparatus connected in parallel to the same electrodes. Here another time constant (0·3—1 sec) is used and if necessary also muscle filters. If possible the ECoG is recorded simultaneously and the two records are compared.

The character of the cerebellar responses to electric stimulation of the sciatic and radial nerves is then determined, as well as that to acoustic and

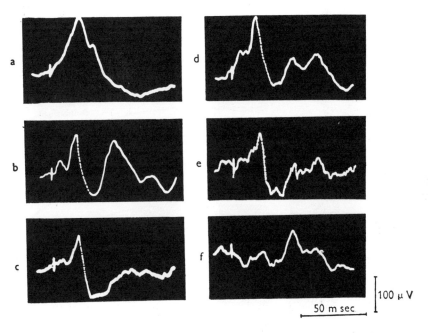

Fig. 269. Examples of typical evoked potentials in the cerebellar cortex. a — tibial nerve — anterior lobe. b — acoustic stimulus — vermis. c — radial nerve — paramedian lobe. d — stimulation of acoustic cortex — anterior lobe. e — stimulation of somatic cortex — vermis. f— visual stimulus — vermis. Positivity of active electrode upward.

optic stimuli and to electrical stimulation of the somatosensory, auditory and visual projection areas of the cortex. Short clicks are used as acoustic stimuli. These are produced by feeding rectangular pulses from the stimulator. into a loud speaker. Light flashes from a stroboscope serve as visual stimuli. By systematically moving the recording electrodes at 2 mm steps the extent of the electrically responding cerebellar cortex is determined for each stimulus. When recording evoked potentials a longer time constant is used (0·3—1·0 sec) and no filtration of fast components is applied. Responses are recorded with single sweeps synchronised to the stimulus and with camera shutter opened.

In the second phase of the experiment, spontaneous and evoked electric activity of the cerebellum are recorded during barbiturate anaesthesia. Fifty mg/kg Penthotal are given to an animal which is beginning to recover from the effect of curare and changes in the electrocerebellogram and electrocorticogram are compared. When anaesthesia has attained a constant level as shown by characteristic spindles in the cerebral cortex (cf. experiment IX C) mapping of cerebellar projection areas is repeated and again the shape of the responses and the extent of their projection are noted.

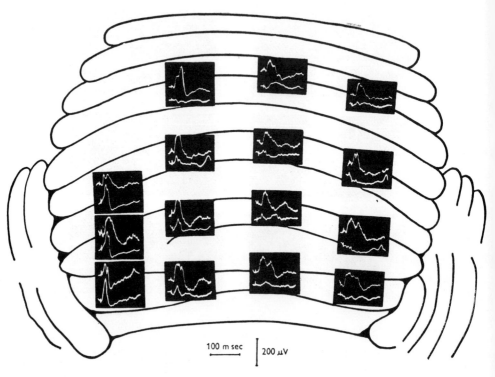

100 m sec   200 μV

Fig. 270. Projection of cerebellar responses to stimulation of the left superficial radial nerve. Recording from anterior lobe of the cerebellum in a curarised cat, before (above) and after (below) administering Pentothal. Positivity of active electrode upward.

Means described in other experiments may also be used to analyse cerebellar activity, e. g. microelectrodes (cf. p. 553), anoxia (cf. p. 561) and application of strychnine (cf. p. 546).

Results: The electrocerebellogram of the curarised unananaesthetised cat is characterised by fast low-voltage activity (Fig. 268), which can sometimes be distinguished only with difficulty from the noise of the amplifier. Such small waves lasting 2—4 msec and having an amplitude of 20—50 μV cannot be found in the ECoG registered in the same way. Under the influence of general anaesthesia this fast activity rapidly disappears. It is closely related to unit discharges of cerebellar cortical neurones (Purkinje cells and cells of the granular layer — Brookhart et al. 1950).

The ink recording of the electrocerebellogram with a longer time constant shows that this fast activity is superimposed on much slower waves that do not differ substantially from ECoG waves. The characteristic effect of barbiturates appears in the electrocerebellogram as well as in the ECoG,

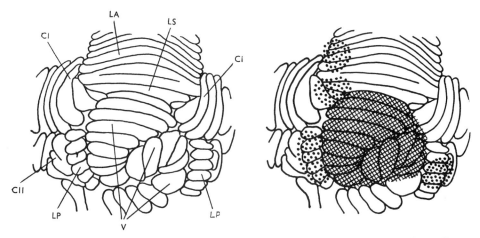

Fig. 271. a) Localisation of responses to different stimuli in the cat cerebellum. Cross-hatched area — auditory and visual projections. Dotted area — somatic projection. b) The main anatomical structures of the cat cerebellum. C I — crus I, C II — crus II, LA — lobus anterior, LP — lobus paramedianus, LS — lobus simplex, V — vermis.

in the form of increased amplitude and appearance of slower frequencies. Cerebellar and cortical potentials are, however, only rarely synchronous. This indicates that the subcortical structures controlling synchronisation of electrical activity of cortical structures of the telencephalon and rhombencephalon are not identical.

Evoked potentials, although showing considerable variability, are mainly surface positive waves lasting 10—20 msec, with an amplitude of 50—100 μV (Fig. 269). This positive wave is very similar to primary responses in the cerebral cortex and is usually followed by a slower and less pronounced negativity. The latent period depends on the structure stimulated. Stimulation of peripheral nerves evokes a cerebellar response with a latency of 10—20 msec, for acoustic stimuli the latency is 10—14 msec, for visual stimuli 40—50 msec and after stimulation of the cerebral cortex the response appears after 4—10 msec. The main positive deflection is sometimes preceded for 4—6 msec by a smaller (20—50 μV) positive wave or waves. The interpretation of the cerebellar potentials is not easy — mainly because of the complex convolution of the cerebellar surface into deep gyri and sulci. It appears that the small positive waves preceding the main component of the response are due to activation of the granular layer by arriving afferent impulses (Morin et al. 1957). The positive phase of the positive-negative response is an expression of arrival of afferent impulse into the cerebellar cortex and of postsynaptic activation of the granular layer. The negative phase, on the other hand, is due to activity of superficial elements of the cerebellar cortex (dendrites of Purkinje cells).

The projection of various afferent systems into the cerebellum is not strictly specific. Particularly in the unanaesthetised animal, the projection areas overlap considerably. Barbiturates, on the other hand, narrow the projection areas. Fig. 270 shows examples of the distribution of potentials evoked by stimulation of the left superficial radial nerve before and after Pentothal administration. The extent of the projection of other afferent systems according to Snider and Stowell (1944), Hampson (1949), Combs (1954), Morin et al. (1957) and our own results is shown diagramatically in Fig. 271.

Conclusion: Despite a certain structural similarity between the cerebral and cerebellar cortex, the latter is characterised by very intense fast spontaneous activity, the nature of which is not yet understood (Bremer 1958). The extensive overlapping of the primary responses in the cerebellar cortex indicates a diffuse distribution of the afferent inflow mediating convergence of impulses of heterogenous modalities.

D. Electric activity of the hippocampus

Problem: Record the spontaneous and evoked activity in the hippocampus of the rabbit.

Principle: The rhinencephalon is an important part of the telencephalon differing from other structures in the forebrain phylogenetically and morphologically (for literature see Gastaut and Lammers 1960). It is formed by:

1) bulbus olfactorius, tractus olfactorius, tuberculum olfactorium,
2) gyrus intralimbicus, which among others includes the hippocampus and gyrus dentatus, induseum griseum, striae longitudinales lateralis and medialis,
3) gyrus fornicatus composed of the gyrus hippocampi and gyrus cinguli.

Cytoarchitectonically the rhinencephalic cortical structures are characterised by a preponderance of pyramidal elements and the absence of granular layers. Very favourable conditions for studying electrical activity of different parts of neurones in extracellular recordings are thus created, especially in the hippocampus (allocortex). Ramón y Cajal (1909) and Lorente de Nó (1934) gave a detailed description of the morphology of the hippocampus. Starting from the ependymal surface facing the lateral ventricle, the following layers are found in the cornu Ammonis: the alveus, a 100—200 μ layer of tangential fibres formed partly by axons of pyramidal cells and partly by afferent fibres. A layer of basal dendrites (200—300 μ) follows and a layer of pyramidal cell bodies forming a narrow band (70—100 μ) parallel to the enpendymal surface. The apical dendrites at first have only few branches (250 μ), then, however, they form a dense net reaching into the preceding layer (600—700 μ). By stimulating afferent fibres ending on different parts of pyramidal neurones it is possible to observe the isolated response of dendrites and cell bodies and to compare the relationship between structure and response (Cragg and Hamlyn 1955, Euler et al. 1958). By studying the shape and latency of the responses to stimulation of different rhinencephalic structures, it is possible to clarify the functional and anatomical relations between them (Green and Adey 1956).

A special relation to the ECoG can be found in records of hippocampal activity registered with the conventional EEG apparatus in an unanaesthetised rabbit. The arousal reaction in the cortex is accompanied by very characteristic theta waves — regular waves of a frequency of 4—6/sec in the hippocampus (theta waves — Jung and Kornmüller 1939, Green and Arduini 1954). Transition to slow sleep EEG pattern, on the other hand, is accompanied by appearance of irregular asynchronous activity in the hippocampus. For review of literature see Green (1964).

Object: Rabbit, weight 2—3 kg.

Apparatus: A two channel D. C. or A. C. amplifier with an input resistance of 5 MΩ and a double beam cathode-ray oscilloscope. Overall sensitivity at least 100 μV/cm. A stimulator giving square wave pulses or condenser discharges synchronised with the time base of the oscilloscope. An isolation unit. A time marker. An EEG ink-writing apparatus with at least four channels.

Other requirements: Surgical instruments, trephines, dental drill. A stereotaxic apparatus with equipment for fixation of the head of the rabbit. Steel semimicroelectrodes with a tip diameter of 30—40 μ pointed electrolytically from the original diameter of 0·2 mm. Wick Ag-AgCl electrodes for corticography. Concentric needle electrodes for recording spontaneous activity of the hippocampus. 10% Pentothal. A respiration pump. A vacuum pump and glas pipettes for sucking off brain tissue. Physostigmine salicylate.

Procedure: The trachea of the rabbit is exposed under light Pentothal anaesthesia and tracheal cannula is introduced. After exposing and cleaning the skull bones an extensive craniotomy is performed above both hemispheres and the head is fixed in the holder of the stereotaxic apparatus. The dura is reflected and the larger pial vessels of one hemisphere are sealed with a cautery or with local application of $AgNO_3$. Pia is cut at first and the cortex lying above the lateral ventricle is then removed by suction using a fine pipette and minimal negative pressure. The hippocampus then appears at the bottom of the lateral ventricle. Finally the spinal cord is severed at CI-CII and artificial respiration is given.

By the microdrive of the stereotaxic electrode carrier, one steel needle electrode is advanced to a depth of 0·8 mm below the ependymal surface and fixed there. It is connected to the output unit of the stimulator. The second stimulating electrode (with a large surface) is fixed to the muscles or skin of the neck.

Another electrode is then fixed into the electrode carrier and inserted into the hippocampus parallel to the stimulating electrode, 1·5 to 2·0 mm from the latter. The reference electrode is placed on the cut skin. The needle electrode is pushed down in 0·1 mm steps, and at each position a recording is made of responses to a threshold stimulus and a stimulus about three times as strong. We attempt to decrease stimulation artefacts first by changing the mutual position of the stimulating or reference electrodes.

During the next phase of the experiment the connection of the electrodes is reversed. The fixed electrode (0·3 mm deep) is connected to the amplifier input and the movable electrode to the stimulator output. Stimulation is again performed at 0·1 mm intervals with an intensity producing a threshold reaction when the stimulating electrode is at a depth of 0·1 mm.

Finally, without changing the position of the fixed recording electrode, we attempt to determine with this electrode the response to stimulation of other rhinencephalic structures — e. g. area entorhinalis, fornix, gyrus cinguli etc. A similar experiment may be performed in another part of the hippocampus with the fixed electrode in another position.

If the state of the animal is satisfactory and if it has quite recovered from the effect of Pentothal, the arousal reaction can be demonstrated in the other, as yet untouched, hemisphere. Otherwise another rabbit must be used for this experiment. Here no craniotomy is performed. Only small trephine openings are made for introducing the deep needle electrodes and cortical screw electrodes. Using the stereotaxic apparatus, concentric recording electrodes are pushed down into a region of the hippocampus symmetrical to the exposed cornu Ammonis. Wick Ag-AgCl electrodes for recording the ECoG are placed on the exposed surface of the hemisphere. Spontaneous activity of the hippocampus is registered for a few minutes together with the spontaneous ECoG. When the recording from the cortex shows typical slow sleep activity, sensory stimuli (tactile stimulation of the skin of the cheek, acoustic stimuli, olfactory stimulation — blowing of aromatic substances into the nostrils) are used to produce an arousal reaction, and the electric response in the cerebral cortex and hippocampus is studied. Finally 1 mg/kg physostigmine salicylate is injected intraperitoneally and EEG is observed for 20 minutes.

Results: Fig. 272B shows, in a monopolar recording from various depths, the response to stimulation of a point 2 mm away and 0·8 mm deep. Fig. 272A also shows diagramatically the individual layers of the hippocampal cortex. The shortest latency of the response is at a depth of 0·8 mm. This indicates a tangential course of nerve fibres from the stimulation locus to the apical dendrites at the recording site. The response is characterised by a nega-

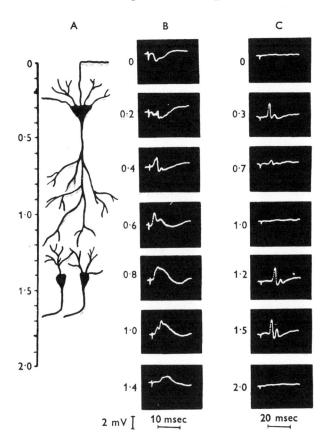

Fig. 272. Hippocampal responses to electrical stimulation. A — diagram of pyramidal cells in the hippocampus and g. dentatus, scale in mm. B — response from different depths below the dorsal surface of the hippocampus to stimulation of a point 0·8 mm deep. C — responses from a depth of 0·3 mm to stimulation at different depths (figures on side). Negativity of active electrode upward.

tive wave lasting for about 15 msec, which is well in evidence, particularly with weaker stimuli. It is registered as a positive wave in superficial leads (in the region of basal dendrites and axons), evidently because is does not pass beyond the level of cell bodies. If this wave attains a certain amplitude one or more spikes are formed on it. These have an amplitude several times larger and a duration of 2—3 msec. Their spread can be followed down to the layer of basal dendrites. Maximum amplitude is reached immediately above the region of cell bodies. It then decreases, evidently because of the scatter of axons in the alveus. In the region of apical dendrites, the negative spike disappears in slow potential waves.

Similar results are obtained when moving the stimulating electrode (Fig. 272C). The response with the shortest latency is obtained when stimulating at a depth corresponding to the position of the recording electrode. From deeper or more superficial positions the latent period is longer. When using slightly suprathreshold stimuli, hippocampal responses disappear when the stimulating electrode reaches a depth of about 1 mm. If, however, it is pushed further down, the response again reappears when stimulation begins to reach the gyrus dentatus (depth 1·2—2·0 mm).

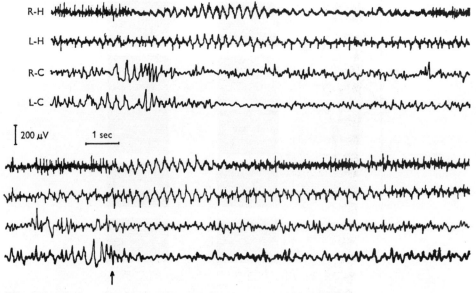

Fig. 273. Bipolar recording of cortical and hippocampal activity in the unanaesthetised rabbit. R-H, L-H — recordings from right and left hippocampus. R-C, L-C — recordings from right and left hemisphere. Top: spontaneous arousal. Bottom: arousal evoked by tactile stimulus (arrow).

Characteristic reponses in the hippocampus may be obtained by stimulating nearly all rhinencephalic structures. From the latent periods at different depths, one may conclude whether termination of corresponding afferent axons is on apical or basal dendrites. The responses to stimulation of the fornix, fimbriae and gyrus cinguli are most easily elicited. The hippocampus also responds to stimulation of peripheral nerves and to acoustic stimuli but responses are obtained only if the intervals between stimuli are very long (Green and Adey 1956).

A characteristic record of spontaneous hippocampal activity and ECoG is shown in Fig. 273. From it the reciprocal relationship between the two structures is evident. While sleep spindles occur in the cortex, the activity of the hippocampus is relatively asynchronous. In the intervals between cortical spindles, however, it is strikingly regular (4—6/sec, 100—200 μV). The theta activity is even more pronounced during the arousal reaction characterized by desynchronization in the EEG. As it is best developed during paradoxical sleep, it may be regarded as the rhinencephalic counterpart of neo-cortical desynchronization. After i. p. injection of physostigmine (Brücke et al. 1957) the theta waves are nearly sinusoidal and last without interruption for minutes. In a freely moving animal theta activity accompanies orienting re-action and approach behavior (Grastyán 1959). At present it is not possible to give a satisfactory explanation of its physiological significance, particularly with respect to the activity of the cerebral cortex.

Conclusion: A study of electrical activity of the hippocampus permits a better understanding of electrical phenomena in different parts of the neurone and contributes to the elucidation of physiological relationships between the rhinencephalon and the other parts of the telencephalon.

E. Nonspecific subcortical influences on the cerebral cortex

Problem: Record the electrical phenomena evoked in the cerebral cortex by stimulation of nonspecific systems in the medulla oblongata, mesencephalon, thalamus and the caudate nucleus.

Principle: Electric stimulation (100—300/sec) of some subcortical structures changes the slow sleep type of ECoG (slow waves, spindles — cf. p. 516) into the waking type simultaneously throughout the cortex in an animal under light chloralose anaesthesia or in the unanaesthetised "encéphale isolé" preparation. Electrocorticography and the behaviour of the animal show that it is aroused as after stimulation of peripheral sense organs. Moruzzi and Magoun (1949), Lindsley et al. (1949, 1950), French and Magoun (1952), French (1952), French et al. (1953), Narikashvili (1953), Segundo et al. (1955) and others using stimulation and extirpation, determined the extent of the corresponding areas of the brain mainly in the ventromedial reticular formation of the medulla oblongata, the reticular formation of the mesencephalon, sub-thalamus (zona incerta, corpus Luysi, Forrel's field) and in nonspecific nuclei of the thalamus.

Slow stimulation (6—10/sec) of nonspecific thalamic nuclei in the cat under barbiturate anaesthesia produces a special type of generalised ECoG reaction (Dempsey and Morison 1942a,b, 1943, Morison et al. 1943, Morison and Dempsey 1942). The response can be found over the whole cortex especially in the associative areas, its latency is 20—35 msec and it is predominantly surface negative. If stimulation of the same intensity lasts for several seconds, the responses show a gradual increase (hence the term "recruiting response") and then decrease in amplitude ("waxing and waning"). The work of Jasper (1949, 1954), Jasper et al. (1946, 1952a, 1955), Starzl et al. (1951, 1952) and Verzeano et al. (1953) exactly defined those areas in the thalamus that produce this type of reaction when stimulated: n. centralis lateralis, n. paracentralis, n. parafascicularis, n. suprageniculatus, centrum medianum, nucleus centralis medialis, n. submedius, n. rhomboideus, n. reticularis thalami, n. ventralis anterior. EEG synchronizing structures in the lower brain stem were described by Batini et al. (1959), Magni et al. (1959) and Magnes et al. (1961).

The recruiting response is similar in many respects to spontaneous barbiturate or sleep spindles (Morison et al. 1943) characteristic for the EEG of anaesthetised or drowsy animals. Spindles may also be evoked by stimulation of the nonspecific thalamus and especially of the head of the caudate nucleus by single electric shocks (caudate spindles — Shimamoto and Verzeano 1954, Jung and Tönnies 1950).

The relationship between the bulbo-ponto-mesencephalic activating system and the diffuse thalamic system is not quite clear. According to some (Starzl et al. 1951) these are two components of the same complex. Others (Gellhorn 1952, 1953a, b, Gellhorn et al. 1954) maintain that two opposing systems exist, an activating (reticulohypothalamic) and an inhibiting (purely diencephalic). For review of literature see Buser (1957), Rossi and Zanchetti (1957), Bremer (1961).

Object: Cat weighing 3—4 kg.

Apparatus: Ink writing EEG apparatus. A cathode-ray oscilloscope that can be connected to the output of the EEG channels. A stimulator giving rectangular pulses with a frequency of 0·5 to 500 cycles/sec. An isolation unit for the stimulator. A time marker.

Further requirements: Surgical instruments, trephines, dental drill. Stereotaxic apparatus for the cat. Screw electrodes for corticography (diameter 2 mm). Bipolar needle electrodes for subcortical stimulation (steel needles, diameter 0·1—0·2 mm, length 5 cm, distance between the tips 1 mm).

Procedure: 1) Encéphale isolé preparation (Bremer 1935, 1936a,b). The cat is anaesthetised with ether. First the subcutaneous and later also deep tissues of the neck are infiltrated with 1—2% novocaine. After introducing a tracheal cannula, the cat is fixed in the stereotaxic head holder with the head elevated and a long sagittal skin incision is made from the frontal bones to about 3—4 cm behind the occipital protuberance. The neck muscles are dissected away from the occipital region of the skull and the atlas. The occipital joint is exposed and the atlanto-occipital membrane is cut between the occipital bone and the atlas. Through the opening thus formed a blunt leukotomy knife is inserted and the spinal cord is severed. Spinal shock may be prevented by chilling the cord or by local application of anaesthetics at the level of transsection. Care is taken not to injure the spinal arteries so as to preserve normal blood supply to the brain base. Immediately after the spinal cord is transsected artificial respiration is applied (17—20/min). It is suitable to use a thermostatic animal board and to introduce a cannula connected to a mercury manometer into the femoral artery for measuring the blood pressure. If the last falls below 80 mmHg, continuous i. v. infusion of adrenaline is given.

The cortical screw electrodes are fixed to the exposed skull bones in the usual way (experiment IX B). The bipolar needle electrodes are inserted according to the stereotaxic atlas into the bulbopontine reticular formation and fixed to the bone with phosphate cement.

Following all surgical procedures, administration of ether is interrupted. Only an hour later, when the last effect of ether has disappeared is the actual experiment commenced. The type of EEG activity is decisive, as only in a drowsy state can a clear cut arousal be obtained. If sleep pattern is pronounced, the reticular formation is stimulated (100—300 cycles/sec, 0·5—1·0 msec, 2—5 sec). A voltage of 0·5 V is used first, and the stimulus intensity is increased to 5 V gradually, with successive stimuli separated by about 1 minute intervals. That threshold stimulus voltage is sought which will produce a pronounced arousal reaction in the cortex. With suprathreshold stimuli, the duration of EEG desynchronisation after cessation of stimulation is observed. We attempt to determine the extent of the reticular area from which the ECoG can be activated by moving the electrodes vertically or inserting them in other frontal or parasagittal planes.

In experiments with recruiting responses and synchronized spindles the same preparation is used. The cortical electrodes remain in place without change. The metal frame of the stereotaxic apparatus serves as reference electrode. The paired stimulating electrodes (of the same shape as when stimulating the reticular formation) are inserted according to the coordinates of the stereotaxic atlas into the region of the nucleus ventralis anterior and nucleus centralis medialis, or into the head of the caudate nucleus. Stimulation is commenced several minutes after introducing the subcortical electrodes. Trains of stimuli of 2 sec duration (frequency 6—9/sec, pulse duration 0·5—1·0 msec, voltage 2—10 V) are used for eliciting the recruiting response. As far as possible the response is recorded simultaneously from several cortical electrodes. By changing the localisation of the stimulating electrodes, the depth and, if possible, other coordinates of the thalamic structures giving the recruiting response are determined. Single shocks (0·1 msec, 10—20 V) to the caudate nucleus are used to elicit the spindle response.

If two stimulators are available, we may attempt to stimulate simultaneously thalamic structures producing a recruiting response and the reticular activating system. When the recruiting response is pronounced the reticular activating system is stimulated for 2—3 sec with a stimulus producing a good arousal response. Changes in cortical potentials are observed.

The following experiments are further suggested:

a) Registration of cortical response to stimulation of the reticular formation by single stimuli or stimuli of low frequency (up to 10/sec).
b) The effect of high frequency stimulation (100—300/sec) of the nonspecific thalamus on the sleep EEG pattern.
c) The effect of stimulating the reticular formation on primary cortical responses produced by stimulating peripheral nerves or receptors.

Results: Periods of slow sleep activity alternating with periods of desynchronization can be seen in the spontaneous ECoG of the "encéphale isolé" preparation. Since the movements of the animal are restricted to the facial

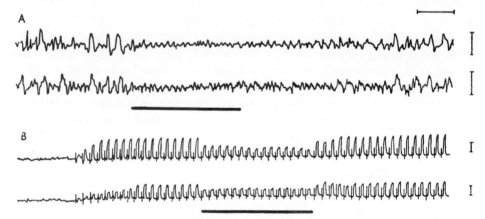

Fig. 274. The effect of reticular stimulation (300/sec, 1 msec, 3 V) on sleep activity in the cerebral cortex (A) and on the recruiting response (B). Cat, encéphale isolé preparation. Calibration: 1 sec, 200 μV.

and eye muscles, it is sometime difficult to decide whether the desynchronized EEG corresponds to the arousal or to the paradoxical sleep (see chapter IX B). The two conditions can be easily distinguished according to the state of pupils, which are mydriatic during arousal and extremely miotic during paradoxical sleep, with occasional phasic dilations synchronized with the rapid eye movements (Berlucchi et al. 1964). Since the preparation receives afferent impulses from cranial nerves it can be aroused by sensory stimulation particularly by acoustic and trigeminal stimuli. It, therefore, happens sometimes that nociceptive stimuli due to the surgery and fixation of the head in the stereotaxic head holder maintain the preparation in a state of wakefulness. Usually local anaesthesia of all sites that might be a source of pain and a small dose of chloralose (5—10 mg/kg i. p.) produce the required level of sleep activity.

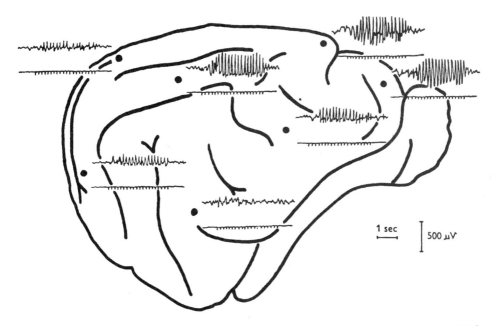

Fig. 275. Recruiting response in different areas of the cerebral cortex in the cat (Dial anesthesia). Stimulation of homolateral centrum medianum.

Fig. 274A shows the effect of stimulating the reticular formation in the left half of the midbrain on the ECoG of both hemispheres in a drowsy animal. Both hemispheres are activated equally. With threshold stimuli (decrease in stimulus voltage or frequency), however, activation may be homolateral. More intense stimuli produce fast low activity persisting for a longer time after termination of stimulation. Different threshold values correspond to different localisations of the stimulating reticular electrodes so that it is possible to map the extent of the areas giving optimum results.

The effect of slow stimulation of the nonspecific thalamic nuclei is shown in Fig. 275. A gradual increase in amplitude of responses and their diffuse projection are characteristic signs. The extent of cortical projection as well as the shape and latency of the response is, of course, considerably dependent upon the localisation of the thalamic electrodes. Hence, in addition to well defined generalisation, we find cases with predominating frontal or occipital distribution of the responses. The response is usually bilateral but more pronounced on the side of the stimulus. In monopolar recordings, increasing cortical responses are surface negative and attain their maximum amplitude in associative areas of the cortex, in the cat, particularly in the suprasylvian gyrus and the sensorimotor cortex. When stimulating the nucleus ventr. ant.,

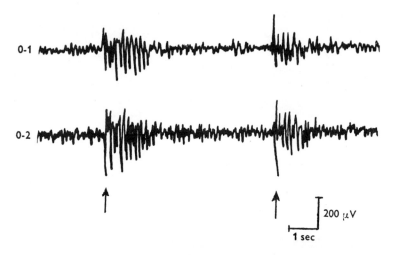

Fig. 276. Spindles evoked in both suprasylvian gyri in the encéphale isolé cat by single shock stimulation of the caudate nucleus.

the latency is about 4—5 msec. Stimuli acting on the centrum medianum or n. centralis med. produce a response only after 20—40 msec.

A single electric shock to the head of the caudate nucleus elicits a positive-negative evoked response (latency about 10 msec) followed after 200—300 msec by a train of high voltage oscillations (frequency 7—10/sec, duration 1—2 sec), which can be recorded from the same cortical region as the recruiting response. An example is given in Fig. 276.

If a stimulus with a frequency of 200 cycles/sec is introduced into the thalamic electrodes giving a recruiting response, a typical arousal reaction is elicited. Slow stimulation (6—9/sec), on the other hand, with electrodes placed in the reticular formation of the mesencephalon does not produce a recruiting response nor is usually a clear cut cortical response observed.

Stimulation of the reticular activating system affects in the same way the so called augmenting response, a reaction similar in character to the recruiting response but produced by stimulation of specific thalamic nuclei (Morison and Dempsey 1942, Gauthier et al. 1956) and limited to the corresponding projection area of the cortex.

Primary cortical responses to stimulation of peripheral nerves are either not fundamentally changed by stimulation of the reticular activating system (Moruzzi and Magoun 1949) or are decreased (Hernández-Péon and Hagbarth 1955). Cortical primary responses evoked by electrical stimulation of the thalamic relay nuclei are, on the contrary, facilitated by reticular stimulation (Bremer and Stoupel 1959, Dumont and Dell 1960). Associative secondary responses, on the other hand (Albe-Fessard and Rougeul 1955, 1956, Buser and Borenstein 1956), which are especially pronounced in the suprasylvian gyrus after large doses of chloralose, disappear almost completely on reticular stimulation.

Conclusion: The nonspecific system of the thalamus and the bulbomesencephalic reticular activating system represent complex polysynaptic structures controlling the functional state of the cerebral cortex and especially of its associative regions.

F. Unit activity of reticular neurons

Problem: Make an extracellular recording of spontaneous activity of reticular neurons and of its changes caused by application of sensory stimuli and by direct electrical or chemical stimulation. Show convergence of stimuli onto reticular cells and compare their relative effectiveness.

Principle: Individual nerve cells represent variable electric dipoles. The electric field that extends around them in the external environment can be examined with an electrode having a tip diameter of about $1\ \mu$. Since in $1\ mm^3$ of nervous tissue there are about 10^4 neurons (Bok 1959) it may often happen that the tip of the electrode is affected by the electric fields of several units. Hence successful recording requires good isolation of the activity of one cell the action potential of which must be considerably higher than the discharges of other elements. Optimum recordings can be obtained from large cells that produce up to 100 times larger extracellular current flow than small neurons or thin axons (Tasaki et al. 1954). By suitably moving the microelectrode a position may be found where the action potential of a given unit is maximal while the amplitude of other unit potentials is minimal.

Electrodes gradually inserted into nervous tissue reveal the presence of spontaneously active neurons that continuously generate irregular trains of output activity. Other, so called "silent" neurons start firing only if stimulated or injured by the approaching tip of the microelectrode. The random sequence of impulses of spontaneous activity can be significantly influenced by the action of external stimuli which either increase or decrease their frequency. In particular, reticular neurons react intensively to stimuli acting on different regions of the surface of the body or to vestibular or auditory stimuli and hence we are fully justified in terming such neurons polysensory. The reaction sometimes appears as a very conspicuous and characteristic change, often, however, it is lost in the fluctuations of spontaneous activity and then statistical analysis has to be applied to detect it.

In addition to external stimuli a number of central influences also affect the activity of reticular neurons, particularly those mediated by cerebelloreticular and corticoreticular connections. The activity of a given neuron can, of course, be affected not only synaptically but also by immediate electrical or chemical stimulation. For this purpose we use either the recording electrode itself or (when using multibarrelled capillary electrodes) parallel microelectrodes. Chemical stimuli (electrolytes) are applied iontophoretically.

Object: Rat weighing 200 g.

Apparatus: A double beam cathode-ray oscilloscope with an A. C. amplifier (with possibility to cut off frequencies below 200 c/sec, total amplification at least 50 μV/cm). An input cathode follower with an input resistance above 1000 MΩ and a grid current below 10^{-11}A. A stmulator giving single rectangular pulses and permitting in connection with relay circuits the application of nonelectric stimuli. A radio frequency isolation unit for the stimulator. A sensitive galvanometer (10^{-10}A/mm/m). A time marker. Calibrator. A two channel tape recorder.

Other requirements: Surgical instruments, circular trephine (diameter 5 mm). Stereotaxic apparatus for the rat with vertical microdrive for introducing the microelectrode (minimum sensitivity 100 μ per revolution). A respiratory pump (open system, 60/min). A tracheal cannula for the rat. D-tubocurarine. Ether. Novocain 1%. Metal (steel or tungsten) microelectrodes. Capillary microelectrodes filled with 3м-KCl or saturated Na-glutamate. A pulling machine for preparation of glass capillary electrodes and the set up for filling them (see p. 208). Electrodes for stimulating the sciatic nerve. A tone generator and miniature earphones for applying sound stimuli. An electromagnetic valve and compressed air for stimulating the snout with puffs of air.

Procedure: The sciatic nerve of the rat is exposed under light ether anaesthesia and a circular trephine opening is made above the cerebellum (5 mm in diameter). Within the opening the dura mater is carefully removed. After tracheotomy and introduction of the tracheal cannula novocain is infiltrated into all the exposed wound edges and pressure points and the animal is fixed in the stereotaxic apparatus. Ether application is ceased, 2 mg/kg D-tubocurarine is applied and artificial respiration (60/min) is started. Pressure is regulated in such a way that respiratory movements are evident but not too intensive. If necessary the temperature of the rat is checked and regulated by a D. C. heater. Also heart rate monitoring usually proves indispensable.

The rat is earthed via a large Ag-AgCl electrode placed on a skin flap or on an exposed muscle. There is no danger of muscle artefacts since curare has been applied. The microelectrode is fixed to the micromanipulator of the stereo-taxic apparatus and is connected via a screened flexible cable to the input of the cathode follower. This connection is kept as short as possible. The screen-ing of the input cable is connected to the cathode of the cathode follower (not to earth, see p. 156). Metal electrodes of low resistance (below 0·3MΩ) can be connected directly to the input of the preamplifier if the latter's input impedance

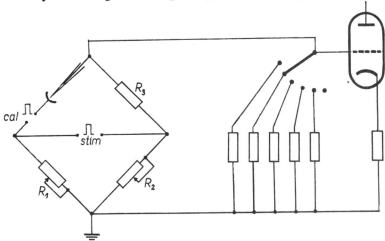

Fig. 277. A bridge circuit for elimination of the stimulation artefact when stimulating and recording with the same microelectrode. For details see text.

is larger than 10MΩ. The microelectrode is than placed by means of the controls of the stereotaxic apparatus above the exposed cerebellar surface in the frontal plane AP 10, about 0·5 to 1·5 mm laterally from the midline. The moment when the tip touches the cerebellar surface cannot usually be observed directly but it can be followed electrically. As long as the electrode is above the brain surface the input of the cathode follower is open and there is noise with a consider-able 50/sec component in the recording. As soon as the microelectrode touches the brain surface there is a sudden decrease in resistance and noise. Since the electrode may be destroyed when touching the brain (broken or clogged) it is of advantage to monitor continuously its resistance by introducing periodically rectangular pulses (1 mV, 1 msec) into the circuit of the indifferent electrode. These impulses appear on the oscilloscope as soon as the electrode touches the brain surface. Their amplitude V is related to the initial amplitude of the calibrating signal V_c as is the resistance of the microelectrode R_m to the total resistance of the input circuit of the cathode follower $R_m + R_g$:

$$V : V_c = R_m : (R_m + R_g)$$

If it is possible to change the grid resistance R_g in exponentially increasing steps from 1 to 100 MΩ (Fig. 277) then the value R_g can roughly be determined for which $V = V_c/2$. Under these conditions (as follows from the above equation) $R_g = R_m$.

Using the microdrive of the stereotaxic apparatus the microelectrode is slowly inserted into the brain tissue. As long as the tip is in the region of the cerebellum (to a depth of 5 mm) a high frequency high voltage activity is recorded and several active units usually are within the range of the microelectrode. At a depth of 5—6 mm the recording suddenly changes: after a short interval of complete silence (the electrode is passing through the IVth ventricle) well isolated units with an amplitude of up to several mV appear on a relatively silent background. For prolonged recordings neurons somewhat lateral to the trajectory of the microelectrode are most suitable: as the microelectrode is gradually lowered the amplitude of the spikes first increases to a maximum and then decreases again and this whole change occurs, depending on the size of the cell within a range of 100 μ. These neurons are less endangered by possible movements of brain tissue, due mainly to changes in blood pressure, which are a particularly disturbing factor near the exposed brain surface. In the deep parts near the base of the brain, on the other hand, such movements are only slight. Other ways of limiting undesirable movements of brain tissue are used particularly in larger animals (hermetic closure of skull together with the electrode, pneumothorax, decompression etc. — Frank and Becker 1964).

After finding a well isolated unit with a spike frequency of 2—10/sec its reaction to external stimuli is tested. A 1 sec time base is set on the oscilloscope and is triggerred by a synchronizing pulse of the stimulator. Stimuli are applied about 300 msec later so that for each sweep resting activity is obtained during the first third of the record and activity influenced by the stimulus during the remaining two thirds. In this way it is easy to recognize whether the stimulus has caused an increase or a decrease in the frequency of the neuron studied. A more detailed analysis is performed from a photographic recording (a continuously moving film with a speed of at least 2 cm/sec) or from a tape recording which is easily obtained by connecting the resistance input of a conventional tape recorder in parallel with the plates of the oscilloscope (via a small condenser of about 100 pF). In both cases the second channel is used to record the synchronizing pulses without which results could not be evaluated. More detailed instructions on how to evaluate unit activity are given in chapter X G.

The stimulus is either electrical stimulation of the sciatic nerve (rectangular impulse, 1 msec) or nonelectrical stimuli gated by the stimulator. In the latter case a positive rectangular impulse lasting about 0·5 sec is preset on the stimulator and led to an electromechanical relay which controls a sound stimulus or opens an electromagnetic valve allowing a current of air to stimulate the trigeminal region.

A neuronal reaction can also be induced by direct electrical stimulation or by microelectrophoresis. The output of the stimulator is connected through a galvanometer and a series resistance R_1 (100 MΩ) to a microelectrode (Fig. 278) which is connected to the input of the cathode follower via a small condenser C_2 (50—100 pF). Possible noise at the stimulator output is removed by connecting a large condenser C_1 (10 μF) in parallel to the stimulator. This

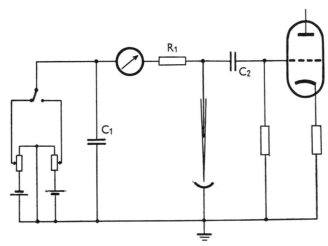

Fig. 278. Arrangement used in experiments with microelectrophoresis. On the left: batteries and potentiometers controlling the braking and injecting currents. For details see text.

condenser allows long polarizing pulses to pass but filters off short irregularities. When using this simple circuit brief spikes are obtained at the input of the cathode follower when switching the polarization on and off. Their duration depends mainly on the value of C_2. Even though such stimulation artefacts are easily distinguished from action potentials in more rapid recordings, they can nearly completely be suppressed if the stimulating current is applied using a bridge circuit shown in Fig. 277. Unit activity is recorded through the left part of the bridge where the microelectrode is connected against the indifferent electrode connected to earth via a calibrator and a small variable resistance R_1. The right part of the bridge serves for stimulation. It consists of a constant resistance R_3 (100 MΩ) connected to the microelectrode and a variable resistance R_2 connected to earth. The stimulator (which must be well insulated from earth) is connected to the horizontal diagonal of the bridge. The bridge is balanced by setting the variable resistors R_1 and R_2 so that

$$R_m : R_3 = R_1 : R_2 .$$

If this condition is fulfilled the potential at the input of the cathode follower (connected to the vertical diagonal of the bridge) remains unchanged against earth during application of the electric stimulus. This makes it possible to use D. C. amplification in the whole recording system.

After termination of the experiment the position of the metal microelectrodes is marked by microlesions (D.C. current of about 10 µA for 1—2 min). It is much more difficult to mark the localization of glass microelectrodes unless these are filled with staining solutions (carmin, ferricyanide). A simple method is to squash the glass capillary with a strong pinsette at the point where it enters the brain. The animal is then injected with 10% formaline and fixed with the tip of the microelectrode remaining in the brain. During fixation a relatively well visible track is formed where the electrode was and this can be distinguished, even after removal of the capillary, up to a distance of several tens of µ from the tip itself so that the localization of the electrode can be roughly determined.

Results: The action potentials of individual nerve cells recorded with an extracellular microelectrode usually appear as negative or positive-negative waves sometimes continuing in a further small positivity (Fatt 1957a, b, Hubel 1960, Bishop et al. 1962a, b). The presence of an active neuron is first indicated by negative potentials, the amplitude of which gradually increases as the electrode approaches the active membrane (Fig. 279). The initial positive wave increases even more rapidly and may become the most pronounced part of the recording when the electrode is very close to the membrane of the

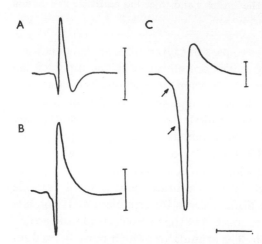

Fig. 279. Changes of the unit potential shape as the microelectrode approaches the neuron. Calibration: 2 msec, 500 µV. The arrows indicate the inflection points on the ascending branch of the action potential. Positivity downward.

neuron. As long as the electrode is at some distance from the cell surface it is affected during the action potential mainly by the current flow into the region of the soma. Nearer to the cell surface the somatic membrane starts to play a role also as a current source for synaptic depolarization of dendrites and also for spikes arising in the region of the initial segment. This is reflected in the positive component of the unit potential in the ascending branch of which sometimes several discrete steps (Fig. 279C) can be distinguished. The positive wave is followed by a negativity when the somatic membrane is depolarized. When the somatic membrane is injured by the microelectrode, the electrode is in the region of the current source also at the height of the action potential — this is a sort of transient state between extra- and intracellular recording, characterized by purely positive waveforms. With fine microelectrodes similar responses may also be recorded from axons. These are suddenly appearing positive potentials with a very steep rising phase which are probably recorded when the electrode tip is in direct contact with the axon or with its myeline sheath (Hubel 1960, Bishop et al. 1962a, b.).

For an evaluation of the reactions of reticular neurons the frequency of action potentials is, of course, more important than their shape and amplitude. Spontaneous activity is usually about 10/sec but there are units that maintain a frequency of about 50/sec for a long time. Activity of too high a frequency and regularity often indicates cell injury. The amplitude of reticular units is not stable but usually varies with changes in the position of the microelectrode in relation to the nerve cell (e. g. during breathing) or with larger changes in frequency (in a train of spikes occurring rapidly one after the other the amplitude gradually decreases).

Reticular neurons usually are highly reactive to nociceptive and somatosensory stimuli. Stimulation of the sciatic nerve induces a pronounced increase (more rarely decrease) in the frequency of discharges in most neurons and this lasts for up to several seconds. Trigeminal stimuli have a similar effect (particularly stimulation of whiskers by air puffs). Acoustic stimuli are much less effective. By determining the response of a single neuron to different stimuli we ascertain the degree of convergence, the relative effectiveness of different stimuli and their interaction (Scheibel et al. 1955, Mancia et al. 1957, Bell et al. 1964). Examples of some typical reactions are shown in Fig. 280. It must be emphasized that the statistical character of the responses makes it necessary to use statistical methods for evaluating the different reactions.

Polarization of a nerve cell may cause a rise or fall in the frequency of unit activity (Fig. 281A). The final effect is determined by the change in membrane potential induced by the polarizing current. If the distance between the electrode and the cell surface is large, only a small part of the stimulating current passes through the intracellular space. When using a positive pulse the current enters the region of the soma and causes hyperpolarization of the membrane but depolarization in the more distant parts of the neuron where it again leaves the cell to enter the extracellular space. A negative pulse, on the other hand, causes depolarization of the somatic membrane. If the electrode is very close to the cell body a considerable part of the current passes through the intracellular space and leaves it via the somatic membrane. Here conditions are very similar to intracellular application of the current — the positive pulse decreases the positive potential of most of the membrane while the negative pulse causes hyperpolarization of the membrane. If depolarization in the most sensitive region of the neuron attains the threshold value an action potential is generated with subsequent repolarization. The membrane potential again attains the discharge level at a rate depending on the intensity the applied current which determines the output frequency of impulses. Action potentials are generally formed according to the principle of a relaxation oscillator (a dripping water tap, a thyratron stimulator — see p. 107). Excessive depolarization leads to supraliminal (Wedensky) inhibition after a short discharge of rapidly decreasing action potentials.

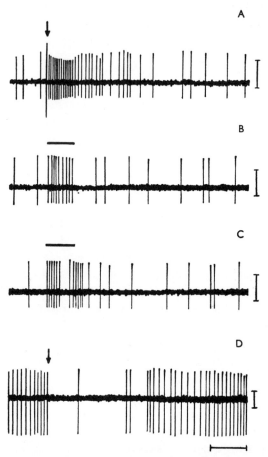

Fig. 280. Excitatory reactions of a reticular unit to sciatic nerve stimulation (arrow — A), to an acoustic stimulus (200 c/sec, horizontal bar — B) and to a trigeminal stimulus (air puff on the snout, horizontal bar — C). D — inhibitory reaction of another reticular unit to sciatic nerve stimulation (arrow). Calibration: 1 sec, 200 μV.

Conditions are more complex if the microelectrode is filled with an electrolyte solution, the ions of which can be applied to the cell membrane by electrophoresis and have a chemical effect there. Since about 10^5 Asec is required for the release of one gramequivalent of ions, about 10^{-14} gramequivalents are transported through the tip of the microelectrode for every nanoamperesec of the applied current. The total current measured by the galvanometer is carried both by anions and cations moving in opposite directions. The current fraction carried by anyone ion is characterized by the transport number which depends upon ion mobility and activity and is influenced by the presence of other ions (especially H⁺ and OH⁻) in the solution. Since also the electrical properties of the electrode tip must be taken into account, transport numbers may considerably differ from 0·5. This together with the uncertain geometry of the microelectrode — nerve cell relationship makes even approximate estimation of the concentration obtained in the tissue extremely difficult. Usually currents of 10—100 nanoamps cause after a few seconds (if the substance is

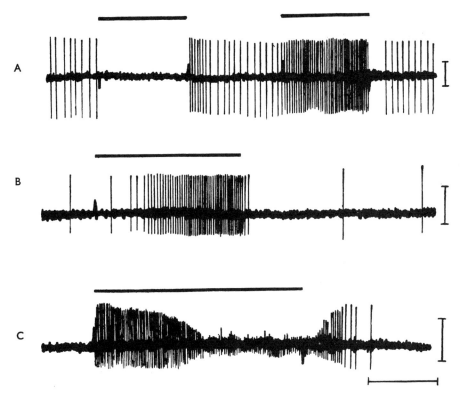

Fig. 281: Reactions of reticular units to polarizing currents (A) and to microelectrophoretic application of glutamate (B, C) through the recording microelectrode. A — the first horizontal bar indicates positive, the second bar negative polarization. B, C — horizontal bars indicate glutamate ejection using 5 and 40 nanoamp currents respectively. Calibration: 5 sec, 200 μV.

effective) pronounced changes in unit activity (Fig. 281B, C). Application of some aminoacids elicits particularly marked effects. In particular, glutamate causes pronounced depolarization of the cell membrane and a corresponding increase in unit firing (Curtis et al. 1960, Curtis 1964, Krnjevic 1964, Salmoiraghi 1964).

It is often not easy to distinguish between the direct and the iontophoretic effects of the current, since the final result may be a combination of both. The

623

iontophoretic effect usually appears after a latent period of up to several sec and is preserved even after the current is switched off. The part played by the current itself can best be evaluated by using multibarrelled electrodes (Curtis and R. M. Eccles 1958) filled with different substances. The current applied via the barrel filled with NaCl has mainly an electric effect, since the electrophoretically injected quantities of Na^+ or Cl^- do not considerably change the concentration of these ions in the extracellular fluid.The same current applied via the barrels filled with active electrolytes has a more pronounced electrophoretic effect in addition to the electric one. When applying drugs electrophoretically it must be borne in mind that substances diffuse out from the open end of the microelectrode even without application of any current. When using ultramicroelectrodes this can be neglected while with electrodes having a lower resistance diffusion can easily be prevented by application of a constant braking voltage which helps to retain ions of the corresponding polarity within the microelectrode (Fig. 278).

Conclusion: Recordings of the activity of individual neurons considerably enlarge the possibilities of electrophysiological analysis of the functions of higher parts of the brain particularly by enabling us to study the whole scale of reactions of individual neurons participating in the average potentials recorded by macroelectrodes. Unit activity also reflects the effect of long lasting stimuli which lead to evoked macroresponses only at their start and termination. Unit recording opens up new perspectives for explaining the principle of coding of information in sensory systems, for obtaining data on the convergence of afferent impulses onto individual neurons and on the mutual interaction of different influences which is the basis of the functional organization of the nervous system. A more general interpretation of the results obtained is, of course, not simple, since it is necessary to predict the characteristics of a complex system on the basis of the behaviour of its individual elements which are connected by complex mutual relationship. It is imperative to use mathematical-statistical methods of analysis to demonstrate and classify the actual reactions of the neurons, to express the incidence of different reactions in the population examined, to demonstrate different forms of the mutual interrelationships of individual elements in the nervous network etc.

G. Simple methods for evaluating unit activity

While classical electrophysiology of higher levels of the brain was mainly based on a study of the resultant activity of millions of neurons, during the last 10 years methods making it possible to follow the behaviour of individual nerve cells have become more and more important. Undoubtedly the key to the understanding of the most complex activity of the brain will be found by clarifying the organisation of the nervous network and the transmission of signals between individual elements of this network.

The classical idea, based mainly on the anatomical structure of the nervous system (Lorente de Nó 1936) assumed that the information concerning a certain, constant external stimulus is encoded in a definite sequence of impulses circulating along closed neuron loops. Direct observations of unit activity, however, have not demonstrated repeating time patterns of impulses. On the contrary, most experimental data indicate that the sequence of impulses at the output of an individual neuron may be characterized as a random process making the coding of information possible mainly by a change in the mean frequency of discharges. Such an assumption is in good agreement with our knowledge of the physiology of receptors where the frequency of impulses carried by afferent nerve fibres is a function of the intensity of the stimulus. The brain may be compared technically to a hybrid computer in which the greater reliability of distant connections of individual analog elements (neurons with gradually changing membrane potential) is attained after analog-to-digital conversion (a certain frequency of discharges corresponds to a certain membrane potential) by discrete signals (impulses carried by nerve fibres). Obviously the exact sequence of impulses is not particularly important but their number per unit time is more decisive. The basic problems of neurophysiological experimentation logically follow from this basic concept: how can the generator of impulse activity be characterized mathematically (e. g. Poissonian process, sine function with Gaussian noise, etc.); how can one distinguish a significant change in this activity from random fluctuations, how do the statistical characteristics of unit activity affect transmission of signals; what are the relationships between input and output signals, between background activity and the response to a constant stimulus, etc. All these problems can only be solved by applying mathematical-statictical analysis mainly based on computer processing of the experimental data (see chapter IIG). Since such computers are relatively expensive and are not available in every neurophysiological laboratory simple methods for the basic analysis of unit activity are described below.

1. Slow changes in unit activity

It is not difficult to study changes in the mean frequency of impulses at 10 sec or longer intervals and to correlate them with long term variations in EEG activity. Unit activity recorded on the screen of a cathode-ray oscilloscope is transferred to the input of a discriminating circuit (the Schmitt trigger circuit, see p. 99). which transforms potentials exceeding the preset threshold amplitude to standard 1 msec long impulses. By adding further logical circuits

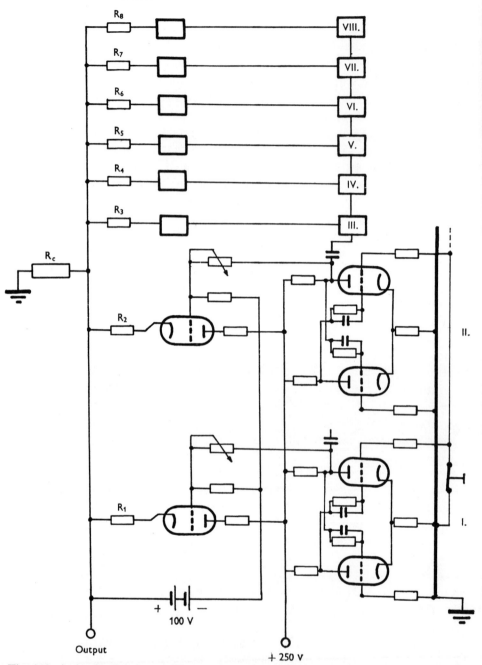

Fig. 282. A simple digital-to-analog converter. The output anodes of the binary counters shown schematically (without the coupling diodes) on the right are connected
through variable resistors with the grids of the corresponding converter valves (left),
blocked by a large negative bias applied between the grid resistors and the common cathode resistor (R_c).

(Schoenfeld and Eisenberg 1962) the conditions of this transformation can further be limited only to potentials not exceeding a certain amplitude (the so called amplitude window) and a certain duration (the so called interval window). Output impulses are led to a simple counter the state of which is read off at

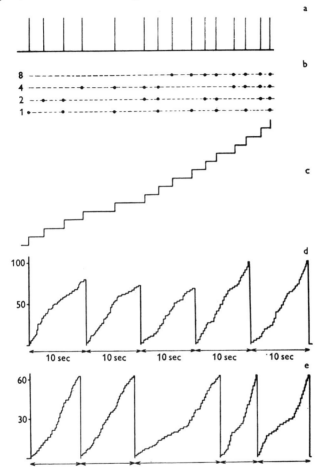

Fig. 283. Different techniques of recording changes of unit activity in time. a — The original sequence of spikes (after shaping). b — Consecutive states of the binary counters counting the activity a. c — Digital-to-analog conversion of the above states. d — Digital-to-analog conversion of a continuing activity with reset of the binary counter at regular ten sec intervals. e — Digital-to-analog conversion of a continuing activity without reset. For details see text.

regular intervals or printed, perforated or photographed. After recording each value the counter is reset to zero and counting is again started.

Sometimes it is, of course, more desirable to obtain the record of unit activity directly in the form of a curve of mean frequency which can be compared with other analog data. Here we use digital-to-analog conversion which expresses the content of the counter by a proportional output voltage. A simple digital-to-analog convertor (DAC) operating in connection with a 8 bit binary counter is shown in Fig. 282. It consists of 8 triodes, the cathodes of which are connected to a common cathode resistor R_c via resistors $R_1 - R_8$ of varying size. Normally all valves are closed by a large negative grid bias. They are connected through resistors to the output anodes of the individual flip-flops of the counter in such a way, that the positive pulse generated during transition from position 0 to 1 opens the

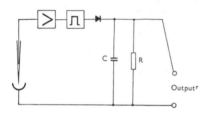

C
R
Output

Fig. 284. A simple integrating circuit for evaluating long lasting changes of unit firing. For description see text.

corresponding valve of the DAC. The intensities of the currents that pass through the individual valves when made conducting are related to corresponding stages of the binary counter and form thus a geometric series 1, 2, 4, 8, 16, 32, 64, 128. When several valves are conducting their currents are summated at the common cathode resistor. The voltage on it is an analog expression of the state of the counter and may be recorded with a suitable D. C. recorder. The counter is either reset to zero at regular intervals so that the recording appears as saw-tooth waves of varying amplitude, the envelope of which corresponds to changes in the mean frequency, or it works continuously and changes in activity can be evaluated from the slope and frequency of the saw-tooth waves. Both types of recordings are shown diagramatically in Fig. 283.

It is easier but less exact to use a simple integrating circuit. Shaped impulses are led through a germanium rectifier to the integrating condenser C which is continuously discharged via a large resistor R (Fig. 284). If the time constant of the circuit ($\tau = RC$) is considerably longer than the average inter-spike interval of the recorded unit activity, then the voltage on the condenser follows the slow changes in the firing rate with a certain delay which again depends upon the time constant τ. The output voltage at time t is approximately equal to the mean value of the input signal during the preceding interval τ.

2. Detection of evoked responses. Correlation with other events

Methods elaborated for the detection of signals in noise are used to decide whether a certain change in unit activity is bound to a given external stimulus. Usually they are based on repeated comparisons of activity before and during application of the stimulus. The simplest technical solution can be realized with a conventional cathode-ray oscilloscope. Action potentials are shaped as described in paragraph 1, and the resulting pulses are used for brightness modulation of the beam (Suzuki et al. 1960, Taira and Okuda 1962, Hind et al. 1963, Rose et al. 1963). The synchronizing pulse of the stimulator triggers the time base of the cathode-ray oscilloscope and only after some delay do we

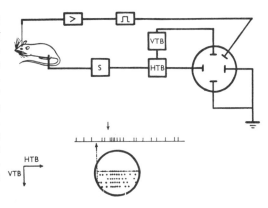

Fig. 285. Photographic plotting of poststimulation histograms. Above: block scheme of the experimental set-up (S — stimulator, VTB-vertical time base, HTB-horizontal time base). Below: example of the unit activity recording. The lower arrow indicates triggering of the horizontal time base, the upper arrow denotes the stimulus application. Display of 5 responses is schematically shown on the oscilloscope screen using the brightness modulation technique.

apply the proper stimulus (when the sweep has passed along about 1/5th of the screen). A voltage is applied to the vertical amplifier which moves the beam one step down for each sweep (as in television). A slow saw-tooth wave or discrete voltage steps produced e. g. by a stepping relay, may be used for this purpose. Up to 100 rows corresponding to individual responses can be placed on the screen if the trace is sufficiently focussed and photographic technique is good. The vertical time base can, of course, also be replaced by continuous movement of the film with a practically unlimited number of rows. Fig. 285 gives an example of a recording of an excitatory reaction. It is evident that after the stimulus more action potentials are found than can be seen in the irregular initial activity. Responses are very variable in the individual rows but the overall picture of more than 30 rows shows the increase convincingly. The results can be expressed quantitatively by using a photometer. For this it is necessary to depict the individual action potentials as distinct points that do not overlap with each other. The slit of the photometer is vertical to the rows and its width is such as to make a sufficient time resolution of changes possible. Post-stimulation histograms can be prepared in this way nearly as exactly as with a digital computer if the photographic process is of good quality and if suitable calibration recordings are used. It is of advantage that the recording contains the whole information which, if necessary, can be used for further analysis of the responses (variation analysis etc.).

The same method can be used for determining the relationship between unit activity and other processes (e. g. strychnine spikes, action potentials of other units, QRS complex of the electrocardiogram etc.), which trigger the time base. On the screen we obtain the distribution of unit activity after the synchronizing process in the chosen interval. When using a two-channel tape recorder with changeable position of the reproduce heads we can also examine the distribution of unit activity in the interval preceding the synchronizing event.

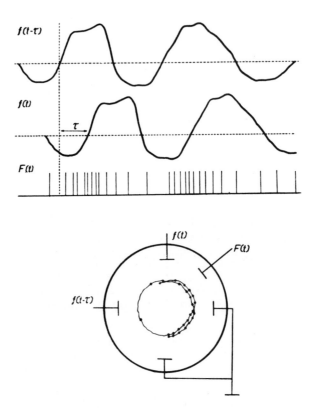

Fig. 286. Relationship between the phase of the periodic EEG activity and occurrence of unit discharges. For details see text.

Sometimes, when determining the relationship between unit activity and rhytmic processes (EEG activity, respiration), it is of advantage to use a different display (Green et al. 1961). The sinusoidal wave form is connected to the vertical and with a phase shift of 90° also to the horizontal amplifier of the cathode-ray oscilloscope so that an irregular circular pattern is obtained on the screen (Lissajou's figures). The phase shifting network for circular display of EEG was described by Kamp (1965). Unit activity is then again used after apropriate shaping for brightness modulation. If activity appears to depend on a certain phase of the compared sinusoidal event the largest number of points is concentrated in one of the quadrants of the screen (Fig. 286). Even though this method makes sensitive qualitative detection of similar relationships possible, the quantitative expression of these is difficult.

In the above examples the time relationship between the frequency of unit activity and other processes was examined. Similar methods can also be used to test far more complex relationships. As an example we may mention

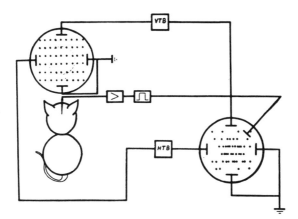

Fig. 287. Application of the brightness modulation technique for automatically plotting the shape of the receptive field of a visually driven unit in a cat. VTB — vertical time base, HTB — horizontal time base. For description see text.

automatic examination of the receptive field of single elements of the optic system in the cat (Suzuki et al. 1960). The screen of the cathode-ray oscilloscope is in the visual field of the examined eye (Fig. 287). It is scanned by a flashing spot of light (30 msec, 18 c/sec) at a rate of 3 cm/sec. The beam of the recording oscilloscope moves synchronously, its brightness, however, is modulated by action potentials recorded from the nervous system of the experimental animal. If the light spot appears in the receptive field of the unit, there is a rise (excitatory reaction) or a fall (inhibitory reaction) of the impulse frequency which appears as a change in the density of points on the recording screen. Of course the effect of the afterdischarge during "off" reactions must be considered. It is possible to limit the recording to "on" or "off" responses by choosing suitable "interval windows".

A similar technique may be used for automatic determination of the relationship between unit activity and stepwise changes of the stimulus qualities (intensity or pitch of tone, intensity or colour of light etc.). Individual stimuli trigger the horizontal time base of the cathode-ray oscilloscope but the Y deviation is proportional to the studied parameter of the stimulus (Suzuki et al. 1960) which is changed along a certain number of steps as determined beforehand.

3. Distribution of inter-spike intervals

In order to evaluate the character of the process which generates unit activity it is important to know the distribution of the intervals between individual action potentials. To a limited extent this problem can also be solved with the cathode-ray oscilloscope supplemented with simple auxillary circuits.

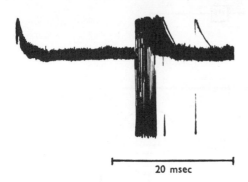

20 msec

Fig. 288. Measuring the interspike intervals with the superposition technique. For details see text.

The simplest is the superposition technique: the time base duration is set to approximately 3 times the average interspike interval and the sweeps are triggered by individual spikes (Fig. 288). The shortest interspike intervals are well represented and so is the modal interval, when the frequency is regular enough. Considerable information is lost, however, since the number of triggered sweeps is only 10—30% of the total number of spikes, and the longer interspike intervals cannot be differentiated from repeated occurrence of spikes in the same sweep.

Some of these drawbacks may be overcome by a somewhat more elaborate technique (Fig. 289). The action potential triggers the horizontal time base. The next action potential is recorded by brightness modulation as a point on that locus of the screen which has just been reached by the sweep. The time base is then reset to the initial position and again triggered. Repeated saw tooth waves are applied to the vertical amplifier so that the recording is spread out along a zone chosen beforehand in order to ensure better utilization of the screen surface. As long as the probability of coincidence of individual points at the same locus of the screen is not too large it is possible to obtain an interval histogram of sufficient exactness by examining the distribution of several hundred to thousand action potentials with a photometer.

Another method makes use of a binary counter in connection with a digital-to-analog converter (see p. 628). The action potential resets the counter to zero and a constant frequency of 1 kc/sec is led to the input of the counter from a generator of impulses. The output of the digital-to-analog converter is connected to the vertical amplifier of the cathode-ray oscilloscope. The next action potential sets off the horizontal time base so that the beam crosses the whole width of the screen during 0·5 msec. Immediately afterwards the counter is reset and again started. With further action potentials a number of sweeps accumulates at those parts of the screen where interspike intervals occur most frequently. This appears in the cummulative photographic recording as a more intensive blackening of the sensitive layer. Even more perfect resolution can

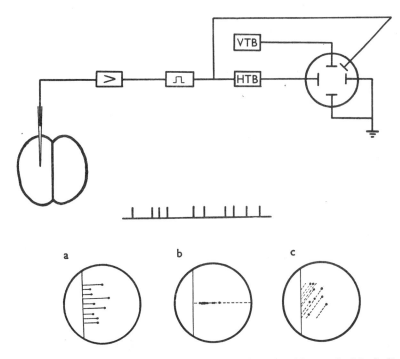

Fig. 289. Plotting interspike histogram with an oscilloscope. Above: the block diagram of the experimental set-up. Middle: example of a sequence of nerve impulses (after shaping). Below: corresponding interspike intervals recorded as abscissae. a — with discrete vertical shifts for each sweep, b — with the vertical time base switched off, c — with a repetitive vertical time base.

be obtained if an optic wedge is placed in front of the screen which then decreases the intensity of the beam from left to right. The point on the screen where the trace disappears varies according to the number of overlapping sweeps. By introducing suitable calibration measurements and by repeated copying of the recording onto high contrast photographic material the interspike distribution can be obtained, the interval being shown on the ordinate and the incidence on the abscissa.

The interspike intervals often form a Poisson distribution (Gerstein and Kiang 1960, Viernstein and Grossman 1961, Herz et al. 1964) characterized by an exponential decrement of the occurrence of longer intervals. Interspike intervals may, however, also show a Gaussian or a bimodal distribution. If spontaneous activity is affected by a regularly repeated stimulus, maxima corresponding to its period are found in the interval histogram.

H. Application of the functional ablation technique in analysing the cortico-subcortical relationships

Problem: Use cortical, hippocampal and thalamic spreading depression to analyse the cortico-subcortical relationships in the generation of spontaneous and evoked EEG activity in rat.

Principle: While the neuronographic techniques described in chapter IX F are based on the recording of responses evoked by remote stimulation, the problems of the functional organization of brain may also be approached from another direction. Activity of a given nervous structure is affected by impulses coming from different sources. Acute elimination of an important afferent influence induces in the partly deafferented structure changes of activity which may reveal the inhibitory or excitatory character of the inactivated input as well as its relative power. Because of the drawbacks of surgical ablation (irreversibility, shock, oedema etc.) the so called functional ablation techniques are being used for the above purpose to an ever increasing extent. Any such technique has to comply with several general demands: it must be easily reproducible and fully reversible; its effects must be uniform and limited to a specific nervous structure only; it must be accompanied by clear-cut electrical signs indicating its extent and duration. While local application of drugs and local cooling can be used for inactivation of restricted areas, transitory elimination of larger cortical or subcortical structures can be achieved by using Leão's spreading depression (Bureš 1959, Bureš and Burešová 1962). This remarkable phenomenon can be evoked not only in the neocortex (for description of cortical depression see chapter IX J), but also in the hippocampus, the caudate nucleus and the thalamus. In all these structures it is characterized by suppression of all kinds of activity due to depolarization of the somato-dendritic membrane of neurons. The maximum depression coincides with development of the negative slow potential and lasts in every point of the depressed structure for 1—2 min, full recovery being attained much later. Changes of activity occurring in structures not directly entered by spreading depression can be considered as consequence of the functional ablation procedures if they are synchronised with the primary depression.

Object: Rat weighing 200 g.

Apparatus: A conventional EEG apparatus with at least four channels. A dual beam cathode-ray oscilloscope. A cathode follower and preamplifier for microelectrode work. A stimulator giving rectangular pulses with a radio frequency isolation unit. A chopper circuit for two-channel DC potential recording (if necessary with symmetrical input cathode followers). An electronic counter. Photographic camera.

Other requirements: Surgical instruments, trephines. A dental drill. Stereotaxic apparatus for rat with electrode holders and a vertical microdrive. Wick calomel cell electrodes. An electrode-cannula assembly. Bipolar silver wire electrode for recording and stimulation. Glass microelectrodes filled with 3M-KCl. Ether, D-tubocurarine, Dial. 1%, 5% and 25% KCl.

Procedure:

1. Spreading depression in deep brain structures

While most techniques used in this chapter were already described in other parts of this book, special attention must be paid to hippocampal and thalamic spreading depressions, which can be evoked by injecting small amounts (0·5 µl or less) of KCl into the appropriate brain region. Since the stimulus evoking spreading depression is applied stereotaxically, its efficiency has to be checked by slow potential recording from the same structure. This is simply achieved by using an electrode-cannula assembly (Fig. 264g) consisting of a 12 mm long piece of injection needle tubing (outer diameter 0·7 mm) and of a glass capillary (about 200−300 µ external diameter). The glass capillary and the steel cannula are arranged in parallel at a distance of 2·0 mm from each other, the latter ending about 1 mm higher than the former. The glass capillary is filled with saline and connected by a salt bridge with a calomel cell electrode. The steel tubing is used as a guide for a well fitting hypodermic needle, which protrudes, when fully inserted, for about 1·5 mm from the orifice of the cannula. Using a microinjection device 0·3−0·5 µl of 25% KCl is injected through the inner needle into a point approximately 2·0 mm distant from the tip of the capillary electrode.

Experiments are performed on anesthetised (Dial 40 mg/kg) or unanesthetised curarized rats. The animal is fixed in the stereotaxic apparatus and the electrode-cannula assembly is introduced into the brain in such a way that the stereotaxic coordinates of the injection and of the capillary electrode are (AP 2·5; R 1·0; V 6·0) and (AP 2·5; L 1·0; V 5·0) for the thalamic spreading depression and (AP 4·0; L 4·0; V 4·0) and (AP 3·0; L 2·5; V 3·5) for the hippocampal spreading depression respectively. The slow potential wave spreading from the point of KCl application is led off by the capillary electrode and recorded in one channel of the EEG apparatus using the chopper technique (see p. 163). The reference electrode is placed on the rostral part of the exposed frontal bones. Other capillary electrodes may be used to record the slow potential change from adjacent structures (caudate, hippocampus, thalamus) while steady potential recording from the exposed cortical surface checks penetration of spreading depression to the neocortex. Further channels can be used for recording the EEG activity or unit activity.

2. Cortico-hippocampal and hippocampo-cortical relationships

Under light ether anesthesia trephine openings are made above both hemispheres according to Fig. 292. After cannulation of the trachea the rat is immobilized with D-tubocurarine (2 mg/kg) and fixed in the stereotaxic apparatus. The electrode-cannula assembly (3-S) is inserted into the left hippocampus while a bipolar electrode (4) is introduced into the right dorsal hippocampus (AP 2·5; R 2·5; V 3·0). Cortical steady potential and EEG is recorded from the

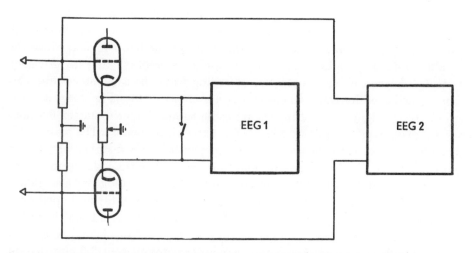

Fig. 290. Separation of the chopper from the electrodes using a symmetrical cathode follower in the input. Channel 1 is used for the slow potential recording and channel 2 for the ECoG recording from the same electrodes.

frontoparietal cortex (trephine openings 1,2). The electrical activity in cortex and hippocampus is recorded first (see chapters IX B and X D). The cortical spreading depression is then evoked by applying a filter paper soaked with 5% KCl onto the trephine opening B. Changes of the spontaneous hippocampogram and of hippocampal reactions to external stimuli are observed in the right hippocampus. KCl is then removed and the trephine opening B is washed with saline. After the EEG in the right hemisphere is normalized again, spreading depression is evoked in the left hippocampus. The injection needle is inserted through the guiding cannula and KCl is applied. Spontaneous EEG and the electrocortical reactions to arousing stimuli are observed while the slow potential wave spreads through the underlying hippocampus. The hippocampal capillary electrode can be used not only for slow potential recording but also for monopolar recording of the hippocampogram. In this case it is necessary to separate the chopper from the electrodes in order to avoid distorting artefacts in the EEG. The same electrode is then simultaneously connected to the input of a simple symmetrical cathode follower and to the input of one of the EEG amplifiers. The chopper is used to short circuit periodically the DC output of the cathode follower which is connected to the input of another EEG channel (Fig. 290).

3. Cortico-thalamic influences

The same procedure is used as in the preceding experiment but the trephine openings are made as indicated in Fig. 294. After the animal is immobilized with D-tubocurarine, dura mater is carefully removed from the trephine opening 1. The glass microelectrode is brought above the exposed cortical surface and connected via a cathode follower and a preamplifier with the cathode-ray oscilloscope. Wick calomel cell electrodes used for both EEG and slow potential recording are applied to the trephine openings 0,3 and 4. They are connected with the EEG apparatus using the arrangement described in the preceding paragraph. Using the microdrive of the stereotaxic apparatus, the microelectrode is slowly inserted into the cortex at a point characterized by sterotaxic coordinates (AP 2·5; L 1·5) to a depth of 1·5—2·0 mm. All precautions described in detail in chapter X F must be observed. After a well isolated unit is found its activity is recorded for several minutes and averaged in thirty second intervals using an electronic counter connected to the output of the oscilloscope amplifier. If the unit frequency is stable enough spreading depression is evoked by applying a filter paper soaked with 1% KCl onto the trephine opening A, while EEG and the slow potential changes are recorded from trephine openings 3 and 4 with the EEG apparatus and the unit activity is counted. After completion of this experiment, the microelectrode is inserted deeper until a unit is found at a depth of 3·5 to 5·5 mm below surface, i. e. in some thalamic nucleus. After several minutes of control recording a new spreading depression is elicited from trephine opening A and the unit activity is again recorded together with the EEG and the slow potential change.

The microelectrode is then removed and two pairs of silver wire electrodes are introduced through trephine openings 1 and 2 into the thalamus to points

(AP 2·5; R, L 1·5; V 4·0—5·0). After recording the spontaneous EEG in both cortex and thalamus for 10—20 min, cortical spreading depression is elicited in the left hemisphere by applying 5% KCl to the trephine opening A and a few minutes later also in the right hemisphere (trephine opening B). The effects of unilateral and bilateral spreading depression on thalamic activity are observed.

After restitution of the initial EEG activity one pair of the thalamic electrodes is connected to the radiofrequency output of the stimulator. Cortical recruiting responses are evoked as described in chapter X E using slightly supra-maximal trains of rectangular pulses (0·2 msec, 7 c/sec, 5 sec) applied at 20 sec intervals. Cortical spreading depression is then elicited in the hemisphere over-lying the stimulating electrodes, while the recruiting responses are recorded in the contralateral hemisphere. The result is compared with the effect of spread-ing depression elicited in the hemisphere contralateral to the thalamic stim-ulation.

4. Reversible deafferentation of the cerebral cortex by thalamic spreading depression

In a rat anesthetized with Dial trephine openings are made as indicated in Fig. 295 and the sciatic nerves are exposed. The animal is fixed into the stereo-taxic apparatus and the electrode-cannula assembly is introduced into the rostral thalamus (see p. 635). Using wick electrodes the cortex exposed by large trephine openings S and V is explored for maximal responses to single shock stimulation of the sciatic nerve or to intense light flashes. Stimuli are applied at regular 1—3 sec intervals. The evoked responses and the spontaneous EEG are monopolarly recorded in 2—4 channels of the EEG apparatus. The former are also recorded with a cathode-ray oscilloscope. A further two channels are used for recording the slow potential changes in the thalamus and in the cortex overlying the site of injection after a stable level of responding is reached, thalamic SD is evoked and its effect on both spontaneous and evoked activity is observed.

Results:

1) Spreading depression evoked in the hippocampus and thalamus (and also in the mammalian caudate nucleus and in the avian striatum) has the same fundamental properties as in the neocortex (p. 568): it is characterized by a slow negative wave the amplitude of which may exceed 20 mV which spreads from the point of KCl application with a velocity of 3 mm/min. When the re-cording electrode is too close to the site of the injection, it may be affected also by the local depolarization process due to accumulation of the injected KCl

in the tissue. Such potentials can be easily recognized by the slow exponential decline corresponding to the gradual removal of K$^+$ ions into the circulation. During the maximal depolarization spontaneous and evoked EEG is suppressed.

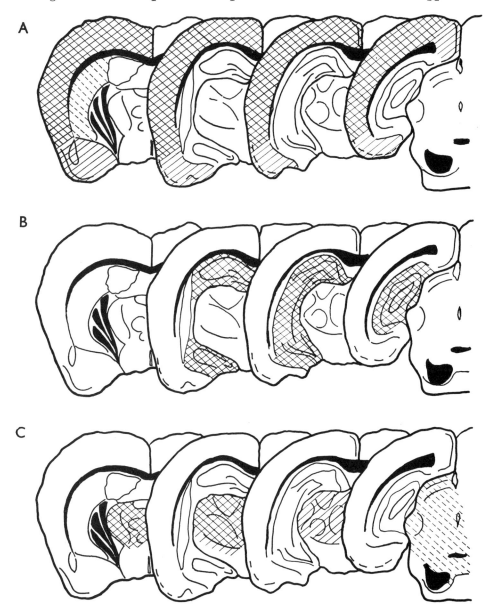

Fig. 291. Extent of the cortical (above), hippocampal (middle) and thalamic (below) spreading depression. Cross-hatching — structures directly invaded by spreading depression; shading — structures entered by spreading depression with a long refractory period; interrupted shading — structures entered by spreading depression irregularly.

The recovery is more rapid in the thalamus than in the cortex. In the hippocampus repolarization is often accompanied by local seizures which regularly occur in unanesthetized animals. Application of spreading depression for functional ablation purpose requires good knowledge of the exact limits of the affected brain structures. First of all it is necessary to check that the injected KCl has not penetrated along the cannula to the adjacent tissue, i. e. to the neocortex with hippocampal injection and to both neocortex and hippocampus with thalamic injection. When spreading depression has been elicited only in the given structure, its extent can be explored by recording from different parts of the same structure and from the adjacent areas. When evaluating the results of such experiments the physical spreading of the negative slow potential to inactive tissue must also be taken into account (space constant is approximately 0·5 mm — Monakhov et al. 1962b).

Results obtained in the rat brain are summarized in Fig. 291 constructed according to the data of Fifková (1964a, b). Neocortical spreading depression affects the whole neocortex, it does not penetrate, however, into the cingular cortex medially and below the rhinal fissure laterally. With a long refractory period it enters the amygdala and in 50% of cases reaches the caudate nucleus. It never invades the hippocampus, thalamus, hypothalamus and the brain stem structures.

Hippocampal spreading depression affects all parts of the hippocampus (dorsal, posterior and ventral) but stops in the subiculum. It enters neither neocortex nor thalamus.

Thalamic spreading depression spreads through the entire thalamus as if it were a homogeneous structure (this is due to the fact that the bands of myelinated fibres separating the individual thalamic nuclei are too narrow to prevent spreading). It spreads across the midline and with some decrement enters also the posterior group of thalamic nuclei and occasionally also the superior colliculus. It does not penetrate into the inferior colliculus, tegmentum, hypothalamus, caudate nucleus and neocortex.

2) Unilateral (and even bilateral) neocortical spreading depression has no marked effect on the spontaneous hippocampogram of unanesthetized curarized rats (Bureš 1959, Weiss and Fifková 1960 — Fig. 292). Theta activity occasionally appears in the hippocampus or may be elicited by external stimuli or by physostigmine application. Similarly hippocampal spreading depression does not influence the EEG activity in the overlying cortex and the slow sleep EEG can be desynchronized by arousing stimuli. Only in the later phases of hippocampal spreading depression characterized by seizure type activity, may some weak paroxysmal discharges appear in the cortical recording. The results indicate that the generators of bioelectrical activity in neocortex and hippocampus are largely independent of each other and that the respective arousal reactions are mediated by parallel independent routes.

3) Although it is more difficult to find spontaneously active neurones in the cerebral cortex than in the reticular formation of rats (see chapter X F) several units are usually encountered during a single penetration. Their activity is usually slower and more irregular than that of the reticular units. After application of KCl the firing rate does not appreciably change until the slow potential reaches the area of recording. At that time a brief high frequency

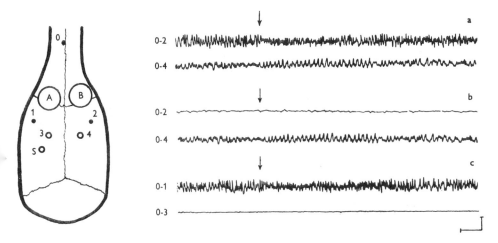

Fig. 292. Cortical and hippocampal electrical activity under control conditions (above) and during ipsilateral cortical (middle) or hippocampal (below) spreading depression. A, B — trephine openings for inducing the cortical spreading depression; S — trephine opening for inserting the hippocampal injection cannula; 1, 2 — cortical electrodes; 3, 4 — hippocampal electrodes; 0 — reference electrode. Calibration: 1 sec, 200μV. For details see text.

discharge occurs which is followed by complete silence lasting for 1—3 min. The recovery is slow, the predepression firing rate being reached after 5—10 min with a single spreading depression wave (Fig. 293 — Burešová et al. 1963). When counting at 30 sec intervals the brief discharge may often be obscured by the following silence, which is the most prominent feature in such recordings. Better results are obtained, therefore, using more dynamic recording techniques (see chapter X G), e. g. recording every tenth impulse (output of the first decade of the electronic counter) in one EEG channel.

Thalamic units react to cortical spreading depression less uniformly. The firing rate of most of them (80% — Bureš et al. 1963) is also considerably decreased but usually some activity (about 30% of predepression frequency) is preserved even at the height of the cortical spreading depression (Fig. 293). In a few cases a pure increase of unit firing may be found as well.

Unilateral cortical spreading depression reduces the spontaneous EEG activity in the underlying thalamus to approximately 60% of the predepression peak-to-peak amplitude (Weiss and Fifková 1961 — Fig. 294), while the EEG in the contralateral thalamus is much less affected. The duration of the thalamic effect corresponds to the depression of cortical activity. When the brief unilateral spreading depression is succeeded by bilateral spreading depression, the amplitude of the thalamogram is further decreased (to 30% of the initial amplitude). Recovery of normal activity proceeds in both cortex and thalamus synchronously.

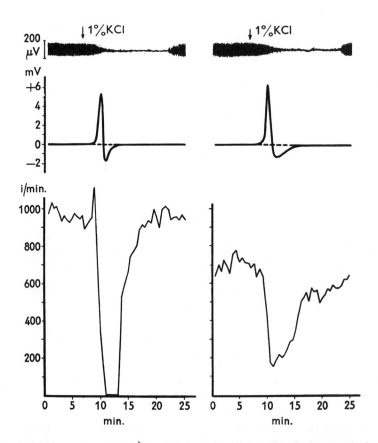

Fig. 293. Activity of cortical (left) and thalamic (right) units during cortical spreading depression. Above: ECoG; middle: slow potential change in the cortex. Below: average firing rate of the units per 30 sec intervals. For details see text.

The reduction of the thalamic EEG is accompanied by concurrent changes of excitability. This is shown in experiments with the recruiting responses (Fig. 294): Spreading depression evoked in the cortex overlying the point of thalamic stimulation causes a clear cut reduction of the recruiting response in both hemispheres, although the stimulus intensity remains constant. By contrast spreading depression evoked in the hemisphere contralateral to the stimulated thalamus induces a slight reduction of the recruiting response only in the depressed cortex (this change is more pronounced in bipolar than in monopolar recording). Greater efficiency of the ipsilateral spreading depression in comparison with the contralateral one indicates that during cortical spreading depression the excitability of ipsilateral nonspecific nuclei is reduced and

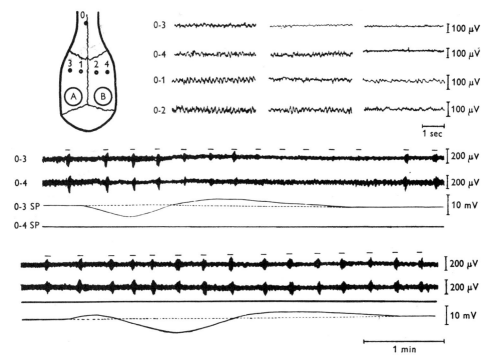

Fig. 294. The effect of cortical spreading depression on the spontaneous EEG activity in thalamus (above: control, unilateral and bilateral SD) and on the recruiting responses evoked by stimulation of the ipsilateral (middle) or contralateral (below) thalamus. 1, 2 — thalamic electrodes; 3, 4 — cortical electrodes; 0 — reference electrode. Horizontal bars indicate thalamic stimulation. For details see text.

that the constant stimulus becomes submaximal or even subthreshold. As a result generation of recruiting reponses is impaired at the site of stimulation and the recruiting potentials are reduced or abolished in all projection areas.

4) During thalamic spreading depression (Bureš et al. 1965) spontaneous EEG activity is depressed at first in the hemisphere ipsilateral to the KCl injection and with a delay of 30—40 sec also in the contralateral cortex. At the height of the thalamic spreading depression EEG becomes irregular, short periods of complete silence alternating with isolated spikes or bursts of slow waves. The maximum effect lasts for 20—30 sec, full recovery being reached

after 90 sec. All these changes occur slightly earlier in the parietal than in the occipital cortex. Since on the side of the injection some of the above effects may be prolonged by the direct KCl influence, examination of the contralateral hemisphere which is affected by the spreading depression process alone gives more consistent results (Fig. 295). Somatosensory evoked responses disappear completely as soon as the underlying thalamus has been invaded by spreading

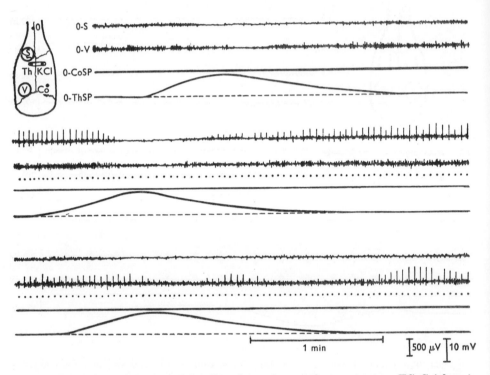

Fig. 295. The effect of thalamic spreading depression on the spontaneous ECoG (above), on the somatosensory (middle) and visual (below) evoked responses. S, V — somatosensory and visual projection areas. Th, KCl — location of the electrode-cannula assembly. Calibration: 500 μV for ECoG, 10 mV for the slow potential recording. For details see text.

depression and are suppressed for a period coinciding with the maximum decrement of the EEG. Their recovery is, however, more rapid, the evoked potentials being very clear cut on the background of partly depressed EEG. Visual evoked responses are affected by thalamic spreading depression in a biphasic way. During the first phase which develops approximately 20 sec after the onset of the somatosensory depression, the visual evoked responses do not completely disappear but their amplitude is significantly reduced (to about 60% of the average predepression amplitude). At the same time the background EEG is reduced. The partial depression of visual evoked responses lasts for approximately 20 sec and is followed by a period of recovery (another 20—30 sec). Only then the second phase starts, characterized by a complete suppression of the visual responses (30 sec), followed by a period of potentiation during which the positive component of the primary response increases up to 170% of its initial amplitude. The depression of EEG and of the somatosensory evoked responses is due to spreading depression invading the nonspecific thalamic nuclei and the ventrobasal complex relaying the sciatic volleys. The long delay of the second phase of visual evoked response suppression is probably due to anatomical separation of the lateral geniculate body from the remaining thalamus by a distinct band of white matter, the lamina medullaris externa, which may considerably slow down the rate of spreading. In experiments in which depression of visual evoked responses is monophasic, the amplitude and latency of depression indicates that spreading depression was either not delayed at the lamina medullaris externa and directly entered the lateral geniculate body (the first and second phase overlapping) or that it did not reach this structure at all (the second phase being absent).

Conclusion: Spreading depression can be used to achieve reversible ablation of several prosencephalic structures. Examination of the remote effects of this intervention may reveal the inhibitory or excitatory character of the eliminated influx, its relative role in the afferentation of the given structure, lateralization of the pathways mediating the observed effects etc. In the experiments described above the functional ablation technique was used to demonstrate the relative independence of cortical and hippocampal activity contrasting with the close cortico-thalamic and thalamo-cortical relationships. When evaluating the remote effects of spreading depression the possibility of tertiary changes must also be considered, since the observed effect is not necessarily due to elimination of direct influence arising in the primarily depressed structure but may be a consequence of secondary changes occurring in other brain regions. Further experiments including elimination of the suspected mediating links are required to decide whether the remote effects are of direct (secondary) or indirect (tertiary) character.

I. Electrophysiological signs of an epileptic seizure

Problem: Record the EEG during an epiletic seizure elicited by electroshock, cardiazol or by sensory stimulation. Demonstrate signs of epileptic irradiation.

Principle: During an epileptic seizure characteristic changes in cerebral electrical activity occur. These differ according to the type of seizure (Penfield

645

ε nd Jasper 1954). Experimental seizures produced in animals usually corres-
рοnd to the human grand mal, characterised by tonic-clonic paroxysms and
a complex of vegetative symptoms. Electrophysiologically, such a seizure
appears as a discharge of fast, high voltage activity affecting the whole cortex
and extensive subcortical regions. At other times, clear cut spike and wave
complexes with a frequency of 2—5/sec are found in the EEG.

An epileptic seizure is an expression of diffuse spreading of activity in
central synapses. This occurs as the result of increased excitability of nervous
structures due to drugs (e. g. cardiazol, picrotoxin) or metabolic effects (e. g.
hypoglycaemia, hypocalcaemia). That region of the brain in which excitability is especially high, and which consequently may become the starting point
of an epileptic seizure through the action of nervous and humoral factors,
is termed a focus. Generalisation of a seizure is due to spreading of this pathological activity into regions of the brain stem and thalamus which are diffusely
connected with the cortex and other parts of the brain.

Reflex audiogenic epilepsy in rats and mice may serve as an example
of focal activation of an epileptic seizure. These are paroxysms occurring
in a certain percentage of animals when strong acoustic stimuli act on them
(jingling of keys, strong buzzers, bells, Galton's whistle). So-called Amantea's
reflex epilepsy permits a detailed analysis of the relationship between the
stimulus, the focus and the mechanism of generalisation. Amantea's epilepsy
(Amantea 1921) develops in a certain percentege of dogs in whom an epileptogenic focus has been formed artificially by applying strychnine to a cortical
projection area. Generalised seizures occur as the result of strong stimulation
of the corresponding sense organ.

The ability to outlast the stimulus and to develop from a certain moment
independently of it are fundamental characteristics of an epileptic seizure.
In addition to automatic activity of centres, this feature may be explained by
reverberating circuits permitting auto-re-excitation of activated nerve centres
until they become exhausted or inhibited. Afterdischarge evoked by direct stimulation of the cerebral cortex (even of the neurally isolated cortex — Burns
1951) or an epileptic seizure produced by electroshock may serve as examples.

Object: Cat weighing 2—3 kg, rat 200 g.

Apparatus: A multichannel ink writing EEG apparatus. A dual beam
cathode-ray oscilloscope with A. C. amplification of 50 μV/cm. A stimulator
giving square wave pulses or condenser discharges synchronised with the
time base of the oscilloscope. A tone generator or a time marker. A loudspeaker.
A regulating autotransformer, a time switch for limiting the electroshock
current to 0·1—1·0 sec.

Other requirements: Surgical instruments, trephines, dental drill. Needle
electrodes for recording the ECoG in rats, wick Ag-AgCl electrodes with

electrode carriers for recording the ECoG in the cat. Animal boards with stereotaxic head holders for the cat and rat. 10% Dial, 1% cardiazol, 1% strychnine.

Procedure:

1) Electroconvulsive shock (ECS). Under light ether anaesthesia the skin of the rat's head is cut along the midline and the skull bones are cleaned. The animal is then tied to the board, the head fixed in the stereotaxic head holder and needle electrodes are inserted into the bone as in experiment IX B. The fixing bars for the meati end in insulated metallic tips serving as electrodes for administering the electroshock current. The area of the meati is treated with EEG paste before the ear plugs are fixed.

The animal is permitted to recover from the anaesthesia. Fixation is then checked and the experiment begins. Electroshock is given after several minutes of control recording. A. C. current is used, the voltage of which can be changed by the regulating autotransformer. It is led to the electrodes across a serial resistance of 50 kΩ. The shock duration is set at $0\cdot2-0\cdot3$ sec with a bipolar switch. Normally both shock electrodes are grounded, and only during use are they connected to the autotransformer.

The switch and the autotransformer are outside the screened chamber and are connected with shielded cables to the shock electrodes. Just before giving a shock, the inputs of the EEG amplifiers are short-circuited and after the shock has ended, they are again disconnected. Even so, the amplifiers are often blocked because of the large input signal. This can be prevented be means of a special anti-blocking device (Baumann et al. 1956).

Usually several seconds immediately following the ECS are lost for recording. If convulsions do not occur, the voltage (measured with a voltmeter in parallel to the autotransformer) is increased by $10-20$ V and, after a pause of 10 min, another shock is given. This is repeated until threshold shock values are obtained. If the threshold is to be determined more exactly, the value of the seizure producing quantity of electricity or the density of this quantity with respect to the size of the head and the brain of the animal is used instead of voltage values.

2) Audiogenic epilepsy in rats. Only animals which, by previous tests, are known to react consistently to epileptogenic acoustic stimuli with a running fit and convulsion are used. The animal is prepared as in paragraph 1. The head is fixed only by the upper jaw and the front teeth. The ears remain free. By dissecting the temporal muscle the bone above the cortical auditory areas is made accessible, and one or two recording electrodes are placed in it. They serve to determine changes in reactibility of the auditory cortex.

When the effect of anaesthesia has completely worn off, EEG recording is commenced. After a short control period, 50 mg/kg of cardiazol is given intraperitoneally. Changes in the ECoG are observed and 5 minutes later an acoustic stimulus is given. This is most simply done by jingling a bunch of keys at a distance of 20—30 cm from the rat's head. If there is no seizure, the stimulus is discontinued after 2 minutes and the experiment is repeated on another animal. In highly sensitive animals, a seizure may be obtained even without a sensitising dose of cardiazol. Since restraint has an inhibitory effect on audiogenic epilepsy (Bureš 1953), the animal must first be adapted to the conditions of the experiment, i. e. it must be repeatedly fixed in this way before the epileptogenic stimulus is applied.

In addition to recording the spontaneous ECoG, an attempt may be made to record responses to short acoustic stimuli (clicks produced by feeding the stimulator discharges into the loudspeaker) in the auditory cortex with a cathode-ray osciloscope. The effect of cardiazol on primary auditory responses and their changes during and after the epileptogenic stimulus are observed.

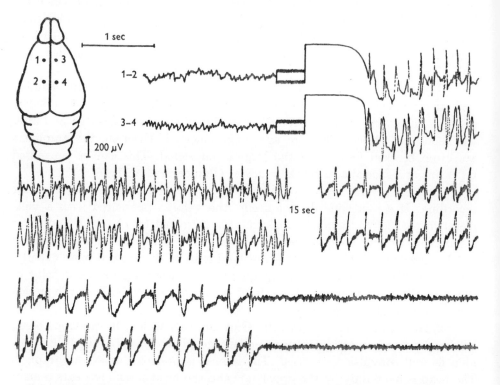

Fig. 296. Epileptic seizure evoked by electroshock in the unanaesthetised rat.

In animals that did not have an epileptic seizure following an acoustic stimulus, the dose of cardiazol is increased to 100—150 mg/kg and an EEG recording of a series of epileptic seizures thus evoked is made.

3) Elements of epileptic irradiation. A cat under Dial anaesthesia (40 mg/kg) is operated as in experiment IX D. The extent of the auditory cortex

is determined in the usual way. Using bipolar stimulating electrodes (ball-tipped silver wires, diameter 0·5 mm, distance 1·5 mm) the motor area of the front limb is determined with rectangular pulses or A. C. current, regulated by an autotransformer. Care is taken to keep the stimulus parameters constant (50 — 100/sec, 3 — 5 sec). Using threshold or slightly suprathreshold stimuli, a narrow projection is obtained. If a two channel oscilloscope is available one

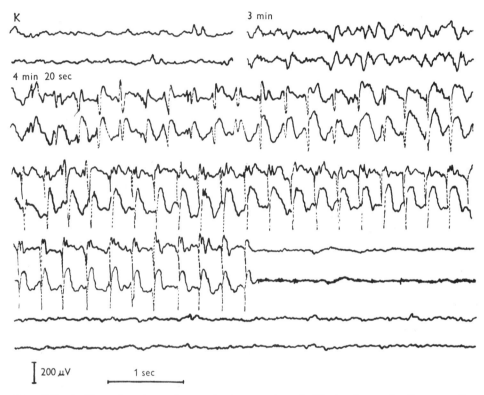

Fig. 297. Audiogenic seizure in a rat given metrazol. Electrodes as in the preceding figure. Figures indicate time following metrazol injection. Epileptogenic stimulus applied after 4 minutes.

electrode is placed onto the motor and the other onto the auditory area, and the experiment is commenced. After a short control recording of the electrical responses to clicks, a filter paper soaked with 0·5% strychnine is applied on the auditory area. When the effect of strychnine is maximal, cardiazol (20 — 40 mg/kg) is given and the development of the responses in the motor and auditory areas is observed for 20 minutes. If responses are clear in the motor areas, the strychnine focus is circumcised with a subpial knife and the effect of this intervention on responses in the motor areas is studied.

Results:

1) Fig. 296 shows a typical EEG recording before and after a suprathreshold electroshock. The motor reaction corresponds to the EEG tracing. After subthreshold stimuli, this is only a jerk of a defensive character which is sometimes followed by a short clonus of facial muscles. Suprathreshold stimuli, on the other hand, produce a typical seizure with a very regular course — initial clonus, followed by a tonic phase which is finally replaced by coarse clonic convulsions of both fore- and hind limbs.

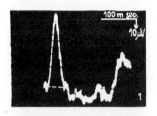

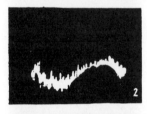

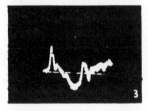

Fig. 298. Primary cortical responses to acoustic stimuli in the rat after metrazol injection (1), during fit (2), and 1 min. afterwards (3). Monopolar recording. Positivity of active electrode upward.

2) Within several minutes after subthreshold doses of cardiazol, pronounced changes in the EEG occur (Fig. 297). The amplitude increases rapidly, synchronous slow waves of high voltage appear, first singly and later in groups. In general there is a change in frequency towards slower rhythms. After 3—5 minutes pathological rhythms predominate in the EEG (amplitude more than 500 μV, frequency about 3/sec). At this stage the epileptogenic stimulus usually does not elicit pronounced EEG changes. Pathological slow rhythms of high amplitude continue to be present, sometimes during the whole duration of the acoustic stimulus. At other times, when manifest convulsions occur, they are replaced by another type of activity. The first sign of seizure is a rapid decrease or even disappearance of the slow waves. This may either occur suddenly or in several successive periods. During the seizure itself, this decrease in activity is even more pronounced, so that in all leads we find nearly iso-electric lines with superimposed low voltage, rapid activity. In some rats a shorter or longer period of the spike and wave activity with a frequency of about 3/sec (Fig. 297) precedes this "electric silence". This type of activity is characteristic mainly for clonic convulsions. The tonic

phase, on the other hand, is usually accompanied by disappearance of electric potentials (Ajmone-Marsan and Marossero 1950).

After an injection of cardiazol, primary responses to an acoustic stimulus, poorly discernible in unanaesthetised rats, begin to increase rapidly and soon attain an iterative character (Bureš 1953). During the epileptogenic stimulus, primary responses to simultaneously applied "clicks" disappear, evidently

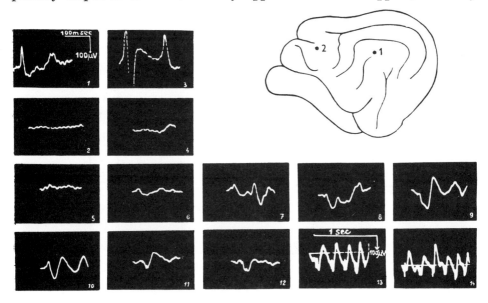

Fig. 299. Spreading of auditory responses into cortical motor areas following metrazol. Responses to acoustic stimulus in the auditory (1) and motor (2) areas. 3, 4 — the same responses after application of strychnine to the auditory area. 5-10 — responses in the motor area at 5 min. intervals after metrazol injection (40 mg/kg i. p.). 11-12 — after circumcision of the cortical strychnine focus (depth 4 mm). 13-14 — recording from the motor area during clonic convulsions.

because they are masked by the stronger epileptogenic stimulus. If convulsions occur the primary responses remain suppressed even after the epileptogenic acoustic stimulus is off and reappear only after 2—3 minutes (Fig. 298).

3) No responses to acoustic stimuli can be recorded in motor areas of the cat's cerebral cortex before or after application of strychnine to auditory areas (Fig. 299). If, however, cardiazol is injected when the effect of strychnine is maximal, first weak and later more and more pronounced potentials of irregular shape and with a longer latency appear in the motor areas. A second wave usually follows the first one with an interval of about 100 msec. Potentials in the motor cortex begin to be accompanied by spasms of the ears, limbs and trunk. With a sufficient dose of cardiazol, repeated acoustic stimuli may produce a clonic convulsion with the characteristic spike and wave activity

of a frequency corresponding to that of clonus. A subpial incision interrupting intracortical connections between the auditory cortex and the other cortical areas usually only produces a certain decrease but not suppression of responses to acoustic stimuli in motor areas. This indicates that cortico-subcortical connections between the strychnine-treated focus in the auditory area and motor areas are involved. Such connections may be regarded as elements of the mechanism of an epileptic seizure (Gastaut and Hunter 1950, Bureš 1953, Hunter and Ingvar 1955).

Conclusion: An experimental epileptic seizure manifests itself in the ECoG by a characteristic high voltage activity indicating synchronous discharge of cortical neurones. An analysis of the paths along which epileptic activity spreads from the focus, shows that subcortical centres are essential in the mechanism of generalisation of the fit. An epileptic seizure is terminated by deep depression of spontaneous and evoked cortical activity.

Appendix I

Stereotaxic atlases for the cat, rabbit and rat

E. Fifková, J. Maršala

Institute of Physiology, Czechoslovak Academy of Sciences, Prague,
Institute of Histology, Faculty of Medicine, Šafarik University, Košice

Three cats, rabbits and six rats were used for determining the stereotaxic coordinates. The animals were exsanguinated under pentothal anaesthesia. The brain was fixed by perfusion through the carotid arteries with saline, followed by a solution of two parts of 96% ethanol and one part of 40% formalin. The soft tissues were dissected away, trephine openings made and the skull was immersed in 10% formalin for an additional 8 days. During this procedure, only insignificant weight changes occur (Flatau 1897, Frontera 1958).

After satisfactory fixation was attained, the skull was rigidly clamped into the head holder of a stereotaxic apparatus and the reference planes necessary for calculation of coordinates were determined (see also chapter IX A).

After additional fixation in ascending alcohol series, the brains were embedded in celloidine. Complete serial frontal sections (50 μ) were made and stained by the method of Nissl with cresylviolet. The plane of section was parallel to the zero frontal plane in all animals used. Three rat brains to be used for frozen sections were fixed by 10% formalin only. Frozen sections 50 μ thick were made. All sections were used and mounted on slides. The sections of one series were covered with glycerol. With this method the shrinkage is kept to a minimum (van Tienhoven and Juhász 1962). The two other series were stained with toluidine blue.

In the cat, the head was rigidly held by steel plugs introduced into the auditory meati, and by bars holding the upper jaw. A plane passing through the auditory meatus and the lower orbital margins is the basal reference plane. A parallel plane lying 10 mm dorsally was used as the horizontal zero plane. Needles were introduced perpendicularly to the basal plane 7 mm laterally from the sagital zero plane at the level of the bregma and 0·5 mm in front of the interaural plane. The latter determine the frontal zero plane. The sagittal coordinates of structures lying rostrally or occipitally from this plane are marked AP+ or AP−. As the interaural line-bregma distance varies considerably (from 16·5−23 mm with the modal class of 20−21 mm, $n = 30$) the frontal zero plane passing through the interaural line was considered to be more exact than that passing through the level of the bregma (used in the first edition). The deep coordinates are determined with respect to the horizontal zero plane: above this level they are marked V+, below it V−. Distances to the left from the sagittal zero plane are marked S, to the right D. All values for stereotaxic coordinates are given in mm. The corresponding enlargement is given in the diagrams.

In the cat the stereotaxic atlas of Jasper and Ajmone-Marsan (1952b) was used as the basic guide for topographic orientation of subcortical structures. The data in the papers of Rioch (1929a, b) and Ingram et al. (1932) were used for the diencephalon, those of Fox (1940) for the septum, basal ganglia and amygdala and those of Taber (1961) for the reticular formation of the brain stem. To make the determination of diencephalic lesions easier for neurophysiologists, the nomenclature of Jasper and Ajmone-Marsan was principally adhered to. Only those structures in which a secondary division can easily be found even with low magnification, were localised more precisely.

In the hypothalamus, the regio preoptica (RPO in the atlas of Jasper and Ajmone-Marsan) was divided into two parts having a different cellular structure. The part lying near to the medial wall of the III ventricle (of its most rostral part) is marked as area praeoptica medialis (APM), the part lying more laterally and composed of disperse cells is called area praeoptica lateralis (APL). It joins the area hypothalamica lateralis caudally.

In the tuberal part of the hypothalamus, Jasper and Ajmone-Marsan distinguished two nuclei, nucleus periventricularis hypothalami and nucleus filiformis. As can be seen from their microphotographs and diagrams, these are two parts of the nucleus paraventricularis. The term "periventricularis" should be reserved for cellular groups directly adjoining the ependyma of the III ventricle, which form in many mammals true periventricular nuclei with typical small round cells. The cytoarchitectonic structure of the nucleus paraventricularis, on the contrary, is characterised by large, intensively-stained cells, clearly different from those of the periventricular nuclei.

In the septal region, described by Jasper and Ajmone-Marsan briefly as "septum" (SPT), the nucleus septalis medialis, nucleus septalis lateralis, nucleus septalis triangularis, nucleus septalis fimbrialis and nucleus fasciculi diagonalis (Broca) were distinguished. The nomenclature of thalamic nuclei remained unchanged; only the nucleus lateralis dorsalis of Jasper and Ajmone-Marsan is termed nucleus lateralis anterior (Kappers 1947). The nuclei of the brain stem reticular formation are organized into three columns according to Brodal et al. (1960): the lateral parvocellular zone, the medial magnocellular zone and the zone of raphe. For different nuclei the nomenclature of Taber (1961) was used.

In the rabbit, the skull is immobilised in the stereotaxic head holder in such a way that the bregma is 1·5 mm higher than the lambda (Sawyer et al. 1954). A plane passing through the bregma and an imaginary point 1·5 mm above the lambda corresponds to the basal horizontal plane. A parallel plane lying 12 mm lower is the horizontal zero plane. Structures above this plane have the depth coordinates marked V+, below it V−. Needles were introduced perpendicularly to the basal plane into brain, 5 mm. to the left of the midline at the level of the bregma and 3·5 mm. and 7·5 mm caudally. The plane passing through the bregma and perpendicular to the above basal plane is the frontal zero plane. Structures lying rostrally to the zero frontal plane have the stereotaxic coordinates AP−, those lying caudally AP+. Distances to the left and right from the sagittal zero plane are marked in the same way as in the cat (SD).

In describing subcortical structures, the stereotaxic atlas of Sawyer et al. was used. In the thalamic region the data of this atlas were elaborated, however. A part of the midline nuclei described by Sawyer at al. briefly as massa intermedia (M) was subdivided according to d'Hollander (1913) into nucleus reuniens and nucleus rhomboidalis rostrally and nucleus centralis medialis caudally. In the medial group of thalamic nuclei, nucleus centralis lateralis (nucleus magnocellularis of d'Hollander) and nucleus paracentralis (nucleus lamellaris of d'Hollander) were further distinguished. The nucleus parafascicularis, which accompanies the tractus habenulointercrulis also belongs to this group and is the most occipital of the medial group of thalamic nuclei (Gurdjian 1927).

In the rat the skull was fixed in a similar way. The bregma, however, was only 1 mm higher than the lambda (Krieg 1946b). The plane perpendicular to the sagittal plane and passing through the bregma and a point 1 mm above the lambda is the basal horizontal reference plane and at the same time the horizontal zero plane. In a group of three rats the needles were inserted into the brain perpendicularly to the basal plane, 3 mm laterally from the sagittal zero plane at the level of bregma and 2·5 and 5·5 mm occipitally. In another group of three animals one pair of needles was introduced 4.5 mm laterally from the sagittal plane at the level of the bregma and a second one 2 mm laterally and 10 mm occipitally. A knife made from a steel wire mounted in the electrode holder was used to make a coronal cut at the level 4·5, 5·0 or 5·5 mm behind bregma (in different animals). The brain was thus divided into two parts from the coronal surface of which the tissue could be sectioned. The frontal zero plane passes through the bregma perpendicularly to the basal plane. The marking of the stereotaxic coordinates with respect to the reference planes is the same as in the cat and rabbit.

The subcortical strucures were described according to the stereotaxic atlas of Krieg (1946b), and according to the paper of Gurdjian (1927). Instead of the nucleus lateralis (Gurdjian 1927), the group of cells lying rostrally in the dorsalateral region of the thalamus is termed nucleus lateralis anterior (Kappers 1947). In describing basal ganglia and nucleus amygdalae the nomenclature of Gurdjian (1928) and Brodal (1947) is used. The different nuclei of the brainstem reticular formation were determined according to the description of Petrovický (1963a, b, 1964).

Abbreviations

A	= aqueductus mesencephali		CO	= locus coeruleus
AAA	= area amygdalaris anterior		COM	= subnucleus compactus
AB	= nucleus basalis amygdalae			nuclei pedunculopontini
AC	= nucleus centralis			tegmenti
	amygdalae		CP	= commissura posterior
ACB	= nucleus accumbens septi		CPYR	= cortex pyriformis
ACO	= nucleus corticalis amygda-		CR	= corpus restiforme
	lae		CRU	= sulcus cruciatus
AD	= nucleus antero-dorsalis		CS	= colliculus superior
AHA	= area hypothalamica		CT	= corpus trapezoides
	anterior		CU	= area cuneiformis
AHL	= area hypothalamica		CUL	= nucleus cuneatus lateralis
	lateralis		CUM	= nucleus cuneatus medius
AI	= nucleus intercalatus		CUS	= area subcuneiformis
	amygdalae			
AL	= nucleus lateralis amygdalae		DA	= nucleus Darkschwitsch
AM	= nucleus anteromedialis		DB	= nucleus fasciculi diago-
	thalami			nalis (Broca)
AMB	= nucleus ambiguus		DBC	= decussatio brachium
AME	= nucleus amygdalae			conjunctivum
	medialis		DE	= nucleus dentatus
AN	= nucleus annularis		DI	= subnucleus dissipatus nuc-
APL	= area praeoptica lateralis			lei pedunculo pontini
APM	= area praeoptica medialis			tegmenti
ARC	= nucleus arcuatus		DPYR	= decussatio pyramidum
AV	= nucleus antero-ventralis		DG	= nucleus dorsalis tegmenti
				Guddeni
BC	= brachium conjunctivum		DS	= decussatio supramammil-
BCI	= brachium colliculi inferioris			laris
BCS	= brachium colliculi superio-		DTD	= decussatio tegmenti
	ris			dorsalis
BP	= brachium pontis		DTV	= decussatio tegmenti
CA	= commissura anterior			ventralis
CC	= corpus callosum			
CCS	= commissura colliculi		E	= epiphysis
	superioris		ECSA	= sulcus ectosylvius anterior
CD	= nucleus caudatus		ECSP	= sulcus ectosylvius
CDP	= caudate-putamen complex			posterior
CES	= nucleus centralis superior		EN	= nucleus entopeduncularis
CEXT	= capsula externa		ENT	= area entorhinalis
CF	= commissura fornicis		EW	= nucleus Edinger-Westphal
CH	= chiasma opticum			
CHA	= commissura habenularum		F	= fornix
CI	= capsula interna		FF	= fimbria fornicis
CIF	= colliculus inferior		FI	= fossa intercruralis
CL	= claustrum		FLM	= fasciculus longitudinalis
CM	= centrum medianum			medialis

FO	=	fasciculus opticus
G	=	nucleus gracilis
GL	=	corpus geniculatum laterale
GLV	=	corpus geniculatum laterale, pars ventralis
GM	=	corpus geniculatum mediale
GP	=	globus pallidus
H	=	hypophysis
HDM	=	nucleus dorsomedialis hypothalami
HIP	=	hippocampus
HL	=	nucleus habenulae lateralis
HM	=	nucleus habenulae medialis
HVM	=	nucleus ventromedialis hypothalami
IAM	=	nucleus inter-antero-medialis
IP	=	nucleus interpeduncularis
IS	=	nucleus interstitialis
IV	=	nucleus interventralis
LA	=	nucleus lateralis anterior
LAT	=	sulcus lateralis
LC	=	nucleus linearis caudalis
LI	=	nucleus linearis inter-medius
LIM	=	nucleus limitans
LL	=	lemniscus lateralis
LM	=	lemniscus medialis
LME	=	lamina medullaris externa
LP	=	nucleus lateralis posterior
LR	=	nucleus linearis rostralis
MD	=	nucleus medio-dorsalis
ML	=	nucleus mammillaris lateralis
MM	=	nucleus mammillaris medialis
MV.	=	radix mesencephalica nervi trigemini
N III	=	nucleus originis nervi oculomotorii

N IV	=	nucleus originis nervi trochlearis
N VI	=	nucleus originis nervi abducentis
N VII	=	nucleus originis nervi facialis
N X	=	nucleus terminationis nervi vagi
N XII	=	nucleus originis nervi hypoglossi
NAM	=	nucleus anteromedialis
NB	=	nucleus reticularis tegmenti pontis Bechterewi
NCAST	=	nucleus commissurae anterioris et striae terminalis
NCL	=	nucleus centralis lateralis
NCM	=	nucleus centralis medialis
NCP	=	nucleus commissurae posterioris
NCT	=	nucleus corpus trapezoides
ND	=	nucleus terminationis dorsalis nervi vestibuli (Bechterew)
NFS	=	nucleus fimbrialis septi
NHD	=	nucleus hypothalamicus dorsalis
NHP	=	nucleus hypothalamicus posterior
NL	=	nucleus terminationis lateralis nervi vestibuli (Deiters)
NLLD	=	nucleus lemnisci lateralis dorsalis
NLLV	=	nucleus lemnisci lateralis ventralis
NM	=	nucleus terminationis medialis nervi vestibuli (Schwalbe)
NO V	=	nucleus originis nervi trigemini
NP	=	nuclei pontis
NPC	=	nucleus paracentralis
NPL	=	nucleus paralemniscalis
NR	=	nucleus ruber
NS	=	nucleus terminationis spinalis nervi vestibuli (Roller)
NS V	=	nucleus tractus spinalis nervi trigemini

NSL	= nucleus septalis lateralis		RSAG	= area retrosplenialis agranularis
NSM	= nucleus septalis medialis			
NT V	= nucleus terminationis nervi trigemini		RSG	= area retrosplenialis granularis
NTOLFLAT	= nucleus tractus olfactorii lateralis		RV	= nucleus reticularis ventralis
NTS	= nucleus triangularis septi			
			S	= stria medullaris
OI	= oliva inferior		SA	= nucleus sagulum
OS	= oliva superior		SCC	= sulcus corporis callosi
			SCH	= nucleus suprachiasmaticus
			SG	= nucleus suprageniculatus
P	= putamen		SGC	= stratum griseum centrale
P XII	= nucleus praehypoglossus		SIL	= sulcus cerebri lateralis
PB	= nucleus parabrachialis		SM	= nucleus submedius
PC	= pedunculus cerebri		SN	= substantia nigra
PF	= nucleus parafascicularis		SO	= nucleus supraopticus
PM	= pedunculus mammillaris		SOD	= nucleus supraopticus diffusus
PP	= nucleus peripeduncularis			
PR	= praetectum		SPF	= nucleus sub-parafascicularis
PT	= nucleus parataenialis			
PTH	= nucleus posterior thalami		SPL	= sulcus splenialis
PU	= pulvinar		SPM	= nucleus supramammillaris
PV	= nucleus paraventricularis		SSPL	= sulcus suprasplenialis
PVA	= nucleus paraventricularis anterior		ST	= stria terminalis
			STH	= nucleus subthalamicus
PVP	= nucleus paraventricularis posterior		SUPS	= sulcus suprasylvius
PYR	= pyramides		TA	= taenia anterior
			TAC	= tuberculum acusticum
R	= nucleus reticularis		THP	= tractus habenulo-peduncularis
RAD	= nucleus raphe dorsalis			
RAM	= nucleus raphe magnus		TMG	= tractus mammillo-tegmentalis
RAN	= sulcus rhinalis anterior			
RAP	= nucleus raphe parvus		TO	= tractus opticus
RD	= nucleus reticularis dorsalis		TMT	= tractus mammillo-thalamicus
RE	= nucleus reuniens			
RG	= nucleus reticularis gigantocellularis		TOLFLAT	= tractus olfactorius lateralis
RH	= nucleus rhomboidalis		TS	= nucleus tractus solitarii
RL	= nucleus reticularis lateralis		TS V	= tractus spinalis nervi trigemini
RP	= nucleus reticularis parvocellularis			
			TTC	= tractus tegmentalis centralis
RPC	= nucleus reticularis pontis caudalis			
			TUBOLF	= tuberculum ofactorium
RPG	= nucleus reticularis paragigantocellularis			
			V	= ventriculus lateralis
RPO	= nucleus reticularis pontis oralis		V III	= ventriculus tertius
			V IV	= ventriculus quartus
RPS	= sulcus rhinalis posterior		VA	= nucleus ventralis anterior

VDM	= nucleus ventralis pars dor-somedialis		VPM	= nucleus ventralis **postero-**medialis
VE	= nucleus ventralis thalami			
VG	= nucleus ventralis tegmenti Guddeni		ZI	= zona incerta
			III.	= nervus oculomotorius
			IV.	= nervus trochlearis
VL	= nucleus ventralis lateralis		V.	= nervus trigeminus
VM	= nucleus ventralis medialis		VI.	= nervus abducens
VPL	= nucleus ventralis postero-lateralis		VII.	= nervus facialis
			VIII.	= nervus statoacusticus

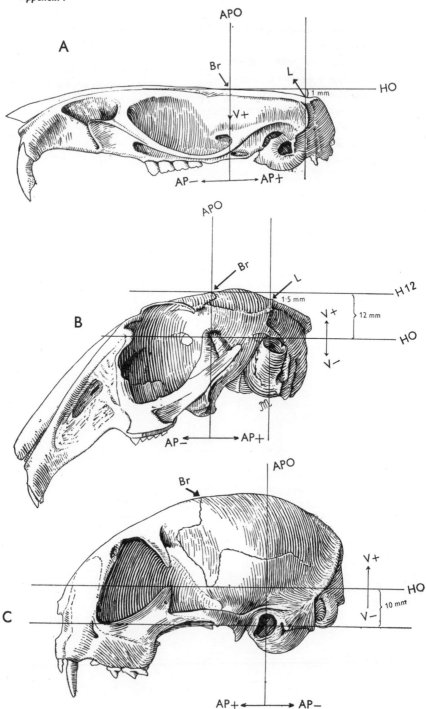

Fig. 300. The skulls of rat (A), rabbit (B) and cat (C) (lateral aspect) illustrating the construction of the horizontal and vertical reference planes. For details see text.

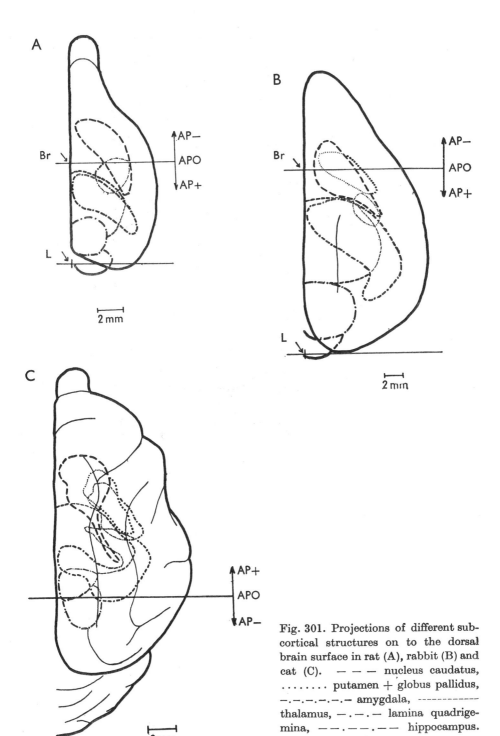

Fig. 301. Projections of different subcortical structures on to the dorsal brain surface in rat (A), rabbit (B) and cat (C). — — — nucleus caudatus, ... putamen + globus pallidus, —.—.—.—.—. amygdala, ------------ thalamus, —.—.— lamina quadrigemina, ——.——.—— hippocampus.

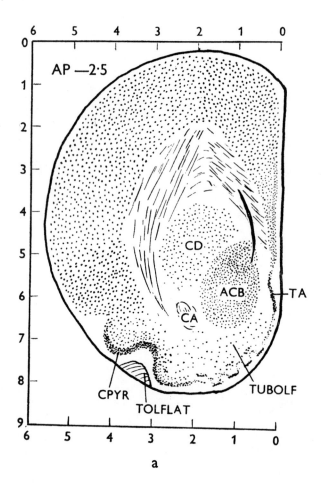

a

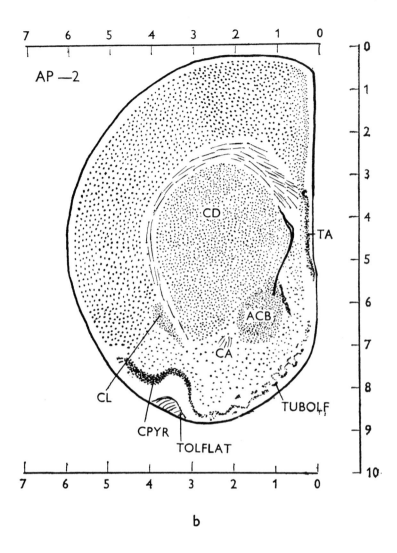

b

Fig. 302a, b. Coronal sections of brain. Rat.

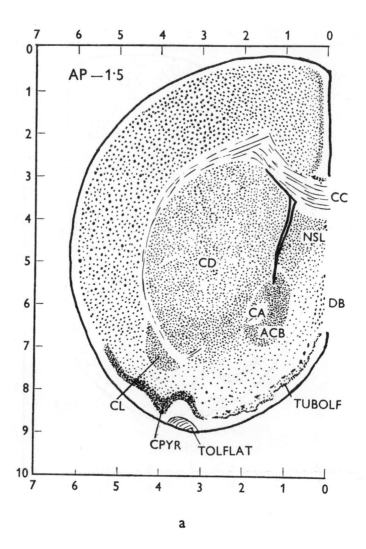

AP —1·5

CC

NSL

CD

DB

CA

ACB

CL

TUBOLF

CPYR

TOLFLAT

a

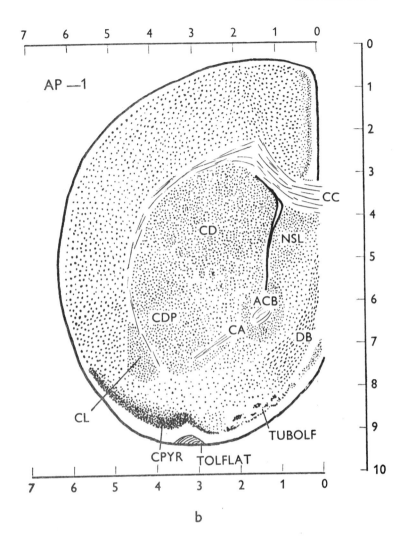

AP —1

CC

CD

NSL

ACB

CDP

CA

DB

CL

TUBOLF

CPYR   TOLFLAT

b

Fig. 303a, b. Coronal sections of brain. Rat.

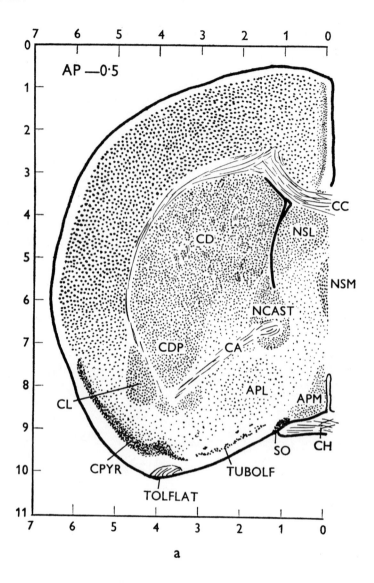

a

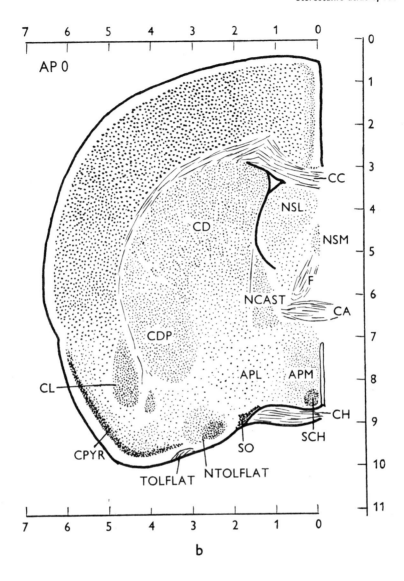

Fig. 304a, b. Coronal sections of brain. Rat.

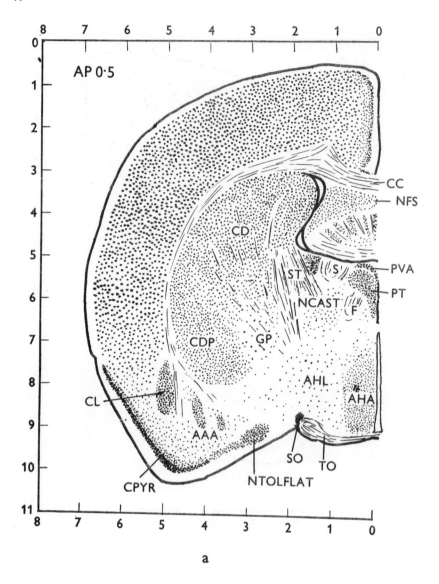

AP 0·5

a

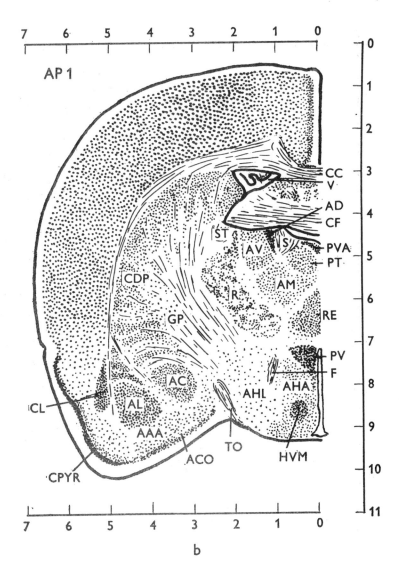

Fig. 305a, b. Coronal sections of brain. Rat.

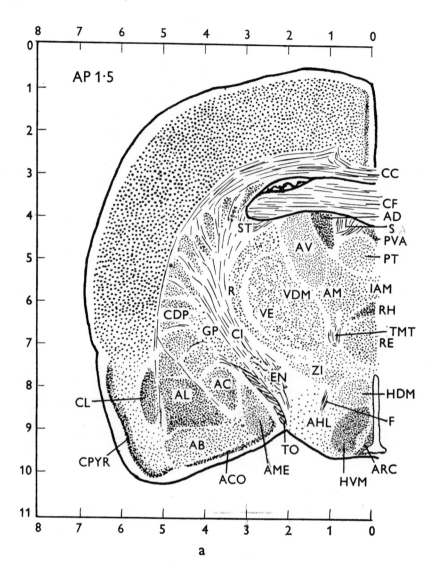

AP 1·5

a

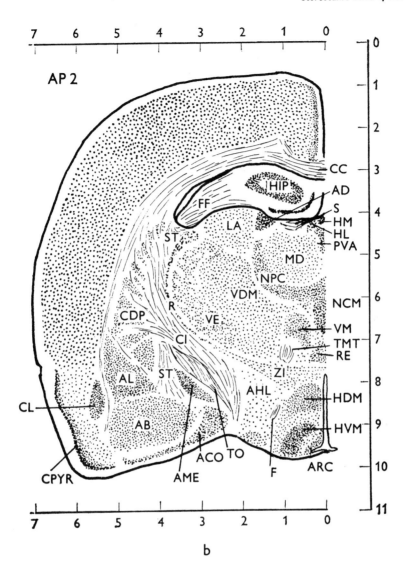

Fig. 306a, b. Coronal sections of brain. Rat.

671

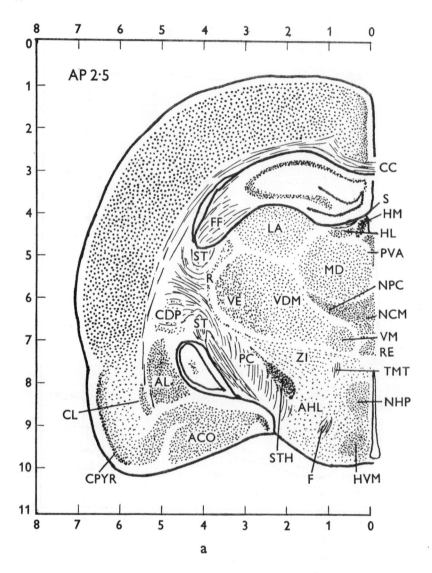

AP 2·5

CC
S
HM
HL
PVA
MD
NPC
NCM
VM
RE
TMT
NHP

FF
LA
ST
R
VE
VDM
CDP
ST
PC
ZI
AL
CL
AHL
ACO
STH
CPYR
F
HVM

a

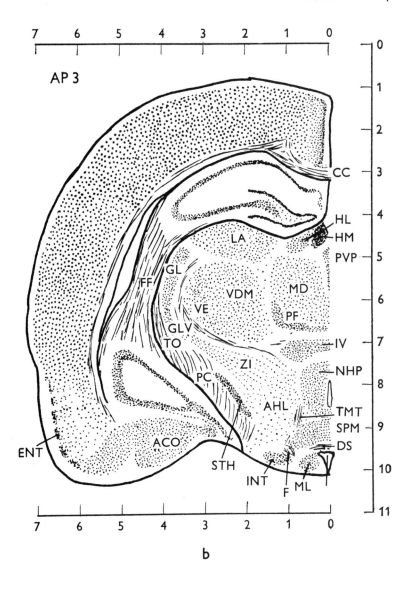

Fig. 307a, b. Coronal sections of brain. Rat.

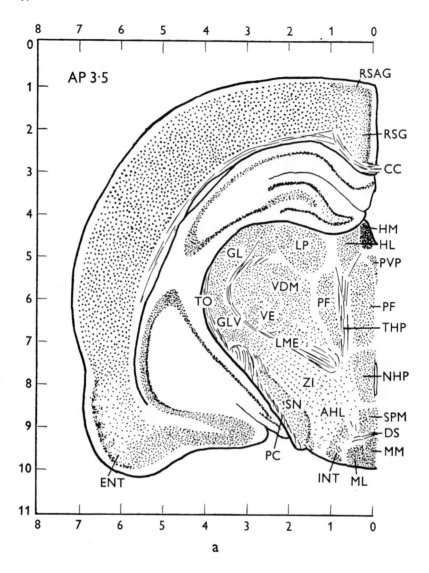

AP 3·5

a

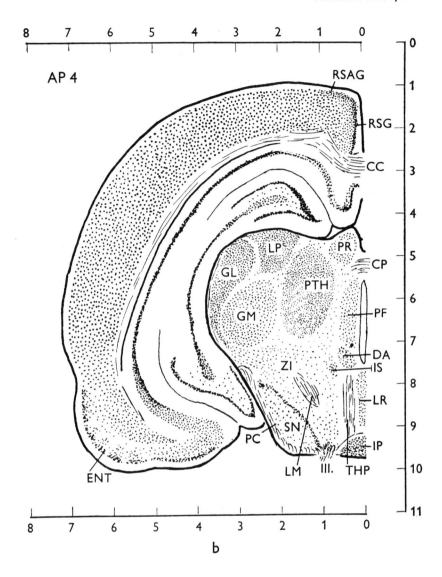

Fig. 308a, b. Coronal sections of brain. Rat.

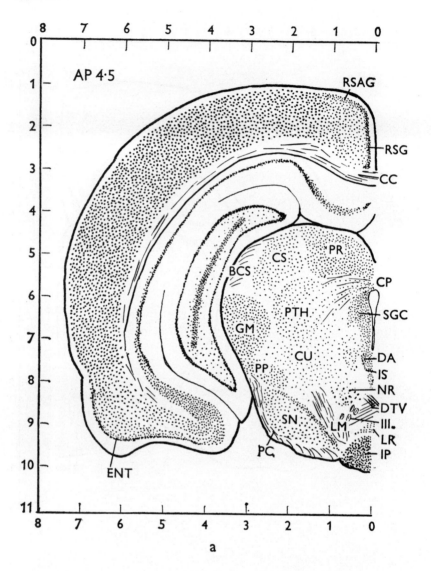

AP 4·5

a

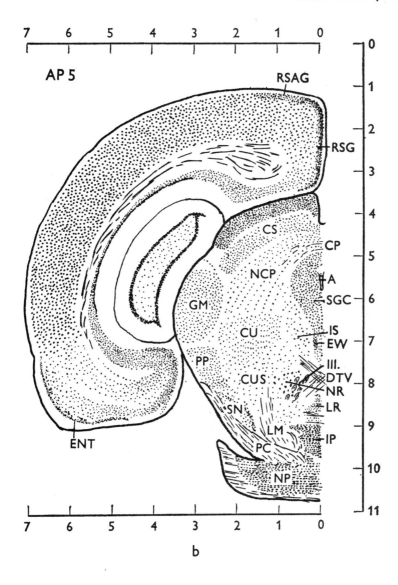

Fig. 309a, b. Coronal sections of brain. Rat.

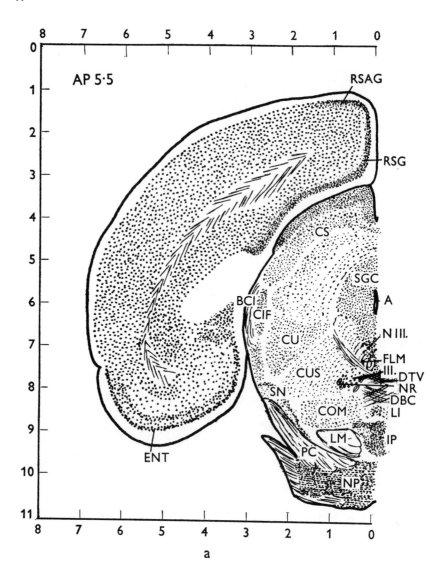

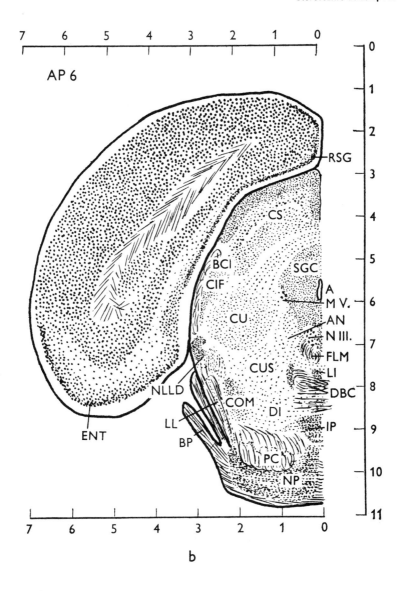

Fig. 310a, b. Coronal sections of brain. **Rat.**

679

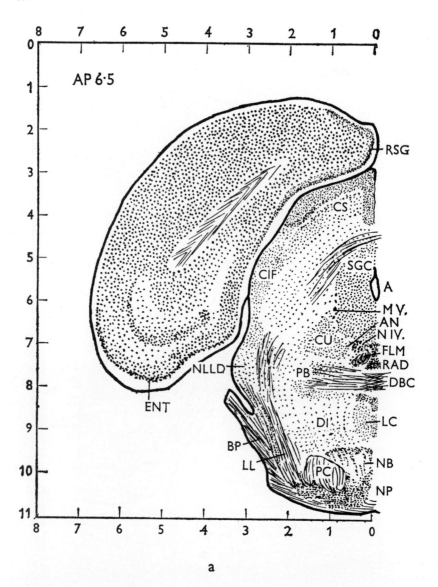

AP 6·5

a

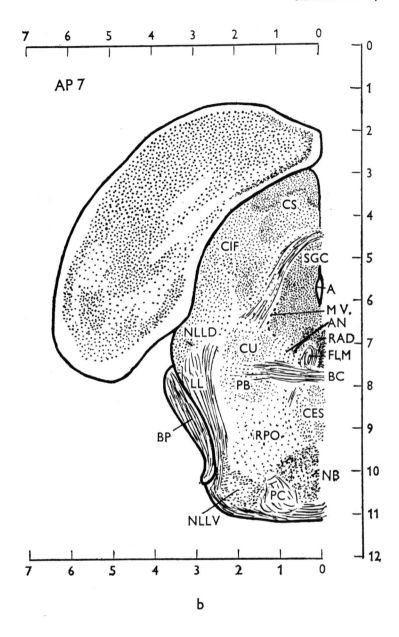

Fig. 311a, b. Coronal sections of brain. Rat.

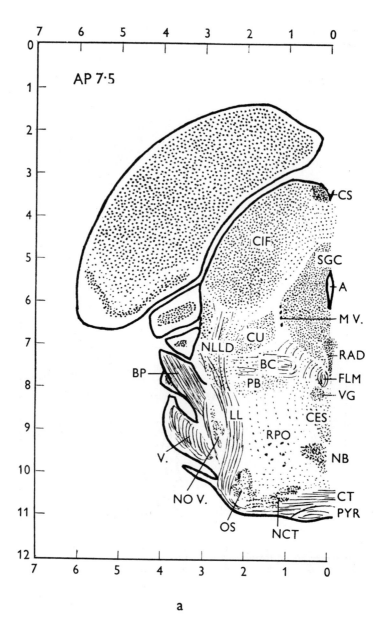

AP 7·5

a

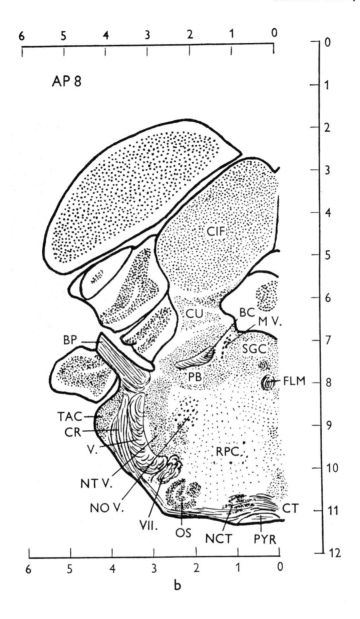

Fig. 312a, b. Coronal sections of brain. Rat.

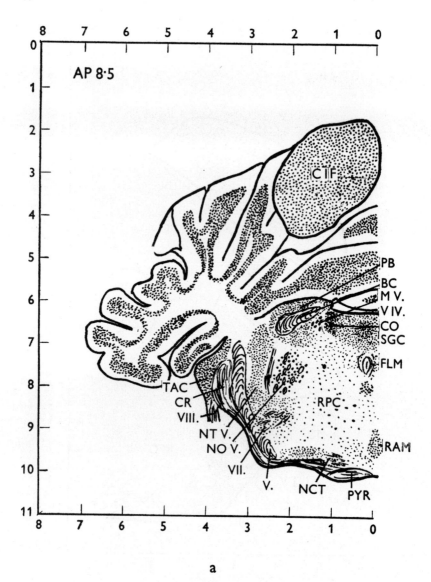

a

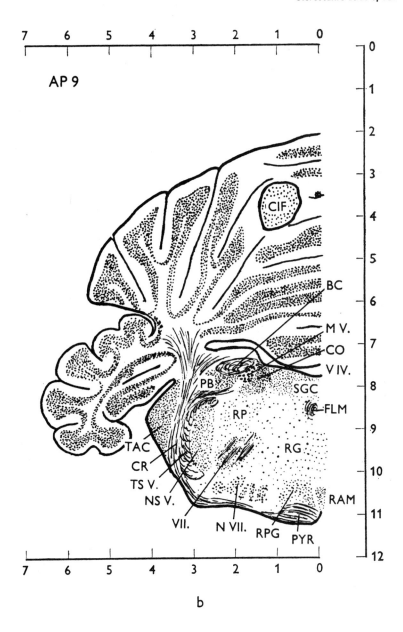

Fig. 313a, b. Coronal sections of brain. Rat.

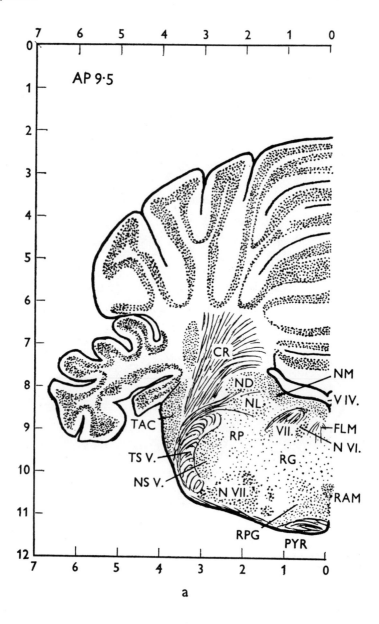

a

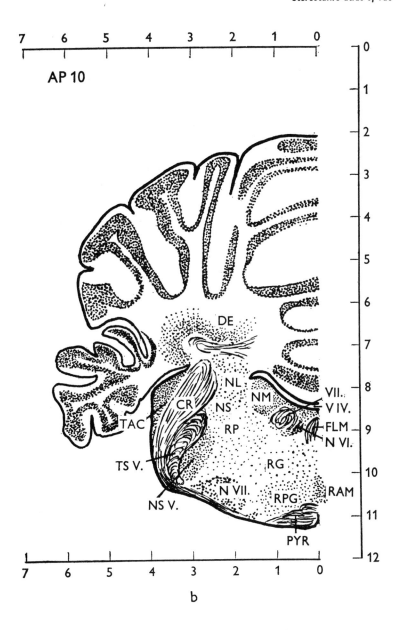

Fig. 314a, b. Coronal sections of brain. Rat.

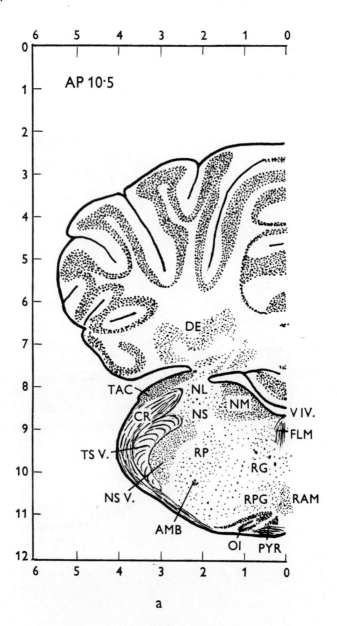

a

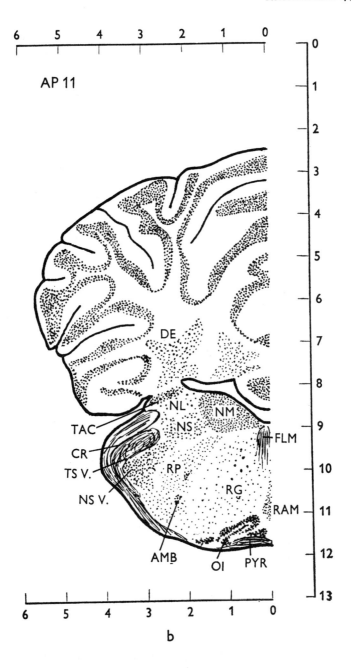

Fig. 315a, b. Coronal sections of brain. Rat.

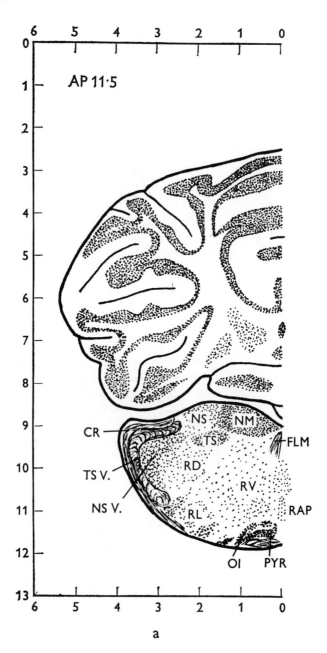

a

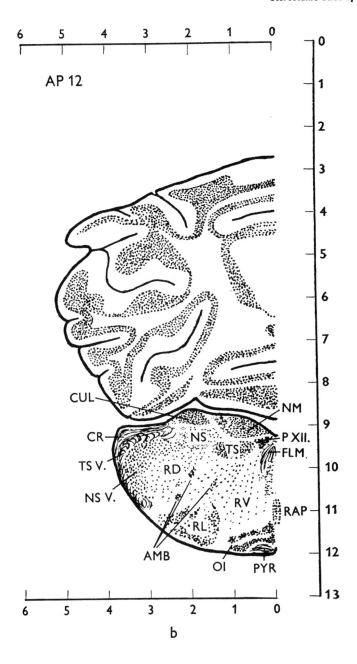

Fig. 316a, b. Coronal sections of brain. Rat.

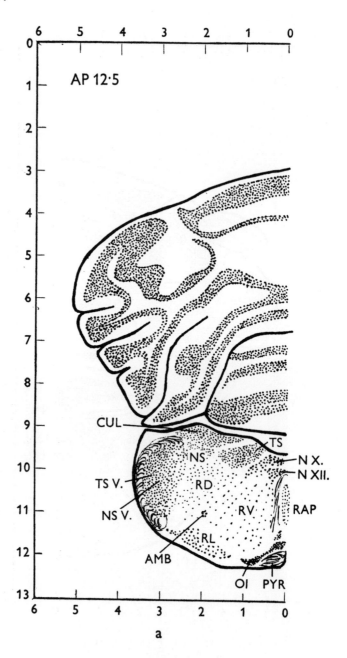

a

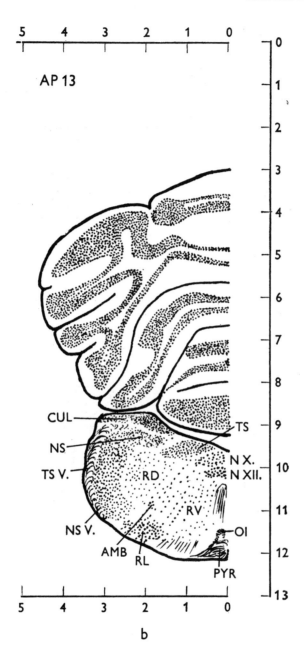

Fig. 317a, b. Coronal sections of brain. **Rat.**

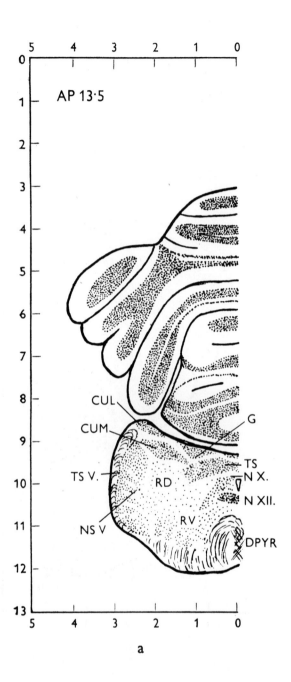

a

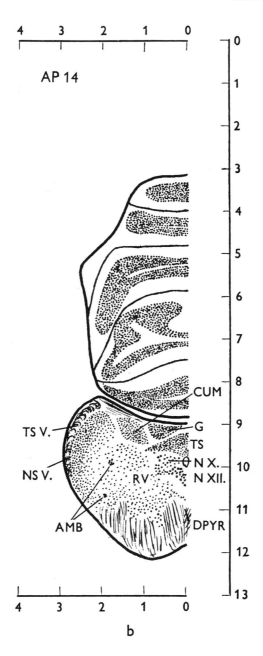

Fig. 318a, b. Coronal sections of brain. Rat.

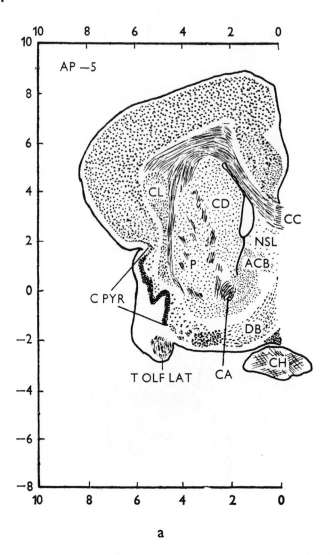

a

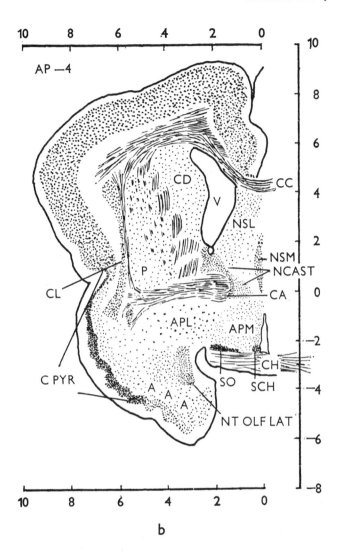

Fig. 319a, b. Coronal sections of brain. Rabbit.

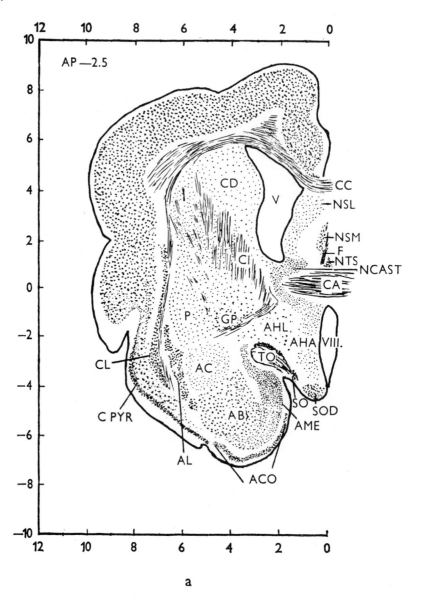

AP —2.5

a

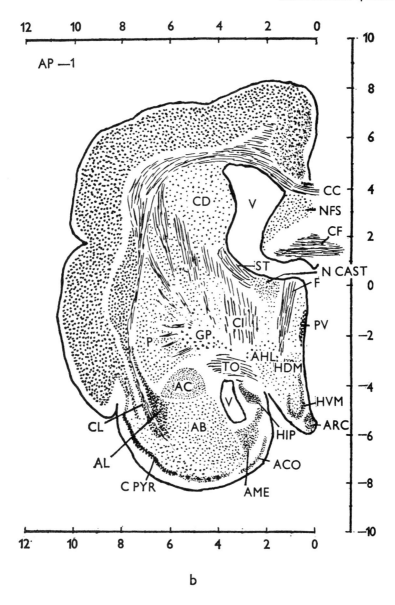

AP —1

Fig. 320a, b. Coronal sections of brain. Rabbit.

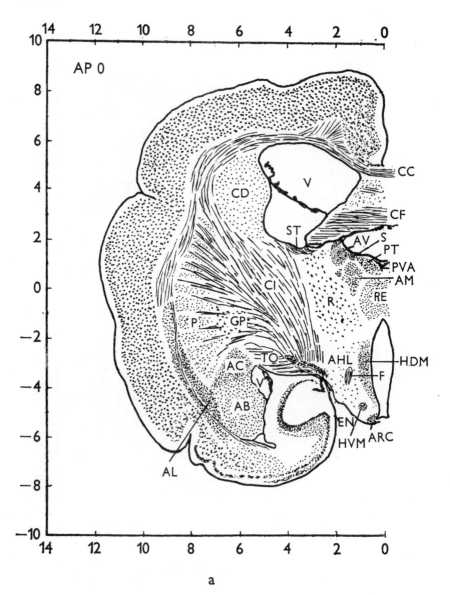

a

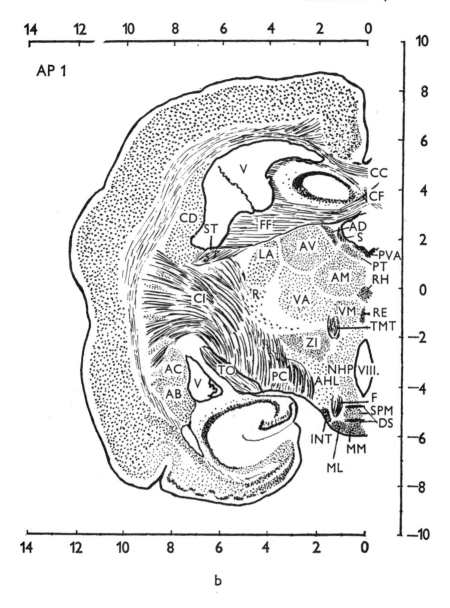

Fig. 321a, b. Coronal sections of brain. Rabbit.

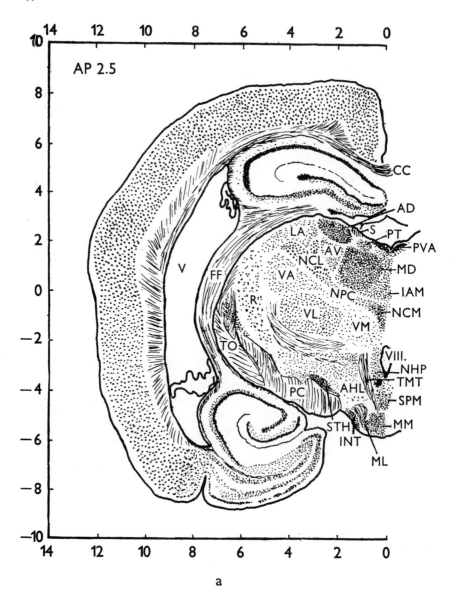

a

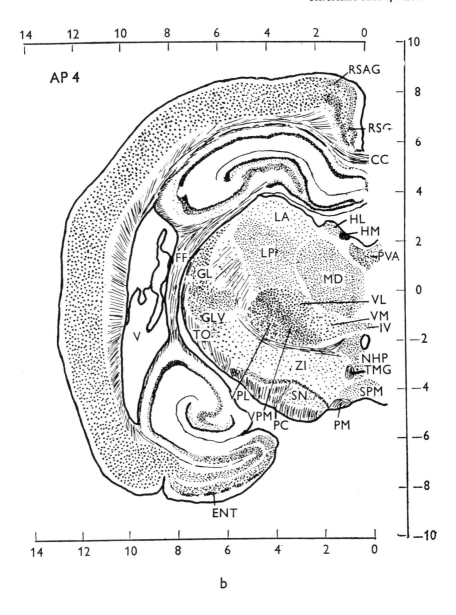

Fig. 322a, b. Coronal sections of brain. Rabbit.

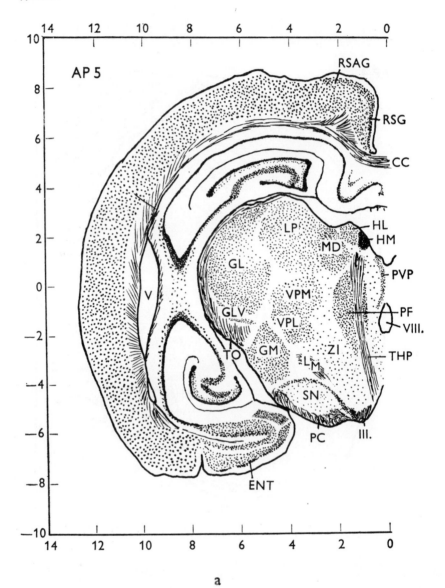

a

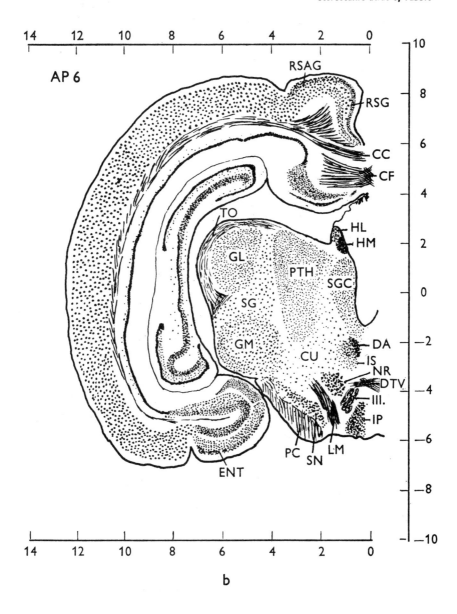

Fig. 323a, b. Coronal sections of brain. **Rabbit.**

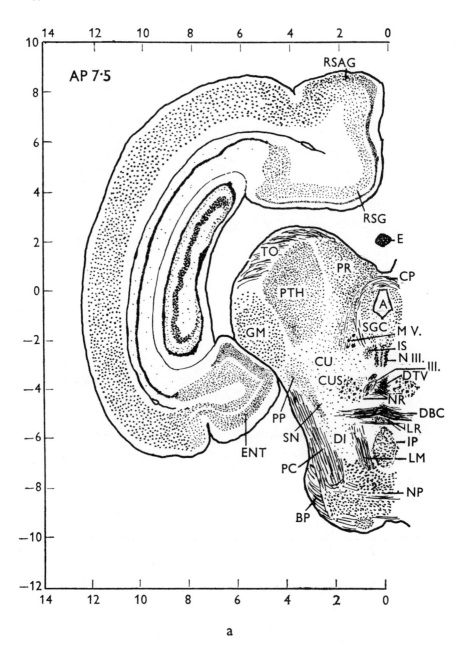

a

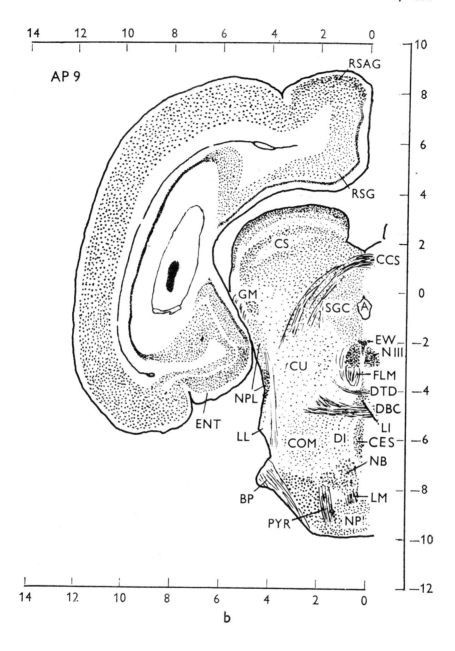

Fig. 324a, b. Coronal sections of brain. Rabbit.

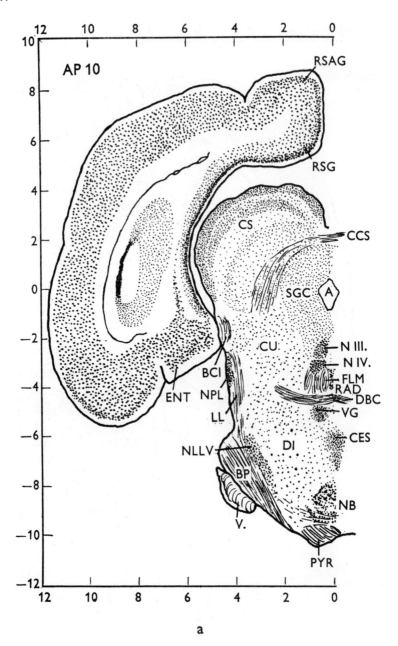

AP 10

a

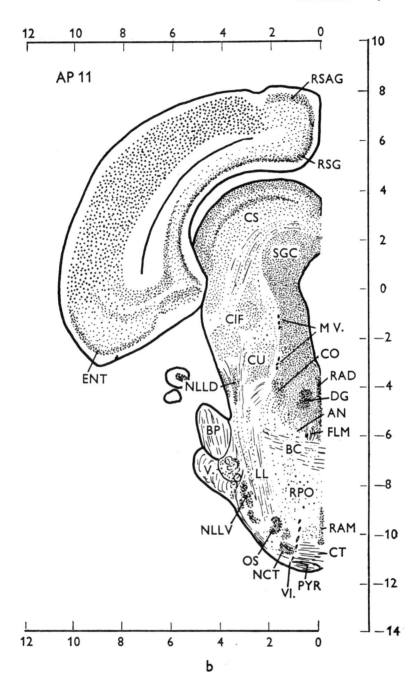

AP 11

RSAG

RSG

CS

SGC

CIF

CU

M V.

CO

RAD

DG

AN

FLM

BC

ENT

NLLD

BP

V.

LL

RPO

NLLV

OS

NCT

VI.

PYR

RAM

CT

Fig. 325a, b. Coronal sections of brain. Rabbit.

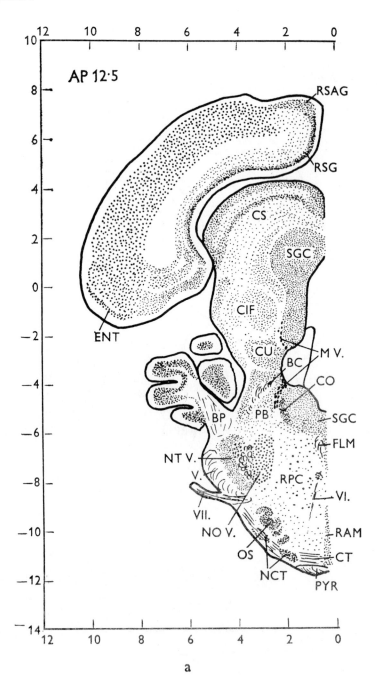

a

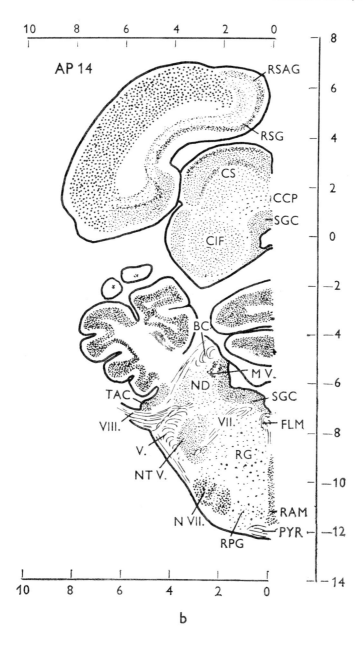

Fig. 326a, b. Coronal sections of brain. Rabbit.

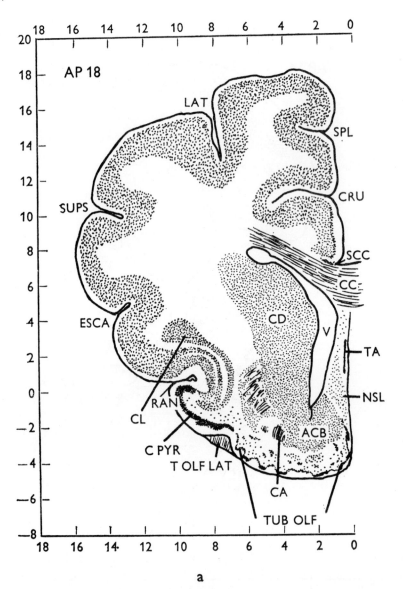

a

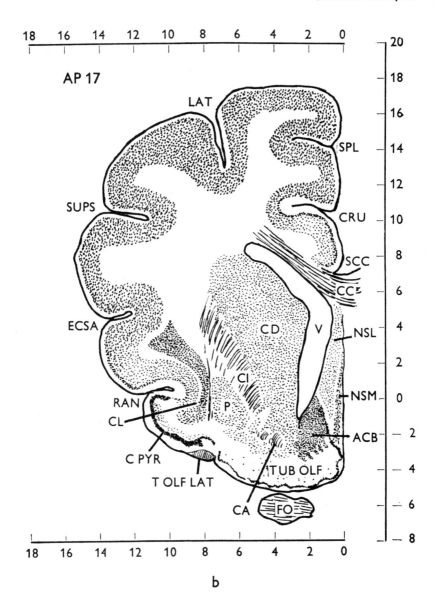

Fig. 327a, b. Coronal sections of brain. Cat.

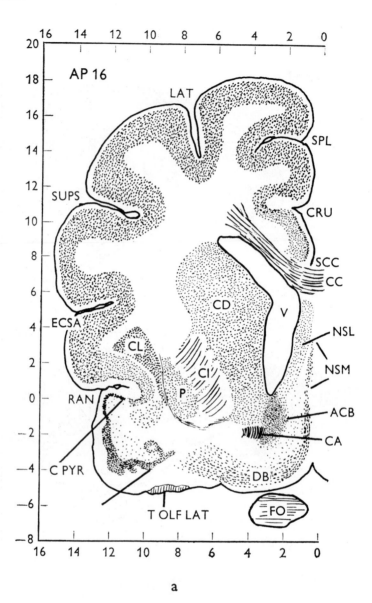

a

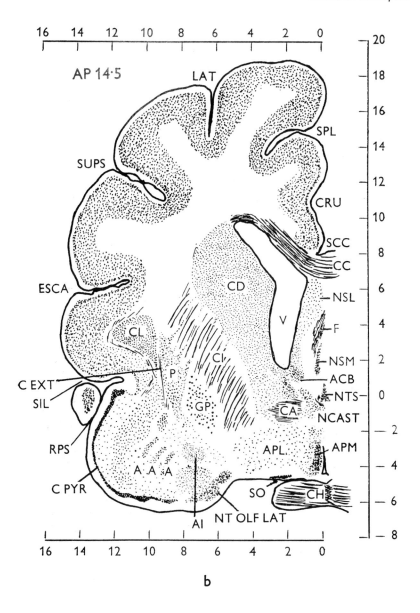

b

Fig. 328a, b. Coronal sections of brain. Cat.

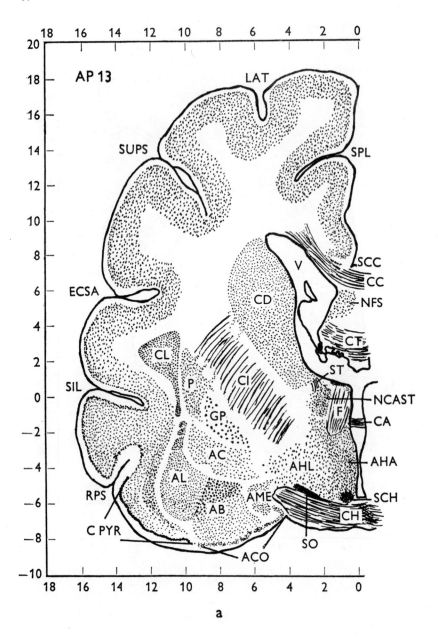

a

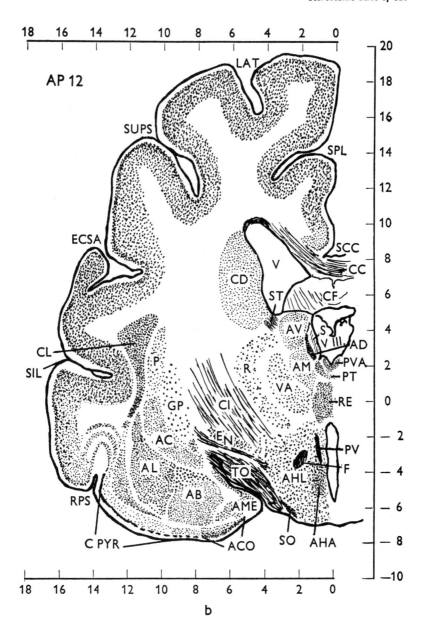

Fig. 329a, b. Coronal sections of brain. Cat.

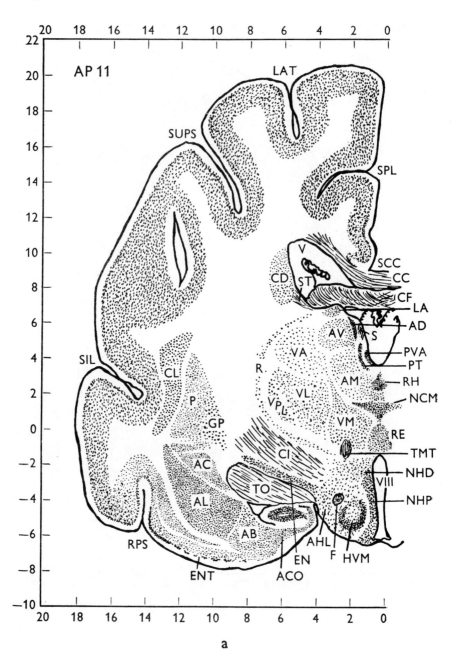

a

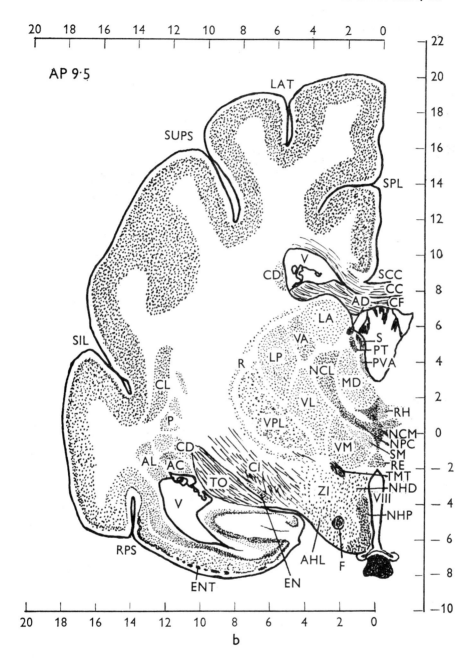

Fig. 330a, b. Coronal sections of brain. Cat.

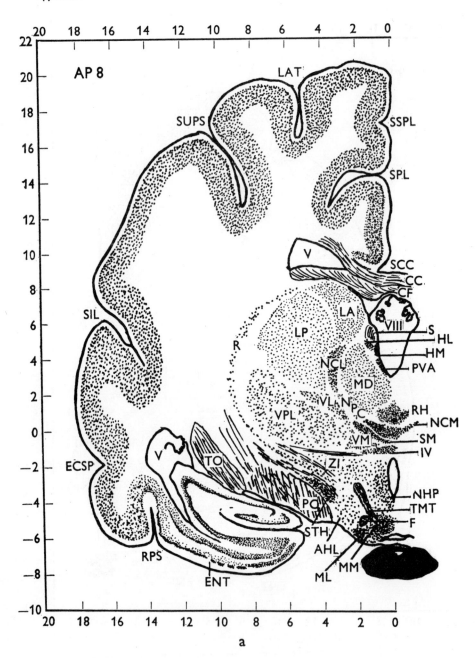

AP 8

a

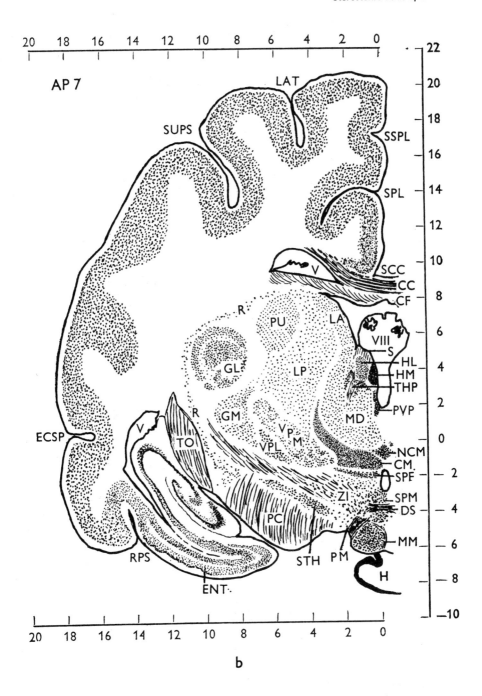

Fig. 331a, b. Coronal sections of brain. Cat.

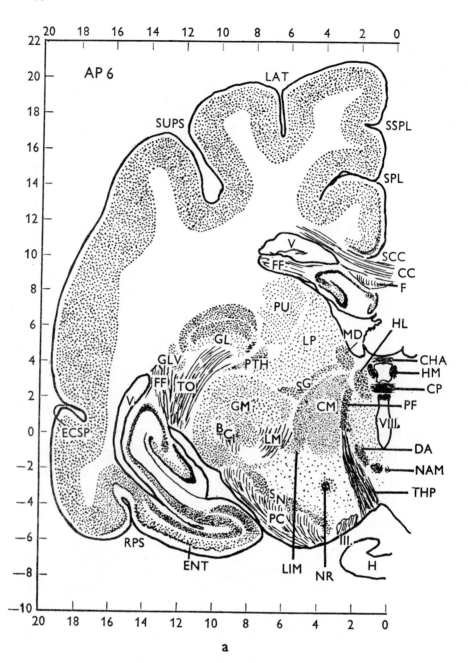

722

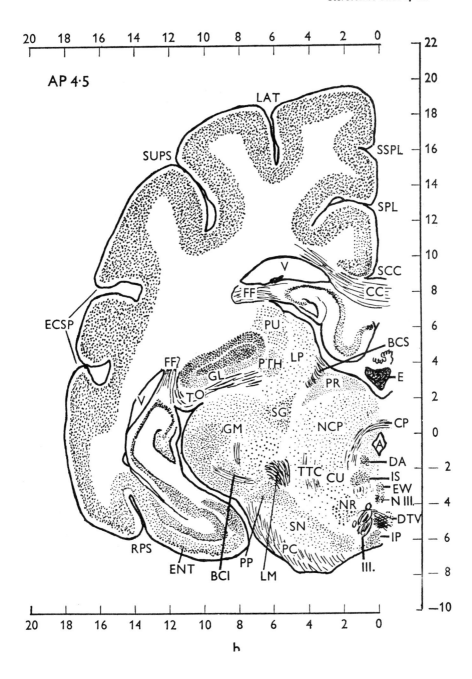

Fig. 332a, b. Coronal sections of brain. Cat.

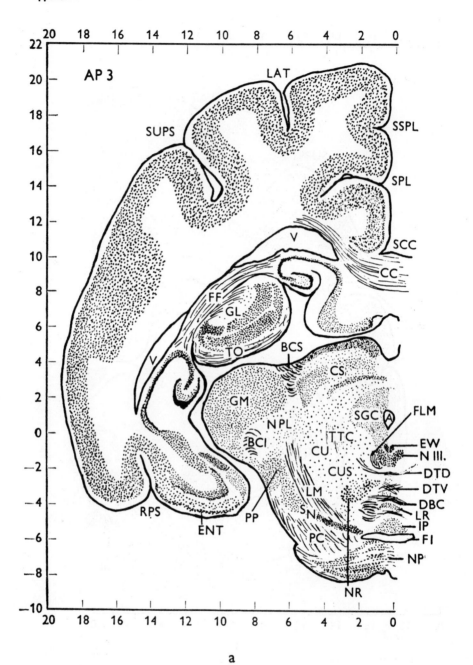

a

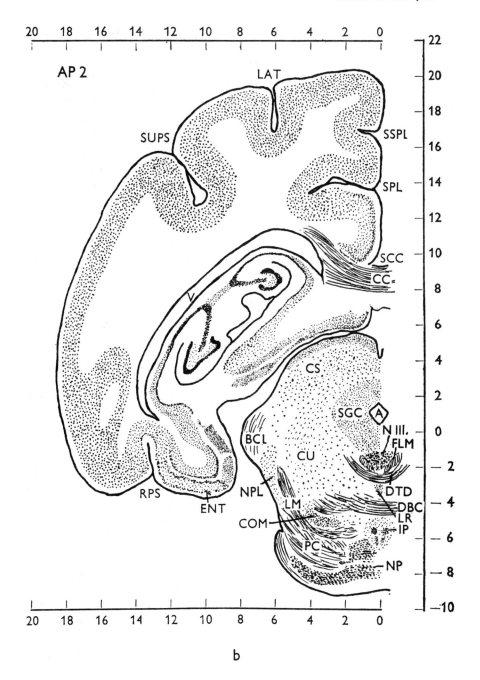

b

Fig. 333a, b. Coronal sections of brain. Cat.

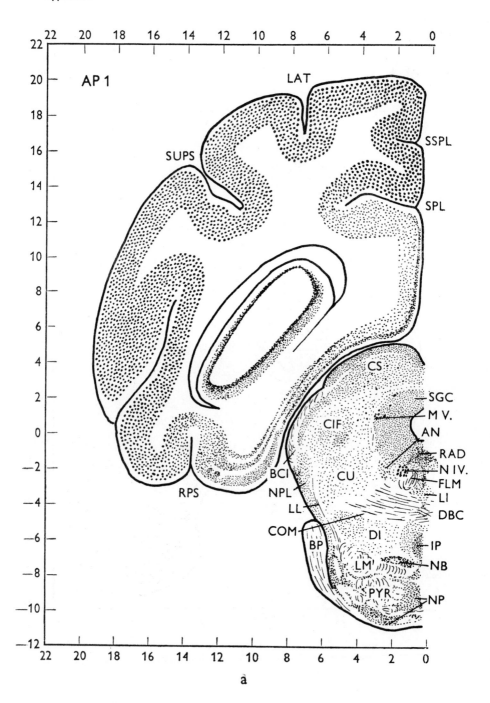

a

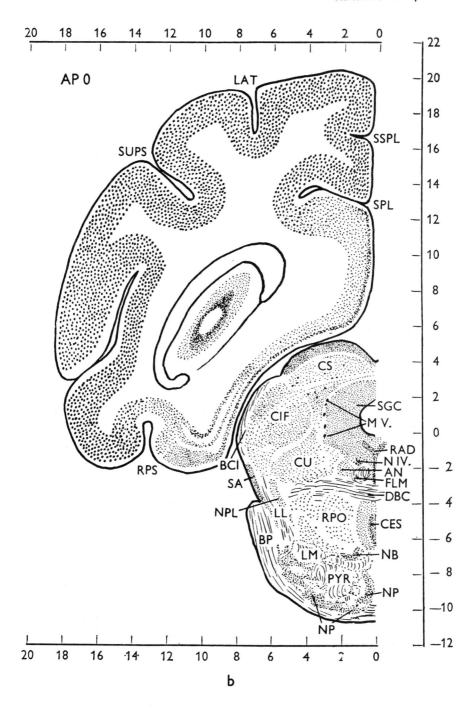

Fig. 334a, b. Coronal sections of brain. Cat.

727

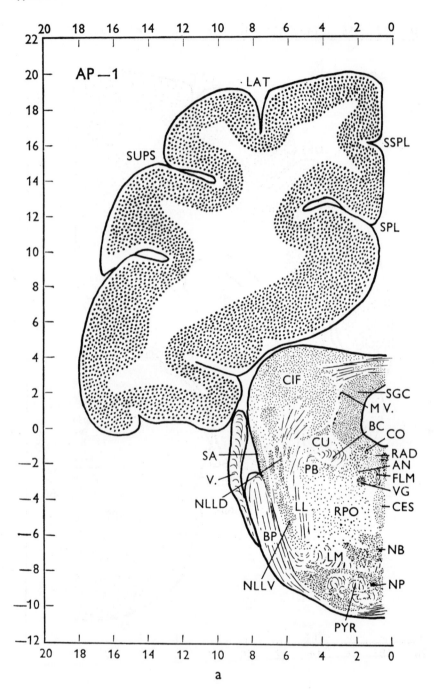

AP—1

a

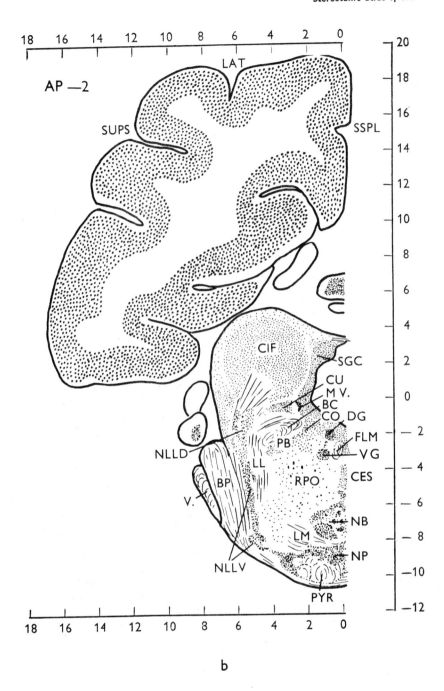

Fig. 335a, b. Coronal sections of brain. Cat.

729

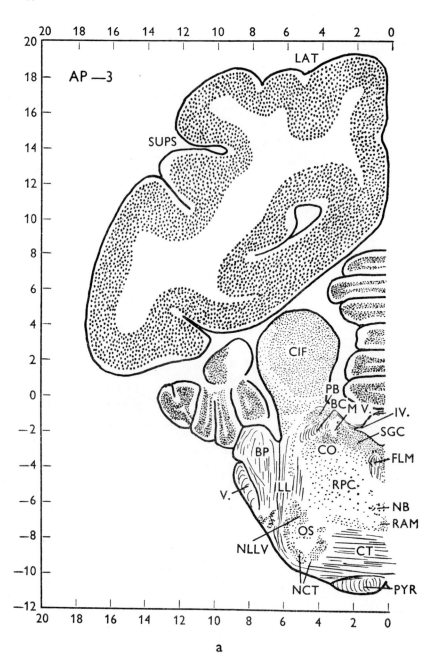

a

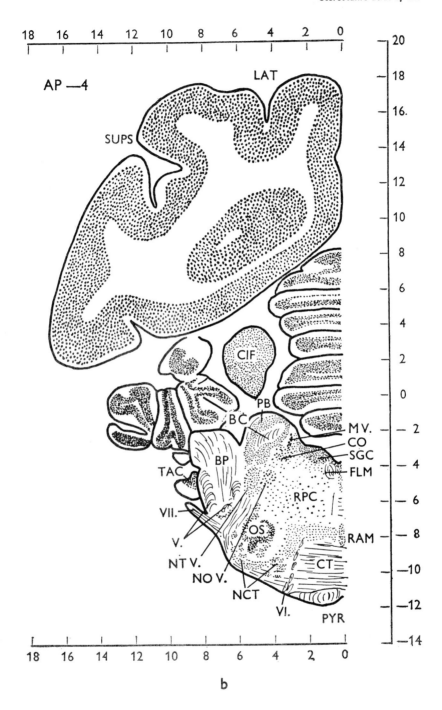

Fig. 336a, b. Coronal sections of brain. Cat.

Appendix II

Mathematical terms

On the advice of the reviewers and other friends some terms of differential calculus, the theory of probability and the theory of stochastic processes are explained below. Since we are not mathematicians and since the appendix must of necessity be short, it is not meant to be complete or mathematically exact, but serves only to illustrate some major points. More detailed data must be sought for in the special literature, e. g. Zeldovich (1963), Loeve (1960), Bartlett (1955).

Function

If one quantity depends on another, it is said to be its function. Independent variables are called the argument of the function (there may be more than one), the dependent variable is termed a function. Variables that only change from one special case to another are termed parameters. For instance in the equation for a circle: $y^2 = r^2 - x^2$, r, the radius, is a parameter.

The value of a function

This is the number or value of a physical quantity which is attained by the function for the given value of the argument. The values of a function or argument are often designated by numerical indices: $y_1 = f(x_1); y_0 = f(5)$ etc.

The limit of a function

This is the quantity which, without limitations, approaches the value of the function if the quantity of the argument, also without limitation, approaches a certain given value. It is written e. g. $\lim_{x \to 2} [(x^2 - 4)/(x - 2)] = + 4$ and reads as: the limit of the given function for x approaching two is four. (As can be seen from the example it is often not sufficient when searching for the limit, to insert the value of the argument into the functional relationship and to calculate the corresponding value of the function i. e. the limit. More complex considerations and procedures are necessary.)

Interval

of the argument or function is a set of values which may be attained by the argument or function. The interval is given by its limits and may be closed with respect to them, i. e. if the limits are its components, i. e. if $x_1 \leqq x \leqq x_2$. It is designated e. g. $\langle x_1 , x_2 \rangle$, where x_1 , x_2 are certain values of x. It may also be open, if x_1 and x_2 are values that may not be attained by x but which may be approached without limitation, i. e. $x_1 < x < x_2$. This is designated (x_1 , x_2). Finally the interval may be closed at one side and open at the other, $\langle x_1 , x_2)$ or $(x_1 , x_2 \rangle$. In statistics, for instance, class intervals must be open at one end and closed at the other.

Differences or increments

of the argument or function are most frequently designated by the Greek letter Δ. Thus $\Delta x = x_2 - x_1$ or $\Delta x = x - x_0$, similarly $\Delta y = y - y_0$ etc.

Derivative function or differential quotient

expresses how the function changes in dependence on unlimited small changes in the argument. If $y = f(x)$ (1)

then $y + \Delta y = f(x + \Delta x)$.

The ratio $\Delta y / \Delta x = [f(x + \Delta x) - f(x)]/\Delta x = \Delta f(x)/\Delta x$ is called the difference quotient and its limit for Δx approaching zero without limitation the differential quotient or derivative

$$\lim_{\Delta x \to 0} \frac{\Delta y}{\Delta x} = \lim_{\Delta x \to 0} \frac{f(x + \Delta x) - f(x)}{\Delta x} = \frac{dy}{dx} = \frac{df(x)}{dx} = d\frac{f(x)}{dx} = y' \qquad (2)$$

The meaning of the differential quotient can be very different but two examples are usually used to explain it:

733

a) Let us take a function $y = f(x)$ and its graphical representation. Let us take two close points A, B of coordinates $A(x_1, y_1)$, $B(x_2, y_2)$ and let us draw a secant, then

$$\frac{y_2 - y_1}{x_2 - x_1} = \frac{\Delta y}{\Delta x} = k = \tan \varphi \tag{3}$$

is the slope of this secant, i.e. the tangent of its angle of intersection with the abscissa. If B approaches A without limitation, the secant changes into the tangent and its slope will no longer depend on Δx and Δy (these are infinitely small) but only on the values x and $y = f(x)$. The dependence of the slope on x can be found by finding the differential quotient:

$$\tan \varphi = k = \lim_{\Delta x \to 0} \frac{\Delta y}{\Delta x} = \frac{dy}{dx} = \frac{df(x)}{dx} \tag{4}$$

thus

$$g(x) = \frac{df(x)}{dx} \tag{4a}$$

is again a function of the same x as is $f(x)$, but another function than $f(x)$. This is found by calculating the corresponding limit of the difference quotient from the above relationships, which, of course, often requires more profound considerations. Rules for calculating this can be found in textbooks of differential calculus.

b) Now let us imagine that the tip of a pencil is led along the curve of the function $y = f(x)$ towards the abscissa at a constant velocity in the direction of the X-axis. Here the rate of movement in the direction of the Y-axis changes in dependence on the slope of the curve, i. e. on $\tan \varphi$, i. e. on dy/dx. Here we have chosen time as the independent variable (argument) and the differential quotient of the given time function thus signifies the rate of change of this function in time. Since time is very often the argument in physical equations the differential quotient often signifies the rate of change, e. g. the rate of change in position, i. e. velocity or some other speed.

Second order differential quotient

Function (4) can be treated in the same way as function (1) and thus we obtain the second order differential quotient of the function $f(x)$

$$y'' = \frac{d(y')}{dx} = \frac{d[df(x)/dx]}{dx} = \frac{d\, df(x)}{(dx)^2} = \frac{d^2 f(x)}{(dx)^2} \tag{5}$$

The significance of the second order differential quotient is the following:

a) geometrical. In the graph of $y = f(x)$, $\dfrac{d^2 f(x)}{(dx)^2}$ indicates the rate of change of the tangent, i. e. its significance is related to the curvature of the curve.

b) physical. The second order differential quotient of a time function signifies the rate of change of velocity, i. e. acceleration. Higher order derivatives rarely are of greater significance.

Differentials and differential equations

From a definition of the derivative we obtain

$$\Delta f(x) = y' \Delta x + \varepsilon \tag{6}$$

where ε is a small value by which the product $\Delta x \cdot d f(x)/d x$ must be corrected, (where Δx is a *finite* increment of x) in order to obtain an exact value for the *finite* increment of the function. Often, of course, we can approximate

$$\Delta f(x) \approx y' \Delta x \tag{7}$$

In order to be able to write an exact equality instead of the approximation, or in order to leave out the correction we often write unlimited small increments d, called differentials, instead of the finite increments Δ, so that

$$dy = df(x) = \frac{df(x)}{dx} dx = y' dx \tag{8}$$

Equations containing differentials of variables or derivative functions are called differential equations. They are very important in physics since often we can only derive the differential equation logically (for limitless small changes), and from this, often by very devious routes, we can calculate the equations for finite values of the argument and function (see integral).

The function of several arguments, partial differential quotient and total differential

Let us take the volume of a quadrangular prism and its dependence on its dimensions x, y, z:

$$V = x \cdot y \cdot z \tag{9}$$

Now the concept derivative ceases to be unambiguous since any of the variables x, y, z may be taken as the argument and hence three different derivatives which may be designated $V'_{(x)}$, $V'_{(y)}$, $V'_{(z)}$ may be calculated, where the index designates the argument that is considered to be variable. The others are taken as constants or parameters. For such derivatives we no longer use the notation dV/dx, but instead of d small modified delta is used:

$$V'_{(x)} = \frac{\partial V}{\partial x} ; \quad V'_{(y)} = \frac{\partial V}{\partial y} ; \quad V'_{(z)} = \frac{\partial V}{\partial z} \tag{10}$$

Of course the equation $\Delta V \approx \partial V$ is valid only if y and z remain really constant. Then $\Delta V \approx \partial V = V'_{(x)}\, \partial x$. If all arguments change at the same time then

$$\Delta V = \frac{\partial V}{\partial x} \Delta x + \varepsilon_x + \frac{\partial V}{\partial y} \Delta y + \varepsilon_y + \frac{\partial V}{\partial z} \Delta z + \varepsilon_z , \tag{11}$$

where ε_x, ε_y, ε_z are the corresponding corrections for the finiteness of changes Δx, Δy, Δz (see above). In order to leave out the corrections we insert limitless small increments d instead of the finite differences Δ. The former are again called differentials; dx, dy, dz are the differentials of the arguments, dV is the differential of their function. Since its value depends on the value of all the differentials of the arguments it is called a total or exact differential. We may now write:

$$dV = \frac{\partial V}{\partial x} dx + \frac{\partial V}{\partial y} dy + \frac{\partial V}{\partial z} dz \tag{12}$$

$$dV = yzdx + xzdy + xydz \tag{13}$$

$$df(x, y, z \,..\,) = \frac{\partial f}{\partial x} dx + \frac{\partial f}{\partial y} dy + \frac{\partial f}{\partial z} dz + \; \ldots \tag{14}$$

The derivatives $\partial f/\partial x$ etc are called partial derivatives or partial differential quotients and are sought for as normal derivatives except for the fact that the other arguments excepting only that according to which we differentiate, are considered to be constant.

Partial derivatives and total differentials are needed e. g. for calculating electronic devices (see p. 148).

Indefinite integral

Let us take the function $y = f(x)$ and let us seek another function $F(x)$ with the property that

$$f(x) = \frac{dF(x)}{dx} , \tag{15}$$

736

i. e. the function $F(x)$ of which the given function $f(x)$ is the derivative. The function $F(x)$ is called the original function or the antiderivative to $f(x)$ or the indefinite integral of the function $f(x)$.

From the equation (15) we do not obtain $F(x)$ but only the differential $dF(x)$

$$dF(x) = f(x) \cdot dx \tag{16}$$

Determination of $F(x)$ from this differential equation is called integration. It is prescribed by the operator $\int$.

$$F(x) = \int dF(x) = \int f(x)\, dx \tag{17}$$

Thus integration is the inverse operation to differentiation. Integral calculus deals with this. If it is at all possible to find the original function for a function $f(x)$, then there is an infinite number of these original functions and they differ by an arbitrary additive constant (hence the term indefinite integral).

Definite integral

Let us take a curve representing the function

$$y = f(x) \tag{18}$$

and on it two points (x_1, y_1) and (x_2, y_2) and let us draw lines parallel to the ordinate through these points. A certain area is included between these two parallel lines, the curve and the abscissa. The size of this area is to be found. Let us divide this area into narrow strips parallel to the ordinate, the length of which is $f(x)$, where x is the distance of the corresponding strip from the origin in the direction of the abscissa and dx is a limitless small width. The area of each strip is then $f(x)\, dx$ and we have to find the sum of the infinitely large number of these limitless narrow strips between x_1 and x_2. For simplicity's sake let us first assume that $x_1 = 0$ and x_2 a continuous variable from zero to b is then designated as x. The total area between $x_1 = 0$ and $x_2 = x$ is evidently some function of x, since its value depends only on the original curve and on the position of the point that bounds it from the right. This function $F(x)$ we do not know but we do know its differential, or the fact that its differential is the product of the differential dx and the function $f(x)$ determining the curve

$$d\,F(x) = f(x)\, dx \tag{19}$$

This is the same equation as (16). The only difference is that we are now not looking for the original function $F(x)$ in general but only for its *value* $F(b)$ for

$x = b$. Hence this integral, although it can be usually solved using $F(x)$, is designated as the definite integral and is written with its limits of integration. If the area to be found is designated as

$$P_{x_1,x_2} = P_{0,x_2} - P_{0,x_1} \text{, then} \tag{20}$$

$$P_{0,x_2} = \int_0^b \mathrm{d}\, F(x) = \int_0^b f(x)\,\mathrm{d}x = F(b) \tag{20a}$$

$$P_{0,x_1} = \int_0^a \mathrm{d}\, F(x) = \int_0^a f(x)\,\mathrm{d}x = F(a) \tag{20b}$$

$$P_{x_1,x_2} = \int_a^b \mathrm{d}F(x) = \int_{x_1}^{x_2} \mathrm{d}F(x) = \int_a^b f(x)\,\mathrm{d}x = F(b) - F(a)\,, \tag{20c}$$

where $F(x)$ is an indefinite integral, i. e. an original function (including an arbitrary constant, see above), $F(b)$ and $F(a)$ are values of this function for $x = b$ and $x = a$ but without an arbitrary constant, which is not important here since the definite integral is the difference between $F(b)$ and $F(a)$ and the additive constants cancel out.

A definite integral is thus determined, if possible, by finding its original function, its two values are calculated and the difference between them is computed.

Despite this narrow relationship between the definite and indefinite integral there is a basic difference between them. The latter is a *function* while the former is the *value* of a function, i. e. *a number*. This is important practically: In many cases it is not possible to find the indefinite integral, even for relatively simple functions, e. g. $\int \sin x/x\,\mathrm{d}x$, yet it is possible to find any arbitrarily exact approximation of a definite integral for any values of the limits of the argument. For important integrals these values are tabulated, see e. g. the probability integral, p. 234.

Multiple integral

Starting from partial derivatives and total differentials it is clear that the total differential $\mathrm{d}f$ of the function $f = f(x_1, x_2, \ldots x_n)$ of several (n) arguments $x_1, x_2, \ldots x_n$

$$\mathrm{d}f = \frac{\partial f}{\partial x_1}\,\mathrm{d}x_1 + \frac{\partial f}{\partial x_2}\,\mathrm{d}x_2 + \ldots + \frac{\partial f}{\partial x_n}\,\mathrm{d}x_n \tag{14}$$

is again a function of these arguments, i. e.

$$\mathrm{d}f = \mathrm{d}f(x_1, x_2, \ldots x_n) = \varphi(x_1, x_2, \ldots x_n)\,\mathrm{d}v \tag{15}$$

where $\mathrm{d}v$ is a limitless small increment of the n-dimensional "volume" (for $n = 2$ the element is an area, for $n = 3$ a volume etc). If x_1, x_2, ... x_n are cartesian coordinates then evidently

$$\mathrm{d}v = \mathrm{d}x_1 \cdot \mathrm{d}x_2 \cdot \ldots \mathrm{d}x_n \tag{16}$$

and

$$\mathrm{d}f = \varphi(x_1, x_2, \ldots x_n)\, \mathrm{d}x_1\, \mathrm{d}x_2 \ldots \mathrm{d}x_n \tag{15a}$$

If now we want to find the value of the function $f(S)$ for the whole n-dimensional region S from this differential equation for the increment of the function in an elementary "volume" $\mathrm{d}v$, this equation must be integrated over the whole of this domain. We seek for this integral as follows:

$$f(S) = \int_S \varphi(x_1, x_2, \ldots x_n)\, \mathrm{d}v \tag{17}$$

using (16) we obtain

$$f(S) = \iint_S \ldots \int \varphi(x_1, x_2, \ldots x_n)\, \mathrm{d}x_n \ldots \mathrm{d}x_2\, \mathrm{d}x_1 =$$

$$= \int_{x_{1a}}^{x_{1b}} \int_{x_{2a}}^{x_{2b}} \ldots \int_{x_{na}}^{x_{nb}} \varphi(x_1, x_2, \ldots x_n)\, \mathrm{d}x_n \ldots \mathrm{d}x_2\, \mathrm{d}x_1 \tag{18}$$

This is called the multiple integral, the integration sign appears n-times to indicate the way of calculating, which is as follows: First we integrate according to argument x_n, i. e. we integrate function φ from x_{na} to x_{nb}, taking all arguments excepting x_n as constants, i. e. writing them without change. Thus we obtain the multiple integral one order lower where $\mathrm{d}x_n$ and x_n are already substituted for by the corresponding integral, i. e. a constant, parameter or function of the same or another argument. This integral is integrated again in the same way, this time according to x_{n-1}, i. e. all arguments, excepting x_{n-1} are considered as constants (i. e. also x_n if it is still contained in the new integral). After n such steps we obtain the *value* of the required multiple integral.

Examples:

a) the volume of a quadrangular prism of sides a, b, c is

$$V = \int_S \mathrm{d}v = \int_0^a \int_0^b \int_0^c \mathrm{d}z\, \mathrm{d}y\, \mathrm{d}x = \int_0^a \int_0^b c\, \mathrm{d}y\, \mathrm{d}x = \int_0^a cb\, \mathrm{d}x = c\, b\, a \tag{19}$$

b) the mass of a quadrangular prism in which density rises in proportion to its altitude z, $\varrho = \varrho_0 z$. The volume element has a mass $dm = dv \cdot \varrho = \varrho_0 z\, dv = \varrho_0 z\, dx\, dy\, dz$. (20)

$$M = \int_S dm = \int_S \varrho_0 z\, dv = \varrho_0 \int_0^a \int_0^b \int_0^c z\, dz\, dy\, dx = \varrho_0 \int_0^a \int_0^b \tfrac{1}{2} c^2\, dy\, dx =$$

$$= \tfrac{1}{2}\, \varrho_0\, a\, b\, c^2 \tag{21}$$

Calculus of probability

If we obtain n different random results of one experiment, of which m events correspond to the expected assumption, then the ratio

$$P(A) = \frac{m}{n} = p_A \tag{22}$$

is called the (apriori) probability of the result A. (This definition is evidently based on intuition only and in exact theory gives rise to a number of inexact results, vicious circles etc. Hence attempts to establish the theory of probability on the base of the theory of sets and to derive it from a few axioms is gaining grounds, see Kolmogorov 1933, 1936.)

The probability of a certain result is evidently $n/n = 1$, of an impossible one 0. The probability of a result B, the opposite of A, $\overline{A}$ (non A) is hence

$$P(B) = P(\overline{A}) = q_A = \frac{n-m}{n} = 1 - p_A = p_B = p_{\overline{A}} \tag{23}$$

The probability of either A_1 or A_2, if these exclude each other (i. e. only one of them occur) is

$$P(A_1 \text{ vel } A_2) = p_1 + p_2 \tag{24}$$

evidently

$$P(A \text{ vel } \overline{A}) = 1 = p_A + q_A \tag{25}$$

If A and B do not exclude each other the following combinations may occur: A, B; $A, \overline{B}$; $\overline{A}, B$; $\overline{A}, \overline{B}$; The probabilities of these events do not only depend on $P(A)$ and $P(B)$ but also on their mutual formal and causal relationships. Here we use the term conditioned or aposteriori probability. Event A occurs if event B has already occurred (hence aposteriori). It is designated $P_B(A)$ or $P(AB)$

$$P_B(A) = \frac{P(A \text{ and } B)}{P(B)} \tag{26}$$

(we assume, of course, that $P(B) \neq 0$), and from the axiom of commutability we obtain

$$P_B(A) = P_A(B)\frac{P(A)}{P(B)} \tag{27}$$

Statistical independency

Two events AB are defined as *statistically independent* if their aposteriori probabilities are the same as the apriori ones, i. e. if $P_B(A) = P(A)$ and $P_A(B) = P(B)$. In this case the following holds:

$$P(A \text{ and } B) = P(A) \quad P(B) \tag{28}$$

and also

$$P(\overline{A} \text{ and } \overline{B}) = P(\overline{A}) \quad P(\overline{B}) \tag{28a}$$

and this may be used to solve problems of joined probability of events, that do not exclude each other.

Event A is usually a set of elementary events A_1, A_2, $A_3 \ldots A_m$. If these events exclude one another then

$$P(A) = \overset{m}{\Sigma} P(A_i) . \tag{29}$$

If $P(A_1) = P(A_2) = \ldots = P(A_m) = p$, then $P(A) = mp$ $\qquad$ (29a)

Let us again take two statistically independent events A and B, that do not exclude each other and let us look for the probability of event C which means the realization of either A or B.

$$P(C) = P(A \text{ vel } B) = 1 - P(\overline{C}) = 1 - P(\overline{A} \text{ and } \overline{B}) = 1 - P(\overline{A}) P(\overline{B}) =$$
$$= 1 - (1 - P(A)) (1 - P(B)) = P(A) + P(B) - P(A) P(B) \quad (30)$$

Hence when seeking for the joined probability of non-exclusive events A and B we must subtract the probability of the simultaneous occurrence of A and B (which would be included twice otherwise) from the sum of probabilities.

When seeking for the probability of more complicated events combination analysis must be used which gives us the necessary data on the numbers m and n. This will be discussed under "distribution of random values".

Stochastic process, random function, random variable, random value

That process (or sometimes also its mathematical model) the event of which depends on chance is called a stochastic process. It can be described and studied only with the use of mathematical statistics, by methods developed from the theory of probability. Good examples of such processes are the thermal movement of molecules, electric noises, interaction of elementary particles of matter etc.

A *random function* is a time function which characterizes the stochastic process. It may hence be said that the values of a random function for every value of their argument are *random variables* (usually the argument is time).

A random variable is any physical (or mathematical) variable that is realized once (and once only) in one "experiment" out of several possible random values.

The distribution of values of random variables

Dependending on the set of these values and the probabilities belonging to them every random variable is characterized by its law of distribution, which can be very different depending on the character of the corresponding stochastic process which generates the random variables. Thus when throwing a dice the number on the upper face is a random variable which may attain values 1, 2, 3, 4, 5, 6 each with a probability of 1/6. Now let us take two dices and let the random variable x_i be the sum of the two numbers on the upper faces. We can construct a table of values (2—12) and their probabilities.

i	1	2	3	4	5	6	7	8	9	10	11
x_i	2	3	4	5	6	7	8	9	10	11	12
m_i	1	2	3	4	5	6	5	4	3	2	1
n	36	36	36	36	36	36	36	36	36	36	36
p_i	$\dfrac{1}{36}$	$\dfrac{1}{18}$	$\dfrac{1}{12}$	$\dfrac{1}{9}$	$\dfrac{5}{36}$	$\dfrac{1}{6}$	$\dfrac{5}{36}$	$\dfrac{1}{9}$	$\dfrac{1}{12}$	$\dfrac{1}{18}$	$\dfrac{1}{36}$

Binomial (or Bernoulli) distribution

The most frequently occurring similar problem is to find the probability for several identical events of the same basic experiment. Let the probability of one positive (successful) result be $P(A) = p$ and the probability $P(\bar{A}) = 1 - p = q$. These probabilities are the same for each of the n experiments, i. e. the experiments are mutually independent. The probability $P_n(k)$ that result A is obtained for n repetitions k times (regardless of sequence) is

$$P_n(k) = \binom{n}{k} p^k q^{n-k} \tag{31}$$

The whole set of events $0A$, $1A$, $2A$, ... kA ... nA with their corresponding probabilities $P_n(0)$, $P_n(k)$, ... $P_n(n)$ is called the distribution of probabilities, in this case the *binomial* (or *Bernoulli*) *distribution* since the individual $Pn(k)$ are the k-th members of the development of $(p + q)^n$ according to the binomial theorem. Since $p + q = 1$

$$\sum_{k=0}^{n} P_n(k) = 1 \tag{32}$$

There is a finite number of n possible events, probabilities are given by n discrete numbers, other events are excluded and thus have a zero probability. The graphic representation of such a distribution is thus a set of sections (parallel to the Y-axis). Their abscissa gives the possible values of the random variable i. e. e. g. the k and the ordinate the probability belonging to them. With increasing n the size of all probabilities is decreased: for one value k designated k_0 (exceptionally for two adjacent k, k_0 and k_0-1), the probability is maximal. This k_0 is a whole number fulfilling the relationship

$$np < k_0 \leqq (n + 1) p \tag{33}$$

Asymptotic formulae of Laplace and Poisson

In order to be able to compare different distributions it is suitable to transform the corresponding graphs so that the abscissa is no longer k but

$$\delta_k = k - k_0 \tag{34}$$

so that k_0 is in the origin of the Y-axis. Thus all distributions for different n will be more or less symmetrically distributed around their k_0 but will still have different heights and widths. Hence we introduce $npq = \sigma^2$ by substitution and denote

$$x_k = \frac{\delta_k}{\sigma\sqrt{2}} \tag{35}$$

and thus distributions for different n (and equal p, q) attain approximately the same width. In order to normalise also their heights we introduce

$$y_k = \sigma P_n(k) \cdot \sqrt{(2\pi)} \tag{36}$$

These transformed distributions have a "line density" proportional to $\sqrt{n}$ and with increasing n their "shapes" approach (unless pq is not very small) a certain limit "shape", for which we can obtain the corresponding distribution $P_n(k)$ for large n, by using the *asymptotic formula of Laplace*, if we, as already mentioned, introduce δ_k and σ into the expression for $P_n(k)$ and if we introduce instead of the factorials their approximation according to the formula of Stirling, apply natural logarihms and expand them into series of which we take the first term only. Then

$$P_n(k) \approx \frac{1}{\sigma\sqrt{(2\pi)}}\, e^{-\delta_k{}^2/2\sigma^2} \tag{37}$$

If p (or q) is very small, it is not permissible to ommit the higher terms in the logarithmic series, but we can use other approximations and reach another *asymptotic formula*, that *of Poisson*, which holds as an approximation for the binomial distribution with a large n and a very small p or q which is hence called the *Poisson distribution*

$$P_n(k) \approx \frac{n^k p^k}{k!}\, e^{-np} \tag{38}$$

Gaussian distribution

If n increases unlimitedly, then in the graphic representation of the probabilities $P_n(k)$ after the transformations described above, the "density" of the ordinates (i. e. their number per unit length in the direction of x) also increases unlimitedly. Thus we move from the discrete distributions (binomial, Poisson) to the *Gaussian distribution* which is the limit of both for $n \to \infty$ and where the above substitutions δ, σ, x, y are not only possible but also necessary. Some of them, of course, in order not to become indefinite expressions, attain a new significance, e. g. y is now the *density of probability*, i. e. the differential quotient (of the limitless small) probability that x will have a value within the limits x_k and $x_k + dx$, to the (limitless small) width of this interval dx. We can no longer speak of the probability that x will have a given value but only that x will be lying between certain limits, which probability is given by the integral Gaussian law as stated in the main text (p. 234).

The above distributions of random variables are not the only ones to be considered for both discrete and random variables.

Every distribution can be more or less characterized by certain parameters calculated from it, the most common being the mode, i. e. the value of the random variable for which the probability attains a maximum (for the binomial distribution this is k_0, for the Gaussian distribution the mean value, for other distributions more maxima may be present), the standard deviation (defined on p. 235), obliqueness for nonsymmetric distributions, this being the measure of asymetry. The less regular the distribution the larger the number of parameters for its exact definition (thus the Gaussian distribution is fully described by two parameters).

Stochastic process characteristics

The characteristics that sufficiently exactly describe the random variable and its distribution do not by far suffice to characterize random functions and stochastic processes. Here we must include time as an argument. First we may consider the distribution of the random variable in fixed individual moments. This gives us the so called *one-dimensional distribution function* of probability at the moment t which is described by the corresponding distribution or the parameters already mentioned, i. e. the *ensemble parameters*. In order to consider the dynamics of the process we search for a connection between the probabilities of the random variable at moment t_1 and the probabilities at moment t_2. This gives us a *two-dimensional distribution* function. For moments t_1, t_2, ... t_n we obtain an *n-dimensional function*. The process is defined the more exactly by the function the higher n. This exactness, however, decreases clarity and as n increases the difficulties of calculation increase rapidly.

Stationarity

A stochastic process is called *stationary* if the distribution function (of any order n) is invariable with shift of the whole process along the time axis. The one-dimensional distribution in this case does not depend on time, the two-dimensional one only on the time difference $t_2 - t_1$, the n-dimensional one only on n-1 time differences $t_n - t_{n-1}$ to $t_n - t_1$ (see *autocorrelative function*). The process is called stationary according to Khinchin (in the broader sense) if at least the one- and two-dimensional distribution functions (and thus also the correlative and autocorrelative coeffcients) are invariable to time shift.

Ergodicity

If certain requirements are fulfilled (e. g. of a one-dimensional Gaussian distribution) and if the process is stationary in the broader sense, it is also ergodic, i. e. every statistic parameter (mean value, standard deviation etc.) obtained from the set of n realizations of a random variable (i. e. every ensemble parameter) (more exactly its limit for $n \to \infty$) is equal to the same parameter obtained from a single realization of the random variable for a longer time interval (i. e. the *time parameter*), or more exactly to its limit for $T \to \infty$. The ergodicity has very great practical significance, as mentioned in the main text. In order to use it, however, the stationary character of the process and the fulfillment of the requirements for which the stationary process is also ergodic, must first be proved. For examples of stationary but not ergodic processes see Levin(1960), for the ergodicity of nonstationary processes see Zheleznov (1959).

References

Ades H. W., 1943: A secondary acoustic area in the cerebral cortex. J. Neurophysiol. 6 : 59—63.

Ades H. W., 1944: Midbrain auditory mechanism in cats. J. Neurophysiol. 7 : 415—424.

Ades H. W., Brookhart J. M., 1950: The central auditory pathway. J. Neurophysiol. 13 : 189—205.

Adey W. R., Carter I. D., Porter R., 1954: Temporal dispersion in cortical response. J. Neurophysiol. 17 : 167—182.

Adey W. R., Kado R. T., Didio J. 1962: Impedance measurements in brain tissue of animals using microvolt signals. Exp. Neurol. 5 : 47—66.

Adey W. R., Kado R. T., McIlwain J. T., Walter D. O., 1966: The role of neuronal elements in regional cerebral impedance changes in alerting, orienting and discriminative responses. Exp. Neurol. 15 : 490—510.

Adrian E. D., 1920: The recovery process of excitable tissues. Part I. J. Physiol. 54 : 1—16.

Adrian E. D., 1928: The basis of sensation. The action of sense organs. Chrisophers. London.

Adrian E. D., 1932: The mechanism of nervous action. Oxford Univ. Press, London.

Adrian E. D., 1935: Discharge frequencies in cerebral and cerebellar cortex. J. Physiol. 83 : 32—33.

Adrian E. D., 1936: The spread of activity in cerebral cortex. J. Physiol. 88 : 127—161.

Adrian E. D., 1940: Double representation of the feet in the sensory cortex of the cat. J. Physiol. 98 : 16—18.

Adrian E. D., 1943: Afferent areas in the cerebellum connected with the limbs. Brain 66 : 289—315.

Adrian E. D., Bronk D. W., 1929: The discharge of impulses in motor nerve fibres. Part II. The frequency of discharge in reflex and voluntary contractions. J. Physiol. 67 : 119—151.

Adrian E. D., Bronk D. W., Phillips G., 1932: Discharges in mammalian sympathetic nerves. J. Physiol. 74 : 115—133.

Adrian E. D., Lucas K., 1912: Summation of propagated disturbances in nerve and muscle. J. Physiol. 44 : 68—124.

Adrian E. D., Matthews B. H. C., 1934: The interpretation of potential waves in the cortex. J. Physiol. 81 : 440—471.

Adrian R. H., 1956: The effect of internal and external potassium concentration on the membrane potential of frog muscle. J. Physiol. 133 : 631—658.

Adrian R. H., 1960: Potassium chloride movement and the membrane potential of frog muscle. J. Physiol. 151 : 154—185.

Ajmone-Marsan C., Marossero F., 1950: Electrocorticographic study of the convulsions induced by Cardiazol. EEG Clin. Neurophysiol. 2 : 133—142.

Aladzhalova N. A. — Аладжадова Н. А., 1954: Об электрических константах коры головного мозга. Доклады Акад. Наук СССР 94 : 1053—1056.

Albe-Fessard D., 1957: Activités de projection et d'association du neocortex des mammiféres. I. Les projections primaires. J. Physiol. (Paris) 49 : 521—588.

Albe-Fessard D., Chagas C., Couceiro A., Fessard A., 1951: Characteristics of responses from electrogenic tissue in *Electrophorus electricus*. J. Neurophysiol. 14 : 243—252.

Albe-Fessard D., Rougeul A., 1955: Activités bilatérales tardives évoquées sur le cortex du chat sous chloralose par stimulation d'une voie somesthésique. J. Physiol. (Paris) 47 : 69—72.

Albe-Fessard D., Rougeul A., 1956: Relais thalamiques d'afférences somesthésiques aboutissant à certaines régions localisées du cortex associatif du chat. J. Physiol. (Paris) 48 : 370—374.

Albe-Fessard D., Stutinsky F., Libouban S.; 1966: Atlas stéréotaxique du diencéphale du rat blanc. Editions CNRS, Paris.

Albe-Fessard D., Szabo T., 1955: Observations sur l'interaction des afférences d'origines péripherique et corticale destinées à l'écorce cérébelleuse du chat. C. R. Soc. Biol. 149 : 457—466.

Alexander J. T., Nastuk W. L., 1953: An instrument for the production of microelectrodes used in electrophysiological studies. Rev. Sci. Instrum. 24 : 528—531.

Alexandrowicz J. S., 1951: Muscle receptor organs in the abdomen of *Homarus vulgaris* and *Palinurus vulgaris*. Quart. J. Micr. Sci. 92 : 163—199.

Alexandrowicz J. S., 1952: Receptor elements in the thoracic muscles of *Homarus vulgaris* and *Palinurus vulgaris*. Quart J. Micr. Sci. 93 : 93—315.

Alvarez-Buylla R., Ramirez de Arellano J., 1953: Local responses in Pacinian corpuscles. Amer. J. Physiol. 172 : 237—244.

Amantea G., 1921: Über experimentelle beim Versuchstier infolge afferenter Reize erzeugte Epilepsie. Pflüg. Arch. Ges. Physiol. 188 : 287—297.

Amassian V. E., Waller H. J., Macy J., 1964: Neural mechanism of the primary somatosensory evoked potential. Annals of the New York Acad. Sci. 112 : 5—32.

Amatniek E., 1958: Measurement of bioelectric potentials with microelectrodes and neutralized input capacity amplifiers. I. R. E. Trans. Med. Electronics P. G. M E. 10 : 3—14.

Ananyev V. M. — Ананьев Б. М., 1956: Электроэнцефалоскоп. Физиол. журнал СССР 42 : 981—987.

Andersen P., Eccles J. C., Schmidt R. F., 1962: Presynaptic inhibition in the cuneate nucleus. Nature 194 : 741—743.

Anderson-Cedergren E., 1959: Ultrastructure of motor end plate and sarcoplasmic components of mouse skeletal muscle fibre. J. Ultrastr. Res., Suppl. 1 : 1—191.

Antipov A. P. — Антипов А. П., 1951: Релаксационный генератор. Вестник ИНИ МПСС 6 : 21.

Araki T., Otani T., 1955: Response of single motoneurons to direct stimulation in toad's spinal cord. J. Neurophysiol. 18 : 472—485.

Araki T., Terzuolo C. A., 1962: Membrane currents in spinal motoneurons associated with the action potential and synaptic activity. J. Neurophysiol. 25 : 772—789.

Arduini A., 1958: The physiology of the thalamocortical connections. Monitore Zool. Ital. 66 : 125—151.

Arduini A., Berlucchi G., Strata P., 1963: Pyramidal activity during sleep and wakefulness. Arch. Ital. Biol. 101 : 530—544.

Artemev V. V. — Артемьев В. В., 1951: Электрическая реакция коры мозга при действии звуковых раздражений у наркотизированных и ненаркотизированных животных. Физиол. журнал СССР 37 : 688—702.

Arvanitaki A., 1938: Les variations graduées de la polarisation des systèmes excitables. Paris, Hermann Cie.

Arvanitaki A., 1942: Effect evoked in an axon by the activity of a contiguous one. J. Neurophysiol. 5 : 89—108.

Ashby W. R., 1956:An introduction to cybernetics. Chapman and Hall, London.

Ashby W. R., 1960: Design for a brain. Chapman and Hall, London.

Atlas D., Ingram W. R., 1937: Topography of the brain stem of the rhesus monkey with special reference to the diencephalon. J. Comp. Neurol. 66 : 265—289.

Autrum H., 1959: Nonphotic receptors in lower forms. Handbook of Physiology. Vol. I., Section 1-Neurophysiology, pp. 369—385, Amer. Physiol. Soc., Washington.

Bailey P., Stein S. N., 1951: A stereotaxic instrument for use on the human brain. In Studies in Medicine: A volume of Papers in Honour of Robert Wood Keeton. Springfield, Thomas pp. 40—49.

Bak A. F., 1958: A unity gain cathode follower. EEG Clin. Neurophysiol. 10 : 745—748.

Baker P. F., Hodgkin A. L., Shaw T. I., 1961: Replacement of the protoplasm of a giant nerve fibre with artificial solutions. Nature 190 : 885—887.

Baker P. F., Hodgkin A. L., Shaw T. I., 1962a: Replacement of the axoplasm of giant nerve fibres with artificial solutions. J. Physiol. 164 : 330—354.

Baker P. F., Hodgkin A. L., Shaw T. I., 1962b: The effects of changes in internal ionic concentrations on the electrical properties of perfused giant axons. J. Physiol. 164 : 355—374.

Balthazar K., 1952: Morphologie der spinalen Tibialis-und Peronaeus-Kerne bei der Katze: Topographie, Architektonik, Axon-und Dendritenverlauf der Motoneurone und Zwischenneurone in den Segmenten L_6 — S_2. Arch. Psychiat. Nervenkr. 168 : 345—365.

Bannister R. G., Sears T. A., 1962: The changes in nerve conduction in acute idiopathic polyneuritis. J. Neurol. Neurosurg. Psychiat. 25 : 321—329.

Barer R., Saunders—Singer A. E., 1951: A low power micromanipulator and microdissector. J. Sci. Instrum. 28 : 65—68.

Barker D., 1948: The innervation of the muscle spindle. Quart. J. Micr. Sci. 89 : 143—186.

Barker D., 1962a: The structure and distribution of muscle receptors. In: D. Barker (editor): Muscle receptors, pp. 227—240. Hong Kong University Press.

Barker D. (editor), 1962b: Muscle receptors. Hong Kong University Press.

Barron D. H., Matthews B. H. C., 1938: The interpretation of potential changes in the spinal cord. J. Physiol. 92 : 276—321.

Bartlett M. S., 1955: An introduction to stochastic processes with special reference to methods and applications. University Press, Cambridge.

Batini C., Moruzzi G., Palestini M., Rossi G. F., Zanchetti A., 1959: Effects of complete pontine transsections on the sleep-wakefulness rhythm: the midpontine pretrigeminal preparation. Arch. Ital. Biol. 97 : 1—12.

Batrak T. E. — Батрак Т. Е., 1958: Методика вживления электродов. Физиол. журнал СССР 44 : 1001—1005.

Baumann C., Kaada B. R., Kristiansen K., 1956: A simple anti-blocking device. EEG Clin. Neurophysiol. 8 : 685—687.

Baumgarten R., 1957: Zur Technik der Mikroableitung am pulsierenden Gehirn. Naturwissenschaften 44 : 22—23.

Bayda L. I., Semenkovich A. A. — Байда Л. И., Семенкович А. А., 1953: Электронные усилители постоянного тока. Государственное энергетическое изд. Москва—Ленинград.

Bayliss L. E., Cowan S. L., Scott Jr., 1935: The action potentials in *Maia* nerve before and after poisoning with veratrine and yohimbine hydrochlorides. J. Physiol. 83 : 439—454.

Békésy G. von, 1952: Micromanipulator with four degrees of freedom. Trans. Amer. Micr. Soc. 71 : 306—310.

Bell C., Sierra G., Buendia N., Segundo J. P., 1964: Sensory properties of neurons in the mesencephalic reticular formation. J. Neurophysiol. 27 : 961—987.

Benjamin R. M., Pfaffmann C., 1955: Cortical localization of taste in albino rat. J. Neurophysiol. 18 : 56—64.

Bennet S. H., 1960: The structure of striated muscle as seen by the electron microscope. In: G. H. Bourne (editor): The structure and function of muscle. Vol. I. Structure. Chapter VI, 137—181.

Beránek R., 1959: Mikromyografie. Nitrobuněčná registrace akčních potenciálů z jediného vlákna lidského kosterního svalu. Čs. fysiol. 8 : 172.

Beránek R., 1965: Nitrobuněčná elektromyografie u člověka. Státní zdravotnické nakladatelství, Praha.

Berger H., 1929: Über das Elektrenkephologramm des Menschen I. Arch. Psychiat. Nervenkr. 87 : 527—570.

Berger H., 1930: Über das Elektronkophalogramm des Menschen II. J. Psychol. Neurol. 40 : 160—179.

Berger W., 1963: Die Doppelsacharosetrennwandtechnik. Eine Methode zur Untersuchung des Membranpotentials und der Membraneigenschaften glatter Muskelzellen. Pflüg. Arch. Ges. Physiol. 277 : 570—576.

Berlucchi G., Moruzzi G., Salvi G., Strata P., 1964: Pupil behavior and ocular movements during synchronized and desynchronized sleep. Arch. Ital. Biol. 102 : 230—244.

Berman A. L., 1961a: Overlap of somatic and auditory cortical response fields in the anterior ectosylvian gyrus of the cat. J. Neurophysiol. 24 : 595—607.

Berman A. L., 1961b: Interactions of cortical responses to somatic and auditory stimuli in anterior ectosylvian gyrus of the cat. J. Neurophysiol. 24 : 608—620.

Bernardis L. L., Skelton F. R., 1965: Stereotaxic localization of supraoptic, ventromedial and mamillary nuclei in the hypothalamus of weanling to mature rats. Am. J. Anat. 116 : 69—74.

Bernhard C. G., 1942: Isolation of retinal and optic ganglion response in the eye of *Dytiscus*. J. Neurophysiol. 5 : 32—48.

Bernstein J., 1902: Untersuchungen zur Thermodynamie der bioelektrischen Ströme. Erster Theil. Pflügers Arch. Ges. Physiol. 92 : 521—562.

Bernstein J., 1912: Elektrobiologie. Braunschweig.

Berry C. M., Karl R. S., Hinsey J. C., 1950: Course of spinothalamic and medial lemniscus pathways in cat and rhesus monkey. J. Neurophysiol. 13 : 149—156.

Bickford R. G., 1950: Automatic electroencephalographic control of general anesthesia. EEG Clin. Neurophysiol. 2 : 93—96.

Bickford R. G., 1951: Use of frequency discriminations in the automatic electroencephalographic control of anesthesia (servo-anesthesia) EEG Clin. Neurophysiol. 3 : 83—86.

Biederman W., 1895: Elektrophysiologie. G. Fisher, Jena.

Birks R., Huxley H. E., Katz B., 1960: The fine structure of the neuromuscular junction of the frog. J. Physiol. 150 : 134—144.

Birks R., MacIntosh F. C., 1957: Acetylcholine metabolism at nerve-endings. Brit. Med. Bull. 13 : 157—161.

Birks R., MacIntosh F. C., 1961: Acetylcholine metabolism of a sympathetic ganglion. Canad. J. Biochem. 39 : 787—827.

Bishop G. H., 1927: The form of the record of the action potential of vertebrate nerve of the stimulated region. Amer. J. Physiol. 82 : 462—476.

Bishop G. H., 1928: The relation between the threshold of nerve response and polarisation by galvanic current stimuli. Amer. J. Physiol. 84 : 417—436.

Bishop G. H., 1929: The reactance of nerve and the effect upon it of electrical currents. Amer. J. Physiol. 89 : 618—693.

Bishop G. H., 1936: Interpretation of potentials led from cervical sympathetic ganglion of rabbit. J. Cell. Comp. Physiol. 8 : 465—477.

Bishop G. H., 1937: La théorie des circuits locaux permet-elle de prévoir la forme du potential d'action? Arch. Internat. Physiol. 45 : 273—297.

Bishop G. H., Clare M. H., 1951: Radiation path from geniculate to optic cortex in cat. J. Neurophysiol. 14 : 497—505.

Bishop G. H., Clare M. H., 1953: Responses of cortex to direct electrical stimuli applied at different depths. J. Neurophysiol. 16 : 1—19.

Bishop G. H., Gilson A. S., 1929: Action potentials from skeletal muscle. Amer. J. Physiol. 89 : 135—151.

Bishop G. H., Heinbecker P., 1932: A functional analysis of the cervical sympathetic nerve supply to the eye. Amer. J. Physiol. 100 : 519—532.

Bishop G. H., O'Leary J. L., 1942a: The polarity of potentials recorded from the superior colliculus. J. Cell. Comp. Physiol. 19 : 289—300.

Bishop G. H., O'Leary J. L., 1942b: Factors determining the form of the potential recorded in the vicinity of the synapses of the dorsal nucleus of the lateral geniculate body. J. Cell. Comp. Physiol. 19 : 315—331.

Bishop G. H., Tharaldson G. E., 1921: An apparatus for microdissection. Amer. Naturalist 55 : 381.

Bishop P. O., 1953: Synaptic transmission. — An analysis of the electrical activity of the lateral geniculate nucleus in the cat after optic nerve stimulation. Proc. Roy. Soc. B 141 : 362—392.

Bishop P. O., Burke W., Davis R., 1962a: The identification of single units in central visual pathways. J. Physiol. 162 : 409—431.

Bishop P. O., Burke W., Davis R., 1962b: The interpretation of the extracellular response of single lateral geniculate cells. J. Physiol. 162 : 451—472.

Bishop P. O., McLeod J. G., 1954: Nature of potentials associated with synaptic transmission in lateral geniculate of cat. J. Neurophysiol. 17 : 387—413.

Blair E. A., Erlanger J., 1933: A comparison of the characteristics of axons through their individual electrical responses. Amer. J. Physiol. 103 : 524—564.

Blobel R., Gonzalo L., Schuckardt E., 1960: Stereotaktisches Vefahren zur Lokalisation der Zwischenhirnkerne beim Meerschweinchen. Endokrinologie. 167—172.

Bogue J. Y., Rosenberg H., 1936: Electrical responses of Maia nerve to single and repeated stimuli. J. Physiol. 87 : 158—180.

Boistel J., Fatt P., 1958: Membrane permeability change during inhibitory transmitter action in crustacean muscle. J. Physiol. 144 : 176—191.

Bok S. T., 1959: Histonomy of cerebral cortex. Elsevier, Amsterdam.

Bonch-Bruyevich A. M. — Бонч-Бруевич А. М. 1955: Применение электронных ламп в экспериментальной физике. Госуд. изд. техннко-теоретический литературы, Москва.

Bose J. C., 1926: The nervous mechanism of plants. Longsmanns, London.

Bovet D., Longo V. G., 1956: Pharmacologie de la substance réticulée du tronc cérébral. XXth Internat. Physiol. Congr. Abstracts of reviews, Brussels pp. 306—329.

Boyd T. E., Ets H. N., 1934: Studies on cold block in nerve. I. Block with and without freezing. Amer. J. Physiol. 107 : 76—84.

Boyle P. J., Conway E. J., 1941: Potassium accumulation in muscle and associated changes. J. Physiol. 100 : 1—64.

Bradley P. B., Elkes J. 1957: The effects of some drugs on the electrical activity of the brain 80 : 77—117.

Brazier M. A. B., 1961 (Editor): Computer techniques in EEG analysis. EEG Clin. Neurophysiol. Suppl. 20, Elsevier, New York.

Brdička R., 1951: Základy fysikální chemie. Přírodověd. vyd. Praha.

Bremer F. 1927 : Sur le mécanisme de la sommation d'influx nerveux. C. R. Soc. Biol. 97 : 1179—1184.

Bremer F., 1935: "Cerveau isolé" et physiologie du sommeil. C. R. Soc. Biol. 118 : 1235—1242.

Bremer F., 1936a: Nouvelles recherches sur le mécanisme du sommeil. C. R. Soc. Biol. Biol. 122 : 460—464.

Bremer F., 1936b: Activité électrique du cortex cérébral dans les états du sommeil et de veille chez le chat. C. R. Soc. Biol. 122 : 464—467.

Bremer F., 1937: L'activité cérébral au cours du sommeil et de la narcose. Contribution à l'étude du mécanisme du sommeil. Bull. Acad. Roy. Med. Belgique 2 : 68—86.

Bremer F., 1952: Analyse oscillographique des réponses sensorielles des écorces cérébrale et cérébelleuse. Rev. Neurol. 87 : 65—92.

Bremer F., 1958: Cerebral and cerebellar potentials. Physiol. Rev. 38 : 357—388.

Bremer F., 1961: Neurophysiological mechanisms in cerebral arousal. In: The Nature of Sleep, edited by G. E. Wolstenholme and M. O'Connor, pp. 30—56. Ciba Foundation Symposium, Little, Brown and Co., Boston.

Bremer F., Bonnet V., 1951: Caractéres généraux de la réponse du cervelet à une volée d'influx afférent. J. Physiol. 43 : 622—664.

Bremer F., Bonnet V., Terzuolo C., 1954: Étude électrophysiologique des aire auditive corticales du chat. Arch. Intern. Physiol. 62 : 390—428.

Bremer F., Dow R. S., 1939: The cerebral acoustic area of the cat. J. Neurophysiol 2 : 308—318.

Bremer F., Gernandt B. E., 1954: A microelectrode analysis of the acoustic response and the strychnine convulsive pattern of the cerebellum. Acta Physiol. Scand. 30 : 120—136.

Bremer F., Stoupel N., 1959: Facilitation et inhibition des potentiels evoqués corticaux dans l'eveil cérébral. Arch. Intern. Physiol. Biochem. 67 : 240—275.

Brinley F. J., 1963: Ion fluxes in the central nervous system. Int. Rev. Neurobiol. 5 : 183—242.

Brinley F. J. ,Kandel E. R., Marshall W. H., 1960: Potassium outflux from rabbit cortex during spreading depression. J. Neurophysiol. 23 : 246—256.

Brock L. G., Coombs J. S., Eccles J. C., 1952: The recording of potentials from motoneurons with an intracellular electrode. J. Physiol. 117 : 431—460.

Brodal A., 1947: The amygdaloid nucleus in rat. J. Comp. Neurol. 87 : 1—17.

Brodal A., Taber E., Walberg F., 1960: The raphe nuclei of the brain stem in the cat. J. Comp. Neurol. 114 : 239—282.

Bronk D. W., 1939: Synaptic mechanism in sympathetic ganglia. J.Neurophysiol. 2 : 380—401.

Bronk D. W., Tower S. S., Solandt D. Y., Larrabee M. G., 1938: The transmission of trains of impulses through a sympathetic ganglion and in its postganglionic nerves. Amer. J. Physiol. 122 : 1—15.

Brookhart J. M., Moruzzi G., Snider R. S. 1950: Spike discharges of single units in the cerebellar cortex. J. Neurophysiol. 13 : 465—486.

Brookhart J. M., Zanchetti A., 1956: The relation between electrocortical waves and responsiveness of the cortico-spinal system. EEG Clin. Neurophysiol. 8 : 427—444.

Browaeys J., 1943: Micromanipulateur à pantographe pour le travail à un grossissement limité. Bull. Soc. Path. Exot. 36 : 69—73.

Brücke F., Sailer S., Stumpf C., 1957: Pharmakologische Beeinflussung der Frequenz der Hippocampustätigkeit während retikulärer Reizung. Arch. Exp. Pathol. Pharmacol. 231 : 267—278.

Bubeník V., 1958: Impulsová technika. SNTL, Praha.

Buchthal F., Clemensen S., 1943: Electromyogram of atrophic muscles in cases of intramedullar affections. Acta Psychiat. Neurol. Kbn. 18 : 377—388.

Buchthal F., Honcke P., 1944: Electromyographical examination of patients suffering from poliomyelitis ant. ac. up to six months after acute stage of disease. Acta Med. Scand. 116 : 148—164.

Buchthal F., Lindhard J., 1937: Direct application of acetylcholine to motor end-plates of voluntary muscle fibres. J. Physiol. 90 : 82—83.

Buchthal F., Lindhard J., 1939: Acetylcholine block of the motor end-plate and electrical stimulation of nerve. J. Physiol. 95 : 59—60.

Buchthal F., Madsen A., 1950: Synchronous activity in normal and atrophic muscle. EEG Clin. Neurophysiol. 2 : 425—455.

Buchthal F., Persson Ch., 1936: A micromanipulative apparatus. J. Sci. Instrum. 13 : 20—23.

Buff H., 1854: Über die Elektricitätserregung durch lebende Pflanzen. Ann. Chem. Pharm. 89 : 76—89.

Büllbring E., 1944: The action of adrenaline on transmission in the superior cervical ganglion. J. Physiol. 103 : 55—67.

Bullock T. H., Hagiwara S., 1957: Intracellular recording from the giant synapse of the squid. J. Gen. Physiol. 20 : 565—578.

Burdon-Sanderson J., 1882: On the electromotive properties of the leaf of *Dionaea* in the excited and unexcited states. Philosophical transactions 173/I : 1—5.

Bureš J., 1953: Experiments on the electrophysiological analysis of the generalisation of of an epileptic fit. Physiol. bohemoslov. 2 : 347—356.

Bureš J., 1956: Some metabolic aspects of Leão's spreading depression. J. Neurochem. 1 : 153—158.

Bureš J., 1957a: The effect of anoxia and asphyxia on spreading EEG depression. Physiol. bohemoslov. 6 : 447—453.

Bureš J., 1957b: The ontogenetic development of steady potential differences in the cerebral cortex in animals. EEG Clin. Neurophysiol. 9 : 121—130.

Bureš J., 1959: Reversible decortication and behavior. In: M. A. B. Brazier (Editor): The Central Nervous System and Behavior. Transaction of the 2nd Macy Conference, J. Macy, Jr. Foundation, N. Y. pp. 207—248.

Bureš J., Burešová O., 1957: Die anoxische Terminaldepolarisation als Indicator der Vulnerabilität der Grosshirnrinde bei Anoxie und Ischämie. Pflüg. Arch. Ges. Physiol. 264 : 325—334.

Bureš J., Burešová O., 1960: The use of Leão's spreading cortical depression in research on conditioned reflexes. EEG. Clin. Neurophysiol., Suppl. 13 : 359—376.

Bureš J., Burešová O., 1962: La dépression envahissante comme instrument de recherche en neurophysiologie. Actualités neurophysiologiques 4 : 107—124.

Bureš J., Burešová O., Fifková E., Rabending G., 1965: Reversible deafferentation of cerebral cortex by thalamic spreading depression. Exp. Neurol. 12 : 55—67.

Bureš J., Burešová O., Weiss T., Fifková E., 1963: Excitability changes in nonspecific thalamic nuclei during cortical spreading depression in the rat. EEG Clin. Neurophysiol. 15 : 73—83.

Burešová O., 1957: Changes in cerebral circulation in rats during spreading EEG depression. Physiol. bohemoslov. 6 : 1—11.

Burešová O., Shima I., Bureš J., Fifková E., 1963: Unit activity in regions affected by the spreading depression. Physiol. bohemoslov. 12 : 488—494.

Burgen A. S. V., Dickens F., Zatman L. J., 1948: The action of botulinum toxin on the neuromuscular junction. J. Physiol. 109 : 10—24,

Burns B. D., 1950: Some properties of the cat's isolated cerebral cortex. J. Physiol. 3 : 50—68.

Burns B. D., 1951: Some properties of isolated cerebral cortex in the unanesthetized cat. J. Physiol. 113 : 156—175.

Burnstock G., Holman M. E., 1961: The transmission of excitation from autonomic nerve to smooth muscle. J. Physiol. 155 : 115—133.

Burnstock G., Holman M. E., 1962: Spontaneous potentials at sympathetic nerve endings in smooth muscle. J. Physiol. 160 : 446—460.

Burr H. S., 1943: An electrometric study of Mimosa. Yale J. Biol. Med. 15 : 823—830.

Burr H. S., Hovland C. I., 1936a: Bio-electric potential gradients in the chick. Yale J. Biol. Med. 9 : 247—258.

Burr H. S., Hovland C. I., 1936b: Bio-electric correlates of development in Amblystoma. Yale J. Biol. Med. 9 : 541—549.

Buser P., 1957: Activités de projection et d'association du néocortex cérébral des mammiféres. Activités d'association et d'élaboration; projection spécifiques. J. Physiol. (Paris) 49 : 589—656.

Buser P., Borenstein P., 1956: Observations sur les réponses corticales visuelles recueillies dans le cortex associatif suprasylvien chez le chat sous chloralose. J. Physiol. (Paris) 48 : 422—424.

Buser P., Borenstein P., 1959: Réponses corticales "sécondaires" à la stimulation sensoriel le chez le Chat curarisé non anesthésié. EEG Clin. Neurophysiol., Suppl.6 : 87—108.

Bush V., Duryee W. R., Hastings J. A., 1953: An electric micromanipulator. Rev. Sci. Instrum. 24 : 487—489.

Butler F., 1965: Applications of metal oxide silicon transistors. Wireless World: 58—61.

Büttner L., 1966: Aufbau und Eigenschaften von Metall-Oxid-Feldeffekttransistoren. Funktechnik, 1966, 163—164 und 203—205.

Byzov A. L. — Бызов А. Л., 1960: Анализ распределения потенциалов и токов внутри сетчатки при световом раздражении. Биофизика, 5 : 284—292.

Byzov A. L. — Бызов А. Л., 1966: Электрофизиологические исследования сетчатки. Изд. Наука, Москва.

Cailloux M., 1943: Un nouveau micromanipulateur hydraulique. Rev. Canad. Biol. 2 : 5—15.

Cajal S. R., 1909: Histologie du système nerveux de l'homme et des vertébrés. Vol. I. Paris, Maloine.

Caldwell P. C., Downing A. C., 1955: The preparation of capillary microelectrodes. J. Physiol. 128 : 31P.

Carpenter D., Engberg I., Lundberg A., 1962a: Presynaptic inhibition in the lumbar cord evoked from the brain stem. Experientia 18 : 450—451.

Carpenter D., Lundberg A., Norrsell U., 1962b: Effect from the pyramidal tract on primary afferents and on spinal reflex actions to primary afferents. Experientia 18:337—338.

Carpenter M. D., Whittier J. R., 1952: Study of methods for producing experimental lesions of the central nervous system with special reference to stereotaxic technique. J. Comp. Neurol. 97 : 73—132.

Caspers H., 1959: Über die Beziehung zwischen Dendritenpotential and Gleichspannung an der Hirnrinde. Pflüg. Arch. Ges. Physiol. 269 : 157—181.

Caspers H., 1962: Die Veränderung der corticalen Gleichspannung und ihre Beziehungen zur sensomotorischen Aktivität (Verhalten) bei Weckreizungen am freibeweglichen Tier. Proc. Int. Union Physiol. Sci., XXII Int. Congress, Leiden, Vol. I, part I, pp. 442—447.

Cazard P., Buser P., 1963a: Réponses sensorielles recueillies au niveau du cortex moteur chez le Lapin. EEG Clin. Neurophysiol. 15 : 403—412.

Cazard P., Buser P., 1963b: Modification des réponses sensorielles corticales par stimulation de l'hippocampe dorsal chez le Lapin. EEG Clin. Neurophysiol.15 : 413—425.

Celesia G. G., 1963: Segmental organization of cortical afferent areas in the cat. J.Neurophysiol. 26 : 193—206.

Cerf J., Libert Ch., 1955: Neurostimulateur multiple à l'impulsions rectangulaires et sorties isolées. EEG Clin. Neurophysiol. 7 : 433—437.

Chambers R., 1921: A simple apparatus for micromanipulating under the highest magnifications of the microscope. Science 54 : 411—413.

Chang H. T., 1950: The repetitive discharge of corticothalamic reverberating circuit. J. Neurophysiol. 13 : 235—258.

Chang H. T., 1951a: Dendritic potential of cortical neurons produced by direct electrical stimulation of the cerebral cortex. J. Neurophysiol. 14 : 1—21.

Chang H. T., 1951b: An observation on the effect of strychnine on local cortical potentials. J. Neurophysiol. 14 : 23—28.

Chang H. T., 1952: Cortical response to stimulation of lateral geniculate body and the potentiation there of by continuous illumination of retina. J. Neurophysiol. 15 : 5—26.

Chang H. T., 1953: Cortical response to activity of callosal neurones. J. Neurophysiol. 16 : 117—131.

Chang H. T., 1955a: Cortical response to stimulation of medullary pyramid in rabbit. J. Neurophysiol. 18 : 332—352.

Chang H. T., 1955b: Activation of internuncial neurons through collaterals of pyramidal fibers at cortical level. J. Neurophysiol. 18 : 452—471.

Chernigovsky V. N. — Черниговский В. Н., 1956: О корковом представительстве внутренных органов. Доклады на XX. международном конгрессе физиологов в Брюселле, 384—389.

Child C., 1929: The physiological gradients. Protopl. 5 : 447.

Clare M. H., Bishop G. H., 1954: Properties of dendrites: apical dendrites of the cat cortex. EEG Clin. Neurophysiol. 3 : 449—464.

Clare M. H., Landau W. M., Bishop G. H., 1961: The cortical response to direct stimulation of the corpus callosum in the cat. EEG Clin. Neurophysiol. 13 : 21—33.

Clark G., 1939: The use of the Horsley-Clarke instrument on the rat. Science 90 : 92.

Clarke R. H., Henderson E. E., 1911: Atlas of photographs of sections of the frozen cranium and brain of the cat (*Felis domestica*). J. Psychol. Neurol. 18 : 391—409.

Clarke R. H., Henderson E. E., 1920: Atlas of photographs of the frontal sections of the cranium and brain of the rhesus monkey (*Macacus rhesus*). John Hopkins Hosp. Rep., Special volume 163—172.

Clarke, R. H., Horsley V., 1906: On a method of investigating the deep ganglia and tracts of the central nervous system (cerebellum). Brit. Med. J. 2 : 1799—1800.

Cobb W., Sears T. A., 1956: The superficial spread of cerebral potential fields. Some evidence provided by hemispherectomy. EEG Clin. Neurophysiol. 8 : 717—718

Cohn R., 1956: Laminar electrical responses in lateral geniculate body of cat. J. Neurophysiol. 19 : 317—324.

Cohen M., Grundfest H., 1954: Thalamic foci of electrical activity initiated by afferent impulses in cat. J. Neurophysiol. 17 : 193—207.

Cole K. S., Antosiewicz H. A., Rabinowitz P., 1955: Automatic computation of nerve excitation. J. Soc. Indust. Appl. Math., 3 : 153—172.

Cole K. S., Curtis H. J., 1939: Electrical impedance of the squid giant axon during activity. J. Gen. Physiol. 22 : 649—670.

Cole K. S., Hodgkin A. L., 1939: Membrane and protoplasma resistance in the squid giant axon. J. Gen. Physiol. 22 : 671—687.

Collins R. L., 1964: Photographic elimination of transients (PET): A simple technique for retrieving bioelectrical signals from noise. Proc. Soc. Exp. Biol. Med. 117 : 724—726.

Combs C. M., 1954: Electro-anatomical study of cerebellar localization. Stimulation of various afferents. J. Neurophysiol. 17 : 123—142.

Conway E. J., 1957: Nature and significance of concentration relations of potassium and sodium ions in skeletal muscle. Physiol. Rev. 37 : 84—132.

Cooke P. M., Snider R. S., 1954: The electrocerebellogram as modified by afferent impulses. EEG Clin. Neurophysiol. 6 : 415—423,

Coombs J. S., Curtis D. R., Eccles J. C., 1957a: The interpretation of spike potentials of motoneurones. J. Physiol. 138 : 231—198.

Coombs J. S., Curtis D. R., Eccles J. C., 1957b: The generation of impulses in motoneurones. J. Physiol. 139 : 232—250.

Coombs J. S., Curtis D. R., Eccles J. C., 1959: The electrical constants of the motoneurone membrane. J. Physiol. 145 : 505—528.

Coombs J. S., Eccles J. C., Fatt P., 1955a: The electrical properties of the motoneurone membrane. J. Physiol. 130 : 291—325.

Coombs J. S., Eccles J. C., Fatt P., 1955b: The specific ionic conductances and the ionic movements across the motoneuronal membrane that produce the inhibitory postsynaptic potential. J. Physiol. 130 : 326—373.

Cooper R., Winter A. L., Crow H. J., Walter W. Grey, 1965: Comparison of subcortical, cortical and scalp activity using chronically indwelling electrodes in man. EEG Clin. Neurophysiol. 18 : 217—228.

Coppée G., 1939: Les voies auditives au niveau dela moelle allongée. Topographie des voies chez le chat et la lapin. C. R. Soc. Biol. 130 : 1364—1366.

Cort J. H., 1957: Nový stereotaxický přístroj pro přesné určení bodů v centrální nervové soustavě krysy, králíka, kočky, morčete a malého psa. Čs. fysiol. 6 : 530—532.

Couteaux R., 1944: Nouvelles observations sur la structure de la plaque motrice et interpretation des rapports myo-neuráux. C. R. Soc. Biol. 138 : 976—978.

Couteaux R., 1960: Motor end-plate structure. In: G. H. Bourne (Editor): The structure and function of muscle, vol. 1, Academic Press, New York — London.

Cragg B. G., Hamlyn L. H., 1955: Action potentials of the pyramidal neurones in the hippocampus of the rabbit. J. Physiol. 129 : 608—627.

Creed R. S., Denny-Brown D., Eccles J. C., Liddell E.G. T., Sherrington C. S., 1932: Reflex Activity of the Spinal Cord. Oxford Univ. Press, London.

Cremer M., 1909: Die allgemeine Physiologie des Nerven. Naget's Handb. Physiol. Mensch. 4 : 793—796.

Cress R. M., Harwood T. R., 1953: Activity in the medulla elicited by electrical stimulation of the posterior column of the spinal cord in the cat. Anat. Rec. 115 : 389.

Curtis D. R., 1964: Microelectrophoresis. In: Nastuk W. L. (Editor): Physical techniques in biological research. Vol. V. — Electrophysiological Methods. Part A, pp. 144—192, Academic Press, New York — London.

Curtis D. R., Eccles R. M., 1958: The excitation of Renshaw cells by pharmacological agents applied electrophoretically. J. Physiol. 141 : 435—445.

Curtis D. R., Phillis J. W., Watkins J. C., 1960: The chemical excitation of spinal neurones by certain acidic amino acids. J. Physiol. 150 : 656—682.

Curtis H. J., 1940: Intercortical connections of corpus callosum as indicated by evoked potentials. J. Neurophysiol. 3 : 407—413.

Curtis H. J., Cole K. S., 1938: Transverse electric impedance of the squid giant axon. J. Gen. Physiol. 21 : 757.

Curtis H. J., Cole K. S., 1940: Membrane action potentials from the squid giant axon. J. Cell. Comp. Physiol. 15 : 147—157.

Curtis H. J., Cole K. S., 1942: Membrane resting and action potentials from the squid giant axon. J. Cell. Comp. Physiol. 19 : 135—144.

Dale H. H., 1937: Transmission of nervous effects by acetylcholine. Harvey Lectures 32 : 229—245.

Dale H. H., Feldberg W., 1934: Chemical transmission at motor nerve endings in voluntary muscle. J. Physiol. 81 : 39P.

Dale H. H., Feldberg W., Vogt M., 1936: Release of acetylcholine at voluntary motor nerve endings. J. Physiol. 86 : 353—380.

Dalton J. C., 1957: Effects of external ions on membrane potential of a lobster giant axon. J. Gen. Physiol. 41 : 529—542.

Dalton J. C., 1959: Effects of external ions on membrane potentials of a crayfish giant axon. J. Gen. Physiol. 42 : 971—982.

Dawson G. D., 1947: Cerebral responses to electrical stimulation of peripheral nerve in man. J. Neurol. Neurosurg. Psychiat. 10 : 134—140.

Dawson G. D., 1951: A summation technique for detecting small signal in a large irregular background. J. Physiol. 115 : 2P—3P.

Dawson G. D., 1954: A summation technique for the detection of small evoked potentials. EEG Clin. Neurophysiol. 6 : 65—84.

Dawson G. D., 1956: The relative excitability and conduction velocity of sensory and motor nerve fibres in man. J. Physiol. 131 : 436—451.

Dawson G. D., Scott J. W., 1949: The recording of nerve action potentials through skin in man. J. Neurol. Neurosurg. Psychiat. 12 : 259—267.

Dawson G. D., Pitman J. R., Wilkie D. R., 1960: A low-capacitance, low output impedance stimulus coupler. Proceedings of the Physiological Society, 1—37.

Degelman J., 1956: A coupling circuit for the Bickford integrator. EEG Clin. Neurophysiol. 8 : 693—694.

Del Castillo J., Katz B., 1955: On the localization of acetylcholine receptors. J. Physiol. 128 : 157—181.

Del Castillo J., Katz B., 1956: Biophysical aspects of neuromuscular transmission. Progr. Biophys. 6 : 121—170.

Del Castillo J., Katz B., 1957: A study of curare action with an electrical micro-method. Proc. Roy. Soc. B. 146 : 339—356.

Delgado J. M. R., 1955: Evaluation of permanent implantation of electrodes within the brain. EEG Clin. Neurophysiol. 6 : 637—644.

Delgado J. M. R., 1962: Chronic implantation of chemitrodes in the monkey brain. Proc. Int. Union Physiol. Sci. vol. II., XXII Int. Congress, Leiden, Abstracts of Communications No. 1090.

Delgado J. M. R., 1964: Electrodes for extracellular recording and stimulation. In: W. L. Nastuk (Editor): Physical Techniques in Biological Research, vol. V., Electrophysiological Methods. Part. A, pp. 88—143, Academic Press, New York-London.

DeLucchi M. R., Garoutte B., Aird R. B., 1962: The scalp as an electroencephalographic averager. EEG Clin. Neurophysiol. 14 : 191—196.

Dempsey E. W., Morison R. S., 1942a: The production of rhytmically recurrent cortical potentials after localized thalamic stimulation. Amer. J. Physiol. 135 : 293—300.

Dempsey E. W., Morison R. S., 1942b: The interaction of certain spontaneous and induced cortical potentials. Amer. J. Physiol. 135 : 301—308.

Dempsey E. W. Morison R. S., 1943: The electrical activity of a thalamocortical relay system. Amer. J. Physiol. 138 : 283—296.

Denny-Brown D., Pennybacker J. B., 1938: Fibrillation and fasciculation in voluntary muscle. Brain 61 : 311—334.

Derbyshire A. J., Rempel B., Forbes A., Lambert E. F., 1936: Effect of anesthetics on action potentials in cerebral cortex of the cat. Amer. J. Physiol. 116 : 577—596.

Dern H., Walsh J. B., 1963: Analysis of complex waveforms. In: Nastuk W. L. (Editor): Physical Techniques in Biological Research, vol. VI, Electrophysiological Methods, Part B, pp. 99—218. Academic Press, New York — London.

De Robertis E., 1958: Submicroscopic morphology and function of the synapse. Exp. Cell Res., Suppl. 5 : 347—369.

Desmedt J. E., 1962: Auditory-evoked potentials from cochlea to cortex as influenced by activation of the efferent olivocochlear bundle. J. Ac. Soc. Amer. 34 : 1478—1496.

D'Hollander F., 1913: Recherches anatomiques sur les couches optiques. Le Névraxe 14—15: 470 (cit. by Rioch 1929).

Dickinson C. J., 1950: Electrophysiological technique. Offices of electronic engineering, London

Digby P. S. B., 1965: Semi-conduction and electrode processes in biological material. I. Crustacea and certain soft-bodied forms. Proc. Roy. Soc. B. 161 : 504—525.

Di Palma J. R., Mohl R., Best W. Jr., 1961: Action potential and contraction of *Dionaea muscipula* (Venus Flytrap). Science 133 : 878—879.

Donaldson P. E. K., 1958: Electronic apparatus for biological research. Butterworths Scientific Publications, London.

Dondey M., Albe-Fessard D., Le Beau J., 1962: Premières applications neurophysiologiques d'une méthode permettant le blocage électif et reversible de structures centrales par réfrigération localisée. EEG Clin. Neurophysiol. 14 : 758—763.

Dow R. S., 1938: The electrical activity of the cerebellum and its functional significance. J. Physiol. 94 : 67—86.

Dow R. S., 1939: Cerebellar action potentials in response to stimulation of various afferent connections. J. Neurophysiol. 2 : 543—555.

Dow R. S., 1942: Cerebellar action potentials in response to stimulation of the cerebral cortex in monkeys and cats. J. Neurophysiol. 5 : 122—136.

Dow R. S., Anderson R., 1942: Cerebellar action potentials in response to stimulation of proprioceptors and exteroceptors in the rat. J. Neurophysiol. 5 : 363—371.

Du Bois-Reymond E., 1841, 1849: Untersuchungen über thierische Electrizität. Vol. 1, 1841, Vol. 2, 1849, Reiner, Berlin.

Du Bois D., 1931: A machine for pulling glass micropipettes and needles. Science 73 : 345—345,

Dudel J., 1965: The mechanism of presynaptic inhibition at the crayfish neuromuscular junction. Pflüg. Arch. Ges. Physiol. 284 : 66—80.

Dudel J., Kuffler S. W., 1961a: The quantal nature of transmission and spontaneous miniature potentials at the crayfish neuromuscular junction. J. Physiol. 155 : 514—529.

Dudel J., Kuffler S. W., 1961b: Mechanism of facilitation at the crayfish neuromuscular junction. J. Physiol. 155 : 530—542,

Dudel J., Kuffler S. W., 1961c: Presynaptic inhibition at the crayfish neuromuscular junction. J. Physiol. 155 : 543—562.

Dumont S., Dell P., 1960: Facilitation réticulaire des mécanismes visuels corticaux. EEG Clin. Neurophysiol. 12 : 769—796.

Dusser de Barenne J. G., McCulloch W. S., 1936a: Some effects of local strychninization on action potentials of the cerebral cortex of the monkey. Trans. Amer. Neurol. Ass. 62 : 171.

Dusser de Barenne J. G., McCulloch W. S., 1936b: Functional boundaries in the sensory-motor cortex of the monkey. Proc. Soc. Exp. Biol. N. Y. 35 : 329—331.

Dusser de Barenne J. G., McCulloch W. S., 1939: Physiological delimination of neurones in the central nervous system. Amer. J. Physiol. 127 : 620—628.

Dusser de Barenne J. G., Garol H. W., McCulloch W. S., 1941: Physiological neuronography of the cortico-striatal connections. Res. Publ. Ass. Nerv. Ment. Dis. 21 : 246—266.

Dvořák J., Koryta J., Boháčková V., 1966: Elektrochemie. Academia, Praha.

Eccles J. C., 1935a: The action potential of the superior cervical ganglion. J. Physiol. 85 : 179—205.

Eccles J. C., 1935b: Slow potential waves in the superior cervical ganglion. J. Physiol. 85 : 464—501.

Eccles J. C., 1936: Synaptic and neuro-muscular transmission. Ergebn. Physiol. 38 : 339—444.

Eccles J. C., 1937: Synaptic and neuro-muscular transmission. Physiol. Rev. 17 : 538—555.

Eccles J. C., 1943: Synaptic potentials and trasmission in sympathetic ganglion. J. Physiol. 101 : 465—483.

Eccles J. C., 1944: The nature of synaptic transmission in a sympathetic ganglion. J. Physiol. 103 : 27—54.

Eccles J. C., 1946: Synaptic potentials of motoneurons. J. Neurophysiol. 9 : 87—120.

Eccles J. C., 1948: Conduction and synaptic transmission in the nervous system. Ann. Rev. Physiol. 10 : 93—116.

Eccles J. C., 1951: Interpretation of action potentials evoked in the cerebral cortex. EEG Clin. Neurophysiol. 3 : 449—464.

Eccles J. C., 1953: The neurophysiological basis of mind: the principles of neurophysiology. Clarendon Press: Oxford.

Eccles J. C., 1957: The physiology of nerve cells. The Johns Hopkins Press, Baltimore.

Eccles J. C., 1959: Neuron Physiology. Introduction. In: Handbook of Physiology, Sec. 1. Neurophysiology, vol I. Amer. Physiol. Soc. Washington.

Eccles J. C., 1963: Presynaptic and postsynaptic inhibitory actions in the spinal cord. In: G. Moruzzi (editor): Brain mechanisms, Elsevier, Amsterdam.

Eccles J. C., 1964a: The physiology of synapses. Springer-Verlag, Berlin-Göttingen-Heidelberg.

Eccles, J. C., 1964b: Presynaptic inhibition in the spinal cord. Progr. Brain Res. 12 : 65 – 89.

Eccles J. C., Eccles R. M., Ito M., 1964a): Effects of intracellular potassium and sodium injections on the inhibitory postsynaptic potential. Proc. Roy. Soc. B 160 : 181 – 196.

Eccles J. C., Eccles R. M., Ito M., 1964b): Effects produced on inhibitory postsynaptic potentials by the coupled injections of cations and anions into motoneurons. Proc. Roy. Soc. B 160 : 197 – 210.

Eccles J. C., Eccles R. M., Magni F., 1961: Central inhibitory action attributable to presynaptic depolarization produced by muscle afferent volleys. J. Physiol. 159 : 147 – 166.

Eccles J. C., Fatt P., Koketsu K., 1954: Cholinergic and inhibitory synapses in a pathway from motor-axon collaterals to motoneurones. J. Physiol. 126 : 524 – 562.

Eccles J. C., Katz, B., Kuffler S. W., 1941: Nature of the "end-plate potential" in curarized muscle. J. Neurophysiol. 4 : 362 – 387.

Eccles J. C., Kostyuk P. G., Schmidt R. F., 1962a: Central pathways responsible for depolarization of primary afferent fibres. J. Physiol. 161 : 237 – 247.

Eccles J. C., Kuffler S. W., 1941: Initiation of muscle impulses at neuro-musclar junction. J. Neurophysiol. 4 : 402 – 417.

Eccles J. C., Macfarlane W. V., 1949: Actions of anti-cholinesterases on end-plate potential of frog muscle. J. Neurophysiol. 12 : 59 – 80.

Eccles J. C., Magni F., Willis W. D., 1962: Depolarization of central terminals of group I afferent fibres from muscles. J. Physiol. 160 : 62 – 93.

Eccles J. C., Malcolm J. L., 1946: Dorsal root potentials of the spinal cord. J. Neurophysiol. 9 : 139 – 160.

Eccles J. C., McIntyre A. K., 1951: Plasticity of mammalian monosynaptic reflexes. Nature 167 : 466 – 468.

Eccles J. C., O'Connor W. J., 1939: Responses which nerve impulses evoke in mammalian striated muscle. J. Physiol. 97 : 44 – 102.

Eccles J. C., Rall, W., 1951: Effects induced in a monosynaptic reflex path by its activations. J. Neurophysiol. 14 : 353-376.

Eccles J. C., Schmidt R. F., Willis W. D., 1962: Presynaptic inhibition of the spinal monosynaptic reflex pathway. J. Physiol. 161 : 282 – 297.

Eccles J. C., Schmidt R. F., Willis W. D., 1963a: Depolarization of central terminals of group Ib afferent fibers of muscle. J. Neurophysiol. 26 : 1 – 10.

Eccles J. C., Schmidt R. F., Willis W. D., 1963b: Depolarization of central terminals of cutaneous afferent fibers. J. Neurophysiol. 26 ; 646 – 656.

Eccles R. M., 1952a: Action potentials of isolated mammalian sympathetic ganglia. J. Physiol. 117 : 181 – 195.

Eccles R. M., 1952b: Responses of isolated curarized sympathetic ganglia. J. Physiol. 117 : 196 – 217.

Eccles R. M., 1955: Intracellular potentials recorded from a mammalian sympathetic ganglion. J. Physiol. 130 : 572 – 584.

Eccles R. M., 1963: Orthodromic activation of single ganglion cells. J. Physiol. 165 : 387 – 391.

Eccles R. M., Libet B., 1961: Origin and blockade of the synaptic responses of curarized sympathetic ganglia. J. Physiol. 157 : 484 – 503.

Eckert B., Zacharová D., 1954: The lability of the excitor and inhibitor nerve fibers of the claw of the crayfish. Physiol. bohemoslov. 3 : 191 – 197.

Eckert B., Zacharová D., 1957: Unterschiede zwischen dem erregenden und dem hemmenden Axon des Öffnermuskels der Krebsschere. Physiol. bohemoslov. 6 : 39—48.

Edwards C., Ottoson D., 1958: The site of impulse initiation in a nerve cell of a crustacean stretch receptor. J. Physiol. 143 : 138—148.

Edwards C., Terzuolo C. A., Sashizu Y., 1963: The effect of changes of the ionic environment upon an isolated crustacean sensory neuron. J. Neurophysiol. 26 : 948—957.

Eidelberg E., Saladias C. A., 1960: A stereotaxic atlas for cebus monkeys. J. Comp. Neurol. 115 : 103—123.

Erlanger J., Gasser H. S., 1937: Electrical signs of nervous activity. Philadelphia.

Ernst E., 1957: Über den heutigen Stand des Erregungsproblems. Acta Physiol. Hung. 11, suppl: 9—11.

Euler C. von, Green J. D., Ricci G., 1958: The role of hippocampal dendrites in evoked responses and after-discharges. Acta Physiol. Scand. 42 : 87—111.

Eyzaguirre C., 1957: Functional organization of neuromuscular spindle in toad. J. Neurophysiol. 20 : 523—542.

Eyzaguirre C., Kuffler S. W., 1955a: Processes of excitation in the dendrites and in the soma of single isolated sensory nerve cell. J. Gen. Physiol. 39 : 87—119.

Eyzaguirre C., Kuffler S. W., 1955b: Further study of soma, dendrite and axon excitation in single neurons. J. Gen. Physiol. 39 : 121—153.

Fadiga E., Pupilli G. C., 1964: Teleceptive components of the cerebellar function. Physiol. Rev. 44 : 433—486.

Falk G., Fatt P., 1964: Linear electrical properties of striated muscle fibres observed with intracellular electrodes. Proc. Roy. Soc. B 160 : 69—123.

Falk G., Gerard R. W., 1954: Effect of micro-injected salts and ATP on the membrane potential and mechanical response of muscle. J. Cell. Comp. Physiol. 43 : 393—403.

Falkovich S. E. — Фалькович С. Е., 1961: Прием радиолокационных сигналов на фоне флюктуационных помех. Изд. ,,Советское радио'', Москва.

Fatt P., 1957a: Electrical potentials occurring around a neuron during its antidromic activation. J. Neurophysiol. 20 : 27—60.

Fatt P., 1957b: Sequence of events in synaptic activation of a motoneuron. J. Neurophysiol. 20 : 61—80.

Fatt P., 1959: Skeletal neuromuscular transmission. In: Handbook of Physiology, Section 1, Neurophysiology, vol. 1, pp. 199—213, Amer. Physiol. Soc., Washington.

Fatt P., Gingsborg B. L., 1958: The ionic requirements for the production of action potentials in crustacean muscle fibres. J. Physiol. 142 : 516—543.

Fatt P., Katz B., 1951: An analysis of the end-plate potential recorded with an intracellular electrode. J. Physiol. 115 : 320—370.

Fatt P., Katz B., 1952a: The electric activity of the motor end-plate. Proc. Roy. Soc. B. 140 : 183—186.

Fatt P., Katz B., 1952b: Spontaneous subthreshold activity at motor nerve endings. J. Physiol. 117 : 109—128.

Fatt P., Katz B., 1953a: The electrical properties of crustacean muscle fibres. J. Physiol. 120 : 171—204.

Fatt P., Katz B., 1953b: Distributed "end-plate potentials" of crustacean muscle fibres. J. Exp. Biol. 30 : 433—439.

Fessard A., 1936: Propriétés rythmiques de la matière vivante. I. Nerfs myélinisés. II. Nerfs non myélinisés. Paris.

Fifková E., 1964a: Spreading EEG depression in the neo-, paleo- and archicortical structures of the brain of the rat. Physiol. bohemoslov. 13 : 1—15.

Fifková R., 1964b: Leão's spreading depression in the thalamic nuclei of rat. Experientia 20 : 635—637.

Flatau E., 1897: Beitrag zur technischen Bearbeitung des Centralnervensystems. Anat. Anz. 30 : 323—329.

Fleckenstein A., Hille H., Adam W. E., 1951: Aufhebung der Kontraktur-Wirkung depolarisierender Katelektronika durch Repolarisation im Anelektrotonus. Pflüg. Arch. Ges. Physiol. 253 : 264.

Flood T., 1951: Wireless Eng. 28, No. 335, 231.

Florey E., Florey E., 1955: Microanatomy of the abdominal stretch receptors of the crayfish (*Astacus fluviatilis*) J. Gen. Physiol. 39 : 69—85.

Florian J., 1928: Ein Hebelmikromanipulator. Z. Wiss. Mikroskopie 45 : 460—471.

Foŋbrune P. de, 1932: Nouveau micromanipulateur et dispositif pour la fabrication des microinstruments. C. R. Acad. Sci. 195 : 603—706.

Forbes A., Battista A. F., Chatfield P. O., Garcia J. P., 1949: Refractory phase in cerebral mechanisms. EEG Clin. Neurophysiol. 1 : 141—175.

Forbes A., Merlis J. K., Henriksen G. I., Burleigh S., Jiusto J. H., Merlis G. L., 1956: Meaurement of the depth of barbiturate narcosis. EEG Clin. Neurophysiol. 8 : 541—558.

Forbes A., Morison B. R., 1939: Cortical response to sensory stimulation under deep barbiturate narcosis. J. Neurophysiol. 2 : 112—118.

Fourier J. B. J., 1822: Théorie analytique de la chaleur, Paris.

Fox C. A., 1940: Certain basal telencephalic centers in the cat. J. Comp. Neurol. 72 : 1—62

Frank G. B., 1960: Maximum activation of the contractile mechanism in frog's skeletal muscle by potassium depolarization. J. Physiol. 154 : 345—353.

Frank K., 1959a: Basic mechanisms of synaptic transmission in the central nervous system. I. R. E. Trans. Med. Electron. ME. 6, 85—88.

Frank K., 1959b: Identification and analysis of single unit activity in the central nervous system. Handbook of Physiology, Sec. 1, Neurophysiology, vol. I. Amer. Physiol. Soc., Washington.

Frank K., Becker M. C., 1964: Microelectrodes for recording and stimulation. In: W. L. Nastuk (Editor), Physical Techniques in Biological Research. Vol. V. Electrophysiological Methods. Part A, pp. 22—87. Academic Press, New York.

Frank K., Fuortes M. G. F., 1955: Potentials recorded from the spinal cord with microelectrodes. J. Physiol. 130 : 625—654.

Frank K., Fuortes M. G. F., 1956a: Unitary activity of spinal interneurones of cats. J. Physiol. 131 : 424—435.

Frank K., Fuortes M. G. F., 1956b: Stimulation of spinal motoneurones with intracellular electrodes. J. Physiol. 131 : 451—470.

Frank K., Fuortes M. G. F., 1957: Presynaptic and postsynaptic inhibition of monosynaptic reflexes. Fed. Proc. 16 : 39—40.

Frankenhaueuser. B., 1951: Limitations of method of strychnine neuronography. J. Neurophysiol. 14 : 73—79.

Franzini-Armstrong C., 1964: Fine structure of sarcoplasmic reticulum and transverse tubular system in muscle fibers. Fed. Proc. 231 : 887—895.

French J. D., 1952: Brain lesions associated with prolonged unconsciousness. Arch. Neurol. Psychiat. 68 : 727—740.

French J. D., Magoun H. W., 1952: Effects of chronic lesions in central cephalic brain stem of monkeys. Arch. Neurol. Psychiat. 68 : 591—604.

French J. D., Verzeano M., Magoun H. W., 1953: A neural basis of the anesthetic state. Arch. Neurol. Psychiat. 69 : 519—529.

Freygang W. H., 1958: An analysis of extracellular potentials from single neurons in the lateral geniculate nucleus of the cat. J. Gen. Physiol. 41 : 543—564.

Freygang W. H., Landau W. M., 1955: Some relations between resistivity and electrical activity in the cerebral cortex of the cat. J. Cell. Comp. Physiol. 45 : 377—392.

Frontera J. G., 1958: Evaluation of the immediate effects of some fixatives upon the measurements of the brain of Macaques. J. Comp. Neurol. 109 : 417—438.

Fuortes M. G. F., Frank K., Becker M. C., 1957: Steps in the production of motoneuron spikes. J. Gen Physiol. 40 : 735—752.

Furshpan E. J., Potter D. D., 1959a: Transmission at the giant synapses of the crayfish. J. Physiol. 145 : 289—325.

Furshpan E. J., Potter D. D., 1959b: Slow post-synaptic potentials recorded from the giant motor fibre of the crayfish. J. Physiol. 145 : 326—335.

Furukawa T., Fukami Y., Asada Y., 1963: A third type of inhibition in the Mauthner cell of goldfish. J. Neurophysiol. 26 : 759—774.

Furukawa T., Furshpan E. J., 1963: Two inhibitory mechanisms in the Mauthner neurons of goldfish. J. Neurophysiol. 26 : 140—176.

Furusawa K., 1929: The depolarization of crustacean nerve by stimulation or oxygen want. J. Physiol. 67 : 325—342.

Gaffey C. T., Mullins L. J., 1958: Ion fluxes during the action potential in *Chara*. J.Physiol. 144 : 505—524.

Galambos R., Rose E., Bromiley B., Hughes R., 1953: Microelectrode studies on medial geniculate body of cat. II. Response to clicks. J. Neurophysiol. 16 : 359—380.

Galleotti G., 1904: Concerning the EMF which is generated at the surface of animal membranes on contact with different electrolytes. Z. Physik. Chem. 49 : 542—562.

Galvani L., 1791: De viribus electricitatis in motu muscularis commentarius. In: Debononiensi scientarium et artium instituto atque academia commentarii, Bologna 7 : 363—418.

Gasser H. S., 1938: Electrical signs of biological activity. J. Appl. Physics 9 : 88—98.

Gasser H. S., Graham H. T., 1933: Potentials produced in the spinal cord by stimulation of the dorsal roots. Amer. J. Physiol. 103 : 303—320.

Gasser H. S., Grundfest H., 1939: Axon diameter in relation to the spike dimension and conduction velocity in mammalian A fibres. Amer. J. Physiol. 127 : 393—414.

Gastaut H., Hunter J., 1950: An experimental study of the mechanism of photic activation in idiopathic epilepsy. EEG Clin. Neurophysiol. 2 : 263—287.

Gastaut H., Lammers H. J., 1960: Anatomie du rhinencéphale. Les grandes activités du rhinencéphale, vol. I. Masson et Cie, Paries.

Gastaut H., Roger Y., Corriol. J., Naquet R., 1951: Etude électrographique du cycle d'excitabilité cortical. EEG Clin. Neurophysiol. 3 : 410—428.

Gauthier C., Parma N., Zanchetti A., 1956: Effect of electrocortical arousal upon development and configuration of specific evoked potentials. EEG Clin.Neurophysiol. 8 : 237—243.

Gellhorn E., 1952: Experimental contribution to the duplicity theory of consciousness and perception. Pflüg. Arch. Ges. Physiol. 255 : 75—92.

Gellhorn E., 1953a: The hypothalamic-cortical system in barbiturate anesthesia. Arch. Int. Pharmacodyn. 93 : 434—442.

Gellhorn E., 1953b: Physiological foundations of neruology and psychiatry. Univ. of Minnesota Press, Minneapolis.

Gellhorn E., Koella W. P., Ballin H. M., 1954: Interaction in cerebral cortex of acoustic or optic with nociceptive impulses: the problem of consciousness. J. Neurophysiol. 17 : 14—21.

Gerard R. W., Marshall W. H., Saul L. J., 1936: Electrical activity of the cat's brain. Arch. Neurol. Psychiat. 36 : 675—738.

Gerstein G. L., Kiang N. Y. S., 1960: An approach to the quantitative analysis of electro-physiological data from single neurons. Biophys. J. 1 : 15—28.

Gibbs, F. A., Davis H., Lennox W. G., 1935: The electroencephalogram in epilepsy and in conditions of impaired consciousness. Arch. Neurol. Psychiat. 34 : 1133—1148.

Gilliatt R. W., Sears T. A., 1958: Sensory nerve action potentials in patients with peripheral nerve lesions. J. Neurol. Neurosurg. Psychiat. 21 : 109—119.

Glass H., 1933: Effect of light on the bioelectric potentials of isolated *Elodea* leaves. Plant Physiol. 8 : 263.

Glasstone J., 1949: An introduction to electrochemistry. D. van Nostrand Comp. N. Y.

Goldman D., 1950: The clinical use of the "average" reference elèctrode in monopolar recording. EEG Clin. Neurophysiol. 2 : 211—216.

Goldman D. E., 1943/44: Potential, impedance and rectification in membranes. J. Gen. Physiol. 27 : 37.

Goldman S., Vivian W. F., Chi-Kuang Chien, Bowes H. N., 1948: Electronic mapping of the activity of the heart and the brain. Science 108 : 720—723.

Goldring S., O'Leary J. L., Holmes T. G., Jerva M. J. ,1961: Direct response of isolated cerebral cortex of cat. J. Neurophysiol. 24 : 633—650.

Goldring S., O'Leary J. L., Lam R. L., 1953: Effect of malononitrile upon the electrocorticogram of the rabbit. EEG Clin. Neurophysiol. 5 : 395—400.

Gosling W., 1964: Circuit application of field effect transistors. Brit. Commun. Electronics: 702—704.

Gozzano M., 1936: Bioelektrische Erscheinungen bei der Reflexepilepsie. J. Psychol. Neurol. 47 : 24—39.

Grafstein B., 1956: Mechanism of spreading cortical depression. J. Neurophysiol. 19 : 154-171.

Grafstein B., 1959: Organization of callosal connections in suprasylvian gyrus of cat. J. Neurophysiol. 22 : 504—515.

Grafstein B., 1963: Postnatal development of the transcallosal evoked response in the cerebral cortex of the cat. J. Neurophysiol. 26 : 79—99.

Graham J., Gerard R. W., 1946: Membrane potentials and excitation of impaled single muscle fibres. J. Cell. Comp. Physiol. 28: 99—117.

Granit R., 1933: The components of the retinal action potential and their relation to the discharge in the optic nerve. J. Physiol. 77 : 207—240.

Granit R., 1947: Sensory mechanisms of the retina. Oxford University Press, London.

Granit R., 1955: Receptors and sensory perception. Yale Univ. Press.

Granit R., 1962: Some problems of muscle-spindle physiology. In: D. Barker (Editor): Muscle receptors. Hong Kong University Press, pp. 1—12.

Granit R., Therman P. O., 1938: The "slow potentials" associated with excitation and inhibition in the excised eye. J. Physiol. 93 : 9 P.

Grastyán E., 1959: The hippocampus and higher nervous activity. In: M. A. B. Brazier (editor) Central Nervous System and Behavior. Trans. Second Macy Conf., Josiah Macy, Jr. Foundation, New York pp. 119—205.

Gray E. G., 1957: The spindle and extrafusal innervation of a frog muscle. Proc. Roy. Soc. B. 146 : 416—430.

Gray E. G., 1962: A morphological basis for pre-synaptic inhibition. Nature 193 : 82—83.

Gray E. G., 1963: Electron microscopy of presynaptic organelles of the spinal cord. J. Anat. 97 : 101—106.

Gray J. A. B., 1959a: Initiation of impulses at receptors. In: Handbook of Physiology, Section 1, Neurophysiology, vol. I., Amer. Physiol. Soc., Washington.

Gray J. A. B., 1959b: Mechanical into electrical energy in certain mechanoreceptors. Progr. Biophys. 9 : 285—324.

Gray J. A. B., Sato M., 1953a: Receptor potentials in Pacinian corpuscles. J. Physiol. 122 : 27—28.

Gray J. A. B., Sato M., 1953b: Properties of the receptor potential in Pacinian corpuscles. J. Physiol. 122 : 610—636.

Gray J. A. B., Sato M., 1955: The movement of sodium and other ions in Pacinian corpuscles. J. Physiol. 129 : 594—607.

Gray T. S., 1954: Applied electronics. — Wiley, N. Y.

Green J. D., 1964: The hippocampus. Physiol. Rev. 44 : 561—808.

Green J. D., Adey W. R., 1956: Electrophysiological studies of hippocampal connections and excitability. EEG Clin. Neurophysiol. 8 : 245—262.

Green J. D., Arduini A., 1954: Hippocampal electrical activity in arousal. J. Neurophysiol. 17 : 533—557.

Green J. D., Maxwell D. S., Petsche H., 1961: Hippocampal electrical activity. III. Unitary events and genesis of slow waves. EEG Clin. Neurophysiol. 13 : 854—867.

Grenell R. G., 1953: Central nervous system resistance. J. Comp. Neurol. 99 : 117—128.

Griswold D. M., 1964: Understanding and using the MOS-FET. Electronics 37 : 66—70.

de Groot J., 1959a: The rat forebrain in stereotaxic coordinates. Trans. Royal Neth. Acad. Sci. 52 : No 4.

de Groot J., 1959b: The rat hypothalamus in stereotaxic coordinates. J. Comp. Neurol. 113 : 389—400.

Grundfest H., 1950: Biological requirements for the design of amplifiers. Proc. I. R. E. 38 : 1018—1028.

Grundfest H., 1957a: The mechanism of discharge of the electric organs in relation to general and comparative electrophysiology. Progress in Biophysics and Biol. Chemistry 7 : 3—85.

Grundfest H., 1957b: Electrical inexcitability of synapses and some of its consequences in the central nervous system. Physiol. Rev. 37 : 337—361.

Grundfest H., 1957c: Excitation triggers in post-junctional cells. In: T. H. Bullock (editor): Physiological triggers and discontinuous rate processes. Amer. Physiol. Soc., Washington.

Grundfest H., 1959: Synaptic and ephaptic transmission. In: Handbook of Physiology, Section 1, Neurophysiology, vol. 1, Chapter V, pp. 147—197. Amer. Physiol. Soc., Washington.

Grundfest H., 1960a: Central inhibition and its mechanisms. In: E. Roberts (editor): Inhibition in the nervous system and γ-aminobutyric acid, pp. 47—65. Pergamon Press, New York.

Grundfest H., 1960b: Functional specification for membranes in excitable cells. Fourth int. neurochem. symposium. pp. 378—402, Pergamon Press, Oxford.

Grundfest H., 1961a: Ionic mechanisms in electrogenesis. Ann. N. Y. Acad. Sci. 94 : 405—457.

Grundfest H., 1961b: Excitation by hyperpolarizing potentials. A general theory of receptor activities. In: E. Florey (editor): Nervous Inhibition. p. 326. Pergamon Press, London.

Grundfest H., Campbell B., 1942: Origin, conduction and termination of impulses in the dorsal spino-cerebellar tract of cats. J. Neurophysiol. 5 : 275—294.

Grundfest H., Kao C. Y., Altamirano M., 1954: Bioelectric effects of ions microinjected into the giant axon of *Loligo*. J. Gen. Physiol. 38 : 245—282.

Grundfest H., Nachmansohn D., 1950: Increased sodium entry into squid giant axons at high frequencies and during reversible inactivation of cholinesterase. Fed. Proc. 9 : 53—63.

Guld Ch., 1962: Cathode follower and negative capacitance as high input impedance circuits. Proceedings of the IRE, 1912—1927.

Gulyaev P. I., Zhukov E. K., — Гуляев П. И., Жуков Е. К., 1948: Методы электрофизиологических исследований. Изд. Ленинградского Государственного Унив. им. А. А. Жданова, Ленинград.

Gurdjian E. S., 1927: The diencephalon of the albino rat. J. Comp. Neurol. 43 : 1—114.

Gurdjian E. S., 1928: The corpus striatum of the rat. J. Comp. Neurol. 45 : 249—281.

Gutmann E. (Editor) 1962: The denervated muscle. Publishing House of the Czechoslov. Acad. Sci., Prague.

Haapenen L., 1952: A direct coupled amplifier for electrophysiological investigations. Acta Physiol. Scand. 29 suppl. 106: 157—160.

Haapenen L., Hyde J., Skoglund C. R., 1952: The application of radiofrequency coupled input stages and stimulators to neurophysiological experiments. Acta Physiol. Scand. 29 suppl. 106 : 394—409.

Hagiwara S., Tasaki I., 1958: A study of the mechanism of impulse transmission across the giant synapse of the squid. J. Physiol. 143 : 114—137.

Hampson J. L., 1949: Relationships between cat cerebral and cerebellar cortices. J. Neurophysiol. 12 : 37—50.

Hansen M. W., 1938: A new micromanipulator. J. Roy. Micr. Soc. 18 : 250—260.

Hardy W. B., 1894: On some histological features and physiological properties of the postesophageal nerve cord of the *Crustacea*. Phil. Trans. B, Part I. 185 : 83—117.

Harrison F., 1938: Modifications in technique for use of Horsley-Clarke stereotaxic instrument. Arch. Neur. Psych. 40 : 563—565.

Hartline H. K., Graham C. H., 1932: Nerve impulses from single receptors in the eye. J. Cell Comp. Physiol. 1 : 277—295.

Harwood T. H., Cress R. H., 1954: Activity in medulla elicited by electrical stimulation of posterior funiculus of spinal cord in cat. J. Neurophysiol. 17 : 157—166.

Hayne R. A., Meyers R., 1950: An improved model of a human stereotaxic instrument. J. Neurosurg. 7 : 463—466.

Heilbrunn L. V., 1956: The dynamics of living protoplasm. Acad. Press, New York.

Heinbecker P., Bartley S. H., 1940: Action of ether and nembutal on the nervous system. J. Neurophysiol. 3 : 219—236.

Helstrom C. W., 1960: Statistical theory of signal detection. Pergamon Press.

Henček M., Zachar J., 1965: The electrical constants of single muscle fibres of the crayfish (*Astacus fluviatilis*). Physiol. bohemoslov. 14 : 330—342.

Hering E., 1882: Beiträge zur allgemeinen Nerven- und Muskelphysiologie. IX. Über die Nervenreizung durch den Nervenstrom. Sitzungsber. k. Akad. Wissensch., Math.-naturwissensch. Cl., Wien 85 : 237.

Hermann L., 1879: Handbuch der Physiologie, F. C. W. Vogel, Leipzig.

Hermann L., Gendre A., 1885: Über eine electromotorische Eigenschaft des bebrüteten Hühnereies. Pflüg. Arch. Ges. Physiol. 35 : 34.

Hernández-Peón R., Hagbarth K. E., 1955: Interaction between afferent and cortically induced reticular responses. J. Neurophysiol. 18 : 44—55.

Herz A., Creutzfeldt O., Fuster J. M., 1964: Statistische Eigenschaften der Neuronaktivität im ascendierenden visuellen System. Kybernetik 2 : 61—71.

766

Hess W. R., 1932: Beiträge zur Physiologie des Hirnstammes. I. Die Methodik der lokalisierten Reizung und Ausschaltung subkortikaler Hirnabschnitte. Thieme, Leipzig.

Hewlett M. G. T., 1951: An electronic trigger mechanism. EEG Clin. Neurophysiol. 3 : 513—516.

Hill A. V., 1936: Excitation and accomodation in nerve. Proc. Roy. Soc. B. 119 : 305—355.

Hill A. V., 1937—1938: Methods of analysing the heat production of muscle. Proc. Roy. Soc. B. 124 : 115—136.

Hill A. V., 1949: The abrupt transition from rest to activity in muscle. Proc. Roy. Soc. B 136 : 399—420.

Hill A. V., 1950: A discussion on muscular contraction and relaxation: their physical and chemical basis. Proc. Roy. Soc. B 137 : 40—85.

Hill A. V., 1951a: The transition from rest to full activity in muscle: the velocity of shortening. Proc. Roy. Soc. B 138 : 329—338.

Hill A. V., 1951b: The earliest manifestations of the mechanical response of striated muscle. Proc. Roy. Soc. B 138 : 339—348.

Hill A. V., 1958: The priority of the heat production in a muscle twitch. Proc. Roy. Soc B 148 : 397—407.

Hill D. K., 1950: The effect of stimulation on the opacity of a crustacean nerve trunk and its relation to fibre diameter. J. Physiol. 111 : 283—303.

Hind J. E., 1953: An electrophysiological determination of tonotopic organization in auditory cortex of cat. J. Neurophysiol. 16 : 475—489.

Hind J. E., Goldberg J. M., Greenwood D. D., Rose J. E., 1963: Some discharge characteristisc of single neurons in the inferior colliculus of the cat. II. Timing of the discharges and observations on binaural stimulation. J. Neurophysiol. 26 : 321—341.

Höber R., 1945: Physical chemistry of cells and tissues. Blakiston Company, Philadelphia.

Höber R., 1946: The membrane theory. Ann. N. Y. Acad. Sci. 47 : 381—394.

Hodes R., 1948: Electromyographic study of defects of neuromuscular transmission in human poliomyelitis. Arch. Neurol. Psychiat. 60 : 457—473.

Hodgkin A. L., 1937a: Evidence for electrical transmission in nerve. Part I. J. Physiol. 90 : 183—210.

Hodgkin A. L., 1937b: Evidence for electrical transmission in nerve. Part II. J. Physiol. 90 : 211—232.

Hodgkin A. L., 1938: The subthreshold potentials in a crustacean nerve fibre. Proc. Roy. Soc. B 126 : 87—121.

Hodgkin A. L., 1948: The local electric changes associated with repetitive action in a nonmedullated axon. J. Physiol. 107 : 165—181.

Hodgkin A. L., 1951: The ionic basis of electrical activity in nerve and muscle. Biol. Rev. 26 : 339—409.

Hodgkin A. L., 1958: Ionic movements and electrical activity in giant nerve fibres. Proc. Roy. Soc. B 148 : 1—37.

Hodgkin A. L., 1963: Conduction of the nervous impulse. Liverpool University Press.

Hodgkin A. L., Horowicz P., 1957: The differential action of hypertonic solutions on the twitch and action potential of a muscle fibre. J. Physiol. 136 : 17P.

Hodgkin A. L., Horowicz P., 1959: The influence of potassium and chloride ions on the membrane potential of single muscle fibres. J. Physiol. 148 : 127—160.

Hodgkin A. L., Huxley A. F., 1939: Action potential recorded from inside a nerve fibre. Nature 144 : 710—711.

Hodgkin A. L., Huxley A. F., 1945: Resting and action potentials in single nerve fibres. J. Physiol. 104 : 176—195.

Hodgkin A. L., Huxley A. F., 1952: A quantitative description of membrane current and its application to conduction and excitation in nerve. J. Physiol. 117 : 500—544.

Hodgkin A. L., Huxley A. F., Katz B., 1949: Ionic currents underlying activity in the giant axon of the squid. Arch. Sci. Physiol. 3 : 129—150.

Hodgkin A. L., Huxley A. F., Katz B., 1952: Measurements of current voltage relations in the membrane of the giant axon of *Loligo*. J. Physiol. 116 : 424—448.

Hodgkin A. L., Katz B., 1949: The effect of sodium ions on the electrical activity of the giant axon of the squid. J. Physiol. 108 : 37—77.

Hodgkin A. L., Keynes R. D., 1955: Active transport of cations in giant axons from *Sepia* and *Loligo*. J. Physiol. 128 : 28—60.

Hodgkin A. L., Rushton W. A. H., 1946: The electrical constants of a crustacean nerve fibre. Proc. Roy. Soc. B 133 : 444—479.

Hoefer P. F. A., 1942: Physiology of motor innervation in ischemias. Res. Publ. Ass. Nerv. Ment. Dis. 21 : 502—528.

Holubář J., 1964a: Mechanisms of the primary cortical response (PCR) of the somatosensory area in rats. Physiol. bohemoslov. 13 : 385—396.

Holubář J., 1964b: Some characteristics of the direct cortical response (DCR) in rats. Physiol. bohemoslov. 13 : 488—495.

Horsley V., Clarke R. H., 1908: The structure and functions of the cerebellum examined by a new method. Brain 31 : 45—124.

Howarth J. V., 1958: The behaviour of frog muscle in hypertonic solutions. J. Physiol. 144 : 167—175.

Howland B., Lettvin J. Y., McCulloch W. S., Pitts W., Wall P. D., 1955: Reflex inhibition by dorsal root interaction. J. Neurophysiol. 18 : 1—17.

Hruška V., 1952: Počet grafický a graficko-mechanický. Přírodovědecké nakl. Praha.

Hubel D. H., 1957: Tungsten microelectrode for recording from single units. Science 125 : 549—550.

Hubel D. H., 1960: Single unit activity in lateral geniculate body and optic tract of unrestrained cats. J. Physiol. 150 : 91—104.

Hughes J. R., 1958: Post-tetanic potentiation. Physiol. Rev. 38 : 91—113.

Hume D. M., Ganong W. F., 1956: A method for accurate placement of electrodes in the hypothalamus of the dog. EEG Clin. Neurophysiol. 8 : 136—140.

Hunt E., O'Leary L., 1952: Form of the thalamic response evoked by peripheral nerve stimulation. J. Comp. Neurol. 57 : 491—513.

Hunter J. Ingvar D. H., 1955: Pathways mediating Metrazol induced irradiation of visual impulses: an experimental study in the cat. EEG Clin. Neurophysiol. 7 : 39—60.

Hursh J. B., 1939a: Conduction velocity and diameter of nerve fibres. Amer. J. Physiol. 127 : 131—139.

Hursh J. B., 1939b: The properties of growing nerve fibers. Amer. J. Physiol. 127 : 140—153.

Huxley A. F., 1957a: Local activation of striated muscle from the frog and the crab. J. Physiol. 135 : 17—18P.

Huxley A. F., 1957b: Muscle structure and theories of contraction. Progr. Biophys. 7 : 255—318.

Huxley A. F., 1959: Ion movements during nerve activity. Ann. N. Y. Acad. Sci. 81 : 221—246.

Hydén H., 1959: Biochemical changes in glial cells and nerve cells at varying activity. Proc. 4th Int. Congress Biochemistry 3 : 64—89. Pergamon Press, London.

Ingram W. R., Hannett F. I., Ranson S. W., 1932: The topography of the nuclei of the diencephalon of the cat. J. Comp. Neurol. 55 : 333—394.

Jansen J., 1957: Afferent impulses to the cerebellar hemispheres from the cerebral cortex and certain subcortical nuclei. Acta Physiol. Scand. 41 suppl. 143 : 1—99.

Jarcho L. W., 1949: Excitability of cortical afferent systems during barbiturate anesthesia. J. Neurophysiol. 12 : 447—457.

Jasper H. H., 1949: Diffuse projection systems: the integrative action of the thalamic reticular system. EEG Clin. Neurophysiol. 1 : 405—419.

Jasper H. H., 1954: Functional properties of the thalamic reticular system. In: Brain Mechanisms and Consciousness. Blackwell, Oxford.

Jasper H. H., Ajmone-Marsan C., 1952a: Thalamocortical integrating mechanisms. A. Res. Nerv. Ment. Dis. Proc. 30 : 493—512.

Jasper H. H., Ajmone-Marsan C., 1952b: A stereotaxic atlas of the diencephalon of the cat. Nat. Research Council of Canada. Ottawa.

Jasper H. H., Droogleever-Fortuyn J., 1946: Experimental studies on the functional anatomy of petit maal epilepsy. Res. Publ. Ass. Nerv. Ment. Dis. 26 : 272—298.

Jasper H. H., Naquet R., King E. E., 1955: Thalamocortical recruiting response in sensory receiving areas in the cat. EEG Clin. Neurophysiol. 7 : 99—114.

Jefferson A., 1954: Aspects of the segmental innervation of the cat's hind limb. J. Comp. Neurol. 100 : 569—596.

Jiménez-Castellanos J., 1949: Thalamus of the cat in Horsley-Clarke coordinates. J. Comp. Neurol. 91 : 307—330.

Johnson G. E., 1924: Giant nerve fibres in crustaceans with special reference to *Cambarus* and *Palaemonetes*. J. Comp. Neurol. 36 : 323—372.

Johnston D. L., 1947: Electroencephalograph amplifier. Wireless Engineer 24 : 231—242.

Jouvet M., 1962: Recherches sur les structures nerveuses et les mécanismes responsables des différentes phases du sommeil physiologique. Arch. Ital. Biol. 100 : 125—206.

Jung R., Kornmüller A. E., 1939: Eine Methodik der Ableitung lokalisierter Potentialschwankungen aus subcorticalen Hirngebieten. Arch. Psychiat. Nervenkr. 109 : 1—30.

Jung R., Tönnies J. F., 1950: Hirnelektrische Untersuchungen über Entstehung und Erhaltung von Krampfentladungen: Die Vorgänge am Reizort und die Bremsfähigkeit des Gehirns. Arch. Psychiat. Nervenkr. 185 : 701—735.

Kaminir L. B. — Каминир Л. Б., 1956: Усилители постоянного тока. Биофизика 1 : 729—740.

Kamp A., 1965: A 90-degree phase difference network for circular display of the EEG on a cathode ray tube. EEG clin. Neurophysiol. 19 : 96—97.

Kandel E. R., Spencer W. A., Brinley F. J., 1961: Electrophysiology of hippocampal neurons. 1. Sequential invasion and synaptic organization. J. Neurophysiol. 24 : 225—242.

Kanevsky Z. M., Finkelshteyn M. N. — Каневский З. М., Финкельштейн М. Н., 1963: Флуктуационная помеха и обнаружение импульсных радиосигналов. Государственное энергетическое изд., Москва.

Kao C. Y., 1960: Postsynaptic electrogenesis in septate giant axons. II. Comparison of medial and lateral giant axons of crayfish. J. Neurophysiol. 23 : 618—635.

Kappers C. U. A., 1947: Anatomie comparée du système nerveux. Haarlem - Paris.

Kaptsov N. A. — Капцов Н. А., 1950: Электрические явления в газах и вакууме. Гос. энерг. изд. Москва—Ленинград.

Kaptsov N. A. — Капцов Н. А., 1956: Электроника. Гос. изд. техникотеоретической литер., Москва.

Katz B., 1937: Experimental evidence for a nonconducted response of nerve to subthreshold stimulation. Proc. Roy. Soc. B 124 : 244—276.

Katz B., 1939: Electric excitation of nerve. Oxford Univ. Press, London.

Katz B., 1948: The electric properties of the muscle fibre membrane. Proc. Roy. Soc. B 135 : 506—534.

Katz B., 1949: The efferent regulation of the muscle spindle in the frog. J. Exp. Biol. 26 : 201—217.

Katz B., 1950: Depolarization of sensory terminals and the initiation of impulses in the muscle spindle. J. Physiol. 111 : 261—282.

Katz B., 1958: Microphysiology of the neuromuscular junction. J. Hopkins Hosp. Bull. 102 : 275—312.

Katz B., 1962: The transmission of impulses from nerve to muscle, and the subcellular unit of synaptic action. Proc. Roy. Soc. B 155 : 455—477.

Katz B., Schmitt O. H., 1939: Excitability changes in a nerve fibre during passage of an impulse in an adjacent fibre. J. Physiol. 96 : 9—10P.

Katz B., Schmitt O. H., 1940: Electric interaction between two adjacent nerve fibres. J. Physiol., 97 : 471—488.

Katz B., Thessleff S., 1957: A study of the "desensitization" produced by acetylcholine at the motor end-plate. J. Physiol. 138 : 63—80.

Keinath G., 1934: Die Technik elektrischer Messgeräte. München.

Kemp E. H., Coppée G. E., Robinson E. H., 1937: Electric responses of the brain stem to unilateral auditory stimulation. Amer. J. Physiol. 120 : 304—315.

Kemp E. H., Robinson E. H., 1937: Electric responses of the brain stem to bilateral auditory stimulation. Amer. J. Physiol. 120 : 316—322.

Kempinsky W. H., 1951: Cortical projection of vestibular and facial nerves in cat. J. Neurophysiol. 14 : 203—210.

Kennard D. W., 1958: Glass microcapillary electrodes used for measuring potential in living tissues. In: P. E. K. Donaldson (editor): Electronic Apparatus for Biological Research, Chapt. 35, Academic Press, New York — Butterworths Scientific Publications, London.

Keynes R. D., 1949: The movements of radioactive ions in resting and stimulated nerve. Arch. Sci. Physiol. 3 : 165—175.

Keynes R. D., Lewis D. R., 1951: The sodium and potassium content of cephalopod nerve fibres. J. Physiol. 114 : 151—182.

Khinchin A. I., — Chintschin A. J. 1933: Zu Birkhoffs Lösung des Ergodenproblems. Math. Ann., 107.

Khinchin A. J. — Хинчин А. Я., 1938: Теория корреляции стационарных случайных функций. Успехи математических наук, вып. 5.

Khinchin A. J., 1949: Mathematical foundations of statistical mechanics. Dover, New York.

Kibjakov A. W., 1933: Über humorale Übertragung der Erregung von einem Neuron auf das andere. Pflüg. Arch. Ges. Physiol. 232 : 432—443.

Kimura D., 1962: Multiple response of visual cortex of the rat to photic stimulation. EEG Clin. Neurophysiol. 14 : 115—122.

Kireyev V. A. — Киреев Б. А., 1951: Курс физической химии. Москва.

Kirchner L. B., 1955a: On the mechanism of active sodium transport across the frog skin. J. Cell. Comp. Physiol. 45 : 61—87.

Kirschner L. B., 1955b: The effect of atropine and the curare on the skin of *Rana esculenta*. J. Cell. Comp. Physiol. 45 : 89—102.

Kishimoto U., 1964: Current voltage relations in *Nitella*. Jap. J. Physiol. 14 : 515—527.

Klein B., 1898: Zur Frage über die elektrischen Ströme in Pflanzen. Berichte Deutsch. Bot. Ges. 16 : 335—346.

Klensch H., 1954: Einführung in die biologische Registriertechnik. Thieme, Stuttgart.

Koefoed-Johnsen V., Ussing H. H., 1956: Nature of the frog skin potential. Abstr. Comm. XXth int. Physiol. Congr. p. 511.

Koefoed-Johnsen V., Ussing H., Zerahn K., 1953: The origin of the short-circuit current in the adrenaline stimulated frog skin. Acta Physiol. Scand. 27 : 38—48.

Koella W. P., 1959: Some functional properties of optically evoked potentials in cerebellar cortex of cat. J. Neurophysiol. 22 : 61—77.

Kogan A. B. — Коган А. Б., 1949: Электрофизиологическое исследование центральных механизмов некоторых сложных рефлексов. Изд. А.М.Н. СССР, Москва.

Kogan A. B. — Коган А. Б., 1952: Методика хронического вживления электродов для отведения потенциалов и раздражения мозга. Изд. А.М.Н. СССР, Москва.

Kohlrausch F., 1943: Praktische Physik. Teubner, Leipzig 1955.

Kolmogorov A. N. (Kolmogorow A. N.), 1933: Grundbegriffe der Wahrscheinlichkeitsrechnung, Ergebnisse d. Mathematik 2(3), Berlin.

Kolmogorov A. N. — Колмогоров А. Н., 1936: Основные понятия теории вероятностей. ОНТИ, Москва.

Kolmogorov A. N. — Колмогоров А. Н., 1941: Интерполирование и екстраполоривание стационарных последовательностей. Изв. АН СССР, сер. матем., 5 (1) 1 : 3—14.

König J. F. R., Klippel R. A., 1963: The rat brain. A stereotaxic atlas of the forebrain and lower parts of the brain stem. Williams and Wilkins, Baltimore.

Kononov V. N., — Конов В. Н., 1960: Симметрические триггеры на плоскостных полупроводниковых триодах. Гос. Энерг. Издат., Москва-Ленинград.

Kopac M. J., 1929: A micromanipulator for biological investigations. Trans. Amer. Micr. Soc. 48 : 438—442.

Kopac M. J., 1964: Micromanipulators: principles of design, operation and application. In: W. L. Nastuk (Editor): Physical Techniques in Biological Research. Vol. V. V. Electrophysiological Methods. Part A, pp. 191—233. Academic Press, New York.

Korn G., Korn T., 1955 — Корн Г., Корн Т., 1955: Электронные модулирущие устройства, Москва.

Kostyuk P. G. — Костюк П. Г., 1956: О месте возникновения электротонических потенциалов в спинномозговых корешках при раздражении мышечных нервов. Физиол. журнал СССР 9 : 800—810.

Kostyuk P. G. — Костюк П. Г., 1959: Двухнейронная рефлекторная дуга. Государств. изд. мед. лит., Москва.

Kostyuk P. G. — Костюк П. Г., 1960: Микроэлектродная техника. ИАН УССР, Киев.

Kozhevnikov V. A., — Кожевников В. А., 1956: Стабильный усилитель постоянного тока. Биофизика 1 : 292—295.

Kozhevnikov V. A., Meshchersky R. M. — Кожевников В. А., Мещерский Р. М., 1963: Современные методы анализа электроэнцефалограммы. Государств. изд. мед. литературы, Москва.

Kravitz E. A., Kuffler S. W., Potter D. D., van Gelder N. M., 1963a: Gamma-aminobutyric acid and other blocking compounds in crustacea. II. Peripheral nervous system. J. Neurophysiol. 26 : 729—738.

Kravitz E. A., Kuffler S. W., Potter D. D., 1963b: Gamma-aminobutyric acid and other blocking compounds in crustacea. III. Their relative concentrations in separated motor and inhibitory axons. J. Neurophysiol. 26 : 739—751.

Krekule I., 1965: Přehled zařízení pro detekci vyvolané odpovědi zprůměrňováním. Čs. fysiol. 6 : 490—498.

Krieg W. J. S., 1946a: Connections of the cerebral cortex. I. The albino rat. A. Topography of the cortical areas. J. Comp. Neurol. 84 : 221—275.

Krieg W. J. S., 1946b: Accurate placement of minute lesions in the brain of the albino rat. Quart. Bull. Northwestern U. Med. Sch. 20 : 199—208.

Křivánek J., 1962: Concerning the dynamics of the metabolic changes accompanying cortical spreading depression. Physiol. bohemoslov. 11 : 383—391.

Křivánek J., Bureš J., 1960: Ion shifts during Leão's spreading cortical depression. Physiol. bohemoslov. 9 : 494—503.

Krnjević K., 1964: Microiontophoretic studies on cortical neurons. Int. Rev. Neurobiol. 7 : 41—98.

Krnjević K., Miledi R., 1958: Acetylcholine in mammalian neuromuscular transmission. Nature 182 : 805—806.

Kuffler S. W., 1943: Specific excitability of the end-plate region in normal and denervated muscle. J. Neurophysiol. 6 : 99—110.

Kuffler S. W., 1948: Physiology of neuro-muscular junction: electrical aspects. Fed. Proc. 7 : 437—446.

Kuffler S. W., Eyzaguirre C., 1955: Synaptic inhibition in an isolated nerve cell. J. Gen. Physiol. 39 : 155—184.

Kuffler S. W., Gerard R. W., 1947: The small-nerve motor system to skeletal muscle. J. Neurophysiol. 10 : 383—394.

Kuffler S. W., Hunt C. C., 1952: The mammalian small-nerve fibres: a system for efferent nervous regulation of muscle spindle discharge. Res. Publ. Ass. Nerv. Ment. Dis. 30 : 24—47.

Kuffler S. W., Hunt C. C., Quilliam J. P., 1951: Function of medullated fibres in mammalian ventral roots: efferent muscle spindle innervation. J. Neurophysiol. 14 : 29—54.

Kurela C. A., — Курела Ц. А., 1958: Метод изготовления внутриклеточных микроэлектродов. Биофизика 3 : 243—245.

Kwassow D. G., Naumenko A. I., 1936: Störungen in der isolierten Leitung der Impulse im durch hypertonische Lösungen und Austrocknung alterierten Nervenstamm. Pflüg. Arch. Ges. Physiol. 237 : 576—584.

Lance J. W., Manning R. L., 1954: Origin of the pyramidal tract in the cat. J. Physiol. 124 : 385—399.

Landau W. M., 1951: Comparison of different needle leads in EMG recording from a single site. EEG Clin. Neurophysiol. 3 : 163—168.

Landau W. M., 1956: An analysis of the cortical response to antidromic pyramidal tract stimulation in the cat. EEG Clin. Neurophysiol. 8 : 445—456.

Lapicque L., 1926: L'excitabilité en fonction du temps; la chronaxie, sa signification et sa mesure. Paris.

Laporte Y., Lorente de Nó R., 1950: Potential changes evoked in a curarized sympathetic ganglion by presynaptic volleys of impulses. J. Cell. Comp. Physiol. 35 Suppl. 2 : 61—106.

Lau E., Kind E. H., Roose G., 1958: Photometry of photographic plates by means of blackening relief interference microscopy. Monthly technical review, 2 : 118—122.

Lawson J. L., Uhlenbeck G. E., 1950: Threshold signals. New York.

Leão A. A. P., 1944: Spreading depression of activity in the cerebral cortex. J. Neurophysiol. 7 : 359—390.

Leão A. A. P., 1947: Further observations on the spreading depression of activity in the cerebral cortex. J. Neurophysiol. 10 : 409—419.

Leão A. A. P., 1951: The slow voltage variations of cortical spreading depression of activity. EEG Clin. Neurophysiol. 3 : 315—321.

Leão A. A. P., Ferreira M. H., 1953: Alteracão da impedancia eléctrica no decurso da depressão alastrante da atividade do córtex cerebral. Anais da Academia Brasileira de Ciencias 25 : 259—266.

Leão A. A. P., Morison R. S., 1945: Propagation of spreading cortical depression. J. Neurophysiol. 8 : 33—45.

Leksell L., 1945: The action potential and excitatory effects of the small ventral root fibres to skeletal muscle. Acta Physiol. Scand. 10 : suppl. 31—84.

Lele P. P., 1962: A simple method for production of trackless focal lesions with focused ultrasound: Physical factors. J. Physiol. 160 : 494—512.

Le Messurier D. H., 1948: Auditory and visual areas of the cerebral cortex of the rat. Fed. Proc. 7 : 70.

Lennox M. A., 1956: Geniculate and cortical responses to colored light flash in cat. J. Neurophysiol. 19 : 271—279.

Levin B. R. — Левин Б. Р., 1960: Теория случайных процессов и ее применение в радиотехнике. Издат. ,,Советское радио'', Москва.

Levin A., 1927: Fatigue, retention of action current and recovery in crustacean nerve. J. Physiol. 63 : 113—129.

Levy C. K., Loeser J. D., Koella W. P., 1961: The cerebellar acoustic response and its interaction with optic responses. EEG clin. Neurophysiol. 13 : 236—242.

Lezin Yu. S. — Лезин Ю. С., 1963: Оптимальные фильтры и накопители импульсных сигналов. Издат. ,,Советское радио'', Москва.

Li C. L., Chou S. N., 1962: Cortical intracellular synaptic potentials and direct cortical stimulation. J. Cell. Comp. Physiol. 60 : 1—16.

Li C. L., Cullen C., Jasper H. H., 1956a: Laminar microelectrode studies of specific somatosensory cortical potentials. J. Neurophysiol. 19 : 113—130.

Li C. L., Cullen C., Jasper H. H., 1956b: Laminar microelectrode analysis of cortical unspecific responses and spontaneous rhythms. J. Neurophysiol. 19 : 131—143.

Libet B., 1962a: Slow synaptic responses in sympathetic ganglia. Fed. Proc. 21 : 345.

Libet B., 1962b: Slow excitatory and inhibitory synaptic responses in sympathetic ganglia. Proc. XXII Int. Physiol. Congr. vol. II, p. 809.

Libouban-Letouze S., 1964: Etude électrophysiologique des structures cérébrales du rat blanc. Thèses à la faculté des sciences de l'Université de Paris.

Libouban S., Oswaldo-Cruz E., 1958: Quelques observations relatives aux activités évoquées et spontanées du cerveau du Rat blanc. J. Physiol. (Paris) 50 : 380—383.

Liley A. W., North K. A. K., 1953: An electrical investigation of effects of repetitive stimulation on mammalian neuromuscular junction. J. Neurophysiol. 16 : 509—527.

Lillie R. S., 1923: Protoplasmic action and nervous action. Chicago University Press.

Lilly J. C., 1954: Instantaneous relations between the activities of closely spaced zones in the cerebral cortex. Amer. J. Physiol. 176 : 493—504.

Lilly J. C., Cherry R. B., 1954: Surface movements of click responses from acoustic cerebral cortex of cat: leading and trailing edges of a response figure. J. Neurophysiol. 17 : 521—532.

Lim R. K. S., Lin Chan-Nao, Moffelt R. L., 1960: A stereotaxic atlas of the dog's brain. Charles C. Thomas, Springfield.

Linderholm H., 1952: Active transport of ions through frog skin with special reference to the action of certain diuretics. Acta Physiol. Scand. 27 Suppl. 97 : 1—144.

Lindsley D. B., Bowden J., Magoun H. W., 1949: Effect upon the EEG of acute injury to the brainstem activating system. EEG Clin. Neurophysiol. 1 : 475—486.

Lindsley D. B., Schneider L. H., Knowles W. B., Magoun H. W., 1950: Behavioral and EEG changes following chronic brain stem lesions in the cat. EEG Clin. Neurophysiol. 2 : 483—498.

Ling G., Gerard R. W., 1949: The normal membrane potential of frog sartorius fibres. J. Cell. Comp. Physiol. 34 : 383—396.

Livanov M. N., Ananyev V. M. — Ливанов М. Н., Ананьев В. М., 1955: Электрофизиологическое исследование пространственного распределения активности в коре головного мозга кролика. Физиол. журнал СССР 41 : 461—469.

Lloyd D. P. C., 1937: The transmission of impulses through the inferior mesenteric ganglia. J. Physiol. 91 : 296—313.

Lloyd D. P. C., 1939a: The excitability states of inferior mesenteric ganglion cells following preganglionic activation. J. Physiol. 95 : 464—475.

Lloyd D. P. C., 1939b: The origin and nature of ganglion afterpotentials. J. Physiol. 96 : 118—129.

Lloyd D. P. C., 1941: A direct central inhibitory action of dromically conducted impulses. J. Neurophysiol. 4 : 184—190.

Lloyd D. P. C., 1943: Reflex action in relation to the pattern and peripheral source of afferent stimulation. J. Neurophysiol. 6 : 111—119.

Lloyd D. P. C., 1946: Facilitation and inhibition of spinal motonerons. J. Neurophysiol. 9 : 421—438.

Lloyd D. P. C., 1949: Post-tetanic potentiation of response in monosynaptic reflex pathways of the spinal cord. J. Gen. Physiol. 33 : 147—170.

Lloyd D. P. C., 1951: Electrical signs of impulse conduction in spinal motoneurones. J. Gen. Physiol. 35 : 255—288.

Lloyd D. P. C., 1952: Electrical manifestations of action in neurones. In: The Biology of Mental Health and Disease, p. 135—161. Paul B. Höber, New York.

Loève M., 1960: Probability Theory. D. van Nostrand Company Inc., Princeton, New Jersey.

Lorente de Nó R., 1934: Studies on the structure of the cerebral cortex. II. Continuation of the study of the Ammonic system. J. Psychol. Neurol. 46 : 113—177.

Lorente de Nó R., 1936: Transmission of impulses through cranial motor nerve nuclei. J. Neurophysiol. 2 : 402—464.

Lorente de Nó, R., 1947: A study of nerve physiology. Studies from the Rockefeller Institute for Medical Research, vol. 131—132, New York.

Lucas K., 1912: On a mechanical method of correcting photographic records obtained from the capillary electrometer. J. Physiol. 44 : 225—242.

Lund E. J., 1923: Normal and experimental delay in the initiation of polyp formation in *Obelia* internodes. J. Exp. Zool. 37 : 69—87.

Lund E. J., Kenyon W. A., 1927: Relation between continuous bioelectric currents and cell respiration. I. Correlation potentials in growing root tips. J. Exp. Zool. 48 : 333 to 357.

Lundberg A., 1952: Adrenaline and transmission in the sympathetic ganglion of the cat. Acta Physiol. Scand. 26 : 252—263.

Luparello T. J., Stein M., Park C. D., 1964: A stereotaxic atlas of the hypothalamus of the guinea pig. J. Comp. Neurol. 122 : 202—218.

MacInnes D. A., 1939: The principles of electrochemistry. Reinhold Publishing Corp, New York.

MacNichol E. F., 1962: Negative impedance electrometer amplifiers. Proceedings of the IRE, 1909—1911.

Magladery J. W., MacDougal D. B. Jr., 1950: Electrophysiological studies of nerve and reflex activity in normal man: I. Identification of certain velocity in peripheral nerve fibers. Bull. Johns Hopkins Hosp. 86 : 265—290.

Magladery J. W., McDougal D. B., Jr., Stoll J., 1950: Electrophysiological studies of nerve and reflex activity in normal man. II. Bull. John Hopkins Hosp. 86 : 291—320.

Magnes J., Moruzzi G., Pompeiano O., 1961: Electroencephalogram synchronizing structures in the lower brain stem. In: G. E. W. Wolstenholme and M. O'Connor (Editors): The Nature of Sleep. Churchill, London, pp. 57—78.

Magni F., Moruzzi G., Rossi G. F., Zanchetti A., 1959: EEG arousal following inactivation of the lower brain stem by selective injection of barbiturate into lower brain stem circulation. Arch. Ital. Biol. 97 : 33—46.

Magoun H. W., 1958: The waking brain. Charles C. Thomas, Springfield.

Malliani A., Rudomin P., Zanchetti A., 1965: Contribution of local activity and electric spread to somatically evoked potentials in different areas of the hypothalamus. Arch. ital. Biol. 103 : 119—135.

Mancia M., Mechelse K., Mollica A., 1957: Microelectrode recording from the mesencephalic reticular formation in the decerebrate cat. Arch. Ital. Biol. 95 : 110—119.

Markus J., Zeluff V., 1948: Handbook of industrial electronic circuits. McGraw Hill, New York — Toronto — London.

Marrazzi A. S., 1939: Adrenergic inhibition at sympathetic synapses. Amer. J. Physiol. 127 : 738—744.

Marsh G., 1928: Relation between continuous bioelectric currents and cell respiration. IV. The origin of electric polarity in the onion root. J. Exp. Zool. 51 : 309.

Marsh G., Beams H. W., 1946: In vitro control of growing chick nerve fibres by applied electric currents. J. Cell. Comp. Physiol. 27 : 139.

Marsh G., Beams H. W., 1952: Electrical control of morphogenesis in regenerating *Dugesia tigrina*. J. Cell. Comp. Physiol. 39 : 191—210.

Marshall W. H., 1941: Observations on subcortical somatic sensory mechanism of cats under Nembutal anesthesia. J. Neurophysiol. 4 : 25—43.

Marshall W. H., 1949: Excitability cycle and interaction in geniculate-striate system of cat. J. Neurophysiol. 12 : 277—288.

Marshall W. H., 1959: Spreading cortical depression of Leão. Physiol. Reviews 39 : 239—278.

Marshall W. H., Talbot S. A., Ades H. W., 1943: Cortical response of the anaesthetized cat to gross photic and electrical afferent stimulation. J. Neurophysiol. 6 : 1—15.

Mathews A. P., 1903: Electrical polarity in the hydroids. Amer. J. Physiol. 8 : 294.

Matteucci C., 1845: Electrophysiological researches. Philos. Trans. Part II.

Matthews B. H. C., 1931a: The response of a single end organ. J. Physiol. 71 : 64—110.

Matthews B. H. C., 1931b: The response of a muscle spindle during active contraction of a muscle. J. Physiol. 72 : 153—174.

Matthews B. H. C., 1933: Nerve endings in mammalian muscle. J. Physiol. 78 : 1—33.

Mayer R. F., 1963: Nerve conduction studies in man. Neurology (Minneap.) 13:1021—1030.

Mayorov F. V. — Майоров Ф. В., 1957: Электронные цифровые вычислительные устройства (элементы и схемы). Гос. энерг. издат., Москва—Ленинград.

McLennan H., 1963: Synaptic transmission. W. B. Saunders Co., Philadelphia.

McNeil E., Gullberg J. F., 1931: A new micromanipulator. Science. 74 : 460—470.

Merril L. C., Slater T. L., 1956: Linear sweep-voltage generators and precision amplitude comparator using transistors. Electrical Communication, Sept. v. 33, No. 3.

Meschersky R. M. — Мещерский Р. М., 1960: Методика микроэлектродного исследования. Медгиз, Москва.

Meshchersky R. M. — Мещерский Р. М., 1961: Стереотактический метод. Медгиз, Москва.

Meshchersky R. M. — Мещерский Р. М., 1965: О преимуществах моноролярной регистрации для анализа вызванных потенциалов. Ж. высш. нерв. деят. 15 : 755—760.

Meyerovich L. A., 1948: Základy impulsové techniky. Technickovědecké vydavatelství, Praha.

Meyerovich L. A., Zelichenko J. C. — Меерович Л. А., Зелнченко Л. Г., 1953: Импульсная техника, Советское радио, Москва.
(German translation: Mejerowitch L. A., Selitschenko L. G., 1959: Impulstechnik. VEB Verlag Technik, Berlin).

Mickle W. A., Ades H. W., 1952: A composite sensory projection area in the cerebral cortex of the cat. Amer. J. Physiol. 170 : 682—689.

Mickle W. A., Ades H. W., 1953: Spread of evoked cortical potentials. J. Neurophysiol. 16 : 609—633.

Mikhaylov V. V. — Михайлов В. В., 1958: Универсальный прибор для электрофизиологических исследований. Биофизика 3 : 516—518.

Monachov K. K., Fifková E., Bureš J., 1962a: Vertical distribution of the slow potential change of spreading depression in the cerebral cortex of the rat. Physiol. bohemoslov. 11 : 269—276.

Monakhov K. K., Fifková E., Bureš J., 1962b: Steady potential field of hippocampal spreading depression. J. Cell. Comp. Physiol. 59 : 155—161.

Monk C. B., 1961: Electrolytic dissociation. Academic Press, London — N. Y.

Monnier A. M., Dubuisson M., 1931: Etude à l'oscillographe cathodique des nerfs pedieux de quelques arthropodes. Arch. Int. Physiol. 34 : 25—57.

Monnier C., 1930: On the relation between the shape of stimulus and the shape of the nerve action potential. Amer. J. Physiol. 93 : 675.

Monnier M., Gangloff H., 1961: Atlas for stereotaxic brain research on the conscious rabbit. Elsevier, Amsterdam.

Monnier M., Laue H., 1953: Technique de dérivation des activités électriques corticales et sous-corticales pendant la stimulation du diencéphale chéz le lapin. Helv. Physiol. Pharmacol. Acta 11 : 73—80.

Moore J. W., 1963: Operational amplifiers. In: Nastuk W. L. (Editor), Physical Techniques in Biological Research, vol. VI. Electrophysiological Methods, Part B, pp. 77—79, Academic Press, New York — London.

Moore J. W., Gebhart J. H., 1962: Stabilized wide-band potentiometric preamplifiers. Proceedings of the IRE, 1928—1941.

Morey G. W., 1938: The properties of glass. Reinhold, New York.

Morin F., 1952: Afferent projection to the midbrain tegmentum in the cat. Anat. Rec. 112 : 63.

Morin F., 1953: Afferent projections to the midbrain tegmentum and their spinal course. Amer. J. Physiol. 172 : 483—496.

Morin F., Catalano J. V., Lamarche G., 1957: Wave form of cerebellar potentials. Amer. J. Physiol. 188 : 263—273.

Morison R. S., Dempsey E. W., 1942: A study of thalamo-cortical relations. Amer. J. Physiol. 135 : 281—292.

Morison R. S., Finley K. H. Lothrop G. N., 1943: Influence of basal forebrain areas on the electrocorticogram. Amer. J. Physiol. 139 : 410—416.

Moruzzi G., Magoun H. W., 1949: Brain stem reticular formation and activation of the EEG. EEG clin. Neurophysiol. 1 : 455—473.

Mountcastle V., Henneman E., 1949: Pattern of tactile representation in thalamus of cat. J. Neurophysiol. 12 : 85—99.

Mütze K., 1961: Brockhaus ABC der Optik. Brockhaus, Leipzig.

Nachmansohn D., 1961: Chemical factors controlling nerve activity. Science 134 : 1962—1968.

Nakao K., Nakanishi T., Tsubaki T., 1965: Action potentials recorded by coaxial needle electrodes in Ringer's solution. EEG clin. Neurophysiol. 18 : 412—414.

Narikashvili S. P. — Нарикашвили С. П., 1953: Влияние раздражения подкорковых образований на электрическую активность коры больших полушарий. АН ГССР Тр. Ин-та физиологии 9 : 133—154.

Nasonov G., Aleksandrov V. — Насонов Г., Александров В., 1940: Реакция живого вещества на внешнее раздражение. Москва—Ленинград.

Nastuk W. L., 1953: The electrical activity of the muscle cell membrane at the neuro-muscular junction. J. Cell. Comp. Physiol. 42 : 249—272.

Nastuk W. L., 1957: Membrane potential changes at a single muscle end-plate produced by transitory application of acetylcholine with an electrically controlled microjet. Fed. Proc. 12 : 102.

Nastuk W. L., 1959: Some ionic factors that influence the action of acetylcholine at the muscle end-plate membrane. Ann. N. Y. Acad. Sci. 81 : 317—327.

Nastuk W. L. (Editor), 1963 and 1964: Physical Techniques in Biological Research, vol. VI. Parts A and B. Academic Press, New York and London.

Nastuk W. L., Hodgkin A. L., 1950: The electrical activity of single muscle fibres. J. Cell. Comp. Physiol. 35 : 39—74.

Neeteson P. A., 1955: Elektronenröhren in der Impulstechnik. Philips, Eindhoven.

Nekrasov M. M. — Некрасов М. М., 1965: Микроминиатуризация и микроэлектроника на нелинейных сопротивлениях. Изд. „Советское радио", Москва.

Nelepets V. V., Nelepets V. S. — Нелепец В. В., Нелепец В. С., 1960: Импульсные режимы в радиотехнических цепях. Военное издат. министерства обороны СССР, Москва.

Nernst W., 1908: Zur Theorie des elektrischen Reizes. Pflüg. Arch. Ges. Physiol. 122 : 275—314.

Nernst W., 1926: Theoretische chemie. Enke, Stuttgart.

Nobili C. L., 1825: Über einen neuen Galvanometer. J. Chem. Phys. 45 : 249.

Noell W. K., 1951: The effect of iodacetate on the vertebrate retina. J. Cell. Comp. Physiol. 37 : 283—308.

Noell W. K., 1952a: The impairment of visual cell structure by iodacetate. J. Cell. Comp. Physiol. 40 : 25—55.

Noell W. K., 1952b: Azide-sensitive potential difference across the eye-bulb. Amer. J. Physiol. 170 : 217—238.

Nüsslein G., 1952: Gleichspannungsverstärker. Berlin.

Oerstad H. Ch., 1820: Experimenta circa effectum conflictus electrici in acum magneticum. Copenhagen.

Ochs S., 1956: The direct cortical response. J. Neurophysiol. 19 : 513—523.

Ochs S., 1962: The nature of spreading depression in neural networks. Intern. Rev. Neurobiol. 4 : 1—69.

Offner F. F., 1950: The EEG as potential mapping: the value of the average monopolar reference. EEG Clin. Neurophysiol. 2 : 215—216.

Offner F. F., 1954: The tri-phasic action potential. EEG Clin. Neurophysiol. 6 : 507—508.

Oikawa T., Spyropoulos C. S., Tasaki I., Teorell T., 1961: Methods for perfusing the giant axon of Loligo pealii. Acta Physiol. Scand. 52 : 195—196.

O'Leary J. L., Bishop G. H., 1938: The optically excitable cortex of the rabbit. J. Comp. Neurol. 68 : 423—478.

O'Leary J. L., Goldring S., 1964: D-C potentials of the brain. Physiol. Rev. 44 : 91—125.

Olszewski J., 1952: The thalamus of the Macaca mulatta; an atlas for use with the stereotaxic instrument. Karger, Basel.

Oomura Y., Ooyama H., Yoneda K., 1967: Miniaturized high input impedance preamplifier. Physiology and Behavior 2 : 93—95.

Ordway F., 1952: Techniques for growing and mounting small single crystals of refractory compounds. J. Res. Nat. Bureau Stand. 48 : 152—162.

Osterhout W. J. V., 1936: Electrical phenomena in large plant cells. Physiol. Rev. 16 : 216—237.

Osterhout W. J. V., 1958: Studies of some fundamental problems by the use of aquatic organisms. Ann. Rev. Physiol. 20 : 1—12.

Ottoson D., 1956: Analysis of the electrical activity of the olfactory epithelium. Acta Physiol. Scand. 35, suppl. 122.

Ottoson D., 1964: The effect of sodium deficiency on the response of the isolated muscle spindle. J. Physiol. 171 : 109—118.

Ottoson D., 1965: The action of calcium on the frog's isolated muscle spindle. J. Physiol. 178 : 68—79.

Overton E., 1902: Beiträge zur allgemeinen Muskel- und Nervenphysiologie. Pflüg. Arch. Ges. Physiol. 92 : 346—386.

Palade G. E., Palay S. L., 1954: Electron microscope observations of interneuronal and neuromuscular synapses. Anat. Rec. 118 : 335—345.

Parris S. R., Staar D. A., 1960: Highly accurate phantastron delay circuit. Electronics, Oct. ,2.

Patton H. D., Amassian V. E., 1954: Single- and multiple-unit analysis of cortical stage of pyramidal tract activation. J. Neurophysiol. 17 : 345—362.

Pavlov I. P. — Павлов И. П., 1926: Лекции о работе больших полушарий головного мозга. Полное собрание трудов. Т. 4. Москва—Ленинград 1949.

Pavlov B. V. — Павлов Б. В., 1958: Стереотактические аппараты и их применение при экспериментальных повреждениях подкорковых образований (по материалам иностранной литературы). Физиол. журнал СССР 44 : 897—900.

Pélegrin M., 1959: Machines à calculer électroniques, arithmétiques et analogiques. Dunod, Paris.

Penfield W., Jasper H. H., 1954: Epilepsy and the functional anatomy of the human brain. Little Brown, Boston.

Perl E. R., Casby J. U., 1954: Localization of cerebral electrical activity: the acoustic cortex of cat. J. Neurophysiol. 17 : 429—442.

Peters C. A. F., 1858: Beschreibung eines auf Altonaer Sternwarte aufgestellten galvanischen Registrierapparates für Durchgangsbeobachtungen nebst Vergleichung einiger an demselber bestimmten Personal-Differenzen mit solchen, die auf gewöhnliche Weise gefunden sind. Astronomische Nachrichten 49 : 1—32.

Petráň M., — Петрань М., 1952: Улучшенный способ осциллографии на движущийся фильм. Physiol. bohemoslov. 1 : 167—172.

Petrovich N. T., Kozyrev A. V. — Петрович Н. Т., Козырев А. В., 1954: Генерирование и преобразование электрических импульсов. Изд. ,,Советское радио", Москва.

Petrovický P., 1963a: Formatio reticularis medullae oblongatae in rat. Acta Univ. Carol. Med. 9 : 733—740 (in Czech).

Petrovický P., 1963b: Formatio reticularis pontis in rat. Acta Univ. Carol. Med. 9 : 741—749 (in Czech).

Petrovický P., 1964: Formatio reticularis mesencephali in rat. Acta Univ. Carol. Med. 10 : 423—453 (in Czech).

Petsche H., Marko A., 1954: The photocell toposcope, a simple method to determine field distribution of electrical activity of the brain. EEG Clin. Neurophysiol. 6 : 521.

Pflüger E., 1859: Physiologie des Electrotonus, Berlin, Hirschwald.

Phillips C. G., 1956a: Intracellular records from Betz cells in the cat. Quart. J. Exp. Physiol. 41 : 58—69.

Phillips C. G., 1956b: Cortical motor threshold and the thresholds and distribution of excited Betz cells in the cat. Quart. J. Exp. Physiol. 41 : 70—84.

Phillips C. G., 1964: Experiments on single neurones within the central nervous system of vertebrates. In: The Scientific Basis of Medicine. Ann. Reviews, Chapt. VI: 81—101.

Polissar M. J., 1954: Physical chemistry of cell irritability and of the nerve impulse. In: Kinetic Basis of Molecular Biology, John Wiley, New York.

Popov P. A. — Попов П. А., 1964: Расчет транзисторных усилителей звуковой частоты. Издат. ,,Энергия", Москва, Ленинград.

Porter K. R., Palade G. E., 1957: Studies on the endoplasmic reticulum. III. Its form and distribution in striated muscle cells. J. Biophys. Biochem. Cytol. 3 : 269—300.

Porter R., 1955: Antidromic conduction of volleys in pyramidal tract. J. Neurophysiol. 18 : 138—150.

Porter R., Sanderson J. H 1964: Antidromic cortical response to pyramidal tract stimulation in the rat. J. Physiol. 170 : 355—370.

Pravdicz-Neminski W. W., 1925: Zur Kenntnis der elektrischen und der Innervations-vorgänge in den functionellen Elementen und Geweben des tierischen Organismus. Elektrocerebrogramm der Säugetiere. Pflüg. Arch. Ges. Physiol. 209 : 326—382.

Purpura D. P., 1955: Further analysis of evoked "secondary discharge". A study in re-ticulocortical relations. J. Neurophysiol. 18 : 246—260.

Purpura D. P., Grundfest H., 1956: Nature of dendritic potentials and synaptic mechan-isms in cerebral cortex of cat. J. Neurophysiol. 19 : 573—595.

Ranck J. B., 1963: Analysis of specific impedance of rabbit cerebral cortex. Exp. Neurol. 7 : 153—174.

Ranck J. B., 1964: Specific impedance of cerebral cortex during spreading depression and analysis of neuronal, neuroglial and interstitial contributions. Exp. Neurol. 9 : 1—16.

Ranson S. W., 1934: On the use of the Horsley-Clarke stereotaxic instrument. Psychiat. Neurol. Bladen 38 : 534—543.

Rashevsky N., 1933: Outline of a physico-mathematical theory of excitation and inhibi-tion. Protoplasma 20 : 42—56.

Rashevsky N., 1948: Mathematical Biophysics, University of Chicago Press, Chicago.

Rayport M., 1957: Anatomical identification of somatic cortical neurones responding with short latencies to specific afferent volleys. Fed. Proc. 16 : 104.

Reger J. F., 1958: The fine structure of neuromuscular synapses of gastrocnemii from mouse and frog. Anat. Rec. 130 : 7—24.

Rehm W. S., 1943: Positive injury potentials of the stomach. Amer. J. Physiol. 140 : 720—725.

Rehm W. S., 1946: Evidence that the major portion of the gastric potential originates between the submucosa and mucosa. Amer. J. Physiol. 147 : 69—77.

Reichel H., 1960: Muskelphysiologie. Springer Verlag, Berlin.

Reighard J., Jenkins H. S., 1923: Anatomy of the cat. H. Holt, New York.

Rein M., 1940: Ein Drehbügelgalvanometer. Pflüg. Arch. Ges. Physiol. 243 : 557.

Reinert G. G., 1939: Ein neuer Mikromanipulator für Arbeiten bis T = 2.500×. Arch. Exp. Zellforsch. 22 : 681—685.

Reinoso-Suárez F., 1961: Topographischer Hirnatlas der Katze für experimental-physiologische Untersuchungen. E. Muck, Darmstadt.

Renshaw B., 1946: Central effects of centripetal impulses in axons of spinal ventral roots. J. Neurophysiol. 9 : 191—204.

Rexed B., 1952: The cytoarchitectonic organization of the spinal cord in the cat. J. Comp. Neurol. 96 : 415—495.

Rheinberger M., Jasper H. H., 1937: The electrical activity of the cerebral cortex in the unanesthetized cat. Amer. J. Physiol., 119 : 186—196.

Rice S. O., 1945: Mathematical analysis of random noise. BSTJ 24, No 1.

Rieker G., Bolte H. D., Bubnoff M., 1963: Ein Verfahren zur Messung von Einzelfaserpotentialen menschlicher Musckelzellen in situ. Pflüger. Arch. Ges. Physiol. 277 : 231-235.

Rindt K., Kretzer K. (Editors), 1954—1957: Handbuch für Hochfrequenz- und Elektro-Techniker I-V, Verlag für Radio-Foto-Kinotechnik, Berlin.

Rioch D. Mc. K., 1929a: Studies on the diencephalon of carnivora. Part I. The nuclear configuration of the thalamus, epithalamus and hypothalamus of the dog and cat. J. Comp. Neurol. 49 : 1—119.

Rioch D. Mc. K., 1929b: Studies on the diencephalon of carnivora. Part II. Certain nuclear configuration and fiber connections of the subthalamus and midbrain of the dog and cat. J. Comp. Neurol. 49 : 121—153.

Robertson J. D., 1955: Recent electron microscope observations on the ultrastructure of the crayfish median-to-motor giant synapse. Exp. Cell. Res. 8 : 226—229.

Robertson J. D., 1956: The ultrastructure of a reptilian myoneural junction. J. Biophys. Biochem. Cytol. 2 : 381—394.

Robertson J. D., 1957: In: Sjötsrand and Rhodin. Electron Microscopy. p. 197. Almquist and Wiksell, Stockholm.

Robertson J. D., 1960: Electron microscopy of the motor end-plate and the neuromuscular spindle. Amer. J. Phys. Med. 39 : 1—43.

Robertson J. D., 1961: Ultrastructure of excitable membranes and the crayfish median—giant synapse. Ann. N. Y. Acad. Sci. 94 : 339—389.

Rocha-Miranda C. E., Oswaldo-Cruz E., Neyts F. L. K., 1965: Stereotaxically oriented macrotome: a device for blocking the brain in rectangular and polar coordinates. EEG clin. Neurophysiol. 19 : 98—100.

Rohlíček V., 1964: A device for polarity alternation of pulses for biological stimulation. Med. Electron. Biol. Eng. 2 : 439—441.

Roldán E., Weiss T., Fifková E., 1963: Excitability changes during the sleep cycle of the rat. EEG Clin. Neurophysiol. 15 : 775—785.

Romanes G. J., 1951: The motor cell columns of the lumbo-sacral spinal cord of the cat. J. Comp. Neurol. 94 : 313—364.

Romanes G. J., 1964: The motor pools of the spinal cord. Progr. Brain Res., vol. 11, Organization of the spinal cord. Elsevier, Amsterdam.

Rose J. E., Galambos R., 1952: Microelectrode studies on medial geniculate body of cat. I. Thalamic region activated by click stimuli. J. Neurophysiol. 15 : 343—357.

Rose J. E., Greenwood D. D., Goldberg J. M., Hind J. E., 1963: Some discharge characteristics of single neurons in the inferior colliculus of the cat. I. Tonotopical organization, relation of spike-counts to sound intensity and firing patterns of single elements. J. Neurophysiol. 26 : 294—320.

Rosenblith W. A., 1962: Processing Neuroelectric Data. MIT Press, Cambridge, Mass.

Rosene H. F., 1935: Proof of the principle of summation of cell E. M. F.s. Plant Physiol. 10 : 209—224.

Rosenzweig M. R., Rosenblith W. A., 1953: Responses to successive auditory stimuli at the cochlea and the auditory cortex. Psychol. Monogr. 67 : 1—26.

Rossi G. F., Zanchetti A., 1957: The brain stem reticular formation. Anatomy and physiology. Arch. Ital. Biol. 95 : 199—438.

Rothenberg M. A., 1950: Studies on permeability in relation to nerve function. II. Ionic movements across axonal membranes. Biochem. Biophys. Acta 4 : 96—106.

Roytbak A. I. — Ройтбак А. И., 1955: Биоэлектрические явления в коре больших полушарий. Изд. АН ГССР Тбилиси.

Roytbak A. I. — Ройтбак А. И., 1956: Первичные ответы коры больших полушарий нормальных животных. Труды Ин-та физиологии им. И. С. Бериташвили 10 : 131 —137.

Rubio R., Zubieta G., 1961: The variation of the electric resistance of microelectrodes during the flow of current. Acta Physiol. Latinoamer. 11 : 91—94.

Rumler G., Švarc G., 1960: Amplitudový analyzátor impulsů s obrazovkou a fotometrickým klínem. Čs. čas. fys. 107.

Rushton W. A. H., 1934: A physical analysis of the relation between threshold and interpolar length in the electric excitation of medullated nerve. J. Physiol. 82 : 332—352.

Rushton W. A. H., 1937: Initiation of the propagated disturbance. Proc. Roy. Soc. B 124 : 210—243.

Salmoiraghi G. C., 1964: Electrophoretic administration of drugs to individual nerve cells. Neuropsychopharmacology 3 : 219—231.

Sandow A , 1944: Studies on the latent period of muscular contraction. Method. General properties of latency relaxation. J. Cell. Comp. Physiol. 24 : 221—231.

Sandow A., 1945: The effect of activity on the latent period of muscular contraction. Ann. N. Y. Acad. Sci. 46 : 153—162.

Sandow A., 1948: Transverse latency relaxations of muscle stimulated with massive transverse shocks. Fed. Proc. 7 : 107—117.

Sandow A., 1952: Fundamental mechanics of skeletal muscle contraction. Amer. J. Phys. Med. 31 : 103—125.

Sandow A., 1955: Contracture responses of skeletal muscle. Amer. J. Phys. Med. 34 : 145—160.

Sandow A., 1964: Potentiation of muscular contraction. Archives of Physical Medicine and Rehabilitation 45 : 62—81.

Saunders M. G., 1954: A circuit to synchronise time marker, time base and stimulator. EEG Clin. Neurophysiol. 6 : 327—328.

Sawyer C. H., Everett J. W., Green J. D., 1954: The rabbit diencephalon in stereotaxic coordinates. J. Comp. Neurol. 101: 801—824.

Schaefer H., 1936: Über den "Reizeinbruch" bei Registrierung von Aktionsströmen mit Oszillographen. Pflüg. Arch. Ges. Physiol. 237 : 717—721.

Schaefer H., 1940, 1942, Elektrophysiologie. I, II, Franz Deuticke. Wien,

Schaefer H., 1941: Beiträge der elektrischen Reizung und Registrierung. Pflüg. Arch. Ges. Physiol. 244 : 475—483.

Schaefer H., Haas P., 1939: Über einen lokalen Erregungsstrom an der motorischen Endplatte. Pflüg. Arch. Ges. Physiol. 242 : 364—381.

Scheibel M., Scheibel A., Mollica A., Moruzzi G., 1955: Convergence and interaction of afferent impulses on single units of reticular formation. J. Neurophysiool. 18 : 309—331.

Schlag J., Brand H., 1958: An analysis of electrophysiological events in cerebral structures during ether anesthesia. EEG Clin. Neurophysiol. 10 : 305—324.

Schmidt R. F., 1963: Pharmacological studies on the primary afferent depolarization of the toad spinal cord. Pflüg. Arch. Ges. Physiol. 277 : 325—346.

Schmidt R. F., 1964: The pharmacology of presynaptic inhibition., Progress in Brain Research 12 : 119—131.

Schmitt O., Dubbert D. R., 1949: Tissue stimulators utilising radiofrequency coupling. Rev. Sci. Inst. 20 : 170—173.

Schoenfeld R. L., Eisenberg L., 1962: Multifunction instantaneous display counter. IRE Transactions on Bio-Medical Electronics, BME-9, 108—112.

Schouten S. L., 1934: Der Mikromanipulator. Z. Wiss. Mikroskopie 51 : 421—515.

Schricker J. L., O'Leary J. L., 1953: Observations on thalamic and cortical responses to peripheral nerve volley. EEG Clin. Neurophysiol. 5 : 279—290.

Schriever H., 1931: Über Einschleichen von Strom. I. Mitteilung. Versuch der Aufstellung allgemeiner Gesetzmässigkeiten. Z. Biol. 91 : 173—195.

Schwartz S., (Editor) 1960: Selected Semiconductor Circuits Handbook. Willey, New York and London.

Scott B. I. H., McAulay A. L., Ieyes P., 1955: Correlation between the electric current generated by a bean root growing in water and the rate of elongation of the root. Austral. J. Biol. Sci. 8 : 36.

Segal J., 1956: Bemerkungen zur Theorie der Erregung. In: Internationales Symposium über den Mechanismus der Erregung. S. 12—25. Deutscher Verlag der Wissenschaften, Berlin.

Segundo J. P., Naquet R., Buser P., 1955: Effects of cortical stimulation on electrocortical activity in monkeys. J. Neurophysiol. 18 : 236—245.

Servít Z., 1958: Evoluční pathologie epilepsie. Publishing House of the Czechoslov. Acad. Sci., Prague.

Shanes A. M., 1958: Electrochemical aspects of physiological and pharmacological action in excitable cells. I. II. Pharmacol. Rev. 10 : 59—164; 165—273.

Shats S. J. — Шац С. Я., 1963: Транзисторы в импульсной технике. Судпромгиз, Ленинград.

Shaw J. C., Roth M., 1955: Potential distribution analysis. II. A theoretical consideration of its significance in terms of electrical field theory. EEG Clin. Neurophysiol. 7 : 285—292.

Sherrington C. S., 1890: Notes on the arrangement of some motor fibres in the lumbosacral plexus. J. Physiol. 13 : 621—772.

Sherrington C. S., 1894: On the anatomical constitution of nerves of skeletal muscles, with remarks on recurrent fibres in the ventral spinal nerve root. J. Physiol. 17 : 211—258.

Sherrington C. S., 1906: Integrative action of the nervous system. Yale University Press, New Haven.

Shimamoto T., Verzeano M., 1954: Relations between caudate and diffusely projecting thalamic nuclei. J. Neurophysiol. 17 : 278—288.

Shipton H. W., 1949: An electronic trigger circuit as an aid to physiological research. J. Brit. I. R. E. 4 : 374—383.

Shockley W., 1952: Unipolar "field-effect" transistor. Proc. IRE 40 : 1365—1376.

Shterk M. D. — Штерк М. Д., 1964: Расчет и проектирование импульсных устройств на транзисторах. Издат. Советское радио, Москва.

Sibaoka T., 1962: Excitable cells in mimosa. Science 137 : 226.

Siler W., King K., 1963: Fundamentals of digital and analog computers. In: Nastuk W. L. (Editor): Physical Techniques in Biological Research, vol. VI, pp. 1—75, Academic Press, New York and London.

Simpson R., Valenstein E. S., 1963: An inexpensive swivel-device for electrical stimulation of unrestrained animals. EEG Clin. Neurophysiol. 15 : 900—901.

Škoda J., 1839: Abhandlung über Perkussion und Auskultation. J. G. von Mösle's Witwe, Wien.

Smirenin B. A. — Смиренин Б. А., 1950: Справочник по радиотехнике. Гос. энерг. издат., Москва—Ленинград.

Smirnov G. D. — Смирнов Г. Д., 1953: О подвижности нервных процессов в центральном и периферическом отделах зрительного анализатора. Ж. высшей нервной деятельности 3 : 941—951.

Smith O. A., Bodemer Ch. N., 1963: A stereotaxic atlas of the brain of the golden hamster (*Mesocricetus auratus*). J. Comp. Neurol. 120 : 53—64.

Smith R. A., 1959: Semiconductors. Cambridge Univ. Press, Cambridge (Mass).

Smolov et al. — Смолов В. Б., Лебедев А. Н., Сапожников К. А., Дубинин Я. И., Смирнов Н. А., Бодунов В. П., Угрюмов Е. П., Яценко В. П. — 1964: Вычислительные машины непрерывного действия. Издательство „Высшая школа", Москва.

Snider R. S., Eldred E., 1951: Electro-anatomical studies on cerebro-cerebellar connections in the cat. J. Comp. Neurol. 95 : 1—16.

Snider R. S., Niemer W. T., 1961: A stereotaxic atlas of the cat brain. The University of Chicago Press.

Snider R. S., Stowell A., 1944: Receiving areas of the tactile, auditory and visual systems in the cerebellum. J. Neurophysiol. 7 : 331—358.

Solandt D. Y., 1936: The measurement of "accomodation" in nerve. Proc. Roy. Soc. B 119 : 355—379.

Soubies-Camy H., 1961: Les Techniques Binaires et le Traitement de l'Information.

Spiegel E. A., Wycis H. T., Marks M., Lee A. J., 1947: Stereotaxic apparatus for operation on the human brain. Science 106 : 349—350.

Stämpfli R., 1952: Demonstration der Messung des Membranpotentials einzelner markhaltiger Nervenfasern. Helv. Physiol. Acta 10 : 41—42.

Stämpfli R., 1954: A new method for measuring membrane potentials with external electrodes. Experientia 10 : 508—509.

Starzl T. E., Magoun W. H., 1951: Organization of the diffuse thalamic projection system. J. Neurophysiol. 14 : 133—146.

Starzl T. E., Taylor C. W., Magoun H. W., 1951: Collateral afferent excitation of reticular formation of brain stem. J. Neurophysiol. 14 : 479—496.

Starzl T. E., Whitlock D. G., 1952: Diffuse thalamic projection system in monkey. J. Neurophysiol. 15 : 449—468.

Steiner J. E., Sulman F. G., 1961: Implanted permanent electrode for corticographical observations in wake laboratory animals. EEG Clin. Neurophysiol. 13 : 287—288.

Stellar E., Krause N. P., 1954: New stereotaxic instrument for use with the rat. Science 120 : 664—666.

Šterc J., Dvořáček J., 1964: Appareil stéréotaxique universel pour animaux de laboratoire. J. Physiol. (Paris) 56 : 853—858.

Susskind A. F., 1958: Notes on analog-digital conversion techniques. Willey, New York.

Sutin H., Campbell B., 1955: A model of cortical activity. Nature 175 : 338—339.

Sutton D., Miller J. M., 1963: Implanted electrodes: cable coupler for elimination of movement artifact. Science 140 : 988—989.

Suzuki H., Ochs S., 1964: Laminar stimulation for direct cortical responses from intact and chronically isolated cortex. EEG Clin. Neurophysiol. 17 : 405—413.

Suzuki H., Taira N., Motokawa K., Kikuchi R., Tanaka I., 1960: Spectral response curves and receptive fields of pre- and postgeniculate fibres of cat. In: Electrical activity of single cells, pp. 39—52. Igakushoin, Hongo, Tokyo.

Svaetichin G., 1951: Electrophysiological investigations on single ganglion cells. Acta Physiol. Scand., Suppl. 86 : 1—57.

Svaetichin G., 1954: The cone action potential. Acta physiol. scand. 29, Suppl. 106:565—600.

Swank R. L., Brendler S. L., 1951: The cerebellar electrogram: effects of anesthesia, analeptics and local novocaine. EEG Clin. Neurophysiol. 3 : 207—212.

Szabo T., Albe-Fessard D., 1954: Répartition et caractéres des afférences somesthésiques et d'origine corticale sur le lobe paramédian du cervelet du chat. J. Physiol. (Paris) 46 : 528—531.

Szentágothai J., 1957: A központi idegrendszer mélyen fekvö részein végzett kisérleti ·beavatkozások módszerei. A "stereotaxis" elvén alapuló müszerek és alkalmazásuk. A kisérleti orvostudomány vizsgáló módszerei III. 19—126. Akadémiai Kiadó, Budapest.

Taber E., 1961: The cytoarchitecture of the brain stem of the cat. I. Brain stem nuclei of cat. J. Comp. Neurol. 116 : 27—70.

Taira N., Okuda J., 1962: Sensory transmission in visual pathway in various arousal states of cat. Tohoku J. Exp. Med. 78 : 76—97.

Talbot S. A., Woolsey C. N., Thompson J. M., 1946: Visual areas I and II of cerebral cortex of rabbit. Fed. Proc. 5 : 103.

Tasaki I., 1953: Nervous transmission. Thomas, Springfield.

Tasaki I., 1956: Initiation and abolition of the action potential of a single node of Ranvier. J. Gen. Physiol. 39 : 377—395.

Tasaki I., 1959: Conduction of the nerve impulse. In: Handbook of Physiology, Sec. 1, Neurophysiology, vol. I, Chapter III pp. 75—121. Amer. Physiol. Society, Washington.

Tasaki I., Polley E. H., Orrego F., 1954: Action potentials from individual elements in cat geniculate and striate cortex. J. Neurophysiol. 17 : 454—474.

Tasaki I., Watanabe A., Takenaka T., 1962: Resting and action potential of intracellularly perfused squid giant axon. Proc. Nat. Acad. Sci., Wash. 48 : 1174—1184.

Tauc L., 1960: Evidence of synaptic inhibitory actions not conveyed by inhibitory postsynaptic potentials. In: E. Roberts (Editor): Inhibition in the nervous system and gamma-aminobutyric acid, pp. 85—89. Pergamon Press, New York.

Taylor G. W., 1942: The correlation between sheath birefringence and conduction velocity with special reference to cat nerve fibres. J. Cell. Comp. Physiol. 20 : 359—369.

Taylor R., 1953: Appendix to Jenerick H. and Gerard R. W., 1953.

Terman P. P., 1943: Radio-engineers handbook. McGraw-Hill, New York — London.

Therman P. O. 1941: Transmission of impulses through the Burdach nucleus. J. Neurophysiol. 4 : 153—166.

Thewlins J. (Editor): Encyclopedic dictionary of physics, vol. 1—9, Pergamon Press, London.

Thompson J. M., Woolsey C. N., Talbot S. A., 1950: Visual areas I and II of cerebral cortex of rabbit. J. Neurophysiol. 13 : 277—288.

Thompson R. F., Johnson R. H., Hoopes J. J., 1963a: Organization of auditory somatic sensory and visceral projection to association fields of cerebral cortex in the cat. J. Neurophysiol. 26 : 343—364.

Thompson R. F., Smith H. E., Bliss D., 1963b: Auditory, somatic sensory and visual response interactions and interrelations in association and primary cortical fields of the cat. J. Neurophysiol. 26 : 365—378.

Thurlow W. R., Gross N. B., Kemp E. H., Lowy K., 1951: Microelectrode studies of neural auditory activity of cat. J. Neurophysiol. 14 : 289—304.

Tiegs O. W., 1953: Innervation of voluntary muscle. Physiol. Rev. 33 : 90—144.

Tienhoven A. van, Juhász L. P., 1962: The chicken telencephalon, diencephalon and mesencephalon in stereotaxic coordinates. J. Comp. Neurol. 118 : 185—198.

Tindal J. S., 1965: The forebrain of the guinea pig in stereotaxic coordinates. J. Comp. Neurol. 124 : 259—266.

Tishenko A. M. et al.—Тишенко А. М., Лебедев В. М., Штерк М. Д., Климушев Б. Я., Федоринин А. А., Егорычев В. И. : 1964: Расчет и проектирование импульсных устройств на транзисторах. Издат. ,,Советское радио'', Москва.

Titaev A. A. — Титаев А. А., 1938: Потенциал желудка и его физиологическое значение. Бюлл. Э. Б. М. 5 : 63.

Toenies J. F., 1938: Reflex discharge from the spinal cord over the dorsal roots. J. Neurophysiol. 1 : 378—390.

Toenies J. F., 1939: Conditioning of afferent impulses by reflex discharges over the dorsal roots. J. Neurophysiol. 2 : 515—525.

Transistor mannual 1962 (6th edition). General Electric Comp., Syracuse, New York.

Trnka Z., Dufek M., 1958: Elektrické měřicí přístroje. SNTL, Praha.

Troshin A. S. — Трошин А. С., 1956: Проблема клеточной проницаемости. Изд. АН СССР, Москва.

Tsarev B. — Царев Б., 1952: Расчет и конструирование электронных ламп. Гос. энерг. издат., Москва—Ленинград.

Tschachotin S., 1912: Eine Mikrooperationvorrichtung. Z. Wiss. Mikroskopie 29 : 188—190.

Tunturi A. R., 1945: Further afferent connections to the acoustic cortex of the dog. Amer. J. Physiol. 144 : 389—394.

Tunturi A. R., 1946: A study on the pathway from the medial geniculate body to the acoustic cortex in the dog. Amer. J. Physiol. 147 : 311—319.

Tunturi A. R., 1950: Physiological determination of the arrangement of the afferent connections to the middle ectosylvian auditory area in the dog. Amer. J. Physiol. 162 : 489—502.

Tutenmacher L. I. — Тутенмахер Л. И., 1962: Электронные информацинно-логические машины. Академия наук СССР, Москва.

Ungar G., 1957: Les phénomènes chimiques de l'excitation cellulaire, leur role dans le mécanisme de l'excitation. J. Physiol. (Paris) 49 : 1235—1277.

Uspenskaya V. L. — Успенская В. Л., 1951: Биоэлектрический потенциал фотосинтеза. ДАН СССР 78 : 259—262.

Ussing H. H., 1949: The active ion transport through the isolated frog skin in the light of tracer studies. Acta Physiol. Scand. 17 : 1—37.

Ussing H., 1952: Some aspects of the applications of tracers in permeability studies. Adv. in Enz. 13 : 21—65.

Ussing H. H., Zerahn K., 1951: Active transport of sodium as the source of electric current in the short-circuited isolated frog skin. Acta Physiol. Scand. 23 : 110—127.

Valley G. E. Jr., Wallman H., 1948: Vacuum tube amplifiers. Mc Graw-Hill, N. York.

Van Harreveld A., 1936: A physiological solution for freshwater crustaceans. Proc. Soc. Exp. Biol. Med., 34 : 428—432.

Van Harreveld A., 1957: Changes in volume of cortical neuronal elements during asphyxiation. Amer. J. Physiol. 191 : 233—242.

Van Harreveld A., 1966: Brain tissue electrolytes. Butterworths, Washington.

Van Harreveld A., Ochs S., 1956: Cerebral impedance changes after circulatory arrest. Amer. J. Physiol. 187 : 180—192.

Van Harreveld A., Ochs S., 1957: Electrical and vascular concomitants of spreading depression. Amer. J. Physiol. 189 : 159—166.

Van Harreveld A., Stamm J. S., 1953a: Effects of pentobarbital and ether on the spreading cortical depression. Amer. J. Physiol. 173 : 164—170.

Van Harreveld A., Stamm J. S., 1953b: Cerebral asphyxiation and spreading cortical depression. Amer. J. Physiol. 173 : 171—175.

Vastola E. F., 1955: Steady potential responses in the cat lateral geniculate body. EEG Clin. Neurophysiol. 7 : 557—567.

Venchikov A. I. — Венчиков А. И., 1954: Биоэлектрические потенциалы желудка. МЕДГИЗ, Москва.

Veratti E., 1902: Ricerche sulla fine struttura della fibra muscolare striata. Memorie de Reale Instituto Lombardo 19 : 87, No. 10 of Series III. English transl.: Investigations on the fine structure of striated muscle fiber. J. Biophys. Bioch. Cytol. 10, Suppl. Part II. 1—69, 1961.

Verhaart W. J. C., 1964: A stereotaxic stlas of the brain stem of the cat. Part I. II. Van Gorcum Assen.

Verzeano M., 1951: Servo-motor integration of the electrical activity of the brain and its application to the automatic control of narcosis. EEG Clin. Neurophysiol. 3 : 25—30.

Verzeano M., Lindsley D. B., Magoun H. W., 1953: Nature of recruiting response. J. Neurophysiol. 16 : 183—195.

Viernstein L. J., Grossman R. G., 1961: Neural discharge patterns in the transmission of sensory information. Information Theory 20 : 262—269.

Vladimirsky K. V. — Владимирский К. В., 1951: О синхронном фильтре. Журналь эксп. и теор. физики 21 : 3—10.

Vodolazsky L. A. — Водолазский Л. А., 1952: Техника клинической электрографии. МЕДГИЗ, Москва.

Vogel J., Kryšpín J., 1956: Eine neue Methode zur Messung des Ohmschen Widerstandes von Geweben. Physiol. bohemoslov. 5 : 381—384.

Volokhov A. A., Shilyagina N. N. — Волохов А. А., Шилягина Н. Н., 1965: Определение стереотаксических координат подкорковых образований мозга у развивающихся животных. Ж. высш. нерв. деят. 15 : 176—184.

Volokhov A. A., Shilyagina N. N. — Волоход А. А., Шилягина Н. Н.,1966: Стереотаксический атлас мозга кроликов молодого возраста. Ж. высш. нерв. деят. 16 : 145—184.

Vorontsov D. S., Emchenko A. U. — Воронцов Д. С., Емченко А. У., 1947: Электрический потенциал куриного яйца, как показатель его способности к развитию. Науч. записки Т. 2, вып. 2 : 101.

Wall P. D., 1958: Excitability changes in afferent fibre terminations and their relation to slow potentials. J. Physiol. 142 : 1—21.

Waller A. D., 1900: Action électromotrice de la substance végétale consécutive à l'excitation lumineuse. C. R. Soc. Biol. 52 : 342—343.

Waller J. C., 1925: Plant electricity. I. Photo-electric current associated with activity of chlorophyl in plants. Ann. Bot. 39 : 515.

Waller J. C., 1929: Plant electricity. II. Towards an interpretation of the photoelectric currents of leaves. New Phytologist 28 : 291—302.

Walter W. G., 1936: The location of cerebral tumours by electroencephalography. Lancet 2 : 305—308.

Walter W. G. 1956: In: Cobb W. A., Greville G. D., Heppenstall M. E., Hill D., Walter W. G., Whitteridge D.: Electroencephalography. Macdonald, London.

Walter W. G., Shipton H. W., 1951: A new toposcopic display system. EEG Clin. Neurophysiol. 3 : 281—292.

Walzl E. M., Mountcastle V., 1949: Projection of vestibular nerve to cerebral cortex of the cat. Amer. J. Physiol. 159 : 595.

Watanabe A., 1926: A new device of micromanipulator. Bot. Magaz. 40 : 115—121.

Watanabe A., Grundfest H., 1961: Impulse propagation at the septal and commisural junctions of crayfish lateral giant axons. J. Gen. Physiol. 45 : 267—308.

Webb R. E., 1965: Field effect transistor for biological amplifiers. Electronic Engineering 37 : 803—805.

Wedensky N., 1883: Über die telephonischen Erscheinungen im Muskel bei künstlichem und natürlichem Tetanus. Arch. Anat. Physiol. pp. 313—325.

Wedensky N. E. — Введенский Н. Е., 1884: Телефонические исследования над электрическими явлениями в мышцах и нервах. Тр. СПб общ. естество-исп. т. 15, вып. 1.

Wedensky N., 1892: Des rélations entre les processes rythmiques et l'activité fonctionnelle de l'appareil neuro-musculaire excité. Arch. Physiol. Norm. Pathol. 4 : 50—59.

Wedensky N., 1903: Die Erregung, Hemmung und Narkose. Pflüg. Arch. Ges. Physiol. 100 : 1—102.

Weidmann S., 1952: The electrical constants of Purkinje fibres. J. Physiol. 118 : 348—360.

Weiss T., Fifková E., 1960: The use of spreading depression to analyse the mutual relationship between the neocortex and the hippocampus. EEG Clin. Neurophysiol. 12 : 841—850.

Weiss T., Fifková E., 1961: Bioelectric activity in the thalamus and hypothalamus of rats during cortical spreading depression. EEG Clin. Neurophysiol. 13 : 734—744.

Werigo B., 1883: Die secundären Erregbarkeitsänderungen an der Cathode eines andauernd polarisierten Froschnerven. Pflüg. Arch. Ges. Physiol. 31 : 417.

Werigo B., 1901: Die depressive Kathodenwirkung, ihre Erklärung und ihre Bedeutung für Elektrophysiologie. Pflüg. Arch. Ges. Physiol. 84 : 547—618.

Whieldon J. A., Van Harreveld A., 1950: Cumulative effects of minimal cortical stimulations. EEG Clin. Neurophysiol. 2 : 49—57.

White W. D., Ruvin A. E., 1957: Recent advances in the synthesis of comb filters, IRE National Convent Record, March, P-2, p. 186 — 199.

Whitfield I. C., 1959: An introduction to electronics for physiological workers. MacMillan, London.

Whitlock D. G., Arduini A., Moruzzi G., 1953: Microelectrode analysis of pyramidal system during transition from sleep to wakefulness. J. Neurophysiol. 16 : 414—429.

Whittembury G., 1964: Electrical potential profile of the toad skin epithelium. J. Gen. Physiol. 47 : 795—808.

Whittier J. R., Mettler F. A., 1949: Studies on the subthalamus of the rhesus monkey. I. Anatomy and fiber connections of the subthalamic nucleus of Luys. J. Comp. Neurol. 90 : 281—317.

Wiener N., 1948: Cybernetics or Control and Communication in the Animal and the Machine. John Wiley and Sons, New York.

Wiener N., 1949: Extrapolation, interpolation and smoothing of stationary time series. The technology press and John Wiley and Sons, New York.

Wiersma C. A. G., Furshpan E., Florey E., 1953: Physiological and pharmacological observations on muscle receptor organs of the crayfish, *Cambarus Clarkii* G. J. Exp. Biol. 30 : 136—150.

Winkler C., Potter A., 1911: An anatomical guide to experimental researches on the rabbit's brain. Amsterdam, Verslays.

Winkler C., Potter A., 1914: An anatomical quide to experimental research on the cat's brain. Amsterdam, Verslays.

Winsbury G. J., 1956: Machine for the production of microelectrodes. Rev. Sci. Instrum. 27 : 514—516.

Woodbury J. W., 1952: Direct membrane resting and action potentials from single myelinated nerve fibres. J. Cell. Comp. Physiol. 39 : 323—339.

Woodbury J. W., Brady A. J., 1956: Intracellular recording from moving tissues with a flexible mounted ultramicroelectrode. Science 123 : 100—101.

Woodbury J. W., Patton H. D., 1952: Electrical activity of single spinal cord elements. Cold Springs Harb. Symp. Quant. Biol. 17 : 185—188.

Woolsey C. N., 1947: Pattern of sensory representation in the cerebral cortex. Fed. Proc. 6 : 437—441.

Woolsey C. N., 1958: Organization of somatic sensory and motor areas of the cerebral cortex. In: H. F. Harlow and C. N. Woolsey (Editors) Biological and Biochemical Bases of Behaviour, pp. 63—81. Univ. Wisconsin Press, Madison.

Woolsey C. N., 1960: Organization of cortical auditory system. A review and synthesis. In: G. Rasmussen and W. F. Windle (Editors). Neural mechanisms of the auditory and vestibular systems. Chapter 12. Ch. Thomas, Springfeld.

Woolsey C. N., Chang H. T., 1948: Activation of the cerebral cortex by antidromic volleys in the pyramidal tract. Res. Publ. Ass. Nerv. Ment. Dis. 27 : 146—161.

Woolsey C. N., Le Messurier D. H., 1948: Pattern of cutaneous representation in rat's cerebral cortex. Fed. Proc. 7 : 137—138.

Woolsey C. N., Walzl E. M., 1942: Topical projection of nerve fibers from local regions of the cochlea to the cerebral cortex of the cat. Johns Hopkins Hosp. Bull. 6 : 315—344.

Woolsey C. N., Wang G. H., 1945: Somatic sensory areas I and II of the cerebral cortex of rabbit. Fed. Proc. 4 : 79.

Woronzow D. S., 1924: Über die Einwirkung des konstanten Stromes auf den mit Wasser, Zuckerlösung, Alkali- und Erdalkalichloridlösungen behandelten Nerven. Pflüg. Arch. Ges. Physiol. 203 : 300—318.

Woronzow D. S., 1925: Über die Einwirkung des konstanten Stromes auf den alterierten Nerven. III. Einwirkung des konstanten Stromes auf den mit Alkali-, Säure-, Zinkchlorid-, Eisenchlorid- und Aluminiumchlorid-Lösungen behandelten Nerven. Pflüg. Arch. Ges. Physiol. 210 : 672—688.

Wreschner M., 1934: Elektrokapillarität (Kapillar-Elektrometr). Abderhaldens Hbch. d. biolog. Arbeitsmethoden, II, 3, 1, p. 3339.

Wurtz R. H., 1966: Steady potential fields during sleep and wakefulness in the cat. Exp. Neurol. 15 : 274—292.

Yakovchuk N. S., Chelnokov V. E., Geifman M. P., — Яковчук Н. С., Челноков В. Е., Гейфман М. П., 1961: Плоскостные транзисторы, Судпромгиз, Ленинград.

Yakovlev K. P. — Яковлев К. П., 1953: Математическая обработка результатов измерений. Гостехиздат, Москва.

Yoffe A. F. — Иоффе А. Ф., 1957: Физика полупроводников (2. изд.), Москва—Ленинград.

Young J. Z., 1939: Fused neurons and synaptic contacts in the giant nerve fibres of cephalopods. Phil. Tr. Roy. Soc. London, Ser. B 229 : 465—475.

Zadeh L. A., Ragazzini J. R., 1952: Optimum filters for the detection of signals in noise. Proc. IRE 40, 1223.

Zachar J. — Захар Й., 1955: Замечания по поводу методики электрофизиологический регистрации неимпульсных форм нервной сигнализации. Physiol. Bohemoslov. 4 : 335—343.

Zachar J. — Захар Й., 1956: Изменения медленного потенциала в очаге парабиоза и его окружении. Physiol. Bohemoslov. 5 : 20—25.

Zachar J., Zacharová D. 1961: Subthreshold changes at the site of initiation of spreading cortical depression by mechanical stimuli. EEG clin. Neurophysiol., 13 : 896—904.

Zachar J., Zacharová D., 1965: The effect of calcium ions on potassium contractures in single muscle fibres of the crayfish. XXIIIth Internat. Congress Physiol. Sci., Abstr. No 803, Tokyo.

Zachar J., Zacharová D., 1966: Potassium contractures in single muscle fibres of the crayfish. J. Physiol., 186 : 596—618.

Zachar J., Zacharová D., Henček M., 1964a: Membrane potential of the isolated muscle fibre of the crayfish (Astacus fluviatilis). Physiol. bohemoslov. 13 : 117—128.

Zachar J., Zacharová D., Henček M., 1964b: The relative potassium and chloride conductances in the muscle fibre membrane of the crayfish. Physiol. bohemoslov., 13 : 129—136.

Zachar J., Zacharová D., Kuncova M. J., Henček M., 1964c: The effect of γ-aminobutyric acid on membrane conductance of the isolated crayfish muscle fibre (in Slovak). Čs. fysiol. 13 : 512—513.

Zacharová D., 1957: The recovery cycle of excitability of the inhibitor and excitor fibre. Physiol. bohemoslov. 6 : 143—17.

Zacharová D., Zachar J., 1963: The inhibition of potassium contractures in isolated muscle fibres after Ca ions withdrawal (in Slovak). Čs. fysiol., 12 : 208—209.

Zacharová D., Zachar J., 1965: Contractions in single muscle fibres with graded electrogenesis. Physiol. bohemoslov., 14 : 401—411.

Zacharová D., Zachar J., 1967: The effect of external calcium ions on the excitation-contraction coupling in single muscle fibres of the crayfish. Physiol. bohemoslov., 16 : 100—200.

Zavolokin A. K. — Заволокин А. К., 1962: Последовательные преобразователи непрерывных величин в числовые эквиваленты. Гос. энерг. издат., Москва—Ленинград.

Zeldovich J. B. — Зельдович Я. Б., 1963: Высшая математика для начинающих и ее приложения к физике. Гос. из. Физико-математической литературы, Москва.

Zhdanov G. M. — Жданов Г. М., 1952: Телеизмерение. Част 1. Гос. энерг. издат., Москва—Ленинград.

Zheleznov N. A. — Железнов Н. А., 1959: Некоторые вопросы спектрально-корреляционной теории нестационарных сигналов. Радиотехника и электроника 4 (3) 151.

Ziel A. van der, 1954: Noise. Prentice-Hall, N. York.